Proceedings of the

16th International Conference

on

Defects in Semiconductors

ICDS 16

Proceedings of the

16th International Conference

on

Defects in Semiconductors

Part 1

Lehigh University, Bethlehem, Pennsylvania
22 - 26 July 1991

Edited by
Gordon Davies, Gary G. DeLeo and Michael Stavola

Trans Tech Publications
Switzerland — Germany — UK — USA

This work was partially supported by U.S. Air Force Grant AFOSR-91-0217 issued by the U.S. Air Force Office of Scientific Research.

This work was partially supported by the Department of The Army Grant DAAL03-91-G-0113 issued by the Army Research Office. The views, opinions, and/or findings contained in this report are those of the authors and should not be construed as an official Department of the Army position, policy, or decision, unless so designated by other documentation.

This work relates to Department of Navy Grants N00014-91-J-1761 and N00014-91-J-1912 issued by the Office of Naval Research. The United States Government has a royalty-free license throughout the world in all copyrightable material contained herein.

ISBN 0-87849-628-9
(Pts. 1, 2 & 3)

Volumes 83-87 (Pts. 1, 2 & 3) of
Materials Science Forum
ISSN 0255 - 5476

Distributed in the Americas by

Trans Tech Publications
c/o Gower Publishing Company
Old Post Road
Brookfield VT 05036
USA

Fax: (802) 276 - 3837

and worldwide by

Trans Tech Publications Ltd
Segantinistr. 216
CH-8049 Zürich
Switzerland

Fax: (++41) 1 342 05 29

16th International Conference
on
DEFECTS IN SEMICONDUCTORS

Lehigh University
Bethlehem, Pennsylvania, USA

22-26 July 1991

ICDS - 16 SPONSORS

Air Force Office of Scientific Research
Bandgap Technology Corporation
Defense Advanced Research Projects Agency
Eastman Kodak Company
Ford Motor Company
Fujitsu Laboratories Ltd.
IBM
Kawasaki Steel Corporation
Lehigh University
Mobil Solar Energy Corporation
NEC Research Institute
Philips Laboratories
Rome Laboratory, Solid State Sciences
Sematech
Sherman Fairchild Endowment
Solar Energy Research Institute
Sumitomo Electric Industries, Ltd.
United States Army Research Office
United States Navy Office of Naval Research
Wright Laboratory Materials Directorate
Xerox Corporation

ICDS - 16 COMMITTEES

International Advisory Committee

R.N. Bhargava, U.S.A.
J.C. Bourgoin, France
J.W. Corbett, U.S.A.
G.G. DeLeo, U.S.A.
G. Ferenczi, Hungary
H.G. Grimmeiss, Sweden
E.E. Haller, U.S.A.
L.C. Kimerling, U.S.A.
H. Kukimoto, Japan
J.M. Langer, Poland
M. Lannoo, France
J.R. Leite, Brazil
R.C. Newman, U.K.
S.T. Pantelides, U.S.A.
G.-G. Qin, P.R. China
H.J. Queisser, F.R.G.
M.K. Sheinkman, U.S.S.R.
M. Stavola, U.S.A.
A.M. Stoneham, U.K.
K. Sumino, Japan
G.D. Watkins, U.S.A.
E.R. Weber, U.S.A.

Program Committee

G.A. Baraff, U.S.A.
S.G. Bishop, U.S.A.
B. Clerjaud, France
G. Davies, U.K.
G.G. DeLeo, U.S.A.
R.J. Jaccodine, U.S.A.
H. Kukimoto, Japan
J.M. Langer, Poland
P.M. Mooney, U.S.A.
S.T. Pantelides, U.S.A.
P.M. Petroff, U.S.A.
M. Stavola, U.S.A.
G.D. Watkins, U.S.A.
E.R. Weber, U.S.A.

Local Organizing Committee

M. Stavola, co-chair
G.G. DeLeo, co-chair
B.W. Benson
W.B. Fowler
S.J. Pearton
G.D. Watkins

PREFACE

The 16th International Conference on Defects in Semiconductors, like previous conferences in this series, was a productive and exciting meeting of research workers who are advancing the fundamental science of defects in semiconductors. However, for the first time, ICDS was held as an independent conference and no longer as a satellite of the International Conference on the Physics of Semiconductors. The organizers of ICDS-16 were concerned that this conference might not be as successful as past defect conferences without the draw of the big semiconductor physics meeting. These concerns were proved to be unjustified. There were more that 400 abstracts submitted, 335 attendees registered, 259 papers presented, and 24 countries represented.

The participation of many delegates from the Eastern Block at a conference in the United States was especially satisfying. The number of Soviet delegates in attendence demonstrated how well international relations have progressed in recent years.

The energy and enthusiasm of the participants seemed boundless. Attendence was excellent at sessions that began as early as 8:30 AM. Scientific discussion dominated meal times. Poster sessions were still well attended when they had to be forcibly closed at 11 PM. Productive scientific discussion then continued at the Tally-Ho Tavern until the early morning hours. We are aware of some delegates who had been up until 3 AM and still found the 9 AM sessions too exciting to miss.

These proceedings document our understanding of the many topics of current interest in the field of defects in semiconductors. Important advances are discussed in roughly 20 invited papers. Over 200 contributed papers describe the breadth of the activity in the field. Among the work presented are resolutions of long standing puzzles, numerous new ideas on old and new problems, and presentations of work at the periphery of the field that may warrant our future attention. Many of the experiments and ideas presented are new and/or controversial and will lead to continued activity and expansion of the field.

We wish success to the future organizers of conferences in the ICDS series and hope that they will enjoy organizing the meeting and editing its proceedings as much as we have.

Michael Stavola

Gary G. DeLeo

Gordon Davies

ACKNOWLEDGEMENTS

The outstanding success of the 16th International Conference on Defects in Semiconductors was due to the efforts of many people. The excellent scientific program was due to the thoughtful consideration of the International Advisory Committee and the Program Committee as well as to the authors and speakers who have presented their work. The following session chairs were responsible for ensuring that the meeting ran smoothly and for helping to referee these proceedings:

W.B. Fowler	G.D. Watkins	R.C. Newman	J.-M. Spaeth
B. Clerjaud	M.L.W. Thewalt	L.C. Kimerling	C.A.J. Ammerlaan
H.P. Gislason	K. Sumino	J.R. Leite	B. Monemar
S.J. Pearton	P.M. Mooney	J.M. Langer	T.A. Kennedy
M.H. Nazaré	C. Van de Walle	H. Kukimoto	G. Baraff
J. Weber	D. Mathiot	A.M. Stoneham	J. Farmer
E. Weber	S.G. Bishop	M. Mizuta	N.M. Johnson
E. Lightowlers	J. Chadi		

Many members of the Lehigh University staff helped provide the excellent local services which we often take for granted. Jessica Dunlap, Bobbie Stahr, and Jefferson Huang saw to the housing needs of the participants. Michael Rosenberg directed the catering services which we all enjoyed. Gail Vollman organized the bus transportation which made it possible for us to utilize the spacious Lehigh campus. Craig Wood and Dotty Stoudt helped manage the conference finances and Linda Cope was our interface to the sponsoring agencies.

We are grateful for the efforts of our graduate students who acted as airport guides, excursion coach captains, projectionists, local drivers, etc. and who were responsible for much of the personal attention the participants received during the week of the conference. We do not know what we would have done without the help of the Physics Department secretaries, Louise McCormick, Barbara Szymanski, and especially Lois Groff who worked hard for several months before and during the meeting week to help type, process, and organize the various conference materials.

The financial contributions of the many industrial and government sponsors of ICDS-16 listed on the next page supported the expenses of the meeting and enabled the organizers to provide the highest quality speakers.

We express our sincere thanks to all those who have contributed to the success of ICDS-16.

CONTENTS — PART 1

1. HYDROGEN IN ELEMENTAL HOSTS

2. TRANSITION METAL IMPURITIES IN ELEMENTAL HOSTS

CONTENTS — PART 2

7. HYDROGEN IN COMPOUND SEMICONDUCTORS

9. TRANSITION METAL IMPURITIES IN COMPOUND SEMICONDUCTORS

10. DONORS IN COMPOUND SEMICONDUCTORS

11. EL2 AND ANTI-SITE RELATED DEFECTS

13. GROWTH DEFECTS

CONTENTS — PART 3

14. NEW TECHNIQUES

17. HETERO-EPITAXY AND STRAINED LAYERS

18. DISLOCATIONS

19. SUPERLATTICES

20. DEFECTS AT SURFACES AND INTERFACES AND IN LOW-DIMENSIONAL STRUCTURES

21. PROCESSING-INDUCED DEFECTS

22. EFFECTS OF DEFECTS ON DEVICES

Materials Science Forum Vol. 83-87 (1992) pp. 1-8

MICROSTRUCTURE OF HYDROGEN AND DOPANTS IN HYDROGENATED AMORPHOUS SILICON

J. B. BOYCE and S. E. READY
Xerox Palo Alto Research Center, Palo Alto, California 94304, USA

ABSTRACT

Nuclear magnetic resonance has been successfully applied to the study of the microstructure of hydrogenated amorphous silicon and related materials. Here we briefly review results from these measurements on (1) the H distribution and bonding and (2) the local bonding structure of dopants in hydrogenated amorphous silicon. These results for amorphous silicon are compared and contrasted with what is known about the corresponding structures in crystalline silicon.

1. Introduction

Nuclear magnetic resonance (NMR) has provided an invaluable probe of the local atomic bonding and microstructure in hydrogenated amorphous silicon (a–Si:H) and related materials [1–11]. Here we review some of these results, with emphasis on the comparison of the microstructures determined for a–Si:H with the corresponding structures known or proposed for crystalline Si (c–Si). The detailed discussions of the measurements and results can be found in Ref. 1 and the references cited therein.

2. Hydrogen Microstructure of a–Si:H

In electronic–grade, plasma–deposited a–Si:H, most of the hydrogen is bonded to the Si in a dilute phase, that gives rise to a narrow NMR line, and in a clustered phase, that gives rise to a broad NMR line [1–3]. The total hydrogen content is found to be $\approx$ 10 atomic %, with $\approx$ 4 atomic % in the narrow line and $\approx$ 6 atomic % in the broad line. In addition, a small fraction of the H (~1%) is in the form of H_2 molecules (~ 0.1 atomic %). These molecules reside in small internal voids and can be observed directly in the NMR spectrum at low temperatures [8] or indirectly in the temperature variation of the H nuclear spin–lattice relaxation time which exhibits a minimum near 40K [2, 6].

The narrow line is Lorentzian in shape with a full width at half maximum (FWHM) of $\approx$ 3 kHz and corresponds to isolated Si–H units randomly distributed throughout the amorphous Si network. The 3 kHz linewidth yields an estimate [2, 5, 6] of the mean spacing between the H atoms in this dilute phase of between 5.5 and 7.7Å, depending on the specific assumptions made.

The origin of the broad line, which is Gaussian with FWHM $\approx$ 25 kHz, is less certain, but it does arise from regions with more clustered H. Various types of structures have been considered. Hydrogenated monovacancies and divacancies are ruled out since their linewidth is too large ($\sim$ 100 kHz). SiH_2, SiH_3 and polysilane chains give line widths that are too small but could yield a 25 kHz line if clustered. Also SiH_2 molecules have been directly observed in the NMR spectrum [9] but only in poor quality films prepared on room temperature substrates and not for electronic–grade material prepared on heated substrates. These results indicate that SiH_2 molecules contribute substantially to the broad line in low–grade a–Si:H but can be ruled out as a major contributor to the broad line in good material. Hydrogenated internal surfaces are, therefore, a preferred candidate since internal voids and platelets are known to exist. Multiple–quantum NMR experiments [3] indicate that these clusters consist of 5–7 H atoms. However, no specific structural model of the clusters or hydrogenated internal surfaces has been proposed.

A specific model, based on the results of theoretical calculations [12] on hydrogen clusters in crystalline silicon, has been proposed [13] for the clustered H responsible for the broad NMR line in a–Si:H. It is based on the H_2^* complex shown in Fig. 1a. A Si–Si bond is broken and one H bonds at the bond–center site to a Si atom which relaxes away from the original bond. A second H bonds in an antibonding site to the other Si atom which relaxes toward the antibonding site, putting it in the plane of the neighboring three Si atoms. This complex is neutral, ESR–inactive, lower in energy by 1.2 eV than two H atoms inserted into a Si–Si bond, and only 0.4 eV higher in energy than molecular H_2 in a T_d site. Since crystalline and amorphous Si are similar locally, such a complex may also be appropriate for a–Si:H. However, the H–H distance at 3.38Å is too long to account for the broad NMR line, giving FWHM = 5 kHz. If two or more H_2^*'s complex as first neighbors to one another [13], as shown in Fig. 1b, the shortest H–H distance becomes 2.26Å. The line width for a pair of H_2^* near neighbors is 17.4 kHz, still somewhat small. The width for four neighboring H_2^*'s is 21.4 kHz and for an infinite H_2^* cluster is 30 kHz. These values are close to the linewidth of the broad line. Thus higher level complexes of H_2^*, containing eight or more H atoms, can account for the broad NMR line and provide a specific model for the clustered phase of H in a–Si:H. It also accounts for the fact that, as more H is added to a–Si:H, the linewidth does not increase. Going from $(H_2^*)_4$ to $(H_2^*)_\infty$ has only a small effect on the linewidth. We should mention, however, that this model is based on the crystalline case, and distortions in a–Si could modify the distances and thereby the predicted linewidth. Also, even in the crystalline case, distortions of the lattice will modify the H–H

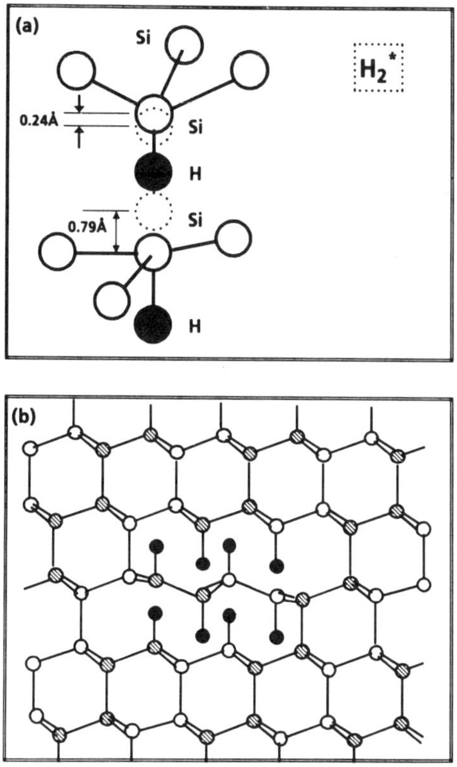

Fig. 1. (a) Structure of the H_2^* diatomic hydrogen complex for crystalline Si. (b) A higher order complex consisting of four H_2^*'s that can account for the broad NMR line.

distance as the cluster size changes. Nonetheless, these distance should not vary significantly so that an $(H_2^*)_n$ cluster provides the correct mean spacing to account for the broad NMR line as well as other measured properties of a-Si:H [13].

When a–Si:H is cooled to low temperatures, the NMR spectrum evolves into a three–component line: the central line with two width components, as observed at room temperature and due to the H bonded to Si, plus a Pake doublet [1, 4, 8] with a splitting of 175 $\pm$10 kHz. The observed doublet is due to orientationally ordering of molecular H_2 or a slowing down of the molecular tumbling rate to less than the 175 kHz splitting. As temperature increases, the Pake doublet collapses due to the averaging by the rapid molecular tumbling and the signal due to H_2 is transferred to the broad line. As temperature increases further, the solid H_2 melts and the signal becomes motionally narrowed and contributes to the narrow line. This has a measurable effect on the signal fraction in the two central line components in samples containing a large amount of H_2, i. e., $n(H_2)/n(H) \geq 0.05$, and a small effect for samples with small amounts of H_2, i. e., $n(H_2)/n(H) \leq 0.05$. The latter case occurs for as-deposited a-Si:H prepared by plasma-enhanced CVD where typically $n(H_2)/n(H) \approx 0.01$ or $n(H_2) \approx 0.1$ atomic %. This H_2 content can be increased by up to a factor of 10 by annealing, by modifying the preparation parameters to make microcrystalline Si and by adding Ge to make Ge-Si alloys. These latter materials have more and larger voids, containing larger quantities of non-bonded molecular H_2 [1, 4]. In all cases, however, the presence of a small number of H_2 molecular relaxation centers on the surfaces of voids in the a-Si host accounts for the NMR spin-lattice relaxation which exhibits a characteristic minimum in the vicinity of 40-50K [1, 2]. So molecular H_2 is readily observed to be present in measurable amounts in a-Si:H. This is to be contrasted with the situation in c-Si where no direct, spectroscopic observation of H_2 has been made to date, even though H_2 in a T_d site is theoretically the the most stable configuration for H in c-Si.

3. Local Bonding Structure of Dopants in a-Si:H

The details of the local bonding structure of dopants in crystalline and amorphous silicon play a key role in determining the electrical properties of these semiconductors. The specific parameters that are pertinent are the number of neighbors (three-fold or four-fold coordinated), the type of neighbors (Si, dopant, hydrogen), the local order (random or clustered), and the near-neighbor distances. The dopants in crystalline silicon occupy four-fold coordinated substitutional sites. When H is added to the c-Si, passivation of the dopants occurs [14-21]. Several theoretical dopant-hydrogen bonding structures have been proposed to account for the passivation [17, 22-24]. In contrast, the local environment of dopants in hydrogenated amorphous silicon (a-Si:H) is not known. The doping efficiency is quite low, leading to the suggestion that the majority of the dopants are three-fold coordinated rather than four-fold. This is a likely structure, but the location of H is unspecified. Hydrogen may also play a role in determining the doping mechanism and efficiency in a-Si:H because of its high H content ($\approx$ 10 atomic %) and since the local structure of the amorphous and crystalline phases are similar. We now discuss the NMR spin echo double resonance (SEDOR) results that have been used to shed some light on the structure of the dopant-H complex in compensated, p-type, and n-type a-Si:H [10, 11].

First, consider the B-P local structure for a compensated a-Si:(H,B,P) sample, containing 2 atomic % B and P. From the SEDOR data, two structurally inequivalent B sites are determined to be present: $40 \pm 15\%$ of the B with a B-P distance of 2.1 ± 0.3Å and $60 \pm 15\%$ of the B with a P at a mean distance of about 3.8 ± 0.3Å. The first has its B-P distance close to that in BP (1.96Å), showing that clustering occurs for 40±15% of the pairs. This is well beyond the random value of 8%, indicating a chemical affinity of the B and P in a–Si. The conclusion then is that B and P tend to cluster in compensated a–Si:H.

Second, the results for the p–type material indicates that 40±15% of the B atoms have a hydrogen near neighbor located at a distance of 1.6±0.3Å. This is close to the B–H bonding distances found in boranes (1.2–1.4Å) and agrees with the theoretically derived B–H distance (1.63Å) in c–Si [22, 23]. Thus the data would indicate that about half of the B atoms have a hydrogen atom as a near neighbor. In contrast, the data for the compensated material indicate that there is no short B–H distance and neither the random nor clustered case apply. Possibly the clustering of the B and P in the compensated material prevents significant B–H bonding.

This determined structure for about half of the B dopants in p–type a–Si:H is very close to proposals for the B–H structure in passivated p–type c–Si [22, 23], shown in Fig. 2. For this crystalline case, it is found that the total–energy minimum configuration corresponds to the hydrogen residing in a bond–center site between the boron and a Si neighbor. The B relaxes by 0.47Å from the ideal four–fold site toward the three–fold site. This plus some Si

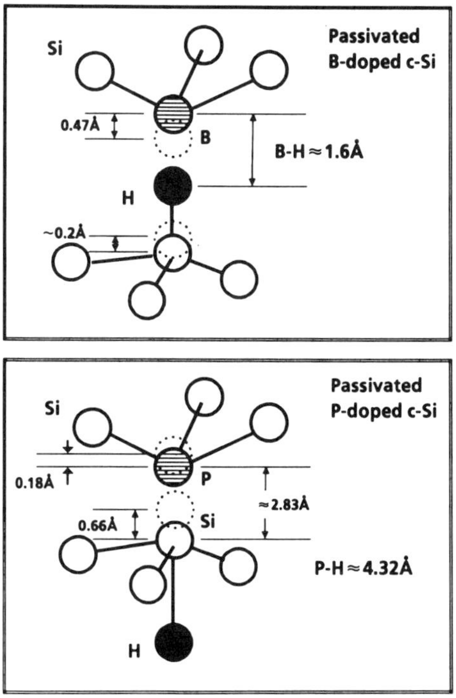

Fig. 2. The theoretically determined H–passivation structures for B–doped and P–doped crystalline Si.

relaxation provides a B–H distance of about 1.6Å, as does a cluster calculation (1.63Å) [22]. Both are in reasonable agreement with the NMR experimental value of 1.6±0.3Å. Ion channeling results on c–Si arrive at this same atomic configuration but with a smaller B relaxation of 0.28Å [18]. Also, infrared and Raman studies [19–21] of c–Si find the H atom in the B–H complex bonded to a Si atom, not the B atom, consistent with Fig. 2. In a–Si:H, on the other hand, B–H modes are observed in materials with high B content [14]. This implies that a significant fraction of the B atoms have an H directly bonded to it. Both structures, the H–B complex of c–Si (Fig. 2) and a direct B–H bond, are consistent with the SEDOR results and are indistinguishable from these measurements.

Third, the P–H results for n–type samples yield a closest P–H distance of 2.6±0.3Å for half the P and 4.1±0.3Å for the other half. (Similar P–H distances are also obtained in the compensated material.) 2.6Å is too large to be attributed to direct P–H bonding which is typically 1.4Å. This result is in qualitative agreement with the structure proposed for passivated n–type crystalline silicon [17, 23, 24]. For this crystalline case, it is found that the total–energy minimum position for H is at the antibonding site of a Si nearest neighbor of a substitutional P atom, as shown in Fig. 2. The direct bonding of H to a P has a significantly higher total energy than the bonding to a neighboring Si. The detailed structure of this model, however, places the H 4.3Å from the P, significantly further than the 2.6Å measured for a–Si:H. This distance is closer to the second P–H distance determined experimentally, namely, 4.1±0.3Å. The c–Si calculations found that the other three antibonding sites that are closest to the P are less favorable than this further site. The amorphous case, however, has significantly more H, and the bulk of this H provides the fourth bond to a Si that is three–fold coordinated to other Si. It ties up the Si dangling bond and is not thought to provide a fifth bond to the Si. A specific model [13] that does yield the experimentally determined distance is the structure of Fig. 2 plus a near–neighbor H_2^* of Fig. 1a. The H in the antibonding site of the H_2^* is 2.45Å from the P. So the SEDOR data are consistent with the picture in which about half of the P has a neighboring H_2^* (P–H=2.45Å) and the other half are in the theoretically–derived passivated structure for c–Si (P–H=4.3Å, Fig. 2). Whether the half of the P that have a neighboring H_2^* also are in the passivated structure of Fig. 2 cannot be determined from the SEDOR data since the 2.45Å distance dominates the signal decay so that a second 4.3Å distance makes a negligible contribution.

All these local structural results indicate that H may play a role in determining the doping efficiency of a–Si:H, as it does in c–Si. Approximately half of the B in p–type and half of the P in n–type a–Si:H have local structures that are consistent with the H–dopant complexes that passivate dopants in c–Si. Substantial difference, however, do exist. For example, in p–type a–

Si:H, a number of B atoms have an H directly bonded to it, whereas in p–type c–Si the H bonds to a Si next to the B. The SEDOR results merely indicate that half the B have an H in the first neighbor shell. Nonetheless, the fact that the local structures of the amorphous and crystalline phases are similar indicates that H most likely plays a role in passivation. The substantial relaxations that do occur in these H–dopant complexes place the dopants in a highly distorted environment, essentially three–fold coordinated. This then is consistent with the single NMR results that most of the dopants are three-fold coordinated [10, 11] and with models of doping that rely on three–fold coordination [25]. Nonetheless, detailed models will have to include the role of H in doping efficiency in a–Si:H and consider the effects of different "three–fold" configurations such as BSi_3 and BSi_3H.

ACKNOWLEDGMENTS

The authors are pleased to acknowledge the significant input of W. B. Jackson, N. M. Johnson, R. E. Norberg, M. Stutzmann, C. C. Tsai, and S. B. Zhang. This work was supported in part by SERI.

REFERENCES

1. J. B. Boyce and S. E. Ready, Physica B 170, 305 (1991).
2. P. C. Taylor, in *Semiconductors and Semimetals*, Vol 21, Part C (Academic Press, New York, 1984) p. 99.
3. J. A. Reimer and M. A. Petrich, in *Advances in Disordered Semiconductors – Amorphous Silicon and Related Materials*, ed. H. Fritzsche (World Scientific Publishing Co., Singapore, 1989) p. 3.
4. J. B. Boyce, S. E. Ready, M. Stutzmann, and R. E. Norberg, J. Non–Cryst. Solids, 114, 211 (1989).
5. J. A. Reimer, R. W. Vaughn, and J. C. Knights, Phys. Rev. B 24, 3360 (1981).
6. W. E. Carlos and P. C. Taylor, Phys. Rev. B 26, 3605 (1982).
7. D. J. Leopold, J. B. Boyce, P. A. Fedders, and R. E. Norberg, Phys. Rev. B 26, 6053 (1982).
8. J. B. Boyce and M. Stutzmann, Phys. Rev. Lett. 54, 562 (1985).
9. J. B. Boyce, in *Hydrogen in Disordered and Amorphous Silicon*, eds. G. Bambakidis and R. C. Bowman (NATO Advanced Study Inst. Proc. Rhodes, Greece, 1985) p. 101.
10. J. B. Boyce and S. E. Ready, Phys. Rev. B 38, 11008 (1988).
11. J. B. Boyce and S. E. Ready, in *Advances in Disordered Semiconductors – Amorphous Silicon and Related Materials*, ed. H. Fritzsche (World Scientific Publishing Co., Singapore, 1989) p. 29.
12. K. J. Chang and D. J. Chadi, Phys. Rev. Lett. 62, 937 (1988).
13. W. B. Jackson and S. B. Zhang, in *Transport, Correlation and Structural*

Defects, ed. H. Fritzsche (World Scientific Publishing Co., Singapore, 1990) p. 63; W. B. Jackson and S. B. Zhang, Physica B 170, 197 (1991).
14. For a review, see *Hydrogen in Semiconductors*, eds. J. I. Pankove and N. M. Johnson (Academic Press, New York, 1991).
15. J. I. Pankove, D. E. Carlson, J. E. Berkreyheiser and R. O. Wance, Phys. Rev. Lett. 51, 2224 (1983).
16. N. M. Johnson, Phys. Rev. B 31, 5525 (1985).
17. N. M. Johnson, C. Herring and D. J. Chadi, Phys. Rev. Lett. 56, 769 (1986).
18. A. D. Marwick, G. S. Oehrlein and N. M. Johnson, Phys. Rev. B 36, 4539 (1987); A. D. Marwick, G. S. Oehrlein, J. H. Barrett and N. M. Johnson, Mat. Res. Soc. Symp. Proc., Vol 104 (Materials Research Society, Pittsburgh, 1988) p. 259.
19. M. Stutzmann, Phys. Rev. B 35, 5921 (1987); C. P. Herreo, M. Stutzmann, and A. Breitschwerdt, Phys. Rev. B 43, 1555 (1991).
20. M. Stavola and S. J. Pearton, in Ref. 14, p. 139.
21. N. M. Johnson, C. Doland, F. Ponce, J. Walker, and G. Anderson, Physica B 170, 3 (1991).
22. G. G. DeLeo and W. B. Fowler, Phys. Rev. B 31, 6861 (1985); G. G. DeLeo and W. B. Fowler, Ref. 14, p. 511.
23. K. J. Chang and D. J. Chadi, Phys. Rev. Lett. 60, 1422 (1988).
24. S. B. Zhang and D. J. Chadi, Phys. Rev. B 41, 3882 (1990).
25. R. A. Street, J. Non–Cryst. Solids 77&78, 1 (1985).

Materials Science Forum Vol. 83-87 (1992) pp. 9-14

ELASTIC ENERGY LOSS DUE TO THE REORIENTATION OF H⁻ AROUND B IN SILICON

G.CANNELLI,[1] R.CANTELLI,[2] M.CAPIZZI,[3] F.CORDERO,[4] A.FROVA,[3]
M. STUTZMANN,[5] and F. TREQUATTRINI[4]

[1]Dipartimento di Fisica, Università di Perugia, Perugia, Italy
[2]Dipartimento di Energetica, Università "La Sapienza", Rome, Italy
[3]Dipartimento di Fisica, Università "La Sapienza", Rome, Italy
[4]Istituto di Acustica "O.M. Corbino", C.N.R., Rome, Italy
[5]MPI, Institut für Festkörperphysik, Stuttgart, Germany

ABSTRACT

Measurements of elastic energy dissipation and elastic modulus have been carried out in Si:B before and after hydrogen incorporation by both ion-gun irradiation and plasma exposure. Experiments were performed at three widely separated frequencies (2.4, 13, and 32 kHz). Both types of H-treated samples present a peak due to a thermally activated relaxation process at 130 K (2.4 kHz). The activation energies deduced from both the peak shift with frequency and the peak width have the same value, W = (0.22±0.01)eV. This indicates that the observed process is characterized by a single relaxations time, whose pre-exponential factor is $\tau_0 = (8\pm4\times10^{-14})$s.

The value measured for the activation energy is in excellent agreement with the one estimated from infrared spectroscopy and logarithmic decrement measurements, and with the theoretical estimate for H trapped in bond-center position between neighboring B and Si atoms. The absence of broadening of the dissipation peak demonstrates that the interaction between the boron trapping centers is still negligible at the B concentrations presently used (up to 6×10^{19} atoms/cm^3).

Measurements on highly B doped samples H-treated by plasma exposure show, an additional peak at 210 K, whose rather unusual features will be illustrated.

1. INTRODUCTION

The anelastic relaxation technique supplies the simultaneous measurements of elastic energy loss and elastic constants in solids. Dissipation may be caused by different mechanisms like classical or quantum atomic diffusion, phase transformations, phonon or electron interactions. The various absorption effects give rise to peaks which are in general well separated on the temperature scale, according to their characteristic diffusional or thermodynamic parameters, so that they can be studied selectively.

The technique can be applied in the low frequency range (Hz and kHz) by resonant frequency methods, or in the high frequency range (MHz) by ultrasound attenuation. By the first method lower background dissipations and hence higher sensitivities are attained.

The diffusion of hydrogen in semiconductors has been studied only in recent years and most of decisive questions like mobility regimes, structure of H complexes, site occupancy, and isotope influence, are still open. The presence of B in crystalline Si increases the H solubility, due to the trapping effect played by the dopant and gives rise to B-H complexes. It has been shown that H occupies bond-center (BC) positions between neighboring B and Si atoms [1], and that H is not immobile in the complex but performs jumps between sites coordinated with the B trapping center [2,3]. The activation energy for reorientation of the B-H complex was theoretically estimated to be 0.2 eV [1]. Experimental results yielded 0.19 eV by infrared absorption measurements under simultaneous application of a uniaxial stress to the sample [2], and 0.23 eV by the amplitude logarithmic decrement measurements of a vibrating circular membrane [4], whose outer silicon frame, however, induced a stress between 17 and 50 Mpa.

Most recently, the dynamics of hydrogen at low temperature was studied by anelastic relaxation measurements on bulk SiB_xH_y, without application of external stresses other than the modulating one introduced by the sample vibration [5]. The elastic energy dissipation peak appearing around 130 K was attributed to the stress-induced reorientation of H around B. The peak was studied as a function of vibration frequency and H concentration, and the activation energy for the jump and the prefactor were derived; information was also extracted concerning the nature of the H jump and the elastic interactions.

In the present paper we report results on the mobility parameters of hydrogen in Si:B, and on a new peak at about 210 K, whose features are rather anomalous.

2. EXPERIMENT

The samples were five rectangular bars (43x5x0.4 mm^3) of Si; three of them (samples 1-3) contained 10^{19} B atoms/cm^3 and two (samples 4 and 5) $6x10^{19}$ B atoms/cm^3. Samples 1 and 2 were H-loaded in their central portion (~3 cm) by ion-gun irradiation at impinging doses of $7x10^{18}$ H atoms/cm^2 at 300 ºC for 3 h (sample 1, irradiated for three hours), and $7x10^{17}$ H atoms/cm^2 at 300 ºC for 19 min (sample 2, irradiated for 19 min). Samples 3 and 5 were left in the "as received" state as reference. Sample 4 was loaded with H in a remote plasma atmosphere at 150 ºC, as described in detail elsewhere [6,7]. The H penetration depths were of the order of a few micrometers and their concentrations in the H containing layers were expected to be comparable to those of boron.

Elastic energy loss and resonant frequency measurements were carried out between 60 and 300 K, by electrostatically exciting flexural vibration modes of the bars. The samples were horizontally suspended on two nodal lines by thin wires; three modes were excited (2.4, 13, and 32 kHz), which correspond to a frequency variation by a factor 13.

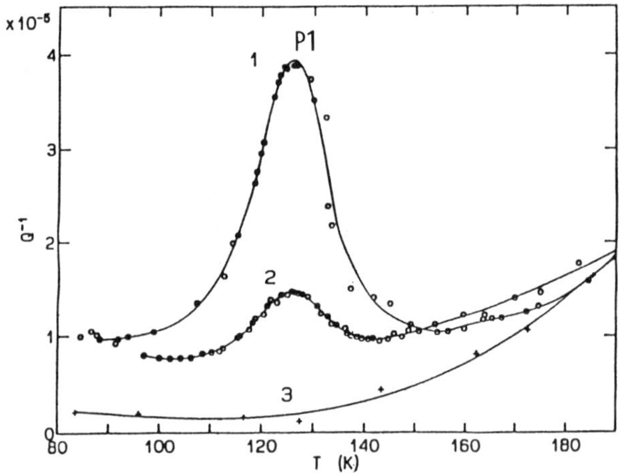

Fig. 1 Elastic energy dissipation of three samples of Si doped with 10^{19} at.B/cm^3, measured at 2.4 kHz: 1) sample 1, irradiated at $7x10^{18}$ H at./cm^2; 2) sample 2, irradiated at $7x10^{17}$ H at./cm^2; 3) sample 3, unhydrogenated.

3. RESULTS AND DISCUSSION

- Relaxation of the B-H complex.

Figure 1 shows the elastic energy dissipation curve as a function of temperature for samples 1 and 2. For comparison, the dissipation of sample 3, not loaded with hydrogen, is also shown. While the dissipation in the unhydrogenated sample decreases monotonically down to 80 K, the H containing samples display a well pronounced maximum at about 130 K. The peak height increases with increasing H content while its peak temperature remains nearly unchanged. The peak appears to be stable, because the cooling curves, measured from room temperature with a rate of ~0.4 K, are retraced on heating even after aging several hours at 80 K.

A detailed analysis of the present relaxation process has been carried out in Ref.5, and the main results are as follows: i) the peak is thermally activated with an activation energy $W = (0.22\pm0.01)$ eV, and has been attributed to the reorientation of H around B; ii) the obtained value of W is in excellent agreement with the theoretical estimate for H trapped in BC position [1] and with experimental results from infrared spectroscopy [2] and elastic energy absorption [4]; iii) the relaxation process is characterized by a single relaxation time, whose pre-exponential factor is $\tau_0 = (8\pm4)\times10^{-14}$ s; this value is similar to that resulting from Ref. 4, but 3 orders of magnitude higher than that derived from infrared measurements under uniaxial stress [2]; iv) the elastic interactions between the H-B complexes do not appear to affect the H jumps within experimental error, and an upper limit of less than 10 meV is set for the random shifts of the H site-energies due to such interactions. As a consequence of that, the cause of the anomalous isotope effect observed for the jump rate of H around B [2], should be searched in a mechanism different from the strain coupling between complexes.

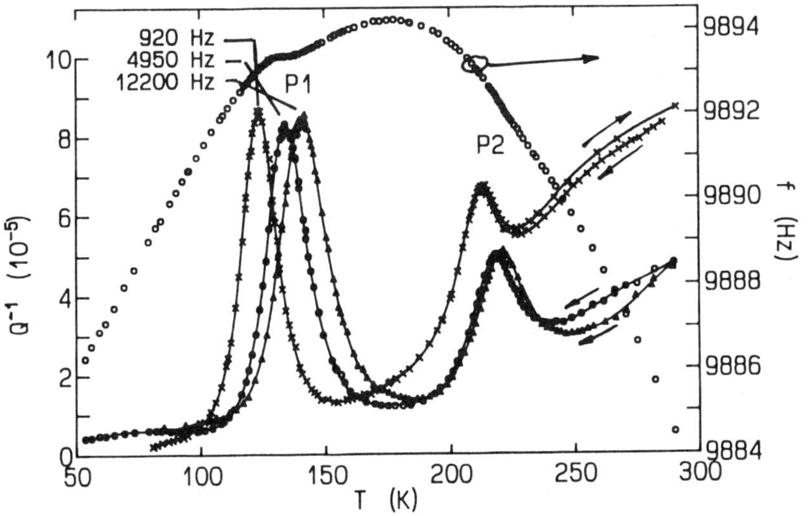

Fig. 2 Elastic energy dissipation and resonant frequency of the Si sample 4 doped with 6×10^{19} B atoms/cm^3 and hydrogenated in a plasma atmosphere. The three curves correspond to the 1st, 2nd and 5th vibration mode.

- Other cluster configurations?

The effects of H in Si:B at higher concentrations of H and B have also been examined. The relaxation spectrum of sample 4, measured at three different frequencies after hydrogenation, is presented in Fig.2. The peak at lower temperature (here labelled P1) is the same of Fig.1, with the same values of W and $\tau(T)$. The elastic modulus, which is proportional to the square of the vibration frequency f, has a maximum around 180 K; its inflection in correspondence to the dissipation peak is the so called "modulus defect" predicted by the general theory of relaxation [8].

The additional peak around 210 K (labelled P2) presents some instability, because the heating curve does not retrace the preceding cooling one. In addition, while the 3rd vibrational mode has the peak shifted with respect to the 1st one, very little shift occurs for the 5th mode. These features might be indicative of an instability of the possible complex involved, and/or of a phase transformation.

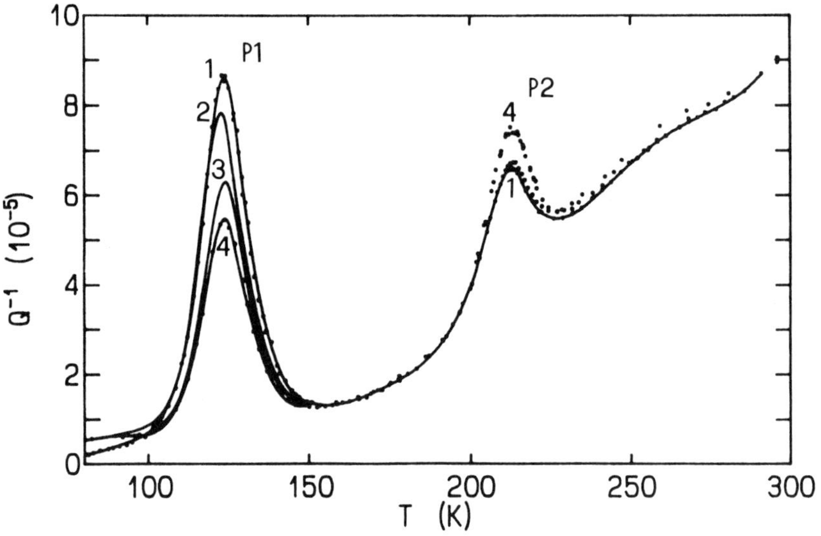

Fig. 3 Q^{-1}curves of sample 4. After subsequent thermal treatments at 250 °C in 100 mbar air, the height of P1 decreases and that of P2 increases.

In the attempt to understand the nature of peak P2, and to study the role played by H (if any) in the process, a sequence of annealing treatments at 250 °C was undertaken on sample 4. Vibrational spectroscopy measurements [9] in samples treated in a similar way have suggested that, at 250 °C, formation of complexes more stable than the B-H pair (e.g. H_2 molecules) may become competitive with effusion of H from the sample. Sample 4 was then first subjected to a thermal treatment at 230 °C for 10 min (including the temperature rise time) in an air atmosphere of 100 mbar: after this annealing the height of P2 slightly increased whilst that of P1 remained unchanged (result not shown here). Instead, three subsequent annealings at 250 °C for 30 min in air as above, caused a progressive decrease of the height of P1 (by 15% each step) and concurrently an enhancement of that of P2 (by smaller amounts), as shown in Fig. 3.

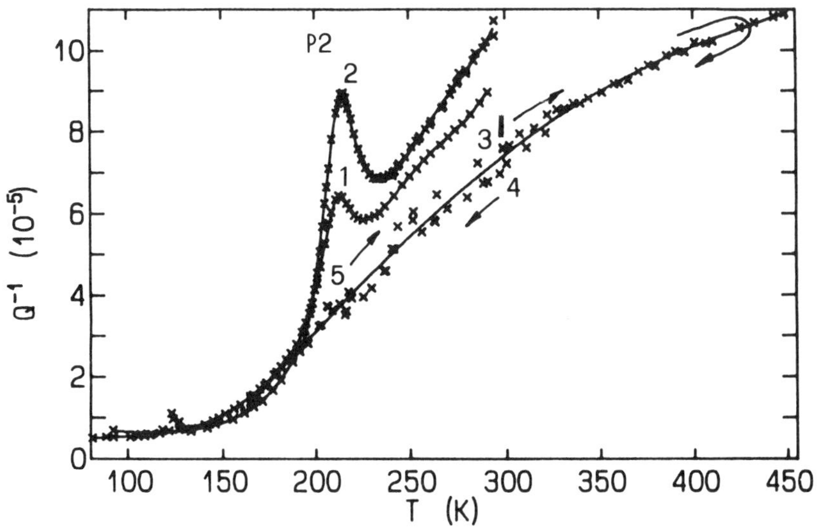

Fig. 4 Elastic energy dissipation of unhydrogenated Si sample 5 doped with 6x19[19] B at./cm[3]. Curve 1: cooling measurement in the "as received" state. Curve 2: cooling after annealing at 600 °C in vacuum for 90 min. Curve 3, 4, 5: Cycle (300-450-160-300) K after cooling of curve 2.

The correlation between the decrease of P1 with the increase of P2 upon annealing might suggest that P2 is the signature of the formation of H_2 molecules. However, this assignement can be ruled out on the ground that peak P2 is also present in the unhydrogenated sample 5, as shown in fig. 4, curve 1: the absence of peak P1 in the Q^{-1} curves confirms that in sample 5 the B-H population is absent, within the limit of experimental error. Therefore, peak P2 seems not to be directly related to H.

A thermal treatment of sample 5 at 600 °C in vacuum for 90 min remarkably increased the height of peak P2, as seen from curve 2. However, the subsequent measurement cycle (300-450-160-300) K revealed, surprisingly, the disappearance of the peak, besides a steadily lower background dissipation (curves 3-5). This low temperature-induced suppression of P2 was not observed in the hydrogenated sample 4 :this indicates that H might have the effect of stabilizing this peak.

Because a dissipation peak at about 220 K has been reported in silicon implanted with oxygen [10], the question arises whether the annealing in air at 250 °C of sample 4 or in vacuum at 600 °C of sample 5, caused oxygen incorporation and consequent formation of O complexes responsible for P2. However, as P2 disappeared after aging at 80 K, this possibility is unlikely, although an involvement of oxygen already present from the manifacturing process might not be excluded.

Finally, the fact that a weak, poorly reproducible P2 peak has been observed in samples 1 and 2, which have a lower B concentration with respect to sample 4, also seggests an involvement of B atoms in this complex.

In conclusion, peak P1 is a single time relaxation process attributed to the stress-induced reorientation of H in BC positions around B. Peak P2 displays very unusual features and might be due to some unstable complexes or to a phase transformation. Its existence is pointed out here, but further experiments are needed to understand its nature.

REFERENCES

1. P.J.H. Denteneer, C.G. Van de Walle, and S.T. Pantelides, Phys. Rev. B **39**, 10809 (1989).
2. M. Stavola, K. Bergman, S.J. Pearton, and J. Lopata, Phys. Rev. Lett. **61**, 2786 (1988).
3. C.P. Herrero and M. Stutzmann, Solid State Comm. **68**, 1085 (1988).
4. B.S. Berry and W.C. Pritchet, J. Appl. Phys. **67**, 3661 (1990).
5. G. Cannelli, R. Cantelli, M. Capizzi, C. Coluzza, F. Cordero, A. Frova, and A. Lo Presti, submitted to Phys. Rev. B.
6. C.P. Herrero, M. Stutzmann, and A. Breitschwerdt, Phys. Rev. B **43**, 1555 (1990).
7. M. Stutzmann, Phys. Rev. B **35**, 5921 (1987).
8. A.S. Nowick and B.S. Berry, "Anelastic Relaxation in Crystalline Solids" (Academic Press, New York, 1972).
9. M.Stutzmann, W.Beyer, L.Tapfer, and C.P.Herrero, Physica B **170**, 240 (1991).
10. B.S. Berry, The Physics Teacher **21**, 435 (1983)

laterials Science Forum Vol. 83-87 (1992) pp. 15-20

KINETICS OF IN-H AND CD-H COMPLEXES IN SILICON

M. GEBHARD, B.VOGT, AND W. WITTHUHN
Physikalisches Institut der Universität Erlangen-Nürnberg,
8520 Erlangen, Germany

Abstract

In-H and Cd-H complexes are studied by perturbed angular correlation (PAC) spectroscopy, using the probe atom $^{111}In/^{111}Cd$ in hydrogenated silicon. The thermal reactivation kinetics of passivated In is investigated by direct observation of the fraction of close In-H pairs. From second order kinetics the dissociation energy $E_{In\text{-}H}$ = 1.46 eV has been deduced. The internal kinetics of hydrogen in the Cd-H complex is monitored directly by the relaxation of the PAC signal. These measurements give evidence for a localized motion of hydrogen and deuterium in the Cd-H and Cd-D complex, respectively, with an activation energy of E_A=0.21eV.

1. Introduction

Hydrogen-related complexes in semiconductors have been studied extensively [1]. Formation of close acceptor-hydrogen complexes causes the well known acceptor passivation phenomena. Here, especially the annealing kinetics and the internal dynamics of these complexes are of interest. Most of the studies are related to the annealing kinetics of B-H complexes [2-4]. For the In-H complex C/V measurements have found a dissociation energy of 1.42±0.05 eV given by a first order kinetics [5]. In this contribution we report on the isothermal annealing kinetics of the In-H complex observed by PAC spectroscopy to determine the order of kinetics and the dissociation energy. A hydrogen atom in next vicinity to the probe atom $^{111}In/^{111}Cd$ produces an electric field gradient (EFG) at the probe atom. These hydrogen induced EFG's are detected by the PAC spectroscopy at the ^{111}Cd probe level, which is populated by the electron capture decay of ^{111}In. Therefore trapping and detrapping of the hydrogen are governed by the In, whereas the EFG is detected at the Cd. Dynamical processes by fluctuating EFG's are thus observed at the Cd. Concerning the internal kinetics of hydrogen and deuterium in the Cd-H and Cd-D complex we have investigated the thermally activated motion of H and D indicated by the relaxation of the PAC spectra.

2. Experimental procedures

In this study samples of crystalline CZ-silicon with $[In]=1.2\cdot10^{16}cm^{-3}$ or $[B]=3...5\cdot10^{18}cm^{-3}$ were used. For PAC measurements the samples were implanted with $^{111}In^{+}$ ions. The energy was 400 keV and the dose was about $5\cdot10^{12}cm^{-2}$. The calculated implantation depth beneath the surface is 160 nm and a standard deviation of 45 nm leads to a maximum In concentration of about $5\cdot10^{17}cm^{-3}$. The samples were furnace annealed at 800 K, 1000 K, and 1200 K for ½ h to remove the radiation damage and to electrically activate the In impurities. The hydrogenation or deuteration was carried out in a dc glow-discharge tube at 400 K up to 540 K for ½ h at a pressure of 5 mbar. The samples were positioned 4 cm downstream off the plasma, in order to avoid plasma induced defects.

The observed coincidence spectra are combined by a standard procedure to the PAC time spectra : $R(t) \approx \Sigma_i f_i \cdot G_i(t)$. The fraction f_i corresponds to the relative number of probe atoms, which have formed the same complex. In hydrogenated silicon up to three different hydrogen specific complexes, according to three different structural and/or electronic configurations of the complexes have been observed. The three complexes are characterized by the frequency $\nu_Q=e\cdot Q\cdot V_{zz}/h$ with the values ν_{Q1} = 270 MHz (at 100 K), ν_{Q2} = 348 MHz (at 295 K) and ν_{Q3} = 483 MHz (at 100 K) [7,8], with V_{zz} the largest component of the diagonalized EFG tensor and Q the nuclear quadrupole moment. The perturbation factor $G_i(t)$ for static and for dynamic hyperfine interaction is described in detail in refs. [9] and [10], respectively.

3. Annealing kinetics of In-H complexes

In this contribution the formation and dissociation of In-H complexes has been investigated by an isothermal annealing program. The hydrogenated In doped sample was isothermally annealed at T_A = 408 K. The PAC measurements were always performed at room temperature. We observed the complexes characterized by ν_{Q2} and ν_{Q3}. For quantitative analysis the sum f of the fractions of these two complexes is discussed. Figure 1a shows the logarithmic plot of this sum versus annealing time t_A. After a fast initial decrease within about 180 min, there is an only slow decrease of f for longer times. The data are analyzed in the framework of a model proposed by Sah et al. [2]. This model has been sucessfully applied to the annealing kinetics of acceptors in oxidized Si after keV electron irradiation or avalance electron injection [3,2] and also to the recovery kinetics of B acceptors in Si [4]. The model involves the following two important processes:

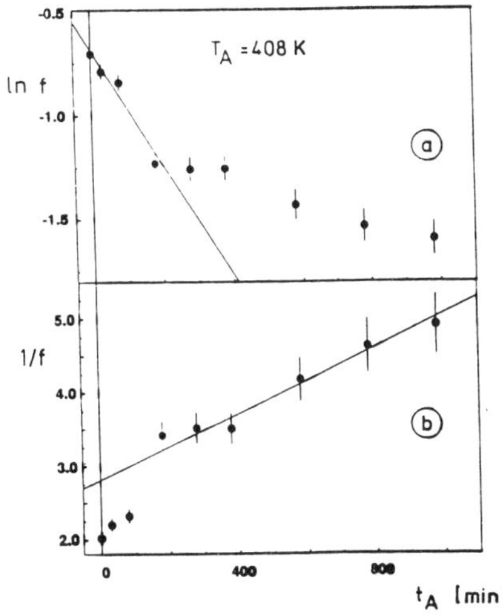

Fig. 1: Sum f of the fractions of the ν_{Q2} and ν_{Q3} complex observed in an In doped sample as function of the annealing time: a) logarithmic plot b) inverse of f. The solid lines represent fits of equation (1) and (2) to the data.

Formation and dissociation of the In-H complex: In + H ↔ In-H;

Formation and dissociation of H molecules: H + H ↔ H_2.

Sah et al. assume in their model that the emission and capture rates of hydrogen at the acceptor are larger than the hydrogen molecule formation rate. The acceptor-hydrogen processes influence the reactivation rate at least in the initial annealing period. Therefore the thermal reactivation behavior can be separated into two distinct phases:

(i) It is assumed, that the atomic hydrogen concentration H remains constant in the initial annealing period with $t_A \leq t_0$. This leads to a first order differential equation for the acceptor-hydrogen complex. With f denoting the fraction of the In-H complex relativ to the substitutional In concentration the solution is:

$$f(t_A) = (f_0 - f_\infty)\cdot\exp\{-[\nu_{In}+\sigma_{In}\cdot H(0)]\cdot t_A\} + f_\infty \quad , \qquad (1)$$

where ν_{In} is the dissociation rate of the In-H complex, σ_{In} is the capture rate for hydrogen at the acceptor, and f_0 and f_∞ are the initial and steady state fractions of the In-H complex, respectively. Therefore an initial exponential decay of the fraction f is expected.

(ii) For longer annealing times ($t_A > t_0$) the In-H formation and dissociation process is in a quasi steady state and hydrogen molecule formation and dissociation governs the reactivation process. If the capture rate σ_H of the molecule formation is large compared to the corresponding emission rate and the final H concentration is assumed to be zero, the differential equation of the In-H complex is of second order and the solution is:

$$1/f = 1/f_0 + \nu_{In} \cdot (\sigma_H/\sigma_{In}) \cdot t_A \quad , \qquad (2)$$

where $\nu_{In} \cdot (\sigma_H/\sigma_{In})$ is the second order annealing rate.

The data of figure 1a confirm the feature of Sah's model of two distinct phases: the fraction decreases fast in the initial annealing period ($t_A \leq t_0$) and slowly for times $t_A > t_0$. In figure 1b the inverse of the In-H fraction is plotted versus the annealing time. The solid line represents the fit of equation (2) to the data recorded for times $t_A > t_0 = 180$ min. The second order annealing rate is determined from the slope of the line to $\nu_{In} \cdot (\sigma_H/\sigma_{In}) = (3.7 \pm 0.4) \cdot 10^{-5}\ s^{-1}$. The assumption made in the model provides $\sigma_H/\sigma_{In} \leq 1$. Indeed, Sah et al. yield in their work on avalance electron injected samples in a large temperature region an almost constant rate of $\sigma_H/\sigma_B \approx 0.5$ for B doped and $\sigma_H/\sigma_{Ga} \approx 0.7$ for Ga doped samples, respectively [2]. With the assumption $\sigma_H/\sigma_{In} \approx 0.5$ and an attempt frequency $\nu_0 = 8.4 \cdot 10^{13}\ s^{-1}$ [5] the thermal activation energy of hydrogen emission at the In acceptor $E_{In\text{-}H} = -k \cdot T_A \cdot \ln(\nu_{In}/\nu_0)$ results to $E_{In\text{-}H} = 1.46 \pm 0.03$ eV. This is in excellent agreement with the energy given by the first order kinetics in reverse bias annealed samples obtained by C/V measurements $E_{In\text{-}H} = 1.42 \pm 0.05$ eV [5]. However, one has to keep in mind that the parameters ν_0 and σ_H/σ_{In} are uncertain within one order of magnitude. The reduction of each of the parameters by a factor of 10 would reduce the energy $E_{In\text{-}H}$ by 0.1 eV. PAC measurements of an isochronal annealing program under the assumption of $\sigma_{In} = 0$ and $\nu_0 = 10^{13}\ s^{-1}$ yield an upper limit of the energy of $E_{In\text{-}H} = 1.3$eV [6].

4. Localized motion of H in Cd-H complexes

Stavola et al. have found by the relaxation of stress-induced dichroism a thermally activated motion of hydrogen in the B-H complex with an activation energy of $E_A = 0.19$ eV [11]. For the acceptor B calculations of Denteneer et al. result in a barrier height of 0.2 eV for the motion of H about B from bond-center site to bond-center site [12]. Peale et al. conclude from their work on stress dependence of Be-H and Be-D complexes, that these complexes either tunnel or rotate between equivalent equilibrium orientations along <111> directions [13].

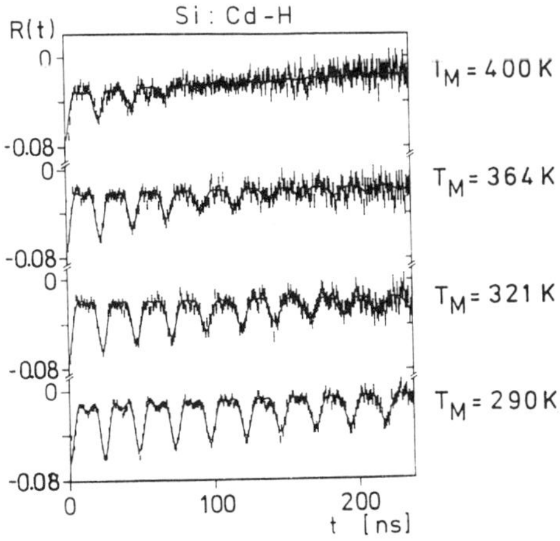

Fig. 2: R(t) spectra recorded at different measuring temperatures for a hydrogenated sample doped with $[B] = 5 \cdot 10^{18} cm^{-3}$.

In the investigation of the internal kinetics of Cd-H complexes by PAC we found direct evidence of a locally restricted motion of hydrogen and deuterium at Cd acceptors (for details see ref.[14]). This study has been restricted to the Cd-H complex with the smallest frequency ν_Q = 270 MHz, which is observed especially in heavily boron doped samples. The relaxation of the modulation in the PAC spectra yields information about this locally restricted motion.

Figure 2 shows the PAC spectra of a hydrogenated Si sample ($[B] = 5 \cdot 10^{18} cm^{-3}$) at different measuring temperatures. At T_M=290 K the modulation is slightly damped. For increasing temperatures an increasing relaxation is visible. The same behavior is observed for Cd-D complexes. The relaxation of the PAC signal has been taken into account by an exponential factor, depending on the fluctuation rate ν of H/D in the Cd-H/D complex [10].

In the present case the PAC spectroscopy allows the observation of fluctuation rates between some 100kHz and 10MHz. In figure 3 the rate ν, which is the inverse of the lifetime of H/D in one of the four crystallographically equivalent sites (see insert), is plotted versus reciprocal temperature. The filled circles correspond to hydrogenated- and the open squares to deuterated samples. The slopes of the lines gives an activation energy of 0.21±0.03 eV for the locally restricted motion of H as well as of D. In the whole

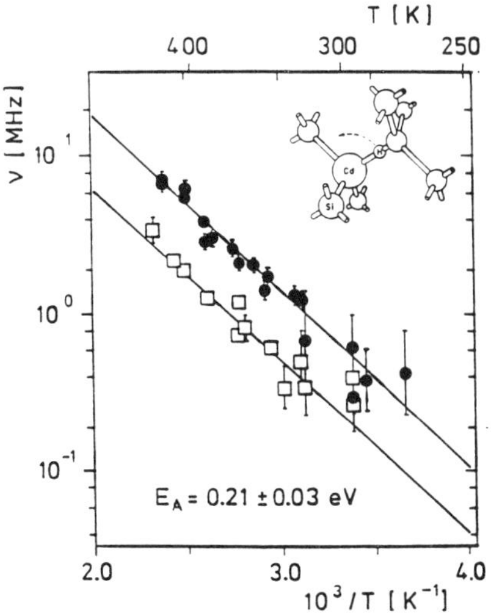

Fig. 3: Arrhenius plot of the fluctuation rate ν. Filled circles correspond to hydrogenated, the open squares to deuterated samples. The insert illustrates the H/D motion about the Cd.

temperature range the fluctuation rate of D is smaller than of H. The attempt frequencies are determined to be in the order of 10^9 s^{-1}, which is about 4 orders of magnitude smaller than a typical phonon frequency. In the discussion of a classical motion versus a thermally activated tunneling process [15] the low fluctuation rates favours the interpretation in terms of a tunneling process.

Acknowledgements

The In implantations were kindly performed by Dr. M. Uhrmacher (Universität Göttingen, II. Physikalisches Institut). This work was financially supported by the Bundesminister für Forschung und Technologie.

References

[1] J.I. Pankove, in: "Hydrogen in Semiconductors", eds. J.I. Pankove and N.M. Johnson (Academic Press, 1990).
[2] C.T. Sah, S.C.S. Pan and C.C.H. Hzu, J. Appl.Phys. 57, 5148 (1985).
[3] S.C.S. Pan and C.T.Sah, J. Appl. Phys 60, 156 (1986).
[4] T. Zundel, A. Mesli, J.C. Muller, and P. Siffert, Appl. Phys. A 48, 31 (1989).
[5] T. Zundel and J. Weber, Phys. Rev B 39, 13549 (1989).
[6] M. Deicher, G. Grübel, R. Keller, E. Recknagel, N. Schultz, H. Skudlik, and Th. Wichert, Mater. Sci. For. 38-41, 1045 (1989).
[7] Th. Wichert, H. Skudlik, M. Deicher, G. Grübel, R. Keller, E. Recknagel and L. Song, Phys. Rev. Lett. 59, 2087 (1987).
[8] M. Gebhard, N. Achtziger, A. Baurichter, D. Forkel, B. Vogt and W.Witthuhn, Physica B 170, 320 (1991).
[9] H.H. Rinneberg, Atomic Energy Rev. 172, 477 (1979).
[10] H. Winkler and G. Gerdau, Z. Phys. 262, 363 (1973).
[11] M. Stavola, K. Bergman, S.J. Pearton, and J. Lopata, Phys. Rev. Lett. 61, 2786 (1988).
[12] P.J.H. Denteneer, C.G. Van de Walle, and S.T. Pantelides, Phys. Rev. 39, 10809 (1989).
[13] R.E. Peale, K. Muro, and A.J. Sievers, Phys. Rev. B 41, 5881 (1990).
[14] M. Gebhard, B.Vogt, and W.Witthuhn, to be published.
[15] C.P. Flynn and A.M. Stoneham, Phys. Rev. B 1, 3966 (1970).

Materials Science Forum Vol. 83-87 (1992) pp. 21-26

DIFFUSION AND DRIFT OF HYDROGEN IN Si AND GaAs

A. W. R. LEITCH*, T. ZUNDEL, Th. PRESCHA, and J. WEBER
Max-Planck-Institut für Festkörperforschung, Heisenbergstr. 1, 7000 Stuttgart 80, Federal Republic of Germany

ABSTRACT

The ability of charged hydrogen to drift in Si and GaAs has been used to study two properties: Firstly, the effective hydrogen diffusion coefficient D_{eff} is accurately determined in B-doped Si for temperatures in the range from 60°C to 140°C . The values of D_{eff} rigorously satisfy the Arrhenius equation $N_A D_{eff} = P_0\, exp(-E_P/kT)$ with $E_P = 1.28$ eV and $P_0 = 1 \times 10^{20}$ $cm^{-1}s^{-1}$. The diffusion process in the studied temperature regime is entirely trap-limited and shows no dependence on the diffusivity of the free hydrogen. Secondly, by drifting the H away from the donors in hydrogenated n-type GaAs, the reactivation of the passivated donors is studied. The thermal dissociation of the donor-hydrogen complex obeys first-order kinetics, with a dissociation energy $E_{SeH} = 1.52$ eV.

1. Introduction

The diffusion and drift of hydrogen in both Si and GaAs is a subject of considerable current interest.[1,2] When in a charged state, hydrogen readily drifts under the influence of an applied electric field at elevated temperatures.[3,4,5,6] The electric field is established in the semiconductor by applying a reverse bias to a fabricated Schottky diode. By means of capacitance-voltage (C-V) profiling measurements using the same Schottky diode, the electrical activity of the carriers within the semiconductor can be monitored. This *in situ* technique can therefore be used to yield valuable information regarding not only the drift and diffusion of hydrogen, but also the dissociation kinetics of the dopant-hydrogen complexes formed after hydrogenation.

In this paper we have used this technique to investigate two relevant issues pertaining to hydrogen in Si and GaAs: Firstly, the effective hydrogen diffusion coefficient D_{eff} is accurately determined in B-doped Si for temperatures in the range from 60°C to 140°C . Secondly, the kinetics governing the thermal reactivation of the passivated donors in hydrogenated n-type GaAs is determined.

2. Diffusion of H in B-doped Si

An RF (13.6 MHz) driven plasma was used to introduce hydrogen into B-doped, (100) float-zone Si (0.47 Ωcm or 10 Ωcm), at 130°C for 2 h. Ti Schottky contacts (1 mm diam) were then evaporated onto the hydrogenated surface. The Ohmic contact consisted of an In-Ga alloy scratched onto the back surface. C-V measurements at room temperature were used to profile the net electrically active boron concentration, $N_I(x) = N_A - [BH] - [H^+]$, where N_A is the boron concentration in the nonhydrogenated control sample, [BH] and [H^+] the concentrations of electrically inactive BH-complexes and untrapped, positively charged hydrogen, respectively. Because the concentration of [H^+] is always much smaller than [BH], as will be shown in Ref. 7, we can obtain the profile of [BH] from $N_A - N_I(x)$. The annealing experiments were performed in the dark and under ambient atmosphere.

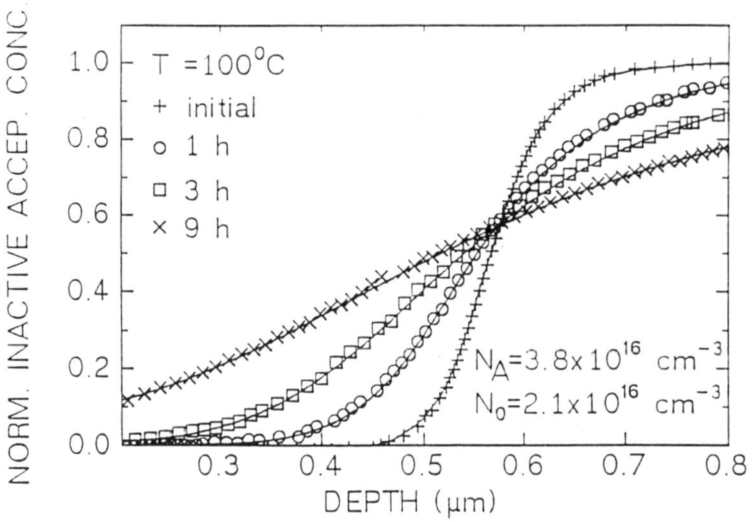

FIGURE 1: Normalized inactive boron profiles after zero bias annealing for increasing times t_D at 100°C. The symbols N_A and N_0 denote the total boron concentration and the normalization factor. The solid lines represent the fit to Eqs. (1) and (2) with $D = 1 \times 10^{-12}$ cm²/s and $R = 3.5$ nm.

The H plasma treatment of the Si successfully passivates the B-acceptors to a depth of $\simeq 2$ μm. We then generate a sharp H-concentration gradient in the *bulk*, by annealing under reverse bias ($V_R = 9$ V) at 140°C for 10 h followed by 100°C for 24 h. The applied electric field causes the thermally dissociated hydrogen, which will be in the positively charged state, to drift deeper into the material, where new BH complexes are formed.[4] The resulting profile $N_A - N_I(x)$ (Fig. 1, curve labelled "initial") is characterized by a layer (x <0.5 μm) free of BH-complexes and a uniform BH-complex concentration ($N_0 = 2.1 \times 10^{16}$ cm^{-3}) in the region $x > 0.7$ μm. The BH concentration strongly increases in the transition region 0.5 μm $< x <$ 0.7 μm which coincides with the edge of the space charge region ($W = 0.57$ μm) in the biased diode. This gradient defines a well defined initial state for subsequent diffusion from the bulk towards the surface.

The diffusion experiment is initiated by annealing at 100°C under *zero* bias for a time t_D, after which the sample is rapidly cooled to room temperature for the *C-V* measurement. The BH-profiles are shown in Fig. 1 for different values of t_D. As t_D increases, the BH-concentration progressively decreases in the region $x > 0.57$ μm. In the region $x < 0.57$ μm an increase of [BH] occurs; the sharp initial profile is thus smoothened. All profiles cross at the point defined by $x = 0.57$ μm and $[\mathrm{BH}]/N_0 = 0.57$.

The profiles in Fig. 1 are explained in the following way: The BH complexes partially dissociate, and the hydrogen diffuses toward the defect free surface layer where it is trapped by the free boron atom. The free H-concentration satisfies the modified diffusion equation

$$\frac{\partial[H]}{\partial t} = D\frac{\partial^2[H]}{\partial x^2} - \frac{\partial[BH]}{\partial t} \tag{1}$$

where D is the H-diffusion coefficient in the absence of a trap (intrinsic diffusivity). The term $\partial[BH]/\partial t$ accounts for the trapping and detrapping of the H-atom at the boron acceptor and is given by

$$\frac{\partial[BH]}{\partial t} = \sigma(N_A - [BH])[H] - \nu[BH], \tag{2}$$

where σ is the capture rate of H at the B-atom and ν the dissociation frequency of the BH-complex. The parameter σ is related to the collision radius R and the diffusion coefficient D according to $\sigma = 4\pi RD$. The dissociation frequency ν has been accurately determined in Ref. 4 and is given by the Arrhenius equation $\nu = \nu_0\, exp(-E_{BH}/kT)$ with $\nu_0 = 2.8 \times 10^{14}$ s^{-1} and the dissociation energy of the BH complex $E_{BH} = (1.28 \pm 0.03)$ eV.

The profiles [H](x) and [BH](x) are calculated by solving Eqs. (1) and (2) numerically. As boundary conditions, we assume that the medium is impermeable to diffusion at the planes $x = W(0\ \text{V}) \simeq 0.15\ \mu$m and $x = x_0$ with $x_0 = 2.5\ \mu$m . Further details are given in Ref. 7. Using the total hydrogen concentration $H_T(T)$=[H]+[HB] calculated from Eqs. (1) and (2), we fit the experimental profiles of Fig. 1 by adjusting the two parameters D and R. The profiles calculated for $D = 1 \times 10^{-12}$ cm^2/s and $R = (3.5\pm0.5)$ nm (Fig. 1 solid lines) are in very good agreement with the experimental curves. It is impossible to obtain a good fit for D lower than about 4×10^{-13} cm^2/s. For D larger than 1×10^{-12} cm^2/s the calculated profiles show a negligible dependence on D but are very sensitive to changes in R. Therefore, the fit provides only a *lower limit* D_l of the diffusivity D which is in the range from 4×10^{-13} cm^2/s to 1×10^{-12} cm^2/s. The fit provides a constant value of R [$R = (3.5\pm0.5)$ nm] for any value of D larger than D_l.

An analytical solution of Eqs. (1) and (2), which supports the numerical results, allows us to define an effective diffusion coefficient D_{eff}, given by[7]

$$D_{eff} \approx \frac{\nu}{4\pi RN_A}, \tag{3}$$

which does *not* depend on the intrinsic diffusivity D, i.e. the diffusion process is entirely controlled by trapping and detrapping of the H atom at the boron acceptor. D_{eff} can be accurately determined from the experimental profiles (Fig. 1), by measuring the slope at $x = W$. Equation (3) predicts that D_{eff} is inversely proportional to N_A and is thermally reactivated with an activation energy $E_{BH} = 1.28$ eV. In order to verify this equation, the diffusion experiment was also performed on the same material at T=120°C and T=140°C, yielding qualitatively similar results to those in Fig. 1. We have plotted the product $P = N_A D_{eff}$ as a function of $1/T$ in Fig. 2 for both $N_A = 3.8 \times 10^{16}$ cm^{-3} and $N_A = 1.4 \times 10^{15}$ cm^{-3}. The values of P are well fitted to the Arrhenius equation $P = P_0\, exp(-E_P/kT)$ with a prefactor $P_0 = (1^{+1}_{-0.5}) \times 10^{20}$ cm^{-1}s^{-1} and an activation energy $E_P = (1.28 \pm 0.03)$ eV.

Thus D_{eff} is inversely proportional to N_A and shows the same temperature dependence as the BH-complex dissociation frequency ν. Under the experimental conditions of this work ($T \leq$140°C, $N_A \geq 1 \times 10^{15}$ cm^{-3}), the diffusion process is entirely controlled by the trapping and detrapping of H at the boron atom. In contrast to previous studies,[8,9,10] the method used in this present work allows us to separate the hydrogenation treatment from the diffusion experiment. If thermal equilibrium prevails, the free electron concentration is negligible, and H is predominantly in the positive charge state [11]. Due to the coulomb interaction between H^+ and B^-, the trapping of H at the boron atom is very effective, in agreement with the quite large collision radius R=3.5 nm.

3. Drift of H in Se-doped GaAs

To study the kinetics governing the thermal reactivation of the passivated donors in hydrogenated n-type GaAs, we used Se-doped ($n = 1 \times 10^{17}$ cm^{-3}) GaAs, grown by metalorganic

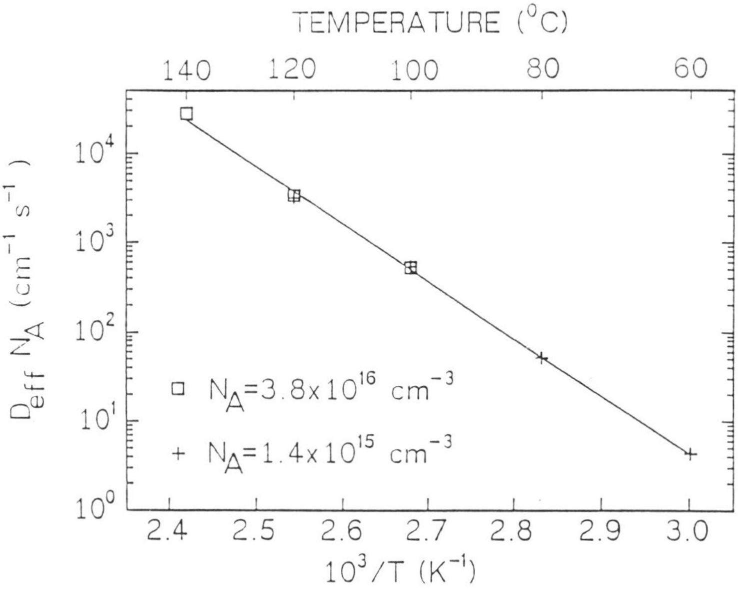

FIGURE 2: Arrhenius plot of $P = N_A D_{eff}$. The solid line represents the fit to the equation $P = P_0\, exp(-E_P/kT)$ with $P_0 = 1 \times 10^{20}$ $cm^{-1}s^{-1}$ and E_P=1.28 eV.

vapor phase epitaxy (MOVPE) on undoped semi-insulating GaAs. Au-Ge/Ni Ohmic contacts were first formed on corner sections of the epilayer surface, after which hydrogen was introduced in the material using a DC plasma system[12] for 2 h at 180°C . Au Schottky contacts (0.5 mm diam) were then evaporated onto the exposed surfaces.

Fig. 3(a) shows a series of carrier concentration profiles, $N_I(x)$, after annealing at 172°C for increasing times while applying a reverse bias $V_r = 2$ V across the Schottky contact. After each annealing time t, the temperature was rapidly decreased by immersing the sample holder in liquid nitrogen, while still applying the reverse bias. For increasing annealing times t, there is a progressive reactivation of the Se donors within the high-field region ($x \leq 0.15$ μm) of the space charge layer. After 200 min, the majority of the Se donors closest to the Schottky contact have been reactivated, leading to a restoration of the original carrier concentration of the GaAs. By comparison, we find an increased passivation of the Se donors in the lower and zero field regions ($0.2 < x < 0.3$ μm).

The results seen in Fig. 3(a) are adequately described in terms of the drift of a negatively charged hydrogen species: Annealing the passivated GaAs at 172°C results in the thermal dissociation of the electrically neutral SeH complex according to $(SeH)^0 \rightarrow Se^+ + H^-$. Considering the high electric-field region ($x < 0.15$ μm) of the reverse-biased Schottky diode, the released H^- will drift deeper into the material, away from the Schottky contact. Thus in the region $x > 0.2$ μm, the increase in the local concentration of H^- results in the increased passivation of the Se donors. This will readily occur, provided there are sufficient trapping centres (ionized Se donors) for the drifting H^-. The profiles of Fig. 3(a) therefore provide further evidence of the presence of H^- in n-type GaAs.[13]

Similar to the case of the BH complex in B-doped Si, the formation and dissociation of the SeH complex may be described by an expression having the same form as Eq. (2). Provided the field is sufficiently high in the region $x < 0.15$ μm for the local hydrogen concentration to be negligible, and provided there is no indiffusion of hydrogen from the surface region, Eq. (2)

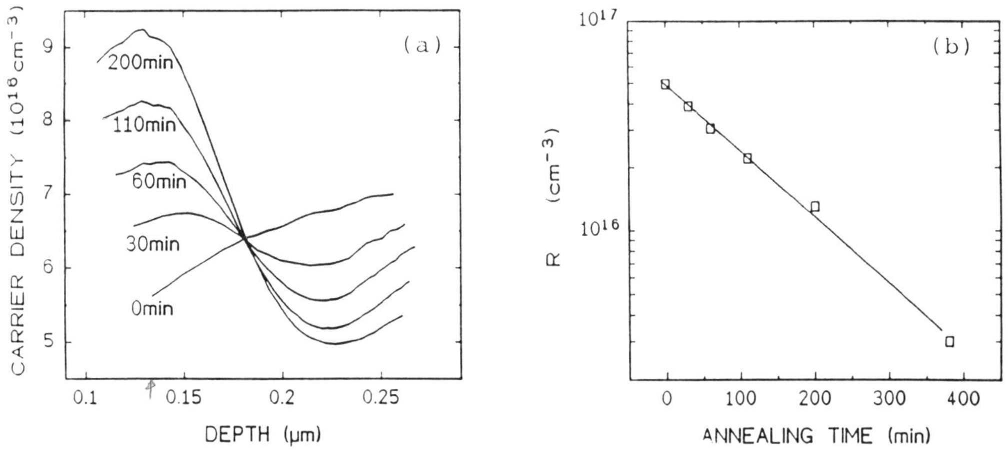

FIGURE 3: Reverse bias ($V_r = 2$ V) annealing of hydrogenated Se-doped GaAs at 172°C, for increasing times. (a) Carrier concentration profiles; and (b) analysis of the annealing kinetics.

predicts that the dissociation of the SeH complex will obey first-order kinetics. In order to quantitatively analyze the reactivation of the SeH complexes in Fig. 3(a), we have measured the inactive donor concentration $R(t) = N_D - N_I(t)$ at $x = 0.13$ μm. The exponential time dependence seen in Fig. 3(b) confirms that the reactivation of the Se donors follows first-order kinetics, and satisfies the equation $R(t) = R_0 exp[-\nu t]$. By carrying out the annealing experiment at various temperatures within the range 154 - 191°C, we have shown in Ref. 13 that the reactivation of the Se donors always follows first-order kinetics over the entire annealing time, with a dissociation energy $E_{SeH} = 1.52 \pm 0.05$ eV and attempt frequency $\nu_0 = 2 \times 10^{13}$s^{-1}.

In comparison, when annealing under *zero* bias conditions, the absence of the electric field results in a second-order reactivation process being measured. To illustrate this, we show in Fig. 4(a) a series of profiles measured for a similar Se-doped hydrogenated GaAs sample, after annealing at 172°C with zero bias applied to the Schottky contact. The reactivation of the Se donors is seen to be much slower compared with annealing at the same temperature

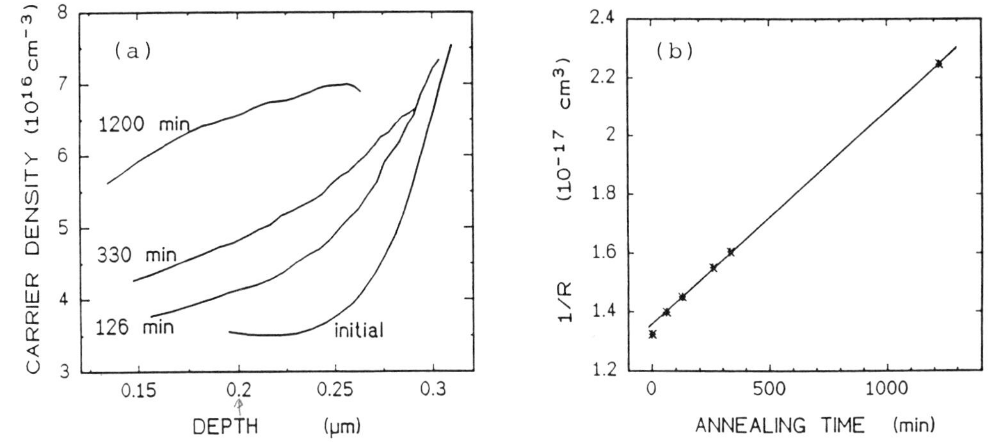

FIGURE 4: Zero bias annealing of hydrogenated Se-doped GaAs at 172°C, for increasing times. (a) Carrier concentration profiles; and (b) analysis of the annealing kinetics.

under reverse bias conditions. By measuring the inactive donor concentration $R(t) = N_D - N_I(t)$ at $x = 0.2$ μm, we obtain a linear relationship only when plotting $R^{-1}(t)$ versus t, as seen in Fig. 4(b).

The results of Fig. 4(b) indicate that the reactivation process can be described by the second-order equation:

$$\frac{\partial R}{\partial t} = -r'R^2, \tag{4}$$

where the second-order annealing parameter $r' = 1.2 \times 10^{-21}$ cm^3/s. This suggests that the reactivation process under zero-bias annealing conditions is rate-limited by the combination of two hydrogen atoms in a molecular species. Such an effect has also been measured when annealing under illumination at temperatures as low as 100°C.[14]

We are grateful to H.J. Queisser for his interest and support throughout this work. We acknowledge stimulating discussions with M. Stutzmann and the technical assistance of W. Heinz and W. Krause.

References

[*] Present Address: Department of Physics, University of Port Elizabeth, P.O. Box 1600, Port Elizabeth 6000, Republic of South Africa.

[1] S. J. Pearton, J. W. Corbett, T. S. Shi, Appl. Phys. A **43**, 153 (1987).

[2] J. Chevallier, B. Clerjaud and B. Pajot, in *Hydrogen in Semiconductors*, edited by J. I. Pankove and N. M. Johnson, Semiconductors and Semimetals Vol. 34 (Academic, New York, 1991), p. 447.

[3] A. J. Tavendale, D. Alexiev, and A. A. Williams, Appl. Phys. Lett. **47**, 316 (1985).

[4] T. Zundel and J. Weber, Phys. Rev. B **39**, 13349 (1989).

[5] A. W. R. Leitch, Th. Prescha, and M. Stutzmann, Appl. Surf. Sci. **50**, 390 (1991).

[6] A. J. Tavendale, S. J. Pearton, A. A. Williams, and D. Alexiev, Appl. Phys. Lett. **56**, 1457 (1990).

[7] T. Zundel and J. Weber, to be submitted.

[8] T. Zundel, A. Mesli, J.C. Muller, and P. Siffert, Appl. Phys. A **48**, 31 (1989).

[9] C.P. Herrero, M. Stutzmann, A. Breitschwerdt, and P.V. Santos, Phys. Ref. B **41**, 1054 (1990).

[10] C. P. Herrero, M. Stutzmann, and A. Breitschwerdt, Mat. Res. Soc. Symp. **163**, 395 (1990).

[11] T. Zundel and J. Weber, Phys. Rev. B **43**, 4361 (1991).

[12] M. Stutzmann and C. P. Herrero, in *The 20th International Conference on the Physics of Semiconductors, Thessaloniki, Greece, 1990,* edited by E. M. Anastassakis and J. D. Joannopoulos (World Scientific, Singapore, 1990), p. 783.

[13] A. W. R. Leitch, Th. Prescha, and J. Weber, Phys. Rev. B **44**, 1375, (1991).

[14] A. W. R. Leitch, Th. Prescha, and J. Weber, accepted for publication in Phys. Rev. B.

Materials Science Forum Vol. 83-87 (1992) pp. 27-32

NEW TRAPS FOR H^0 IN BORON- AND PHOSPHORUS-DOPED Si

Levente KORPAS,† James W. CORBETT,† and Stefan K. ESTREICHER‡

†Physics Department, SUNY, Albany, NY 12222 (USA)

‡Physics Department, Texas Tech University, Lubbock, TX 79409-1051 (USA)

ABSTRACT

We have calculated the equilibrium geometries, electronic structures, and relative stabilities of complexes involving substitutional boron or phosphorus and two H interstitials in Si. Trapping of a second interstitial H at an already H-passivated B or P impurity is energetically preferred over the formation of H^*, H_2^*, or H_2 far away from the dopant. The lowest-energy structures are very likely to be strong traps for (at least) one additional H interstitial.

1. INTRODUCTION

Hydrogen is a highly mobile interstitial impurity in Si and other semiconductors. It interacts strongly with the lattice as well as with lattice defects and impurities. These interactions are for the most part of covalent nature and profoundly affect the properties of the material. In particular, hydrogen passivates the electrical activity of shallow acceptors and donors (for recent reviews, see refs.1-3).

A number of experimental and theoretical studies[1-3] of the {B,H} and {P,H} pairs in Si have led to a general agreement on the ground state properties of these pairs. In the equilibrium configuration, {B,H} has an essentially 3-fold coordinated B atom which overlaps weakly with an Si−H bond. H is located near the bond-centered (*BC*) site, between B and one of its Si nearest neighbors (NNs): B $\cdots$ H−Si. In the {P,H} complex, P is also 3-fold coordinated, but H is bound to a host Si atom at the antibonding (*AB*) site, forming P $\cdots$ Si−H.

However, there is evidence from SIMS[4,5] and hydrogenation[6] data that the concentration of hydrogen substantially exceeds that of the B or P dopants up to a depth of 1 μm or more from the surface of the sample, implying the existence of new trapping centers for H. Real time hydrogenation data[6] show that the density of these traps correlates with the dopant concentration, suggesting that the already passivated {B,H} or {P,H} complexes are trapping centers for additional H interstitials. This trapping occurs at much greater depths than the ones normally associated with platelets.[5]

We have investigated the properties of the {B,H,H} and {P,H,H} complexes and calculated the depths of these traps for H^0 relative to traps in intrinsic material, such as the *BC* site,[7] the H_2^* configuration,[8,9] or the H_2 molecule. The calculations were performed near the *ab-initio* Hartree-Fock (HF) level (PRDDO[10]) in molecular clusters[11] containing 35, 44, and 48 host atoms. PRDDO is an all-electron, self-consistent, and parameter-free method which reproduces accurately the results of *ab-initio* HF calculations at a fraction of the cost. In the case of substitutional P, a set of $3d$ orbitals was included on P and on each of its four Si NNs. In the case of B, only minimal basis set results are available at this time, but we are in the process of double-checking our calculations with larger basis sets.[12] Unrestricted (UHF) wavefunctions were used throughout.

Geometry optimizations, normally including first and second NN relaxations around the complex, were carried out for many possible structures of the two complexes, about a dozen in each case. For some configurations, geometry constraints were used to lower the number of degrees of freedom (15 in some cases).

In the following, the *zero of the energy* corresponds to atomic H far outside the cluster containing the complex under study. A positive energy means that it *costs* energy to take a H atom far away from the cluster and place it in a given optimized configuration inside the cluster. The energy is negative when one *gains* that energy. Our energies are therefore dissociation energies for one H from a given complex, with the refence point corresponding to atomic H far outside the cluster. With this convention, the energy of H at the *BC* site[13] in otherwise perfect Si is $0.25\,eV$. Note that this number corresponds to first and second NNs optimized around H at the *BC* site.

2. THE {B,H,H} COMPLEX

Preliminary results for {B,H,H} have already been published[14] and can be summarized as follows.

(a) The lowest-energy configuration, with energy $-0.61eV$, is shown in Fig. 1. The two H interstitials bridge symmetrically two of the four B–Si bonds. The two Hs form strong bonds with the Si atoms, and only weakly overlap with the B atom. The latter ends up mostly 2-fold coordinated to its remaining 2 Si NNs. The odd electron is localized entirely in a non-bonding orbital on the B atom. Because of the presence of this odd electron, of a two-fold coordinated B atom, and of lattice strain, this complex will almost certainly attract at least one more H interstitial. How strong this new trap really is will be determined in later calculations.

(b) We explored 11 other possible configurations. All of them have energies higher than (or equal to) that of H at the *BC* site (see §1). The lowest one is the {B,H} with the second H at a *BC* site far away ($0.25\,eV$), then are a variety of configurations with the second H at various *BC* or *AB* sites ($0.80\,eV$ to $1.80\,eV$), then several configurations with the second H at various tetrahedral interstitial (T) sites around {B,H}. The highest energies correspond to substitutional B and an

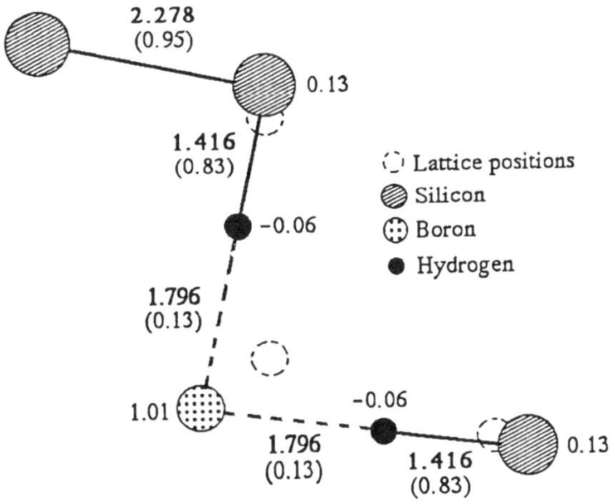

FIGURE 1: Lowest-energy configuration of the {B,H,H} complex. The dashed circles show the perfect lattice sites. The bond lengths are given in Å, and the degrees of bonding are in parenthesis. The atomic spin populations are given next to each atom. The energy of this configuration is $-0.61\ eV$.

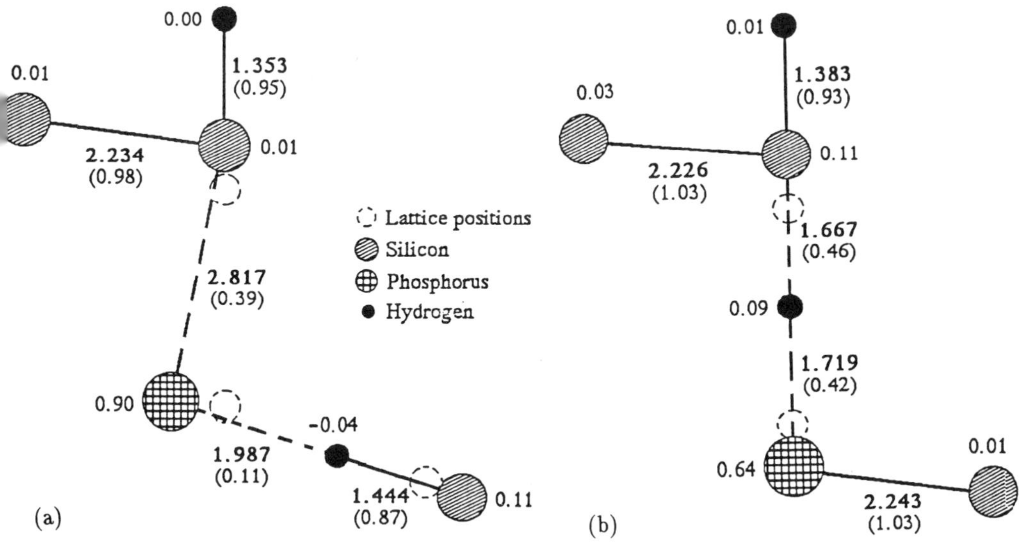

FIGURE 2: Lowest-energy configurations of the {P,H,H} complex. The dashed circles show the perfect lattice sites. The bond lengths are given in Å, and the degrees of bonding are in parenthesis. The atomic spin populations are given next to each atom. The energies of these configurations are $-1.51\ eV$ for (a) and $-1.14\ eV$ for (b).

H_2 molecule at a neighboring T site (3.79 eV) or at a far-away T site (4.03 eV).

Thus, in the case of {B,H,H}, there is only one favorable configuration. It is nearly 0.9eV more stable than trapping the second H at a BC site, and over 4 eV more stable than the formation of an H_2 molecule.

3. THE {P,H,H} COMPLEXES

We found 4 configurations with negative energies (see §1).

(a) The lowest one, with −1.51 eV, is shown in Fig. 2a. It has one H at the AB site of one Si and the other H bridging a Si−P bond involving another Si. In this configuration, P is almost 2-fold coordinated. The odd electron is localized almost entirely in a non-bonding orbital on P. As was the case above (§2a), this configuration is likely to be a trap for a third H interstitial.

(b) The second-lowest configuration, with −1.14 eV, is similar to the one above, but the same Si atom is bound to both H interstitials, forming the linear structure P−H−Si−H (Fig. 2b). The odd electron is more delocalized than was the case above, though most of it remains on P.

(c) Two more configurations have negative energies, although they are substantially higher than (a) or (b) above. Further studies will be required to determine with certainty whether these are true local minima of the potential energy surface (PES) rather than saddle points. The first of these configurations (−0.60 eV) has the two H interstitials symmetrically located, each at the AB site of a Si atom adjacent to P. Again P is 2-fold coordinated and carries the odd electron. The second configuration (−0.17 eV) has one H AB to Si and the second AB to P, all along the same trigonal axis.

(d) Finally, seven more configurations were optimized and are in the range 0.25 eV to 3.32 eV. The highest energy again corresponds to substitutional P and an H_2 molecule.

4. DISCUSSION

(a) In hydrogen passivated, $p-$ or $n-$type Si, the deepest traps for any excess H are the {B,H} and {P,H} pairs themselves. The addition of one H leads to the formation of the {B,H,H} and {P,H,H} complexes shown in Figs. 1 and 2a, respectively. While it *costs* 0.25 eV to form bond-centered H in Si, one *gains* 0.61 eV to form {B,H,H} and 1.51 eV to form {P,H,H}. These results are consistent with the experimental observation[15] that, for equal doping concentrations, more H interstitials are trapped at shallow depths in $n-$type than in $p-$type Si.

(b) The formation of H_2 molecules is ruled out: Substitutional B or P and the H_2 molecule in the most favorable location is the highest-energy configuration of the dozen or so we studied in each case, typically $5\,eV$ above the ground state. As is discussed elsewhere,[12,14] we find that the H_2^* complex is only some $1\,eV$ more stable than the H_2 molecule in Si.

(c) In a recent paper,[16] Stutzmann and Herrero have reported new Si−H stretching modes in samples in which the concentration of H exceeded that of the B dopants. This finding appears to be consistent with our results. Finally, our calculations disagree with a recent model proposed by Bonapasta.[17]

Acknowledgements: JWC acknowledges partial support from the Solar Energy Research Foundation, from IBM, and from the Mobil Foundation. The work of SKE was supported by the grant D-1126 from the Robert A. Welch Foundation and by a grant from the Advanced Technology Program of the State of Texas.

REFERENCES

[1]S.J. Pearton, J.W. Corbett, and T.S. Shi, Appl. Phys. A **43**, 153 (1987).

[2] *"Hydrogen in Semiconductors"*, ed. J.I. Pankove and N.M. Johnson, (Academic Press, 1991).

[3]Report the Panel *"Fundamental Issues in Hydrogen-Defect Interactions"*, sponsored by the US DoE, to be published in Rev. Mod. Phys.

[4]J.W. Corbett, J.L. Lindström, S.J. Pearton, and A.J. Tavendale, Solar Cells **24**, 127 (1988); J.W. Corbett, S.J. Pearton, and M. Stavola, Proc. Int. Conf. Sci. Technol. Defect Control in Semic., ed. K. Sumino, (North-Holland, Amsterdam, 1990), vol.I, p.53; J.T. Borenstein, D. Angell, and J.W. Corbett, MRS Symp. Proc. **138**, 209 (1989).

[5]N.M. Johnson, F.A. Ponce, R.A. Street, and R.J. Nemanich, Phys. Rev. B **35**, 4166 (1987); S.J. Jeng, G.S. Ohrlein, and G.J. Scilla, Appl. Phys. Lett. **53**, 1755 (1988).

[6]C.H. Seager and R.A. Anderson, MRS Symp. Proc. **163**, 431 (1990); C.H. Seager, R.A. Anderson, and D.K. Brice, J. Appl. Phys. **68**, 3268 (1990).

[7]T.L. Estle, S.K. Estreicher, and D.S. Marynick, Hyp. Int. **32**, 637 (1986); S.K. Estreicher, Phys. Rev. B **36**, 9122 (1987).

[8]K.J. Chang and D.J. Chadi, Phys. Rev. Lett. **62**, 937 (1989) and Phys. Rev. B **42**, 7651 (1990); Z.B. Zhang, W.B. Jackson, and D.J. Chadi, Phys. Rev. Lett. **65**, 2575 (1990).

[9]P. Deak and L.C. Snyder, Rad. Effects and Def. in Sol. **111-112**, 77 (1989).

[10]T.A. Halgren and W.N. Lipscomb, J. Chem. Phys. **58**, 1569 (1973); D.S. Marynick and W.N.

Lipscomb, Proc. Nat. Acad. Sci. (USA) **79**, 1341 (1982).

[11]S.K. Estreicher, Phys. Rev. B **37**, 858 (1988) and references therein.

[12]L. Korpas, J.W. Corbett, and S.K. Estreicher, to be published.

[13]C.H. Chu and S.K. Estreicher, Phys. Rev. B **42**, 9486 (1990), table IV.

[14]L. Korpas, J.W. Corbett, and S.K. Estreicher, Superlattices and Microsctructures (in press).

[15]See e.g., Tables 2 and 4 in the first of the papers in Ref.4.

[16]M. Stutzmann and C.P. Herrero, Proc. 20th ICPS, ed. E.M. Anastassakis and J.D. Joannopoulos (World Scientific, Singapore, 1990), vol.3, p.783.

[17]A.A. Bonapasta, preprint.

Materials Science Forum Vol. 83-87 (1992) pp. 33-38

HYDROGEN-INDUCED PLATELETS IN SILICON: SEPARATION OF NUCLEATION AND GROWTH

N. M. JOHNSON[1], C. HERRING[1,2], C. DOLAND[1], J. WALKER[1], G. ANDERSON[1], and F. PONCE[1]
1 Xerox Palo Alto Research Center, Palo Alto, CA 94304, USA
2 Department of Applied Physics, Stanford University, Stanford, CA 94305, USA

ABSTRACT

It is demonstrated that the growth of hydrogen-induced platelets in *n*-type single-crystal silicon can be controlled independently of the nucleation process. The results demonstrate that the suppression of platelet formation at high hydrogenation temperatures (e.g., >250°C) is a consequence of the suppression of platelet nucleation and that platelets nucleated at lower temperatures (e.g., 150°C) grow markedly at the higher temperatures. The platelets, though somewhat reminiscent of the Guinier-Preston zones long familiar in certain metallic alloys, have novel traits that suggest that short-range interatomic forces may be sufficiently anisotropic to produce a two-dimensional phase.

1. Introduction

In addition to the numerous, detailed demonstrations of the ability of hydrogen to passivate defects and impurities in single-crystal silicon (e.g., Ref. 1), it has also been shown that hydrogen, diffused into silicon at moderate temperatures, can itself generate extended structural defects.[2,3] These defects are considered to be unrelated to either plasma or radiation damage because they can be introduced with a remote hydrogen plasma system[3] but not with a remote plasma of helium, argon, or oxygen.[4] Similarly, both the crystalline perfection of the starting material and the exclusion of any intentional radiation damage during hydrogenation argue against these defects being due to hydrogen decoration of either pre-existing or plasma-induced defects. Thus, it has been proposed that the chemical reactivity of isolated, interstitial hydrogen can, under certain circumstances, result in the generation of hydrogen-stabilized defects in the silicon lattice. The topic of hydrogen-induced defects is of immediate technological interest since similar extended defects occur after reactive ion etching of silicon wafers[5,6] and in low-temperature epitaxially grown silicon.[7]

Some general features of the extended defects and their formation conditions have thus far been determined. The defects appear within ~0.1 µm of the exposed surface, are planar in shape, and are aligned predominantly along {111} crystallographic planes.[2-4] The density of these platelets is found to depend on the type and concentration of the dopant impurities. For similar dopant concentrations and identical hydrogenation conditions, *n*-type silicon contains higher densities of platelets than *p*-type material, and the platelet density increases monotonically with donor concentration. Raman spectroscopy provides direct evidence that the platelets arise from the coordinated formation of Si-H bonds. The predominant spectral feature is a broad peak centered near 2140 cm^{-1} (at room temperature), which systematically correlates with the presence of the platelets and which shifts to ~1570 cm^{-1} in deuterated specimens.[2,4] Both the vibrational frequency and the isotopic shift are characteristic of Si-H bonds and suggestive of a higher-order Si-H complex (i.e., SiH_n, n=2,3).[8] Still lacking is quantitative information on the kinetics of the generation and dissociation of the platelets. In this paper we demonstrate that the nucleation and growth stages can be clearly distinguished and identify hydrogenation conditions appropriate for the study of each of these processes individually.

2. Kinetics of Platelet Formation

The work to be reported here was undertaken to test the hypothesis that the platelets are regions of a metastable two-dimensional phase. Such a picture is suggested by atomic-resolution electron micrographs,[2,9] where platelets that are not too densely packed or too large (e.g., ≤ 10 nm in diameter) appear when viewed edge-on as two-dimensional regions where adjacent silicon planes are pushed appreciably apart over the seemingly uniform platelet area. The platelet phase must be presumed to be, if sufficiently extended, of lower free energy than the supersaturated solid solution it replaces, though it is doubtless in turn less stable than other possible phases which may be harder to nucleate (e.g., phases, like silicon plus gaseous hydrogen, that cannot exist over a large region in lattice registry with the silicon host). In such case platelet formation might well occur by an initial nucleation, appreciable over only limited ranges of temperature and concentration, followed by relatively rapid growth of those nuclei that have surpassed a certain critical size.

A simple Volmer-type theory[10] suffices to illustrate the qualitative behavior of the nucleation and growth processes in the present study. Consider a roughly circular nucleus of radius R, on a particular lattice plane. The free energy change ΔF due to its formation out of the supersaturated solution contains a "bulk" term AR^2, with $A<0$ below the saturation temperature T_S, a positive term BR due to the perimeter energy, and another positive term $B'R/n(R)$ due to the elastic strain energy.[11] The combination of these terms yields $\Delta F(R)$ that has a maximum at some radius R_m (for $T<T_S$), and just as in the more familiar three-dimensional problem[10] both R_m and $\Delta F(R_m) \equiv \Delta F_m$ become large as $T \rightarrow T_S$. This increase of ΔF_m could account for the failure of platelets to form in typical plasma-hydrogenation experiments above ~250°C, though platelets once formed do not anneal out appreciably until considerably higher temperatures.

The picture just described suggests that if nuclei were formed at a temperature T_1 near the minimum of $\Delta F_m/(kT)$, and if some of these were allowed to grow at T_1 until $R>R_m(T_2)$, where $T_1<T_2<T_S$, then increasing T to T_2, in the continued presence of plasma products, should encourage rapid growth of these nuclei, even though T_2 may be high enough to suppress platelet formation in virgin material. In this study we exposed specimens of single-crystal silicon to monatomic deuterium (or hydrogen) in a remote hydrogen plasma system.[3] The silicon wafers were *n*-type ($[P]=8\times10^{17}$ cm^{-3}), float-zone grown, and of (100) orientation; results are also presented from (111)-oriented, Cz-grown material. The hydrogenations were performed in two steps. The specimens were first hydrogenated at $T_1 = 150$°C (20 min) and then over a range of higher temperatures T_2 (60 min). Control specimens received only the individual steps.

The dramatic effect of the two-step hydrogenation schedule on platelet formation is demonstrated in Fig. 1 (for $T_2=275$°C) with bright-field micrographs from cross-sectional transmission electron microscopy (TEM). Small platelets (diameters $\leq$ 10 nm) are discernable within ~50 nm of the surface after the 150°C step (Fig. 1(A)) but are undetectable for the 275°C step (Fig. 1(B)). The two-step treatment (Fig. 1(C)) produces large platelets with nominal diameters approaching 100 nm which are predominantly situated in the {111} planes.

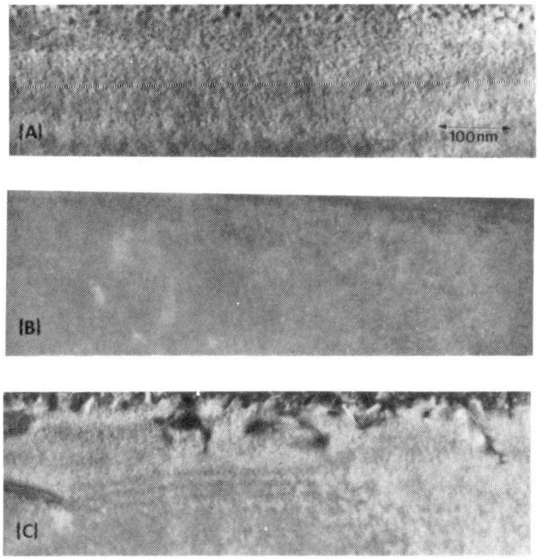

Fig. 1. Cross-sectional TEM micrographs, viewed in a <110> projection, of *n*-type silicon after three different deuteration treatments: (a) D, 150°C, 20 min, (b) D, 275°C, 60 min, and (c) D, 150°C, 20 min *plus* D, 275°C, 60 min.

The incorporation of deuterium in the near-surface layer during the two-step hydrogenation schedule is shown in Fig. 2 (for $T_2=275°C$) with spatial depth profiles from SIMS. The profiles display the same dependence on deuteration schedule as the platelets (Fig. 1). For quantitative comparison of the profiles, the areal density of D in the platelet layer was obtained by depth integration after subtraction of the residual deuterium, which was estimated from the average concentration of D between 200 and 300 nm below the surface and which probably consists mostly of PD complexes[12] and D_2* complexes. (The latter are the more mobile and less stable of two tentatively identified[13] diatomic hydrogen species in silicon). For single-step deuterations, the areal densities of total D are as follows: after the 150°C deuteration the density is 2.0×10^{14} cm^{-2}, and after 275°C it is 1.9×10^{13} cm^{-2}. For the two-step schedule the integrated density is 2.3×10^{15} cm^{-2}. Thus, the concentration of deuterium within ~100 nm of the exposed surface correlates with the generation of platelets.

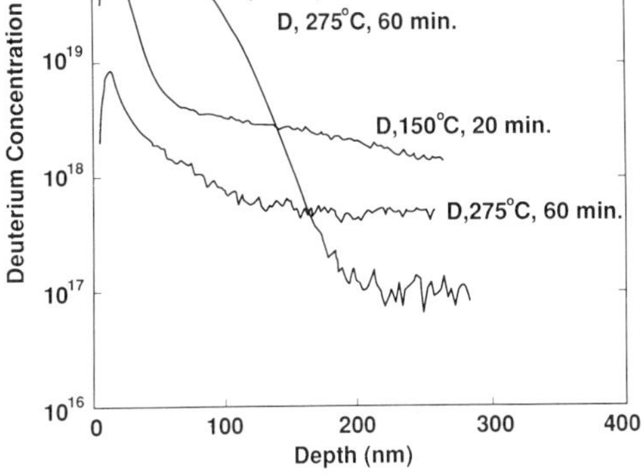

Fig. 2. Depth profiles of deuterium in the near-surface region of *n*-type silicon after three different deuteration treatments: (a) D, 150°C, 20 min, (b) D, 275°C, 60 min, and (c) D, 150°C, 20 min *plus* D, 275°C, 60 min.

Fig. 3. Dependence of the areal density of deuterium on the temperature T_H ($\equiv T_2$) of the second step in the two-step deuteration treatment. The deuteration temperature during the first step was T_1 = 150°C (20 min). Also shown is the deuterium density for single-step deuterations at T_H.

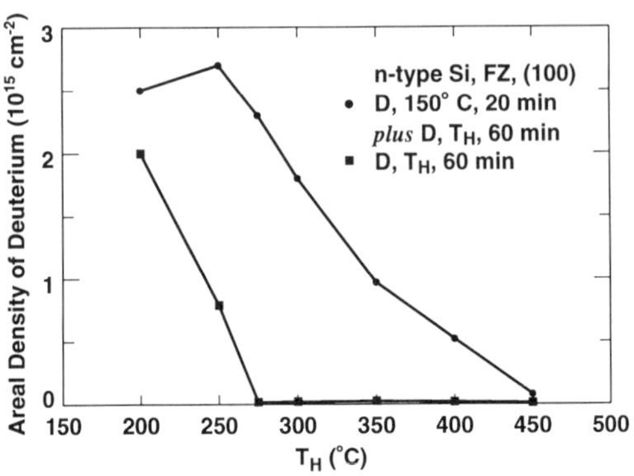

The dependence of the areal density of D on T_2 is shown in Fig. 3. In specimens receiving only the T_2 step, the platelet density is greatly attenuated for temperatures above 250°C. With the two-step schedule, platelets nucleated at 150°C continue to grow for T_2 up to ~450°C, above which thermal dissociation becomes competitive.

Local vibrational modes created by the two-step treatment are shown in Fig. 4 from Raman spectroscopy. After hydrogenation the predominant spectral feature is a broad band centered near 2130 cm^{-1}, with a poorly resolved band(s) in the range of 2060-2090 cm^{-1}. A vacuum anneal at 300°C introduces a resolvable band near 2090 cm^{-1}, which is attenuated by a subsequent anneal at 400°C. Anneal temperatures >450°C (~60 min) are required to appreciably diminish the intensity of the band at 2130 cm^{-1}.

Stress measurements by the optical lever technique[14] were used to directly established that the hydrogen-induced platelets produce compressive stress in the silicon wafer.[15]

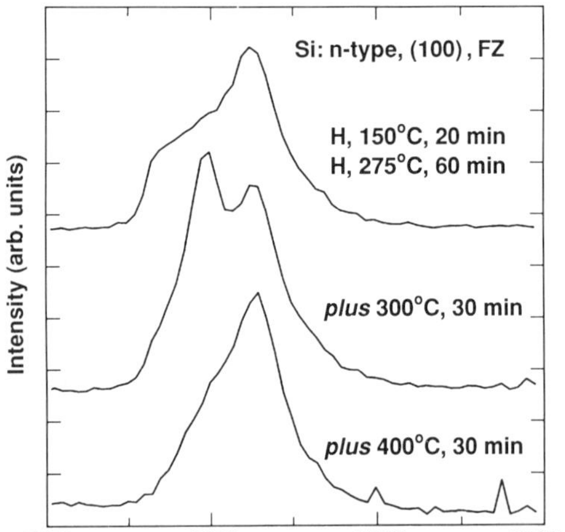

Fig. 4. Raman spectra for *n*-type silicon after a two-step hydrogenation treatment and subsequent vacuum anneals: (a) H, 150°C, 20 min plus H, 275°C, 60 min, (b) plus an anneal at 300°C, 30 min, and (c) plus a further anneal at 400°C, 30 min. For clarity the spectra have been offset vertically.

3. Microscopic Structure of Platelets

While much remains to be learned about the processes involved in the nucleation and growth of platelets, new insights can be presented on their microscopic structure. The platelets are, of course, very reminiscent of the long-studied Guinier-Preston (GP) zones that are precursors of nucleation of new phases in certain metallic alloys.[16] However, there are important differences as well as similarities between the two. While the hydrogen-induced platelets involve an interstitial solid solution rather than the substitutional configurations of the GP zones, the new phase is in both cases coherent with the host matrix. When GP zones adopt a planar shape, as in the Al-Cu system, their shape can plausibly be attributed to an attempt to optimize the strain energy due to mismatch of atomic sizes. By contrast, the present TEM data (e.g., Fig. 5) suggest that platelet growth involves a strong anisotropy in the interatomic interactions. A possible microscopic model with this feature has been proposed by Zhang and Jackson[17] from pseudopotential total energy calculations. These authors suggested an atomic structure for the Si-H platelets that is based on a highly anisotropic interaction of the building blocks of a platelet, which are metastable pairs of hydrogen interstitials (H_2*). With such a model it is conceivable, though not yet proven, that thickening of the initial platelet configuration (i.e., bilayers of H_2*) into a three-dimensional phase by addition of new layers at the middle of a platelet might be energetically disadvantageous. In such case the two-dimensional growth of platelets to sizes of tens of nm could well occur. Our present TEM data do not yet fully suffice to establish long-range two-dimensionality of the platelets, but they are strongly suggestive. Viewed edge-on, the platelets show up as regions where [110] channel transmission of electrons appears enhanced; since sample preparation alone produces sizable fluctuations in this transmission, evident in the mottled appearance of the platelet-free regions of Figs. 1 and 5, it is not surprising that some irregularity is seen in the traces of the edge-on platelets. However, in most cases the fluctuation in the apparent center line of a platelet is no more that an atom spacing or so above a fixed {111} plane, and typically does not increase over the width of a platelet, even when this width is many tens of nm. However, occasional shifts in the center line of a platelet trace might possibly result from correlated nucleation at neighboring sites. We conclude that the case for a basically two-dimensional phase seems quite strong.

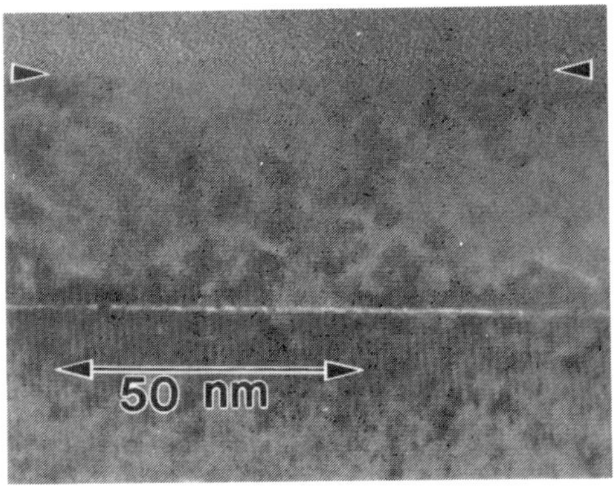

Fig. 5. High-resolution lattice image of a platelet, viewed in a <110> projection, from cross-sectional TEM. The silicon was *n*-type ($[P]=7\times10^{17}$ cm^{-3}) and of (111)-orientation. Platelets were generated with a two-step hydrogenation treatment: H, 150°C, 20 min plus H, 275°C, 60 min. The arrow heads at the top of the micrograph mark the surface of the silicon.

4. Concluding Remarks

Our results demonstrate that platelet formation is suppressed at high hydrogenation temperatures (e.g., >250°C) specifically as a consequence of the suppression of platelet nucleation and that platelets nucleated at lower temperatures (e.g., 150°C) grow dramatically at these higher temperatures, with nominal diameters approaching 100 nm. Platelet generation involves the precipitation of a two-dimensional silicon hydride phase from a supersaturated solution of hydrogen in silicon, with features that are in some ways reminiscent of Guinier-Preston zones that appear in certain metallic alloys. But unlike the latter, which typically grow into three-dimensional nuclei as they enlarge, platelets grow to diameters of many tens of nm while retaining a remarkable two-dimensionality. This suggests that the Si-H phase involves a highly anisotropic interaction between migrating hydrogen and platelets which sustains the planar configuration.

ACKNOWLEDGMENT

This work was supported partially by the Office of Naval Research and partially by the Solar Energy Research Institute.

REFERENCES

1. Hydrogen in Semiconductors, vol. eds. J. I. Pankove and N. M. Johnson, Semiconductors and Semimetals Vol. 34, (Academic, New York, 1991), Chaps. 2-7.
2. N. M. Johnson, F. A. Ponce, R. A. Street, and R. J. Nemanich, Phys. Rev. B 35, 4166 (1987).
3. N. M. Johnson, in Hydrogen in Semiconductors, vol. eds. J. I. Pankove and N. M. Johnson, Semiconductors and Semimetals Vol. 34 (Academic, New York, 1991), Chap. 7.
4. N. M. Johnson, C. Doland, F. A. Ponce, J. Walker, and G. Anderson, Physica B 170, 3 (1991).
5. H. P. Strunk, H. Cerna, and E. G. Mohr, Inst. Phys. Conf. Ser. No. 87: Sect. 1, 457 (1987).
6. A.-J. Jeng, G. S. Oehrlein, and G. J. Scilla, Mat. Res. Soc. Symp. Proc. 104, 247 (1988).
7. C. C. Tsai, G. B. Anderson, and R. Thompson, Mat. Res. Soc. Symp. Proc. 192, 475 (1990).
8. G. Lucovsky, R. J. Nemanich, and J. C. Knights, Phys. Rev. B 19, 2064 (1979).
9. F. A. Ponce, N. M. Johnson, J. C. Tramontana, and J. Walker, IOP Conf. Ser. No. 87: Sect. 1, 49 (1987).
10. e.g., J. W. Christian, The Theory of Transformations in Metals and Alloys, Second Edition (Pergamon, New York, 1981) Part I, Sects. 46, 47.
11. This logarithmic term does not appear in the expressions usually encountered for the strain energy of "disks" idealized as oblate spheroids. Its origin is most easily seen if one represents the distortion outside the platelet as that due to a dislocation loop in an elastic continuum, with a "Burger's vector" normal to its plane.
12. N. M. Johnson, C. Herring, and D. J. Chadi, Phys. Rev. Lett. 56, 2224 (1986).
13. N. M. Johnson and C. Herring, Inst. Phys. Conf. Ser. 95, 415 (1989).
14. e.g., R. W. Hoffman, Physics of Thin Films 3 , eds. G. Hass and R. E. Thun (Academic, New York, 1966), p. 211.
15. N. M. Johnson, K. S. Stevens, and J. Walker, unpublished.
16. For reviews see: A. Guinier, Solid State Physics 9, eds. F. Seitz and D. Turnbull (Academic, New York, 1959), p. 294; J. B. Cohen, ibid. 39, eds. H. Ehrenreich and D. Turnbull (Academic, New York, 1986), p. 131.
17. S. B. Zhang and W. B. Jackson, Phys. Rev. B 43, 12142 (1991).

Materials Science Forum Vol. 83-87 (1992) pp. 39-44

AN ISOTHERMAL PUMPING MECHANISM FOR THE INTRODUCTION OF HIGH HYDROGEN CONCENTRATIONS IN p^+ LAYERS IN SILICON

A.D. MARWICK, M. WITTMER AND G.S. OEHRLEIN
IBM Research Division, T.J. Watson Research Center, P.O. Box 218,
Yorktown Heights, NY 10598, USA

ABSTRACT
Anomalously high hydrogen concentrations were found in highly-doped buried p^+ layers (B or Ga doped, $1 - 7 \times 10^{19}/cm^3$) following isothermal exposure to atomic hydrogen at 200°C. Hydrogen/acceptor concentration ratios of 5-8, depending on the exposure conditions, were obtained. The hydrogen distribution was similar in shape to the dopant profile, which was produced by implantation and rapid annealing, but was in general more sharply peaked. After hydrogenation at 200°C the Ga-doped samples were found to contain extended defects. These were identified as stacking faults of intrinsic (vacancy) type with a Burger's vector of $1/3\langle\bar{1}11\rangle$. The introduction of H in concentrations exceeding the acceptor concentrations is ascribed to a three-step "pumping" process: thermally-activated breakup of HA pairs, followed by trapping of some of the H^+ at neutral clusters, probably extended defects, then finally in-diffusion of new H^+ to compensate the exposed A^-. It is suggested that the H clusters and the extended defects may contain hydrogen-generated point defects.

1. Introduction

Hydrogen concentrations greatly in excess of the equilibrium solubility can be introduced into silicon by exposure to a source of high fugacity, such as a hydrogen plasma. We should expect that this hydrogen will tend to form precipitates of some kind. In addition, it is well established that hydrogen atoms can be trapped at defects of all kinds, including impurity atoms, and there has been extensive study of dopant hydrogen pairs.[1] However, it should be emphasized that, except for a narrow range of experimental conditions, most of the hydrogen atoms introduced into silicon by plasma hydrogenation do not form pairs with dopant atoms. A near-surface layer always forms which contains a high hydrogen concentration and extended defects, in both n-type[2, 3] and p-type[4, 5] material. It is thought that at least some of the hydrogen is associated with these defects. Clear evidence for this has been found in boron-doped silicon, where hydrogen bubbles decorate pre-existing damage following hydrogenation at high plasma pressures.[5] It is accepted that some of the hydrogen in the near-surface layer is in the form of clusters, generically referred to as "H_2", but little is understood about these, other than that the hydrogen they contain has a relatively low mobility. There is certainly no direct evidence that the clusters are interstitial hydrogen molecules. Similar clusters may be formed when A-H pairs are annealed, as the hydrogen released from the acceptors also has a rather low mobility.[6]

Thus, much remains to be established about the majority of the hydrogen in the silicon, i.e. hydrogen that is not associated with dopant atoms. In this paper we describe some new information about the formation and properties of layers containing high hydrogen concentrations. We find that such regions can be formed well below the surface, in buried p^+ layers, during plasma treatment at 200 °C. Atomic-resolution techniques (channeling and high-resolution electron microscopy) are used to investigate the state of the hydrogen and the silicon lattice. A brief report of this work was published previously.[7] Note that we use the term "hydrogen" to include both the ^1H and ^2H isotopes of hydrogen.

2. Experiment

Silicon (100) wafers (p-type, 10 ohm.cm) were implanted with 100 keV boron to doses of 3×10^{14}, 10^{15} and 3×10^{15} ions/cm^2, or with 10^{15} 100 keV Ga, then annealed at 1000°C for 10 s (Ga implants) or 650°C for 30 s (Ga implants). The resulting dopant profiles are sharply peaked, as seen in Figs. 1 and 2. The samples were exposed to atomic deuterium from a plasma source by placing them on a hot stage built into the powered electrode of an RF diode etcher, with a shield which protected them from bombardment by energetic particles from the plasma. Gas pressures of 400 mtorr and 100 mtorr were used for the boron and gallium-implanted samples, respectively, for 3 hours. Figures 1 and 2 show SIMS profiles of the ^{2}H in these samples. In the boron case, data is shown for the three different implant doses and for hydrogenation temperatures of 150°C and 200°C. The gallium data is for a hydrogenation temperature of 200°C. After hydrogenation the samples were examined by cross-sectional transmission electron microscopy, SIMS and ion channeling.

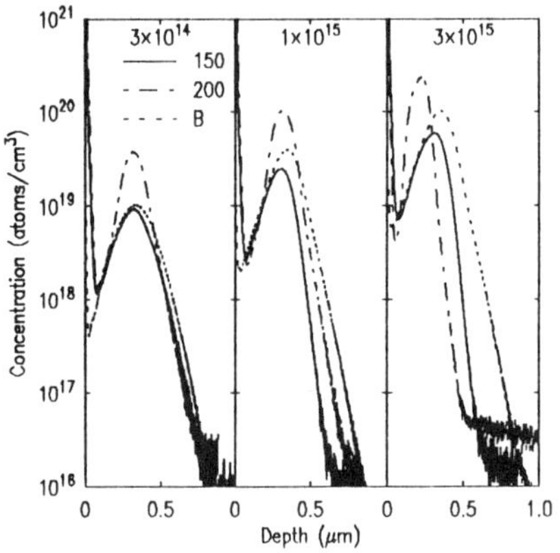

Figure 1: SIMS profiles of boron (dashed line) and deuterium in B-implanted samples, for doses of 3×10^{14}, 1×10^{15} and 3×10^{15} B/cm^2, and hydrogenation temperatures of 150°C and 200°C.

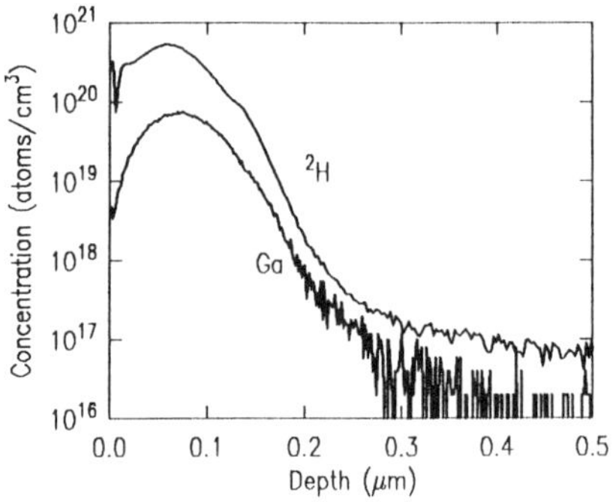

Figure 2: SIMS profiles of Ga and ^{2}H in a sample hydrogenated at 200°C.

3. Results

Hydrogen depth profiles

The deuterium profiles in Figures 1 and 2 show that for hydrogenation at 150°C and the lowest B implant dose a close match between the boron profile and the ^{2}H profile is obtained, as previously observed.[8] This is attributed to the formation of boron-hydrogen pairs. For higher boron doses the development of the ^{2}H profile is incomplete, in the sense that not enough hydrogen has been incorporated to equal the total boron concentration, but the boron and hydrogen concentrations are equal where their profiles overlap. After 200°C hydrogenation, on the other hand, the deuterium concentration greatly exceeds the boron or gallium concentration, although the shape of the deuterium profile still reflects that of the acceptors. For boron doses of 3×10^{14} and 10^{15} /cm^2, and for the Ga dose (10^{15}

/cm²), the ^{2}H peak at 200°C is close to the peak at 150°C and hence to the acceptor peak. A small amount of deuterium trapping at the "end-of-range" loops (see below) causes the shoulder at 120 nm depth in Figure 2.

The build-up of hydrogen in the p^+ layer can be qualitatively explained[7] if we postulate a multi-step "pumping" process which operates during the hydrogenation at 200°C. The key hypothesis is that H^+ in the passivated layer has a small probability of being converted into an immobile neutral form, probably as a cluster involving other hydrogen atoms and atomic defects. The removal of the H^+ in this way perturbs the electrical neutrality of the passivated p^+ layer, which therefore attracts more H^+ to re-establish complete compensation. Thus, the hydrogen concentration in the p^+ layer gradually increases. In equilibrium at 200°C, the passivated layer contains an appreciable concentration of H^+ ions formed by thermal dissociation of AH pairs. These are free to diffuse until re-captured by an A^- ion or a hydrogen cluster. The fact that the shapes of the hydrogen and boron (or gallium) profiles are similar shows that the hydrogen diffuses predominantly as the H^+ ion, with a negligible contribution from neutral H^0, as will be discussed later.

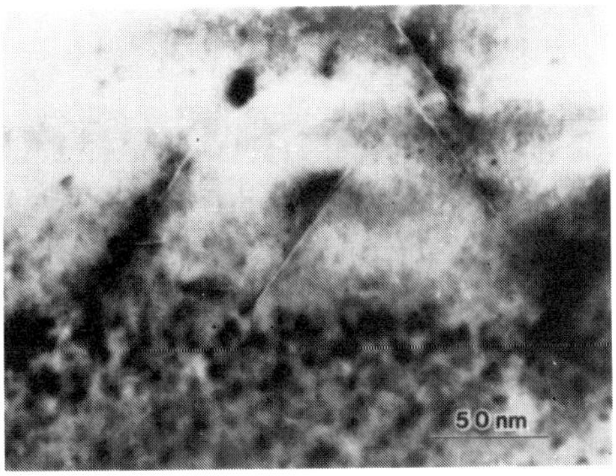

Figure 3: Cross-section electron micrograph of a Ga-implanted sample hydrogenated at 200°C, showing stacking faults. The surface of the sample is at the top of the micrograph, and the band of interstitial loops marking the original amorphous/crystalline interface is at the bottom.

For the highest boron dose, 3×10^{15} /cm², both the depth of maximum penetration of ^{2}H and the depth of its peak concentration were less at 200°C than at 150°C. The implication is that incoming hydrogen is trapped before reaching the passivation front, i.e. conversion of hydrogen from BH pairs to the immobile form is faster than the input rate. Thus, in this case dissolution of the hydrogen in the sample appears to be the rate limiting step. Since this sample contains about 5×10^{15} H/cm² in all, we deduce that the dissolution rate is approximately 5×10^{11}H/cm².s at 200°C. At 150°C, on the other hand, the flux is clearly lower than this, because the total hydrogen in the sample is now less than the amount of boron. Thus the amount of hydrogen incorporated in each sample by the plasma treatment appears to be temperature dependent.

A related point is that the amount of excess hydrogen in the samples of Figure 1 is approximately proportional to the amount of boron they contain. This strongly suggests that the process leading to the growth of the hydrogen excess obeys first order kinetics. We have previously suggested [7] that the following overall equation can summarize the process:

$$(HA)^0 + e^- + nH^0_{imm} \rightarrow A^- + (n+1)\mathrm{H}^0_{imm}. \qquad (1)$$

Here A means an acceptor atom, and nH^0_{imm} represents an immobile neutral cluster containing n hydrogen atoms. Equation (1) is clearly first order. However, it does not address the question of the nucleation of the hydrogen clusters. Observations of high-order kinetics in the annealing of BH pairs may reflect this nucleation stage.[9]

Microstructure

Previous studies[4, 5] have shown that implantation damage in boron-implanted material acts as a trap for hydrogen, and that small bubbles can be found decorating the interstitial loops in such material after exposure to hydrogen plasma under the same conditions as in the present work (400 mtorr and 200°C). Thus, we can expect that the boron-implanted samples used in the present work contain such bubbles after their 200°C treatment. This raises the question of whether the pre-existing damage is necessary for the development of the hydrogen excess.

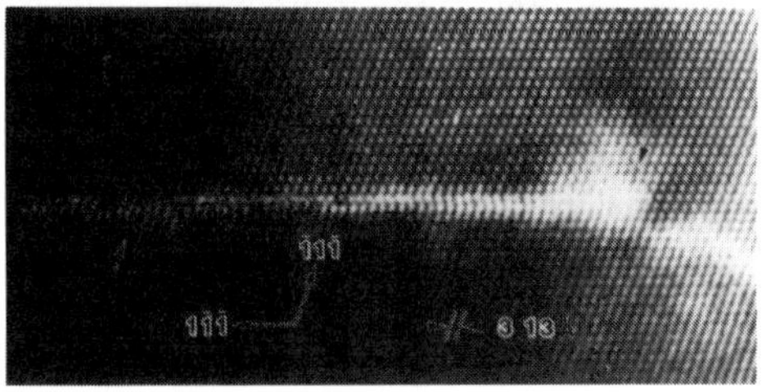

Figure 4: High resolution micrograph of a stacking fault. The arrow indicates the position of the peripheral dislocation.

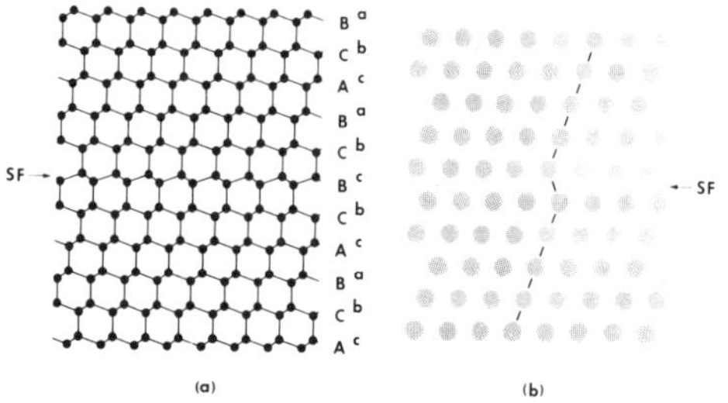

Figure 5: Diagrams showing how a stacking fault with a missing {111} plane gives rise to the contrast observed in Figure 1: (a) the Si lattice, with a missing plane; (b) schematic showing how the high-resolution image of this defect is formed.

Gallium-implanted samples do not contain visible defects after implantation and annealing, except for a thin layer of "end-of-range" interstitial loops at the original amorphous/crystalline interface. However, it was found[7] that after hydrogenation at 200°C the Ga-implanted layer contained Frank loops enclosing stacking faults (SF) which had a Burger's vector of $1/3(\bar{1}11)$. Figure 3 shows a cross-section micrograph of such a sample. Defects similar to stacking faults have been previously found to correlate with the presence of high near-surface hydrogen concentrations in plasma-exposed p-type samples,[4, 5] but in the present case the layer in which the defects were found is thicker, corresponding to the greater depth of the hydrogen-enriched layer. The observed SFs are of intrinsic (vacancy) type. Figure 4 is a high-resolution micrograph of one such defect, while Figure 5 shows a diagram of its structure, demonstrating that the observed contrast is the result of a missing plane of atoms. Hydrogenation-induced defects in n-type silicon have a different structure.[2, 3]

Thus both boron and gallium-implanted material show indications of the presence of extra lattice defects following hydrogenation at 200°C. In the boron case, the presence of hydrogen bubbles necessarily implies vacancy agglomeration, while at the same time expansion of pre-existing interstitial loops[5] accounts for the corresponding interstitials, at least qualitatively. In the Ga-implanted sample, vacancy-type defects are represented by the intrinsic Frank loops, but the corresponding interstitials are not accounted for. Most probably, the interstitials were lost to the surface and to the band of end-of-range loops. The difference in the B and Ga-implanted samples is presumably largely due to the absence of pre-existing defects in the Ga case, although the lower plasma pressure no doubt also played a role, as no bubbles were resolved in the boron case at low pressure either.[5]

Ion channeling, using the low-energy nuclear reaction $^{2}H(^{3}He, ^{4}He)^{1}H$ to detect deuterium nuclei, was previously used to investigate the site of deuterium atoms in the 200°C-hydrogenated Ga-implanted sample.[7] The signal from the nuclear reaction showed no channeling effect, which indicates an effectively random position for the deuterium nuclei relative to the silicon lattice. This means that the deuterium atoms did not occupy just one or two high-symmetry sites in the lattice, as the signature of these in the channeling data would have been clear. Thus the majority of the "excess" hydrogen cannot be in forms which would give such simple lattice sites, such as H_2 molecules at T-sites or <111> Si-H bonds in either hydrogen-vacancy clusters or the extended defects.

4. Discussion

The "pumping" effect we have described is the dominating effect in the evolution of the hydrogen profile at 200°C. The temperature dependence of this effect is quite strong. A study with furnace-annealed boron-implanted samples[10] shows that after hydrogenation at 250°C the amount of hydrogen trapped in the p^+ layer is much smaller than at 200°C. Qualitatively, this can be understood as being a consequence of the reduced stability of BH pairs at the higher temperature. However, just how this affects the various steps subsumed into Equation (1) is not yet understood.

A complication in understanding the temperature dependence of the hydrogen buildup is that the total amount of hydrogen incorporated in the samples was strongly dependent on the doping level, as well as on the hydrogenation temperature. Figure 1 shows that, as already mentioned, the amount of hydrogen scaled roughly with the boron dose. Specifically, at 150°C the amount of hydrogen incorporated in the sample implanted with 3×10^{15} boron/cm^2 is much greater than in the 1×10^{15} /cm^2 sample. Even so, the latter sample could not have been saturated with hydrogen, as the total hydrogen it contained was much less than the amount of boron. This seems to be evidence that the amount of hydrogen entering the sample from the plasma is related to the number of available trap sites it contains, or in other words that during the passivation there is a local equilibrium set up near the surface in which hydrogen entering from the plasma is balanced by untrapped hydrogen leaving the sample.

These ideas, together with the observed scaling of the excess hydrogen production with boron concentration, imply that the net hydrogen incorporation rate is controlled by processes occurring in the passivated p^+ layer. The limiting rate can be estimated from the data on 200°C hydrogenation of the 3×10^{14} B/cm^2 sample, since in this case the supply of hydrogen from the plasma seems to have been enough to ensure that the sample controlled the uptake rate. The mean rate of formation of excess hydrogen in this sample was 1.9×10^{15} /cm^3.s at the peak of the boron profile during the hydrogenation. If we suppose that the rate-limiting reaction was one in which an H^+ was incorporated into a cluster of some sort, then we can estimate the cluster density. The incorporation rate is given by $4\pi r_C n_C D_+ n_+$, where r_C, n_C are the radius and number density of the clusters, and D_+, n_+ are the diffusion coefficient and concentration of H^+. The concentration of H^+ is expected to be determined by equilibrium with BH pairs according to the expression[11] $n_+^2 = (n_B/\Omega)\exp(-\Delta\overline{E}_{BH}/kT)$, where n_B is the boron concentration, Ω the atomic volume of Si, and $\Delta\overline{E}_{BH}$ is the binding energy of BH, 0.87 eV.[11] Thus $n_+ \approx 1.6 \times 10^{16}$ /cm^3 at 200°C. Using for D_+ the Wan Wieringen and Warmolz[12] value of 10^{-7} cm^2/s, and assuming $r_C = 3$Å, we find that the density of cluster sites needed to account for the rate of formation of excess hydrogen is 3×10^{12} /cm^3. This low value shows that it is quite feasible for all the hydrogen clustering to occur at extended defects; it is not necessary to assume the existence of atomic-sized clusters in order to explain the kinetics.

There is evidence for the participation of lattice defects in the passivation process, at least for high hydrogen concentrations. The observation of vacancy-type defects (bubbles or Frank loops) in high-hydrogen p-type silicon indicates that vacancies are generated in the high-hydrogen layer. This is predicted to be an exothermic process,[13] and thus we expect hydrogen-vacancy clusters to form and to be the precursors of extended defects. However, on the basis of the channeling result only a small proportion of the hydrogen can be in such simple defects, which points again to the likely importance of larger defect clusters in high-hydrogen material.

The similarity of the shape of the excess hydrogen depth profile to the acceptor profiles in Figs. 1 and 2 shows that the diffusing hydrogen must be H^+ if the sinks for the mobile hydrogen are on the scale of the extended defects, since diffusion of neutral hydrogen over such distances would result in appreciable broadening of the profile.[7] If the predominant sinks were not the extended defects, but some denser population of sub-microscopic clusters, then one could suppose that the diffusing species is neutral. However, the most economical hypothesis at present is that the extended defects are the dominant sinks.

Conclusions

In both boron and gallium-implanted p^+ layers large concentrations of hydrogen can be built up by exposure to atomic hydrogen at 200°C. The amount of hydrogen incorporated in the samples appears to be determined by quasi-equilibrium between hydrogen in the plasma and in the sample. The shapes of the hydrogen profiles indicate that hydrogen diffuses as H^+, with a negligible contribution of neutral species. The build-up of hydrogen can be accounted for if H^+ ions can be incorporated in extended defects during the plasma treatment. Frank loops are generated in gallium-implanted samples. These are of vacancy character and we tentatively attribute their formation to agglomeration of hydrogen-vacancy complexes.

References

[1] *Hydrogen in Semiconductors*, eds. Jacques I. Pankove and Noble M. Johnson, (Academic Press, New York, 1991).
[2] N. M. Johnson, F. A. Ponce, R. A. Street, and R. J. Nemanich, Phys. Rev. B **35**, 4166 (1987).
[3] F. A. Ponce, N. M. Johnson, J. C. Tramontana, and J. Walker, Microscopy of Semiconducting Materials, Oxford 1987. Inst. Phys. Conf. Ser. **87**, 49 (1987).
[4] S. J. Jeng and G. S. Oehrlein, Appl. Phys. Lett. **50**, 1912 (1987).
[5] S. J. Jeng, G. S. Oehrlein, and G. J. Scilla, Appl. Phys. Lett. **53**, 1735 (1988).
[6] M. Stutzmann, W. Beyer, L. Tapfer, and C. P. Herrero, Physica B **170**, 240 (1991).
[7] A. D. Marwick, G. S. Oehrlein, and M. Wittmer, Appl. Phys. Lett **59**, 198 (1991).
[8] N. M. Johnson, Phys. Rev. B **31**, 5525 (1985).
[9] T. Zundel and J. Weber, Phys. Rev. B **43**, 4361 (1991).
[10] S. J. Jeng, G. S. Oehrlein, and G. J. Scilla. (unpublished)
[11] C. Herring and N. M. Johnson, in *Hydrogen in Semiconductors*, edited by J. I. Pankove and N. M. Johnson (Academic Press, New York, 1991), p. 225.
[12] A. Van Wieringen and N. Warmoltz, Physica **22**, 849 (1956).
[13] C. G. Van de Walle, P. J. Denteneer, Y. Bar-Yam, and S. T. Pantelides, Phys. Rev. B **39**, 10791 (1989).

Materials Science Forum Vol. 83-87 (1992) pp. 45-50

HYDROGEN EFFUSION FROM MONOCRYSTALLINE B-DOPED SILICON.

D. BALLUTAUD[1], P. de MIERRY[1], J.C. PESANT[1], R. RIZK[1], A. BOUTRY-FORVEILLE[2] and M. AUCOUTURIER[1]
[1]Laboratoire de Physique des Solides, CNRS, 1, Place Aristide Briand 92195 Meudon cedex (France).
[2]Laboratoire d'Electrochimie Interfaciale, CNRS, 1 Place Aristide Briand 92195 Meudon cedex (France).

ABSTRACT

Deuterium effusion spectra of crystalline boron-doped silicon ($1x10^{19}$ and $2.5x10^{17}$ $at.cm^{-3}$) are analysed in the light of their dependence on the silicon surface and the doping level. Using chemical etching of the plasma damaged surface layers due to hydrogen (deuterium) plasma and secondary ion mass spectroscopy profiling, we show that the effusion rate seems to be limited by the surface desorption. We suggest also the assignment of the various effusion peaks to specific hydrogen configurations.

1. Introduction

There has been during the last decade a great amount of hydrogen effusion experiments performed on hydrogenated (deuterated) amorphous silicon, all of which aiming at elucidating the role of such species in reducing the density of uncoordinated or "dangling" bonds in this material [1,2]. Since it has been established recently that hydrogen in crystalline semiconductors and especially in silicon deactivates the dopants, [3] analogous effusion experiments have been realized on p and n type crystalline silicon [4]. In our recent simulation works of experimental deuterium diffusion profiles, based on the existence in silicon of neutral and charged hydrogen species (H^+ and H^0 in p-type silicon, H^- and H^0 in n-type silicon), we have deduced the diffusivities, the bonding energies and the hydrogen donor and acceptor level in the gap [5,6]. In order to provide complementary information on the stability of the different hydrogen configurations in the lattice, we report in this paper the preliminary results of deuterium effusion experiments performed on monocrystalline boron-doped silicon, which are analysed in connection with secondary ion mass spectroscopy (SIMS) profiling.

2. Experimental

The monocrystalline FZ boron-doped silicon samples ($1x10^{19}$ and $2.5x10^{17}$ cm^{-3}) were deuterated by a RF plasma (1 mbar, 0.12 W/cm^{-2}) during 40 hours at 180°C. Effusion spectra were measured by a quadrupole mass spectrometer coupled to an evacuated quartz tube (10^{-10} mbar) which contained the deuterated samples. The spectra were obtained either through linear (20°C/min) or isothermal heating of the quartz tube which was submitted to a nearly constant pumping speed by a judicious adjustment of the ionic pump output valve. All the samples had the same surface of 1 cm^2 and the low basic pressure has allowed to detect a low rate of outgassed deuterium. The deuterium profiles were obtained by SIMS using a CAMECA IMS4f apparatus. The primary ion was Cs^+ and the absolute concentration was determined by calibration with deuterium implanted reference samples.

3. Results and discussion.

Fig. 1 shows the deuterium SIMS profiles obtained for the deuterated silicon samples doped with boron to concentrations of $2.5x10^{17}$ and $1x10^{19}$ cm^{-3}. These profiles which have been simulated in recent works [5,6] exhibit a plateau characteristic of the hydrogen-dopant interaction and a shoulder between this plateau and the region of the plasma-damaged near-surface layer. This shoulder is particularly pronounced for the medium doped sample ($2.5x10^{17}$ cm^{-3}) and has been attributed in our simulation studies [5,6] to the diffusion of neutral species. The effusion spectra obtained for these samples during linear heating at a rate of 20°C/min are represented on Fig. 2 (curves a and b). While one observes for the higher doped sample a predominant effusion peak at about 350°C, followed by two shoulders near 380°C and 500°C, the spectrum of the lower doped material reveals only two large desorption bands: a large one with a long tail starting at T< 200°C, with a maximum near 370°C, followed by a more important one at about 510°C.

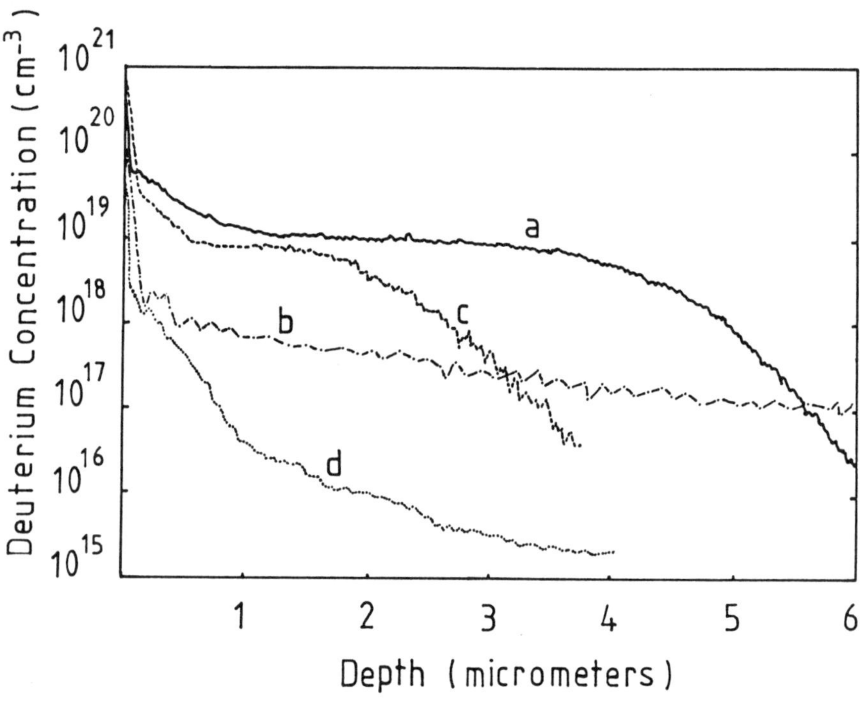

Fig. 1 :

SIMS deuterium diffusion profiles. The curves a and b show respectively the deuterium profiles for silicon samples doped with boron to concentrations of $1x10^{19}$ and $2.5x10^{17}$ cm^{-3} , both deuterated at 180°C during 40 hours. The curves c and d show respectively the deuterium profile after the isothermal effusion during 40 min of the $1x10^{19}$ cm^{-3} doped sample at 220 and 400°C.

The comparison between the two spectra indicates that the total amount of the dissolved deuterium decreases with the dopant concentration, in agreement with the corresponding SIMS profiles. In fact, the drastic decrease in the low temperature region (T<400°C) is, on the one hand, partially compensated by an appreciable increase in the high temperature region (T>450°C) and, on the other hand, accompanied by an overlapping of the two substructures observed for the $1x10^{19}$ cm^{-3} doped material.

The structure of the effusion spectrum observed in the highly B-doped silicon recalls that reported by Stutzmann and Brandt [4] for the same dopant concentration. In a quite similar approach to that reported for p-type hydrogenated amorphous silicon [1], these authors attribute the low and high

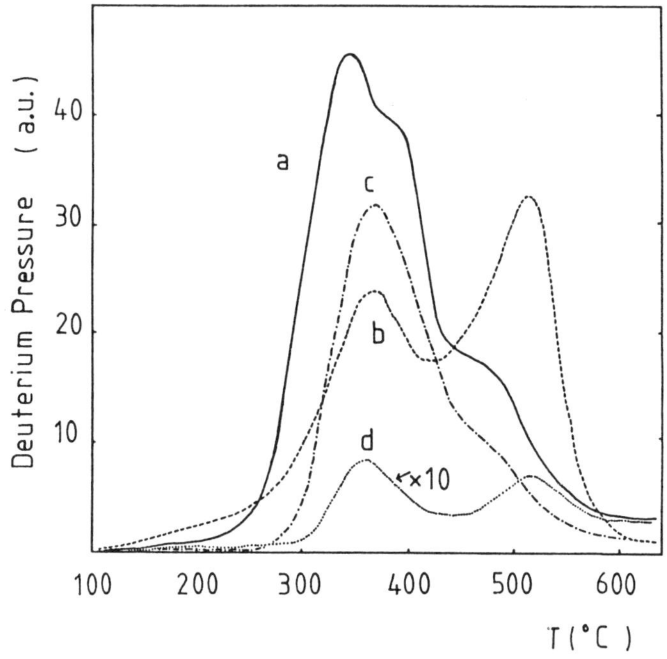

Fig. 2:

Curves a and c show respectively the deuterium effusion spectra of the $1x10^{19}$ cm^{-3} doped sample before and after a 5000 Å chemical etching from the surface.
Curves b and d show the corresponding spectra for $2.5x10^{17}$ cm^{-3} before and after a 5000 Å chemical etching respectively.

temperature peaks to clustered and isolated Si-H bonds respectively. Accordingly, as a prelude to the desorption process, the formation of H_2 molecules from neighboring Si-H bonds occurs at low temperature and is responsible for the 350°C peak, while the isolated bonds lead to this formation at ~ 500°C. Although this interpretation could be applicable in our case, we would like to bring some insight into the origin of these two phases and attempt to identify the role of the plasma damaged layers where the deuterium accumulates in a high concentration. For this purpose, we have proceeded to a chemical etching of about 5000 Å of the deuterated surface. The effusion spectra obtained on these etched samples are reported on Fig. 2 (curves c and d). From the comparison of each couple of spectra, (a, c and b, d) for each dopant level, it appears clearly that the relative intensities of the peaks are greatly modified after the etching process. For the $1x10^{19}$ cm^{-3} doped sample, the removal of a 5000 Å layer leads to an almost total disappearance of the low temperature peak (350°C) and to an important decrease of the shoulder at 500°C. However, the resolved substructure at T ~380°C remains almost unchanged. As for the medium doped sample, the effusion spectrum performed subsequently to the etching process reveals a drastic decrease of the two peaks, keeping in mind that the amount of Si-D...B complexes is expected to remain unchanged.

These observations seem to indicate that the clustered phase is mainly located in the deuterium accumulation region just beneath the surface where defects and traps due to plasma damage are in a high concentration. Moreover, as the high temperature peak (~ 500°C) attributed to isolated Si-D bonds decreases also after etching, more drastically for the $2.5x10^{17}$ cm^{-3} case, it seems that this phase originates from the quenched neutral hydrogen. Indeed, since these neutral species are found to be at the diffusion temperature in a relatively higher concentration in the $2.5x10^{17}$ cm^{-3} doped material than that in the $1x10^{19}$ cm^{-3} one [5], and as we have argued elsewhere [6] that they could quench in some diatomic hydrogen complex [7], this configuration could then be at the origin of the isolated Si-D bonds responsible for the effusion peak near 500°C.

The remaining question concerns the structural origin of the mid-temperature effusion peak (370-380°C). According to the effect of the etching procedure, it is tempting to attribute this band to the desorption of deuterium atoms released from the Si-D...B complexes. This is supported by the agreement between the ratio of the peak areas at this temperature in the etched samples and that of their doping levels. We have assumed here that the large peak at 370°C in the non-etched $2.5x10^{17}$ cm^{-3} doped sample consists of a predominant desorption band of the clustered hydrogen mixed to a small peak due to the outgassed D atoms from D-dopant complexes. However, it is established now that these D-dopant complexes usually dissociate starting from temperatures around 160°C [8]. Indeed, the SIMS profile performed after annealing at 220°C for 40 min and shown in Fig. 1 (curve c) indicates clearly a reduction of the plateau depth accompanied by a slight enhancement of the deuterium accumulation in the subsurface region. After annealing at 400°C for 40 min, the SIMS profile (Fig. 1, curve d) reveals the disappearance of the plateau (i.e. a complete dissociation of the complexes) and the persistence of a hydrogen shoulder. This shoulder recalls that we have suspected to be due to the quench of neutral species in molecular complex configurations [6] and which we assumed above to effuse at about 500°C. These observations clearly indicate that the effusion process of the deuterium atoms is rate-limited by the surface desorption.

Finally, we have performed isothermal annealing at 400°C on etched samples during 3 min and 20 min (Fig.3) .The SIMS profiles obtained subsequently show a slight accumulation of deuterium under the surface indicating that the surface effect on effusion rate is independent of the presence of defects and traps due to plasma.

To confirm the surface effect and the assignment of the various effusion peaks to the different hydrogen configurations, further investigations are necessary. A quantitative study, combining effusion, SIMS and infrared spectroscopy techniques, on the influence of the deuteration temperature, the chemical nature of the dopant and the heating rate, is in progress and will be reported in a subsequent paper.

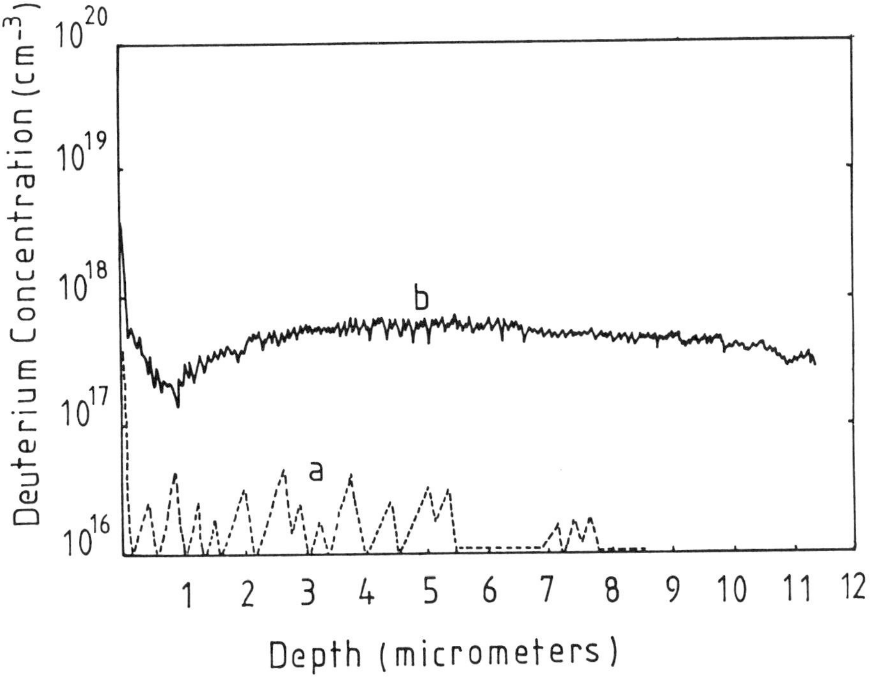

Fig. 3:

SIMS deuterium profiles after isothermal effusion at 400°C during 20 min (curve a) and 3 min (curve b).

5. REFERENCES

1. W. Beyer, in Tetrahedrally-Bonded Amorphous Semiconductors, eds. by D. Adler, and H. Fritzsche, (Plenum, New York, 1985), p.129.
2. W. Beyer, Physica B, 170, 105 (1991).
3. Hydrogen in Semiconductors, eds. J.I. Pankove and N.M. Johnson, Semiconductors and Semimetals, 34 (Academic Press, N.Y. 1991).
4. M. Stutzmann and M.S. Brandt, J. Appl. Phys. , 68, 1406 (1990).
5. R. Rizk, P. de Mierry, D. Ballutaud, M. Aucouturier, and D. Mathiot, Physica B, 170, 129 (1991).
6. R. Rizk, P. de Mierry, D. Ballutaud, M. Aucouturier and D. Mathiot, to be published in Phys. Rev. B.
7. K.J. Chang and D.J. Chadi, Phys. Rev. Lett., 62, 937 (1989).
8. T. Zundel and J. Weber, Phys. Rev. B, 43, 4361 (1991).

Materials Science Forum Vol. 83-87 (1992) pp. 51-56

EFFECT OF MULTIPLE TRAPPING ON HYDROGEN DIFFUSION IN SILICON

JEFFREY T. BORENSTEIN[1], JAMES W. CORBETT[2] AND STEPHEN J. PEARTON[3]

1 Mobil Solar Energy Corporation, 4 Suburban Park Drive, Billerica, MA 01821
2 Physics Department, SUNY at Albany, Albany NY 12222
3 AT&T Bell Laboratories, 600 Mountain Avenue, Murray Hill NJ 07974

ABSTRACT

The multiple trapping of hydrogen atoms at shallow acceptors during plasma hydrogenation of silicon is shown to explain several features of the resulting H diffusion profile. In heavily-doped p-type silicon, the plateau in the hydrogen profile at the acceptor concentration is preceded by a high concentration plateau near the surface. We demonstrate that this plateau can be explained by successive aggregation of hydrogen atoms at acceptors, and that the multiple trapping process appears to terminate after the agglomeration of eight to twelve H atoms. The reaction cross-section for multiple trapping is much lower than for the first capture, consistent with ionic capture of H^+ at acceptors followed by neutral multiple capture. Hydrogen dimer formation, dominant in high resistivity Si, may still be important in high-concentration near-surface profiles seen in heavily-doped silicon.

INTRODUCTION

Investigations of hydrogen diffusion in crystalline silicon have uncovered a wide array of hydrogen reactions in the lattice [1-7]. Most studies continue to support the notion that unencumbered hydrogen will diffuse at the rate determined by the early workers van Wieringen and Warmoltz (VWW) [8]. Yet extrapolation of the VWW diffusion coefficient to temperatures employed during standard plasma hydrogenation treatments yields a diffusion rate of several microns per minute, a rapid rate of penetration never experimentally observed. Kinetic studies [1,9-12] have shown that the in-diffusion depth is severely limited by reactions involving hydrogen capture, with the dominant reaction being capture at shallow dopant sites during the deactivation process. Even in undoped silicon hydrogen in-diffusion is retarded; the shape of the resulting diffusion profile can be explained by invoking the formation of hydrogen dimer complexes [9]. High concentrations of hydrogen near the surface after plasma hydrogenation have been attributed to these and other complexes.

Recently, Tulchinsky et. al. [13] proposed an explanation for the high levels of near-surface hydrogen in p-type silicon. Typical SIMS profiles show that the hydrogen concentration falls off exponentially with increasing depth, eventually reaching a plateau at the boron concentration. These workers demonstrated that the exponential decay can be explained by unsaturable trapping of hydrogen at the boron sites. One of the most striking features of the exponentially-decaying near-surface H profiles is the appearance of a second plateau, about one order of magnitude above the plateau at the doping level. This second plateau often appears as a protrusion or inflection point superimposed on the exponential decay profile. In the present work we demonstrate that the near-surface plateau may be caused by eventual saturation of the trapping sites.

HYDROGEN DEACTIVATION OF BORON - PLATEAU FORMATION

The hallmark feature of hydrogen profiles in passivated p-type silicon is the appearance of a plateau in H concentration at the boron doping level. Beyond this plateau, the hydrogen concentration falls off exponentially. Plateau formation is a direct result of bimolecular reaction kinetics controlling the deactivation process, and occurs whenever the boron trap concentration exceeds the hydrogen concentration.

For free, unencumbered diffusion within a lattice, the diffusion equation results in a complementary error function solution. In the presence of boron trapping sites, the kinetics are modified to include a reaction term, shown below:

$$\frac{\delta[H]}{\delta t} = D\frac{\delta^2[H]}{\delta x^2} - k_1 [H] [B] \qquad (1)$$

and

$$\frac{\delta[B]}{\delta t} = - k_1[H] [B] \qquad (2)$$

Here [H] is the hydrogen concentration, D the diffusion coefficient, and k_1 the forward reaction rate for H capture:

$$k_1 = 4 \pi R D, \qquad (3)$$

with R the capture radius for the reaction. The electrically active boron concentration [B] initially has the value $[B_0]$, equal to the doping concentration, and drops as the acceptor sites are passivated.

Numerical solutions of Eqns. (1) and (2) result in predicted hydrogen profiles such as the one shown in Figure 1. Here we overlay the corresponding experimental profile obtained for the conditions used: 1 h hydrogenation at 125 °C for 1 Ω-cm B-doped FZ-Si. The model predicts the existence of the experimentally-observed plateau; the extent of this plateau is controlled by the product of the hydrogen surface concentration, normally set at 10^{14} cm^{-3}, and the diffusion coefficient [9]. This product must be lowered by about one order of magnitude from the value determined for high resistivity silicon in Ref. [9]; the multiple trapping mechanism described later resolves this apparent discrepancy. The capture radius is determined by the hydrogen charge state [9]. For positively-charged H species the capture radius will take on the value for ion pairing.

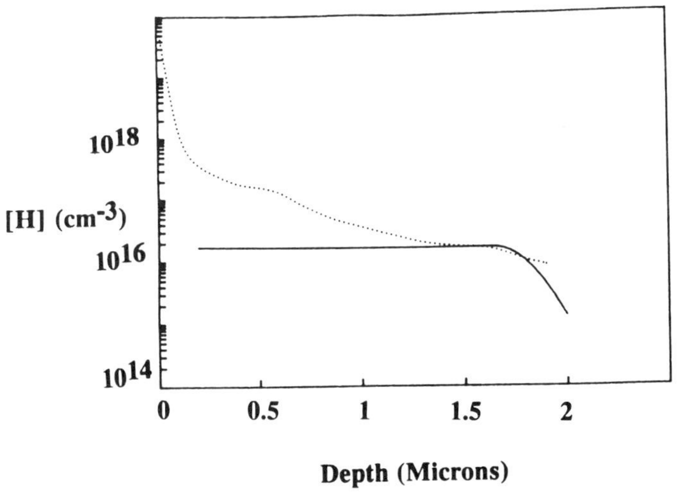

Figure 1. Single trapping model (solid line) vs. experiment (dotted line) for 1 Ω-cm B-doped FZ-Si plasma hydrogenated at 125 °C for 1 h.

Several points regarding the assumptions used to fit the data in Figure 1 are worthy of mention. The bimolecular reaction does not account for hydrogen dissociation from boron sites, which can be significant [14] but adds complexity to the analysis without changing the fundamentals of the reaction process. The drift field induced by the boron deactivation process [10] has also been excluded. Models which retain all of these effects [11] provide excellent fits to the data but make physical insights difficult. Finally, the experimental profile clearly diverges from the simple model near the surface. As described in the next section, infinite trapping of hydrogen at dopant sites results in a profile which, like the experimental near-surface profile, falls off exponentially.

INFINITE TRAPPING

Hydrogen profiles taken after passivation treatments in highly-doped p-type silicon exhibit exponential decay near the front surface. Previous work [13] has demonstrated that this results from unsaturable trapping; the reaction-diffusion equation governing the free hydrogen concentration can be written

$$\frac{\delta[H]}{\delta t} = D\frac{\delta\ [H]}{\delta x^2} - 4\pi D[H]\left(R_1[B] + R_2[B\cdot H] + R_3[B\cdot H\cdot H] + \ldots\right) \quad (4)$$

where R_2 is the capture radius for the second trapping reaction and [B·H] is the concentration of boron with one hydrogen, etc. If we assume that the capture radii for hydrogen-boron trapping reactions are all equal, so that $R_i = R_1$, we have

$$\frac{\delta[H]}{\delta t} = D\frac{\delta^2[H]}{\delta x_2} - 4\pi D R_1[B]\ [H]' \quad (5)$$

where $[B_0]$ is the total boron concentration, which obeys the following relation throughout the hydrogenation process:

$$[B_0] = [B] + [B{\cdot}H] + [B{\cdot}H{\cdot}H] + \ldots.. \quad (6)$$

In the quasi-steady-state regime, the time derivative in Eqn. (5) is equal to zero, and the equation has the solution

$$[H] = [H_0] \exp \{-\surd(4\pi R_1[B_0])\}. \quad (7)$$

In order to construct the in-diffusion profile, an expression for the total concentration of trapped hydrogen $[H_T]$ is required. The second term on the right side of Eqn. (5) corresponds to the time derivative of trapped H concentration:

$$\delta[H_T] = 4 \pi R_1 D [B_0] [H] \delta t. \quad (8)$$

Inserting Eqn. (7) into Eqn. (8), we have the solution for trapped hydrogen, given by

$$[H_T] = 4 \pi R_1 [B_0] D t [H_0] \exp \{-\surd(4\pi R_1[B_0]) x\}. \quad (9)$$

Since the trapped hydrogen concentration far exceeds the free hydrogen concentration in the region of interest, Eqn. (9) represents the approximate near-surface H profile in the case of unsaturable trapping. A typical profile obtained using the unsaturable trapping model is shown in Figure 2, contrasted with the experimental profile from Fig. 1. The observed decay region is reproduced, but two differences arise. The main plateau at the boron concentration is not predicted by the unsaturable trapping model, nor is the rounded plateau which appears as an inflection point in the near-surface exponential decay profile. In the next section we argue that these two features can be explained by assuming that the multiple trapping reaction ceases after a limited number of captures.

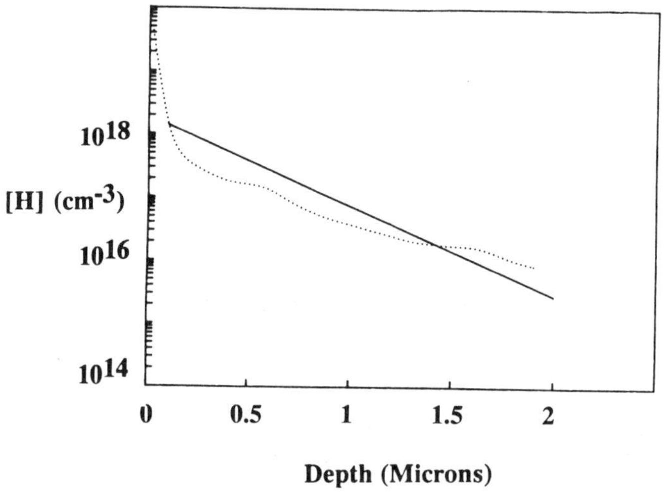

Figure 2. Model calculations for unsaturable trapping (solid line) vs. experiment (dotted line).

MULTIPLE TRAPPING

In order to obtain results for the reaction-diffusion equations for the case of limited multiple capture of hydrogen atoms at boron sites, a numerical solution must be calculated. The pseudo-Crank-Nicholson approach used here has been described previously [15]. By modifying Eqns. (4) and (8) appropriately, we find the set of coupled equations

$$\frac{\delta[H]}{\delta t} = D\frac{\delta\ [H]}{\delta x^2} - 4\pi D[H]\left\{R_1[B]+R_2[B{\cdot}H]+\cdots+R_n[B{\cdot}H(n)]\right\} \quad (10)$$

$$\frac{\delta[H_T]}{\delta t} = 4\pi D\ [H]\ \{R_1[B]+R_2[B{\cdot}H]+\cdots+R_n[B{\cdot}H(n)]\} \quad (11)$$

Here the notation $[B{\cdot}H(n)]$ represents a boron complex with **n** hydrogen atoms. In order to reduce the number of free parameters, the capture radii for all multiple trapping reactions are made equal: $R_i\ (i\geq 2) = R_2$. In addition, the VWW diffusion coefficient, hydrogen surface concentration, and ionic capture radius R_1 discussed earlier have been invoked. Thus the only free parameters are the multiple capture radius R_2 and the size of the largest complex $[B{\cdot}H(n)]$.

Numerical solutions to Eqns. (10) and (11) have been obtained for various values of **n** and R_2. For large **n** values ($n\geq 30$) the predicted hydrogen profile approaches an exponential decay, as expected. For smaller **n** values, the profile contains a discernable plateau at high concentration near the surface. The predicted hydrogen profile for n=10 is plotted against the experimental data in Figure 3. The value of R_2 has been set at 0.25 Angstroms, much smaller than the initial ionic radius, and suggestive of a geometry-dependent capture process.

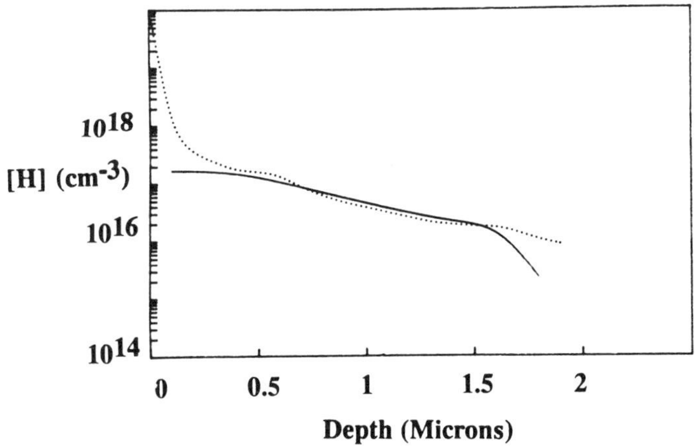

Figure 3. Model calculations for multiple trapping of up to ten H atoms (solid line) vs. experiment (dotted line).

The calculations for multiple trapping find that the hydrogen concentration at the near-surface plateau rises as the size of the largest complex increases. When the value of **n** is set to values of between 8 and 12, the model finds the higher plateau at a concentration at about ten times the boron concentration, most nearly approximating the experimental data. The average cluster size in the first two microns contains approximately four hydrogen atoms, with the larger clusters forming nearer to the surface. Some details of the experimental hydrogen profile are not reproduced by the model calculation, suggesting that additional effects must be included in order to fully describe the observed behavior. The high-concentration region above the multiple trapping plateau may be due to the formation of hydrogen dimers [4], or may be caused by additional multiple captures with much smaller capture cross-sections.

CONCLUSIONS

The features of hydrogen in-diffusion profiles obtained after plasma hydrogenation of p-type crystalline silicon can be explained invoking the principle of multiple trapping of hydrogen atoms at boron acceptor sites. The kinetic model shows that the multiple trapping process appears to terminate with the capture of eight to twelve hydrogen atoms.

REFERENCES

1. J.W. Corbett, J.L. Lindstrom, S.J. Pearton and A.J. Tavendale, Solar Cells 24, 127 (1988).
2. S.J. Pearton, J.W. Corbett and T.S. Shi, Appl. Phys. A 43, 153 (1987).
3. S.J. Pearton, J.W. Corbett and J.T. Borenstein, Physica B 170 85 (1991)
4. S.J. Pearton, M. Stavola and J.W. Corbett, Rad. Eff. and Defects in Solids 111-112 [1-2], 323 (1989).
5. C.G. Van de Walle, Y. Bar-Yam and S.T. Pantelides, Phys. Rev. Lett. 60 2761 (1988).
6. K.J. Chang and D.J. Chadi, Phys. Rev. B 40 11644 (1989).
7. C.H. Seager, R.A. Anderson and J.K.G. Panitz, J. Mater. Res. 2, 96 (1987).
8. A. Van Wieringen and N. Warmoltz, Physica 22, 849 (1956).
9. J.T. Borenstein, D. Tulchinsky and J.W. Corbett, Mat. Res. Soc. Proc. 163 633 (1990).
10. C.H. Seager and R.A. Anderson, Mat. Res. Soc. Proc. 138, 209 (1989). p. 197.
11. D. Mathiot, Phys. Rev. B 40, 5867 (1989).
12. J.P. Kalejs and S. Rajendran, Appl. Phys. Lett., 55, 2763 (1989).
13. D.A. Tulchinsky, J.W. Corbett, J.T. Borenstein and S.J. Pearton, Phys. Rev. B 42 11881 (1990).
14. T. Zundel and J. Weber, Phys. Rev. B 39 13549 (1989).
15. J.T. Borenstein, D. Angell and J.W. Corbett, Mat. Res. Soc. Proc. 138, 209 (1989).

Materials Science Forum Vol. 83-87 (1992) pp. 57-62

LONG-RANGE CORRELATIONS AND HYDROGEN DIFFUSION IN SILICON*

JAMES W. CORBETT, AND IGOR V. VERNER
The Institute for the Study of Defects in Solids, Department of Physics, University at Albany-SUNY, Albany, N.Y. 12222, USA

ABSTRACT

It is shown that the depth profiles with plateaus observed, in some cases, in the diffusion of hydrogen into semiconductors, e.g. silicon, may be explained by including terms in the diffusion equations that describe the different ways for interactions of hydrogen with impurity atoms. Such plateau-type profiles may be formed due to the loss of stability in the system with the local interactions and diffusion exchange between neighbor volumes and due to the appearance of long-range correlations in this case. It is shown that the behavior of the system depends on the dimensionality of the random walk through which the hydrogen diffuse. It is also shown that changes of temperature and concentration of impurity alter the parameters of the random walk, and accordingly this change the character of the diffusion processes.

1. INTRODUCTION

Recently interest in the problem of hydrogen in semiconductors, such as silicon, germanium and some III-V compound semiconductors increased due to an importance of this impurity in the field of modern microelectronics, because hydrogen can change the properties of semiconductors and this fact is extremely important in submicron electronic technology. On the other hand, hydrogen is a very interesting impurity in itself, and only now are some aspects of its behavior in the matrix of elemental semiconductors, like silicon, starting to be understood.

Hydrogen can be incorporated into semiconductors in various ways: by implantation, by plasma-treatment, by wet-etching or electrolysis, during the melt-growth of semiconductor crystals and during annealing in the atmospheres containing hydrogen or hydrogen-containing molecules, etc. [1] There is a wide range of changes in the properties of semiconductors related to presence of hydrogen. There are passivation of the electrical activity of defects and impurity states, reconstruction of surface dangling bonds and defects at interfaces and also reconstruction of the dislocations due to the changes of the dislocation atmosphere, suppression of swirl defects formation during the melt growth of semiconductors, etc. In the field of research of defects in semiconductors hydrogen also plays an important role, because it can be used for masking of defects, for example, for studies of defects by electron paramagnetic resonance [2].

Of all the problems concerning the behavior ofhydrogen in semiconductors matrix which are not yet solved, diffusion processes and the depth profiles associated with hydrogen diffusion in semiconductors exhibit a special interest. A variety of profiles have been observed experimentally for the system "silicon + hydrogen". Under some conditions, the profile is a standard complementary error function, erfc, whose behavior is describable on the basis of Fick's law. However, in some experimental works the diffusion profile has the opposite curvature to the erfc, and in many cases the diffusion profile exhibits a constant plateau (in some experiments, more than one plateau) [3,4]. The existence of such peculiarities are contradictions to the Fick's laws, because the classical diffusion cannot describe such things as plateaus, breaks or parts that deviate from the standard erfc-type curve. The effect of plateau formation in the hydrogen depth profile strong depends on temperature and concentration of other impurity atoms such as boron or phosphorus [3,5]. Several theoretical models have been proposed as an explanation for this phenomenon [4,5-7]. Some of these models were based on the concept of interactions and the formation of complexes of hydrogen with other impurity atoms or amongst other hydrogen atoms. Diffusion based on the process of complex formation is different from classical diffusion. There are indications for trapping of much more than one hydrogen atom and accordingly a model for impurity trapping at unsaturable traps was proposed by Tulchinsky *et al.* [6]. In this paper the authors suggest that platelets may act as unsaturable traps as far as influencing the diffusion profile. They also showed that the diffusion of hydrogen in a percolating lattice must be different than that in a simple cubic lattice. Based on this model, the effective dimensionality of the random walk was found to be infinite. Capizzi and Mattiga [7] proposed a theoretical model that takes into account both neutral and ionized hydrogen diffusion, the

*Supported in part by the Solar Energy Research Institute, IBM Corporation, and Mobil Foundation.

latter enhanced by a built-in electric field associated with the hydrogen doping gradient. This model neglects trapping at impurities and suggests that the formation of molecules and/or trapping plays a minor role in hydrogen diffusion at high temperature. However, none of these models can adequately give the complete picture of the observed behavior. In particular, they cannot explain from a single point of view the temperature and dopant dependencies of the plateau formation process and also the existence, in some cases, of more than one plateau in the depth profiles.

2. A KINETIC MODEL AND LONG-RANGE CORRELATIONS IN THE "HYDROGEN + IMPURITY" SYSTEM OF A SEMICONDUCTOR

In the present work, we considered the process of hydrogen diffusion from the following point of view. It is possible to assume that the nonlinear character of hydrogen diffusion observed in experiments is caused by nonlinear interactions between hydrogen atoms, impurity atoms and self-defects of semiconductor. Such nonlinear interactions under condition where neighboring volumes of the semiconductor are involved in the exchange of matter (diffusion) can be a reason for the formation of feed-backs in the impurity-defects subsystem of the semiconductor. These feed-backs may stimulate a loss of stability in the system. In that case, the fluctuations of concentration will not immediately return to the steady state. These fluctuations can be very large, and according to the transport (diffusion) processes these fluctuations can be transported to the neighboring volumes and then only slowly disappear when diffusion is unable to support the fluctuation transport. However, in some cases, such fluctuations may even grow. As a results, in a system that consists of objects, even with interactions of a short-range nature , long-range correlations may arise. Plateau formation in the hydrogen depth profile may be caused by these correlations. The correlation length of such fluctuations depends on the character of the diffusion processes and on the dynamics of the interactions between objects from which the system consists, i.e. on local quasichemical processes.
Studies of hydrogen in silicon have shown what an important role the interaction of hydrogen with impurities plays in diffusion processes and in the formation of the resulting depth profile [4-6]. We shall consider some scheme of interactions that are based on the assumption that hydrogen can be trapped and immobilized at an impurity site with a total concentration $[X_t]$. In this scheme we will be include the possibility of the trapping of more then one hydrogen atom at an impurity atom, the possibility of different ways for this process to occur and the possibility of dissociation of complexes. For a qualitative analysis, we will consider the formation of complexes up to a complex including two hydrogen atoms on one impurity center. The qualitative results will not change in the case when we include the possibility of the trapping of multiple hydrogens, but the analytical investigation of the problem will become very difficult. Schematically, all acts of interaction which we will study here can be introduced in the following symbolic forms:

$$[\mathbf{H}]+[\mathbf{H}] \underset{\beta_1}{\overset{\alpha_1}{\longleftrightarrow}} [\mathbf{H}\cdot\mathbf{H}]\,, \tag{1}$$

$$[\mathbf{H}]+[\mathbf{X}] \underset{\beta_2}{\overset{\alpha_2}{\longleftrightarrow}} [\mathbf{H}\cdot\mathbf{X}]\,, \tag{2}$$

$$[\mathbf{H}\cdot\mathbf{X}]+[\mathbf{H}] \underset{\beta_3}{\overset{\alpha_3}{\longleftrightarrow}} [\mathbf{H}\cdot\mathbf{X}\cdot\mathbf{H}]\,, \tag{3}$$

$$[\mathbf{H}\cdot\mathbf{H}]+[\mathbf{X}] \underset{\beta_4}{\overset{\alpha_4}{\longleftrightarrow}} [\mathbf{H}\cdot\mathbf{H}\cdot\mathbf{X}]\,, \tag{4}$$

$$[\mathbf{H}] \xrightarrow{\alpha_5} \mathbf{S}\,,\ [\mathbf{H}\cdot\mathbf{H}] \xrightarrow{\alpha_6} \mathbf{S}\,, \tag{5}$$

where [H] is the atomic concentration of hydrogen; [X] is the atomic concentration of impurity; [H·H], [X·H], [H·X·H] are the concentrations of di-hydrogen and complexes of one and two hydrogen atoms with the impurity, respectively; $\alpha_1,\ldots,\alpha_6$, $\beta_1,\ldots,\beta_4$ are the kinetic coefficients; and S is the symbol for the sinks. Last two reactions (5) describe the loss of [H] and [H·H] on sinks (or, in general, all first-order reactions of [H] and [H·H] that are not included in remaining scheme). The kinetic coefficients which describe the forward reactions (capture rate constants) can be expressed in the general form for objects a and b as:

$$\alpha_{\mathbf{n}} = 4\pi(\mathbf{D_a} + \mathbf{D_b})\mathbf{R}(\mathbf{a},\mathbf{b}) \quad (6)$$

with D_a and D_b - the diffusion coefficients of objects a and b, respectively, and R(a,b) is the reaction radius for a specific reaction. For reaction (1) $\alpha_1 = 2\pi(\mathbf{D_H})\mathbf{R}(\mathbf{H},\mathbf{H})$. In the last reactions two α_5, α_6 are also proportional to the corresponding diffusion coefficients (D_H and D_{HH}, respectively), to the radius of reactions and to the concentration of sinks.
The dissociation rates also can be estimated from:

$$\beta_{\mathbf{n}} = \nu_{\mathbf{n}} \exp(-\frac{\mathbf{E_a}}{\mathbf{kT}}), \quad (7)$$

where $\nu_{\mathbf{n}}$ is the vibrational frequency associated with the dissociation of complex, and E_a is the energy associated with that dissociation.
The kinetic equations will be equations for each of the reactant species are as follow:

$$\frac{\partial[\mathbf{H}]}{\partial \mathbf{t}} = -\alpha_1[\mathbf{H}]^2 + \beta_1[\mathbf{H}\cdot\mathbf{H}] - \alpha_3[\mathbf{H}]\cdot[\mathbf{X}] + \beta_3[\mathbf{X}\cdot\mathbf{H}] - \alpha_2[\mathbf{H}]\cdot[\mathbf{X}\cdot\mathbf{H}] + \\ +\beta_2[\mathbf{H}\cdot\mathbf{X}\cdot\mathbf{H}] - \alpha_5[\mathbf{H}] + \mathbf{D_H}\frac{\partial^2[\mathbf{H}]}{\partial \mathbf{R}^2} \quad (8)$$

etc, * * *

$$\frac{\partial[\mathbf{H}\cdot\mathbf{X}\cdot\mathbf{H}]}{\partial \mathbf{t}} = +\alpha_2[\mathbf{H}]\cdot[\mathbf{X}\cdot\mathbf{H}] - \beta_2[\mathbf{H}\cdot\mathbf{X}\cdot\mathbf{H}] + \alpha_4[\mathbf{H}\cdot\mathbf{H}]\cdot[\mathbf{X}] - \\ -\beta_4[\mathbf{H}\cdot\mathbf{X}\cdot\mathbf{H}] \quad (9)$$

The total concentration of impurity species $[X_t]$ at any time t is given by:

$$[\mathbf{X_t}] = [\mathbf{X}] + [\mathbf{X}\cdot\mathbf{H}] + [\mathbf{H}\cdot\mathbf{X}\cdot\mathbf{H}] \quad (10)$$

We will assume that the mobility of the impurity related defects is less than the mobility of pure hydrogen atoms and the di-hydrogen complexes. For the last two objects, we will assume that the diffusion of the single hydrogen atom is much faster than the complex [H·H]. For the following analysis it is better to introduce equations (8)-...-(9) and condition (10) in the form of dimensionless variablesas follow:

$$\frac{\partial \mathbf{z}_1}{\partial \tau} = -\vartheta\delta_1\mathbf{z}_1 - \vartheta\mathbf{z}_1 + \beta\mathbf{z}_2 - \gamma_3\mathbf{z}_1\mathbf{z}_3 + \frac{\gamma_3}{\theta_3}\mathbf{z}_4 - \\ -\gamma_2\mathbf{z}_1\mathbf{z}_4 + \frac{\gamma_3}{\theta_2}\mathbf{z}_5 + \mathbf{d}_1\frac{\partial^2\mathbf{z}_1}{\partial \mathbf{l}^2}, \quad (11)$$

$$\frac{\partial \mathbf{z}_2}{\partial \tau} = +\vartheta\delta_1\mathbf{z}_1 - \beta\mathbf{z}_2 - \gamma_4\mathbf{z}_2\mathbf{z}_3 + \frac{\gamma_3}{\theta_4}\mathbf{z}_5 - \mathbf{z}_2 + +\mathbf{d}_2\frac{\partial^2 \mathbf{z}_2}{\partial \mathbf{l}^2}, \tag{12}$$

$$\varepsilon\frac{\partial \mathbf{z}_3}{\partial \tau} = (-\theta_4\mathbf{z}_1\mathbf{z}_3 + \chi_2\mathbf{z}_4 - \delta_3\theta_4\mathbf{z}_2\mathbf{z}_4 + \mathbf{z}_5), \tag{13}$$

$$\varepsilon\frac{\partial \mathbf{z}_4}{\partial \tau} = (\theta_4\mathbf{z}_1\mathbf{z}_3 - \chi_2\mathbf{z}_4 - \delta_2\theta_4\mathbf{z}_1\mathbf{z}_4 + \chi_1\mathbf{z}_5), \tag{14}$$

$$\varepsilon\frac{\partial \mathbf{z}_5}{\partial \tau} = (\delta_2\theta_4\mathbf{z}_1\mathbf{z}_4 - \chi_1\mathbf{z}_5 + \delta_3\theta_4\mathbf{z}_2\mathbf{z}_4 - \mathbf{z}_5), \tag{15}$$

and

$$\mathbf{z}_3 + \mathbf{z}_4 + \mathbf{z}_5 = 1 \tag{16}$$

In the equations (18)-(23) the dimensionless variables and parameters given by:

$$\begin{gathered}
\mathbf{z}_1 = \frac{\alpha_3}{\alpha_5}\cdot[\mathbf{H}],\ \mathbf{z}_2 = \frac{\alpha_3}{\alpha_5}\cdot[\mathbf{H}\cdot\mathbf{H}],\ \mathbf{z}_3 = \frac{[\mathbf{X}]}{[\mathbf{X_t}]},\ \mathbf{z}_4 = \frac{[\mathbf{X}\cdot\mathbf{H}]}{[\mathbf{X_t}]},\\
\mathbf{z}_5 = \frac{[\mathbf{H}\cdot\mathbf{X}\cdot\mathbf{H}]}{[\mathbf{X_t}]},\ \tau = \alpha_6\mathbf{t},\ \mathbf{d}_1 = \frac{\mathbf{D_H}}{\alpha_6\mathbf{L}^2},\ \mathbf{d}_2 = \frac{\mathbf{D_{H\cdot H}}}{\alpha_6\mathbf{L}^2},\ \varepsilon = \frac{\alpha_6}{\beta_4},\\
\delta_1 = \frac{\alpha_1}{\alpha_3}, \delta_2 = \frac{\alpha_2}{\alpha_3}, \delta_3 = \frac{\alpha_4}{\alpha_3},\ \gamma_2 = \frac{\alpha_2[\mathbf{X_t}]}{\alpha_6},\ \gamma_3 = \frac{\alpha_3[\mathbf{X_t}]}{\alpha_6},\\
\gamma_4 = \frac{\alpha_4[\mathbf{X_t}]}{\alpha_6},\ \theta_2 = \frac{\alpha_5}{\beta_2}, \theta_3 = \frac{\alpha_5}{\beta_3}, \theta_4 = \frac{\alpha_5}{\beta_4},\ \beta = \frac{\beta_1}{\alpha_6},\\
\chi_1 = \frac{\beta_2}{\beta_4}, \chi_2 = \frac{\beta_3}{\beta_4}, \vartheta = \frac{\alpha_5}{\alpha_6}.
\end{gathered} \tag{17}$$

In this brief paper we cannot introduce all algebraic steps for searching the solutions of (11)-(15) with (16). In the case when $\varepsilon \to 0$ these equations have three positive stationary-states for Z_1 and Z_2. Detailed discussion of the behavior of stationary-states solutions and detail of the stability analysis will be published separately. Here we only want to say that straightforward but very lengthy algebraic calculations showed that solutions can be unstable under some critical combination of parameters such as temperature and the total concentration of impurity atoms. For a qualitative understanding of the behavior of our system, it is only necessary to know this fact (fact of instability). When the system becomes unstable as a result of nonlinear interactions between hydrogen and impurity atoms, the fluctuations in the concentration become much larger than in the stable situation and they can "grow" due to the effect of positive feed-backs described by reactions (3) and (4). According to the diffusion process (exchange between tneighboring volumes), the fluctuations can be transported to the nearest unstable region where they can play the role of new perturbations, they can increase due to the existence of positive feed-backs, and the transport of the fluctuations can be continued until diffusion is able to transmit the perturbations. The resulting length of such correlations can be approximately estimated as follow:

$$L_{corr} \approx \sqrt{\frac{D_H}{\tilde{\gamma}}}, \tag{18}$$

where D_H is the diffusion coefficient of diffused objects (in our case it is the hydrogen atoms) and $\tilde{\gamma}$ is the coefficient which characterize the slowest step of the dynamic processes described by reactions (1)-(5). In terms of the above introduced mechanism, the plateau formation process in the depth profile of hydrogen in silicon is the results of the transport of concentration fluctuations. If the fluctuation of concentration arose at some "point" then it can move a distance $L_{corr.}$ and then the system will return to its initial state. But if the system is unstable, new fluctuation can appear. The formation of such fluctuations and plateaus for hydrogen diffusion does not occur in the near-surface region, where the concentration of hydrogen atoms is up to two order large than the concentration of impurity atoms, because in this region other types of reactions occur which do not bring about instability, as they do at deeper depths. This is the reason for the sharp decreasing of the hydrogen concentration after $L_{corr.}$. For some estimation of $L_{corr.}$, we can use the values of the parameters from the literature [3-6]. If we assume that $\tilde{\gamma}$ is of the same order of magnitude that the time of impurity-hydrogen complexes dissociation, then for diffusion coefficient of $D=5.4.10^{-12}$ m^2/sec and a dissociation energy $E_{ij} \approx 0.42$ eV at $T=125^oC$, we have: $L_{corr.} \approx 0.34\mu$. From experimental curves [3] this length can be estimated to be about 0.31 - 0.46 μ.

3. HYDROGEN DIFFUSION TROUGH THE RANDOM WALK

In the above considered system, we are dealing with two different types of processes: local quasichemical processes (1)-(5) and the transport processes, i.e. diffusion. These two processes are both important and necessary, because the first of them "form" (or produce) the fluctuations while the second one is responsible for transportation of the fluctuations, i.e. we need in the transport and we need in the object of this transport. The characteristic time τ_D, during which the heterogeneity in the cell with linear size $\mathcal{L}$ will be destroyed by diffusion, is [8]:

$$\tau_D \approx \mathcal{L}^2 \cdot D^2 \tag{19}$$

We can introduce the size $\mathcal{L}$ through the topological dimensionality (d) of the space in which our system exists:

$$\mathcal{L}^d = \Delta V, \tag{20}$$

where ΔV is the topological volume of the cell. In the final form we have:

$$\frac{1}{\tau_D} = D \cdot \Delta V^{-\frac{2}{d}} \tag{21}$$

According to (21), the magnitude of the diffusion goes down with the decreasing of the topological dimensionality and vice versa. As was shown in [6], impurity atoms in silicon can form fractal lattice. Formation of the fractal lattice is connected with the decreasing of the topological dimensionality. The more compact (dence) lattice has low dimenşionality and the smaller intensity of the diffusion. Therefore, if for some value of d we had long-range correlations in the system, and then after that we start to decrease the dimensionality (for example, by increasing the concentration of dopant), then below some critical value of d (d_C) the correlations will disappear. At the "point" d_C, τ_D must be equal to the characteristic time of the quasichemical processes ($\tau_{q.-ch}$). The time $\tau_{q.-ch}$ is not dependent on the dimensionality d according to the local character of these processes.

Using the condition $\tau_D \approx \tau_{q.-ch}$ we can couple the dimensionality with the internal characteristics of the system and find d_C.

Thus, for $\mathcal{L}\approx 1$ μ, D=5.4.10^{-12} m^2/sec , and $\tau_{q.-ch} \approx \frac{1}{\tilde{\gamma}} \approx 212.8\,\text{sec}$ d_C can be estimated to be ≈1.08. When $d > d_C$ then the long-range correlations arise in the system. When $d < d_C$ then diffusion is insufficient to support correlations between neighboring cells.

It has been shown in experiments that for high levels of dopant concentration plateaus formation is suppressed. From the above introduced point of view, this means that for high level of dopant the "random walk" has a large density, low value of d and low intensity of diffusion. With the decreasing of the dopant level, the dimensionality of the fractal lattice will increase and intensity of the diffusion will also increase. When the value of dopant concentration reachs the value corresponding to d_C , long-range correlations arise in the system. For example in p-type Si this happens when the dopant level is about 0.1 Ω-cm. A similar effect takes place when the temperature changes, because increasing the temperature results in an increasing of the diffusion coefficient (see equations (21)). This fact also explains the difference between the length of the plateaus for different temperatures for the same level of impurity concentration. According to (18), we have in this case:

$$L_{corr}(T_1) > L_{corr}(T_2),\ T_1 > T_2\,. \tag{22}$$

4. SUMMARY

In summary, we propose that the process of plateau formation observed in the depth profiles, in some cases, for the diffusion of hydrogen into semiconductors, e.g. silicon, may be explained and described by including terms in the diffusion equations which describe the different ways for interacting hydrogen atoms with other impurity atoms, like boron. We argue that such plateau-type profiles occur due to the loss of stability in the system as a result of interactions and diffusion exchange between neighbor volumes and due to the appearance of long-range correlations in this case. We showed that the behavior of the considered system depends on the dimensionality of the fractal lattice through which the hydrogen diffuse. The value of the critical dimensionality, above which the effect of long-range correlation occur, depends on the type of interaction, on the temperature and on the concentration of impurity atoms. Changes in the parameters of the system (temperature, concentration of impurity) alter the characteristics of the random walk through which hydroigen diffuse and, accordingly, change the character of the diffusion process. As a result, these processes are visualized on the observed experimental hydrogen depth profiles obtained for different concentrations of impurity and for different temperature regimes. Some numerical estimations, which follow from the above introduced model (for example, plateaus length), yield a good correspond to what was found experimentally. As a next step in the modeling of hydrogen diffusion in semiconductors it is necessary to include interactions due to the charge states and the drifting internal electric field.

REFERENCES

1. *Hydrogen in Semiconductors,* eds. Jacques I. Pankove, and Noble M. Jonson (Academic Press, Inc., Boston, San Diego, New York, London, Sydney, Tokyo, Toronto, 1991).
2. S.J. Pearton, M. Stavola, and J.W. Corbett, in: Defects in Semiconductors, ed. G. Ferenczi (Trans Tech, Switzerland, 1989) p. 25.
3. J.W. Corbett, J.L. Lindstrom, S.J. Pearton, and A.J. Tavendale, Solar Cells, 24, 127 (1988).
4. J.W. Corbett, J.L. Lindstrom, L.C. Snyder, and S.J. Pearton, Mater. Res. Soc. Symp. Proc., 104, 229 (1988).
5. J.T. Borenstein, D. Angell, and J.W. Corbett, Mat. Res. Soc. Symp. Proc., 138, 209 (1989).
6. D.A. Tulchinsky, J.W. Corbett, J.T. Borenstein, and S.J. Pearton, Phys. Rev.B, 42, 881 (1990).
7. M. Cappizzi, and A. Mattigi, Appl. Phys. Lett, 50, 918 (1987).
8. G. Nicolis, and I. Prigogine (eds.): Exploring Complexity: An Introduction (W.H. Freeman and Company, New York, 1989)

Materials Science Forum Vol. 83-87 (1992) pp. 63-68

CHARGE STATES OF DONOR-HYDROGEN PAIRS IN Si: A FRAGILE BALANCE

Stefan K. ESTREICHER,† Carleton H. SEAGER,‡ and Robert A. ANDERSON‡

†Physics Department, Texas Tech University, Lubbock, TX 79409-1051 (USA).

‡Sandia National Laboratories, Division 1112, Albuquerque, NM 87185 (USA).

ABSTRACT

The debonding rates of H-passivated P and As in silicon have been observed to be very sensitive to the concentration of majority and minority charge carriers. A theoretical study of the stable and metastable configurations of the {P,H} and {As,H} pairs in the 0 and +1 charge states has been carried out at the near-*ab-initio* Hartree-Fock level. These calculations show that the lowest-energy configuration in the 0 charge state is the highest-energy configuration in the +1 charge state and vice-versa. This bistability of donor-hydrogen pairs implies that H cannot remain in place upon change of charge state, whether $0 \longrightarrow +1$ or $+1 \longrightarrow 0$. Quantitative differences between the P and the As cases are qualitatively consistent with the observed differences in the temperature dependence of the debonding rates of {P,H} and {As,H}.

1. INTRODUCTION

The passivation by atomic hydrogen of the electrical activity of shallow acceptors and donors has been a subject of intense research in the past ten years. A number of properties of the ground states of (electrically neutral) acceptor-hydrogen and donor-hydrogen pairs in Si have been theoretically predicted and experimentally observed (for recent reviews, see Refs.1-3). It is now apparent that there are several fundamental differences in the interactions of H with shallow acceptors and donors.

(i) Hydrogen passivation of shallow acceptors is very efficient,[4] while that of donors is much more difficult to achieve.[5] In the case of B, more than 99% of the acceptors near the surface can be passivated. This passivation probably results from the reaction $B^- + H^+ \longrightarrow \{B,H\}^0$, i.e., H^+ experiences a long-range Coulomb attraction to B^-. On the other hand, there is no consensus (among theoreticians as well as among experimentalists) as to the existence of H^- in Si. However, whether it exists or not, there is clearly not enough H^- to passivate all the P^+ centers, and the passivation reaction is probably of the type $P^+ + H^0 + e^- \longrightarrow \{P,H\}^0$. This implies a much smaller (non-coulombic) capture cross section. Note that there is considerable evidence[6,7] for the coexistence of H^+ and H^0 from ex-

periments performed in the depletion region of $p-$type diodes. These same species should be present in similar ratios in n-type diode depletion regions.

(ii) H-acceptor pairs are substantially more stable than H-donor pairs. The reactivation (debonding of H leaving an electrically active dopant) is an activated process which requires temperatures of at least $100\,^{\circ}C$ (in reverse bias) in the case of {B,H}. In equilibrium, {P,H} pairs break up at lower temperatures, and in illuminated Si, this can occur at $300\,^{\circ}K$. Further evidence of a weaker interaction between H and P than H and B is provided by infra-red absorption data. The Si$-$H stretching vibration is near $1875\,cm^{-1}$ in the case of {B,H} (Ref.7), with H near the bond-centered (BC) site: B$\cdots$H$-$Si. On the other hand, the dominant Si$-$H stretching mode in the case of donors is much lower,[8] near $1560\,cm^{-1}$, indicating a much weaker Si$-$H interaction. In this case, H is at the antibonding (AB) site of a Si atom: P$\cdots$Si$-$H. It should be noted that in addition to the donor-independent $1560\,cm^{-1}$ line, a second, donor-dependent, Si$-$H stretching vibration is observed:[9] It is $1647\,cm^{-1}$ for P, $1661\,cm^{-1}$ for As, and $1671\,cm^{-1}$ for Sb. The configurations corresponding to these modes have yet to be identified.

(iii) The debonding rate of donor-hydrogen complexes is highly sensitive to the concentration of majority and minority charge carriers. Recent investigations of the processes involved have revealed a variety of phenomena which suggest that charge state effects are important in determining the stability of the donor-hydrogen pairs. Seager and Anderson[10] have shown that, for concentrations of P ranging from $6 \cdot 10^{14}$ to $6 \cdot 10^{16}\,cm^{-3}$, the dissociation rate of {P,H} pairs is (a) inversly proportional to the concentration of electrons in equilibrium, and (b) proportional to the population of excess holes injected with forward bias or light. They have also obtained evidence that a fraction of {P,H} pairs can be converted to a positive charge state after anneals in depletion. Much of this data can be understood if the dissociation of donor-hydrogen pairs proceeds only after conversion to a positive charge state.

We report here a study of the properties of the {P,H} and {As,H} complexes in the 0 and +1 charge states. The calculations have been carried out near the *ab-initio* Hartree-Fock (HF) level, with the method of PRDDO.[11] This method contains no adjustable parameters, and has recently been used[12] to study the properties of five acceptor-H and donor-H complexes. The host crystal is represented by clusters containing 7 and 43 host atoms (in the case of P), and 7 and 17 host atoms (in the case of As). The surface dangling bonds are saturated with H atoms as discussed in Ref.13. The basis set includes valence d orbitals on Si and P, while As has only $1s$, $2sp$, $3spd$, and $4sp$ orbitals. It would be preferable to include a set of *valence* d orbitals ($4d$) on As as well. However, the current version of the code does not allow multiple sets of d functions on the same atom.

2. BISTABILITY OF {P,H} AND {As,H}

Figure 1 shows the equilibrium geometries of the $\{P, H\}^0$, $\{P, H\}^+$, $\{As, H\}^0$, and $\{As, H\}^+$ complexes in the {110} plane in Si. In addition to bond lengths, the figures contain the degrees of bonding[14] associated with each bond. The differences between the 0 and +1 charge states of the donor-hydrogen complexes involve not only the geometry of the ground state (shown in the figure) but the metastable states as well.

Neutral charge state: In the lowest-energy configuration, Figs.1(a) and 1(c), H is at the *AB* site of a Si atom adjacent to the donor and forms a rather strong bond with it. The Si atom is displaced almost to the plane of its 3 nearest neighbors (NNs), and is nearly sp^2−coordinated with these 3 atoms. H attaches to remaining $3p_z$ orbital. There is little overlap between the donor and this Si atom. The donor is only slightly displaced from its perfect substitutional site and is essentially 3-fold coordinated.

The metastable state is 1.14 eV above the ground state in the case of P and corresponds to H at the *AB* site of P. The configuration where H is near the *BC* site, bridging a P−Si bond, is 1.40 eV above the ground state and could be a saddle point of the energy. In the case of As, the configuration where H is *AB* to As involves a 5-fold coordinated donor atom. This particular configuration is always stabilized by the inclusion of valence *d* orbitals which participate to the bonding in that coordination more than when the group V atom is 3-fold or 4-fold coordinated. As mentioned above, we cannot include 4*d* orbitals on As at the present time. However, a study of the gain in energy upon inclusion of valence *d* orbitals for comparable bonding configurations of Si and P gives strong indications that the energy of the *AB* configuration H−As−Si is about 1.5 eV above the ground state. Finally, the *BC* site for H is again the highest inenergy, 2.25 eV above the ground state.

Positive charge state: The lowest-energy configuration is shown in Figs.1(b) and 1(d). H is near the *BC* site, between the donor and a Si atom, and forms a much stronger bond with Si than with the donor. This bond has only about 80% covalent character. There is very little covalent interaction between H and the donor. Most of the odd electron (about 70%) resides on the donor, the rest being on the Si atom.

The metastable state corresponds to H *AB* to the donor, and is 0.69 eV above the ground state in the case of P. In this configuration, H forms a substantial bond with the donor, which is 5-fold coordinated. In the case of As, the metastable state is only 0.37 eV above the ground state, and H is quite close to the tetrahedral interstitial site, where it is 90% atomic. It forms only a weak bond with As, which is essentially 4-fold coordinated. Finally, the configuration where H is *AB* to Si is the highest in energy, 1.36 eV above the ground state in the case of P, but only 0.47 eV in the case of As.

Finally, we also checked a few additional points on the potential energy surface: Substitutional P

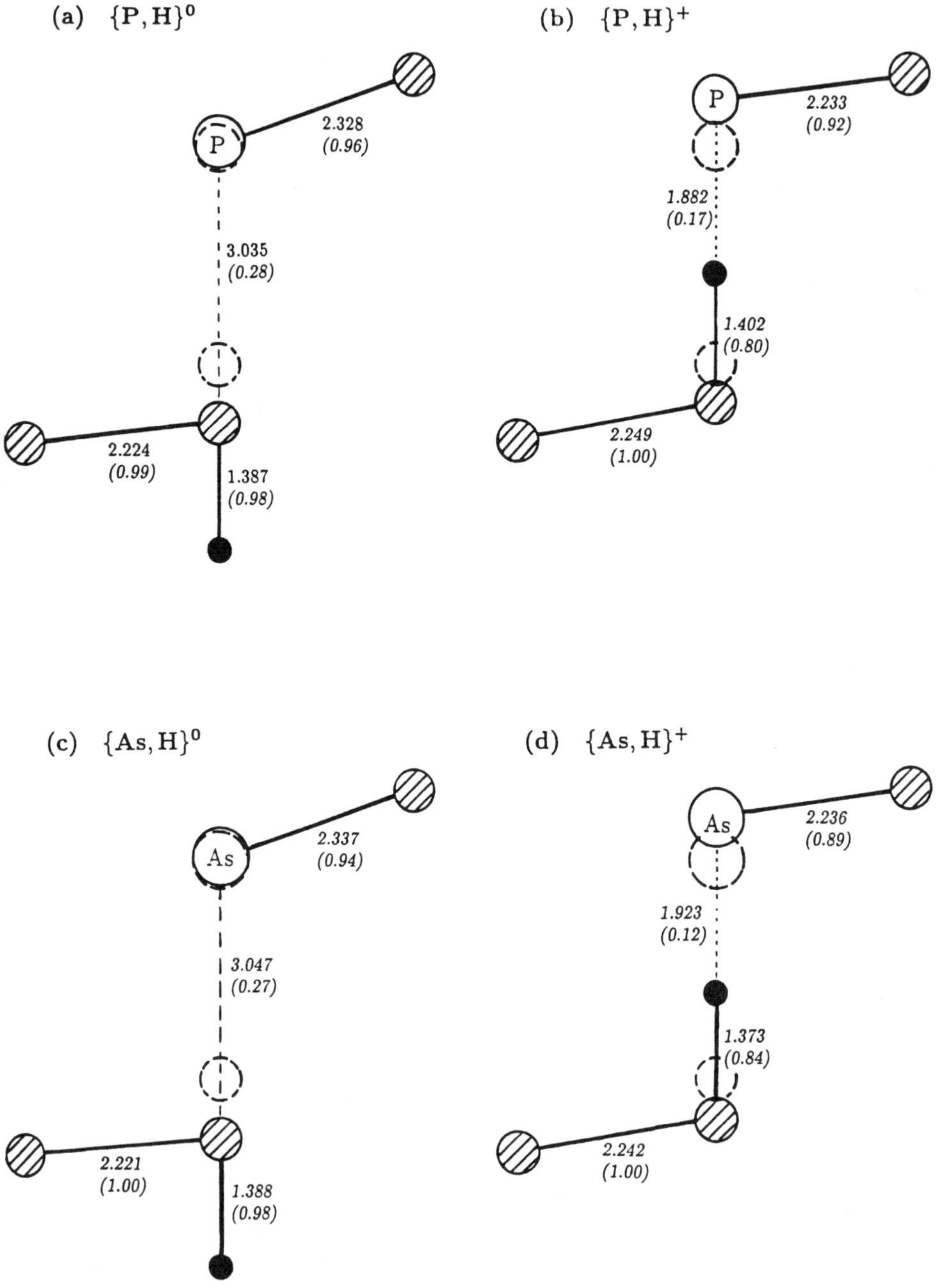

FIGURE 1: Lowest-energy configurations of the $\{P,H\}^0$ (a), $\{P,H\}^+$ (b), $\{As,H\}^0$ (c), and $\{As,H\}^+$ (d) pairs in Si. The bond lengths are in Å, and the degrees of bonding are in parenthesis (see text). Only the three atoms along the $<111>$ direction were optimized. The dashed circles show the perfect lattice sites. The shaded circles are host Si atoms and the full circles are H.

or As with H at 4 different T sites near the dopant, and at a *BC* site between two Si host atoms, with one of them nearest to the donor. In the case of P, these 5 configurations are between 1.37 and 1.64 eV of the minimum-energy configuration, the highest energies corresponding to H the furthest away from P. In the case of As, these energies vary between 0.37 and 0.89 eV.

3. DISCUSSION

The results described above can be interpreted as follows:

Bistability: The change of charge state, whether $0 \longrightarrow +1$ or $+1 \longrightarrow 0$, is accompanied by a complete change in the equilibrium geometry of the complex. The lowest-energy configuration in one charge state coincides with the highest-energy configuration in the other charge state. Going from one charge state to the other forces all the atoms to adjust their positions, with H having to move the most.

In the 0 charge state, H is at the *AB* site of a Si atom, and the Si–H stretching vibration is independent of the donor, which overlaps only very weakly with the Si–H bond (see Figs. 1(a) and 1(c)). In the +1 charge state, H is near the *BC* site, between the donor and a Si atom, and is primarily bound to Si. One would clearly expect this Si–H stretching vibration to be donor *dependent.* A model which involves the *coexistence* of $\{D,H\}^0$ and $\{D,H\}^+$ pairs in the sample (with D being P or As) would be consistent with the observation of a donor-independent and of a donor-dependent Si–H stretching vibrations.

Differences between P and As: Even though the situation is qualitatively the same for both donors, there are a number of quantitative differences that are worth noting.

First, the potential energy surface for H is significantly flatter in the +1 charge state if the donor is As than if it is P. This would suggest that the change of configuration involves more energy in the case of P, and hence that H has a greater probability of being released than in the case of As.

Second, in the +1 charge state, the Si–H degree of bonding (i.e., the covalent character) is a little stronger if the donor is As than if it is P. This indicates that, in the +1 charge state, the Si–H stretching vibration should be a little higher if the donor is As than if it is P. This agrees with the observed trend for the unidentified IR line reported in Ref.8.

Finally, note that the two arguments above are consistent with the experimental observation[9,10] that the debonding of H in As-doped material occurs at higher temperatures (about 25 °C higher) than in P-doped material.

Acknowledgements: The work of SKE was supported by the grant D-1126 from the R.A. Welch Foundation and by a grant from the Advanced Technology Program of the State of Texas. CHS and RAA are supported by the US Department of Energy (contract number DE-AC04-76DP00789).

REFERENCES

[1] S.J. Pearton, J.W. Corbett, and T.S. Shi, Appl. Phys. A **43**, 153 (1987).

[2] *"Hydrogen in Semiconductors"*, ed. J.I. Pankove and N.M. Johnson, Academic Press (NY, 1990).

[3] Report of the Panel *"Fundamental Issues in Hydrogen-Defect Interactions"* sponsored by the US Department of Energy (to be published in Rev. Mod. Phys.)

[4] J.I. Pankove, D.E. Carlson, J.E. Berkeyheiser, and R.O. Wance, Phys. Rev. Lett. **51**, 2224 (1983); J.I. Pankove, R.O. Wance, and J.E. Berkeyheiser, Appl. Phys. Lett. **45**, 1100 (1984).

[5] N.M. Johnson, C. Herring, and D.J. Chadi, Phys. Rev. Lett. **56**, 769 (1986) and **59**, 2116 (1987); N.M. Johnson and S.K. Hahn, Appl. Phys. Lett. **48**, 709 (1986).

[6] C.H. Seager, R.A. Anderson, and D.K. Brice, J. Appl. Phys. **68**, 3268 (1990).

[7] T. Zundel and J. Weber, Phys. Rev. B **39**, 13549 (1989).

[8] M. Stavola, S.J. Pearton, J. Lopata, and W.C. Dautremont-Smith, Appl. Phys. Lett. **50**, 1086 (1987).

[9] K. Bergman, M. Stavola, S.J. Pearton, and J. Lopata, Phys. Rev. B **37**, 2770 (1988); K. Bergman, M. Stavola, S.J. Pearton, and T. Hayes, Phys. Rev. B **38**, 9643 (1988).

[10] C.H. Seager and R.A. Anderson, Sol. St. Comm. **76**, 285 (1990) and MRS Symp. Proc. (fall 1990, in press).

[11] T.A. Halgren and W.N. Lipscomb, J. Chem. Phys. **58**, 1569 (1973); D.S. Marynick and W.N. Lipscomb, Proc. Natl. Acad. Sci. (USA) **79**, 1341 (1982).

[12] S.K. Estreicher, L. Throckmorton, and D.S. Marynick, Phys. Rev. B **39**, 13241 (1989).

[13] S.K. Estreicher, A.K. Ray, J.L. Fry, and D.S. Marynick, Phys. Rev. Lett. **55**, 1976 (1985); S.K. Estreicher, Phys. Rev. B **37**, 858 (1988).

[14] The "degree of bonding" is an index which gives the covalent character of a bond. Its value is 1.00 for a perfectly covalent 2-electron bond, such as H_2 in free space. See D.R. Armstrong, P.G. Perkins, and J.J.P. Stewart, J. Chem. Soc. Faraday Trans. **8**, 838 (1973).

Materials Science Forum Vol. 83-87 (1992) pp. 69-74

RIGID ROTOR IN A TETRAHEDRAL FIELD: AN APPLICATION TO (H,Be) AND (D,Be) IN SILICON

KEVIN R. MARTIN, W. BEALL FOWLER, AND GARY G. DELEO
Department of Physics, Lehigh University, Bethlehem, PA 18015 U.S.A.

ABSTRACT

The energy levels of a rigid rotor in a two-parameter tetrahedral field are calculated. An explicit form for the potential is obtained by using linear combinations of spherical harmonics which transform as the A_1 representation in the group T_d. For the wavefunctions we assume an expansion in spherical harmonics up to and including $\ell = 12$. An attempt is made to interpret the IR spectra of the (H,Be) and (D,Be) complexes in silicon within the framework of this "hindered" rigid rotor model.

1. Introduction

In recent experiments performed by Peale, Muro, and Sievers[1], and by Muro and Sievers[2], the (H,Be) and (D,Be) complexes in silicon have been shown, at low temperatures, to exhibit normal single acceptor lines, each with additional, weaker, blue-shifted replicas. The values of these blue-shifted replicas, 38.8 cm^{-1} for (H,Be) and 16.2 cm^{-1} for (D,Be), are in the correct range for some sort of rotational or tunneling motion. In addition, as the temperature is raised, red-shifted replicas are seen, indicating a complex manifold of levels in the ground state.

Two models have been introduced to explain the spectra of these complexes. The first of these models, here referred to as the tetrahedral tunneling model, was initially introduced by Haller, et al., for a different defect[3]. This tetrahedral tunneling model assumes the Hamiltonian can be written in three parts; $\mathcal{H}_a$, which describes the hydrogenic-like hole acceptor states, $\mathcal{H}_N$, which describes the dynamics of the tunneling nuclei, and V, which couples these two motions. $\mathcal{H}_a$ predicts p-like excited states, so that the coupling between the hole and nuclear motions is assumed negligible for the excited states since p-like functions have zero amplitude at the origin. The complex manifold of levels in the ground state is assumed to result from the coupling of the s-like single acceptor ground state with the nuclear tunneling motion. For $\mathcal{H}_N$, a four dimensional Hilbert space is assumed spanned by four identical functions oriented along different, bonding, <111> directions[3]. Assuming no overlap between these functions and parametrizing matrix elements of $\mathcal{H}_N$, the tetrahedral model predicts two nuclear levels, A_1 and T_2, separated by 4t in energy, where t is the matrix element of $\mathcal{H}_N$ between two different oriented functions[3]. Hence, within this model, the blue-shifted replicas are seen as a transition from the common ground state to the T_2 nuclear tunneling state.

The second of these models, here referred to as the octahedral tunneling model, was very recently introduced by Artacho and Falicov[4]. The basic assumption in the octahedral model is that parity is nearly a good quantum number so that new selection rules are imposed and six identical <100> oriented functions are assumed for the nuclear motion rather than the four <111> functions in the tetrahedral model.

In this paper we propose a more general model, namely that of a rigid rotor hindered by a field of tetrahedral symmetry[5]. Our calculation is reminiscent of one by Devonshire for the octahedral case[6], which has been applied to various tunneling defects in the alkali halides[7,8]. In some sense this hindered rigid rotor model is an extension of the tetrahedral and octahedral models since for the correct choices for the two parameters, the hindered rigid rotor model reduces to either model. Our model allows the investigation of intermediate cases, and it also allows a detailed investigation of assumptions inherent in the other two models.

2. Formulation

The problem we are considering is

$$\left(\frac{L^2}{\hbar^2} + V(\theta,\phi)\right)\psi = W\psi \tag{1}$$

where $W=E/B$ is the dimensionless energy eigenvalue and $B=\hbar^2/2I$ is the rotational constant. V is the static tetrahedral potential. Since the total Hamiltonian is invariant under operations of the group T_d, the eigenfunctions must transform as the irreducible representations of T_d, which are A_1, A_2, E, T_1, and T_2. A specific form for V can be obtained by using symmetry adapted linear combinations of spherical harmonics[9]. We write

$$V(\theta,\phi) = -K(Y_3^{tet} + CY_4^{tet}) \tag{2}$$

$$Y_3^{tet} = \frac{3}{7}\sqrt{\frac{40\pi}{7}}\, i(Y_3^{-2} - Y_3^{2}) \tag{3}$$

$$Y_4^{tet} = \frac{\sqrt{4\pi}}{3}\left(Y_4^0 + \frac{1}{2}\sqrt{\frac{10}{7}}(Y_4^4 + Y_4^{-4})\right) \tag{4}$$

where K and C in eqn. 2 are parameters, and the Y's on the right in eqns. 3 and 4 are the ordinary spherical harmonics. Since K is an overall multiplicative factor, we call K the strength of the potential. Since C adjusts the relative contribution of the $\ell=4$ term to the $\ell=3$ term, we call C the measure of the shape of the potential. It should be pointed out that Y_4^{tet} not only is invariant under the operations of T_d but also the larger group O_h, which has even and odd representations[6]. With C=1.0, eqn. 2 corresponds to the first two non-constant terms in a point-ion expansion of the potential due to the four nearest neighbors of the tunneling nuclei. Table 1 summarizes all the extrema for this potential, with all entries appropriate to positive K.

Table 1

	$0 \leq C < 3\sqrt{3}/7$	$3\sqrt{3}/7 \leq C < 9/7$	$C \geq 9/7$
$V_{<100>}$	Saddle	Minimum	Minimum
$V_{<111>}$	Minimum	Minimum	Maximum
$V_{<\bar{1}\bar{1}\bar{1}>}$	Maximum	Maximum	Maximum

$V_{<100>}$, $V_{<111>}$, and $V_{<\bar{1}\bar{1}\bar{1}>}$ are the values, as a function of K and C, of V for the direction indicated in the subscript. For negative K, the minima and maxima must be interchanged. For C>9/7 and large enough positive K, this hindered rigid rotor model reduces to the octahedral model discussed in the Introduction. For eqn. 1 we assume an expansion in spherical harmonics for the eigenfunctions so that the inner product with $Y_{l'}^{m'}$ becomes,

$$\sum_{lm} a_{lm}[((l)(l+1)-W)\delta_{l'l}\delta_{m'm} + \langle Y_{l'}^{m'}|V(\theta,\phi)|Y_l^m\rangle]=0 \qquad (5)$$

where the a_{lm} are the expansion coefficients and V is as given in eqns. 2-4. Setting the determinant to zero gives a prescription for the eigenvalues W, limited in accuracy by the number of basis functions used. We use in the expansion for the eigenfunctions spherical harmonics up to and including $\ell=12$. Figure 1 shows the dimensionless eigenvalues W for C=0.5 as a function of K, the strength of the potential, and Figure 2 shows the same for C=1.5. For both figures K=0 yields, of course, the eigenvalues of the free rigid rotor. For large |K|, both figs. 1 and 2 exhibit so-called librational structure, but for positive K, the nature of the libration is quite different. In fig. 1 with positive K, a four-fold degenerate ground state has formed because the rotor is trapped in one of the four equivalent <111> (bonding) minima wells, whereas in Fig. 2 for positive K a six-fold degenerate ground state is formed since the minima for C=1.5 are in the six <100> directions. For large negative K, both figs. 1 and 2 clearly show four librational levels, each with a degeneracy that is a multiple of four, required since the minima are in the four $<\bar{1}\bar{1}\bar{1}>$ (anti-bonding) directions. In order to apply this model to the (H,Be) and (D,Be) complexes in silicon, the most appropriate region to look at is in the moderate K

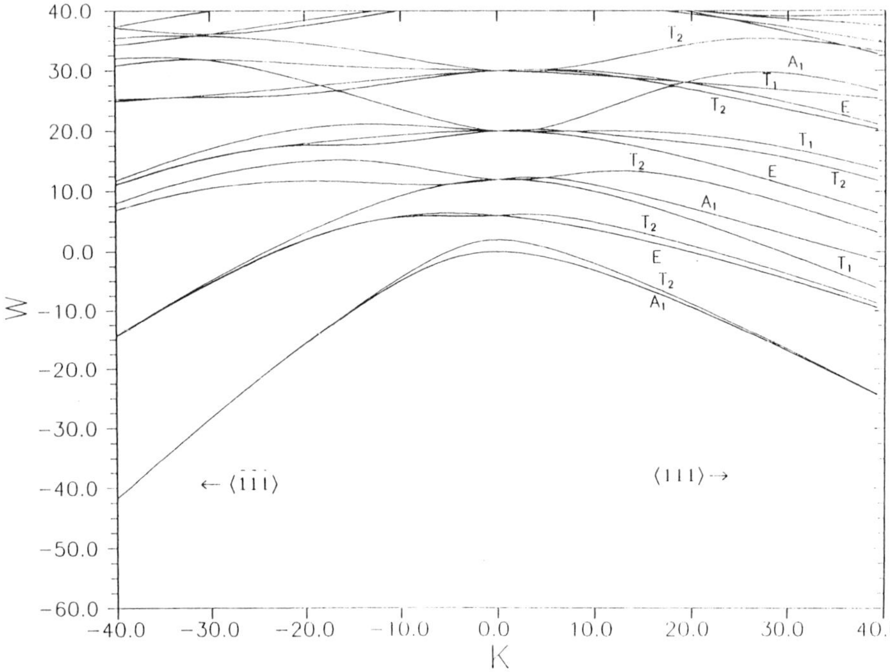

Figure 1. Plot of the dimensionless energy eigenvalues W vs. K for C=0.5. The first 36 eigenvalues are labeled as irreducible representations of T_d. The direction of the minima wells are indicated on the plot.

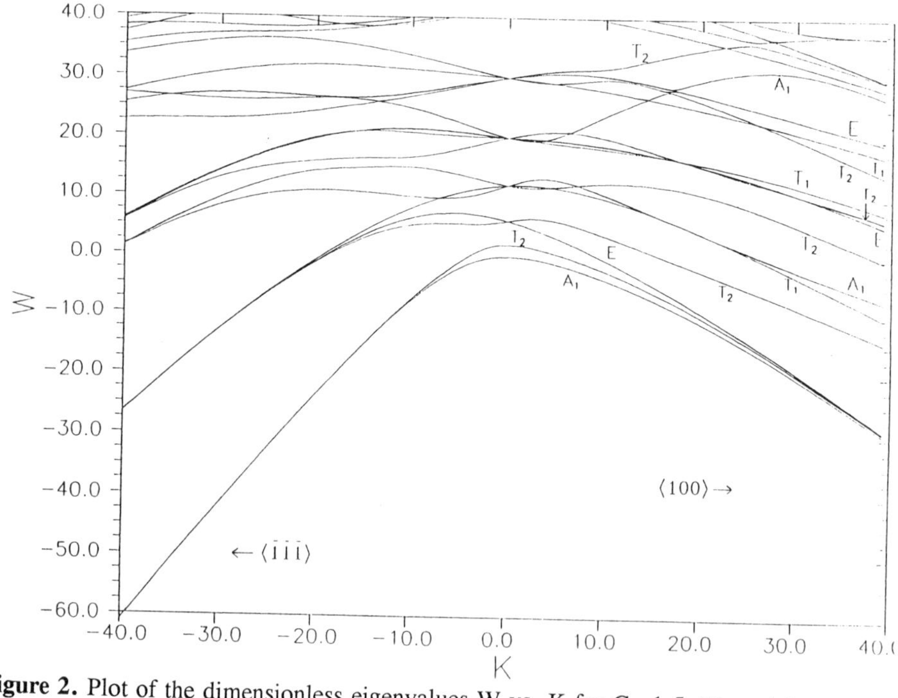

Figure 2. Plot of the dimensionless eigenvalues W vs. K for C=1.5. The minima for positive K are now in the six equivalent <100> directions.

range, which will be referred to as the hindered or tunneling regime.

3. Results and Discussion

In the experiment by Peale, et al., <111> stress was shown to produce the largest shifts in the levels, and they interpreted this as indicating that the system must be tunneling between equivalent <111> directions[1]. Further support for <111> minima comes from Modified Neglect of Diatomic Differential Overlap (MNDO) calculations on a simple, small cluster[10]. The cluster consisted of four silicon atoms situated at the vertices of a tetrahedron with a beryllium atom near the center of the tetrahedron and a hydrogen atom near the beryllium. The silicons were terminated by hydrogens. Two of the silicons and the hydrogen and beryllium were constrained to relax on a [110] plane, with the other two silicons allowed to relax freely. The silicons were observed to not move much from their equilibrium positions, and the minimum configuration was found to be along <111> (bonding) directions. With the assumption of Peale, et al., and the MNDO calculation in mind, we assume that the minima lie in <111> directions and therefore pick $C=0.5$ in the hindered rigid rotor model to see if this model can explain the experiment. Noting that the difference in energy of the $\ell=1$ and $\ell=0$ states in the **free** molecules are 21.0 cm^{-1} and 11.6 cm^{-1} (using $R=1.333$ Å) for (H,Be) and (D,Be) respectively, Fig. 1 shows us the rigid rotor model will have great difficulty in explaining the spectra since the difference of the two lowest levels always decreases from the free value, and the experimental values are 38.8 cm^{-1} and 16.2 cm^{-1} for (H,Be) and (D,Be) respectively[1,2]. This is also the problem of the tetrahedral tunneling model discussed in the Introduction, namely that thc **free** rigid rotor should be the upper limit since the effect of the environment around the defect should "drag" the rotor. One must be careful with Fig. 1, however, since W is in dimensionless units. B, the rotational constant, must multiply W to scale the energy levels correctly. In this sense, B is also a parameter. Obviously, B must be bigger than its free value, which means R, the bond length, must **decrease**. The largest value of R which could account for the experimental separation for (H,Be) with $C=0.5$ is 0.98 Å[5]. However, our MNDO calculation indicates that the bond length of the (H,Be) does not decrease from its free value. For example, when the (H,Be) system was oriented along the bonding directions (the minima) a bond length of 1.78 Å was found, and when oriented along an anti-bonding direction a bond length of 1.40 Å was found. The free (H,Be) in a negative charge state gave a bond length of 1.41 Å. Thus, it is reasonable to assume that the free value for the bond length is a lower limit.

The quasi-parity-conserving octahedral tunneling model proposed by Artacho and Falicov[4] suggests that the first and **third** levels (A_{1g} and E_g, where the "g" subscript should be treated as approximate) are the two levels of interest since new selection rules are implied if parity is nearly a good quantum number. Fig. 2 shows obviously that there must be some positive value of K for which there is agreement with experiment. We find that for $C=2.0$, $K=5.1$ gives the correct spacing for the first and third levels in order to agree with the experimental value of 38.8 cm^{-1} for (H,Be), assuming B is nearly the value of the free molecule.

The existence of <100> minima inherent in the octahedral tunneling model is supported by the local-density calculations of Denteneer, Van de Walle, and Pantelides[11]. Other aspects of this model remain to be investigated, however, including quasi-parity conservation and the predicted effect of stress on the levels (which as stated earlier seemed to support a tetrahedral tunneling model). Our present calculations are designed to resolve these issues.

4. Acknowledgements

The authors wish to thank R. E. Peale and F. S. Ham for their continuing help and very useful comments. This research was supported in part by the U.S Navy Office of Naval Research (Electronic and Solid State Sciences Program), Contract No. N00014-89-J-1223.

4. References

1. R.E. Peale,K. Muro and A.J. Sievers, Phys. Rev. B **41**, 5881 (1990)
2. K. Muro and A. J. Sievers, Phys. Rev. Lett. **57**, 897 (1986)
3. E.E. Haller,B. Joos and L.M. Falicov, Phys. Rev. B **21**,4729 (1980)
4. Emilio Artacho and L.M. Falicov, Phys. Rev. B **43**, 507 (1991)
5. Kevin R. Martin and W. Beall Fowler, Phys. Rev. B **44,** 1092 (1991)
6. A.F. Devonshire, Proc. R. Soc. London, Ser. A **153**, 601 (1936)
7. Judith G. Peascoe, W.R. Fenner and Miles V. Klein, J. Chem. Phys. **60**, 4208 (1974)
8. C.E. Mungan, R.C. Spitzer, J.P. Sethna and A.J. Sievers, Phys. Rev. B **43**, 43 (1991)
9. S.L. Altmann and A.P. Cracknell, Rev. Mod. Phys. **37**, 19 (1965)
10. Vanessa Gates and Gary G. DeLeo (unpublished)
11. P.J.H. Denteneer, C.G. Van de Walle, and S.T. Pantelides, Phys. Rev. Lett. **62**, 1884 (1989)

Materials Science Forum Vol. 83-87 (1992) pp. 75-80

Electron Spin Resonance Study on Hydrogen Passivation of Donors in Silicon

Kouichi Murakami, Shigeru Fujita, and Kohzoh Masuda
Institute of Materials Science, University of Tsukuba
Tsukuba Academic City, Ibaraki 305, Japan

ABSTRACT

It is demonstrated that hydrogen behaviors in phosphorus-doped silicon can be clarified by measuring electron spin resonance (ESR) of phosphorus donor and/or conduction electrons. Efficiency of hydrogen passivation of P donors shows strong donor-concentration dependence. This leads to a conclusion that hydrogen becomes inactive for P-H complex formation in metallic Si heavily doped with P impurities. Significant broadening in motionally narrowed ESR spectra is observed to occur through hydrogen introduction. From donor concentration dependence this broadening is found to come from the hyperfine interaction between P donor and/or conduction electrons and hydrogen nuclear spins. From changes in the ESR intensity and linewidth, we discuss also hydrogen behaviors observed in annealing process in Si.

1. INTRODUCTION

Hydrogen has been one of very important impurities in semiconductors because hydrogen passivates energy levels originated from defects and impurities or produces new levels in semiconductors [1], and because hydrogen terminates almost all dangling bonds at silicon surfaces.[2] Many experimental and theoretical works have been done for the former cases.[1,3-5] In p-type Si, hydrogen state, configurations, and diffusion are relatively well known, as compared with those in n-type Si.[4] Open questions remained are behavior and effect of hydrogen in n-type Si and a possibility of molecule-like hydrogen in semiconductors. Recently, Johnson and Herring [6] reported hydrogen immobilization in Si and Chang and Chadi [5] proposed a new meta-stable diatomic hydrogen complex, i.e., molecule-like hydrogen in Si by ab initio calculations. The formation of the molecule-like hydrogen is considered as a cause for partial passivation of shallow donors.[5]

In order to study these problems, we have investigated hydrogen passivation efficiency for shallow donor, phosphorus (P), and deep donor, off-center nitrogen, in Si, and donor concentration dependence by electron spin resonance (ESR).[7,8] We have utilized ion-implanted P donors with average concentrations ranging from 4×10^{17} to 4×10^{19} /cm^3, as an ESR probe for study on behaviors of hydrogen in Si. In the concentration range from 1×10^{18} to 4×10^{18} /cm^3, the impurity band is formed.[9] Above 4×10^{18} /cm^3 donor electrons are nonlocalized and behave like conduction electrons, since Si shows metallic conduction, whereas below 1×10^{18} /cm^3 donor electrons are localized at a donor atom or donor clusters.[9,10] Therefore, it is expected that the linewidth of motionally narrowed ESR increases due to the Fermi contact hyperfine (hf) interaction between donor and/or conduction electrons and hydrogen nuclear spins above 1×10^{18} /cm^3 if hydrogen concentration becomes higher than some 1×10^{19} / cm^3. Furthermore, even if donor electrons are localized, hydrogen of higher concentrations (an order of 10^{21} /cm^3) will induce the ESR line broadening.

In this study, we report that the P donor concentration shows a great influence on P-H complex formation in n-type Si. We also demonstrated that the line broadening of ESR of P donor and/or conduction electrons is observed by hydrogen plasma treatment and that this is caused by the Fermi contact hf interaction.

2. EXPERIMENTAL

For hydrogen passivation experiments floating-zone (FZ) p-type Si wafers with $5x10^{14}$ B/cm^3 were implanted with P at doses of $2x10^{13}$ to $2x10^{15}$ $/cm^2$ at an energy of 70 keV. The implanted wafers were activated with a furnace annealing of 950°C for 30min to 2 hr. Hydrogen passivation was done with a high frequency (13.56MHz) glow discharge apparatus, which was reported in detail elsewhere.[7,8] Exposure time was varied from 5 to 180 min. For comparison, H-passivation was done also for the off-center nitrogen center in Si that has a deep donor level (E_c-0.33eV).[11] Furthermore we performed annealing experiments at temperatures from 100 to 900°C for passivated P donors to investigate the dissociation process of the P-H complex. In order to know hydrogen profile and diffusion, deuterium, an isotope of low natural abundance, was used for passivation and was detected by secondary ion mass spectrometry (SIMS). This revealed strong correlation between P concentration and D profile, which is described later.

The ESR measurements of shallow donor P and deep donor off-center N in Si were made at temperatures ranging from 4 to 100 K with an X-band spectrometer, using continuous-flow liquid helium cryostats. Since the average concentrations of P are 4.0 x 10^{17} to $4x10^{19}$ $/cm^3$, the ESR spectrum observed becomes a single Lorentzian line due to motional narrowing in addition to exchange narrowing.[9,10] On the other hand, the ESR spectrum of the deep donor off-center N shows hf structure caused by nuclear spin of N because of localization of donor electrons.[11]

3. RESULTS AND DISCUSSION

Figure 1 shows results of hydrogen passivation (a) for the shallow donor P , and of annealing effects (b) and (c) for P donors which were obtained from

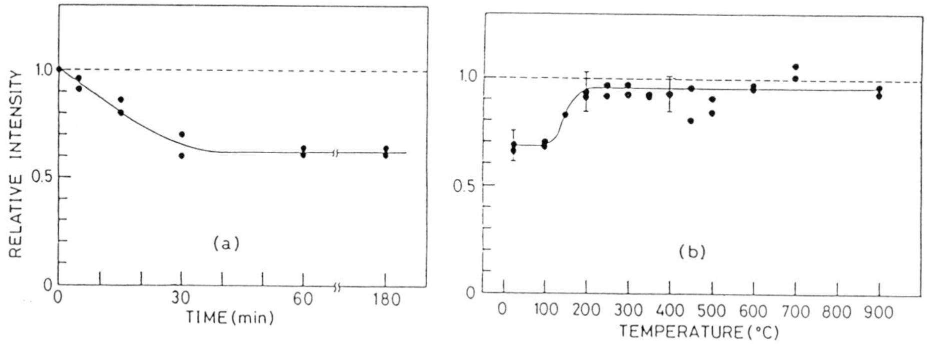

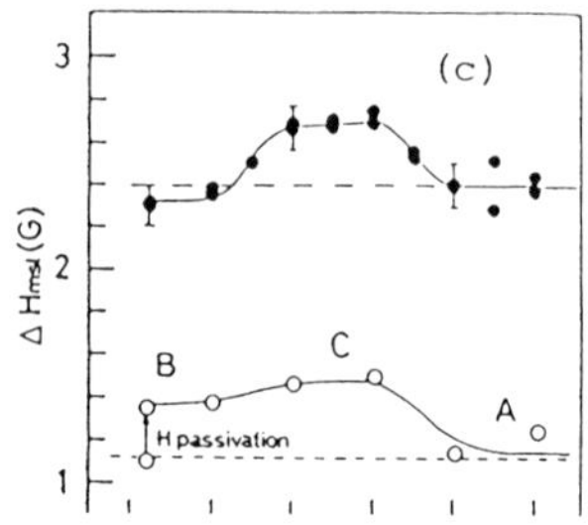

Fig. 1 (a) Relative amount of shallow P donors as a function of passivation time, and annealing behavior of P donors in H-passivated Si,
(b) relative amount of shallow donors and (c) ESR linewidth (● at 100 K and o at 4 K) as functions of annealing temperature

ESR measurements mainly at 100 K. It is found from Fig. 1(a) that the P donors are partially passivated, i.e., 35 %. For the off-center N with a concentration of $1 \times 10^{18}/cm^2$, it was completely passivated at a time of 30 min.[8] It is clear that for comparable hydrogenation conditions, hydrogen passivation of shallow donors is less pronounced than for the deep donor off-center N. The ESR intensity and linewidth were measured at 100 K and 4 K to investigate electrical recovery of donors through dissociation of the P-H complex. The recovery stage for the amount of P donors is seen at 150°C in Fig. 1(b). This recovery temperature is in good agreement with the result obtained by infrared absorption measurement.[12] In Fig. 1(c) the ESR line-width ΔH_{msl} broadens at a limited temperature range between 150 and 350°C. The onset (150°C) of the broadening is coincident with that of the electrical recovery of P donors.

The efficiency of hydrogen passivation of P donors is shown in Fig. 2. This was also obtained from ESR measurements at 4 K for Si samples with various P concentrations. We find for the first time that the passivation efficiency is strongly dependent on the donor concentration under the hydrogen-plasma exposure condition used here, i.e., the efficiency decreases from about 30 % to 0 % as the average donor concentration N_D^* is increased. A critical value above which the efficiency becomes nearly 0 is approximately $2 \times 10^{19}/cm^3$. This concentration corresponds to one at which the Fermi level enters into Si conduction band.[9] This strong P-concentration dependence is inconsistent with the result obtained by Bergman, et. al. [12], suggesting that the hydrogen plasma used and the substrate temperatures are much different each other.

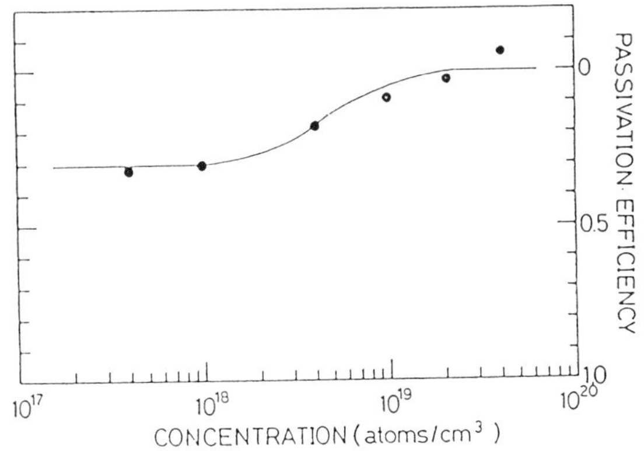

Fig. 2 Efficiency of hydrogen passivation of P donors as a function of N_D^* for Si sample passivated for 1 hr.

Figure 3 shows typical depth profiles of deuterium in Si samples with N_D^* of 0, 4×10^{17}, 4×10^{18}, and $4 \times 10^{19}/cm^3$. The species were exposed to deuterium plasma for 60 min. The arrows in Fig. 3 indicate the position of the peak of P donor profiles. For the Si sample doped with N_D^* of $4 \times 10^{17}/cm^3$, a slight increase of deuterium near the surface region, but there is no significant difference in deuterium profile from undoped p-type Si, as can be seen in Fig. 3 (a) and (b). A peak of deuterium profile is clearly observed just at the peak of P profile for the Si with N_D^* of $4 \times 10^{18}/cm^3$ in Fig. 3(c), whereas a broad plateau is seen at a deeper region for the Si with N_D^* of $4 \times 10^{19}/cm^3$ in Fig. 3(d). It is noted that the concentration of deuterium drastically decreases at the P peak region for the latter Si sample. This suggests that the amount of deuterium trapped by P donors decreases with

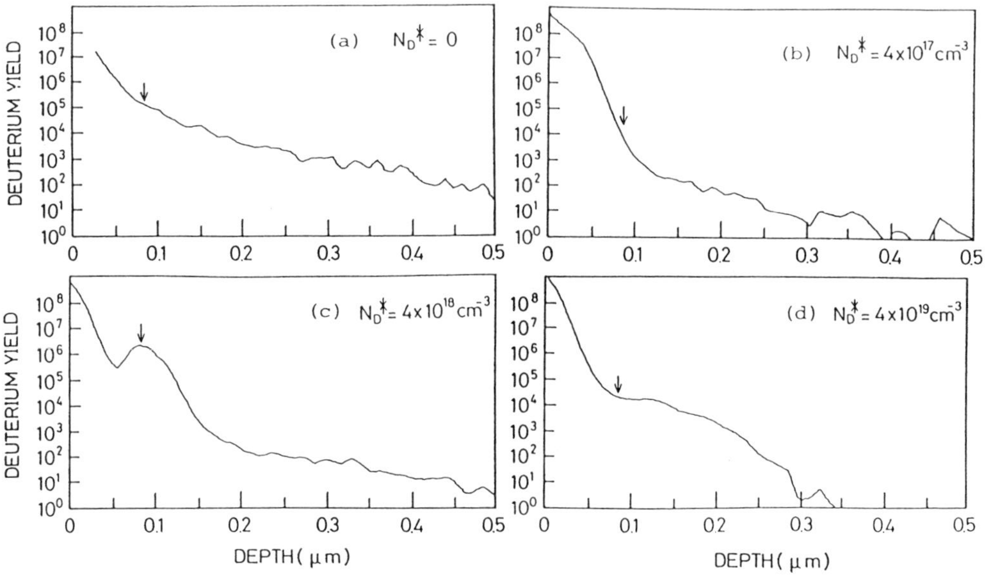

Fig. 3 Typical depth profiles of deuterium in Si with different N_D^*; (a) 0, (b) 4×10^{17}, (c) 4×10^{18}, and (d) 4×10^{19} P/cm^3 after deuterium exposure for 60 min.

increase in donor concentration for highly P-doped Si. This is consistent with the result of donor concentration dependence of hydrogen passivation obtained in Fig. 2. Thus we lead to a conclusion that hydrogen (deuterium) becomes inactive for H-P complex formation in Si heavily doped with P impurities. If an immobile molecule-like hydrogen is efficiently formed in

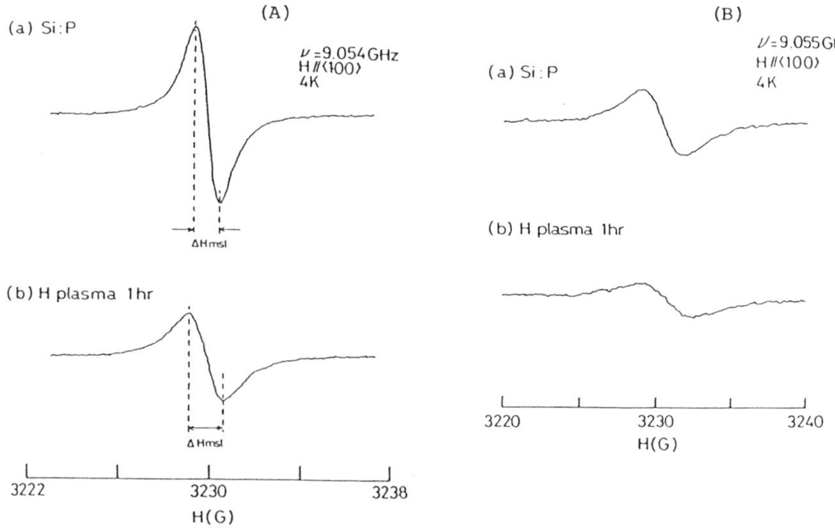

Fig. 4 Typical ESR spectra of P donor and/or conduction electrons with a g value of 1.998 for Si samples with (A) N_D^* of 4×10^{18} and (B) 2×10^{19}/cm^3. (a) and (b) correspond to ESR spectra before and after hydrogen exposure, respectively.

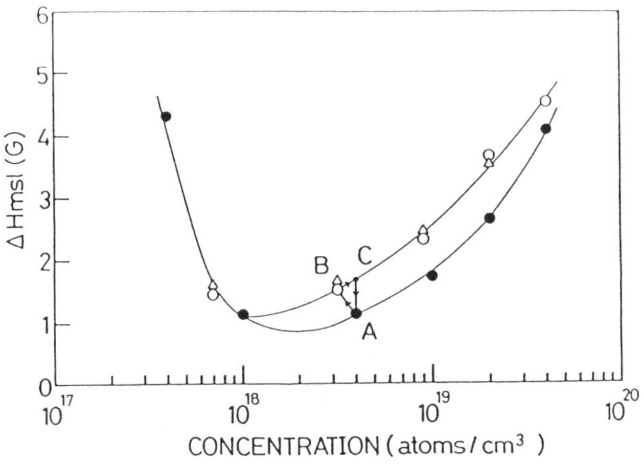

Fig. 5. ESR linewidth of P donor and/or conduction electrons at 4 K vs. N_D^* before (●) and after (o) hydrogen passivation. Signs A, B, and C are thought to correspond to the linewidths A, B, and C in Fig.1(c).

heavily P-doped Si, passivation efficiency will decrease and the deuterium concentrations also will increase in regions of P profile. Since the latter is not realized, however, an immobile molecule-like hydrogen is thought not to be efficiently formed in heavily P-doped Si. The low efficiency of hydrogen passivation is probably caused by screening of the Coulomb interaction between interstitial H^- and substitutional P^+ due to high density conduction electrons or by instability of P-H complex due to the Fermi level effect.

Figure 4 shows typical ESR spectra before and after hydrogen passivation. Significant broadening is observed in the motionally narrowed ESR line due to hydrogen introduction in Si. The ESR linewidths obtained for variously doped Si samples are plotted as functions of N_D^* in Fig. 5, taking decrement of P donors after hydrogen passivation into account. This indicates that hydrogen introduced in Si changes spin relaxation of P donor and/or conduction electrons. The increament of the ESR linewidth is found to become large for Si with higher N_D^* as compared with Si with lower N_D^*. Above 1x10^{18}/cm^3 donor and/or conduction electrons in the impurity band (1-4x10^{18}/cm^3) and the conduction band (higher than 4x10^{18}/cm^3) are delocalized,[9,10] so that they can interact with much more hydrogen nuclear spins in contrast to localized donor electrons below 1x10^{18}/cm^3. The broadening degree increases with increasing N_D^*, except the result of Si doped with N_D^* of 4x10^{19}/cm^3 in which the deuterium concentration was much decreased as shown in Fig.3(d). The ESR linewidth gets narrower by annealing at temperatures above 400°C, as shown in Fig. 1(c). These indicate that the broadening is due to the Fermi contact hf interaction of donor and/or conduction electrons with nuclear spins (I=1/2) of hydrogen. This was recently confirmed by deuterium introduction showing an isotope effect for the ESR broadening, which will be reported elsewhere. From the ESR broadening, hydrogen states will be possible to be analyzed.

Dissociation of the P-H complex occurs at temperatures above 150°C. The ESR line broadening observed from 150 to 350°C in Fig. 1(c) means that hydrogen still remains in the Si surface layer after annealing at these temperatures. This also suggests that an intermediate state yielding the ESR line broadening is formed in the process of dissociation of the P-H complex and is likely to be a state of hydrogen atom or molecule staying just near the P atom. This model will be investigated by ESR of localized donor electrons in Si with much lower P concentrations. In Fig. 1(c) we can see the ESR linewidth begins to decrease at 350°C and reaches the initial value above 400°C. This indicates that H atom or molecule become mobile above 350°C and diffuse away out of or into the Si wafer, as reported on hydrogen atom on the Si surface and hydrogen molecule in Si.[13]

4. CONCLUSIONS

We have first studied hydrogen passivation of P donors in Si and annealing behavior by ESR. The passivation efficiency is found to decrease from 30 to 0 % as the donor concentration increases. This is strongly correlated with hydrogen (deuterium) profiles, indicating that the formation of P-H complex is suppressed by high-density P donors. Broadening of ESR of P donor and/or conduction electrons is observed after hydrogen plasma exposure. This is caused by the Fermi contact hf interaction and will be useful for analysis of hydrogen states in n-type Si. Results of annealing hydrogen-passivated Si indicate that hydrogen is stable in n-type Si at temperatures of 150 to 300°C and that hydrogen diffuses out of or into Si above 350°C.

Acknowledgements

The authors would like to thank Dr. O. Eryu for assistance of ion implantation and SIMS measurement. This work was partly supported by a Grant-in-Aid for Scientific Research (1989 and 1990) on Priority Areas from the Ministry of Education, Science, and Culture, Japan.

References

1. S.J.Pearton, M.Stavola and J.W.Corbett, Materials Science Forum 38-41, 25 (1989).
2. S.S. Iyer, M. Arienzo, and E.de Fresart, Appl. Phys. Lett. 57, 893 (1990).
3. C.G.van de Walle, P.J.H.Denteneer, Y.Bay-Yam and S.T.Pantelides, Phys. Rev. B 39, 10791 (1989).
4. N.M.Johnson, C.Herring and D.J.Chadi, Phys. Rev. Lett. 56, 769 (1986).
5. K.J.Chang and D.J.Chadi, Phys. Rev. Lett. 62, 937 (1989).
6. N.M.Johnson and C.Herring, Phys. Rev. B 38, 1581 (1988).
7. K. Murakami, S. Fujita, and K. Masuda, to be published in Jpn. J. Appl. Phys.
8. K. Murakami, H. Suhara, S. Fujita, and K. Masuda, to be published in Phys. Rev. B.
9. J.D. Quirt and J.R. Marko, Phys. Rev. B 7, 2770 (1973).
10. S. Maekawa and N. Kinoshita, J. Phys. Soc. Japan 20, 1447 (1965).
11. K.Murakami, H.Kuribayashi, and K.Masuda, Jpn. J. Appl. Phys. 27, L1414 (1988) and Phys. Rev. B 38, 1589 (1988).
12. K.Bergman, M.Stavola, S.J.Pearton, and J.Lopta, Phys. Rev. B 37, 2770 (1988).
13. N. Yabumoto, K. Saito, M. Morita, and T. Ohmi, Jpn. J. Appl. Phys. 30, L419 (1991).

Materials Science Forum Vol. 83-87 (1992) pp. 81-86

INTERACTION OF DEUTERIUM WITH INTERNAL SURFACES IN SILICON

S. M. MYERS, D. M. FOLLSTAEDT, H. J. STEIN, AND W. R. WAMPLER
Sandia National Laboratories, Albuquerque, New Mexico 87185 USA

ABSTRACT

The strength of deuterium bonding to the walls of closed cavities within Si was determined in ion-beam experiments. These studies circumvented an inherent indeterminacy in the analysis of external-surface desorption and thereby allowed the Si-H surface bond energy to be quantified for the first time. The bond energy is 2.5 ± 0.2 eV for submonolayer coverages.

1. Introduction

The Si-H bond on the Si surface is central to a number of technologically important reactions involving atomic H, H_2, and more complex H-bearing molecules. Moreover, this surface bond provides an instructive limiting case for the attachment of H to dangling-bond defects in Si; in particular, since steric and volume constraints due to surrounding atoms are essentially absent on the surface, the Si-H bond can adjust freely to minimize energy. Previously, extensive research has elucidated the configurations of chemisorbed H on the (100) and (111) surfaces of Si [1-4] and has determined the activation energies E_D for desorption of H_2 [5-8]; under conditions of submonolayer coverage, these energies are near 2.5 eV. The strength E_B of the Si-H bond has not previously been determined in a definitive way, however, because the relationship between E_B and E_D depends upon the specific path of the desorption reaction, which has still to be established.

The above situation led us to perform new experiments to determine E_B. The key distinguishing feature of our work is that it examined thermal H release from closed internal surfaces instead of the external surfaces investigated previously. In the situation where the bonding surfaces were not directly exposed to external vacuum, heating initially caused the H to be promoted from chemisorption to atomic solution in the Si lattice; then, the H diffused to the external surface and finally desorbed. As we shall argue, this altered path of release allowed, for the first time, a straightforward evaluation of the Si-H bond energy, giving E_B = 2.5 eV.

In interpreting thermal desorption of H_2 from external surfaces, the difficulty is that the process involves not only the breaking of two Si-H bonds but also H-H recombination. The latter step is strongly exothermic, with E_R = 4.5 eV. Depending on the specific reaction path of desorption, the recombination might have either a large or a minimal effect on the observed activation energy E_D. In one extreme scenario which parallels the established behavior of H on metal surfaces, the breaking of the two Si-H bonds and the H_2 formation are concurrent, and one has

$$E_D = 2 \cdot E_B - E_R \ . \tag{1}$$

Equation (1) has often been assumed to hold approximately for Si as well as metals [see, for example, ref. 6]; with E_D = 2.5 eV, this equation gives E_B = 3.5 eV. Alternatively, one might envision a two-step desorption process whereby the rate-controlling break of one Si-H bond is followed by the exothermic reaction H + Si-H → H_2 + Si-. For this situation, $E_D \approx E_B$. In the general case,

$$E_D = 2 \cdot E_B - E_R + E_A \ , \ 0 \le E_A \le E_R - E_B \tag{2}$$

where the undetermined quantity E_A is also equal to the activation energy for H_2 chemisorption on Si. The indeterminacy of Eq. (2) is avoided in the present experiments on thermal release from internal surfaces; here, the promotion from internal chemisorption to solution and the subsequent lattice diffusion are jointly rate-determining, and, with the diffusion activation energy being independently known [9], the promotion energy is readily extracted. We will show that this latter quantity is essentially the desired bond energy, differing from E_B only by being referenced to H in lattice solution rather than H in vacuum.

2. Procedure

Microscopic cavities were formed beneath the surface of (111) Si by ion implanting He at room temperature and then annealing at 973 K for 30 minutes under high vacuum. The He-beam energy of 30 keV and fluence of 1000 nm^{-2} yielded a calculated [10] range of 280 nm and a peak concentration of 8 at.%. Microstructure was characterized by high-resolution TEM.

Deuterium was introduced into the He-implanted Si by D ion implantation at room temperature or by heating in D_2 gas at 873 K and a pressure of 87 kPa (650 torr) for 48 hours. Then, thermal release of the D was induced by temperature ramping in high vacuum at a rate of 2 K/min. The evolution of the system during the ramp was characterized in two ways. First, the areal density of D remaining within the He-implanted layer was determined periodically, without interrupting the ramp, by using the ion-induced nuclear reaction $D(^3He,p)^4He$ in a manner described elsewhere [11]. In addition, the bonding state of the D was characterized through IR vibrational spectroscopy [12] of a succession of specimens where the release process had been halted. The latter measurements provided confirmation that the D was in fact attached to the walls of the internal cavities.

The strength E_B of Si-H bonding at the internal surfaces was extracted from the nuclear-reaction data on D retention versus temperature. This analysis was carried out by applying well established diffusion-trapping theory [11] and taking the trap strength, which is simply related to E_B, to be an adjustable parameter.

3. Results and analysis

The microstructure of Si following He ion implantation and annealing at 973 K is shown in Fig. 1. Cavities are seen with sizes ranging up to about 10 nm. Regions of the specimen

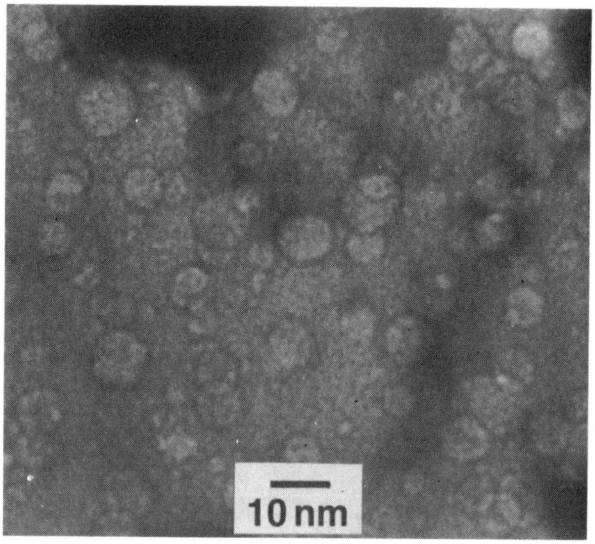

Fig. 1. Bright-field, 300-keV TEM micrograph of cavities within He-implanted Si after annealing at 973 K. The cavities are imaged at slight underfocus to provide contrast.

thinner than the He ion range showed no open volumes, thereby confirming that the cavities were well separated from the surface and not externally vented.

Infrared vibrational spectra are shown in Fig. 2 for He-implanted Si which had been heated in D_2 or H_2 gas at 873 K to the point of saturation uptake. The plotted data reflect local stretching modes of the Si-D and Si-H bonds. Measurements with the ^1H isotope are included to allow direct comparison with published spectra for H chemisorbed on external Si surfaces [1,2] and also H ion-implanted into Si containing no He cavities [12]. The H and D spectra in Fig. 2 are essentially equivalent, being simply shifted by factors of 0.73 in frequency and 0.5 in intensity due to the larger deuteron mass.

The H vibrational spectrum in Fig. 2 conforms closely to that reported for H chemisorbed on the external, (100) surface of Si [2], but it is distinctly different from spectra obtained for H ion-implanted into Si that did not contain He cavities [12]. This supports the assertion that, in our specimens, the H and D were chemisorbed onto the walls of the internal cavities instead of being trapped at implantation damage. When we introduced the D into the cavity-containing Si by room-temperature implantation instead of gas charging, the spectra initially were similar to those expected for damage trapping. With subsequent annealing at increasing temperatures, however, the vibrational spectrum reverted to that of Fig. 2 before the ultimate release of the D from the specimen. This suggests a sequence in which the implanted D initially became attached at random to damage traps, but, with heating, was detrapped and moved to the stronger, cavity-wall traps.

Deuterium retention versus temperature as measured by nuclear-reaction analysis is shown in Fig. 3. The filled circles are from a specimen treated to produce He cavities and then ion implanted with D before temperature ramping. In order to provide a comparison with ion-damage trapping, we have included two additional retention profiles, one for a specimen implanted only with D, and the other for a sample prebombarded with Si ions to simulate the He-ion damage without formation of cavities. The Si-ion bombardment was performed at room temperature, 200 keV, and a fluence of 90 Si/nm², and, as in the case of He injection, vacuum annealing at 973 K followed before the implantation of D. The dashed curve in the figure shows desorption from an external (100) Si surface based on experimentally determined desorption-rate equations [6].

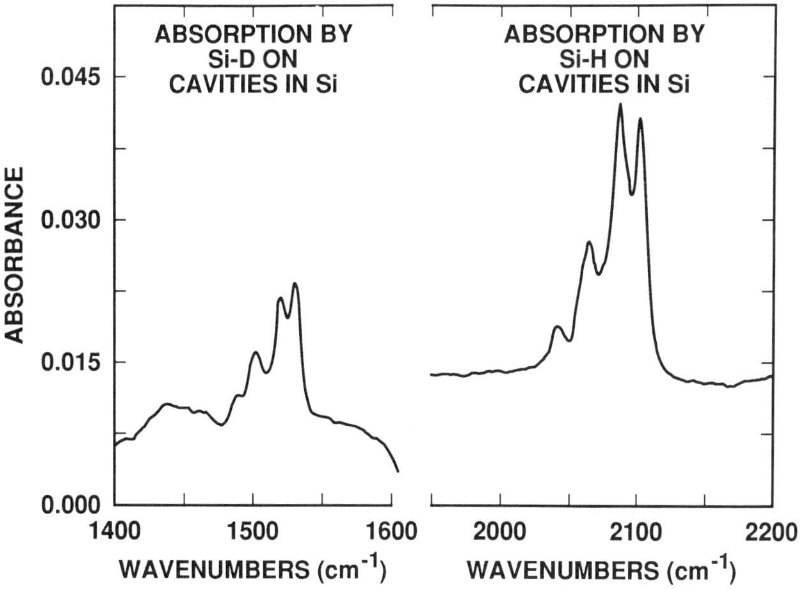

Fig. 2. Room-temperature IR absorption by Si with internal cavities after heating in H_2 or D_2 gas, obtained using internal-reflection plates.

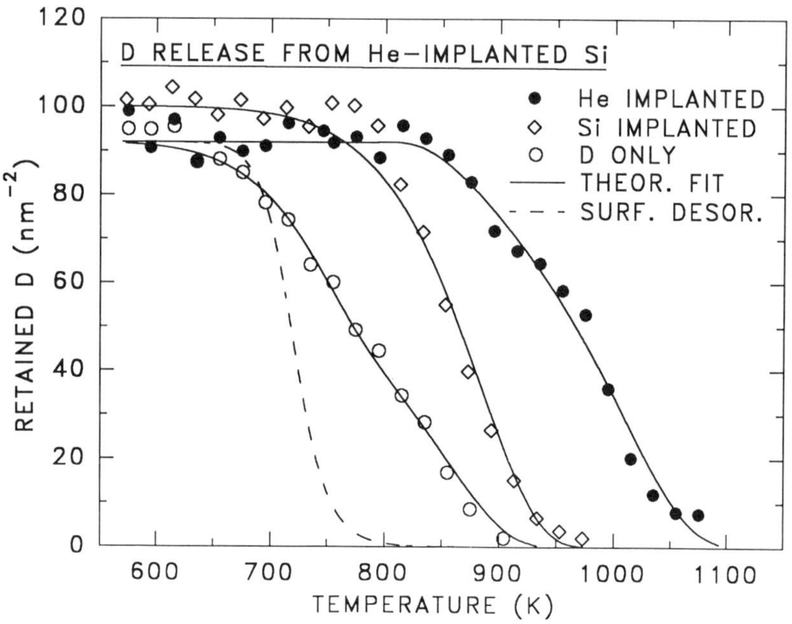

Fig. 3. Deuterium release from Si during temperature ramping for specimens implanted with He (•), or irradiated with Si ions (◊), or implanted only with D (o). The dashed curve shows predicted desorption from an external (100) surface. The solid curves represent fits of diffusion-trapping theory as discussed in the text.

It is apparent from Fig. 3 that the internal cavities in Si produced a distinct, high-temperature release stage near 1000 K. This is indicative of D trapping stronger than that at the implantation damage. Taking into consideration the IR vibrational spectra discussed above, we ascribe this strong binding to the formation of Si-D bonds at the walls of the cavities. The release of D from the internal cavities occurs at a much higher temperature than desorption from the external surface. This reflects the more difficult path for the former process, which requires promotion from the chemisorbed state to solution followed by solid-state diffusion.

The solid curves in Fig 3 represent fitted solutions of diffusion-trapping theory that is discussed in detail elsewhere [11]. Briefly, the formalism is based on a diffusion equation which includes terms for the local exchange of D between solution sites and trap sites. We solved the mathematical problem numerically, taking the diffusion coefficient of atomic H in Si from the literature [9] and adjusting the trap binding energies and areal densities of the D trapping sites to produce agreement with the experimental data. The trap binding energy in this formalism, E_T, is expressed with reference to solution in the crystalline lattice. In the case of implanted Si not containing He cavities, the data were fitted with two types of implantation-damage trap, one with E_{T1} = 1.4 eV and the other with E_{T2} = 1.7 eV. The D-release stage near 1000 K, which is associated with cavity-wall traps, yielded a binding energy of E_{T3} = 2.1 ± 0.2 eV. Because this is the largest binding energy observed in the system, it is ascribed to the more stable, monohydride surface state present at coverages below one monolayer.

In order to extract the Si-H bond strength E_B from E_{T3}, it is simply necessary to take account of the difference between the reference states for the two energies; the first quantity is specified relative to atomic H in vacuum, while the second is relative to H in solution. The conversion can be accomplished using the relation

$$E_B = E_{T3} - E_S + (1/2)\cdot E_R , \tag{3}$$

where E_S = 1.86 eV is the endothermic energy of solution per atom from H_2 gas [9] and E_R = 4.52 eV is the vacuum dissociation energy of H_2. This gives E_B = 2.5 ± 0.2 eV.

4. Implications

The Si-H surface bond strength of 2.5 eV obtained by us is smaller than has often been assumed in the past. Values above 3 eV have been proposed based on an assumption either that E_A in Eq. (1) is small or that the bond strength is similar to those reported for SiH_4 and related molecular species [see, for example, refs. 6 and 8]. It is noteworthy that our new value provides greater consistency with two independent results. First, the observed activation energy for dissociation of Si-H centers at the Si side of the Si-SiO_2 interface is 2.56 ± 0.06 eV [13], very close to our value of the surface bond energy. The second independent observation is that it is difficult to chemisorb H on Si from the molecular gas. This is inconsistent with the previous assumption of a small value for E_A, the activation energy for H_2 chemisorption. When our result E_B = 2.5 eV is combined with E_D = 2.5 eV in Eq. (1), however, one obtains E_A = 2.0 eV, which is quite consistent with the observed resistance to adsorption. A remaining consideration is the existence of larger bond strengths in molecular species containing Si and H, where reported dissociation energies for various Si-H bonds range from 2.6 to 3.9 eV [14]. This difference is not troubling however, since the strength of an Si-H bond can be expected to depend on the other bonding interactions of the Si atom; moreover, such a dependence is confirmed by the large range of bond energies in the molecular species.

It seems probable that the Si-H bond strength obtained in the present work represents an upper limit on the affinity of H for Si-dangling-bond defects within Si. This view derives from the fact that, on the surface, repulsive interactions which would constrain the bond configuration are absent, and as a result the state of lowest energy can be realized; in contrast, when the dangling bond resides on such centers as point defects, dislocations, and grain boundaries, volume and steric constraints are present to varying degrees.

Our value of the Si-H surface bond energy is close to the previously reported activation energies for external-surface desorption, which cluster near 2.5 eV. This leads us to suggest revision of the frequently cited model of desorption in which H-H recombination is concurrent with the breaking of two Si-H bonds and the energetics are given approximately by Eq. (1). As already discussed in the introduction, the observed equality $E_D \approx E_B$ is instead consistent with a two-step process whereby a rate-determining break of one Si-H bond is followed by the exothermic reaction H + Si-H $\rightarrow$ H_2 + Si-.

ACKNOWLEDGEMENTS

The authors are indebted to D. M. Bishop for the temperature ramp experiments and related data analyses and to M. P. Moran for TEM of the cavity microstructure. They also acknowledge valuable discussions with K. L. Brower, T. Klitsner, and J. Y. Tsao. This work was supported by the U.S. Dept. of Energy under Contract No. DE-AC04-76DP00789.

REFERENCES

1. Y. J. Chabal, G. S. Higashi, and S. B. Christman, Phys. Rev. B 28, 4472 (1983).
2. Y. J. Chabal, Surf. Sci. 168, 594 (1986).
3. J. J. Boland, Phys. Rev. Lett. 65, 3325 (1990).
4. K. Mortensen, D. M. Chen, P. J. Bedrossian, J. A. Golovchenko, and F. Besenbacher, Phys. Rev. B 43, 1816 (1991).
5. G. Schulze and M. Henzler, Surf. Sci. 124, 336 (1983).
6. P. Gupta, V. L. Colvin, and S. M. George, Phys. Rev. B 37, 8234 (1988).

7. B. G. Koehler, C. H. Mak, D. A. Arthur, P. A. Coon, and S. M. George, J. Chem. Phys. 89, 1709 (1988).
8. K. Sinniah, M. G. Sherman, L. B. Lewis, W. H. Weinberg, J. T. Yates, and K. C. Janda, J. Chem. Phys. 92, 5700 (1990).
9. A. Van Wieringen and N. Warmoltz, Physica 22, 849 (1956).
10. J. F. Ziegler, J. P. Biersack, and U. Littmark, The Stopping and Range of Ions in Solids (Pergamon Press, New York, 1985).
11. S. M. Myers, W. R. Wampler, and F. Besenbacher, J. Appl. Phys. 56, 1561 (1984).
12. H. J. Stein, J. Electronic Mater. 4, 159 (1975).
13. K. L. Brower and S. M. Myers, Appl. Phys. Lett. 57, 162 (1990).
14. R. Walsh, Acc. Chem. Res. 14, 246 (1981).

Materials Science Forum Vol. 83-87 (1992) pp. 87-92

MEASUREMENTS OF THE DIFFUSION COEFFICIENT OF HYDROGEN IN SILICON MONITORED BY CATALYZED ENHANCED OXYGEN DIFFUSION JUMPS

R.C. NEWMAN, J.H. TUCKER and S.A. MCQUAID.
Interdisciplinary Research Centre for Semiconductor Materials, The Blackett Laboratory, Imperial College, London SW7 2BZ, United Kingdom.

ABSTRACT

Diffusion of hydrogen into Czochralski silicon leads to enhanced oxygen diffusion jumps in a surface region, as monitored by the relaxation of stress-induced dichroism of the 9μm band. There is evidence of a sharp interface up to which dichroism is lost, implying that the hydrogen diffusion is the rate limiting step in the enhancement process for the range 225°C to 350°C. We obtain $D_H=1.7\times10^2\exp(-1.2eV/kT)cm^2s^{-1}$ and make comparisons with the other available data.

1. Introduction

The earliest work on the diffusion of hydrogen into monocrystalline silicon was that of Van Wieringen and Warmoltz [1] who measured the rate of effusion through a cylindrical silicon container heated to a temperature in the limited range 967°C to 1200°C and where hydrogen gas was used as a source. Steady state measurements yielded data for the product $[D_H][H_s]$, where D_H is the diffusion coefficient of hydrogen and $[H_s]$ was its equilibrium concentration. The sum of the diffusion activation energy E_D, and the heat of solution of the hydrogen E_s was reported to be 2.3eV. By introducing temperature and/or pressure transients, it was possible to separate these two quantities, leading to quoted values of E_D=0.48eV (Fig. 1) and E_s=1.86eV. The extrapolated value of $[H_s]$ would be $6\times10^3cm^{-3}$ at 250°C, which is 11 orders of magnitude below the detection limit of secondary ion mass spectrography (SIMS) of $10^{15}cm^{-3}$, even if hydrogen is replaced by deuterium to minimize background effects.

During recent years there has been considerable interest in the behavior of hydrogen at such low temperatures because of observations of the passivation of both shallow and deep defect levels [2,3]. The formation of pair defects such as H-B has been demonstrated directly by infrared (IR) local vibrational mode spectroscopy relating to the stretch modes of the paired hydrogen [3]. In highly doped silicon ($\sim10^{19}cm^{-3}$), the value of [H] rises to that of the acceptor impurity but only to a depth of 1-3μm when in-diffusion occurs from hydrogen in a H-plasma with the sample held at 185°C. However, these measurements have given no direct information about any deeper diffusion at much lower [H] concentrations because of the limitations of the SIMS technique.

The purpose of this paper is to demonstrate that such very low hydrogen concentrations can be detected indirectly because they lead to enhancements in the rate of diffusion jumps of grown-in oxygen impurities measured by the relaxation of stress-induced dichroism of the 9μm absorption band [4]. We then show that such measurements can be used to derive estimates of D_H. These measurements extend our previous limited analyses [5,6] both in their scope and interpretation when a plasma was used as the source of hydrogen and are closely related to our reports [6] of enhanced rates of (a) thermal donor (TD) formation, (b) loss of interstitial oxygen, O_i, from solution and (c) formation of carbon-oxygen complexes in Czochralski (Cz) silicon heated in a H-plasma.

The present paper is in three parts. First, we demonstrate that the introduction of hydrogen into

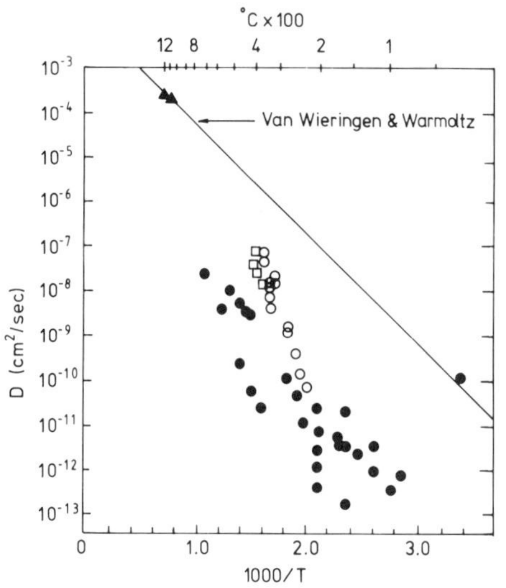

Fig. 1 An Arrhenius plot of the hydrogen diffusion coefficient: ▲ and solid line from Ref. 1; ● other previously published data; □ results from Ref. 17; ○ present results.

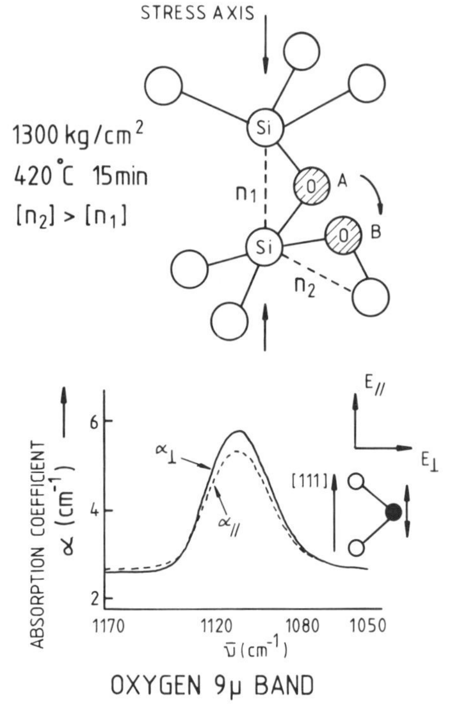

Fig. 2 Oxygen diffusion jumps from sites A to B under uniaxial stress and the resulting dichroism of the 9μm band.

silicon by annealing in hydrogen gas at high temperatures leads to an enhancement in the value of the diffusion coefficient of oxygen, D_{oxy}, during a subsequent heat treatment at a low temperature [7]. We then discuss the diffusion of hydrogen into silicon from an inductively-coupled RF-plasma (40W, 2Torr, 13.56Mz) as a function of heating time. The dichroism in the IR absorption band at 9μm is lost progressively from the external surfaces of the samples, rather than uniformly throughout their volume, leading to estimates of D_H. Finally, the results will be discussed and compared with other data.

2. High temperature hydrogen in-diffusion leading to enhanced values of D_{oxy}

We have shown that heat-treatments of boron doped silicon in hydrogen gas followed by a quench to room temperature led to the formation of H-B pairs in a concentration close to $5x10^{15}cm^{-3}$ which were uniformly distributed throughout the volume of samples 1mm in thickness [7]. It was demonstrated that the hydrogen originated from the external gas source since replacing it with deuterium gas led to the generation of D-B pairs. Similar diffusion of hydrogen into undoped material was also expected but its presence was not detected directly by IR absorption spectroscopy.

Bars of Cz silicon (p-type: $\rho=10\Omega$-cm) containing $[O_i] \sim 10^{18}cm^{-3}$ and cut to dimensions of 2x2x20mm with the long axis parallel to a [111] direction have now been heated in hydrogen gas at temperatures in the range 600°C to 1300°C and then quenched. They were then heated in air for 15min at 420°C while a uniaxial stress of 100MPa was applied along the [111] direction (Fig. 2). The samples were cooled to room temperature (at a rate of 1000°C/h) and the stress was removed. IR measurements of the 9μm band (300K) revealed a dichroism $R=(\alpha_{\perp}-\alpha_{\parallel})/\alpha_{\perp}$ of about 10%, where $\alpha_\perp$ and $\alpha_\parallel$ are the absorption coefficients for incident light polarized perpendicular and parallel to the stress axis respectively. Samples were thinned progressively by grinding and repolishing two opposite surfaces and R was found to be uniform throughout their volume.

Such hydrogen diffused and stressed samples were annealed in air at temperatures in the range 225°C to 350°C and R, which remained uniform throughout the sample, decayed exponentially, with a time constant τ. The oxygen diffusion coefficient was determined from the relation $D_{oxy}=a_o^2/64\tau$, where a_o=5.42Å is the lattice spacing of silicon. The values of D_{oxy} so obtained were greater than those for normal oxygen diffusion but the degree of enhancement decreased as the temperature of the hydrogen pre-diffusion treatment decreased below 800°C (Fig. 3). These observations can be understood since we have shown by means of H-B pairing that the concentration of [H] introduced also decreased with the hydrogen anneal temperature [7,8]. The present measurements, together with a comprehensive set of ancillary measurements discussed elsewhere [6], indicate beyond all reasonable doubt that the enhanced oxygen diffusion is due to the presence of hydrogen impurities. Clearly, D_{oxy} will not be enhanced once [H] falls below some critical value, $[H_{crit}]$, defined as that concentration at which D_{oxy}, due to the enhancement mechanism, is equal to the normal equilibrium value at the same temperature. We write $D_{oxy}=\{D_{oxy}(equil)+D_{oxy}(enh)\}$, where $D_{oxy}(equil)=0.11\exp(-2.51eV/kT)cm^2s^{-1}$ [9] although there is no unique expression for D_{oxy}(enh). The activation energy for the latter quantity is close to 2.0eV [10], but the pre-exponential factor should be proportional to [H].

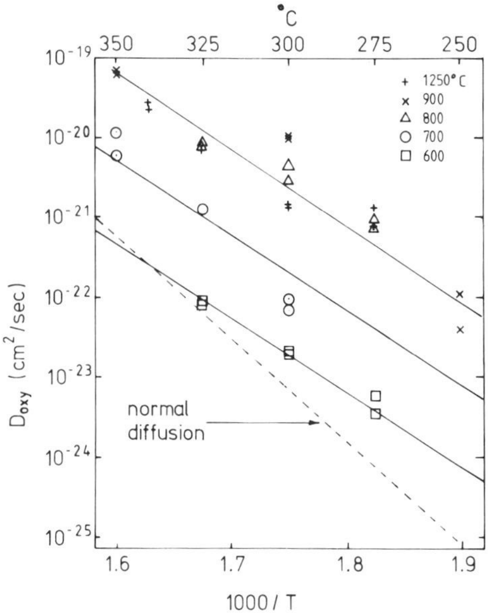

Fig. 3 Arrhenius plots of enhanced D_{oxy} as a function of pre-anneal temperature in H_2(g).

We envisage a process in which hydrogen atoms diffuse rapidly in the silicon matrix and collide at random with the bonded interstitial oxygen impurities [11]. Although there is no evidence that stable O-H pairs are produced, they may exist as a transient species and subsequent dissociation could then lead to an oxygen diffusion jump. The model implies that the H-atoms act only as catalysts and can make collisions with many different O_i atoms, so that a low concentration of hydrogen can produce a relatively large and measurable effect.

The probability P that an H-atom collides with one oxygen atom due to a random walk is $P=4\pi r_c D_H[H]$, where $r_c\sim5$Å is the interaction radius [11]. It is assumed that $D_{oxy}(enh)=Pd^2$, where $d\sim2$Å is the diffusion jump distance of the O_i atom. Such measurements of D_{oxy}(enh) yield values of the product D_H[H], within the approximations inherent in the model. To separate the two quantities it is necessary to devise experimental conditions with a time dependence which modifies the magnitude of either D_H or [H].

3. Low temperature treatments of samples in a plasma

Pre-stressed samples with no other postgrowth heat-treatment, exhibiting a uniform dichroism, were exposed to hydrogen atoms in a H-plasma while they were heated for increasing times at various temperatures in the range 225°C to 330°C. For short times, the hydrogen will diffuse only to a limited depth, up to which D_{oxy}(enh) should be large if the surface concentration of hydrogen $[H_s']$ is significantly greater than $[H_{crit}]$. SIMS measurements show that just below the surface of plasma treated material there is a very high concentration of hydrogen ($\sim 10^{20}cm^{-3}$) which might act as a source for the in-diffusion [12]. In the middle of the plasma-treated samples, where $[H] < [H_{crit}]$, D_{oxy} will have its equilibrium value. As the anneal time t increases, there should be a progressive loss of dichroism extending from the outer surfaces of samples towards the middle and the average measured dichroism should decrease.

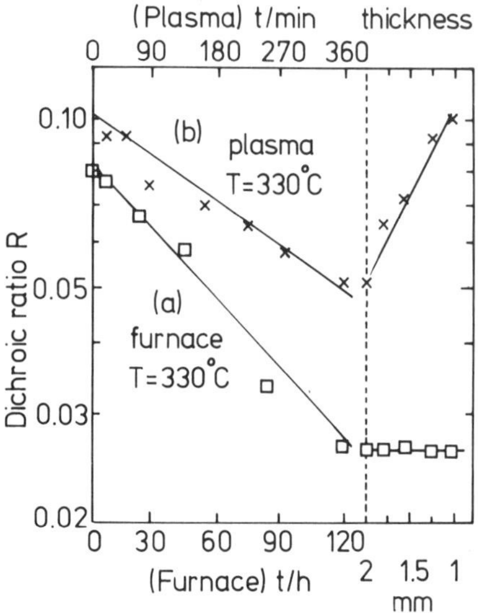

Fig. 4 Measured dichroism (9μm band) as a function of anneal time, followed by thinning of the samples.

Measurements of R (Fig. 4b) showed a decay with time. Some anneals were terminated when R had been reduced to half its original value, R_o. The thickness of such samples was reduced from 2mm to 1mm by progressive grinding and re-polishing as described earlier. Average values of R throughout the remaining thickness increased until they regained their original value. The recovery appeared to be linear (Fig. 4b) with decreasing thickness, implying that $R \sim 0$ in the outer regions but $R \sim R_o$ in the middle region and that the boundary was quite sharp (<0.1mm). Samples annealed in air in a furnace showed an exponential decay of R which remained uniform (Fig. 4a) throughout the thickness.

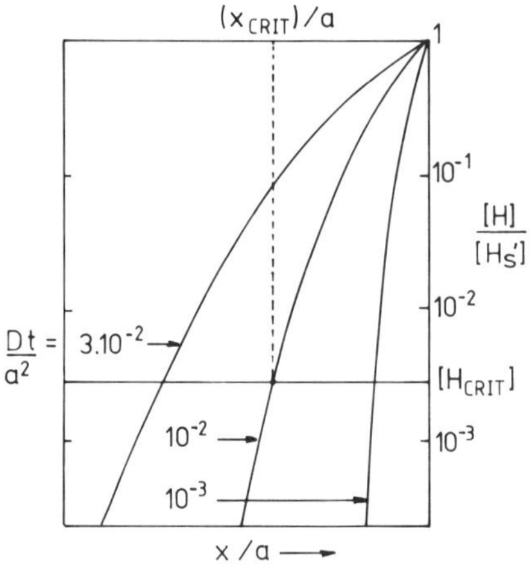

Fig. 5 In-diffused profiles of hydrogen as a function of reduced time showing $[H_{crit}]$ and x_{crit}.

Since there was no other information about the shape of the in-diffused hydrogen profile, it was assumed that it could be fitted to an error function complement, so that $[H]=[H_s']\mathrm{erfc}(x/2\sqrt{D_H t})$. If $[H_{crit}]$ at a depth $x=x_{crit}$ were significantly smaller than $[H_s']$, as implied by the measurements, the gradient d[H]/dx at this depth would be large (Fig. 5). The quantity $\mathrm{erfc}^{-1}[H_s']/[H_{crit}]$ is not then very sensitive to the magnitude of the argument (the ratio $[H_s']/[H_{crit}]$) and may be taken to have the numerical value of 2. In that case, we find that $R=R_o(a-x_{crit})/a$, where 2a is the initial thickness of the sample. A plot of $(1-R(t)/R_o)$ versus $\sqrt{t}$ should be linear with a slope of $4\sqrt{D_H}/a$. Our data can be fitted in this way (Fig. 6) and we obtained the values of D_H shown in the Arrhenius plot in Fig. 1.

4. Discussion

The rates of loss of dichroism in samples given a prior heat treatment in hydrogen gas at 900°C are in agreement with those reported some years ago [7,10]. It has become apparent subsequently that hydrogen was a component of the atmosphere in which these earlier samples were given a pre-treatment [13]. We also note that enhanced oxygen diffusion was found in samples preheated in the presence of iron or copper impurities [14]. The metallic impurities were introduced from hydrated nitrates so that there would again have been a source of hydrogen. As a separate issue, the metals may have catalyzed the formation of atomic hydrogen from the molecular species [15].

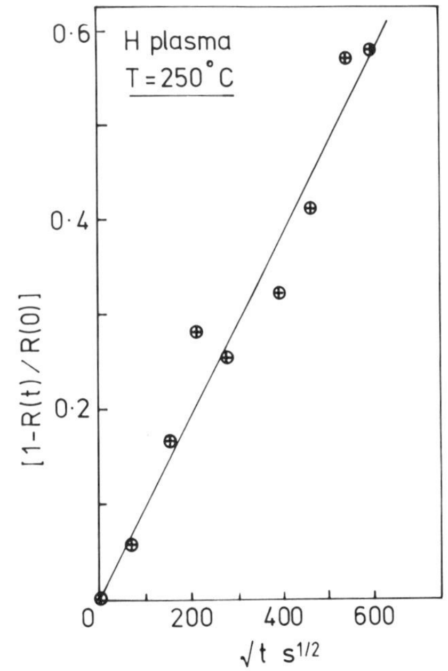

Fig. 6 Loss of dichroism versus $\sqrt{t}$ for a sample annealed in a H-plasma.

The process accounting for the loss of dichroism that we have described for samples heated in a H-plasma involves two basic steps. The first is the diffusion of hydrogen into the sample and the second is the hydrogen catalyzed re-orientation of the interstitial oxygen atom. The depth profiling measurements (Fig. 4b) show unambiguously that the hydrogen in-diffusion is the rate limiting step. Experimentally, we found that the rates of loss of dichroism (spatially uniform) in samples pre-heated at 900°C in hydrogen gas were not radically different from the rates found when hydrogen was diffused into samples from a plasma (giving a non-uniform loss of dichroism) [16]. Consequently it was inferred previously, mistakenly, that the rate limiting step in both cases was that associated with the enhanced diffusion jumps of the oxygen atoms.

The depth profiles (Fig. 4b) show that this interpretation was clearly incorrect for plasma treated samples. The effective concentration of hydrogen present after a 900°C treatment followed by the 420°C treatment may have been significantly smaller than that produced directly in the plasma treatments. Thus a much higher value of D_{oxy}(enh) may have been induced close to the surfaces of the latter samples.

There is recent independent evidence [17] to support our conclusion that the in-diffusion of hydrogen introduced at low temperatures is the rate limiting process for enhanced oxygen diffusion. The latter process leads to an enhanced rate of TD-formation, as shown previously [6]. Since long-range oxygen migration is involved in TD-formation, this process must impose a greater rate limitation than that for single oxygen jumps. Nevertheless, it has been found from spreading resistance measurements that there is an exceedingly sharp front ($<50\mu$m) for enhanced TD-formation in samples annealed in a plasma [17]. The advancement of that front analyzed using the relation $x^2 \sim D_H t$ leads to estimates of the diffusion coefficients also shown in Fig. 1. The data are in broad agreement with the present results but there is a complication because the diffusion took place in p-type Si in our samples but in n-type Si due to TD-formation in the other work [17].

An analysis of our results alone yields $D_H = 1.7 \times 10^2 \exp(-1.2\text{eV}/kT)\text{cm}^2\text{s}^{-1}$. The pre-exponential factor is larger than would be expected for the diffusion of an interstitial species, suggesting that the migration of the hydrogen was trap limited to some extent. The traps may in principle be identified with the boron dopant (10^{15} atom cm^{-3}), hydrogen itself (to form molecules) or the oxygen impurities. The effect of trapping would lead to under-estimates of D_H. Most other measurements of D_H for the temperature range investigated were lower than the values obtained in the present work (see Fig. 1). A detailed review of the earlier work has been given by Herring and Johnson [18]. It is clear that further measurements are required (a) to understand the effect of changing the charge state of hydrogen on its rate of diffusion and (b) to find the relative concentrations of atoms and molecules and/or complexes as a function of temperature.

Acknowledgements

The authors thank the Science and Engineering Research Council (UK) for financial support.

References

1. A. Van Wieringen and N. Warmoltz, Physica 22, 849 (1956).
2. S.J. Pearton, J.W. Corbett and T.S. Shi, Appl. Phys. A45, 153 (1987).
3. M. Stavola and S.J. Pearton, in *Hydrogen in Semiconductors*, Vol. 34 of *Semiconductors and Semimetals*, edited by J.I. Pankove and N.M. Johnson, treatise editors: R.K. Willardson and A.C. Beer, (Academic Press, San Diego, 1991) p139.
4. J.W. Corbett, R.S. McDonald and G.D. Watkins, J. Phys. Chem. Solids 23, 873 (1964).
5. A.R. Brown, R. Murray, R.C. Newman and J.H. Tucker, *Mat. Res. Symp. Proc.* 163, 555 (1990).
6. R.C. Newman, A.R. Brown, R. Murray, A.K. Tipping and J.H. Tucker, *6th Int. Symp. on Silicon Mater. Sci. and Technol.*, in *Semiconductor Silicon*, edited by H.R. Huff, K.G. Barraclough and J. Chikawa, (The Electrochem. Soc., Pennington, N.J., 1990), Vol. 90-7, p734.
7. S.A. McQuaid, R.C. Newman, J.H. Tucker, E.C. Lightowlers, R.A.A. Kubiak and M. Goulding, Appl. Phys. Lett. 58, 2933 (1991).
8. S.A. McQuaid, R.C. Newman and J.H. Tucker, This Volume.
9. R.C. Newman, in *Proc. 20th Int. Conf. Phys. Semicond.*, edited by E.M. Anastassakis and J.D. Joannapoulos, (World Scientific, Singapore, 1990) 1, p525.
10. M. Stavola, J.R. Patel, L.C. Kimerling and P.E. Freeland, Appl. Phys. Lett. 42, 73 (1983).
11. S.K. Estreicher, Phys. Rev. B41, 9886 (1990).
12. R. Murray, A.R. Brown and R.C. Newman, Mat. Sci. and Eng. B4, 299 (1989).
13. L.C. Kimerling, private communication (1990).
14. A.K. Tipping, R.C. Newman, D.C. Newton and J.H. Tucker, in *Defects in Semiconductors III*, edited by H.J. von Bardeleben, Mater. Sci. Forum (Trans Tech. Pub., Switzerland, 1986) 10-12, p887.
15. W.L. Hansen, E.E. Haller and P.N. Luke, IEEE Trans. on Nucl. Sci. NS29, No 1, 738 (1982).
16. A.R. Brown, J.H. Tucker, R.C. Newman and S.A. McQuaid, in *Proc. 20th Int. Conf. Phys. Semicond*, edited by E.M. Annastassakis and J.D. Joannopoulos, (World Scientific, Singapore, 1990) 1, p553.
17. H.J. Stein and S. Hahn, in *Symp. on Defects in Semiconductors*, Washington, (Electrochem. Soc., Pennington, 1991), In the press.
18. C. Herring and N.M. Johnson, in *Hydrogen in Semiconductors*, Vol. 34 of *Semiconductors and Semimetals*, edited by J.I. Pankove and N.M. Johnson, treatise editors: R.K. Willardson and A.C. Beer, (Academic Press, San Diego, 1991), p225.

Materials Science Forum Vol. 83-87 (1992) pp. 93-98

MEASUREMENTS RELATING TO THE SOLUBILITY OF HYDROGEN IN SILICON AT HIGH TEMPERATURES

S.A. MCQUAID[1], R.C. NEWMAN[1] and E.C. LIGHTOWLERS[2]
[1]Interdisciplinary Research Centre for Semiconductor Materials, The Blackett Laboratory, Imperial College, Prince Consort Road, London SW7 2BZ, U.K.
[2]Department of Physics, Kings College London, Strand, London, WC2R 2LS, U.K.

ABSTRACT

The solubility of atomic hydrogen in silicon at high temperatures has been determined in a new way. Boron doped silicon samples were heated at various temperatures in the range 900°C to 1300°C in an atmosphere of hydrogen gas and then cooled rapidly. Concentrations of H-B pairs in the samples were measured by infrared localized vibrational mode spectroscopy. The results indicate that a boron concentration of $\sim 8\text{x}10^{16}\text{cm}^{-3}$ is sufficient to trap the dissolved hydrogen efficiently. It is concluded that the solubility is given by $[H_s]=3.7\text{x}10^{19}\exp(-1.2\text{eV/kT})\text{cm}^{-3}$.

1. Introduction

The earliest estimates of the solubility of hydrogen in silicon, $[H_s]$, were deduced by Van Wieringen and Warmoltz [1] in 1956 from their investigation of hydrogen permeation through silicon grown by the Czochralski (Cz) method. From the pressure dependence of the permeation it was shown that the process was controlled by bulk diffusion in the material and that the species responsible for the permeation was mainly atomic hydrogen. Values of the product of the diffusivity, D_H, and the solubility, $[H_s]$, were determined from steady state permeation measurements over the narrow range of temperatures between 1092°C and 1200°C. They obtained a value of 2.34eV for the sum of the activation energy for diffusion, E_D, and the heat of solution, E_s. Analysis of the time dependence of permeation caused by pressure transients led to estimates of D_H. Values of $[H_s]$ could then be calculated for a hydrogen gas pressure of 1 atmosphere. A subsequent [2] least-squares fit to their data (Fig. 1) led to $[H_s]=4.96\text{x}10^{21}\exp(-1.86\text{eV/kT})\text{cm}^{-3}$ so that at 1200°C $[H_s]\sim 2\text{x}10^{15}\text{cm}^{-3}$. By extrapolating these results, a vanishingly small solubility of $6\text{x}10^{3}\text{cm}^{-3}$ would be expected at 250°C. However, measurable effects believed to be due to the presence of mobile hydrogen in silicon have been reported at such low temperatures [3-5] implying that the real concentration is significantly higher.

In 1968, the solubility of tritium (^{3}H) was measured by Ichimiya and Furiuchi [6] in high resistivity silicon prepared using the floating zone (FZ) technique. Samples were annealed in tritium gas at temperatures in the range 900°C to 1200°C for an hour before being quenched into water. Following the quench, samples were reheated at 1000°C in hydrogen to release the dissolved tritium and values of [^{3}H] were determined from the ionization of the surrounding gas induced by the radioactivity of this isotope. The gas pressure during the initial anneal could not be measured directly and was assumed to be that found for hydrogen under the same conditions. Pressures of around 0.3 atmospheres were used. These measurements were converted to solubilities by scaling the concentrations linearly with pressure up to 1 atmosphere. This procedure seems to be incompatible with the square root dependence of permeation on pressure measured by Van Wieringen and Warmoltz [1]. The values obtained in this later work are given by $[H_s]=6.61\text{x}10^{19}\exp(-1.39\text{eV/kT})\text{cm}^{-3}$ [2] (Fig. 1) so that at 1200°C $[H_s]\sim 1\text{x}10^{15}\text{cm}^{-3}$. Etching the surface prior to the second anneal leading to the measurement of [^{3}H] was necessary since desorption of tritium bonded to the surface led to over-estimates of the concentration dissolved in the material.

We have recently reported other measurements [5] relating to $[H_s]$ at high temperatures. Silicon crystals doped with boron at a level ($\sim 10^{17}cm^{-3}$), much greater than the expected values of $[H_s]$ ($\sim 10^{15}cm^{-3}$), were heated for relatively long times in hydrogen gas in the temperature range 900°C to 1300°C. Following a rapid quench to room temperature, infrared localized vibrational mode (LVM) absorption was detected due to H-B pair defects [7] created by the treatment. It was demonstrated that atomic hydrogen diffused into the material from the gas during the anneal rather than from some other source by annealing a sample in deuterium gas and detecting absorption due to D-B pairs. By establishing a calibration of the integrated absorption (IA) of the H-B stretch mode, concentrations of H-B pairs were determined as a function of anneal temperature. The values obtained in this way were given by $[H\text{-}B] = 5.6x10^{18}exp(-0.95eV/kT)cm^{-3}$.

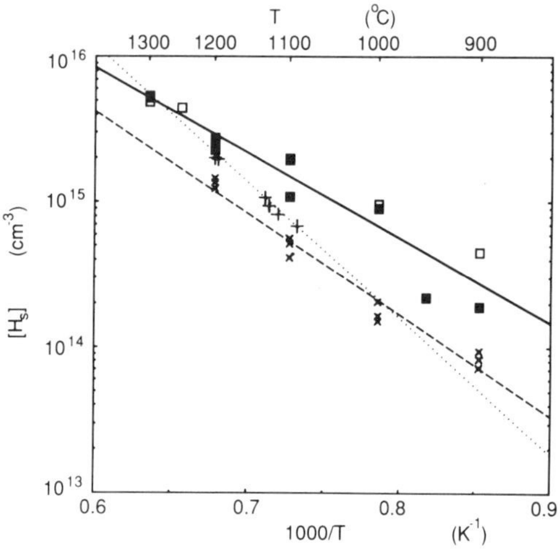

Figure 1. The solubility of atomic hydrogen in silicon (continuous line) fitted to present results (■) and our previous values (□). Measurements of Van Wieringen and Warmoltz (+ and dotted line) and those for ^3H of Ichimiya and Furiuchi (× and dashed line) are also shown.

Since only one boron doping level was used, it was not possible to assess how efficiently the dissolved hydrogen was trapped by the boron atoms during the quench. There may also have been formation of H_2 molecules or larger clusters. Such alternative mechanisms for the loss of isolated atomic hydrogen from solution while the sample was rapidly cooled could not be discounted. Indeed, since there is strong evidence that transient interactions between hydrogen and oxygen occur [3-5,8] at temperatures below 500°C, the presence of oxygen in the Cz material may have affected the proportion of $[H_s]$ which was trapped as H-B pairs. Given these considerations it could only be inferred that $[H_s]$ at the high temperature was greater than or equal to [H-B] following the quench. In this paper we report further measurements made on similarly treated Cz silicon with various doping levels, on boron doped float zone (FZ) refined material, and on n-type phosphorus doped material.

2. Experimental details

Five different single crystal silicon samples were annealed together as a set at each temperature. Three of the samples were prepared from boron doped Cz material with [B] close to $6x10^{16}cm^{-3}$, $9x10^{16}cm^{-3}$ and $4x10^{17}cm^{-3}$ and thicknesses of around 1.5mm, 1.1mm and 0.7mm respectively. A fourth boron doped specimen 1.6mm in thickness was prepared from FZ silicon with $[B] \sim 6x10^{16}cm^{-3}$. The fifth sample in each set was cut from FZ material doped with phosphorus with $[P] \sim 1x10^{17}cm^{-3}$ and a thickness of about 0.4mm. All the crystals were chemically polished and their resistivities were measured using the four point probe technique. The samples were subjected to an RCA cleaning procedure immediately prior to being heated in a similarly cleaned quartz-lined furnace in a continuous flow of hydrogen gas at a pressure of one atmosphere. The heating was terminated by quenching the specimens into silicone oil by allowing them to drop 45cm from the heated zone.

Sets of samples were annealed at each of the temperatures 900°C, 950°C, 1000°C and 1100°C for 2h. Further sets were heated at 1200°C and 1300°C for 90min and 30min respectively. The anneal times were chosen so that the diffusion length was more than four times greater than the maximum sample thickness at the corresponding temperature using values reported by Van Wieringen and Warmoltz [1] for the hydrogen diffusivity. At the highest temperatures excessively long anneal times were avoided in order to prevent roughening of the sample surfaces, presumably caused by a chemical reaction between the silicon and the quartz lining of the furnace.

Sample surfaces were cleaned and polished after the heat treatment and resistivity measurements were again made. Infrared absorption spectra were obtained using a Bruker IFS 113v interferometer operated at a resolution of $0.5cm^{-1}$ with the specimens cooled to around 20K. To remove the background absorption due to both the photoionization of acceptor levels and two-phonon processes, the spectrum of a reference sample prepared from the same untreated material was subtracted from that of the heat treated sample. By completely passivating both boron doped epitaxial and ion implanted silicon in a hydrogen plasma, a calibration of the integrated absorption of the H-B stretch mode was established [5]. Thus a value of IA=1 was found to correspond to $[H\text{-}B]=3.0\times10^{15}cm^{-3}$, allowing the concentrations of H-B pairs to be determined from the LVM spectra.

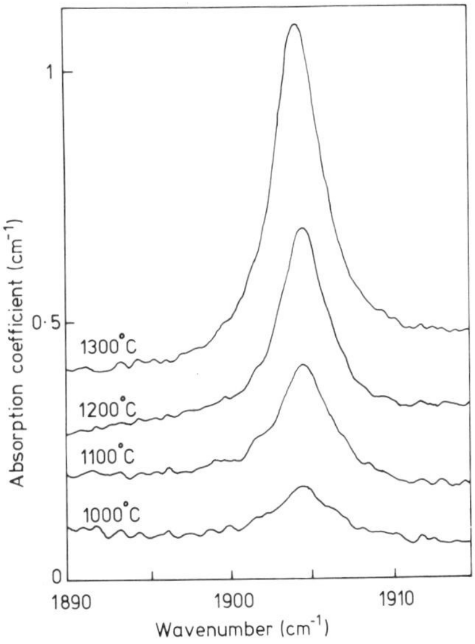

Figure 2. Absorption (at 20K) due to the H-B stretch mode in Cz silicon with $[B]\sim9\times10^{16}cm^{-3}$ after annealing at the indicated temperatures in an atmosphere of hydrogen gas followed by a rapid quench to room temperature.

3. Results

Absorption due to the stretch mode of hydrogen in the H-B complex was strongest for the samples which had been annealed at the highest temperatures (Fig. 2). A slight increase in the resistivity of the samples was measured as a result of the highest temperature (1300°C) treatment, suggesting a drop in the hole concentration of around $1\times10^{16}cm^{-3}$. This reduction is consistent with measured values of $[H\text{-}B]\sim5\times10^{15}$ cm^{-3}. The spectral position of the peak ($1904.3cm^{-1}$) and its full width at half maximum ($\sim3.4cm^{-1}$) were found to be the same for both FZ and Cz material and were independent of the doping level in the range investigated. Similar values for these parameters were previously measured by us for bulk Cz material with $[B]\sim10^{17}cm^{-3}$ passivated to a depth of around 40μm by immersion in a H-plasma at 180°C for 260h. No shift nor broadening of the LVM line was observed for the passivated epitaxial layer ($[B]=3\times10^{18}cm^{-3}$: thickness =5.2$\mu$m) used to calibrate the absorbance. However, hydrogenation of the ion implanted material ($[B]\sim1.5\times10^{20}cm^{-3}$: thickness$\sim$0.7$\mu$m) led to line broadening ($8.4cm^{-1}$) and a peak shift of $2cm^{-1}$ to higher energy.

Absorption due to H-B pairs was detected for all p-type samples which had been quenched from 1200°C and 1300°C (Fig. 3). A depth profile of a sample annealed at 1300°C revealed a constant value (variation $<4\%$) of the pair concentration demonstrating that a) the anneal time was long enough for the hydrogen concentration to reach equilibrium throughout the sample thickness and b) the quench was sufficiently rapid to prevent significant out-diffusion of H. Infrared measurements made on the two most highly B-doped crystals showed significant H-B absorption after anneals at temperatures as low as 900°C (Fig. 1).

No absorption due to the stretch mode of paired hydrogen in H-P complexes [9] was detected in the heat treated n-type material.

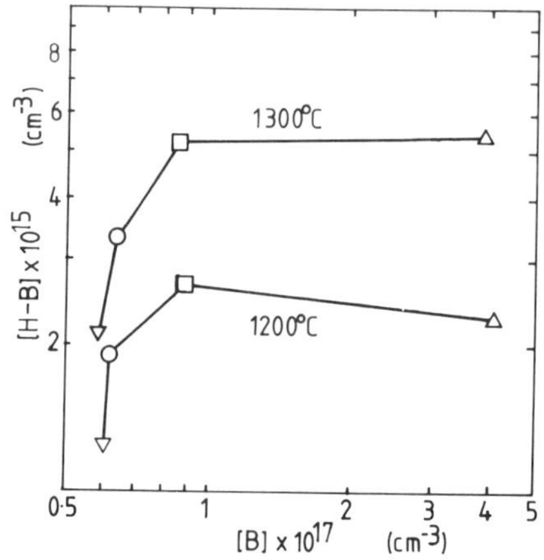

Figure 3. H-B pair concentration as a function of boron concentration in FZ (▽) and Cz (○,□ and △) silicon annealed at the temperatures shown in H_2(g) and quenched.

4. Discussion

The concentration of passivated boron acceptors in the material with low [B] is similar for both float zone and Czochralski material. There is, therefore, no evidence to suggest that the presence of interstitial oxygen in the Czochralski material affects the kinetics of hydrogen trapping by boron.

The rate of loss of isolated hydrogen during a rapid quench due to trapping by boron acceptor ions (present in a concentration $[B^-]$) and self-trapping is expected to be described by

$$d[H]/dt = -K_f[H^+][B^-] - K_f'[H^+][H^0] \qquad (1)$$

where K_f and K_f' are the rate coefficients for the respective forward reactions and where $[H^+]$ and $[H^0]$ represent the concentrations of positively charged and neutral hydrogen atoms respectively. The concentration of dissolved hydrogen would be greater for anneals at the higher temperatures and consequently the second term in equation (1) would be relatively more important. At a given anneal temperature the relative contribution of the first term is expected to increase with increasing $[B^-]$ in the sample.

The concentration of H-B pairs generated by the treatment appears to be independent of the initial boron concentration (Fig. 3) for values of $[B]>8\times10^{16}cm^{-3}$ in the case of anneals at high temperatures. It follows that boron impurities present in such high concentrations act as efficient sinks for $[H_s]$ at or below 1300°C. Under these conditions, the values of the hydrogen solubility $[H_s]$ should be close to measurements of [H-B]. Values obtained in this way are shown in Fig. 1 for the material with initial values of [B] of $9\times10^{16}cm^{-3}$ and $4\times10^{17}cm^{-3}$.

At the high temperatures used in this work the Fermi level would have been close to midgap in even the most highly doped material (the intrinsic carrier concentration in silicon [10] exceeds $4\times10^{17}cm^{-3}$ at temperatures greater than 675°C). Thus the relative concentrations of the different charge states of atomic hydrogen, $[H^+]$, $[H^0]$ and $[H^-]$, during the anneal should be similar to those expected in undoped material. Therefore the hydrogen concentration in the material immediately prior to quenching would have tended to a value close to the intrinsic solubility $[H_s]$ throughout the sample thickness.

As the temperature was reduced rapidly during the quench, the Fermi level would have moved towards the valence band in the p-type material. As a result, the isolated hydrogen atoms would have become predominantly positively charged [11] and therefore would be coulombicly attracted to the negatively charged acceptor ions. At temperatures below 200°C, stable passivated boron [12-14] complexes would have formed in p-type material leading to the measured pair concentrations.

An expression similar to equation (1) is expected to describe the loss of isolated hydrogen in n-type material (substituting H^- for H^+ and P^+ for B^-). The apparent charge of the H-P complex [9] is around half that of the H-B pair [5]. The absence of any detectable concentration of H-P pairs in the n-type material suggests that $[H\text{-}P] < 8 \times 10^{14} cm^{-3}$ even for material annealed and quenched from 1300°C. It would appear then, that phosphorus present in a concentration of $10^{17} cm^{-3}$ is less effective in trapping H^- atoms than the capture of H^+ impurities by the same concentration of boron. Reported estimates of the binding energy of H-P pairs ($\sim$0.5eV) are less than that for H-B complexes (0.87eV) [15] consistent with the present conclusion. Other work has demonstrated the partial passivation of sulphur donors in silicon given a similar treatment [16], i.e., each double donor was paired with only one hydrogen atom to form a single donor. If the interaction was primarily Coulombic then the binding energy of this first hydrogen might be as high as $\sim$1eV (i.e., double that for H-P pairs) so that it would be larger than the corresponding value for H-B pairs. However the binding energy of a second hydrogen atom might be expected to be similar to that between P^+ and H^-. It follows that there is no inconsistency between our observations for n-type material and those reported for sulphur doped silicon.

The solubilities we have determined (Fig. 1) are close to those measured by Van Wieringen and Warmoltz. A least squares fit to the present results and those we obtained previously [5] leads to the Arrhenius expression $[H_s]=3.7 \times 10^{19} \exp(-1.2eV/kT) cm^{-3}$. The heat of solution of $E_s=1.2eV$ we estimate, is closer to that measured by Ichimiya and Furuichi [6] for tritium in silicon (1.39eV) than the value of 1.86eV obtained by Van Wieringen and Warmoltz [1] over the narrow range of temperatures they investigated. Using our estimate of E_s and the sum $(E_s+E_D)=2.34eV$ measured by the latter workers, a value of $E_D \sim 1.1eV$ would be implied. This value may be compared with $E_D=1.2eV$ which we have measured at temperatures in the range 225°C to 350°C. Extrapolating our results to 250°C leads to a value of $[H_s]$ close to $1 \times 10^8 cm^{-3}$ which may reasonably account for effects believed to be due to the presence of mobile hydrogen in silicon at such low temperatures [3-5].

Acknowledgements

The authors thank R.A.A. Kubiak and M. Goulding for providing the boron doped epitaxial and ion implanted material. The assistance of J.H. Tucker in preparing the script was appreciated. This work was supported by the U.K. Science and Engineering Research Council.

References

1. A. Van Wieringen and N. Warmoltz, Physica 22, 849 (1956).
2. C. Herring and N.M. Johnson, in *Hydrogen in Semiconductors*, Vol. 34 of *Semiconductors and Semimetals*, edited by J.I. Pankove and N.M. Johnson, treatise

editors: R.K Willardson and A.C. Beer, (Academic Press, San Diego, 1991), p225.
3. R.C. Newman, J.H. Tucker, A.R. Brown and S.A. McQuaid, J. Appl. Phys. (accepted for publication).
4. R.C. Newman, J.H. Tucker and S.A. McQuaid, This Volume.
5. S.A. McQuaid, R.C. Newman, J.H. Tucker, E.C. Lightowlers, R.A.A. Kubiak and M. Goulding, Appl. Phys. Lett. 58, 2933 (1991).
6. T. Ichimiya and A. Furuichi, Int. J. Appl. Rad. Isotopes 19, 2115 (1968).
7. M. Stavola, S.J. Pearton, J. Lopata and W.C. Dautremont-Smith, Appl. Phys. Lett. 50, 1086 (1987).
8. S.K. Estreicher, Phys. Rev. B41, 9886 (1990).
9. K. Bergman, M. Stavola, S.J. Pearton and J. Lopata, Phys. Rev. B37, 2770 (1988).
10. S.M. Sze, in *Physics of Semiconductor Devices*, (Wiley, New York, 1981) p19.
11. E.E. Haller, Semicond. Sci. Technol. 6, 73 (1991).
12. C.-T. Sah, S.C.-S. Pan and C.C.-H. Hsu, J. Appl. Phys. 57, 5148 (1985).
13. T. Zundel and J. Weber, Phys. Rev. B43, 4361 (1991).
14. S.T. Pantelides, Appl. Phys. Lett. 50, 995 (1987).
15. N.M. Johnson, in *Hydrogen in Semiconductors*, Vol. 34 of *Semiconductors and Semimetals*, edited by J.I. Pankove and N.M. Johnson, treatise editors: R.K. Willardson and A.C. Beer, (Academic Press, San Diego, 1991), p113.
16. R.E. Peale, K. Muro and A.J. Sievers, *Shallow Impurities in Semiconductors*, edited by G. Davies, Mat. Sci. Forum 65-66, 151 (1990).

Materials Science Forum Vol. 83-87 (1992) pp. 99-104

DONOR-HYDROGEN COMPLEXES IN SILICON STUDIED BY MÖSSBAUER SPECTROSCOPY

Z.N. LIANG and L. NIESEN

Laboratorium voor Algemene Natuurkunde, Materials Science Center
Westersingel 34, 9718 CM GRONINGEN, The Netherlands

Abstract

The formation and stability of Sb-H and Sb-D complexes in crystalline Si have been studied by Mössbauer spectroscopy, using the 23.9 keV transition of ^{119}Sn populated in the decay of ^{119}Sb. Hydrogen (or deuterium) was introduced by low energy implantation. More than 80% of the Sb atoms can be associated with hydrogen. The Sb-H complexes dissociate around 440 K. A large reduction of the Mössbauer intensity is observed after introducing of hydrogen and deuterium. This is explained by localized hydrogen tunneling motion in the neighborhood of the excited ^{119}Sn atom.

1. Introduction

Hydrogen passivation of shallow levels in Si has been extensively studied in recent years, especially in the two technologically most important materials, Si and GaAs [1-4]. Great progress has been made in the understanding of hydrogen behavior in p-type Si, notably in boron doped Si [5,6]. However, the knowledge on the behavior of hydrogen in n-type Si is poorer.

In this paper, we present ^{119}Sn Mössbauer results concerning the formation and dissociation of ^{119}Sb-H complexes. Moreover the dynamics of the ^{119}Sn-H and ^{119}Sn-D complexes produced in the decay of ^{119}Sb is investigated by measurements as a function of sample temperature.

2. Experimental

Various doses of 110 keV Sb ions were implanted at room temperature into p-type as well as into n-type Si wafers. The samples were annealed at 900° C for 30 minutes in flowing N_2 gas, then postimplanted at room temperature with 200 eV H_2^+ (or D_2^+). With this procedure the hydrogen is introduced without accompanying lattice damage. The stability of the complexes was investigated by subsequently annealing the samples at stepwise increasing temperatures. ^{119}Sn Mössbauer conversion electron spectra were taken with a Ca^{119}SnO$_3$ resonance counter at room temperature.

3. Results

Fig. 1 shows typical Mössbauer spectra obtained from a sample with a low Sb dose (4×10^{12} Sb cm^{-2}). Before hydrogen implantation, the ^{119}Sb atoms were virtually all at substitutional sites, leading to a single ^{119}Sn Mössbauer line with isomer shift IS = 1.80(3) mm/s, in agreement with that in Ref. [7]. Introduction of hydrogen results in two phenomena shown in Fig. 1b: (1) a new line appears with IS = 2.34(5) mm/s and (2) the total intensity of the Mössbauer spectrum decreases noticeably. The same phenomena are also observed in the deuterium implanted samples. The new line is due to the formation of an Sb(Sn)-H complex [8]. It decreases in intensity after annealing at T_a = 423 K and is absent at 523 K, as shown in Fig. 1c and d. At 523 K the intensity of the spectrum is back to the original situation.

To describe phenomenon (2), we assume that the total intensity of a Mössbauer spectrum should be constant at a sample temperature of 80 K, and equal to I_0, the intensity before hydrogen introduction. By doing so, we neglect the possible difference in vibrational Debye temperatures for the two components observed. This will give a maximum error of about 10% in the total intensity at 80 K. Now for a hydrogen

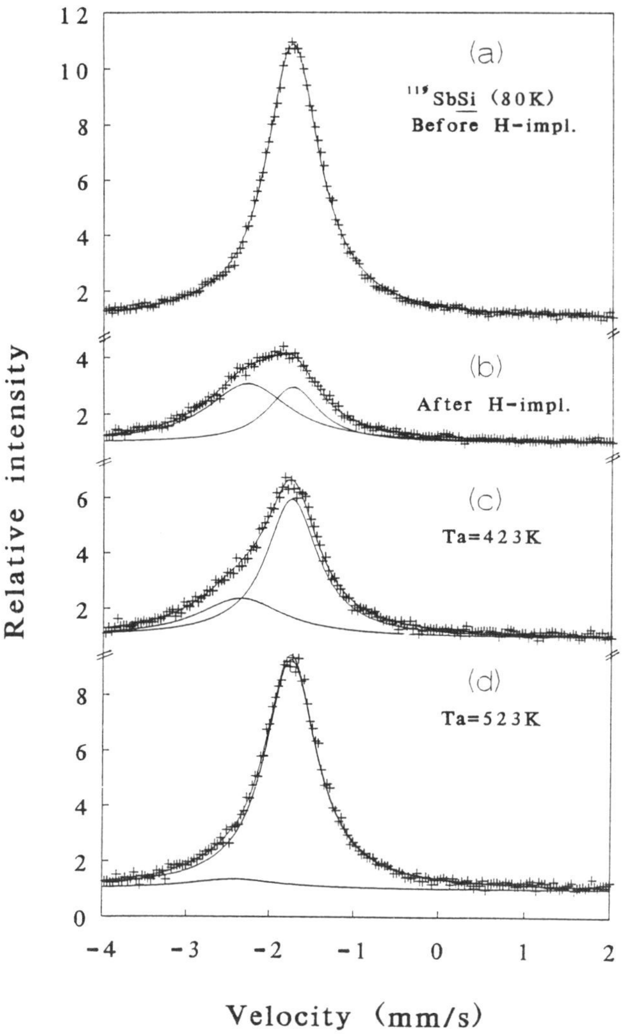

Fig. 1. Mössbauer spectra measured at 80 K: (a) after implantation with 110 keV, $4x10^{12}$ Sb cm^{-2} and annealed at 900° C for 30 minutes in flowing N_2; (b) after introduction of $2x10^{15}$ H cm^{-2} with 200 eV H_2^+ at room temperature; (c) and (d) after annealing at 423 K and 523 K in flowing N_2, respectively. The solid lines are the least-square fitted results.

(or deuterium) implanted spectrum we have two parts: the visible part consisting of the substitutional ^{119}Sn fraction (I_S) and the Sb(Sn)-H complex fraction (I_H), and the "invisible" part given by $I_I = I_0 - I_S - I_H$. We assume that ^{119}Sn atoms belonging to this part have an effective recoilless fraction that is too low to be observed.

The relative intensities of I_H and I_I are given in Fig. 2 as a function of annealing temperature for a hydrogen dose of $2x10^{15}$ cm^{-2} and for Sb doses of $4x10^{12}$ cm^{-2} (Fig.2a) and $1.5x10^{13}$ cm^{-2} (Fig. 2b). One can clearly see that about 80% (2a) and 60% (2b) of the Mössbauer intensity is associated with hydrogen. Furthermore, the relative intensities of I_H and I_I are decreasing dramatically around T_a = 455 K (Fig. 2a) and 425 K (Fig. 2b), corresponding to dissociation energies of 1.46 and 1.36 eV, determined by a simple dissociation model [8]. This result strongly suggests that I_H and I_I are intimately connected.

The area of the Mössbauer line as a function of sample temperature for the ^{119}Sb implanted sample (H-free) is shown in Fig. 3a. A Debye temperature of 225(3) K is derived, in good agreement with earlier results [9]. Fig. 3b and 3c give the areas for

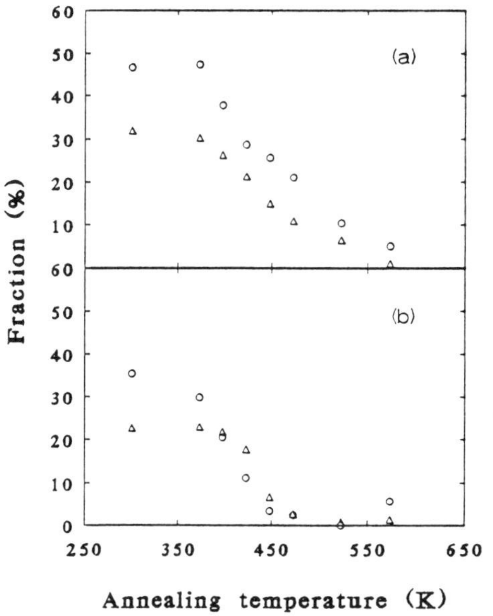

Fig. 2. The fractions of I_H (Δ symbol) and I_I (o symbol) as a function of T_a for a hydrogen dose of $2x10^{15}$ H cm^{-2}: (a) $4x10^{12}$ Sb cm^{-2} and (b) $1.5x10^{13}$ Sb cm^{-2}.

a hydrogenated and deuterated sample, respectively. Here the area of the "invisible" component was derived at each temperature by assuming that the sum of all components should be equal to the fitted area given in Fig. 3a, except for a small correction for 3b where a different detector was used. As is shown in Fig. 3b, the intensity of the

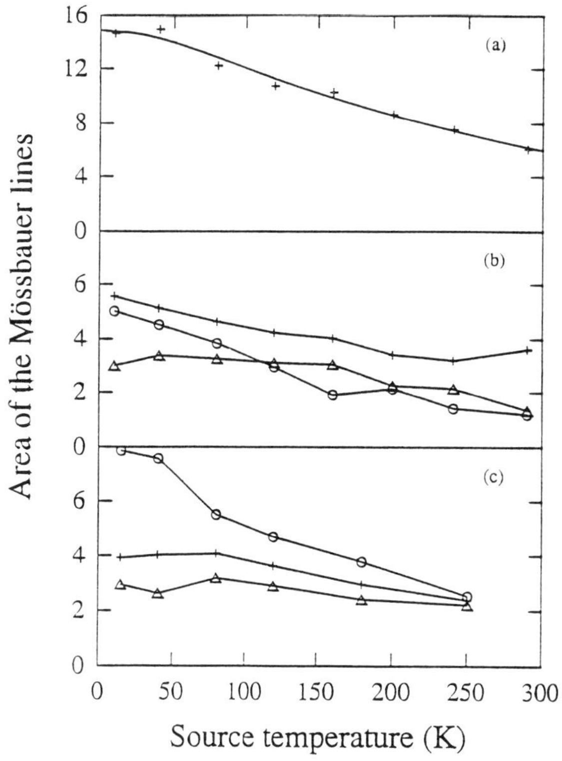

Fig. 3. The area of the Mössbauer lines as a function of the sample temperature: (a) a hydrogen free sample with only ^{119}Sb implanted. The solid line is a least square fit to a Debye model, giving $\Theta_D=225(3)$ K; (b) a sample with $2x10^{15}$ H cm^{-2} and only ^{119}Sb; (c) $2x10^{15}$ D cm^{-2} and $5x10^{12}$ Sb cm^{-2}. I_S is denoted by (+) symbol; I_H by (Δ) and I_I by (o).

substitutional component I_S decreases less steeply than before the introduction of hydrogen and even curls upward around 240 K. The intensities of the other two components show an even more surprising behavior. Below 150 K, I_I grows at the expense of I_H, while at higher temperatures, they both decrease relatively fast. The same trends are observed for the case that D was introduced, the invisible fraction being even somewhat higher. It was verified that all changes were completely reversible. This strange behavior will be discussed later in this paper.

In order to avoid problems from changing surface conditions, Sb-H complex formation as a function of hydrogen dose was studied by taking Mössbauer spectra at 80 K in the UHV implantation set up. Fig. 4 shows the results for a sample with 5×10^{12} Sb cm^{-2}. I_H stays more or less constant as the hydrogen dose increases from 1×10^{14} to 1.5×10^{16} cm^{-2}, while I_I increases steeply at the expense of I_S. The same behavior was found for an Sb dose of 5×10^{13} cm^{-2}.

In another experiment ^{119}Sb was implanted to a dose of 10^{12} cm^{-2} into highly B-doped Si (10^{18} cm^{-3}). The maximum Sb concentration resulting from the implantation was about 10^{17} cm^{-3}, so that the material remained **p-type.** No Sb-H complex line was observed for hydrogen doses up to 2×10^{15} cm^{-2}.

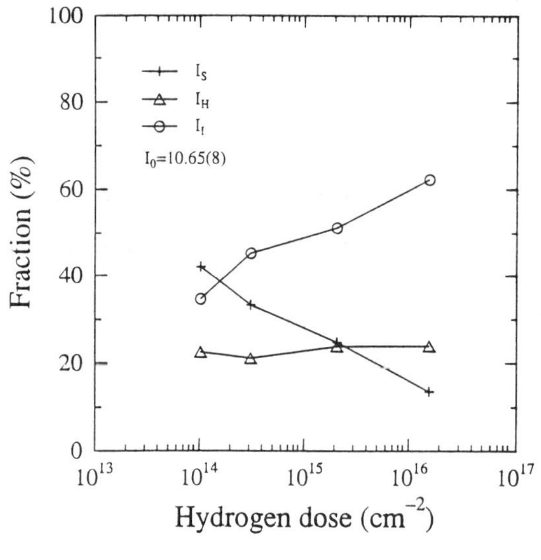

Fig. 4. Fractions of the three components as a function of hydrogen dose for a sample with 5×10^{12} Sb cm^{-2}.

4. Discussion

The Mössbauer results show unambiguously that the donor passivation is related to trapping of hydrogen in the vicinity of Sb donor atoms. Up to 80% of the Sb atoms can be associated with hydrogen. The dissociation energies found from the disappearance of I_H and I_I, 1.36 - 1.46 eV, agree quite well with the value of 1.43 eV found in the infrared studies [10]. In those experiments all donor-H (P, As, Sb) complexes give rise to nearly the same stretching mode and wagging mode, suggesting that the hydrogen atom is not bound directly to the impurity. This supports the AB of Si model [11], *i.e.* the interstitial hydrogen atom is at the antibonding (AB) site of a Si atom which is adjacent to the substitutional donor impurity.

The observed absence of Sb-H complexes after hydrogenation in p-type Si is easily understood by the fact that under these circumstances H^+ is expected to be the stable charge state [12,13]. The Coulomb repulsion between H^+ and Sb^+ will then prevent the formation of Sb-H complexes. This argument implies that in n-type Si Sb-H complex formation is due to trapping of H^- or H^0, in agreement with the fact that hydrogen has a deep donor level in the gap [14].

The disappearance of a large part of the Mössbauer absorption after introduction of hydrogen must be connected with large displacements of the excited ^{119}Sn atom during the emission of the Mössbauer gamma ray. In order to quantify this, we estimate that the recoilless fraction $f = \exp\{-k^2\langle x^2\rangle\}$ of the "invisible" component is at most 5% of that of the visible components. This leads to a lower limit of the rms amplitude of the displacements: $\langle x^2\rangle^{1/2} \geq 0.15$ Å. Such a large number cannot be associated with ordinary lattice vibrations.

According to recent pseudopotential calculations [15,16] the Sb atom in the Sb-H(Si) complex is expected to be surrounded by three Si atoms and a lone electron pair, while the hydrogen atom is at the AB (or T_d) site of the Si atom with the broken Sb-Si bond. This structure is stabilized by the preference of group V atoms for a threefold coordination. When the parent ^{119}Sb atom decays to ^{119}Sn, this structure becomes unstable. The present Mössbauer results can be understood by assuming that at that moment the bound hydrogen atom starts a localized diffusive motion, thereby inducing displacements of the neighboring ^{119}Sn atom of the order of 0.15 Å or more. Calculations show that such atomic relaxations induced by hydrogen are not unreasonable [12]. Moreover, displacements of ^{57}Fe Mössbauer atoms of the order of 0.2 Å due to the motion of neighboring hydrogen atoms have been observed before in a metallic system [17].

In order to explain the data, the hydrogen (or deuterium) motion around the excited nucleus in the ^{119}Sn atom and the response from ^{119}Sn must have the following characteristics:

(i) The hydrogen must be mobile down to 4.2 K on a time scale given by τ_N, the life-time of the excited state (τ_N = 25 ns). This excludes thermal activation and points to hydrogen tunneling. For deuterium we observe the same behavior, the lowest source temperature there being 15 K.

(ii) The hydrogen atom must diffuse locally in the time interval τ_N. One may think of a diffusion path containing the four AB or T_d sites neighboring the Sn atom. If the hydrogen atom would jump away, the neighboring ^{119}Sn atom would contribute to the substitutional component I_S. In fact, a hint of such behavior can be found in Fig. 3b, in which the area of the substitutional line is observed to increase around 240 K. This would correspond to an activation energy for long-range diffusion of about 0.25 eV.

(iii) Localized diffusion or cage motion of Mössbauer atoms between a number of equivalent positions $\mathbf{r}_N$ leads to a reduction of the phase coherence of the emitted radiation. This can be expressed by an effective Debye-Waller factor $f_J = |\sum_N \exp\{-i\mathbf{k}\bullet\mathbf{r}_N\}|^2/N^2$. If the number of positions $\mathbf{r}_N$ is not too large, this expression does not effectively reduce to zero. As we found that the visible component I_H and the invisible one I_I appear and disappear at the same moment, we identify I_H with the remaining Mössbauer intensity given by f_J. This quantity depends on the phase shifts $\mathbf{k}\bullet(\mathbf{r}_i-\mathbf{r}_j)$, *i.e.* on the direction in which the gamma ray is observed. However, we have not found any dependence of the intensity of I_H on the observation direction. This suggests that the position distribution of the Mössbauer atom is not given by a number of fixed points but is more continuous.

The peculiar temperature dependence of the I_H line also suggests that this line and I_I are different aspects of the same phenomenon. The data imply that the displacements of the excited ^{119}Sn nucleus become smaller if T increases. In view of the lack of a structural model for the hydrogen motion we cannot discuss this phenomenon in more detail.

On the other hand, the H-dose dependence shown in Fig. 4 is most easily understood in terms of a different number of hydrogen atoms per complex, *e.g.* one associated with I_H and two with I_I. This would make the situation even more complicated. Although we cannot rule out this possibility, a description in terms of one hydrogen atom per complex seems most appropriate for the time being.

5. Conclusions

The formation of Sb-H and Sb-D complexes in Si is observed via Mössbauer spectroscopy on radioactive ^{119}Sb decaying to ^{119}Sn. Their stability is in agreement with the infrared data on which the current structural model for donor-H complexes is based.

The charge state of hydrogen in n-type Si is H^- or H^0. A localized quantum tunneling of the hydrogen in the neighborhood of the excited ^{119}Sn atom is proposed to explain the large reduction of the Mössbauer intensity after hydrogenation. Theoretical calculations on $Sn\text{-}H_n$ and $Sb\text{-}H_n$ systems would be helpful for the understanding of our observations.

We acknowledge the help of F. Th. ten Broek, J. J. Smit, L. Venema, E. H. du Marchie v. Voorthuysen and the crew of the KVI cyclotron during various stages of the experiments. One of us (Z. N. L.) is grateful to Prof. D. Mo for his encouragement. This work is part of the research program of the "Stichting voor Fundamenteel Onderzoek der Materie" (FOM) with financial support from the "Nederlandse Organisatie voor Wetenschappelijk Onderzoek" (NWO).

References

[1] S. J. Pearton, J.W. Corbett and T.S. Shi, Appl. Phys. A 43, 153(1987).
[2] G. G. DeLeo, W. B. Fowler, T. M Sudol and K. J. O'Brien, Phys. Rev. B 41, 7581(1990).
[3] N. M. Johnson, R. D. Burnham, R. A. Street, and R. L. Thornton, Phys. Rev. B 33, 1102(1986).
[4] P. R. Briddon and R. Jones, Phys. Rev. Lett. 64, 2535(1990).
[5] M. Stavola, K. Bergman, S. J. Pearton and J. Lopata, Phys. Rev. Lett. 24, 2786(1988).
[6] P. J. H. Denteneer, C. G. Van de Walle and S. T. Pantelides, Phys. Rev. B 39, 10809(1989).
[7] A. Nylandsted-Larsen, F. T. Petersen and G. Weyer, J. Appl. Phys. 59, 1908(1986).
[8] Z. N. Liang and L. Niesen, Hyperfine Interactions 60, 749(1990).
[9] J. W. Petersen, O. H. Nielsen, G. Weyer, E. Antoncik and S. Damgaard, Phys. Rev. B 21, 4292(1980).
[10] K. Bergman, M. Stavola, S.J. Pearton and T. Hayes, Phys. Rev. B 38, 9643(1988).
[11] N. M. Johnson, C. Herring and D. J. Chadi, Phys. Rev. Lett. 56, 769(1986).
[12] C. G. Van de Walle, P. J. H. Denteneer, Y. Bar-Yam and S. T. Pantelides, Phys. Rev. B 39, 10791(1989).
[13] K. J. Chang and D. J. Chadi, Phys. Rev. B 40, 11644(1989).
[14] N. M. Johnson and C. Herring, Mat. Sci. Forum 38-41, 961(1989).
[15] P. J. H. Denteneer, C. G. Van de Walle and S. T. Pantelides, Phys. Rev. B 41, 3885(1990).
[16] S. B. Zhang and D. J. Chadi, Phys. Rev. B 41, 3882(1990).
[17] R. Wordel, F. J. Litterst and F. E. Wagner, J. Phys. F 15, 2525(1985).

Materials Science Forum Vol. 83-87 (1992) pp. 105-110

DEPTH PROFILES FOR HYDROGEN-ASSISTED THERMAL DONOR FORMATION IN SILICON

H. J. STEIN[1], S. HAHN[2] AND P. M. RICHARDS[1]
[1]Sandia National Laboratories, P. O. Box 5800, Albuquerque, New Mexico 87185-5800, U. S. A.
[2]Stanford University, Dept. of Matls. Sci. and Eng., Stanford, California 94305, U. S. A.

ABSTRACT

Concentration and depth profiles for thermal donor formation in Si under hydrogen plasma exposure have been investigated by infrared, IR, absorption and spreading resistance probe, SRP measurements. Donors measured by SRP are shown to correlate with oxygen-related TD's detected by IR. Hydrogen-independent precursor nuclei formation at the hydrogenation temperature, and continued hydrogen activity in TD formation after hydrogenation are inferred from effects of extended pre and post heating on TD concentrations and profiles. A model wherein diffusing hydrogen activates precursor sites is presented to explain concentrations and profiles for TD's formed in a hydrogen plasma.

1. Introduction

Studies in the past few years[1,2] have confirmed that interactions of hydrogen with oxygen occur in Czochralski, CZ, Si at temperatures near 400°C. These interactions were shown by hydrogen-assisted thermal donor, TD, formation[1,2] and by enhanced relaxation of stress-aligned interstitial oxygen centers[2]. The interactions are providing new understanding of enhanced oxygen diffusion, and new information on hydrogen diffusion and formation of oxygen-related TD's in Si.

Characteristic infrared, IR, absorption for electronic transitions of oxygen-related TD's provide a definitive measurement for hydrogen-assisted TD formation, and spreading resistance probe, SRP, measurements provide a method for obtaining the depth profile for hydrogen-assisted donor formation. In the present study we compare the penetration depth for hydrogen-assisted donors as measured by SRP to those for TD's determined from IR/polish-back experiments. Preliminary measurements have been made on the effects of pre and post hydrogenation heating on the TD profile. Finally, models are considered to interpret the TD profiles as a function of time in a hydrogen plasma exposure.

2. Experimental Details

Methods and systems for plasma hydrogenation, IR absorption, and SRP measurements have been described previously[1]. Briefly, the hydrogenations were performed in a low power (≈0.03 W/cm^2, 0.5 torr. and 0.5 MHz) plasma generated in a furnace-heated quartz tube, IR absorption measurements were performed at 80 K with 4 cm^{-1} resolution, and the SRP measurements were made after angle lapping to give 10 μm resolution in the donor depth profile measurements. Samples were diced (0.63 x 0.63 cm) from a 10 ohm-cm n-type (100) CZ Si wafer which had been polished one side and given a rapid thermal anneal (800°C/10 sec) to remove pre-existing TD's. Polish back after hydrogen plasma exposure was performed with 6 μm followed by 1 μm diamond paste, and remaining thickness was determined from the intensity of the multiphonon absorption in the measured spectrum. Thicknesses determined in this way are believed more accurate than those measured by a micrometer because rounding of the sample surface occurred during polishing.

3. Experimental Results

A depth profile measured by SRP for TD's in CZ silicon after a 4 hour hydrogen-plasma exposure at 400°C is plotted in Fig. 1. The initial dopant donor concentration has been subtracted from that measured after the exposure. The plotted concentration, therefore, is due to the formation of TD's. The TD concentration is relatively constant in the profile, followed by an abrupt decrease to a background level. While the depth to the abrupt decrease is consistent with that expected for hydrogen diffusion in Si at 400°C, the profile departs markedly from a complementary error function which is illustrated by the dashed line in Fig. 1.

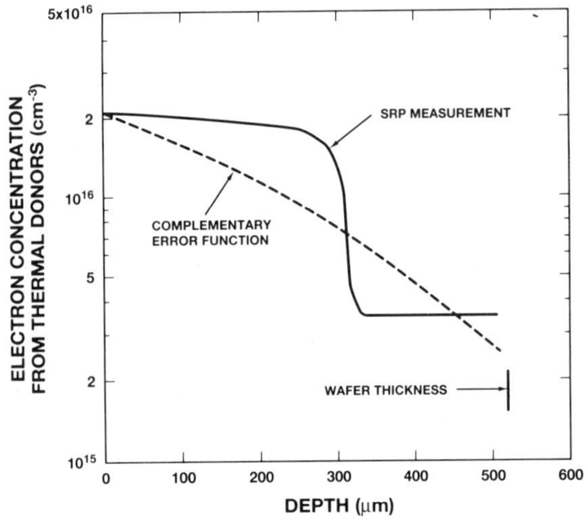

Fig. 1: Thermal donor concentration determined by SRP versus depth into Si after 4 hours in a hydrogen plasma at 400°C (solid line). Plot of complementary error function with depth for 1/3 maximum concentration matched to that for the abrupt decrease in measured concentration (dashed line).

To verify that the donors in CZ silicon formed in the hydrogen plasma are due to oxygen-related TD's, infrared absorption spectra were measured before and after plasma exposure at 400°C. Plotted in Fig. 2 are difference spectra obtained by subtracting a spectrum measured before plasma exposure spectrum from those measured after exposure for 2, 4, and 8 hours. Lattice phonon absorption features are removed from such difference spectra. A small negative going peak near 1100 cm^{-1} in the difference spectra is due to the decrease in interstitial oxygen which occurs upon TD formation[1,2]. Free carrier absorption associated with TD's causes the increase in absorption with decreasing frequency in the spectra. The features labeled A through D in the spectra are characteristic of electronic transitions for double donors of the oxygen-related TD family, thus confirming formation of oxygen-related TD's in a hydrogen plasma.

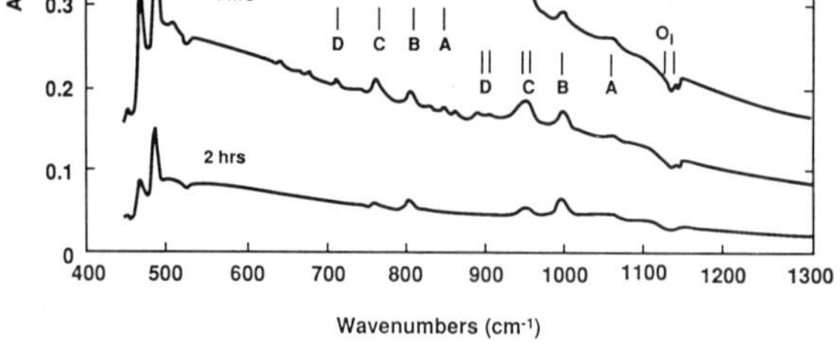

Fig. 2: Absorption by thermal donors after 2, 4 and 8 hours in a hydrogen plasma at 400°C.

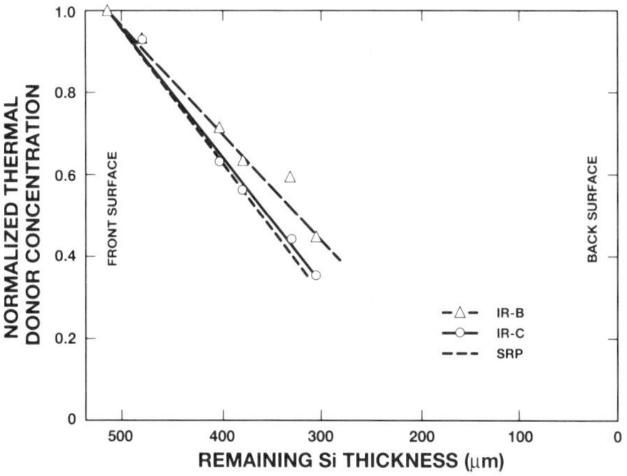

Fig. 3: Normalized integrated thermal donor concentrations from polish-back/IR experiments compared to summation over the concentration profile measured by SRP.

Depth profiles for B and C TD's were obtained from a series of polish-back/IR measurements on samples which had been given a hydrogen plasma exposure for 4 hours at 400° C. Peak intensities versus thickness for the B and C electronic transitions are plotted in Fig. 3. The B member of the TD family penetrates slightly deeper than the C member which is consistent with an A, B, C, etc. sequence of TD formation. Also plotted in Fig. 3 is a normalized integrated donor concentration obtained by summing the TD concentration detected by SRP over the remaining thickness. Agreement between the SRP and IR results for TD's versus remaining thickness indicates that the same donors are measured by both techniques.

A profile such as that shown in Fig. 1 for TD's can be explained by a saturation process[1]. One of us[3] has recently considered three models to describe saturation profiles of reaction products for hydrogen interactions in Si: reversible traps, irreversible traps, and an excitation of sites without strong trapping. In the excitation model, hydrogen activates a site in the vicinity but there is negligible perturbation of the hydrogen diffusion. Activation barrier lowering by hydrogen[4] is an example of a physical process where this model might apply, and for convenience we use its terminology, although the other models give similar results. The pertinent rate equation is therefore

$$\partial C_e/\partial t = \gamma m(C_o - C_e) \qquad (1)$$

where C_e is the measured TD concentration and is a function of depth, x, and time, t. C_o is the saturation TD concentration, γ is an activation rate, and m is the concentration of hydrogen in solution at x and t.

This form of the excitation model, however, cannot explain the effects of exposure time on the TD depth profiles shown in Fig. 4. While the penetration depth follows a square root of time[1], as expected for a diffusion process, the model (Eq. 1) predicts that the formation rate of additional TD's would approach zero as C_e approaches C_o upon saturation. In fact, the hydrogen-assisted TD concentration near the surface increases essentially linearly with time at 400°C. The excitation model has, therefore, been modified to allow C_o to increase with time. With this modification the model can describe hydrogen-assisted TD formation in the presence of an increasing concentration of precursor sites formed uniformly in the wafer independent of the added hydrogen. Precursor nuclei concentration dependence on sample history and annealing has been suggested previously to explain sample-dependent TD formation rates in samples which are otherwise equivalent[5,6].

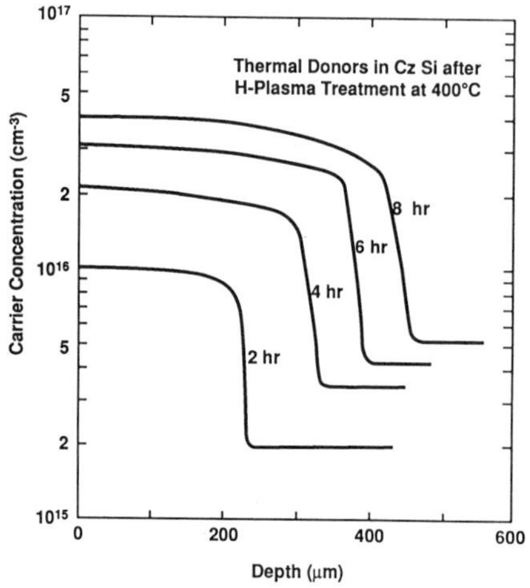

Fig. 4: Thermal donor profiles measured by SRP after 2, 4, 6 and 8 hours in a hydrogen plasma at 400°C.

If precursor nuclei are formed through the depth of the wafer and independent of hydrogen during heating at 400°C, then preheating at 400°C before a hydrogen plasma exposure would be expected to increase the TD formation rate. A TD profile measured after a 6 hour preheat in flowing N_2 followed by a 2 hour hydrogen-plasma exposure at 400°C is depicted by the solid line in Fig. 5. The dashed line in Fig. 5 represents a TD profile after a 2 hour hydrogen-plasma exposure without preheating, and the dot-dash line represents the TD concentration measured for a sample which was given the preheat only. These preliminary results indicate that preheating does increase the formation rate for TD's in the profile, consistent with nuclei formation during preheating. On the other hand, preheating has little effect upon the penetration depth for TD formation. This is in accord with attributing the penetration depth to hydrogen diffusion.

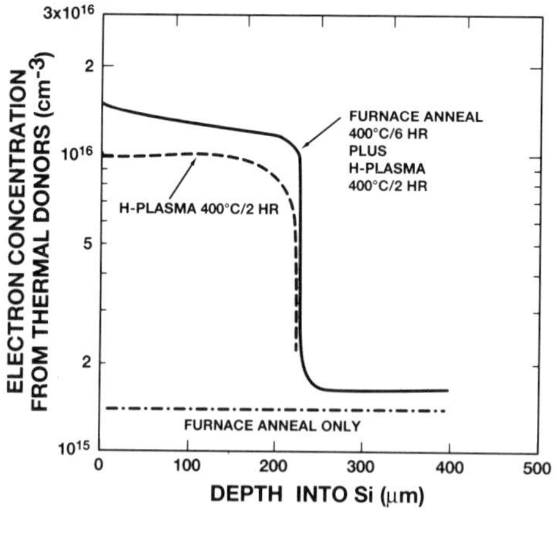

Fig. 5: Thermal donor profile measured by SRP after a 6 hour furnace preanneal at 400°C plus 2 hours in a hydrogen plasma at 400°C (solid line). Profile after hydrogenation with no preanneal (dashed line), and after preanneal only (dash-dot line).

In contrast to TD profiles in plasma-exposed CZ Si, TD profiles which are more characteristic of diffusion have been observed after hydrogen ion implantation[1]. One difference between the ion implantation and plasma exposure experiments was in the cool down rate after hydrogen exposure. Rapid cooldown occurs when the furnace is moved away after a plasma exposure whereas the implanted samples cool slowly in vacuum while mounted on a copper block. If hydrogen remains in the sample upon termination of the hydrogen exposure, and nuclei which can be activated by retained hydrogen continue to increase, then perhaps a diffusion profile will evolve if the sample is heated after termination of the exposure. To examine this possibility, a sample was heated in flowing N_2 at 400°C for 6 hours after a 2 hour hydrogen exposure at 400°C. A TD profile for this sample is shown in Fig. 6. It is apparent that the TD profile has evolved to a more diffusion-like profile during post hydrogenation annealing. This result supports the idea of continued nuclei generation during heating at 400°C. The result further indicates retention of hydrogen (or an interaction product such as oxygen liberated from interstitial sites[4]) within the Si so that the retained species can subsequently interact with nuclei to enhance TD formation. Hydrogen retention, although in low concentration, is the most likely candidate since Newman et al.[7] found hydrogen enhancement of relaxation for stress-aligned interstitial oxygen centers in CZ Si continued after cessation of plasma hydrogenation.

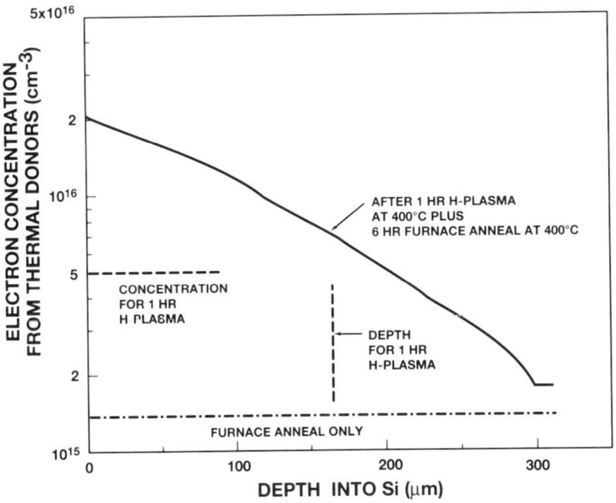

Fig. 6: Thermal donor profile measured by SRP after 1 hour in a hydrogen plasma plus a 6 hour post furnace anneal at 400°C (solid line). Concentration and depth for thermal donors after 1 hour in a hydrogen plasma at 400°C (dashed line), and concentration after furnace anneal only (dash-dot line).

4. Summary and Conclusions

Integrated TD concentrations versus depth determined from IR absorption/polish-back experiments correlate with integrated donor concentrations determined by SRP after hydrogen plasma exposure of CZ Si, thus confirming that the donors measured by SRP are oxygen-related TD's. The TD concentration in the profile after plasma hydrogenation is increased by preheating at the hydrogenation temperature, consistent with hydrogen-independent nuclei formation during the heating. The depth of penetration for TD formation is attributed to hydrogen diffusion and is not significantly affected by preheating. An excitation model combined with continuous formation of nuclei at the hydrogenation temperature is suggested to explain the TD profiles, dependence of the TD concentration on time in the hydrogen plasma, and preheating effects on the TD formation rate. Preliminary results from post hydrogenation heating effects on the TD profiles indicate continued participation of hydrogen, or hydrogen-promoted sites, in TD formation after plasma hydrogenation has ceased.

5. Acknowledgements

Work at Sandia National Laboratories is supported by the U. S. Department of Energy under Contract No. DE-AC04-76DP00789.

6. References

1) H. J. Stein and S. Hahn, Appl. Phys. Lett. 56, 63 (1990); H. J. Stein and S. Hahn in *Defect Control in Semiconductors*, Ed. K. Sumino (Elsevier Sci. Pub., North-Holland, 1990) p. 211; H. J. Stein and S. Hahn, in 1991 ECS Proc. Vol. *Defects in Si*, ed. W. M. Bullis, in press.

2) A. R. Brown, M. Claybourn, R. Murray, P. S. Nandhra, R. C. Newman, and J. H. Tucker, Semicond. Sci. Technol. 3, 591 (1988); R. C. Newman, A. R. Brown, R. Murray, A. Tipping and J. H. Tucker, in ECS Proc. Vol. 90-7 *Semiconductor Silicon/1990*, ed. H. R. Huff and K. G. Barraclough (1990) p. 734.

3) P. M. Richards, J. Appl. Phys. (accepted for publication).

4) S. K. Estreicher, Phys. Rev. B 41, 9886 (1990).

5) B. Y. Mao, J. Lagowski, and H. C. Gatos, J. Appl. Phys. 56, 2729 (1984).

6) V. P. Markevich and L. I. Murin, Phys. Stat. Sol. (a), 111 (1989).

7) R. C. Newman, Private Communication.

Materials Science Forum Vol. 83-87 (1992) pp. 111-118

PASSIVATION OF SHALLOW ACCEPTORS IN Si AND GaAs BY ANNEALING IN H_2

I.A. Veloarisoa, D.M. Kozuch, Michael Stavola, R.E. Peale, and G.D. Watkins, Physics Bldg. 16, Lehigh University, Bethlehem, PA 18015

S.J. Pearton, C.R. Abernathy, and W.S. Hobson, AT&T Bell Laboratories, Murray Hill, NJ 07974

ABSTRACT

In this paper we describe two examples of the passivation of shallow acceptors in semiconductors by annealing in H_2 gas. In the first, a fraction of the shallow acceptors are shown to be passivated throughout the bulk of a Si sample that is several mm thick by annealing in H_2 gas at 1280°C and quenching. In the second, epitaxial GaAs:C grown by metalorganic MBE or MOCVD is shown to be partially passivated near 500°C by H_2 gas that is either in annealing or growth ambients.

I. INTRODUCTION

Most recent work on H in semiconductors [1-3] has focussed on defect passivation in a plasma that contains atomic hydrogen. In the following sections we discuss two new examples where H_2 in growth or annealing ambients leads to significant H incorporation and shallow impurity-H complex formation. (i) Acceptors in Si can by passivated by annealing in H_2 at elevated temperatures (900°C-1280°C) and then quenching rapidly. [4-6] (ii) In GaAs:C epitaxial layers grown by metalorganic MBE (MOMBE) or MOCVD, we have found that annealing in H_2 at 500°C leads to a substantial partial passivation of the C in the layer. Typical growth temperatures and processing-related annealing steps are performed near 500°C; H_2 gas in growth or processing ambients is shown to lead to C-H complex formation.

Although H-passivation from H-containing plasmas has dominated recent work, there are a number of cases where H_2 in growth and annealing ambients has been shown to lead to H incorporation and/or defect passivation. In elemental hosts, Hall [7] and Haller and coworkers [8] have discovered several H-related species in Ge that had been grown in H_2. Si grown in an H_2 atmosphere is also known to contain a number of H-related defects.[9] In experiments performed by the Sievers group,[10,11] annealing Si in H_2 gas led to the formation of impurity-H complexes.

In III-V and II-VI host materials, there are a few examples of H incorporation from H_2 gas of which we are aware. Svob and coworkers diffused D into GaAs [12] and II-VI materials [13] from D_2 gas and detected the D with SIMS. In more recent work, Svob et al. [14] showed with SIMS measurements that H was incorporated into II-VI materials grown by MOCVD from H_2 in the growth ambient. Defect passivation was not demonstrated in these experiments.

Here we focus on the passivation of shallow impurities by hydrogen following growth or annealing in an H_2 containing ambient. In our studies, IR absorption due to H-stretching vibrations is the primary probe of the passivated shallow impurities.

II. PASSIVATION OF SHALLOW ACCEPTORS IN Si

In our experiments, Si samples were sealed in quartz ampules that had been evacuated and then filled with 0.6 atm of H_2 gas at room temperature. The sealed ampules were annealed at temperatures between

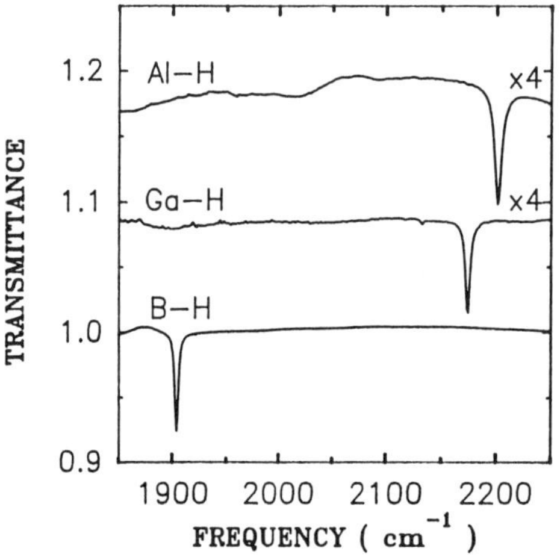

Fig. 1. IR spectra measured near liquid He temperature for acceptor-doped, Si samples that had been annealed in H_2 gas at 1280°C for 30 min. and then quenched to room temperature. The acceptor concentrations are $6x10^{16}$, $1x10^{16}$, and $0.7x10^{16}$ cm^{-3} for the B, Al, and Ga doped samples, respectively.

900°C and 1280°C and then quenched to room temperature by withdrawing the ampule from the furnace and dropping it into ethylene qlycol at room temperature. IR absorption spectra were measured at a resolution of 2 cm^{-1} with a Bomem DA3.16 Fourier transform spectrometer.

IR spectra of bulk, acceptor-doped Si samples (with N_A of the order of 10^{16} cm^{-3}) that had been annealed in H_2 are shown in Fig. 1. The H-stretching features have frequencies and absorption strengths similar to those reported for acceptor-H complexes in samples that were doped with concentrations greater that $10^{18}cm^{-3}$ by ion-implantation and then passivated by exposure to a hydrogen plasma.[15] Absorption strengths in the H_2-soaked, uniformly doped samples that are comparable to thin, heavily doped layers implies acceptor passivation well into the bulk of the H_2-soaked samples. To confirm that the passivation was uniform throughout the sample thickness, we remeasured the IR absorption after grinding 1 mm from the surfaces of an H_2 treated, B-doped sample that was initially 3 mm thick. Vibrational absorption due to B-H centers was observed with a strength consistent with the 1 mm thickness of the thinned sample and a uniform distribution of centers throughout the sample.

To determine [B-H] from the strength of the IR absorption at 1903 cm^{-1} we have reexamined previous results [16] for a B-implanted sample that had been passivated in an H_2 plasma. The calibration is given by,

$$[B\text{-}H]/A = 2.5 \times 10^{15}\ cm^{-1}. \qquad (1)$$

This calibration is close to that determined for Al-H complexes in Si [15] and also to a recent calibration of the B-H absorption by McQuaid et al. [5] With the calibration given by Eq. (1) we find that [B-H] = $4.6x10^{15}$ and $5.0x10^{14}$ cm^{-3} for samples with [B] = $1.7x10^{17}$ and $1.6x10^{16}$ cm^{-3}, respectively. Hence, approximately 3% of the B is passivated by the anneal in H_2 at 1280°C. For the Al doped sample we find [Al-H]=$1.3x10^{15}$ cm^{-3} and that 8% of the Al is passivated. The more complete passivation of Al is consistent with this complex's greater binding energy. [17]

We have explored the role of the temperature and duration of the H_2-soaking treatment on the concentration of B-H complexes that are formed. In Fig. 2 are plotted the concentrations of B-H complexes vs. the reciprocal of the temperature of the H_2 anneal for the samples doped with [B] = $1.7x10^{17}$ cm^{-3}. The open circles are for 30 min. anneals and the closed squares are for 120 min anneals. At 1100°C,

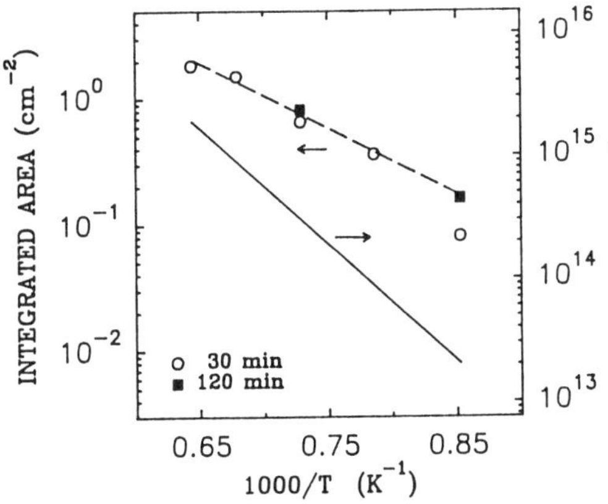

Fig. 2. The integrated absorption coefficient, A, for the H-stretching vibration of the B-H complex vs. the reciprocal of the annealing temperature for Si samples with [B] = 1.7×10^{17} cm^{-3} that were annealed in H_2, and then quenched. The concentration scale on the right is related to A by Eq.(1) in the text. The solid line is the solubility of H in pure Si determined by Van Wieringen and Warmoltz.

increasing the duration of the anneal increases the strength of B-H absorption only slightly. At 900°C, increasing the duration of the anneal increases the B-H absorption by roughly a factor of 2 indicating that H has not fully diffused through the sample during the 30 min anneal. The diffusion constant determined by Van Wieringen and Warmoltz (VWW) [18] is 3.7×10^{15} cm^2/s at 900°C which gives a $\sqrt{Dt}$ diffusion length of 0.26 cm for the 30 min anneal in H_2 which is comparable to the thickness of the sample.

The dashed line shown in Fig. 2 is the best fit to the data. (At 900°C, the data for the 120 min anneal was used.) If we use the calibration of the B-H absorption given in Eq. (1), the dependence of [B-H] upon annealing temperature for our samples is given by

$$[B\text{-}H] = 1.2 \times 10^{19} \exp(-1.04\text{eV}/kT) \text{ cm}^{-3} \qquad (2)$$

Also shown in Fig. 2 is the hydrogen solubility determined by Van Wieringen and Warmoltz [18], $S_H = 2.4 \times 10^{21} \exp(-1.88\text{eV}/kT)$ cm^{-3}. While the B-H concentration is greater than the VWW value by only a factor of 2.7 at 1280°C, the B-H concentration in our samples annealed at 900°C exceeds the VWW value by a factor of 20. The activation energies in Eq.(2) and in the solubility determined by VWW differ by nearly a factor of 2.

When H was introduced into n-type Si doped with P or As in the mid-10^{16} cm^{-3} range by annealing at 1280°C in H_2 gas and quenching, IR absorption bands at 1555 and 1561 cm^{-1} that have been assigned to P-H and As-H complexes [15,19] were not detected. We conclude that shallow donors are not passivated as efficiently as acceptors by the introduction of H into our samples at high temperatures.

Previous studies of H-passivated shallow impurities are made possible, in part, by the high solubility of the dopant. To have a sufficient number of centers in the micron-thick, H-passivated layers that are typical of plasma exposed samples, the dopant concentration must be large. The study of hydrogen-passivated deep impurities has been hampered by their lower solubility which does not lead to a sufficient total number of hydrogen-passivated centers for easy examination by structure sensitive techniques such as IR absorption or EPR. Hence, we expect that there will be renewed interest in techniques by which H can be introduced throughout the bulk of a semiconductor sample. The total number of centers introduced should permit structure-sensitive studies of species with solubilities that are decades smaller than is typical of common dopants.

III. PASSIVATION OF GaAs:C_{As} GROWN BY MOMBE OR MOCVD

In our experiments on the passivation of GaAs:C, several epitaxial layers grown by MOMBE and MOCVD with various growth conditions were examined. MOMBE samples were grown in a Varian Gas-Source Gen II on semi-insulating GaAs substrates.[20] Most samples studied were grown from AsH_3 and trimethylgallium (TMG) that was introduced with He carrier gas. The AsH_3 was introduced through a low pressure cracker heated to 950°C. The growth temperature was 500°C. Immediately after the growth, the substrate heater was switched off and the sample was allowed to cool to 475°C which took 30 s. The AsH_3 flow was stopped and the sample was unloaded. A few MOMBE samples were grown from TMG and a solid As source; for one of these samples a flow of H_2 gas was introduced into the growth chamber.

MOCVD samples were grown from triethylgallium and trimethylarsine following the method of Kobayashi and Inoue.[21] The substrate temperature was 545°C. The metalorganic sources were introduced with hydrogen carrier gas.

Annealing experiments were performed either in an RTA oven with an ambient of forming gas or He or in a muffle furnace in sealed quartz ampules. Characterization methods include Hall effect, SIMS (Cs^+ beam with ion-implanted standards), and IR absorption.

IR spectra for the H-stretching region are shown in Fig. 3 for samples from a MOMBE-grown layer with $N_A = 9.4\times10^{19}$ cm^{-3} that had been annealed in various ambients. In Fig. 3(a) a feature at 2635 cm^{-1} that was previously assigned to the C_{As}-H center by Clerjaud et al. [22] and additional features at 2643, 2651, and 2688 cm^{-1} that also involve C and H [23] are seen in an as-grown layer. We use the denotation C_x-H for these centers.

In previous work it was found that H could be introduced into the GaAs:C layers grown by MOMBE from the TMG or the AsH_3 sources. [23,24] Fig. 3 shows that H_2 in the annealing ambient is also an important source of H in the layers. The spectrum in Fig. 3(b) shows that annealing in H_2 enhances the strength of the C-H stretching feature at 2635 cm^{-1}. This anneal was performed in a sealed ampule; spurious sources of atomic H that might result from the dissociation of H_2 on hot furnace components are unlikely. In Figs.

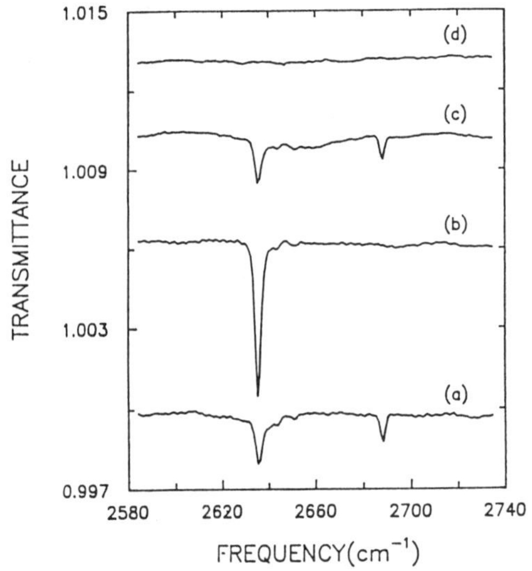

Fig. 3. IR spectra that demonstrate the effect of annealing ambient on C-H centers in GaAs:C grown by MOMBE. Samples were selected from the same wafer with $N_A = 9.4\times10^{19} cm^{-3}$. (a) As-grown. (b) Anneal at 450°C in 0.6 atm H_2 in a sealed ampoule. (c) Anneal for 5 min in forming gas at 600°C in an RTA oven. (d) Anneal for 5 min in He at 600°C in an RTA oven.

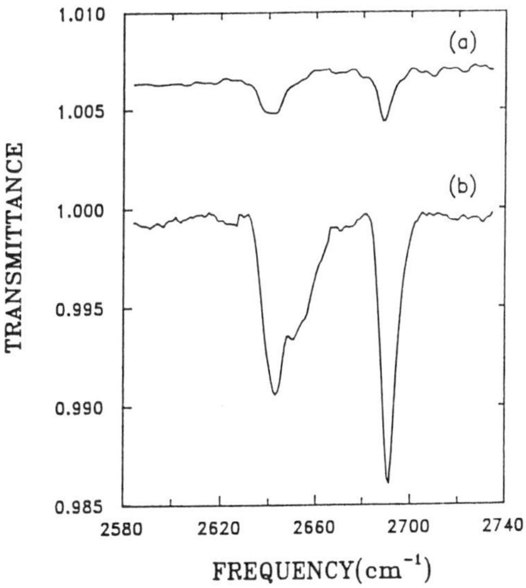

Fig. 4. A comparison of the strength of C-H vibrational features in samples grown by MOMBE with a solid As source. (a) The sample is 0.22 μm thick and was grown without an H_2 flow. (b) The sample is 0.56 μm thick and was grown with an H_2 flow of 10 cc/min.

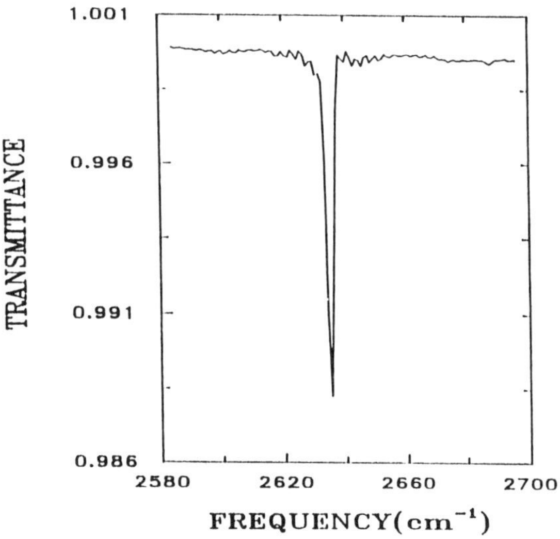

Fig. 5. IR spectrum measured near liquid He temperature for an as-grown GaAs:C layer grown by MOCVD. $N_A = 2 \times 10^{19}$ cm^{-3}.

3(c) and (d), samples are compared following annealing in an RTA oven in forming gas (90% N_2 and 10% H_2) and He gas. An anneal in the inert He ambient at 600°C eliminates the C_x-H centers whereas, following an anneal at 600°C in forming gas, C_x-H centers remain.

To confirm that H_2 present during growth can passivate C acceptors in GaAs:C, we have compared two epilayers grown from TMG and a solid As source. For one epilayer a flow of H_2 (10 cc/min) was introduced into the MOMBE growth chamber. IR spectra of these samples are shown in Figs 4(a) and (b). C-H centers are present in the layer grown from the TMG source and solid As indicating that the TMG alone is a source of H in the layer. Introducing H_2 gas into the chamber enhances the C-H-related IR absorption and hence the concentration of passivated C in the layer. When one accounts for the different layer thicknesses, it is found that the hydrogen content is increased by 2.7 times for the sample grown in an H_2 flow.

An IR spectrum is shown in Fig. 5 for an as-grown, GaAs:C layer grown by MOCVD. The strong feature at 2635 cm^{-1} assigned to C_{As}-H centers is present. The metalorganic sources and H_2 carrier gas are the likely sources of H in the epilayer.

The presence of C_x-H centers in as-grown samples indicates that these defects are stable near the epi-growth temperature. Isochronal (5 min) annealing data are shown in Fig. 6. These anneals were performed in an RTA furnace in a He ambient. The 2635 and 2688 cm^{-1} features are annealed away at temperatures of 800 K (527°C) and 900 K (627°C), respectively. Both centers are stable or marginally stable at the epi-growth temperatures of 500 and 545°C for the MOMBE and MOCVD growths, respectively, and persist to higher temperatures when samples are annealed in H_2-containing ambients. [See Fig. 3(c)].

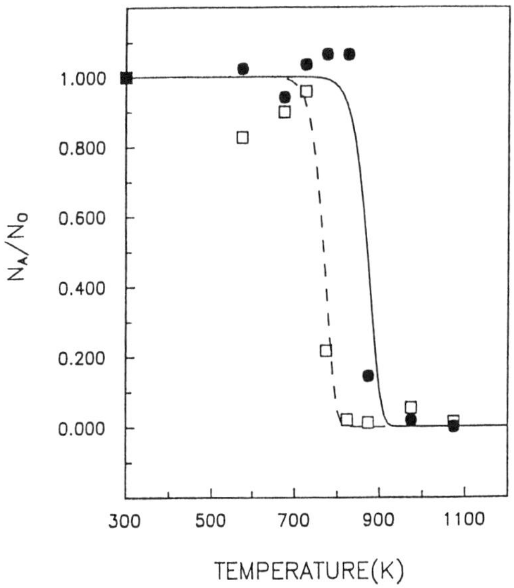

Fig. 6. Isochronal annealing data for the 2636 cm^{-1} (open squares) and 2688 cm^{-1} (closed circles) absorption signals in epitaxial GaAs:C grown by MOMBE. Anneals were performed for 5 min in a He ambient. Here, $N_A = 2x10^{20}$ cm^{-3}.

The stability of the C_{As}-H center (2635 cm^{-1}) is consistent with previous results reported by Clerjaud et al. [25]

We have made an approximate calibration of the concentration of centers that include C and H by making a direct measurement of the H concentration with SIMS. It was assumed that all of the hydrogen in the layers was involved in stable C_x-H complexes and that the different C-H stretching features have the same calibration factor. With these approximations the calibration is

$$[C_x\text{-H}]/A = 1.8x10^{16} \text{ cm}^{-1} \qquad (3)$$

Here [C_x-H] refers to the total concentration of C_x-H centers and A is the integrated absorption coefficient, $A = \int\alpha(\sigma)d\sigma$, in units cm^{-2} for the total H-stretching absorption due to the C_x-H centers.

With the calibration factor given in Eq.(3) we can estimate the concentrations of C_x-H centers in the various epilayers from the strength of the infrared absorption. The as-grown MOMBE layer in Fig. 3(a) has [C_x-H] = $7x10^{18}$ cm^{-3}. Annealing in H_2 [Fig. 3(b)] increases the concentration of passivated carbon to [C_x-H] = $1.2x10^{19}$ cm^{-3} or to roughly 11% of the total [C]. The highest concentrations of C_x-H centers were observed in layers with H_2 in the growth ambient. The sample grown by MOMBE in a flow of H_2 [Fig. 4(b)] has [C_x-H] = $6.9x10^{19}$ cm^{-3}. The MOCVD-grown layer [Fig. 5] has [C_x-H] = $1.4x10^{19}$ cm^{-3}. The fraction of passivated C acceptors varies from 6% to 40% of the total acceptor concentration depending upon the growth method and growth conditions.

IV. CONCLUSION

In recent work, acceptor impurities in Si have been passivated by annealing bulk samples in H_2 at high temperature and quenching.[4-6] H is incorporated at concentrations that are well above the solubility of H determined by Van Wieringen and Warmoltz.[18] Similar treatments did not result in the passivation of shallow donors in n-type Si. It is expected that this bulk passivation technique will prove useful in studies of H-passivated deep levels.

For GaAs:C epitaxial layers, H_2 gas in growth and annealing ambients has been shown to lead to the formation of C_x-H complexes. Up to 40% of the C acceptors are estimated to be passivated in some cases. We presume that the H_2 dissociates at the sample surface and then diffuses into the epilayer where C_x-H centers are readily formed. The C_x-H centers have been shown to be stable or marginally stable at temperatures near 500°C. Hence, the unintentional passivation of C is especially likely because typical epitaxial growth and processing-related annealing steps are performed near 500°C and often in H_2 containing ambients.

The work described in Sec. II was supported by the U.S. Navy Office of Naval Research (Electronics and Solid State Sciences Program) under Contract No. N00014-90-J-1264. The work described in Sec. III was supported by the NSF under Grant No. DMR-9023419.

REFERENCES

1. Hydrogen in Semiconductors, ed. J.I. Pankove, and N.M. Johnson, (Academic, San Diego, 1991).
2. Hydrogen in Semiconductors, ed. M. Stutzmann and J. Chevallier, (North Holland, Amsterdam, 1991).
3. S.J. Pearton, J.W. Corbett, and M. Stavola, Hydrogen in Crystalline Semiconductors, (Springer-Verlag, Heidelberg, 19xx), in press.
4. I.A. Veloarisoa, D.M. Kozuch, R.E. Peale, M. Stavola, and G.D. Watkins, Bull. Am. Phys. Soc. 36, 945 (1991).
5. S.A. McQuaid, R.C. Newman, J.H. Tucker, E.C. Lightowlers, R.A.A. Kubiak, and M. Goulding, Appl. Phys. Lett. 58, 2933 (1991).
6. I.A. Veloarisoa, M. Stavola, D.M. Kozuch, R.E. Peale, and G.D. Watkins, Appl. Phys. Lett., to be published.
7. R.N. Hall, Inst. Phys. Conf. Ser. No. 23, 190 (1975).
8. E.E. Haller, in ref. 1, p. 351; Semicond. Sci. Technol. 6, 73 (1991).
9. G.R. Bai, M.W. Qi, L.M. Xie and T.S. Shi, Solid State Commun. 56, 277 (1985); M.W. Qi, G.R. Bai, T.S. Shi and L.M. Xie, Mat. Lett. 3, 467 (1985) and references contained therein.
10. K. Muro and A.J. Sievers, Phys. Rev. Lett. 57, 897 (1986).
11. R.E. Peale, K. Muro, A.J. Sievers, Shallow Impurities in Semiconductors, ed. G. Davies, (Trans Tech, Switzerland, 1991), p. 141.
12. L. Svob, C. Grattepain, and Y. Marfaing, Appl. Phys. A47, 309 (1988).
13. L. Svob and Y. Marfaing, Shallow Impurities in Semiconductors, ed. G. Davies, (Trans Tech, Switzerland, 1991), p. 181.
14. L. Svob, Y. Marfaing, F. Desjonqueres and R. Druilhe, in ref. 2, p. 550.
15. M. Stavola, S.J. Pearton, J. Lopata and W.C. Dautremont-Smith, Appl. Phys. Lett. 50, 1086 (1987); M. Stavola and S.J. Pearton, in ref. 1, p. 139.
16. M. Stavola, S.J. Pearton, J. Lopata, and W.C. Dautremont-Smith, Phys. Rev. B 37, 8313 (1988).
17. T. Zundel and J. Weber, Phys. Rev. B 39. 13549 (1989).
18. A. Van Wieringen and N. Warmoltz, Physica 22, 849 (1956).
19. K. Bergman, M. Stavola, S.J. Pearton, and J. Lopata, Phys. Rev. B 37, 2770 (1988).
20. C.R. Abernathy, S.J. Pearton, R. Caruso, F. Ren, and J. Kovalchik, Appl. Phys. Lett. 55, 1750 (1989); C.R. Abernathy et al., J. Crystal Growth 105, 375 (1990).
21. T. Kobayashi and N. Inoue, J. Crystal Growth 102, 183 (1990).
22. B. Clerjaud, F. Gendron, M. Krause, and W. Ulrici, Phys. Rev. Lett. 65, 1800 (1990).
23. D.M. Kozuch, M. Stavola, S.J. Pearton, C.R. Abernathy, and J. Lopata, Appl. Phys. Lett. 57, 2561 (1990).
24. K. Woodhouse, R.C. Newman, T.J. de Lyon, J.M. Woodall, G.J. Scilla, and F. Cordone, Semicond. Sci. Technol. 6, 330 (1991).
25. B. Clerjaud, F. Gendron, M. Krause, C. Naud, and W. Ulrici, in ref. 2, p. 417.

Materials Science Forum Vol. 83-87 (1992) pp. 119-124

INTERSTITIAL H AND {H,B}, {H,C}, AND {H,Si} PAIRS IN Si AND Ge

Djordje M. Maric,[†] Peter F. Meier,[†] and Stefan K. Estreicher[‡]

[†]Physics Department, University of Zürich, 8057 Zürich (Switzerland).

[‡]Physics Department, Texas Tech University, Lubbock, TX 79409-1051 (USA).

ABSTRACT

The interactions between interstitial H and substitutional B, C, and Si in crystalline Si and Ge have been studied at the *ab-initio* Hartree-Fock level in several molecular clusters. Large basis sets and corrections for relativistic effects were included. The stable and metastable structures of the {H,B} complex in Ge are compared to the well known {H,B} pair in Si, obtained at the same theoretical level. The formation of {H,C} and {H,Si} pairs in Ge is also studied. The relative depths of these traps for H are compared to the bond-centered and tetrahedral interstitial configurations of hydrogen in perfect Ge.

1. INTRODUCTION

High-purity germanium is the first semiconductor in which hydrogen was observed to alter the electrical activity of substitutional impurities.[1-3] In particular, hydrogen activates[4,5] the normally inactive substitutional C and Si. The lowest concentrations of H and C or Si for which acceptor formation is observed are of the order of $10^{14}\ cm^{-3}$. However, only a small fraction of this total concentration actually forms acceptors. Hydrogen activation of isovalent substitutional impurities has been observed in Ge only, although a complex such as {H,C} could in principle be formed and be electrically active in Si as well. A number of experimental studies[5-8] of the A(H,C) and A(H,Si) acceptors in Ge have established that the substitutional impurity binds H, that the complexes are insensitive to applied stress, and that A(H,C) is stable up to 200 °C, while A(H,Si) dissociates readily at room temperature.

The interpretation of the early optical spectra was consistent with the H interstitial located at a site of T_d symmetry, i.e., at the tetrahedral interstitial (T) site. This led to a tunneling model[7,9] which explained some features of the data. The subsequent empirical calculations[10,11] of the electronic structures of these complexes did not include the evaluation of potential energy surfaces (PES's). Later experiments[12] showed that A(H,C) and A(H,Si) have in fact C_{3v} symmetry. A recent density functional calculation[13] of the PES for various charge states of the {H,Si} pair in Ge has concluded that H is near the bond-centered (BC) site in the 0 and +1 charge states, and is at the antibonding (AB) site of Si, 0.2 Å from the T site, in the −1 charge state. In each case, the calculated binding energies of H are very small, ranging from 0.02 to 0.05 eV.

In addition to the activation of C and Si, hydrogen can also passivate electrically active centers, such as substitutional B. The passivation of B is well known in Si, where nearly 100% of the implanted borons can be passivated by hydrogen at room temperature. Much of the experimental and theoretical work on hydrogen interactions in semiconductors has focused on Si (see Refs.2,3,14). The extraordinary efficiency of shallow acceptor passivation by hydrogen in Si is most probably due to the long-ranged Coulomb attraction between H^+ and B^-. In Ge, although some passivation of B occurs, it is much less efficient. Whether this is due to differences in the microscopic structure or stability of the complex, or to a large difference of capture radii is not clear at present.

In this paper, we report the results of calculations of the PES's and electronic structures of the {H,B}, {H,C}, and {H,Si} pairs in Si and Ge. A few results relevant to isolated interstitial H in Ge are also given. Section II contains a brief discussion of the method and the clusters used. The results for the {H,B}, {H,C}, and {H,Si} pairs in Si and Ge are given in Section III.

2. DETAILS OF THE CALCULATIONS

The results presented below were obtained at the *ab-initio* Hartree-Fock level with split-valence basis sets on all the atoms, and polarization functions on selected atoms: A set of p orbitals for H and a set of d's for the other atoms on the relevant trigonal axis (B, C, Si, or Ge). In the case of Si, all-electron calculations were performed. For Ge, we used model pseudopotentials[15] with relativistic corrections for the core orbitals. In the case of the {H,B} pair, a closed-shell (RHF) wavefunction was used. In the other cases, the calculations were done with unrestricted (UHF) wavefunctions. In this paper, only the neutral charge state of the complexes is considered. Further studies including the negative (−1) charge state will be published shortly.[16]

The following configurations were optimized using gradient-based techniques: H near a BC site (between the substitutional impurity and a host atom), and H at AB sites (AB to the impurity or to a host atom). In the cases of substitutional C and Si, the configurations with H at the two T sites along the trigonal axis containing the impurity were also considered.

The host crystals were represented by two clusters: HXY_7H_{18} and $HXY_{13}H_{24}$, where H is interstitial hydrogen, X is substitutional B, C, or Si, and Y is the host atom (Si or Ge). The former cluster contains two host atom shells around the BC site and the latter contains both the BC and the T sites, each with two complete host atoms shells around them. The calculated equilibrium bond lengths between host atoms in the cluster with no impurity are 2.3854 Å in Si and 2.4827 Å in Ge, i.e., some 0.03 Å longer than the experimental values for crystalline Si (2.3512) and Ge (2.4508). The optimized host-saturator bond lengths are 1.4878 Å (Si–H) and 1.5347 Å (Ge–H).

3. THE RESULTS

1. The {H,B} pair

The lowest-energy configurations of the {H,B} pair in Si and Ge are shown in Fig.1. In both hosts, H is near the *BC* site, strongly bound to the host atom, and only weakly overlapping with the B atom. Within the constraints of C_{3v} symmetry, we found two other local minima of the PES, both much higher in energy. The first has H *AB* to B (+1.54 *eV* in Si, +1.48 *eV* in Ge), and the second has H *AB* to the host atom (+1.83 *eV* in Si and +1.75 *eV* in Ge). A detailed comparison between Si and Ge of the displacements of each atom in all three configurations shows very small differences. For example, in the *BC* configuration, Si moves by 0.162 Å away from the perfect substitutional site, while Ge moves by 0.159 Å. There is also little difference in the details of the electronic structures, as can be seen, e.g., in the bonding indices shown in Fig.1.

In order to further compare the PES's for H near substitutional B in Si and Ge, we calculated two other configurations: H at the *BC* site between two host atoms near B, and H at the *T* site, again near B. For these calculations, we used unrestricted wavefunctions in both the singlet and the triplet states. The results of these calculations show that the PES for H near B in Ge is only slightly flatter than in Si.

Thus, once formed, the {H,B} pairs in Si and Ge are very similar. The stable and metastable structures are nearly identical, and the lowest-energy states are separated from the metastable ones by a substantial amount of energy, which is nearly the same in Si and Ge. Although the PES appears to be a bit flatter in Ge than in Si, it is by an amount too small to explain why shallow acceptors are so easy to passivate in Si and so hard to passivate in Ge. The reason could be the difference in capture radii: In Si, passivation results mainly from the reaction $B^- + H^+ \longrightarrow \{H,B\}^0$, which is characterized by a long-ranged Coulomb capture radius of H^+ by B^-. On the other hand, recent muon spin rotation[17] (μSR) and thermal effusion data[18] suggest that above some 220 °K, the concentration of H^+ in Ge is very low, in contrast to that of H^+ in Si. Thus, hydrogen passivation in Ge could well result only from a reaction such as $B^- + H^0 + h^+ \longrightarrow \{H,B\}^0$, i.e., H has to stumble "accidentally" upon B for passivation to be possible.

2. The {H,C} pair

We found two nearly energetically degenerate configurations for the {H,C} pair in Si and Ge (Fig.2). The lowest-energy one corresponds to H near the *BC* site, bound primarily to a host atom. Most of the unpaired spin density is on C, which is almost sp^2 hybridized. We compared this configuration to

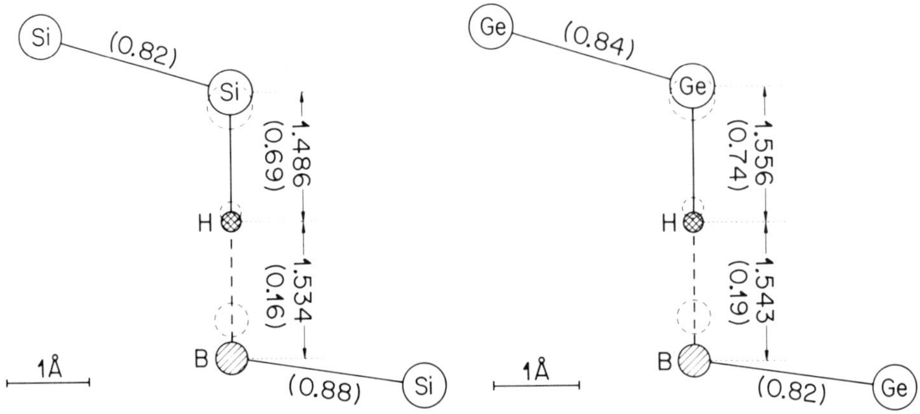

FIGURE 1: Lowest-energy configurations of the {H,B} pair in Si and Ge. The bond lengths are in Å, and the bond indices are given underneath, in parenthesis. The dashed circles show the perfect lattice sites.

{H,C} PAIR

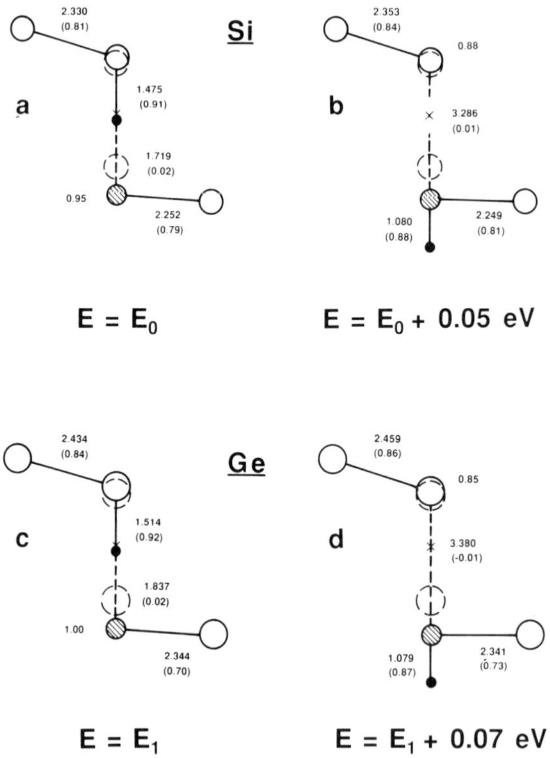

FIGURE 2: Nearly energetically equivalent configurations of the {H,C} in Si and Ge. The bond lengths are in Å, and the bond indices are given underneath, in parenthesis. Atomic spin populations are associated with some atoms. The dashed circles show the perfect lattice sites.

the one with C at the perfect substitutional site and H far outside the clusters. One *gains* 1.42 eV (in the case of Si) and 1.83 eV (in the case of Ge) by forming the {H,C} pair with H near the BC site. By comparison, at the same theoretical level, it *costs* 1.44 eV (for Si) and 1.20 eV (for Ge) to form the BC structures Si–H–Si and Ge–H–Ge, starting with H far outside a cluster, without substitutional C. These numbers show that substitutional C is a strong trap for interstitial H in both hosts.

The second configuration of the {H,C} complex, also shown in Fig.2, corresponds to H AB to C, which is 4-fold coordinated: Three sp^2 hybrids with three host atoms and H attached to the remaining $2p_z$ orbital. The host atom near C along the $< 111 >$ axis remains very near its ideal substitutional site and carries most of the odd electron in a non-bonding orbital.

The first configuration (H near the BC site) is stabilized by having the unpaired electron in a non-bonding orbital on the most electronegative atom (C), while the second configuration (H AB to C) is stabilized by the formation of a C–H bond, which is stronger than an Si–H or a Ge–H bond. That these two very different configurations are within a few hundredths of an eV from each other is probably coincidental, and we see no reason why a similar set of circumstances should occur for other complexes, such as {H,Si} in Ge (see below).

We also investigated the configuration where H is AB to a host atom, and found it to be higher in energy: 0.94 eV in the case of Si, and 1.63 eV in the case of Ge. Finally, the unbound configurations (H at the T site near C or near a host atom at perfect substitutional sites) are much higher in energy (about 3 eV).

We are pursuing geometry optimizations of all these configurations in the (-1) charge state to find out which of them is stabilized the most by the capture of an electron.

3. The {H,Si} pair

The situation is quantitatively very different for three reasons: *(i)* the Si–H bond is weaker than the C–H bond, *(ii)* the odd electron is not as stable in a non-bonding orbital on Si or Ge as it is on C, and *(iii)* the Si–Ge natural bond length is much closer to the Ge–Ge one than is the C–Ge one. The calculations show that four configurations are very close in energy. The most stable one (with energy $= 0$) has H AB to Si (similar to the metastable configuration shown in Fig.2, with C replacing Si). Then come two configurations with the Si and all the host atoms very near their perfect substitutional site and H in a non-bonded configuration at either one of the two T sites along a trigonal Ge–Si axis: $+0.08$ eV at T_{Si} and $+0.15$ eV at T_{Ge}. These two configurations do not result in activation. Finally, the BC configuration (Si–H $\cdots$ Ge) is at $+0.27$ eV. The energy is $+0.64$ eV when H is AB to Ge.

These results show that there is little incentive (0.08 eV) for H to form a bond with Si. When it does, the resulting configuration is rather unstable, in agreement with experimental observations. Note that it *costs* 0.53 eV to bring H from far outside the Ge cluster (containing substitutional Si) and put it in the lowest energy configuration (H–Si $\cdots$ Ge). By comparison, one *gains* 1.83 eV by forming the analogous {H,C} pair (see above). Note that one needs even more energy ($1.44\,eV$) to form the usual BC interstitial in perfect Si, which is the analogous of the {H,Si} pair in Si. More details and results for the -1 charge state will be published shortly.

Acknowledgements: The authors are thankful to E.E. Haller for enlightening discussions.The work of PFM is funded by the Swiss National Science Foundation. The work of SKE has been supported in part by the grant D-1126 from the Robert A. Welch Foundation and by a grant from the Advanced Technology Program of the State of Texas.

REFERENCES

[1]E.E. Haller, MRS Symp. Proc. **46**, 495 (1985).

[2]*Hydrogen in Semiconductors*, edited by J.I. Pankove and N.M. Johnson (Academic Press, New York, 1991).

[3]Report of the Panel *Fundamental Issues in Hydrogen-Defect Interactions*, sponsored by the US Department of Energy, to be published in Rev. Mod. Phys.

[4]R.N. Hall, IEEE Trans. Nucl. Sci. **NS-21**, 260 (1974).

[5]R.N. Hall, Inst. Phys. Conf. Ser. **23**, 190 (1975).

[6]E.E. Haller, W.L. Hansen, and F.S. Goulding, Adv. Phys. **30**, 93 (1981).

[7]E.E. Haller, B. Joos, and L.M. Falicov, Phys. Rev. B **21**, 4729 (1980).

[8]E.E. Haller, Phys. Rev. Lett. **40**, 584 (1978).

[9]L.M. Falicov and E.E. Haller, Sol. St. Comm. **53**, 1121 (1985).

[10]J. Oliva and L.M. Falicov, Phys. Rev. B **28**, 7366 (1983).

[11]J. Oliva, Phys. Rev. B **29**, 6846 (1984).

[12]J.M. Kahn, R.E. McMurray, Jr., E.E. Haller, and L.M. Falicov, Phys. Rev. B **36**, 8001 (1987).

[13]P.J.H. Denteneer, C.G. Van de Walle, and S.T. Pantelides, Phys. Rev. Lett. **62**, 1884 (1989).

[14]S.J. Pearton, J.W. Corbett, and T.S. Shi, Appl. Phys. A **43**, 153 (1987).

[15]W.R. Wadt and P. J. Hay, J. Chem. Phys. **82**, 284 (1985).

[16]Dj.M. Maric, P.F. Meier, and S.K. Estreicher, to be published.

[17]R.L. Lichti, D. Lamp, and T.L. Estle, private communication.

[18]M. Stutzmann, J.B. Chevrier, C.P. Herrero, and A. Breitschwert, Appl. Phys. A (in press).

Materials Science Forum Vol. 83-87 (1992) pp. 125-136

SPECTROSCOPY ON TRANSITION-METAL DEFECTS IN SILICON

M. KLEVERMAN, A. THILDERKVIST, G. GROSSMANN, AND H. G. GRIMMEISS
Department of Solid State Physics, University of Lund
Box 118, S-221 00 LUND, Sweden

ABSTRACT

In Si, the transition metal impurity excitation spectra are dominated by one-particle transitions from a localized many-particle ground state to shallow donor and acceptor like states. In the final states, exchange effects between the shallow particle and the core states may be observed as e.g. for the neutral interstitial Mn impurity. In this paper, some recent results obtained by high-resolution absorption spectroscopy using perturbation techniques i.e. uniaxial stress and Zeeman spectroscopy will be discussed in detail.
The Mn spectrum is due to transitions from the $^4T_1$ ground state to $nS_{3/2}(\Gamma_8)$ shallow-hole states. Exchange interaction between the $^3A_2$ final-state core and the j=3/2 holes results in levels having a total angular momentum of 5/2, 3/2, and 1/2. The splitting of the levels decreases with n due to increasing delocalization of the hole.
The neutral isolated interstitial Fe impurity shows a donor like spectrum dominated by transitions to p-like donor states. The exchange interaction is drastically reduced compared with Mn due to the small overlap between the excited electron and the final core state. Here, three superimposed donor spectra are observed since three final states, derived from the $^4T_1$ ground-state term of Fe^+, are reached in the transitions. The Fe spectrum shows yet another interesting feature in that transitions to states derived from a $^5T_1$ term are observed. Spin unrestricted calculations show an $a_1\uparrow$ resonance close to the conduction band minimum. Transitions to this $a_1\uparrow$ state will then result in a $^5T_1$ term close to the conduction band.

1. Introduction

Transition-metal impurities (TM´s) in silicon form an interesting class of defects due to their partly filled d shells. They generally have energy levels in the band gap for several different charge states and they easily form complexes with other defects e.g. the shallow group-III acceptors. The present understanding of the TM's in silicon is mainly based on EPR and, more recently, also on theoretical calculations. The EPR work was pioneered by Ludwig and Woodbury (LW) [1] about 30 years ago and their model has been successfully applied to many isolated TM centers as well as to TM complexes. Recently, sophisticated calculations [2,3] have shown that the success of the LW model was more or less accidental, its main features, however, are still applicable.
A very important advantage of EPR compared with other methods is that it can chemically identify the constituents and determine the symmetry of the impurity under study. However, it also has some inherent drawbacks, in that it is only applicable to those charge states that are paramagnetic and only probes the ground state(s).
Internal d transitions are rarely observed in silicon but dominate the spectra for most TM´s

in III-V and II-VI compounds. In silicon the spectra, sofar reported, are almost exclusively due to one-particle transitions from the ground state to shallow donor or acceptor like states.

The TM excitation spectra in silicon are usually very weak, although ordinary solid state diffusion experiments, which are usually followed by a rapid temperature quench, result in rather high concentrations. The small intensity must therefore be attributed to correspondingly small optical matrix elements, which may be explained by considering the large difference in spatial extension of the initial and final states. However, the introduction of highly sensitive Fourier spectrometers during the last decade has made it feasible to perform high resolution spectroscopy even for very weak absorption signals. Optical methods therefore provide important information in addition to EPR investigations of the electronic structure of e.g. TM centers and also constitute an important technique for non-paramagnetic centers.

The shallow states are determined by the host material and are usually well described by the effective mass theory [4], EMT, and have thus a general energy-level structure independent of the particular impurity. One would therefore expect to observe shallow like spectra even for very deep impurities although shifted towards higher photon energies due to their considerably larger ground state binding energy. Such spectra have indeed been observed e.g. for the neutral chalcogen double donors in silicon [5,6]. They all show excitation spectra similar to those reported for e.g. the shallow group-V impurity phosphorus, with respect to both the energy splitting between the p-like states and their relative intensities. However, intense transitions are also observed to states in the 1s valley-orbit split manifold which are strictly forbidden in the EMT approximation. This is explained by the inclusion of a considerable deep-level character in the ground state. Furthermore, only the s states have a non-vanishing amplitude at the impurity site and thus experience the localized deep potential. For these states, one would expect considerable central-cell shifts and, for centers having very compact ground states, transitions to s states are likely to dominate the spectrum. One example of such an excitation spectrum is a Ag related donor spectrum which essentially reflects transitions to ns states [7]. Furthermore, interactions between these ns states and the remaining core have in several cases been shown to follow a scaling rule. Accordingly, the effect on an excited state scales as the probability to find the excited particle at the origin, which for a hydrogen like syztem $\sim 1/n^3$, where n is the principal quantum number. Such scaling has been inferred e.g. from the Ag spectrum, in which all observed ns states split accordingly.

Silicon has six equivalent conduction band minima. All EMT ns donor states are therefore six-fold accidentally degenerate. The valley-orbit interaction and the central-cell correction lowers the degeneracy and, for a T_d center, the 1s manifold splits into $1s(A_1)$, 1s(E), and $1s(T_2)$, 1-, 2-, and 3-fold degenerate, respectively. A lower symmetry core and/or spin-orbit interaction may reduce the degeneracy further as well mix different states of the manifold. The group-V donors and the chalcogen double donors have the A_1 state as ground state. The localized potential for the chalcogen donors gives rise to an anti-bonding A_1 state in the gap [8] which may be viewed as the $1s(A_1)$ state, pulled down by the strongly attractive central cell. For other centers, the ground state may have a totally different origin. In Pt doped silicon, a donor like excitation spectrum was investigated in some detail [9]. It was concluded that transitions to <u>all</u> six 1s states were observed in a narrow energy range, presumably close to the 1s EMT position. The ground state was shown to originate

from a t_2 state which was believed to be an antibonding state resulting from interaction between the Pt dt_2 state and a t_2 vacancy-like orbital [10].
As for the donors, shallow acceptor spectra in silicon are also well understood [4] although the small spin-orbit splitting of the silicon valence band complicates the level structure compared with that for GaAs or Ge.
In this paper we will report on recent results for some transition metal impurities in silicon. The discussion will focus on the neutral single interstitial 3d impurities iron and manganese. The experimental methods used are high resolution FTIR studies including perturbation techniques, i.e. Zeeman and uniaxial stress spectroscopy. A variety of interesting physical effects are observed for these centers that considerably increase our understanding of deep defects in silicon in general and of TM´s in particular.

2. Experimental details.

The doping was done by solid state diffusion. After the usual polishing, cleaning, and etching procedures, a diffusion source was obtained by either evaporation of the TM or by merely scraping the metal onto optically inactive surfaces. During the heat treatment, the samples were held in an evacuated quartz ampoule. The diffusion temperature was usually between 1200 - 1250°C and the samples were subjected to a rapid temperature quench by dropping the ampoules into pump oil or ethylene glycole. A special design of the ampoules ensured that they broke and, thus, the samples obtained direct contact with the quench medium immediately. After repolishing or etching to high optical quality, the samples for the uniaxial stress and Zeeman measurements were oriented in a Laue backscattering X-ray setup and cut in the <100>, <110>, and <111> directions. Typical sample dimensions for the stress measurements were 2x2x10 mm^3.
The transmission spectra were obtained by a Bomem DA3-02 FTIR spectrometer equipped with a liquid-nitrogen cooled InSb photodiode detector. The appodized resolution varied between 0.25 - 2 cm^{-1} depending on the specific conditions. A continuous flow He cryostat was used for the stress measurements. In the Zeeman measurements, an Oxford Instruments 7T spectromagnet was used in the Voigt configuration. Suitable polarizers and optical filters were used. In some cases, corrected spectra were obtained by dividing the original spectrum with an artificial background. See the references for further experimental details.

3. Eperimental results and discussion

The Mn_i center

Mn-doped silicon shows, at T=2K, a series of sharp absorption lines, labeled A_1, B_1, C_1, and D_1, between approximately 8150 - 8450 cm^{-1} (Fig. 1). At higher temperatures, excited initial states become thermally populated and all lines are replaced by a group of nearly equally spaced lines. A simple analysis based on the annealing temperature suggested that the fine structure was due to population of excited phonon states in the ground state [11]. In order to establish the role of Mn, an isochronal annealing experiment was performed in

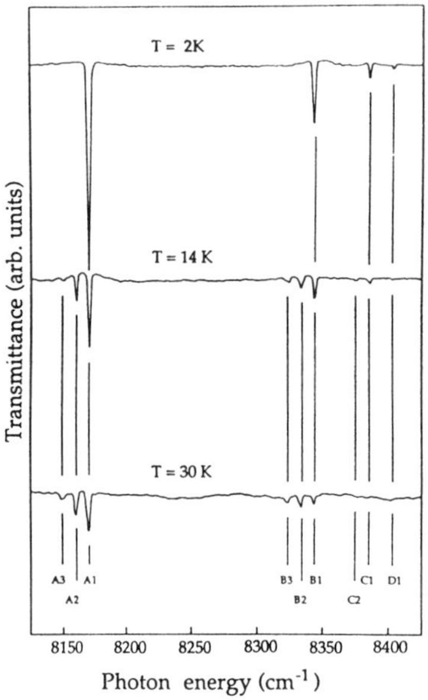

Fig. 1 The excitation spectrum of Mn_i^0 for three different temperatures.

which the integrated intensity of the lowest line was compared with that for the Mn_i^0 EPR signal of the same sample. The result is shown in Fig. 2 and the good aggreement gave strong evidence for the line spectrum being related to Mn_i. This type of experiment is extremely important since it may establish the chemical nature of the spectrum which, in turn, enables a detailed comparison between theory and experiment. One may thus be able to relate e.g. observed splittings or second-order effects to physical quantities such as spin-orbit coupling (s-o) and exchange interaction.

In Fig. 1, the energy spacing between consecutive lines decreases rapidly with increasing photon energy, thus, indicating excitations to shallow states bound by a Coulombic potential. The ionization energy could therefore be estimated by the high energy limit of the line series and a value of about 8450 cm^{-1} (1048 meV) was obtained. The energy position of different charge states of interstitial Mn in Si is known from deep levels transient spectroscopy (DLTS) [12] and a combination of EPR and Hall effect measurements [13]. The Mn^-/Mn^0 level is located about 130 meV below the conduction band. This energy and the estimated ionization energy above add up to a value close to the band gap, indicating that the line spectrum is indeed due to transitions to shallow excited hole states at Mn^0. The excited state of Mn^0 naturally divides into two subsystems: the delocalized hole and the localized final core state. The small probability of finding the hole in the core region ensures that the interaction between these subsystems is small. In a first approximation, therefore, the transitions may be described by $Mn_i^0 + h\nu \rightarrow Mn_i^- + h$. To lowest order, the energy structure observed should thus reflect that of the shallow hole. However, the Mn_i^0 spectrum shows no similarity to the typical shallow acceptor spectrum which is dominated by transitions to excited P-like states. Considering that the Mn_i^0 ground state is very different from that of shallow group-III acceptors it was assumed that the excitations take place to shallow nS-like hole states. In order to test this assumption, the scaling rule mentioned above was used for the observed relative intensities of the zero-phonon lines. Assuming that the electric-dipole matrix elements are determined by the overlap between the nS-states and the localized initial state, the scaling rule should be applicable. In fact, the relative intensity ratio for the lines A_1, B_1, and C_1 scales as 2^{-3}, 3^{-3}, and 4^{-3}, which leads to the level scheme depicted in Fig. 3.

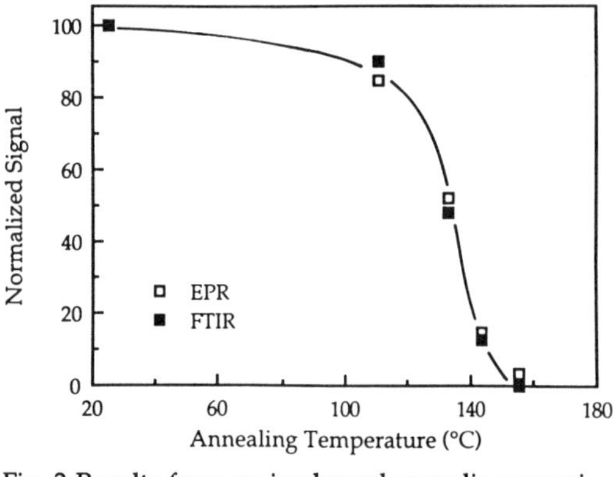

Fig. 2 Results from an isochronal annealing experiment in which the integrated intensity of the A_1 line is compared with that for the Mn_i^0 EPR line.

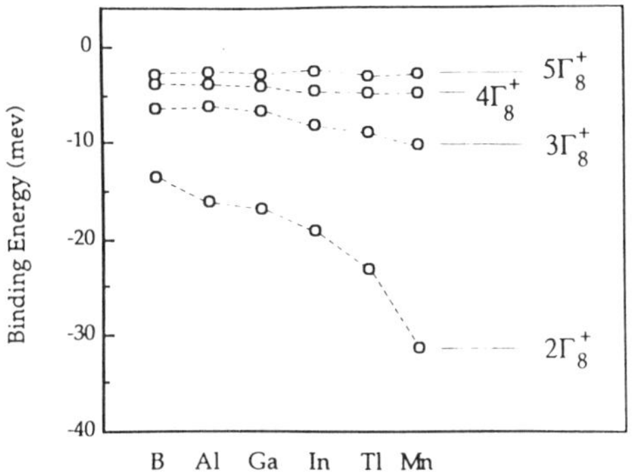

Fig. 3 The proposed excited level scheme for Mn_i^o compared with those for S-like states of B, Al, Ga, In, and Tl in silicon (from Ref. 11).

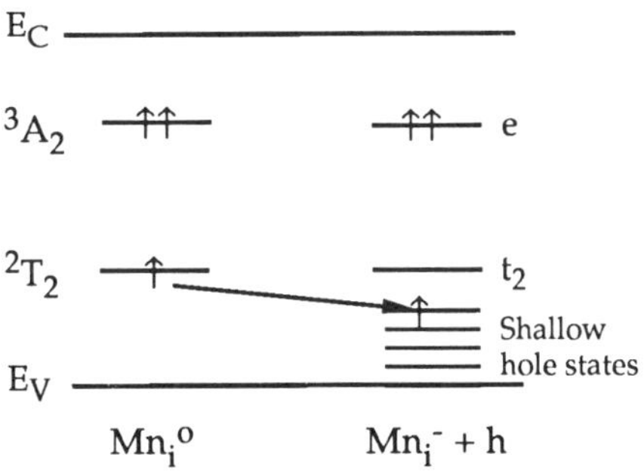

Fig. 4 A schematic picture showing the proposed hole transitions from the t_2 localized hole state to shallow $nS_{3/2}(\Gamma_8)$ states.

Atomic Mn has a $3d^5 4s^2$ configuration and it is assumed that in the silicon crystal the two s electrons are promoted to the d shell leading to a $3d^7$ configuration for Mn_i^o. Calculations show that the d states split into an e and a t_2 level in the gap, with the e level lying at an energy Δ_{CF} above the t_2 level. The configuration with lowest energy is the $e^2 t_2^5$ high-spin configuration. Since the line spectrum is due to hole transitions, it is easier to count the holes instead and we consider an $e^2 t_2^1$ high-spin hole configuration. The two e holes form a $^3A_2$ state and the t_2 hole a $^2T_2$ state. Combining the two subsystems gives a $^4T_1$ state as the ground-state term which is split by the spin-orbit interaction into states having a total effective angular momentum J=1/2, 3/2, and 5/2. EPR measurements have shown the J=1/2 level to be the ground state.

The Mn_i^- ground state configuration is e^2 ($^3A_2$) and the minimum excitation energy results from a transtition of the Mn_i^o t_2 hole to a shallow hole state. In Fig. 4, the hole transitions are depicted schematically.

In Fig. 5 the uniaxial stress result of the Mn_i^o center are presented for F||<100> [14]. As is readily seen, the A_1 and A_2 lines as well as the B_1 and B_2 lines show similar splitting behavior which is also seen for the <110> and <111> directions of the stress. The stress data therefore give additional support for the assumption that the fine structure observed at higher temperatures is caused by population of excited phonon states. The stress splitting is nearly isotropic and, disregarding the small additional splitting of the lowest component of the A lines, similar for the A and B lines. Furthermore, the splitting between the main components is close to that

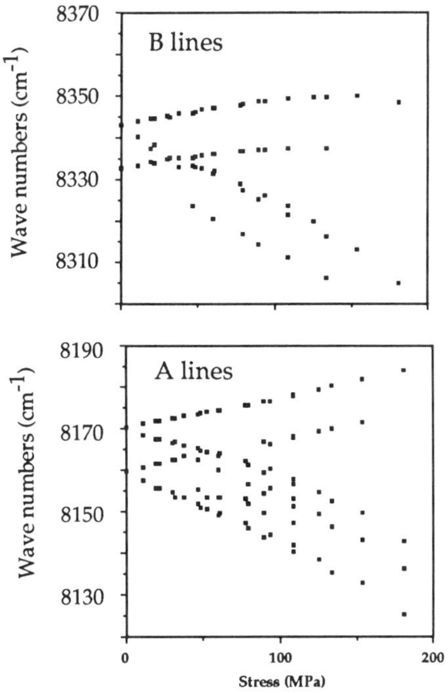

Fig. 5 The uniaxial stress results for the A and B lines for F||<100>.

found for the $1S_{3/2}(\Gamma_8)$ boron ground state in silicon.

Although the overall splitting behavior is almost identical for the A and B lines, the small splitting of the lowereA line component indicates an interesting difference. Assuming T_d symmetry (as observed in EPR for both Mn_i^o and Mn_i^-) we attribute an effective total angular momentum J to the final many-particle state of the transition. The stress split components are at least two-fold degenerate (Kramers degeneracy) and characterized by $|\pm M_J|$. The A line splits into three components which is only possible in if the final state has J=2 (±2, ±1, and 0) or J=5/2 (±5/2, ±3/2, and ±1/2). Since the final core state has S=1 and the hole has half integer spin we attribute a J=5/2 level to the A line final state. This J=5/2 state would result from an exchange interaction between a Γ_8 shallow hole (j=3/2) and the S=1 final core state ($^3A_2$) resulting in levels having J=1/2, 3/2, and 5/2.

In order to describe the uniaxial stress and Zeeman results we employ a simple model [15]. We construct the Mn_i^o ground state by taking into account the exchange interaction between the S=1 ($^3A_2$) state and the (s=1/2) $^2T_2$ hole state as well as the spin-orbit coupling within the $^2T_2$ hole states. The exchange operator is defined by H_{ex}=-2JS·s and the spin-orbit operator by H_{so}=ξL·s, where we attribute an effective angular momemtum L=1 to the T_2 state. The exchange term generates the $^4T_1$ ground state and a $^2T_1$ state. In the calculations, the exchange parameter J was chosen large enough to make the spin-orbit interaction between the $^4T_1$ and a $^2T_1$ terms negligible.

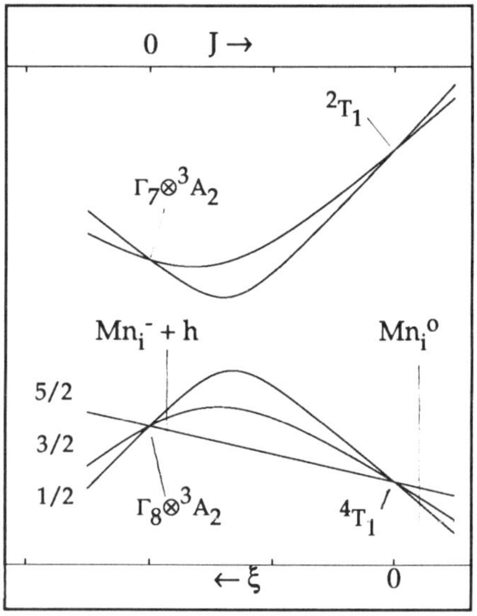

Fig. 6 Our model depicted schematically (see text for details).

We apply the same model to describe the final state (Mn_i^- + h) although with different exchange and spin-orbit parameters. The Γ_8 hole results from the $^2T_2$ state which is spin-orbit split into a Γ_8 and a Γ_7 state. Now, the spin-orbit interaction dominates over the exchange interaction between the hole and the S=1 core.

It should be noted, that it is feasible to construct the initial and final states by only choosing two parameters carefully, the s-o parameter and the exchange parameter.

The level scheme obtained from the model is depicted in Fig. 6 in which the positions of the initial and final states are indicated. The exchange parameter J is positive in both cases which ensures that the spin of the t_2 hole is parallel to that of the $^3A_2$ state leading to a high spin ground state (Hund´s rule). A value of about 500 cm^{-1} has been chosen for the initial state whereas for the A line it is much smaller, only 50 cm^{-1}. The spin-orbit parameter ξ has different signs in the two cases reflecting that the t_2 state has a different origin in the two states. In the ground state the t_2 state is predominantly d like and ξ<0, whereas in the final state the t_2 state is p like and ξ>0 (the sign of ξ refers to an electron picture).

It should be noted that the stress behavior of the ground and the excited state is determined by the t_2 state alone. In T_d symmetry, the stress hamiltonian includes operators of A_1, E, and T_2 symmetry. We choose the stress directions to lie in the plane spanned by the [111] and [110] directions and the stress matrix for the t_2 hole is then given by

$$H_{Stress} = aA_1\sigma_A + eE_\theta\sigma_\theta + t(T_{2\xi}\sigma_\xi + T_{2\eta}\sigma_\eta + T_{2\zeta}\sigma_\zeta)$$

Three parameters, a, e, and t are needed to describe the stress respons, since the direct product $T_2 \otimes T_2 = A_1 + E + T_1 + T_2$ contains all three representations. The matrices are constructed from the L=1 matrices by using the replacement theorem: $\mathbf{A_1} = 1/3(L_x^2 + L_y^2 + L_z^2)$, $\mathbf{E_\theta} = 2L_z^2 - L_x^2 - L_y^2)$, $\mathbf{T_{2\xi}} = (L_yL_z + L_zL_y)$, and cyclic permutation of the indices. The linear combinations of the stress tensor components are defined by $\sigma_A = \sigma_{xx} + \sigma_{yy} + \sigma_{zz}$, $\sigma_\theta = 2\sigma_{zz} - \sigma_{xx} - \sigma_{yy}$, $\sigma_\xi = \sigma_{yz}$, and cyclic permutations of the indices.

In Fig. 7, the results obatined from our stress model are presented and compared with the experimental data. The initial state (J=1/2) only shows a hydrostatic shift when disregarding any second-order effects due stress induced mixing with the 3/2 and 5/2 states. The upper A_1 line component only shows linear response which indicates that the second-order stress effect in the ground state is small. Therefore, only three parameters suffice to describe the stress splitting, with a parameter accounting for the difference in hydrostatic shifts between the final and initial states. The almost perfect fit for the A_1 line is obtained by using a=-0.08 cm^{-1}/MPa, e=0.067 cm^{-1}/MPa, and t=0.42 cm^{-1}/MPa and the spin-orbit parameter ξ is 300 and -17 for the final and initial state, respectively.

In the excited state, a stress induced mixing with the J=1/2 and J=3/2 states leads to the non-linear response observed for the two lower A_1 components. The model we use is a spherical model and the projection of J onto the stress axis remains a good quantum number. The upper stress split component ($\pm M_J = 5/2$) only shows a linear shift, since the two higher lying state do not contain $M_J = \pm 5/2$ and, hence it remains a pure eigenstate irrespective of the applied stress.

In order to further test our model the polarization properties of the stress split lines were investigated in detail. In Fig. 8, the observed and calculated intensities for polarized light perpendicular and parallel to the stress axis are presented. The almost perfect agreement gives further support to our model. It should be noted that no additional parameters were used in these calculations.

The J=1/2 and J=3/2 states are experimentally not observed and our calculations show in fact that the J=5/2 lines contain almost all oscillator strength.

Also a Zeeman study was carried out on the Mn_i^0 excitation spectrum. The Zeeman operator is given by $H_{Zeeman} = \mu_B B(g_S S + g_S s + g_L L)$, where μ_B is the Bohr magneton, g_S the free

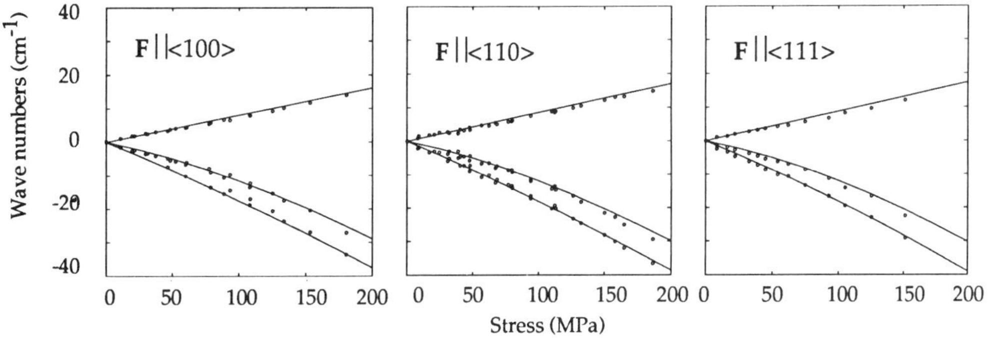

Fig. 7 The experimental stress results for **F**||<100>, <110>, and <111>. The full lines are obtained from our model (see text for details). The wave number origin is at the A_1 line zero stress position.

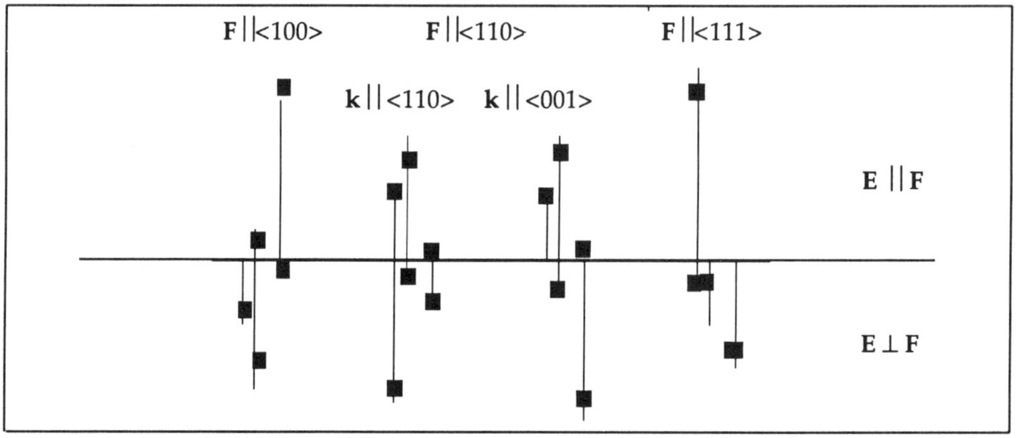

Fig. 8 The observed polarization properties, filled squares, of the stress split A_1 line components compared with our calculations (full lines).

spin g value, and g_L is the orbital g value for the t_2 state. The operators **S** and **s** are the spins for the $^3A_2$ state and the s=1/2 t_2 hole. **L** is the orbital angular momentum operator for an atomic p state representing the t_2 hole.

The g-value for the J=1/2 ground state of Mn_i^o is known from EPR studies to be 3.362 [1] and in order to reproduce this value a value of -0.039 was inferred for g_L. In order to fit the experimental data ,g_L for the excited t_2 hole state was found to be about 1.

The Zeeman results are presented in Fig. 9 together with the theoretically obtained splittings. Also in this case, the calculated polarization properties agreed almost perfectly with those experimentally observed.

It is interesting to note that all Zeeman components show an upward curvature. In our calculation the s-o parameter in the ground state was adjusted to reproduce this non-linear response through mixing effects with the J=3/2 and 5/2 components. It was found that a value for ξ in range from -10 to -20 cm^{-1} gave acceptable results. The energy difference between the J=3/2 and 1/2 state is given by $-3/2\xi$ which results in a value of about 15-30 cm^{-1} for the separation energy. Ludwig and Woodbury [1] actually observed an EPR signal from the Mn_i^o J=3/2 state and a value of about 10 to 20 cm^{-1} has been estimated for its separation from the ground state [16] which is in fair agreement with our estimate from the Zeeman data.

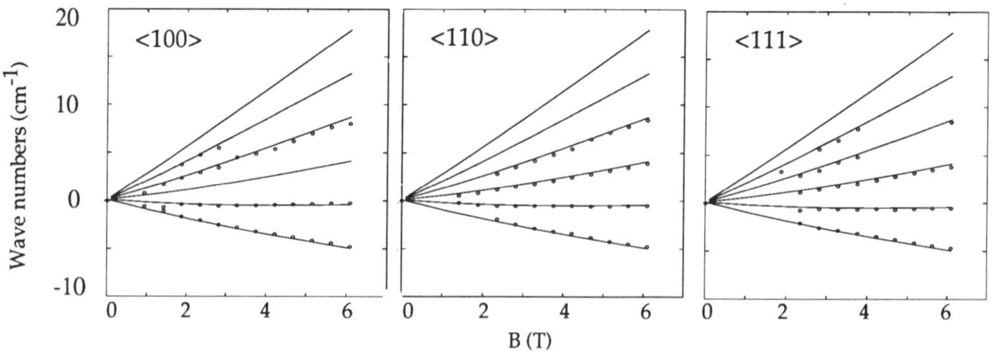

Fig. 9 The Zeeman splitting for the A_1 line. The full lines are obtained from the model described in the text. The energy origin is defined as the energy of the A_1 line at zero field.

The Fe_i center

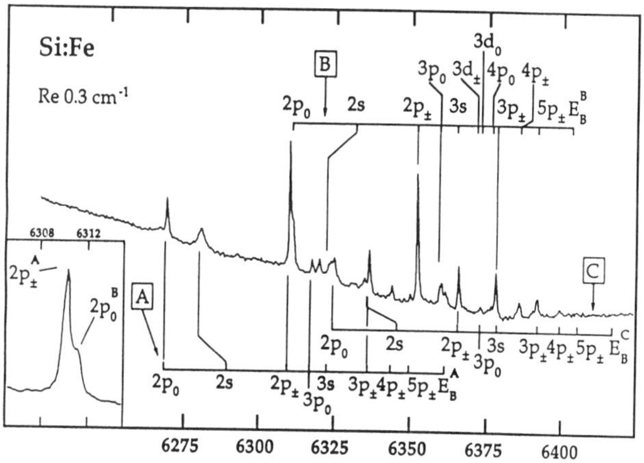

Fig. 10 Donor like excitation spectrum of Fe_i^0. From Ref. 18

The interstitial iron impurity in silicon, Fe_i, has T_d symmetry as deduced by EPR [1]. Iron doped silicon shows a sharp line excitation spectrum [17,18] in the energy range from 6270 and 6390 cm^{-1} as shown in Fig. 10. The same spectrum was also observed in photothermal ionization spectroscopy (PTIS) which also shows excitations to continuum. The photo-ionization cross section could be calculated from the continuum signal and in Fig. 11a this cross section is compared with the cross sections obtained from photo-EPR and photocapacitance. Furthermore, as in the case of Mn, an isochronal annealing experiment was carried out with the result given in Fig. 11b. These two experiments clearly establish that the spectrum shown in Fig. 10 is related to Fe_i^0. The binding energy could be estimated from the onset of the ionization cross-section to be about 6300 cm^{-1} (781 meV) in close agreement with the generally accepted binding energy of the interstitial iron donor. This therefore suggested that the spectrum is due to excitations from the Fe_i^0 ground state to excited shallow donor states. By using similar arguments as for the Mn_i^0 spectrum, the transitions may schematically be described by $Fe_i^0 + h\nu \rightarrow Fe_i^+ + e$.

It was not possible to identify the final excited states (taking all EMT states into account) by using only one series of EMT states. Rather, the spectrum was interpreted as a superposition of <u>three</u> shallow donor series as later confirmed by high resolution measurements. The three series are labeled A, B, and C in Fig. 10.

The relative intensity of the different series did not show any thermalization effects and the origin of the three series had to be attributed the level structure of the final core state.

Fe_i^0 and Fe_i^+ have a $^3A_2$ and a $^4T_1$ ground state term, respectively. The $^4T_1$ ground state term is split into states having J=1/2, 3/2, and 5/2 by the spin-orbit interaction and the three series were ascribed to the possibility that all three final core states could be reached in the transitions. The $2p_0^B$ lines coincides almost exactly with the $2p_\pm^A$ line but the two lines are clearly resolved as shown in the inset, giving additional support for our interpre-

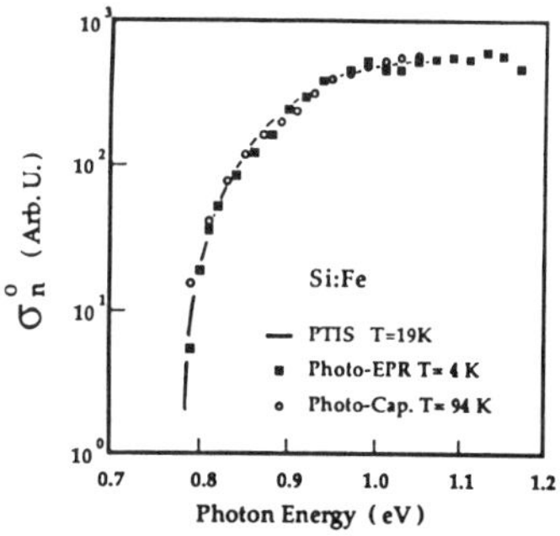

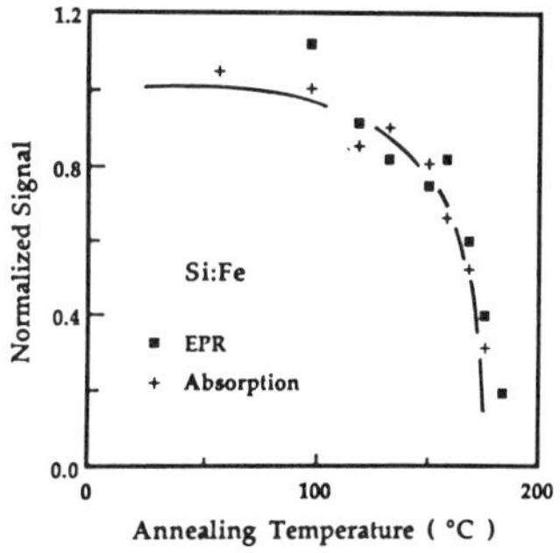

Fig. 11 a) Photoionization spectra obtained from PTIS, photo EPR, and photocapacitance (Ref.17).
b) Results from an isochronal annealing experiment obtained from the intensity of the Fe_i^0 EPR signal and the absorption lines.

tation.
The assignment of the different lines is presented in Fig. 10 and the A, B, and C series correspond to final core levels J=1/2, 3/2, and 5/2 of the $^4T_1$ ground state of Fe_i^+, respectively.
In silicon, the anisotropy of the conduction band minima splits all hydrogenic np states (m_l=0 and ±1) into np_0 and $np_\pm$ states. Only parity and the projection of the angular momentum along the principal axis of the minima remain good quantum numbers.
In Fig. 12, a schematic splitting pattern is depicted according to our model for **B**|| <100>. The Fe_i^o $^3A_2$ (S=1) ground state splits isotropically into three components. Only the lowest initial state Zeeman component is assumed to be populated due to the low experimental temperature (1.9K). The splitting of some excited donor states, $2p_0$, $2p_\pm$, and $3p_0$, is also shown, their spin splitting, however, has been ommited for clarity. To first order, the $2p_0$ and $3p_0$ lines are unaffected by the magnetic field, whereas the $2p_\pm$ line splits into three components. For valleys perpendicular to the field, the $2p_\pm$ state do not split, whereas for valleys parallel to the field it splits into two components. Taking second order effects into account, an avoided crossing behavior between the $2p_\pm$ and $3p_0$ lines is expected which also has been observed for shallow donors [20,21]. In Fig. 13, the result for the $2p_\pm^A$ and $3p_0^A$ is presented for **B**||<110> [19]. As is readily seen, a satisfying agreement is obtained beween the experimental points and the results obtained from the model in Fig. 12. No free parameters were used, the g value for the initial state and the final states were obtained from previuos EPR results [1] and those for the donor states were taken from Ref. 21.
In silicon, shallow donor states show a characteristic stress splitt-

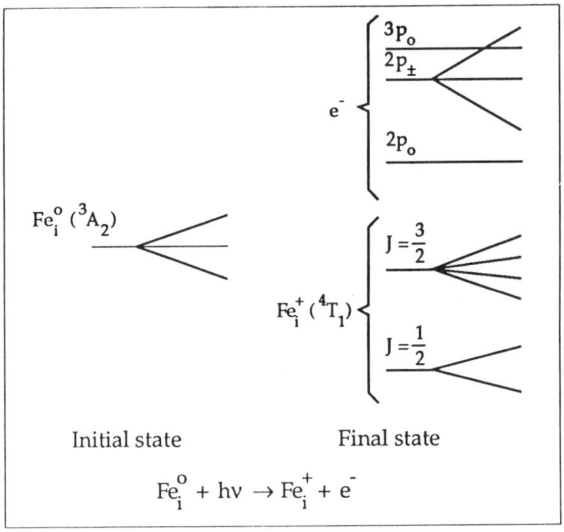

Fig. 12 A schematic Zeeman splitting pattern for the Fe_i^o center (See text for details)

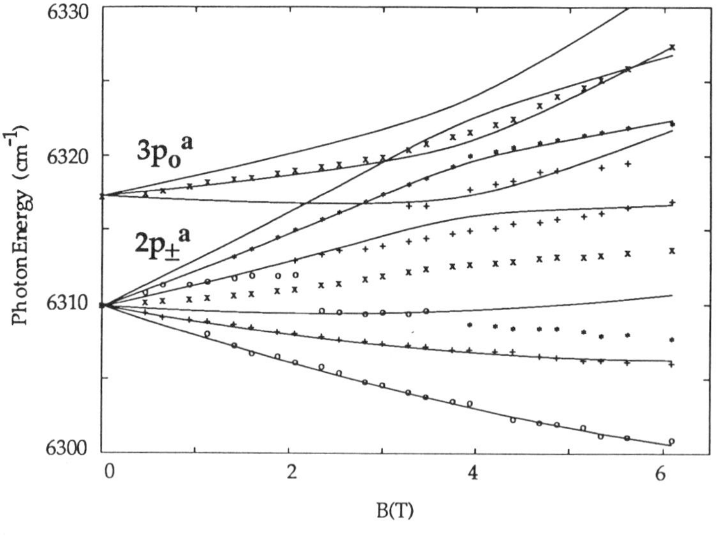

Fig. 13 Zeeman splitting of the $2p_\pm^A$ $3p_0^A$ lines for magnetic field **B**||<110>.

ing pattern reflecting the stress induced shifts of conduction band valleys which depend on the orientaion of their principal axis relative to the stress axis. The excited p-states split as the valleys and are well described by the deformation potential approximation, DPA, whereas the states in the ns manifolds may show a more complicated splitting behavior, in particular for low stress fields. In Fig. 14, the uniaxial stress result for the Fe_i^o donor spectrum is presented. The splitting of the p lines closely follows DPA, which gives a final proof for the assignment of the final states shown in Fig. 10. The $2s^A$ and $3s^A$ lines do not split which is characteristic for the A_1 combination. The s-o splitting of the $^4T_1$ term is in a first approximation determined by the Lande´ interval rule. Surprisingly, the experimentally observed splitting departs considerably from this rule. Also, the calculated relative intensity ratios are in disagreement with this model. Similar effects have in fact been observed for several other centers previously [22] which were attributed to a dynamical Jahn-Teller effect.

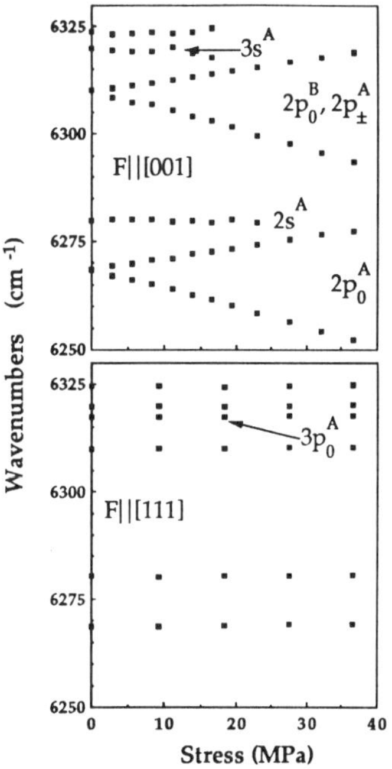

Fig. 14 Uni axial stress reults for the interstitial Fe donor

All our Fe-doped samples show additional lines at 5824.6, 5860.9, and 5862.8 cm^{-1}. They also belong to Fe_i^o since they are always observed with the same relative intensities compared to the donor lines and show the same annealing behavior.

The Zeeman splitting of the 5824.6 cm^{-1} line is presented in Fig. 15. The line shows an isotropic splitting into two prominent components. In addition, a weak component is also observed due to population of the M_S=0 initial state of $^3A_2$. A g value of 3.04 is found for the final state, assuming a J=1 state. A g value of 1.87 is then attributed to the initial state, in fair agreement with that for the Fe_i^o ground state of 2.0699 [1]. The splitting shows some nonlinearity which may very well explain the deviation of our g value, obtained by a linear regression fit, from that obtained by EPR. The relative intensities and polarization properties are also in agreement with a J=1 final state.

The Zeeman splitting of the other two lines is well accounted for by a J=2 level which in T_d symmetry splits into an E state and a T_2 state. The J=1 and J=2 levels could in a natural way be constructed from a s-o split $^5T_1$ term (J=1,2, and 3) of Fe_i^o. In this case, a spin only g value of 3.0035 is predicted for the J=1 state, in close agreement with that experimentally observed.

Spin unrestricted calculations [3] show a high spin $a_1\uparrow$ resonance close to the bottom of the conduction band. When occupied, the $a_1\uparrow$ resonance, exchange coupled to the $^4T_1$ core state of Fe_i^+, results in a $^5T_1$ term, that very well may be found in the band gap [23] close to the conduction band minimum. On the other hand, the $^3T_1$ term, derived from the $a_1\downarrow$ resonance, is expected to appear as a resonance in the band. However, excitations from the $^3A_2$ ground state to the $^5T_1$ levels are spin forbidden but could become allowed via an interaction with $^3T_1$ levels. Transitions to the J=3 level of the $^5T_1$ term were not detected, which is consistent with the fact that the $^3T_1$ term does not contain any J=3 level. The s-o splitting between the J=1 and J=2 levels of the $^5T_1$ term is to first order given by the J=1/2 and J=3/2 splitting of the Fe_i^+ ground state term $^4T_1$ as is experimentally observed.

Conclusions

High resolution spectroscopy studies on TM impurities in Si can provide detailed information on their electronic structure. In conjunction with perturbation spectroscopy, the interplay between spin-orbit and echange effects may be revealed.

Acknowledgements: We gratefully acknowledge financial support from the Swedish Natural Science Research Council.

References

[1] G. W. Ludwig and H. H. Woodbury, Solid State Phys. **13**, 223 (1962)
[2] A. Zunger, Solid State Phys. **39**, 275 (1986), and references therein.
[3] F. Beeler, O. K. Andersen, and M. Scheffler, Phys. Rev. **B41**, 1603 (1990)
[4] A. K. Ramdas and S. Rodriguez, Rep. Prog. Phys. **44**, 1297 (1981)
[5] E. Janzén, R. Stedman, G. Grossmann, and H. G. Grimmeiss, Phys. Rev. **B29** 1907 (1984)
[6] P. Wagner, C. Holm, E. Sirtl, R. Oeder, and W. Zulehner, in: Festkörperprobleme **24**, 191, ed. P. Grosse (Vieweg, Braunsweig, 1984)
[7] J. Olajos, M. Kleverman, and H. G. Grimmeiss, Phys. Rev. **B38**, 10633 (1988)
[8] V. A. Singh, U. Lindefelt, and A. Zunger, Phys. Rev. **B27**, 4909 (1983)
[9] J. Olajos, M. Kleverman, and H. G. Grimmeiss, Phys. Rev. **B40**, 6196 (1989)
[10] G. D. Watkins, Physica **117&118**, 9 (1983)
[11] T. Bever, P. Emanuelsson, M. Kleverman, and H. G. Grimmeiss, Appl. Phys. Lett. **55**, 2541 (1989)
[12] L. C. Kimmerling, J. L. Benton, and J. J. Rubin in *Defects and Radiation Effects in Semiconductors 1980*, Inst. of Physics Conf. Ser. **59**, 217 (1981)
[13] R. Czaputa, H. Feichtinger, and J. Oswald, Solid St. Commun. **47**, 223 (1983)
[14] T. Bever, Diploma work, Lund 1989 (unpublished)
[15] M. Kleverman, T. Bever, B. Bech Nielsen, and H. G. Grimmeiss (unpublished)
[16] F. S. Ham, private communication.
[17] J. Olajos, B. Bech Nielsen, M. Kleverman, P. Omling, P. Emanuelsson, and H, G. Grimmeiss, Appl. Phys. Lett. **53**, 2507 (1988)
[18] A. Thilderkvist, M. Kleverman, G. Grossmann, and H. G. Grimmeiss, Proc. 20th Int. Conf. Physics of Semiconductors, Thessaloniki,1990, E. M. Anastassakis and J. D. Joannopoulos, ed. (World Scientific, Singapore, 1990) p. 581
[19] A. Thilderkvist, G. Grossmann, M. Kleverman, and H. G. Grimmeiss, Mat. Res. Soc. Symp. Proc. **163**, 51 (Boston, 1989)
[20] B. Pajot, F. Merlet, G. Travella, and Ph. Arcas, Can. J. Phys. **50**, 1106 (1972)
[21] B. Pajot, F. Merlet, and G. Travella, Can. J. Phys. **50**, 2186 (1972)
[22] W. Ulrici in *The dynamical Jahn-Teller effect in Localized Systems*, (North-Holland, Amsterdam, 1984) p.439, and references therein.
[23] F. Beeler, private communication.

Materials Science Forum Vol. 83-87 (1992) pp. 137-142

TRANSITION-METAL ACCEPTOR PAIRS IN SILICON

P. EMANUELSSON[1], P. OMLING[1], H. G. GRIMMEISS[1], W. GEHLHOFF[2], J. KREISSL[2], K. IRMSCHER[2] AND U. REHSE[2]
[1]Department of Solid State Physics, University of Lund, Box 118, S-221 00 Lund, Sweden
[2]Forschungsstelle für Kristallzüchtung, Rudower Chaussee 6, D/O-1199 Berlin, FRG

ABSTRACT

Three electron paramagnetic resonance (EPR) spectra are presented. The corresponding defects are identified as different transition-metal acceptor pairs. For two of them (CrIn and MnGa) the electronic structure originates from the $3d^5$ configuration of Cr^+ and Mn^{2+} in a cubic crystal field with a trigonal distortion caused by the acceptor ion. These EPR spectra show very complicated angular dependencies, which can be well described by a spin 5/2 system and a spin Hamiltonian taking both the cubic and trigonal crystal fields into account. The third spectrum shows orthorhombic symmetry and is suggested to originate from a $Fe^{2+}In^-$ pair. This would be the first observation of a pair in silicon with the electronic configuration $3d^6$.

1. Introduction

The pairs between transition metals (TM) and group III acceptors (A) in silicon have been extensively studied during the last years. The electronic properties have been determined from space charge and Hall measurements,[1-5] and electron paramagnetic resonance (EPR) studies have provided information on the identity, symmetry and electronic structure.[6-20]

The EPR investigations started with the comprehensive work by Ludwig and Woodbury,[6] in which the EPR spectra of several different Fe-, Mn- and Cr-acceptor pairs were reported. All pairs showed trigonal symmetry, except the FeIn defect which was found to have orthorhombic symmetry. Ludwig and Woodbury proposed that the pairs consist of a substitutional, negatively charged acceptor with a transition metal ion on the nearest neighbour (nn) or next nearest neighbour (nnn) interstitial position. The nn configuration results in trigonal and the nnn in orthorhombic symmetry. The two models are shown in Fig. 1.

Since the negatively charged acceptor has a closed shell structure, the electronic configuration is essentially determined by the interstitial transition metal ion, i. e. $3d^5$ for the Cr^+A^- and $Mn^{2+}A^-$ pairs, and $3d^7$ for the Fe^+A^- pairs. The role of the acceptor is mainly to produce a crystal field which lowers the symmetry of the interstitial TM.

In recent years it was found that the FeAl, FeGa and FeIn pairs can exist in both the nn and nnn configurations, i. e. one of the two configurations is metastable.[5-10] These new results have motivated further studies on the CrA and MnA pairs in order to investigate whether or not also these pairs show metastable properties, and in order to understand the reason for this exciting physical property.

In this paper we report on the identification of two pairs with the electronic configuration $3d^5$, namely the CrIn and MnGa pairs. In addition, the EPR spectrum of the $Fe^{2+}In^-$ pair is presented, this being the first report on a TM-A pair in silicon with $3d^6$ configuration.

2. Experimental Details

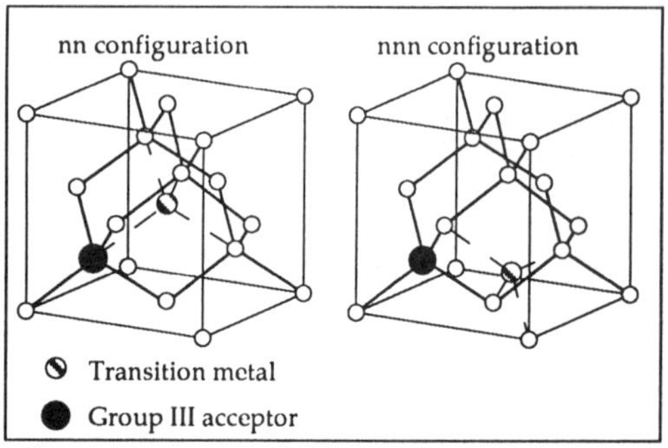

Fig. 1: Nearest neighbour (nn) and next nearest neighbour (nnn) configurations of the transition-metal acceptor pairs.

The samples were prepared from Czochralski grown silicon crystals doped with indium or gallium. The transition metal doping was performed by encapsulating a piece of the metal together with an etched silicon crystal in a quartz ampule. The diffusions took place at a temperature of 1150 to 1200 °C for one hour. After the diffusion, the samples were rapidly quenched. In the case of isotope doping (^{53}Cr and ^{57}Fe) a piece of enriched metal was placed beside the silicon crystal. The EPR measurements were performed using a Bruker ESP 300 spectrometer equipped with a helium gas flow cryostat, and a ZWG ERS 230 spectrometer equipped with a hydrogen finger cryostat. In both spectrometers it was possible to illuminate the samples in situ.

3. Results and analysis

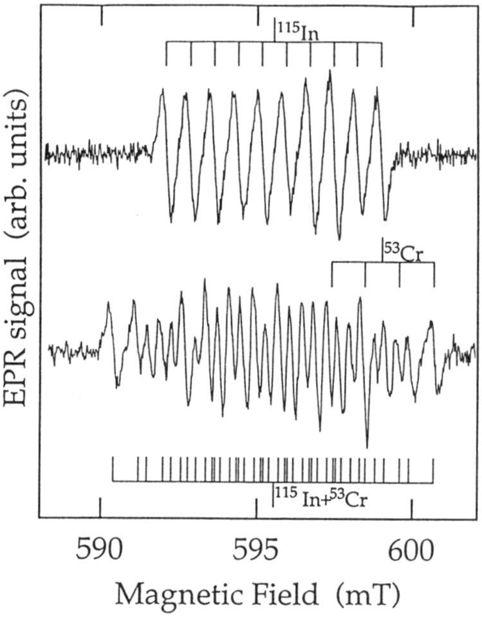

Fig. 2: Hyperfine structure for a sample doped with indium and natural chromium (top) or ^{53}Cr (bottom). Also shown is the calculated hyperfine structure caused by ^{115}In and ^{53}Cr.[13]

In the samples doped with chromium and indium the well known EPR spectrum of interstitial chromium was detected. Upon illumination another spectrum with a complicated hyperfine structure including both allowed and forbidden transitions appeared.[13] In some directions, however, a clear splitting into ten lines proving the incorporation of one indium atom (I=9/2) in the defect was resolved. Chromium consists to 90.45% of isotopes without a nuclear spin and, therefore, does not give rise to any measurable hyperfine structure. However, upon isotope doping with ^{53}Cr (I=3/2), we observed an additional splitting of each line into four components. From this the defect is identified as a chromium-indium pair. The hyperfine structure for samples doped with natural chromium and with ^{53}Cr are both shown in Fig. 2.
The rotation pattern reveals trigonal symmetry but is very complicated and it is therefore difficult to immediately determine the spin. However, in analogy with the other CrA pairs, the spin is anticipated to be 5/2.[6] Neglecting hyperfine interactions, the spin Hamiltonian can be written as:

$$H = H_{FS} + H_Z \quad (1)$$

$$H_{FS} = D\,[S_z^2 - 1/3S(S+1)] - [\,(a-F)/180]\,[35\,S_z^4 - 30S(S+1)S_z^2 + 25\,S_z^2 + 3S^2(S+1)^2 - 6S(S+1)] + (\sqrt{2}/36)a[S_z(S_+^3 + S_-^3) + (S_+^3 + S_-^3)S_z] \quad (2)$$

$$H_Z = g_{||}\mu_B B_z S_z + g_{\perp}\mu_B(B_x S_x + B_y S_y) \quad (3)$$

where all symbols have their usual meaning. The z-axis is the pair axis, and the x-

and y-axes are chosen according to ref. 12. With the magnetic field in the z-axis the energy eigenvalues are given by (neglecting the cubic crystal field parameter):

$$E(M) = D(M^2-35/12) - (a-F)/180\,(35M^4-475/2M^2+2835/16) + g_{||}\mu_B BM \qquad (4)$$

Using Eq. (4), the parameters D, (a-F) and $g_{||}$ can be calculated to: $|D|=0.1075\pm0.0005$ cm^{-1}, $|a-F|=0.0126\pm0.0003$ cm^{-1} and $g_{||}=1.994\pm0.002$. The sign of D/(a-F) is found to be positive.

With the magnetic field in another direction, Eq. (4) cannot be used. Also perturbation theory is impossible to perform in this case since the zero-field splitting and the Zeeman energy is of the same order of magnitude in the magnetic field region where the EPR transitions take place. Instead we made a computer diagonalisation of the spin Hamiltonian, and fitted the calculated rotation pattern to the experimentally observed one by varying the two parameters which could not be determined from Eq. (4), i. e. $g_\perp$ and a. These parameters are determined to: $g_\perp=2.003\pm0.004$ and $a=0.0055\pm0.0010$ cm^{-1}. The calculated rotation pattern is shown in Fig. 3 and the experimental points are indicated.

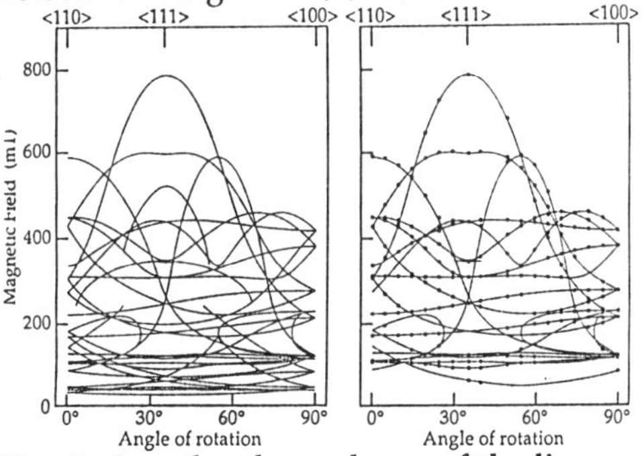

Fig. 3: Angular dependence of the line positions for the CrIn pair obtained at 9.526 GHz. Left: Plot of all possible fine structure transitions. Right: The part of the calculated angular dependence for which experimental data exist is plotted as solid lines. The experimental points are plotted as solid circles.[13]

In the samples doped with manganese and gallium another EPR spectrum was observed.[12] With the magnetic field in the main directions, a sixfold hyperfine splitting is clearly resolved. From this splitting, it can be concluded that one manganese atom (I=5/2, 100% natural abundance) is involved in the defect. A closer inspection of the lines reveals that each Mn hyperfine line is further split into four components. This shows that the defect contains a second atom with nuclear spin 3/2. Gallium consists of two isotopes, ^{69}Ga (60.1% natural abundance) and ^{71}Ga (39.9% natural abundance), both with nuclear spin 3/2. A detailed analysis of the hyperfine structure shows that it is necessary to take the natural abundances and the different nuclear g-values of the galliumisotopes into account in order to explain it. This unambiguously proves the incorporation of one gallium atom, and the defect can therefore be identified as a manganese-gallium pair.

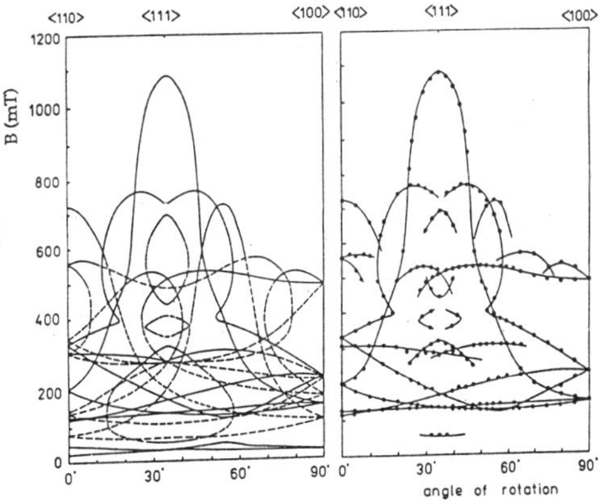

Fig. 4: Angular dependence of the line positions for the MnGa pair obtained at 9.01 GHz. Left: Plot of all possible fine structure transitions. Right: The part of the calculated angular dependence for which experimental data exist is plotted as solid lines. The experimental points are plotted as solid circles.[12]

The fine structure rotation pattern of the MnGa spectrum is very complicated, and reminds slightly of the CrIn spectrum. Since the other MnA pairs, in analogy with the CrA pairs, have a spin 5/2 configuration, we apply the same spin Hamiltonian as in the CrIn case, i. e. the one described by Eqs. (1)-(3). A similar analysis as for the CrIn pair givesthe following spin Hamiltonian parameters: $|D|=0.1773\pm0.0005$ cm^{-1}, $|a-F|=$

0.0025±0.0005 cm^{-1}, $g_{||}$=2.004±0.005 and $g_{\perp}$=2.004±0.005. The sign of D/(a-F) is found to be positive. In Fig. 4 the rotation pattern calculated with these parameters is shown and the experimental points are included. The striking similarity between experimental and calculated data is a strong evidence that the used model is the correct one.

In the samples doped with iron and indium, three EPR spectra are detected. In addition to the well known spectra of the trigonal and orthorhombic neutral FeIn pairs,[6,9,10] a third iron- and indium-related spectrum is detected.

In analogy with the other neutral iron-acceptor pairs (FeB, FeAl and FeGa), also the neutral FeIn pairs can be well described by a substitutional, negatively charged acceptor with a closed shell and a positively charged iron ion at an interstitial position, i. e. $Fe_i^+In_s^-$. The electronic structure of such a pair is therefore expected to be identical with that of the Fe^+ state in a cubic field with a trigonal or orthorhombic distortion. In a cubic crystalline field the ^4F ground state of the Fe^+ with a $3d^7$ configuration is split into the orbital triplets $^4T_1$ and $^4T_2$, and into an orbital singlet $^4A_1$, with the $^4T_1$ lowest in energy.

The combined effects of spin-orbit coupling and the trigonal or orthorhombic fields created by the acceptor split the ground state into six Kramers doublets. The resonances, which take place within the lowest or two lowest doublets can be described with the real spin S=3/2 or approximately with the effective spin S′=1/2 because the zero-field splitting between the ground doublets is always large compared with the microwave photon energy.[10,15,16]

The problem of calculating the g-values have up to now been treated by several authors in a simple approximation by restricting to the twelve basis states of $^4T_1$ using an effective orbital momentum l′=1, and including the admixture of excited states to the magnetism by effective Lande´ factors.[14,18,19] From the fitting of the calculated g-values, the strength of the trigonal and/or orthorhombic crystal fields could be determined. However, taking the measured zero-field splittings into account these simple calculations cannot describe the experimental data satisfactorily. Especially the finding that the zero-field parameter D of the FeB pair is negative[20] cannot be explained within the framework of the used approximations. An improved description of the electronic structure of the Fe^+A^- pairs which takes the different second- and fourth order contributions from the acceptor-induced crystal field distortions as well as the admixtures of excited states via spin-orbit coupling in second order into account is in progress and will be reported elsewhere.

In similarity with the neutral FeIn spectra the new spectrum shows a distinct splitting into ten lines in some directions of the magnetic field, proving that one indium ion (I=9/2) is involved in this centre. Because natural iron consists to 97.8% of isotopes with nuclear spin zero, no hyperfine structure is expected from a possible iron in the defect. However, upon isotope doping with ^{57}Fe (I=1/2) a further splitting of each line into two is indicated. We therefore tentatively identify the spectrum as originating from an iron-indium pair. The angular dependence of the line positions can be described by the phenomenological spin Hamiltonian:

$$H = + g_x'\mu_B B_x S_x' + g_y'\mu_B B_y S_y' + g_z'\mu_B B_z S_z' \tag{5}$$

with S′=1/2 and the effective g-values $g_x' \approx g_y' \approx 0$, $g_z' \approx 14.1$ (x//<100>, y,z//<110>). The rotation pattern and the effective g-values are typical for transitions within the pseudo-doublets for a S=2 manifold in an orthorhombic crystal field. It is therefore not unreasonable to assume that the spectrum is caused by an orthorhombic $Fe^{2+}In^-$ centre. This assumption is supported by the behaviour of the three spectra upon illumination. The intensity of the new spectrum decreases while the intensity of the neutral FeIn spectra increase when the sample is illuminated.

Since the In^- has a closed shell, the spin of such a system is determined by the Fe^{2+}

ion. In a cubic crystal field an orbital triplet, $^5T_2$, resulting from the ^5D term of the free Fe^{2+} ion with a $3d^6$ configuration is the ground state. The orthorhombic field lifts the orbital degeneracy leaving an orbital singlet state with a spin S=2. Neglecting hyperfine structure and fourth order terms, the spin Hamiltonian can be written as:

$$H=D(S_z^2-2)+E(S_x^2-S_y^2)+g_x\mu_B B_x S_x+g_y\mu_B B_y S_y+g_z\mu_B B_z S_z \quad (6)$$

with x//<100> and y,z//<110>. Under the condition $g\mu_B B << D$ the $-2\leftrightarrow 2$ transitions occur at the magnetic field given by:

$$B = (4\, g_z\mu_B \cos\alpha)^{-1}\, [(h\nu)^2-4\Delta^2]^{1/2}\, (\cos\theta)^{-1} \quad (7)$$

where $\Delta=[(D^2+3E^2)^{1/2}-D]$, $\tan\alpha=\sqrt{3}/2\, E/D$ and θ is the angle between the magnetic field and the z-axis.

The calculated angular dependence predicted by Eq. (7) is obeyed within the experimental errors supporting the proposed model. The missing line in the rotation pattern is explained by the fact that $-2\leftrightarrow 2$ transitions only can be induced if the microwave magnetic field has a component parallel with the z-axis.

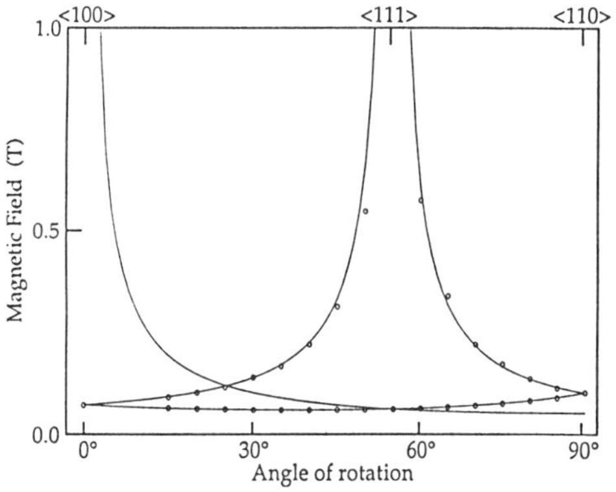

Fig. 5: Angular dependence of the line positions of the $Fe^{2+}In^-$ pair obtained at 9.528 GHz.

4. Discussion

EPR spectra are detected for the transition-metal acceptor pairs with the TM ion in three different electronic configurations, $3d^5$, $3d^6$ and $3d^7$, and in two different charge states, + and 2+. A major problem to solve is to understand why the Fe^+A^- pairs show the metastable properties previously discussed, while the Cr^+A^- and $Mn^{2+}A^-$ pairs apparently do not. One explanation might be the different charge states, the extra charge of the Mn^{2+} compared with the Fe^+ increases the Coulomb attraction and therefore favours the nn configuration. The fact that the Cr^+A^- pairs are only observed in the nn configuration can be explained by that, at least in our experiments on the Cr^+In^- centre, the sample had to be illuminated in order to detect the EPR signal. This means that the defect is in the $(Cr^{2+}In^-)$ charge state during cooling, and if the temperature at the EPR measurements is too low for the Cr atom to jump to the other position, the charge state might be the dominant reason for the lack of metastability. However, the detection of the $Fe^{2+}In^-$ centre in the nnn configuration reported in this paper shows that the charge state cannot be the only reason why CrA and MnA pairs only exist in the nn configuration.

Another reason for the different metastable properties could be the electronic structures of the defects. The ground state of the Cr^+ and Mn^{2+} ions is ^6S which is an orbital singlet state, while the Fe^{2+} and Fe^+ ions have a ^5D and a ^4F term respectively as ground states. Further investigations are necessary to give a clear solution to this problem.

6. Acknowledgements

This work was supported by the Swedish National Science Research Council and the Swedish Board for Technical Development.

REFERENCES

1. H. Lemke, Phys. Status Solidi A 64, 549 (1981); A 71, K215 (1982); A 75, K49 (1983); A 76, 223 (1983).
2. H. Conzelmann, K. Graff and E. R. Weber, Appl. Phys. A 30, 169 (1983).
3. H. Feichtinger, J. Oswald, R. Czaputa, P. Vogel and K. Wünstel, in *Proc. of the 13th ICDS*, eds. L. C. Kimerling and J. M. Parsey Jr. (The Metallurgical Society of AIME, New York, 1984) p. 855.
4. A. Chantre and D. Bois, Phys. Rev. B 31, 7979 (1985).
5. A. Chantre and L. C. Kimerling, in *Defects in Semiconductors*, Vols. 10-12 of *Materials Science Forum*, ed. H. J. von Bardeleben (Trans Tech, Aedermannsdorf, 1986) p. 387.
6. G. W. Ludwig and H. H. Woodbury, in *Solid State Physics*, eds. F. Seitz and D. Turnbull (Academic, New York, 1962) Vol. 13, p. 223.
7. J. J. van Kooten, G. A. Weller, C. A. J. Ammerlaan, Phys. Rev. B 30, 4564 (1984).
8. W. Gehlhoff, K. Irmscher and J. Kreissl, in *New Developments in Semiconductor Physics*, Lecture Notes in Physics Vol. 301, eds. G. Ferenczi and F. Beleznay (Springer, Berlin, 1988) p. 262.
9. P. Omling, P. Emanuelsson, W. Gehlhoff and H. G. Grimmeiss, Solid State Commun. 70, 807 (1989).
10. W. Gehlhoff, P. Emanuelsson, P. Omling and H. G. Grimmeiss, Phys. Rev B 41, 8560 (1960).
11. J. Kreissl, W. Gehlhoff, P. Omling and P. Emanuelsson, Phys. Rev. B 42, 1731 (1990).
12. J. Kreissl, K. Irmscher, W. Gehlhoff, P. Omling and P. Emanuelsson, to be published in Phys. Rev. B 44 (1991).
13. P. Emanuelsson, P. Omling, H. G. Grimmeiss, J. Kreissl and W. Gehlhoff, to be published in Phys. Rev. B.
14. W. Gehlhoff and K. H. Segsa, Phys. Status Solidi B 115, 443 (1983).
15. C. A. J. Ammerlaan, Solid State Phenomena 6&7, 591 (1989).
16. W. Gehlhoff and U. Rehse, Solid State Phenomena 6&7, 257 (1989).
17. W. Gehlhoff, K. Irmscher and U. Rehse, in *Defects in Semiconductors*, Vols. 38-41 of *Materials Science Forum*, ed. G. Ferenczi (Trans Tech, Aedermannsdorf, 1989) p. 373.
18. C. A. J. Ammerlaan and T. Gregorkiewicz, in *New Developments in Semiconductor Physics*, Lecture Notes in Physics Vol. 301, eds. G. Ferenczi and F. Beleznay (Springer, Berlin, 1988) p. 244.
19. C. A. J. Ammerlaan and J. J. van Kooten, Mater. Res. Soc. Proc. 45, 525 (1985).
20. W, Gehlhoff and U. Rehse, to be published.

Materials Science Forum Vol. 83-87 (1992) pp. 143--148

THEORETICAL INTERPRETATION OF EPR MEASUREMENTS ON THE IRON-SHALLOW ACCEPTOR PAIRS IN SILICON

LUCY V.C. ASSALI AND J.R. LEITE
Instituto de Física, Universidade de São Paulo, CP 20516, 01498 São Paulo, SP, Brazil

ABSTRACT

Systematic studies of the electronic structure of the Fe-B, Fe-Aℓ and Fe-Ga pair complexes in silicon are performed. Self-consistent spin-polarized electronic state calculations are carried out for the complexes in both, the trigonal and orthorhombic symmetries. One common trend in the electronic properties of the complexes is that their ground state have a spin $S = 3/2$. This finding explains why the EPR spectra of these defects are well described by a spin quartet ground state. Hyperfine contact energy parameters A are calculated at the Fe and acceptor nuclei and compared with the available EPR data. It is shown that a fairly consistent picture emerges from the covalent model that we propose to describe the physical properties of these complexes.

1. Introduction

Experimental investigations of the 3d transition metal related pair complexes in silicon that involve an acceptor (A) impurity such as B, Aℓ, Ga, or In have been performed since the pioneering electron paramagnetic resonance (EPR) experiments reported by Ludwig and Woodbury thirty years ago.[1-22] Despite the great deal of efforts that have been devoted to the characterization of these pairs complexes in silicon only recently rigorous first-principles calculations have been carried out in order to understand the electronic structure of these complexes.[23] One important achievement of the theory was the demonstration that the covalent effects play a basic role in determining the physical properties of these pairs. However, there are still details of the electronic structure of these defects which require further investigations. In this work we continue our theoretical studies of the electronic structures of the Fe-B, -Aℓ and -Ga pairs in silicon. By comparing our calculations with the reported EPR data we show that the covalent model proposed by us to describe the electronic structure of the pairs is fairly consistent.[23]

The EPR results show that the Fe-A pairs in silicon display both, trigonal and orthorhombic symmetries. The Fe-B pair is an exception to this rule once it is observed only in trigonal symmetry. The trigonal EPR spectra have been reported for the Fe-B pair[1,8], for the Fe-Aℓ pair[7], for the Fe-Ga pair[1] and recently, for the Fe-In pair.[18] The orthorhombic spectra have been ascribed to the Fe-Aℓ pair[7], to the Fe-Ga pair[20] and to the Fe-In pair.[1,22] It has been currently accepted that the trigonal pairs display a ⟨111⟩ symmetry indicating that they may consist of a substitutional acceptor with the iron impurity placed at a nearest-neighbor interstitial site and the orthorhombic pairs display a ⟨100⟩ symmetry consistent with the assumption that the iron impurity is placed at the substitutional acceptor next-nearest-neighbor interstitial site.[1,20] These microscopic atomic configurations for the pairs are assumed as the starting point of our calculations.

The theory used in this work is based on the multiple-scattering Xα formalism within the framework of the Watson-sphere-terminated molecular-cluster model. This approach is well documented and has been used in our previous calculations for pair complexes in silicon.[23] The band edges of the perfect silicon crystal were defined by adopting a 26Si-atom cluster model. In order to simulate the Fe-A pairs, a $25Si + A_sFe_i$ cluster is adopted, were an iron atom is placed at the center of the cluster (T site) and the acceptor atom replaces either a nearest or next-nearest neighbor substitutional site. The self-consistent calculations were carried out by taking into account the spin-polarization effects. In order to account for the core polarization effects on the evaluation of the hyperfine contact fields at the iron and acceptors nuclei, all the electrons were included in the calculations.

2. Results

2.1. Electronic Structure

Figures 1 and 2 summarize the main features of the electronic structure of the Fe-B, -Aℓ and -Ga pairs in silicon. Gap levels and resonances are labelled according to the irreducible representations of the point groups T_d for the isolated impurities and C_{3v} or C_{2v} for the pair complexes. One important feature of our calculations is the finding that the gap states induced by the pairs are Fe_i-like states. The strong localization of the pair gap states within the cluster central atomic sphere emphasizes the Fe_i character of these impurity states. However, it is worth mentioning that the covalent effects play a role once the t_2 impurity state of the acceptor impurity interacts with the Fe_i-derived t_2 resonances and gap levels and induces significative contributions to the pair gap states. For the orthorhombic pairs the $2b_2$ gap states are derived from a strong mixing between the acceptor t_2 impurity state and the Fe_i-3d derived impurity states.

The strong localization of the gap states on the Fe_i impurity gives rise to an interesting common trend in the electronic properties of the complexes. The results depicted in Figs. 1 and 2 show that an ^4A orbital singlet is ascribed to the ground state of all pairs analysed here, implying that a high spin $S = 3/2$ is ascribed to them for both, the ⟨111⟩ and ⟨100⟩ configurations. We have simulated the electronic structure of the complexes for the spin $S = 1/2$ by changing the electron population of the gap levels and carrying out the calculations to the self-consistent limit. In all cases, except for the Fe-Ga pair in ⟨111⟩ symmetry, the calculations lead to hole states below occupied states which indicates that we are dealing with excited electronic states of the complexes. A typical result is shown in Fig. 2(d). Although for the Fe-Ga pair in C_{3v} symmetry we are able to converge the calculations for the $S = 1/2$ case without the presence of a hole state below occupied states (see Fig. 2(b)) the total energy of the cluster indicates that the $S = 1/2$ state is an excited state of the complex.

The results of our calculations that the ground state of the Fe-B, -Aℓ and -Ga pair complexes in silicon is a spin quartet is in agreement with EPR experiments.[22] It is interesting to observe that the analysis of the charge distribution in the cluster regions allow us to conclude that the electronic states of the pairs arise from the covalent interactions involving the Fe_i, the acceptor and the host silicon atoms. Therefore, the covalent effects play an important role in determining the electronic structure of these pairs.

2.2. Hyperfine Contact Fields

The spin-polarized calculations reported here allow us to access the values of the Fermi hyperfine contact energy, A, at the Fe and acceptor nuclei. The values of A have been extracted from EPR experiments for the Fe-B and Fe-Ga pairs. Table I shows the calculated values of A at the Fe nucleus for the Fe-A pair in the ⟨111⟩ and ⟨100⟩ configurations. The entries correspond to the ground ($S = 3/2$) and excited ($S = 1/2$) states of the complex. Table II displays the values of A extracted from the EPR data according to the expression $A = 1/3\, A_{\parallel} + 2/3\, A_{\perp}$. The values of A for the isolated interstitial impurities Fe^0 and Fe^+ are included for comparison.

The agreement between theory and experiment for the Fe-B pair is excellent. For the Fe-Ga pair, even assuming that the measured value corresponds to the $S = 3/2$ ground state, the agreement is very good. There is no clear correlation between the measured values of A for the isolated iron impurity and for the complexes that allows to the conclusion that the iron impurity retains its identity

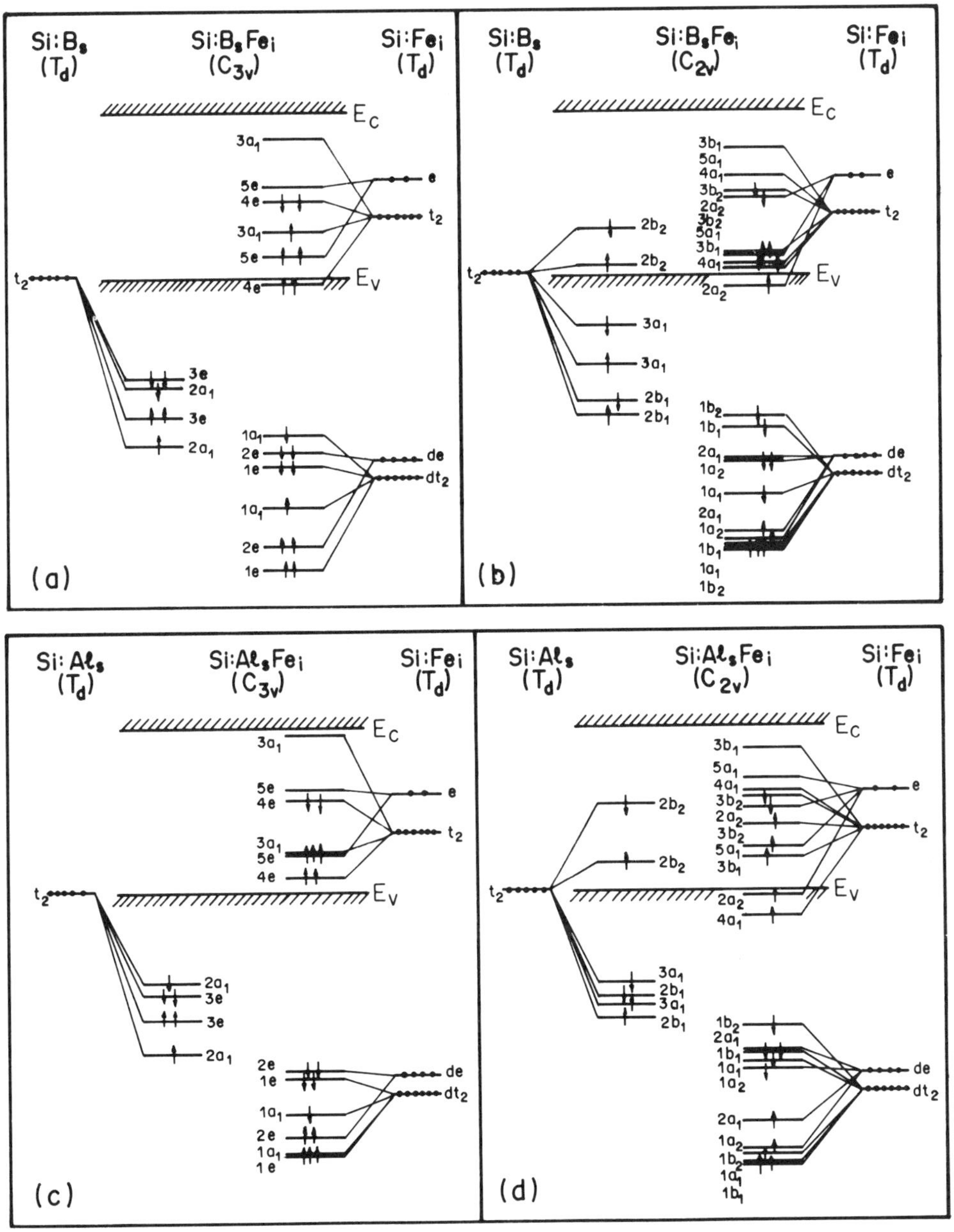

Figure 1. Impurity levels, relative to the band edges, induced by the pair complexes in Si as derived from the interactions between the iron and the acceptor impurities. Fe-B in (a) ⟨111⟩ and (b) ⟨100⟩ symmetries; Fe-Aℓ in (c) ⟨111⟩ and (d) ⟨100⟩ symmetries. The full circles indicate the occupancy of the levels induced by the isolated interstitial iron (Ref. 24) and the isolated substitutional acceptor impurities (Ref. 25) in Si according to spin unpolarized calculations. The arrows indicate the occupancy of the levels with up(↑) and down (↓) spins as derived from the present spin-polarized calculations.

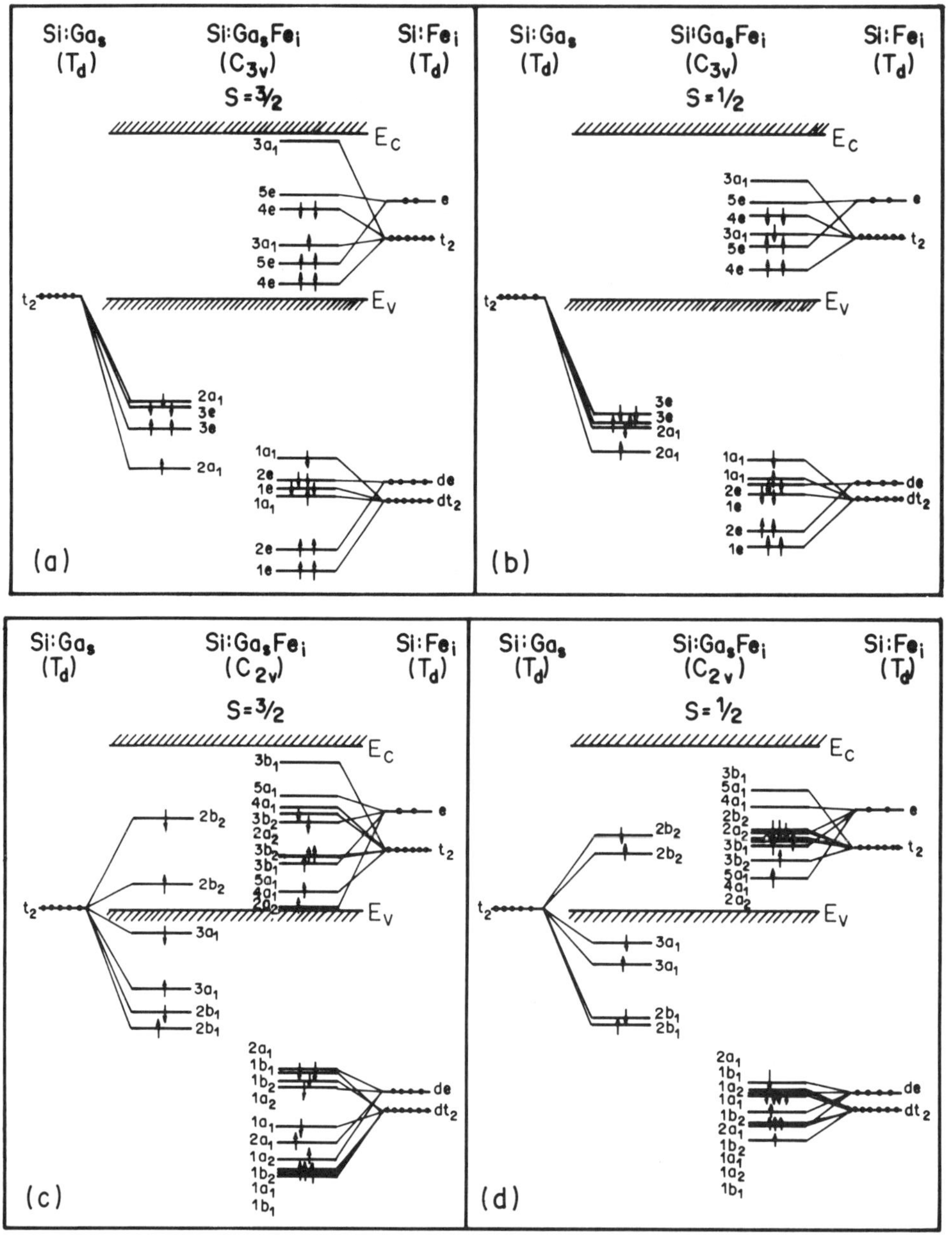

Figure 2. The same as in Fig. 1 for the Fe-Ga pair complex. S indicates the total spin of the complex. (a) $\langle 111 \rangle$ symmetry and S = 3/2 ; (b) $\langle 111 \rangle$ and S = 1/2 ; (c) $\langle 100 \rangle$ and S = 3/2 ; (d) $\langle 100 \rangle$ and S = 1/2 . The open circle in (d) indicates an unoccupied state.

in the complex as Fe^{0} or Fe^{+} . The entries shown in Table I may play an useful role on the interpretation of further EPR and ENDOR experiments performed on these complexes. Table III depicts the calculated values of A at the acceptor nuclei for the Fe-A pairs in silicon. The experimental values of A at the acceptors nuclei for the pairs are shown in Table IV. The agreement between theory and experiment for the ground state (S = 3/2) of the Fe-B pair is again

TABLE I. Calculated hyperfine contact energy parameter A, in units of 10^{-4} cm^{-1}, at the Fe nucleus for the iron-acceptor pairs in Si. S is the total spin of the complex.

	Fe-B	Fe-Aℓ ⟨111⟩	Fe-Ga	Fe-B	Fe-Aℓ ⟨100⟩	Fe-Ga
S = 3/2	5.0	6.2	5.6	8.3	6.2	4.8
S = 1/2	16.6	-0.3	3.8	2.7	5.0	-0.3

TABLE II. Experimental values[(a)] for the hyperfine contact energy parameter A, in units of 10^{-4} cm^{-1}, at the Fe nucleus for the iron-acceptor pairs and for the isolated interstitial iron impurity in Si

	Fe-B	Fe-Ga	Fe^0	Fe^+
A	5.9(6.2)[b]	3.2	6.98	2.99
Symmetry	C_{3v}	C_{3v}	T_d	T_d

[a]Reference 1. [b]Reference 8.

TABLE III. Calculated hyperfine contact energy parameter A, in units of 10^{-4} cm^{-1}, at the acceptor nuclei for the iron-acceptor pairs. S is the total spin of the complex.

	^{10}B	^{11}B	$^{27}Aℓ$ ⟨111⟩	^{69}Ga	^{71}Ga	^{10}B	^{11}B	$^{27}Aℓ$ ⟨100⟩	^{69}Ga	^{71}Ga
S=3/2	0.4	1.3	43.9	122.0	152.0	-0.1	-0.4	4.5	45.1	57.4
S=1/2	15.3	45.6	41.0	-80.0	-100.0	-11.2	-33.3	-56.4	19.5	24.8

TABLE IV. Experimental values for the hyperfine contact energy parameter A, in units of 10^{-4} cm^{-1}, at the acceptors nuclei for the iron-acceptors pairs in C_{3v} symmetry.

^{10}B	^{11}B	^{71}Ga
0.43[a]	1.12[a] (1.15)[b]	9.5[b]

[a]Reference 8. [b]Reference 1.

excellent. However, the calculated value of A for the Fe-Ga pair is significatively larger than the experimental result. It has been argued that the value of A for the acceptor is very small or even absent once the acceptor captures one electron from the Fe atom and stays in a negative charge state, therefore in a closed shell configuration.[7] Our calculations show that this "ionic" point of view is incorrect. The hyperfine contact interaction at the acceptor nucleus is originated from the spin-polarization effects on the acceptor core and valence states caused by the impurity states plus the contribution from the a_1 impurity states themselves. In other words, the Fe-3d derived impurity states play an important role in determining the value of A at the acceptor nuclei. Therefore, the hypothesis that the values of A for the acceptors are small is not supported by our investigations. We expect that the predictions made in Table III should motivate further experimental works on these complexes, leading to a conclusive description of their electronic structure.

3. Conclusions

The analysis of the charge distribution associated to the isolated Fe_i and to Fe-B pair, for example, leads to a typical conclusion related to the pairs electronic structure. Bearing in mind that there are eight electrons occupying the t_2 and e gap levels, we obtain the value 2.56e for the charge associated with these states within the Fe_i sphere of the isolated impurity.[24] The seven electrons occupying the Fe-B gap levels, according to the electronic configuration displayed in Fig. 1(a), give the value 2.57e for the charge within the Fe_i sphere. Therefore, the same amount of charge within the Fe_i atomic sphere is obtained in both cases. Here the Haldane and Anderson mechanism, which is operative if covalent effects play a role, prevents charge transfer out of the Fe_i impurity sphere from taking place by compensating the decrease in one of the gap level occupancy.[23]

Remarkable effects are induced on the B_s t_2 acceptor level when the Fe impurity is placed at the next-nearest-neighbor interstitial site. The t_2 level, which is found at about 0.1 eV above the top of the valence band[25], undergoes a splitting of about 1.47 eV. Two impurity levels of b_2 symmetry are introduced in the band gap. The covalent effects involve now the Coulomb and exchange interactions between the B_s t_2 impurity state and the Fe_i induced resonance and gap states. Similar conclusion is reached for the Fe-Aℓ and Fe-Ga pairs in orthorhombic symmetries.

REFERENCES

1. G.W. Ludwig and H.H. Woodbury, Solid State Phys. **13**, 223 (1962).
2. K. Graff and H. Piepper, J. Electrochem. Soc. **128**, 669 (1981).
3. J. Kreissl and W. Gehlhoff, Phys. Stat. Sol. (b) **112**, 695 (1982).
4. K. Wünstel and P. Wagner, Solid State Commun. **40**, 797 (1981); Appl. Phys. A **27**, 207 (1982).
5. L.C. Kimerling and J.L. Benton, Physica **116**B, 297 (1983).
6. H. Conzelmann, K. Graff, and E.R. Weber, Appl. Phys. A **30**, 169 (1983).
7. J.J. van Kooten, G.A. Weller, and C.A.J. Ammerlaan, Phys. Rev. B **30**, 4564 (1984).
8. W. Gehlhoff and K.H. Segsa, Phys. Stat. Sol. (b) **115**, 443 (1983).
9. K. Wünstel, K.-H. Froehner, and P. Wagner, Physica **116**B, 301 (1983).
10. H. Feichtinger, J. Oswald, R. Czaputa, P. Vogl, and K. Wünstel, J. Electron. Mat. **14**a, 855 (1985).
11. A. Chantre and D. Bois, Phys. Rev. B **31**, 7979 (1985).
12. A. Chantre and L.C. Kimerling, Mat. Sci. Forum **10-12**, 387 (1986).
13. R. Sauer and J. Weber, Physica **116**B, 195 (1983).
14. T.E. Schlesinger and T.C. McGill, Phys. Rev. B **28**, 3643 (1983).
15. C.A.J. Ammerlaan and J.J. van Kooten, Mat. Res. Soc. Symp. Proc. **46**, 525 (1985).
16. L. Samuelson and P. Omling, 18th ICPS, O. Engström, Ed. (World Scientific, Singapore, 1987), Vol. 1, p. 931.
17. S.D. Brotherton, P. Bradley, and A. Gill, J. Appl. Phys. **57**, 1941 (1985).
18. P. Omling, P. Emanuelsson, W. Gehlhoff, and H.G. Grimmeiss, Solid State Commun. **70**, 807 (1989).
19. M. Geddo, B. Pivac, A. Borghesi, A. Stella, S.U. Campisano, and E. Rimini, J. Appl. Phys. **67**, 3538 (1990).
20. W. Gehlhoff, K. Irmscher, and J. Kreissl, Lecture Notes in Phys. **301**, 262 (1988).
21. O.O. Awadelkarim and B. Monemar, J. Appl. Phys. **64**, 6306 (1988).
22. W. Gehlhoff, P. Emanuelsson, P. Omling, and H.G. Grimmeiss, Phys. Rev. B **41**, 8560 (1990).
23. L.V.C. Assali and J.R. Leite, Phys. Rev. B **36**, 1296 (1987); Mat. Sci. Forum **38-41**, 409 (1989); 20th ICPS, E.M. Anastassakis and J.D. Joannopoulos, Eds. (World Scientific, Singapore, 1990), Vol. 1, p. 597.
24. G.G. DeLeo, G.D. Watkins, and W.B. Fowler, Phys. Rev. B **23**, 1851 (1981).
25. L.V.C. Assali and J.R. Leite, unpublished.

Materials Science Forum Vol. 83-87 (1992) pp. 149-154

MAGNETO-OPTICAL PROPERTIES OF FE-AL PAIRS IN SILICON AND THE DISCOVERY OF A NEW TRIGONAL $(Fe_i\text{-}Al_s)^0$ PAIR.

S. GREULICH-WEBER, A. GÖRGER, J.-M. SPAETH, AND H. OVERHOF
Fachbereich Physik, Universität-GH Paderborn, Warburger Str. 100A, W-4790 Paderborn, FRG

Abstract

Iron-Aluminum pairs in Silicon are investigated with conventional and optically detected electron paramagnetic resonance (EPR). For the trigonal and orthorhombic pairs known from previous EPR measurements we found for the first time optical absorption bands by measuring the magnetic circular dichroism of the absorption (MCDA). Direct experimental evidence is presented for the configurational bistability of both pairs by showing that the MCDA of the trigonal configuration can be transformed into that of the orthorhombic configuration by the combined effect of light and temperature. A new trigonal pair is discovered by conventional EPR having the same EPR intensitiy as the known one. Total energy calculations of various $(Fe_i\text{-}Al_s)$ pair configurations show that two trigonal $(Fe_i\text{-}Al_s)^0$ pairs with different $Fe_i\text{-}Al_s$ separations have almost the same binding energy and should occur with the same probability. Fe_i^+ is always on a tetrahedral interstitial site, while Al_s^- is nearest neighbor along <111> in one pair, second nearest neighbor in the other one with one silicon lattice site in between.

Introduction

Pairs of interstitial iron and substitutional acceptors in silicon have recently found particular interest because of their structural transformations and metastability. Chantre and Bois found a bistability for $(Fe_i\text{-}Al_s)$ pairs with deep level transient spectroscopy (DLTS) and associated the two configurations to a trigonal and to an orthorhombic $(Fe_i\text{-}Al_s)$-pair on the basis of an estimate of their different Coulomb binding energies [1]. With EPR both pair symmetries for the neutral $(Fe_i\text{-}Al_s)$ charge state could be identified [2-4]. The configurational bistability, however, could so far not be observed in EPR experiments [3,4]. In this paper we use MCDA to show that the two pairs observed in [2-4] have optical absorption bands. With MCDA-EPR observed for the orthorhombic and trigonal pairs at different energies we further show that under the combined effects of light and cooling the trigonal pairs are transferred into orthorhombic pairs. In conventional EPR we observe two different trigonal pairs (beside the orthorhombic pair), one of which is also seen in MCDA-EPR and is to be identified with 'NL27' [2]. The other trigonal pair (invisible in MCDA-EPR) is reported here for the first time. We propose an identification of this pair on the basis of a self consistent total energy calculation for trigonal pairs where Fe_i and Al_s is altered. Two trigonal pair were proposed so far only by Kimerling [5] for $(Fe_i\text{-}In_s)^0$.

Experimental Results

Magnetic circular dichroism of the absorption and optically detected EPR.
The MCDA measured after diffusing Fe into the Al-doped Si and cooling the sample to 1.5 K in the dark or with light of photon energy below the band gap energy ($h\nu < 1.17$ eV) is shown in Fig.1 (trace (a)). The MCDA is temperature dependent and therefore due to paramagnetic centers. However, when cooling the sample under illumination with light of energy exceeding the band gap energy, changes are observed in the MCDA spectrum (see

Fig.1, trace (b)). The MCDA transition at 0.875 eV and 3 transitions arround 1eV increase by about 70%, while the remainder of the MCDA spectrum decreases by about 40% including the prominent line at 0.78 eV. Apparently, the MCDA is composed of at least two defects, one of which is formed at the expense of the other under the combined effects of light and cooling.The nature of the defects giving rise to the MCDA spectrum could be identified by performing MCDA detected EPR (MCDA-EPR) experiments [6]. Fig.2 shows the MCDA-EPR spectrum measured in the MCDA transition at 0.78 eV. A comparison of the angular dependence of the MCDA-EPR lines with those of the trigonal $(Fe_i\text{-}Al_s)^0$ pair 'NL27' [2] measured conventionally showed that it is identical within experimental error. The measurement of the MCDA-EPR spectrum when setting the photon energy to 0.875 eV yields an EPR spectrum that can be identified with the orthorhombic $(Fe_i\text{-}Al_s)^0$-pair with <100> g-tensor symmetry described previously [2]. The excitation spectra of the MCDA-EPR lines, i.e. the MCDA tagged by EPR [7] show that the sharp transition at 0.78 eV and a broad continuum extending to 0.97 eV belongs to the previously described trigonal $(Fe_i\text{-}Al_s)^0$ pair while the optical transitions at 0.875 eV, 0.997 eV, 1.01 eV and 1.03 eV belong to the orthorhombic pair. The spectra obtained after cooling in the dark and

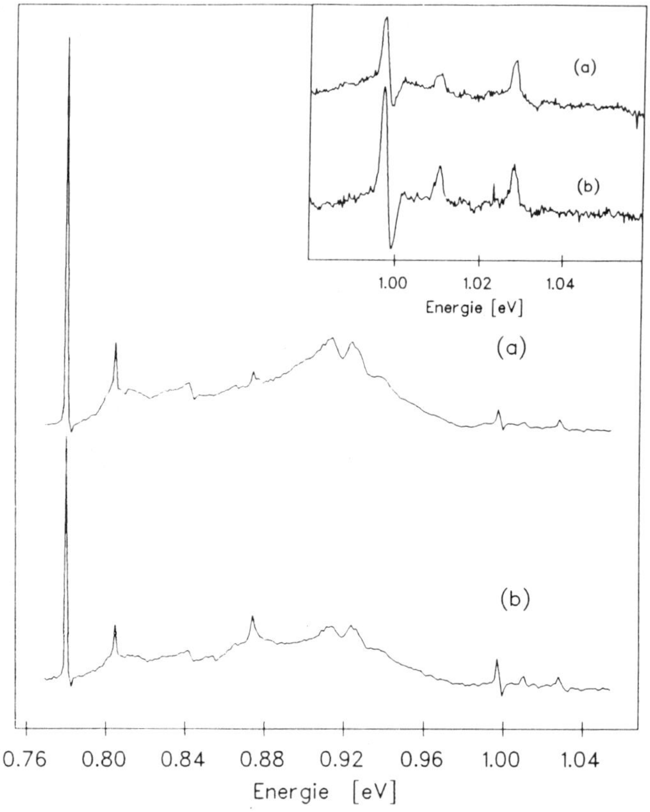

Fig. 1: *MCDA spectrum of Fe diffused Si:Al after cooling the sample in the dark (trace (a)) and after cooling under illumination (trace (b)) with above-band-gap-light (halogen lamp, GaAs-filter) to 1.5 K (B = 1.5 T). The inset shows the spectral part about 1 eV with higher resolution.*

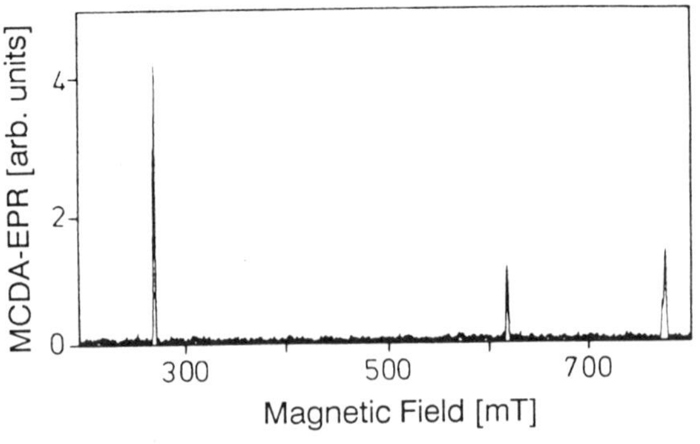

Fig. 2: *MCDA-EPR spectrum of the trigonal $(Fe_i\text{-}Al_s)^0$ pair in Si measured at $h\nu = 0.78$ eV, $B || [111] + 3^o$, ($\nu_{ESR} = 23.52$GHz, $T = 1.5$ K).*

after cooling under above-band-gap illumination (Fig.1) can now be understood: When cooling the sample in the dark both pairs are present. When cooling under light the number of trigonal pairs is reduced while that of the orthorhombic pair is increased. The experiments show directly the structural transformation from one pair configuration to the other. This for the first time confirms the bistability of the two configurations on a microscopic level.
With the MCDA method it is also possible to determine the effective spin of the two pairs [8]. The previous assumptions that at low temperature (our experiments were done at 1.5 K) both pairs have $S' = {}^{1}/2$, was confirmed.

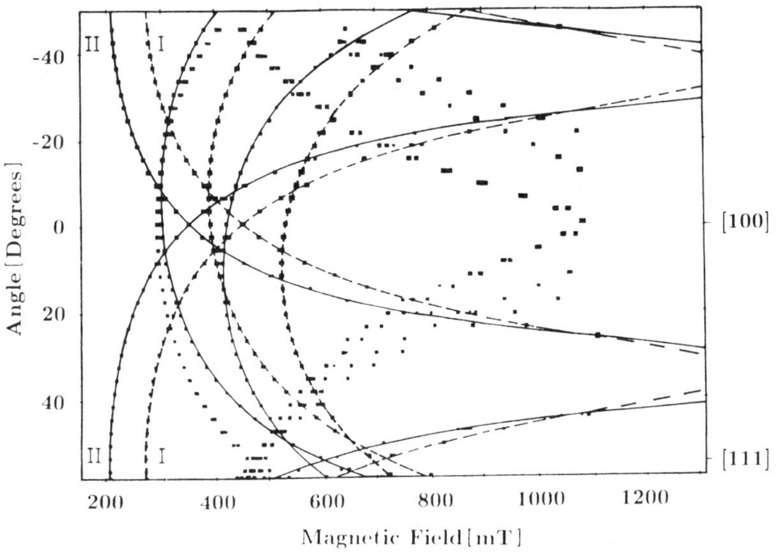

Fig. 3: *EPR angular dependence of $(Fe_i\text{-}Al_s)^0$ pairs in Si for rotation of the magnetic field in a {110} plane measured at 24 GHz and T=6 K. The dots are the experimental field positions of the EPR transitions. The dashed lines are the angular dependence of the trigonal pair I calculated with the g-tensor after [2], the solid lines represent the angular dependence of the trigonal pair II with the g-values as given in table I.*

Electron paramagnetic resonance.
The EPR angular dependence measured conventionally of the same iron diffused Si:Al samples as investigated by optically detected EPR is shown in Fig.3. Dashed and solid lines are the result of an analysis of the EPR angular dependence using the usual spin Hamiltonian with effective spin $S' = {}^{1}/2$. The dots connected with dashed lines belong to the previously described trigonal $(Fe_i\text{-}Al_s)^0$ pair defect [2] (henceforth called trigonal pair I, it is the same as measured in MCDA-EPR). The solid lines in Fig.3 represent the angular dependence of EPR lines not observed up to now, which belong to a new trigonal $(Fe_i\text{-}Al_s)^0$ pair (henceforth called trigonal pair II). All remaining dots are lines from the orthorhombic $(Fe_i\text{-}Al_s)^0$ defect described previously [2]. The g-values of the new trigonal $(Fe_i\text{-}Al_s)^0$ pair (observed also in X band) together with the previously published ones are collected in table I. With conventional EPR the ^{27}Al hyperfine (hf) interactions of both trigonal pairs could be resolved for some orientations. They seem to be essentially isotropic. The ^{27}Al hf interaction of the new trigonal defect is slightly larger than that of the trigonal pair I. The EPR intensities of the two trigonal pairs were found to be the same within the temperature range of 4.2 to 15 K and did not change with additional illumination irrespective of the cooling procedure. Above 15 K both signals vanish. No additional trigonal spectra were observed at higher temperature.

symmetry	g-tensor			aluminium hf interaction $\|A\|$/[MHz]	remarks	Ref.
	$g_\perp$	$g_\parallel$				
trigonal	1.138	6.389		4.6	trigonal I	[2]
trigonal	1.076	8.395		4.9	trigonal II	this work
(<111>-axial)						
	g_x	g_y	g_z			
orthorhombic	1.236	1.612	5.885	3.9	ground state	[2,3,4]
orthorhombic	5.36	2.51	1.73	4.7	excited state	[3,4]
(x,y||<110>, z||<100>)						

Table I: *g-values for the EPR-spectra of $(Fe_i\text{-}Al_s)^0$ pair defects in silicon determined for an effective spin $S'=1/2$. The labels "trigonal I and II" are chosen to distinguish the two trigonal pairs, while "ground and excited state" were proposed by Gehlhoff et al. [3, 4].*

Theoretical calculation of the stability of $(Fe_i\text{-}Al_s)$ pair configurations.

Our calculations have been performed within the general framework of the spin-density functional theory (DFT). Exchange and correlation are treated in the local spin-density approximation (LSDA) of the spin-density functional theory.For the computations we have made use of the selfconsistent linear muffin-tin orbital method in the atomic sphere approximation (LMTO-ASA) ([9,10], see also [11]). In Fig.5 the different trigonal pairs to be discussed are indicated in a {111} plane of the Si lattice. In pair (a) iron and aluminum are separated by a nearest neighbor distance (nnd), for pairs (b) and (c) the separation is two nnd's and for pair (d) we have three nnd. The relative stability of pairs (a) and (c) has been discussed by Kimerling et al., [5] on the basis of a simple ionic bonding model. Electronic structure calculations for trigonal Fe-Al pairs have been published for pair (a) only [12,13].

Total energy calculations for the four pairs (a)-(d) in the three possible charge states are summarized in Fig.6. Here we plot $\Delta E = E[(Fe_i\text{-}Al_s)^{(n)}] - E[Fe_i^+] - E[Al_s^-] + neE_F$ as a function of the Fermi energy E_F. Also plotted is the sum of the total energies for isolated Fe_i^+ and Al_s^- point defects and for isolated Fe_i^0 and Al_s^- point defects, respectively. These lines are

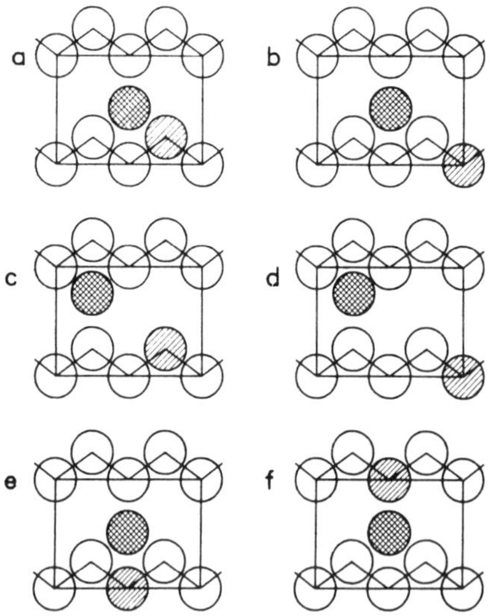

Fig. 5: *Schematic representation of various configurations of $(Fe_i\text{-}Al_s)$ pairs in Si in a {110} plane. The cross line hatched circles represented Fe_i and the hatched circles Al_s. Configuration (e) and (f) are different in a static magnetic field only.*

marked ∞ and can serve as the zero lines for the binding energies of the pairs. For pair (a) we find transition energies $E^{+/0} = E_V + 0.32$ eV and $E^{0/-} = E_V + 0.92$ eV. Similar transition energies are found for pair (b).

In Fig.6 total energies for the neutral pair are given which have been obtained for the 'high-spin' configuration with $S=3/2$, while experimentally an effective spin $S'=1/2$ is observed. According to our calculations the total energy of the "low spin" $S=1/2$ pairs is above that of the $S = 3/2$ pairs by about 0.3 eV. There is no contradiction to the experiment: the $S=3/2$ configuration must be split by the axial field and as long as the zero-field splitting is large enough only a doublet is observed in the EPR experiment. For pairs (a) and (b) the total energies in all three possible charge states of the pairs are very similar and well below the total energies of the dissociated pair. There is no stable Fe_i^{++} point defect and, therefore, pairs with large Fe-Al distance cannot exist in the positive charge stage. We find that in this charge stage pairs (e) and (d) are also nonexistent. The similarity of the total energies of pairs (a) and (b) suggests that we should find both pairs with approximately the same abundance, because at the temperature where Fe_i^+ becomes mobile it is more or less a matter of chance and not of total energy which pair is formed. It is interesting to compare the theoretical contact hf interaction matrix elements with the ^{27}Al nucleus for the different pairs in the neutral charge state: if we use the isotropic value of $g'=2.13$ and approximate the $S'=1/2$ zero-field split system by a simple $S=3/2$ spin system we find A=-7.54 MHz and +7.82 MHz for pairs (a) and (b), respectively, whereas for pairs (c) and (d) the corresponding values are much smaller with -0.64 MHz and +0.24 MHz, respectively. The order of magnitude difference of the data shows that the latter pairs essentially have the hf properties of isolated Fe_i^+ defects. Obviously the spin-density at the different nuclei is dominated by the Fe_i^+ d-electrons and is altered only slightly if a Si atom is replaced by Al.

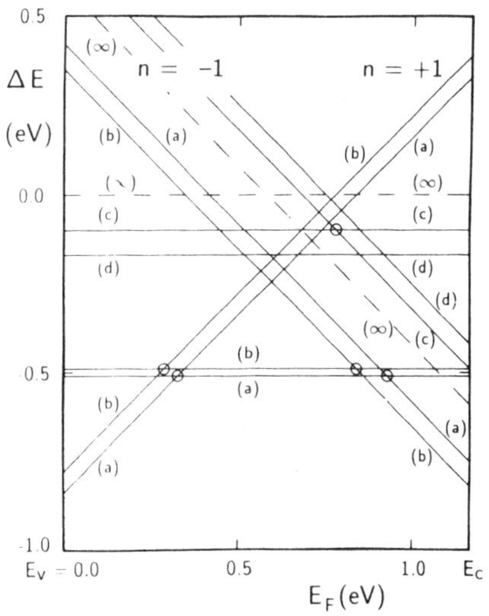

Fig. 6: *Pair formation energy for the $(Fe_i\text{-}Al_s)^{(n)}$ pairs in different atomic positions for the charge states (n) as a function of the Fermi energy. The lines marked ∞ correspond to dissociated pairs. The lines with positive slope belong to $n=+1$, the horizontal one to $n=0$ and those with negative slope to $n=-1$. $E_V=\Delta E=0$.*

Discussion

It is surprising that our new trigonal center has not been observed previously by EPR, in particular since the intensity of EPR lines is the same as that of the known one described earlier [2] within the whole temperature range where the spectra can be observed. The new trigonal center cannot be an excited state of the known trigonal center I since otherwise the relative intensities would be temperature dependent. The effective spin of the trigonal pair I was confirmed by our MCDA experiments to be 1/2, therefore the two trigonal sets of lines cannot be part of a quartet ($S=3/2$) spin system with a small zero field splitting. Hence the new trigonal EPR line set can only come from a trigonal $(Fe_i\text{-}Al_s)^0$ pair with a microscopic configuration different from that of the trigonal pair I. Both trigonal pairs must have the same statistical probabilities for the association of iron and aluminium. Our total energy calculations show

that two pairs, (a) and (b) in Fig.5, have essentially the same pair binding energy of about 0.5 eV. We therefore speculate that the formation of pair (a) and (b) at the temperatures were iron becomes mobile (200 K) is not governed by the small energy differences but simply by geometrical configurations. That would mean that both pairs would be formed with essentially the same probability. For the pair (c) discussed by Kimerling et al. [5] as an alternative to pair (a) we do find a relatively small pair binding energy. According to our calculations formation of this pair would be rather improbable.
Our calculations of the ^{27}Al hf interactions show that their values for configurations (a) and (b) are almost the same with a small enhancement for (b) in very good agreement with experiment. The calculated absolute values are also in good agreement with experiment. However, the calculated hf interaction for configurations (c) and (d) comes out to be one order of magnitude smaller. This confirmes again that pair (c) cannot be the trigonal pair II.
To the best of our knowledge the observations reported here for the $(Fe_i\text{-}Al_s)^0$ pairs are the first MCDA measurements known for defects in silicon. The optical transitions responsible for the MCDA spectrum must be intracenter transitions. Interestingly, in the optically detected EPR only the trigonal pair I was observed. The same crystal was used for EPR and MCDA-EPR measurements. Since for isolated iron point defects we also do not see any MCDA it seems reasonable to assume that the MCDA is somehow connected with the trigonal field caused by the presence of aluminum. In this case the trigonal pair I should be associated with pair (a) in Fig.5 since the Fe-Al pair distance is smallest compared to the other configurations. With our present arguments, however, we have no clearcut distinction between both possibilities.
From experiment we cannot decide whether the trigonal pairs have an effective spin of $S'={}^1/2$ split off from a $S={}^3/2$ state by the crystal field as was found for the orthorhombic centers or whether they have a true low-spin configuration. The lowest state, however, was identified clearly by the MCDA measurements to have $S'={}^1/2$. Our theoretical results strongly suggest a high spin state. This is also in agreement with the observation that the axial field of the Al ligands does not change significantly the distribution of the Fe d-electron density. It is therefore likely to assume that the $S={}^3/2$ configuration of Fe_i^+ is conserved when forming the $(Fe_i^+ \text{-} Al_s^-)$ pair and that the axial field gives rise to $S'={}^1/2$ as the split-off lowest state.

Acknowledgement: We would like to thank Prof. W. Gehlhoff for providing some of the measured Fe diffused crystals and for private communication of his EPR work on bistability of the Fe-Al pairs.

References

1. A.C. Chantre, D. Bois, Phys. Rev. B **31** 7979 (1985)
2. J. J. van Kooten, G.A. Weller, C.A.J. Ammerlaan, Phys. Rev. B **30**, 4564 (1984)
3. W. Gehlhoff, K. Irmscher, J. Kreissl, Lecture Notes in Physics, **301**, 262 (1988)
4. W. Gehlhoff, K. Irmscher, U. Rehse, Material Science Forum, **38-41**, 373 (1989)
5. L.C. Kimerling, M.T. Asom, J.L. Benton, P.J. Drevinsky, C.E. Caefer, Materials Science Forum **38-41**, 141 (1989)
6. L.F. Mollenauer, P. Pan, Phys. Rev. B 6, 772 (1972)
7. F.J. Ahlers, F. Lohse, J.-M. Spaeth, L.F. Mollenauer, Phys. Rev. B **28**, 1249 (1983)
8. A. Görger, B.K. Meyer, J.-M. Spaeth and A.M. Hennel, J. Phys. Semicond. Science and Technol. **3**, 832 (1988)
9. O. Gunnarsson, O. Jepsen, O.K. Andersen, Phys. Rev. B **27**, 7144 (1983)
10. F. Beeler, O. K. Anderson, M. Scheffler, Phys. Rev. B **41**, 1603 (1990)
11. H. Overhof and G. Corradi, this volume
12. M. Scheffler: Festkörperprobleme XXIX, ed. by U.Rössler,(Vieweg Verlag, 1989),p. 231
13. L.V.C. Assali, J. R. Leite: Proc. 20th Conf. on Physics of Semicond., World Scientific Publ. 1990, Vol. **1**, p. 597

Materials Science Forum Vol. 83-87 (1992) pp. 155-160

< 100 > AND < 111 > CONFIGURATIONS OF IRON-ACCEPTOR PAIRS IN SILICON RELATED TO STABLE AND METASTABLE STATES

HIDEKI TAKAHASHI, MASASHI SUEZAWA AND KOJI SUMINO
Institute for Materials Research, Tohoku University, Sendai, 980, Japan

ABSTRACT

The generation process of Fe-acceptor pairs in Si was investigated by means of ESR measurements in order to clarify the relation between their stable/metastable state found by DLTS and the < 100 >/< 111 > configuration found by ESR. The energy differences between the < 100 > and the < 111 > configurations of Fe-Al, -Ga, and -In pairs are determined to be 0.07 ± 0.01, 0.03 ± 0.01, and 0.09 ± 0.01 eV, respectively. These values are in good agreement with those determined by DLTS.

1. Introduction

DLTS studies have shown that a pair of Fe and a group III acceptor, such as Fe-Al, Fe-Ga and Fe-In, in Si has both stable and metastable states and that determined energy differences between the two states are 0.07 [1,2], 0.04 [3] and 0.13 eV [3] for Fe-Al, Fe-Ga and Fe-In pairs, respectively.
On the other hand, ESR studies clarified that Fe-acceptor pairs have two structural configurations with < 111 > trigonal and < 100 > orthorhombic symmetries [4 − 7], except for Fe-B which has only < 111 > trigonal symmetry [8,9].
The purpose of our paper is to clarify the relation between the two configurations and the two states. We propose here a method to obtain the energies related to the stable/metastable states from the analysis of generation kinetics of defect complexes investigated with ESR measurements.

2. Experiment

Crystals used in this experiment are shown in Table 1. All of them were grown by the float-zone method. Specimens were of a rectangularshape, $3.7 \times 3.2 \times 2.7$ mm^3 in size, with edges parallel to the [100], [011] and [011]. After chemical polishing with CP-4, the specimen was sealed into an evacuated quartz capsule together with high purity Fe wire. After being annealed for 2 hours at 1300℃, the specimen was quenched in ice-water by breaking the capsule and was kept in liquid nitrogen until ESR measurement was done since Fe atoms in Si were found to diffuse even at room temperature. The concentrations of Fe and Fe-acceptor pairs were determined from the absorption intensities of ESR at 10 K with a JEOL X-band spectrometer.

3. Experimental results

Table I Crystals used in experiments.

Dopant	Concentration (cm^{-3})	Type	Resistivity ($\Omega \cdot cm$)
Al	$\sim 2 \times 10^{15}$	p	5~7
Ga	$\sim 5 \times 10^{15}$	p	2~3
In	$\sim 3 \times 10^{15}$	p	4~6

The ESR spectra related to Fe^+, Fe^0 and Fe-acceptor pair were observed. In the case of Fe-Al pairs, one absorption line related to trigonal pairs is detected at 1685 G and two lines related to orthorhombic pairs at 1100 and 4500 G when the magnetic field is parallel to the [100] direction. We denote these lines with P_1 (1685), P_2 (1100), and P_2 (4500) in the following.

Fe-Ga and Fe-In pairs also give rise to three absorption lines in the same way as Fe – Al pairs do, but at slightly different positions.

In order to find the temperature range in which the generation or the dissociation of Fe-acceptor pairs takes place, isochronal annealing experiments were conducted. The line intensities of the Fe-acceptor pairs first increase with the increasing temperature and attain the maxima at 100℃ as shown in Fig. 1 for the case of Fe-Al pairs. Then, they decrease with temperature and diminish completely at 200℃. Comparing the temperature dependencies of ESR intensities of Fe^+, Fe^0 and Fe-acceptor pairs, we find that when the intensities of Fe-acceptor pairs attain the maxima at 100℃, the intensities of Fe^+ and Fe^0 show the minima decreasing from the low temperature side. It means that the concentration of Fe atoms decreases on developing Fe-acceptor pairs. Intensities of Fe^+ and Fe^0 increase up to about 150℃. Such increases in the concentrations of Fe^+ and Fe^0 are due to the dissociations of the Fe-acceptor pairs.

In order to follow the generation process of Fe-acceptor pairs in detail, we investigated how the concentrations of Fe-acceptor pairs vary with the time in isothermal annealing. Figure 2 shows the generation behavior of Fe-Al pairs due to isothermal annealing at 40℃. The intensities of Fe-Al pairs increase monotonously with time and attain the constant value after annealing for long time. Essentially similar isothermal behavior is observed with Fe-Ga and Fe-In pairs.

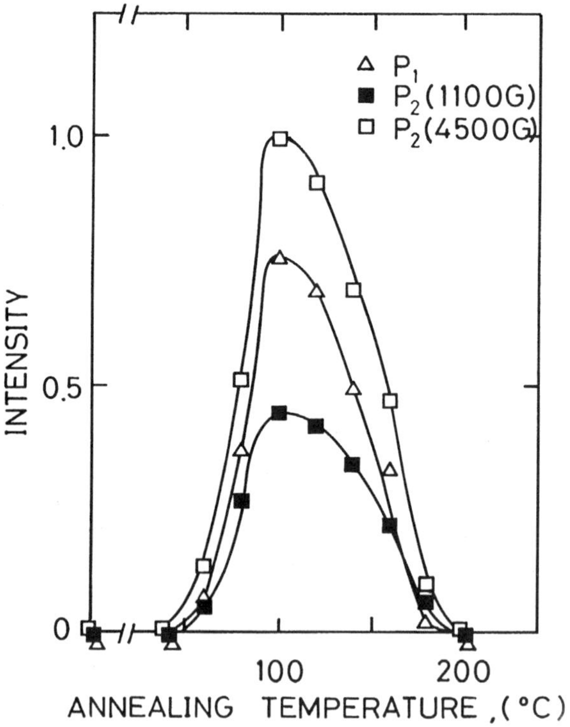

Fig. 1 Variations in the concentrations of Fe-Al pairs in isochronal annealing for 10 min at each temperature.

4. Analysis

In this section, we give the analysis of generation process of

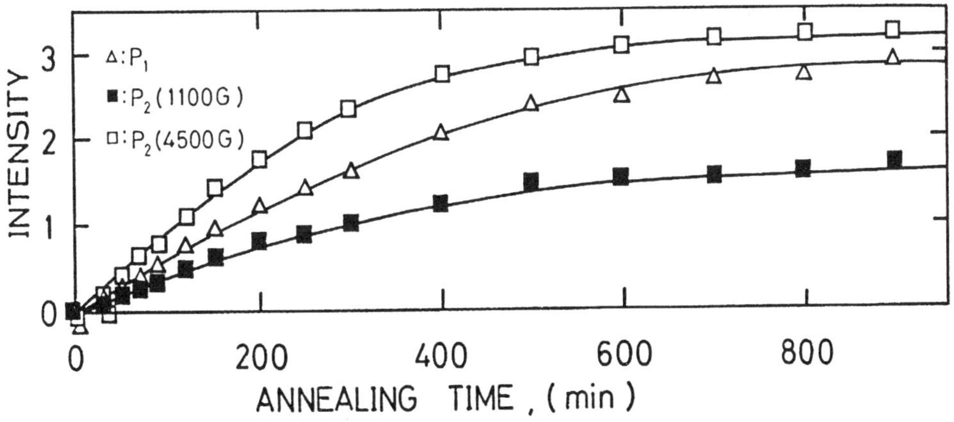

Fig. 2 Generation of Fe-Al pairs in isothermal annealing at 40℃.

Fe-acceptor pairs in silicon. In the case that the annealing temperature are around 40 − 80℃ substitutional acceptor atoms do not diffuse. Therefore, the pairing process is controlled by the diffusion of Fe atoms toward acceptor dopants. We describe the generation of Fe-acceptor pairs with the use of chemical reaction rate theory. The pairing process is expressed as follows :

$$F + A \underset{K_{-2}}{\overset{K_2}{\rightleftarrows}} P_2 \quad , \tag{1}$$

$$P_2 \underset{K_{-1}}{\overset{K_1}{\rightleftarrows}} P_1 \quad , \tag{2}$$

where F, A, P_1 and P_2 denote Fe^+, ionized acceptor, the first and the second nearest neighbor pairs of Fe and acceptor, respectively, together with their concentrations. All acceptors are thought to be ionized because the concentration of Fe is higher than that of acceptors in our case. K_n and K_{-n} (n = 1, 2) are the forward and backward reaction constants, respectively. We can express the time dependencies of the concentrations of P_1 and P_2 by following equations :

$$dP_1/dt = K_1 P_2 - K_{-1} P_1 , \tag{3}$$

$$dP_2/dt = K_2 F A - K_{-2} P_2 - K_1 P_2 + K_{-1} P_1 , \tag{4}$$

where

$$F = F_0 - P_1 - P_2, \quad A = A_0 - P_1 - P_2 \ , \tag{5}$$

$$F A = F_0 A_0 - (F_0 + A_0)(P_1 + P_2) + (P_1 + P_2)^2 \ . \tag{6}$$

F_0 and A_0 are the initial concentrations of Fe^+ and acceptors. Solutions of eqs.(3) and (4) are obtained under following assumptions : First, the term of $(P_1+P_2)^2$ term in eq.(6) is neglected since it is small in comparison with other terms. Second, K_{-2} is taken to be zero. Then, the ratio P_2/P_1 is obtained to be

$$P_2/P_1 = K_{-1}/K_1 + [ab/K_1(b-a)]\,[\exp(-at) - \exp(-bt)] \,/\, [1-\exp(-at)], \tag{7}$$

where

$$a = K_2\,[F_0 + A_0], \qquad b = K_1 + K_{-1}\,. \tag{8}$$

The first term on the right hand side of eq. (7) is a constant which depends only on the temperature, and the second term is a decreasing function of t that goes to zero as $t \to \infty$ as the system approaches the thermal equilibrium. In the equilibrium

$$P_2/P_1 = K_{-1}/K_1\,. \tag{9}$$

This ratio is given by the following equation if there is a energy difference of ΔE between the states P_1 and P_2.

$$P_2/P_1 = \exp(-\Delta E/kT)\,. \tag{10}$$

Therefore, we can determine the magnitude of ΔE from the dependence of this ratio on the annealing temperature for various kinds of Fe-acceptor pairs. Figure 3 shows P_2/P_1 for Fe-Al pairs in isothermal annealing at 40℃. In the beginning stage of annealing the ratio decreases with the time and becomes constant at long annealing time. This behavior coincides with what is expected from eq. (7).

Figure 4 shows the dependencies of the equilibrium values of P_2/P_1 for Fe-Al, Fe-Ga and Fe-In pairs on the temperature. From the slopes of the experimental straight lines in this figure, we recognize that the energy of P_1 state is lower than that of P_2 state in the case of Fe-Al and Fe-Ga pairs,

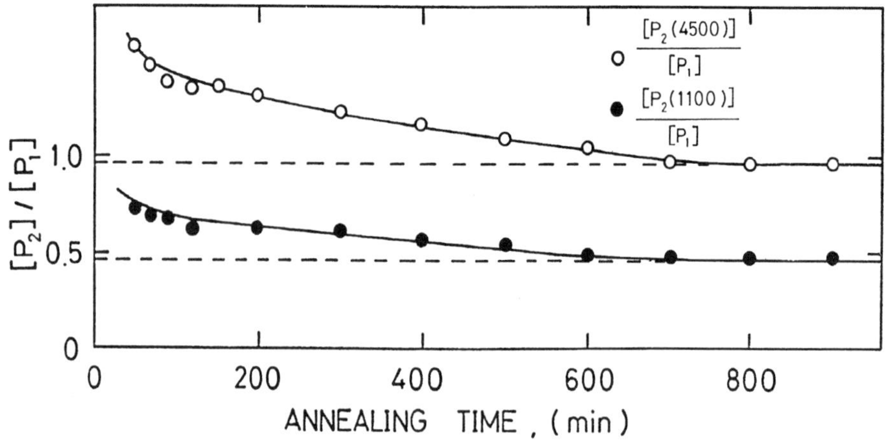

Fig. 3 Variations of P_2/P_1 related to Fe-Al pairs in isothermal annealing at 40℃.

and, in contrast, the energy of P_2 is lower than that of P_1 in the case of Fe-In pairs. The energy differences ΔE between P_1 and P_2 states of Fe-Al, Fe-Ga and Fe-In pairs are determined to be 0.07 ± 0.01, 0.03 ± 0.01 and -0.09 ± 0.01 eV, respectively. These values are in good agreement with the energy differences between the stable and the metastable states determined with DLTS measurements. Thus, we have succeeded in determining the relation between the stable/metastable states reported from DLTS measurements and the trigonal / orthorhombic symmetry structures reported from ESR measurements for Fe-acceptor pairs in Si. The trigonal structure is stable and the orthochombic structure metastable for Fe-Al and Fe-Ga pairs while the reverse holds for Fe-In pairs. The reason for such difference may be attributed to the strain effect originating from the difference in the bond lengths of the acceptors [10].

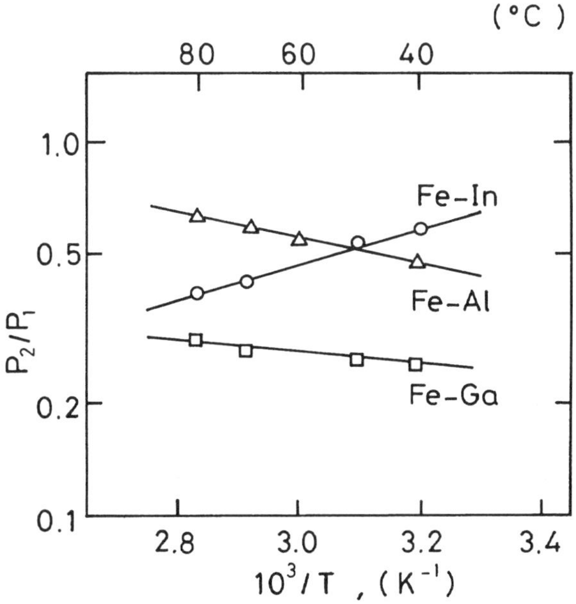

Fig. 4 Dependence of equilibrium values of P_2/P_1 for Fe-Al, Fe-Ga, and Fe-In pairs on temperature.

The generation process of trigonal Fe-acceptor pairs in isothermal annealing is observed to be well described with the following equation :

$$P_1 \propto 1 - \exp(-J_1 t). \qquad (11)$$

This coincides with an approximate solution of eq.(3) and (4), in which J_1 is equal to $K_1 + K_{-1}$. As shown above, the difference in the activation energies between K_1 and K_{-1} is rather small. So, we may take the activation energy of J_1 to be equal to that of K_1. The activation energy of K_1 is determined by fitting eq.(11) to the P_1 versus t relations measured at various annealing temperatures. Such activation energy of K_1 is interpreted to be the height of potential barrier for interstitial Fe^+ for a jumping from the second nearest to the first nearest interstitial site adjacent to an acceptor atom. We determined the values of the potential barrier for Fe-Al, Fe-Ga and Fe-In pairs to be 0.75 ± 0.01, 0.72 ± 0.01 and 0.69 ± 0.01eV, respectively.
The activation energy of the diffusion of Fe^+ in the Si lattice can be determined from the temperature dependence in the generation behavior of P_2. This quantity is determined to be 0.68 ± 0.01eV. Small discrepancies between this and the potential barrier between the first nearest and the second nearest interstitial site determined above may show that the interaction extends between a Fe atom and an acceptor atom located even at the second nearest site.

5. Conclusion

The following conclusions are obtained from the present work.

(1) The trigonal configuration of Fe-Al and Fe-Ga pairs and the orthorhombic configuration of a Fe-In pair are stable while the orthorhombic configuration of Fe-Al, Fe-Ga pairs and the trigonal configuration of a Fe-In pair are metastable. The energy differences between these two states are 0.07 ± 0.01, 0.03 ± 0.01 and 0.09 ± 0.01eV for Fe-Al, Fe-Ga and Fe-In pairs, respectively. These values are in good agreement with the results obtained with DLTS, 0.07, 0.04 and 0.13eV.

(2) The heights of potential barriers for interstitial Fe^{+} for jumping from the second nearest to the first nearest interstitial site adjacent to acceptors, Al, Ga and In are 0.75 ± 0.01, 0.72 ± 0.01 and 0.69 ± 0.01eV, respectively.

(3) The activation energy for diffusion of Fe^{+} in Si lattice is 0.68 ± 0.01eV.

References

[1] A.Chantre and D.Bois : Phys. Rev. B31 (1985) 7979.

[2] A.Chantre : *Mat. Res. Soc. Symp. Proc.* Vol. 104 (1988) 37.

[3] A.Chantre and L.C.Kimerling : *Materials Science Forum*, ed. J. van Bardeleben (Trans Tech Publ., 1986) Vol.10 – 12, p387.

[4] J.J. van Kooten, G.A.Weller and C.A.J.Ammerlaan : Phys. Rev. B30 (1984) 4564.

[5] W.Gehlhoff, K.Irmscher and J.Kreissl : *New Developments in Semiconductor Physics* (Springer, Berlin, 1988) 262.

[6] P.Omling, P.Emanuelsson, W.Gehlhoff and H.G.Grimmeiss : *Materials Science Forum*, ed. G.Ferenczi (Trans Tech Publ., 1989) Vol.38 – 41, p445.

[7] P.Omling, P.Emanuelsson, W.Gehlhoff and H.G.Grimmeiss : Solid States Commun. 70 (1989) 807.

[8] W.Gehlhoff and K.H.Segsa : Phys. Stat. Sol. (b) 115 (1983) 443.

[9] M.Suezawa and K.Sumino : *Defect Control in Semiconducters*, ed. K.Sumino (North-Holland, 1990) Vol.1, p335.

[10] L.C.Kimerling, M.T.Asom, J.L.Benton, P.J.Drevinsky and C.E.Caefer : *Materials Science Forum*, ed. G.Ferenczi (Trans Tech Publ., 1989) Vol.38 – 41, p141.

Materials Science Forum Vol. 83-87 (1992) pp. 161-166

COPPER RELATED DEFECT REACTIONS IN P-TYPE SILICON

A. MESLI and T. HEISER
CRN (IN2P3), Laboratoire PHASE (UPR n° 292), BP 20, F-67037 Strasbourg Cedex, France

ABSTRACT

We report on copper related defect reactions in p-type silicon. Diffusion of Cu followed by fast quenching yields Cu-related defects and acceptor passivation. One of the observed defects, labeled 'D', is shown to be unstable below room temperature. Reaction kinetic analysis reveal a correlation between the binding energy of dopant-copper pairs and the activation energy of the unstable defect. We show that its metastable behavior, can be understood as a pairing reaction between D and free copper atoms, the DCu complex being electrically inactive. The available concentration of mobil copper depends on the acceptor-copper pairing and explains the dopant dependent association frequency.

1. Introduction

Recently, passivation of acceptors in silicon after chemomechanical polishing at room temperature in a copper containing chemical solution was observed.[1] This passivation could be explained by a pairing mechanism involving the dopant and a fast diffusing X defect.[1] The extrapolation to high temperatures of its diffusion coefficient was shown to be higher than all known impurity diffusion coefficients in silicon.[2] Prescha et al.[3] and Prigge et al.[4] found however a strong correlation between copper and the X defect. Very recently, using the PAC technique, Keller et al.[5] identified the X defect as interstitial copper. As a major consequence of its high diffusivity, copper is unstable at room temperature and is expected to be involved in numerous defect reactions. In this paper we report observations of a Cu - related complex revealing a metastable behavior. Its complete electrical characterization helps us to understand the mechanism underlying the observed metastability.

2. Experimental conditions

The samples used in this study were prepared from boron and gallium-doped silicon, respectively float zone (FZ) and Czochralski (CZ)-grown. Doping levels are in the range 0.4 - 1.5 10^{15} cm^{-3}. Preliminary DLTS characterization did not detect levels up to $10^{11} cm^{-3}$. After cleaning and etching the substrates, 2000 Å of high purity (6N) copper was evaporated on both sides. The samples were then annealed in argon atmosphere at 1330 K for 30min. Afterwards, they were quenched in liquid nitrogen (LN2), chemically etched and re-stored in LN2 before and after Schottky barrier realization. The properties of the unstable defects and their transformation kinetics were studied using C-V and DLTS measurements.[6] Possible non-uniformity of the dopant is taken into account.

3. Experimental results

Capacitance-voltage measurements on the diodes show a large neutralization of the boron and gallium through the whole material, as illustrated in figure1. The electrical inactivity starts to disappear above 200 K under reverse voltage. The high electric field, directed toward the bulk, sweeps out the neutralizing mobile Cu_i^+ and increases the passivated dopant concentration in the neutral region. It appears that five minutes at 200 K are enough to completely reactivate the boron in the high field region, while 240 K are needed to reach the same result in the Ga doped silicon within the same space of time. Hence, the Ga_s-Cu_i binding energy is higher compared to the B_s-Cu_i complex, in agreement with the data obtained after chemomechanical polishing.[3] The reverse reaction or passivation is possible below 300 K if we switch off the bias.

The behavior of copper can be monitored through the dissociation - association reaction given by :

$$A_s^-Cu_i^+ \Longleftrightarrow A_s^- + Cu_i^+ \qquad (1)$$

where A_s^- represents the substitutional dopant atom. This reaction is governed by a simple Coulombic attraction between two oppositely charged ions. When a high electric field is applied, the complex is dissociated and reaction (1) is driven to the right, depleting the superficial region from Cu_i^+. The distribution of this impurity becomes nonuniform (see figure 1) and the reverse reaction (annealing under 0V) is influenced by a long range diffusion process. Mathematically, reaction (1) is described by a system of two coupled equations :

$$\frac{\partial[\mathrm{Cu}]}{\partial t} = D_f \frac{\partial^2[\mathrm{Cu}]}{\partial x^2} - \frac{d[\mathrm{ACu}]}{dt} \qquad (2)$$

$$\frac{d[\mathrm{ACu}]}{dt} = c_a([A_o]-[\mathrm{ACu}]) - \frac{[\mathrm{ACu}]}{\tau_a} \qquad (3)$$

where $c_a = 4\pi R_a D_f [\mathrm{Cu}]$ is the capture rate of Cu_i^+ by the acceptor atom or association frequency, $[A_o]$ the total acceptor concentration and $1/\tau_a$ the dissociation frequency. The pairing depends on the Cu_i^+ diffusion coefficient D_f, the capture radius R_a of the fixed dopant A^- and the available concentrations [Cu] and $[A_o]$. This system of coupled equations allowed Zundel et al.[1] to determine the diffusion coefficient of Cu_i^+ at low temperature.
Under high electric field, the region under investigation is depleted from Cu^+ and the first term in equation (3) can be neglected. A first order dissociation reaction is thus expected and experimentally observed. The dissociation frequency of the pair Ga_s-Cu_i is found to be 1.5 10^{13} exp (- 0.73 eV/KT), which is very close to the value found by Prescha et al.[3] for Ga_s - X. The activation energy of 0.73 eV is higher by about 0.1 eV to the corresponding value[3] for B_s- Cu_i as expected from figure 1.

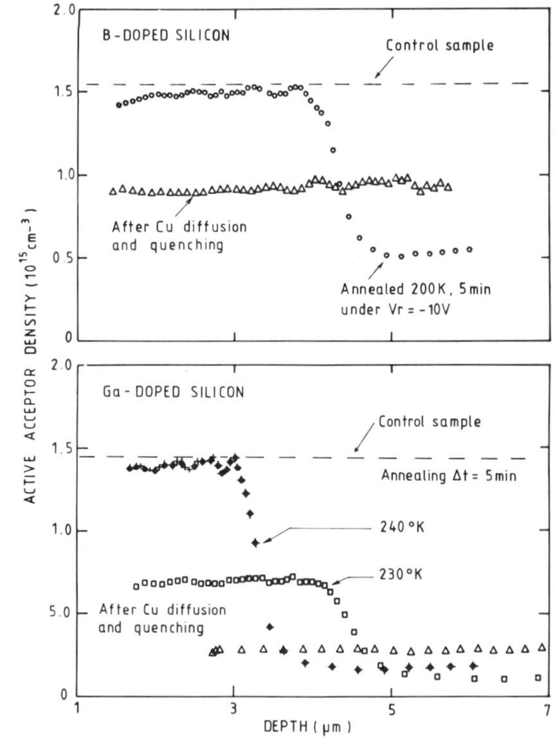

Figure 1 : Electrically active acceptor densities as a function of the reverse bias annealing conditions.

In the course of studying a possible role of additional defects, DLTS measurements have been done. Typical spectra observed in both kinds of samples are shown in figure 2. Two hole traps labelled Ha (E_v + 0.085 eV) and Hb (E_v + 0.210 eV) are observed. Control samples quenched in the same conditions but without intentionnal contamination with copper do not show these centers. This establishes their copper related nature and dopant independence. Oxygen does not seem to play any role in the generation of these defects as well. The first level has already been seen by Graff et al.[7] and was later identified by photoluminescence by Weber et al.[8] as the Cu-Cu pair. This level has a donor like character with a constant capture cross section of 3.5 10^{-15} cm^2 and is annealed above 420 K.

The deeper level Hb, which will retain our attention below, has a concentration of 5 10^{13} cm^{-3} in boron doped silicon. Its density is systematically twice in gallium doped materials. The junction electric field enhanced emission rate, consistent with the Poole-Frenkel[9] effect, specifies its acceptor-like character (-/0). The hole capture cross section is found to be 1.5 10^{-15} cm^2. The pecularity of this defect is its metastable behavior. Actually, the DLTS signal is only seen if the sample is cooled down with an applied reverse bias (full curve). When cooled down without bias (dashed curve), it cannot be detected. Moreover, annealing kinetics show the change from the electrically active to the electrically inactive configuration, and vise versa, to follow an exponential law. One could thus be tempted to relate these observations to a charge state driven configurational bistability as already reported for other defects.[10] However, such a model cannot explain all the observations and should be excluded for two main reasons. i) In gallium doped silicon, the level appears (cooling with reverse bias) and disappears (cooling at zero volt) in a temperature range (260 K-290 K) where it always lies below the Fermi level. The resulting invariant charge state should however forbid a charge state driven metastability. ii) The metastable behavior is no longer observed when the sample is kept at room temperature for several days. These results suggest, as alternative mechanism, a simple pairing reaction between an unknown defect D, responsible for the Hb level in the gap, and the fast diffusing Cu_i^+, the DCu_i^+ complex being electrically inactive. Within this description, the same reasoning as above can be applied, i.e. a thermally activated dissociation-association reaction according to a Coulombic interaction between two oppositely charged ions :

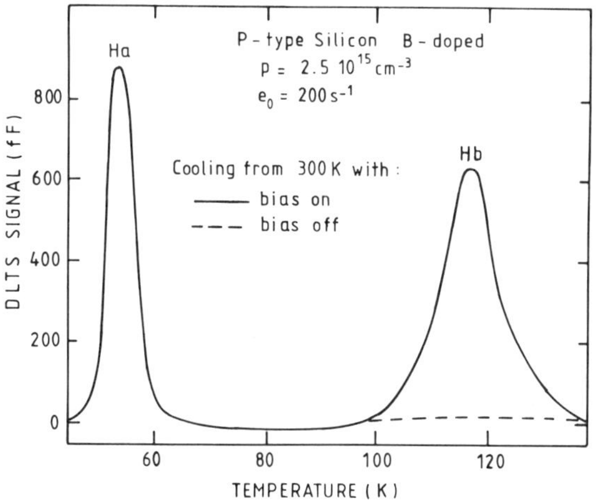

Figure 2 : DLTS spectrum showing the Cu-related defects.

$$D^-Cu_i^+ \quad \Longleftrightarrow \quad D^- + Cu_i^+ \tag{4}$$

As for the acceptor-copper pairing, the equation describing reaction (4) is given by :

$$\frac{d[DCu]}{dt} = c_d (D_o - [DCu]) - \frac{[DCu]}{\tau_d} \tag{5}$$

where all the parameters have the same meaning as above, only the index has been changed reminding their relation to the D defect. Equation (2) is still valid as the D concentration is well below the doping level. Again, under a high electric field, the thermal dissociation of $D^-Cu_i^+$ is followed by a drift toward the bulk of Cu_i^+, leading to the observed first order behavior. More surprising is the first order kinetics obtained when reaction (4) is driven to the left (0V). Indeed, similarly to the acceptor-copper pairing described above, we have to consider the diffusion process as a possible rate limiting factor. Below, a careful analysis of the results will help us to explain this unexpected behavior.

The isothermal annealing data shown in figure 3 are consistent with the relations :

$$1/\tau_d = 1.55\ 10^{14}\ \exp(-0.970\ eV/KT) \qquad \text{(both samples)} \tag{6}$$

$$c_d = 7.50\ 10^{12}\ \exp(-0.790\ eV/KT) \qquad \text{(B doped silicon)} \tag{7}$$

$$c_d = 4.60\ 10^{13}\ \exp(-0.905\ eV/KT) \qquad \text{(Ga doped silicon)} \tag{8}$$

The dissociation frequency $1/\tau_d$ of the pair D^-Cu^+ is dopant independent but the capture rate of Cu_i^+ by D^-, $c_d = 4\pi R_d D_f [Cu]$, differs from boron to gallium doped material.
On the basis of the spectroscopic data described above, and the unique dissociation frequency of the D^-Cu^+ complex in both kinds of doped silicon, we can expect that neither boron nor gallium are involved in the observed defect. However, the examination of the capture rates given by relations (7) and (8) can hardly exclude a role of the dopant in the association kinetics.

To account for these two experimental facts, we must clearly define the role of the different reacting particles. They are of two types : i) the fixed charges, boron respectively gallium, and the defect D, ii) the mobile copper which has a concentration gradient from the surface to the bulk before pairing takes place (see figure 1). The latter is part of both reactions (1) and (4), which do interfere. The association kinetics D^- + Cu_i^+ could be numerically solved by finding a solution of the three coupled equations (2), (3) and (5) with appropriate boundary conditions. This approach seems however rather complicated and even useless. Indeed, the exponential decrease of electrically active D concentration, observed experimentally, indicates that some reasonable simplifications can be made.

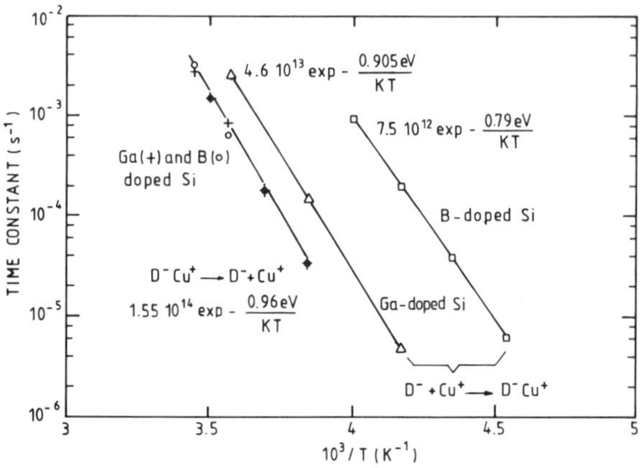

Figure 3 : dissociation and association frequencies of DCu in both B and Ga-doped silicon.

When cooling down the sample under reverse bias, the level Hb is not detected in the neutral region, as shown in figure 4. This indicates that equilibrium of reaction (4) corresponds to a negligible amount of free D defects. Consequently, when we switch off the bias, the dissociation of $D^-Cu_i^+$ ($1/\tau_d$) is negligible and reaction (4) is driven only to the left. To account for the exponential decay of the defect density, the association frequency c_d should be time invariant which leads to a constant density of free Cu. Moreover, an experimental trend is observed between $A^-Cu_i^+$ dissociation and $D^-Cu_i^+$ association barrier. The higher the binding energy of the $A^-Cu_i^+$ complex, the lower the $D^-Cu_i^+$ association rate. This would suggest that the rate limiting factor in the passivation of the defect is the availability of free Cu_i^+. The profiles of 'D' during destruction at ≈ 285 K and V_r = 0 V, not shown here, exhibit a parallel decrease of the density as the annealing time increases.

The following picture can account for all these requirements : Reaction (1), supposed at equilibrium in the neutral region, provides free Cu_i^+ which, when the bias is turned off, diffuses toward the surface and constitutes a limiting factor for reaction (4). Due to the large diffusivity of Cu_i^+, reaction (1) reaches very quickly the equilibrium in the previously depleted region. Indeed, in the temperature range considered for $D^-Cu_i^+$ association (240 K in boron doped Si and 270 K in gallium doped Si), the profiles of figure 1 are flat after only 5 minutes at zero bias, which is very short compared to the time required for a complete $D^-Cu_i^+$ association. Thus, the Cu_i^+ diffusion as a limiting factor is excluded. According to these arguments, we can write d[ACu]/dt = 0 during reaction (4). The substitution of [Cu], extracted from (3), into equation (5) yields the DCu complex association rate :

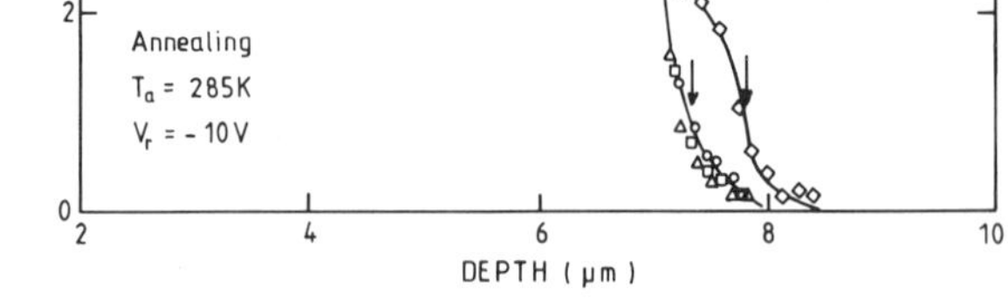

Figure 4 : 'D' defect profiles under different annealing conditions. The arrows show the limits of the depletion region.

$$c_d = \frac{R_d}{R_a} \frac{1}{\tau_a} \frac{[ACu]}{[A_o] - [ACu]} \qquad (9)$$

where $1/\tau_a$ is the acceptor-copper dissociation frequency. [ACu] is extracted from the C-V profiles in the bulk. Relation (9) agrees well with our experimental results. A linear shape is indeed obtained from the plot of c_d versus $[ACu]/(A_o-[ACu])$ as shown in figure 5. The slope is very close to the value given by relation (8) when we take R_d/R_a to be unity and independent on the nature of the dopant. Considering the attraction between D^-, respectively A^- and Cu^+ to be governed by long range coulombic forces, R_d and R_a are much higher than the corresponding ionic radius and should indeed be independent on the defect itself.

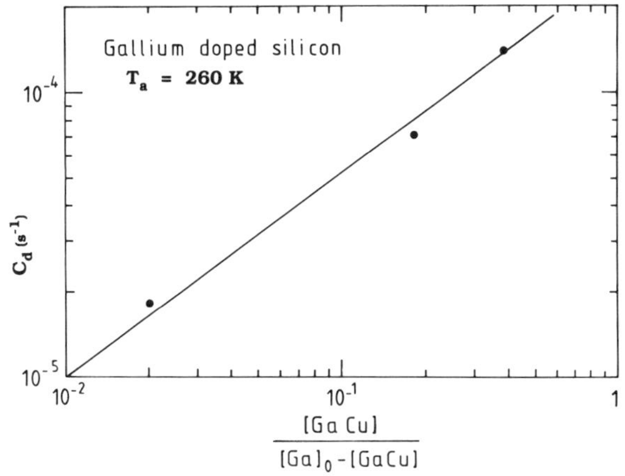

Figure 5 : DCu association rate as a function of the residual fraction of GaCu.

It remains however that we do not find the expected activation energies (0.69 eV in boron and 0.73 eV in gallium-doped silicon) but slightly higher values of about 0.15 eV. This deviation is due to a non negligible loss of free Cu_i^+ during isothermal annealing measurements. Indeed, the association kinetics data in figure 3 were plotted considering $[ACu]/([A_o]-[ACu])$ as constant. However, from C-V measurements, it could be seen that, from one isothermal annealing experiment to the next, this quantity was slightly varying. This loss of available copper followed roughly an exponential decrease with increasing annealing temperature, yielding an apparent energy barrier around 0.17 eV. Although the only physical meaning of this barrier is the irreversible loss of Cu during long time strorage, it is high enough to account for the above deviation. The model proposed here is supported by the fact that after long time storage at room temperature, when a complete irreversible depassivation of acceptors occurs, the source of free Cu vanishes and the defect looses its apparent reversible metastability. Moreover, when the annealing experiments are carried out at intermediate times, the [ACu] and thus the [Cu] concentrations reach lower critical values, below which the assumed simplifications do not hold. The kinetics deviates considerably from a first order behavior as the Cu_i concentration can no more be considered as constant during reaction (4).

Taking into account the sample preparation procedure, thermodynamic properties of copper in silicon as well as the above results, we can tentatively describe the origin of the D defect. Considering that interstitial copper has a pronounced room temperature instability, it vanishes during quenching mostly into precipitates in agreement with the solubility data.[12] In our fast quenching procedure a fraction is trapped by acceptors and by the generated D-defect which is by itself Cu related.The concentration of the defect is systematically twice in gallium doped substrates and in general, the fraction of passivated acceptors is higher in gallium than in boron doped silicon, according to the higher binding energy of the Ga-Cu complex. This would suggest that at a low density scale, the solubility of Cu_i^+ is increased in the presence of gallium. Thus, during the quenching, it is likely that a higher fraction escapes from gallium to form the observed D-defect. It follows that the dopant intermediates the defect formation. This would explain the difficulty to detect the D-center in n-type silicon[11] prepared in the same manner. As no pairing may occur in this material, the defect has at most a very small concentration. It seems reasonable to consider that the Cu precipitation generates clusters of increasing size, including the elementary Cu-Cu pair as well as the D-center. In the latter, binding with other structural defects is not excluded.

4. Conclusion

Copper is reported to be involved in defect reactions in silicon. The high diffusivity of Cu_i^+ allows pairs formation with p-type dopants as well as acceptor-like defects, all unstable well below room temperature. Neither dopants nor oxygen are found to be involved in the complex. The Hb level described here behaves similarly to the already reported[13] PtX center. The major differences with this last system is that the pair PtX is electrically active and the attraction between the two components are not pure Coulombic. Finally, in the light of our observations, we believe that the bistability attributed to the gold-copper complex detected by Czaputa et al.[14] has to be reconsidered. It can probably be understood within the framework of low bounded pairs. Numerous other unstable pairs including copper are expected from this work and references 13 and 14. These systems could be very fruitful in studying the influence of trapping phenomena on the diffusion of interstitial copper. It could lead to a better understanding of the large discrepancy between measured Cu diffusion coefficients at high[2] respectively low[3] temperatures.

Acknowledgments

The authors would like to thank Professors W. Zulehner and C.A.J. Ammerlaan for providing samples with controlled doping nature and concentrations. A. Mesli also thanks Dr.J. Weber and T. Prescha for the facilities of DLTS analysis below liquid nitrogen and fruitfull discussions on their pionnering work on the X defect. This work has been supported by G.C.I.S.

References

[1] T. Zundel, J. Weber and B. Benson, Appl. Phys. Lett. 53, 1426 (1988).

[2] R. H. Hall and J. H. Racette, J. Appl. Phys. 35, 379 (1964).

[3] T. Prescha, T. Zundel and J. Weber in Defects in silicon 1989, ed. by C.A.J.Ammerlaan, A.Chantre and P.Wagner, E-MRS Symposia Proceedings (North-Holland 1989) p.79.

[4] H. Prigge, P. Gerlach, P. O. Hahn, A. Schnegg and H. Jacob, J. Electrochem. Soc. 138, 1385 (1991).

[5] R. Keller, M. Deicher, W. Pfeiffer, H. Skudlik, D. Steiner and Th. Wichert, Phys. Rev. Lett. 65, 2023 (1990).

[6] A.Mesli, E. Courcelle, T. Zundel and P. Siffert, Phys. Rev. B36, 8049 (1987).

[7] K. Graff and H. Pieper, in Semiconductors Silicon 1981, Proceedings of the 4th International Symposium on Silicon Material and Technology, ed. by H. R. Huff and R. J. Kriegler (The Electrochemichal Society, Pennington, New Jersey, 1981), 81-5, p. 331.

[8] J. Weber, H. Bauch and R. Sauer, Phys. Rev. B 25, 7688 (1982).

[9] J. Frenkel, Phys. Rev. 54, 647 (1938).

[10] A. Chantre, Appl. Phys. A 48, 3 (1989).

[11] A. Mesli and T. Heiser to be published.

[12] E. R. Weber, Appl. Phys. A 30, 1 (1983).

[13] T. Prescha, Thesis Max Planck Institute of Stuttgart 1990 (Germany).

[14] R.Czaputa, Appl. Phys. A 49, 431 (1989).

Materials Science Forum Vol. 83-87 (1992) pp. 167-172

INTERACTION OF A COPPER-INDUCED DEFECT WITH SHALLOW ACCEPTORS AND DEEP CENTERS IN SILICON

TH. PRESCHA and J. WEBER
Max-Planck-Institut für Festkörperforschung, Heisenbergstr. 1, 7000 Stuttgart 80, Federal Republic of Germany

ABSTRACT

Copper incorporation into p-type Silicon generates the fast diffusing X defect. Several, electrically inactive complexes AX with acceptors A=B, Al, Ga, In are investigated. The X defect interacts also with deep defects (Pt, Au) by forming new, electrically active centers. The behavior of the PtX defect under zero and reverse bias annealing reveals features typical for metastable impurity complexes. However, the profiles of the defect concentrations can only be explained by a simple dissociation of the PtX center.

1. Introduction

Copper diffusion or polishing in a special, copper containing slurry generates the X defect in p-type Si [1,2]. The X defect exhibits a high diffusivity at room temperature and forms readily neutral complexes AX by passivating the shallow acceptors A=B, Al, Ga, In [3,4]. In the electric field of a reverse biased Schottky diode the X defect drifts like a positively charged particle which allows us to determine precisely the dissociation energy of the AX complexes [3,4]. Detailed information about the binding mechanism and the symmetry are achieved by the study of the vibrational properties of the BX complex [5] and the perturbed $\gamma\gamma$ angular correlation (PAC) of the InX complex [6,7]. From the obvious dependence of the X defect generation on the Cu contamination of the samples, an identification of the X defect with interstitial Cu was made [2,4,8]. However, an unambiguous proof of the Cu incorporation in the AX complexes is still missing.

In this paper we present new complexes of the X defect with the deep centers Pt and Au in p-type silicon. Contrary to the AX complexes the PtX and AuX centers are electrically active. The PtX complex exhibits in the space charge region of a Schottky diode an annealing similar to bistable defects. However, from the defect profiles a total dissociation of the PtX defect and a drift of the X defect in the electrical field of the space charge region explains correctly the annealing behavior.

2. Experimental

Silicon samples doped with B, Al, Ga, In in concentrations of 10^{14} to 10^{16} cm^{-3} are used in our experiments. The X defect is generated by Cu indiffusion or chemomechanical polishing according to Ref. [2]. The samples are stored in liquid nitrogen to prevent the decay of the X defect. The PtX complexes were generated in Pt diffused (850°C , 2h) boron doped wafers by chemomechanical polishing. Titanium contacts are evaporated to provide Schottky diodes. The ohmic contacts consist of an In-Ga alloy which is scratched onto the backside of the sample. Capacitance-Voltage (CV) profiling and DLTS measurements are performed.

3. Investigation of the Acceptor-X Complexes

The dissociation of the AX complexes is studied for all acceptors A=B, Al, Ga and In. In the following, we will describe the experiments by using the AlX pair as an example. The X defect

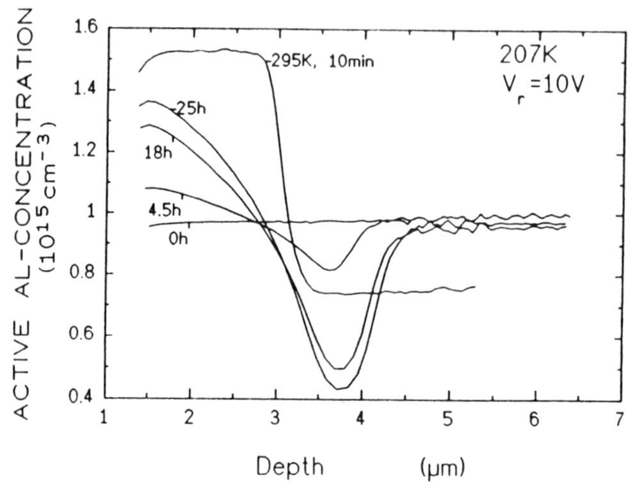

FIGURE 1: Concentration profiles of the electrically active aluminum concentration in a copper diffused sample after annealing under reverse bias (V_r = 10V). The final annealing is performed at a higher temperature to restore the original Al concentration.

is introduced into an Al doped sample ([Al] = 1.5 $\times 10^{15}$ cm^{-3}) by copper indiffusion at 850°C for 5min and subsequent quenching to room temperature in ethylene glycol. A reverse voltage of V_r=10V is applied at T = 207K to the Schottky diode for a certain time. Prior to the CV measurement the temperature is decreased to 77K by immersing quickly the sample holder into liquid nitrogen. The profiles of the electrically active Al concentration are given in Fig.1. Without annealing at 207K the electrically active acceptor concentration $N_A = 0.95 \times 10^{15}$cm^{-3} is uniform and lower than the total acceptor concentration $N_A = 1.5 \times 10^{15}$cm^{-3} of the original Al doped sample (Fig. 1: curve labeled 0h). The annealing under reverse bias causes a reactivation of the Al acceptor in the region 1.5μm$< x <$ 3μm. Reactivation under the above conditions is very slow and needs several days to destroy all AlX complexes in the space charge region. At higher temperatures (T=295K) complete reactivation in the space charge region ($x <$ 3μm) occurs already after 10min. An increased passivation of Al acceptors is detected in the region 3μm$< x <$ 4.5μm directly below the space charge region.

The change of the concentration profiles under the annealing conditions can be understood by the drift of the positively charged X defect. The AlX complexes dissociate and release the X defect according to AX $\rightarrow$ A$^-$ + X$^+$. The electric field of the space charge region drifts the positive X defect to the field free region adjacent to the space charge layer. In this region an increased AX formation occurs. The increase in the active acceptor concentration at x = 1.5μm follows first order kinetics indicating a negligible recapture of X defects by acceptors. The decay of the inactive acceptor concentration, which is identical with the AX complex concentration [AX] can be described by $[AX] = [AX](t = 0)\exp(-\nu t)$ with a dissociation frequency of $\nu = 4.5 \times 10^{-4}$s^{-1}. We determined the temperature dependence of ν by measuring the dissociation kinetics at different temperatures. Figure 5 summarizes the temperature dependence of the dissociation frequencies for all measured AX complexes.

4. X Defect in Pt-Diffused Samples

The X defect forms a complex with the Pt donor. Figure 2 shows DLTS-spectra of a Pt-diffused B-doped sample, which is chemomechanically polished to introduce the X defect. Before the polishing only the Pt donor ($E_t = E_v$ + 320meV) with a peak temperature of 155K is observed [Fig. 2(a)]. The signal height of the Pt donor decreases after the introduction of the X defect. A new DLTS peak of a donor like defect called PtX ($E_t = E_v$ + 560meV) appears at 242K. Annealing the sample at temperatures around 300K under reverse bias decreases the

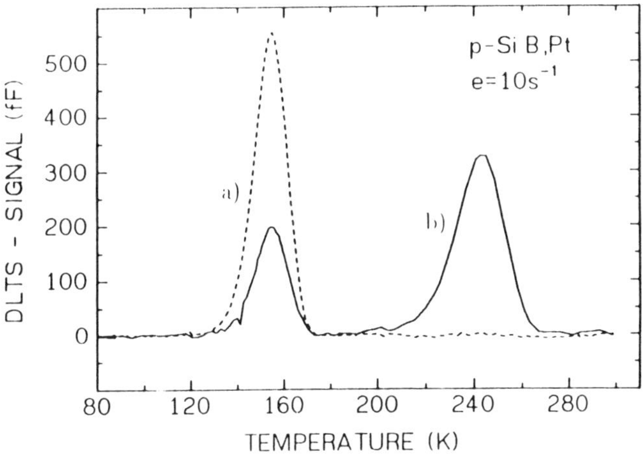

FIGURE 2: DLTS spectra of (a) Pt diffused B-doped silicon, (b) same sample as (a) after additional chemomechanical polishing.

PtX concentration and the Pt-donor signal increases. The sum of both DLTS-signals remains constant under all annealing conditions.

In section 6 we will discuss that the only explanation for the annealing behavior of the PtX complex is the dissociation of the X defect from the electrically active PtX complex: $PtX \rightarrow Pt + X$. The positively charged X defect drifts in the electrical field towards the zero field region below the space charge region. The dissociation kinetics of the PtX complex under applied reverse voltage is studied by measuring the Pt donor and the PtX concentration at different annealing times and annealing temperatures. The bias voltage during the DLTS measurements was smaller than the bias during annealing. Under these conditions the PtX complex is in the neutral charge states during annealing. The resulting dissociation frequencies for the PtX^0 defect at different temperatures are given in Figure 5.

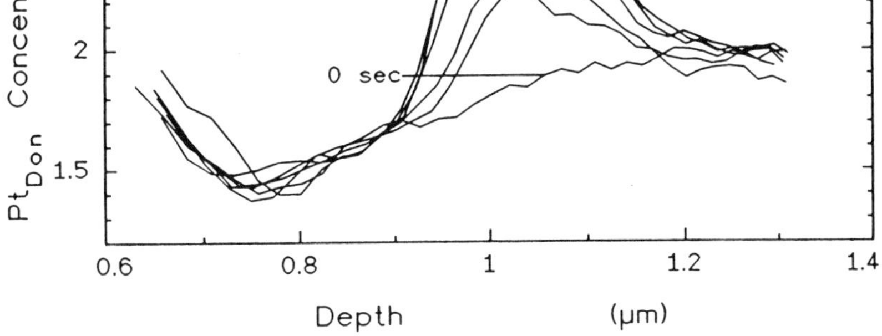

FIGURE 3: Profiles of the Pt donor in a Pt diffused and chemomechanically polished sample as a function of the reverse bias annealing time at 300K.

5. Concentration profiles of the Pt donor and PtX defect

Profiling the defect concentrations directly by DDLTS gives new insight into the PtX dissociation process. A Pt diffused B-doped sample is chemomechanically polished and annealed at $T = 300$K under a mediate reverse bias $V_r = 4$V. Figure 3 displays the change of the Pt-donor concentration profile with annealing. In the depth $0.9\mu m < x < 1.2\mu m$ the Pt-donor concentration increases within 120 sec to an equilibrium value. In the region $x < 0.9\mu m$ no increase in the Pt-donor concentration is detected. The region $x < 0.9\mu m$ coincides with the region which was accessible to the DLTS measurements reported on in the previous chapter. The zone where the strong changes appear, is just before the end of the space charge region $x = 1.2\mu m$. The zone extends in the space charge region to the depth, where the PtX defect is still occupied with a hole under the annealing conditions (PtX^+). The extend of this zone is marked λ in Fig. 4. A profile of the PtX defects shows that the increase in the Pt-donor concentration coincides with the decrease of the PtX concentration within the accuracy of the measurements. The dissociation of the PtX^+ defect is much faster compared to the PtX^0 defect in the previous chapter (see Fig. 5).

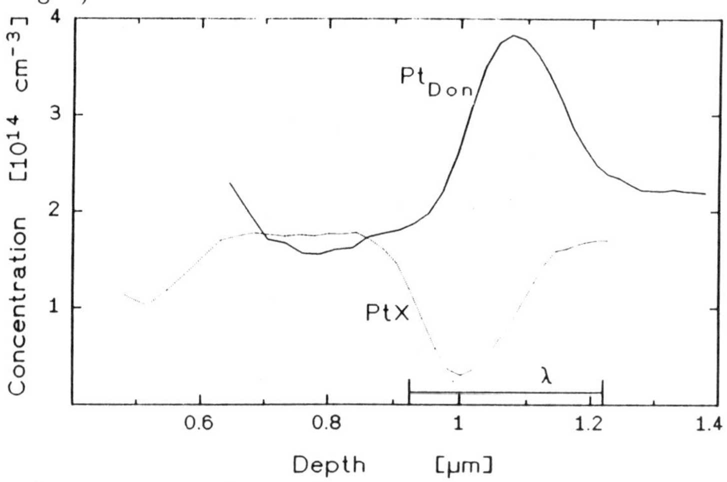

FIGURE 4: Concentration profiles for the Pt-donor and PtX defect after reverse bias annealing. λ marks the region at the end of the space charge region, where the PtX is still occupied by a hole.

6. Discussion

The X defect passivates shallow acceptors by forming stable AX complexes. The drift of the X defect in the electric field of the space charge region reveals the positive charge of the free defect. The dissociation of all X defect complexes follows first order reaction kinetics. The PtX defect shows annealing characteristics (Fig. 2) which are similarily detected in Au doped samples. After Cu contamination a new AuX defect is formed. In Ref. [9] the annealing of this AuX defect under reverse bias is attributed to a bistable behavior of the AuX defect. Our analysis of the annealing of the PtX defect in the λ region gives however evidence for a dissociation process and contradicts any bistable properties of the PtX complex. The conversion of bistable configurations (e.g. AlFe pairs [10]) is controlled entirely by the charge state of the defect. The PtX defect is positively charged in the λ region and the bulk during the annealing process and for a bistable defect the same annealing behavior would be expected in the two regimes. However, the PtX^+ defect changes only concentration in the λ region. Necessary for the first order reactivation of the Pt donor is the electric field in the λ region, which drifts the X^+ after dissociation of the PtX complex into the bulk of the sample. Our results on Au doped samples exclude the possibility of a bistable Au defect and support also the model proposed for the annealing of the PtX defect.

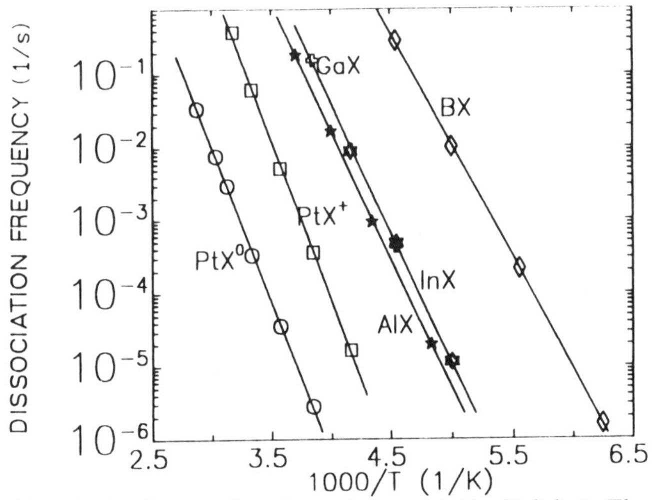

FIGURE 5: Dissociation frequencies of complexes with the X defect. The frequencies are well described by an Arrhenius law.

For the PtX complex two different dissociation processes can be determined. In the space charge region, where the PtX complex is still in the neutral charge state (PtX^0) under reverse bias annealing a slow dissociation is measured. However, in the λ region where the complex has trapped a hole (PtX^+) a fast dissociation occurs. The resulting dissociation frequencies are displayed in Figure 5.

All dissociation frequencies ν are thermally activated and follow the Arrhenius law $\nu(T) = \nu_0 \exp(-E_D/kT)$. Here ν_0 is the attempt frequency for dissociation and E_D the activation energy for dissociation. The values for the different complexes are listed in Table 1.

The PtX^0 and the PtX^+ defects show identical dissociation energies but differ strongly in the attempt frequencies ν_0. We are able to interpret the relative stability of the PtX^0 complex under the assumption that the PtX complex is an effective hole trap. The distribution of the different PtX charge states in the space charge region is governed by the equation

$$\frac{\partial[PtX^+]}{\partial t} = e_e[PtX^0] - e_h[PtX^+]$$

e_e and e_h are the emission rates for electrons and holes, respectively. The total PtX concentration is given by

$$[PtX] = [PtX^0] + [PtX^+]$$

which leads under thermal equilibrium to

$$[PtX^+] = \frac{\alpha}{1+\alpha}[PtX] \quad \text{with } \alpha = \frac{e_e}{e_h}.$$

The binding energy of the neutral PtX^0 complex should be much larger compared to the PtX^+ defect, due to the coulombic attraction of the Pt^- and the Cu^+ in this complex. Therefore the dissociation of the complex will occur preferentially via the PtX^+ complex.

	B	Al	Ga	In	PtX^0	PtX^+
$E_D(eV)$	0.61± 0.02	0.70± 0.02	0.71± 0.02	0.69± 0.02	0.85± 0.03	0.86± 0.03
$\nu_0(s^{-1})$	$2.6 \cdot 10^{13}$	$2.0 \cdot 10^{12}$	$8.0 \cdot 10^{12}$	$2.6 \cdot 10^{12}$	$7.6 \cdot 10^{10}$	$2.4 \cdot 10^{13}$

Table 1: Dissociation energies and attempt frequencies for complexes between the defect X and acceptors or Pt donor.

$$\frac{\partial[PtX]}{\partial t} = \frac{\partial[PtX^+]}{\partial t} = -\nu(PtX^+)\frac{\alpha}{1+\alpha}[PtX] = -\nu_{eff}[PtX]$$

For a hole trap is $\alpha \ll 1$ and the attempt frequency reduces to

$$\nu_{eff} = \nu_0(PtX^+)\alpha \exp\left(-\frac{E_D(PtX^+)}{kT}\right)$$

For the midgap PtX level the temperature dependence of α cancels and the PtX^0 complex will have the same dissociation energy as the PtX^+ complex. The effective attempt frequency $\nu_{eff} = \nu_0(PtX^+)\alpha$ is reduced compared to $\nu_0(PtX^+)$ by α, which can be estimated from the DLTS measurements to be $\alpha < 10^{-1}$.

It is a pleasure to thank H.J. Queisser for his interest in the work and T. Zundel for helpful discussions. We appreciate the technical assistance of W. Heinz and W. Krause.

References

[1] F.G. Vieweg-Gutberlet and P.F. Siegesleitner, in H.R. Huff and E. Sirtl (eds.) *Semiconductor Silicon 1977*, Electrochemical Society, Princeton, NJ, 1977, p. 387.

[2] H. Prigge, P. Gerlach, P.O. Hahn, and A. Schnegg: Ext. Abstracts 89-1, The Electrochemical Society, Inc., Pennington, NJ (1989) p. 372.

[3] T. Zundel, J. Weber, B. Benson, P. O. Hahn, A. Schnegg and H. Prigge, Appl. Phys. Lett. **53**, 1426 (1988).

[4] Th. Prescha, T. Zundel, and J. Weber, H. Prigge and P. Gerlach, Materials Science and Engineering, **B4**, 79 (1989).

[5] J. Hage, H. Prigge and P. Wagner, Appl. Phys. **A 50**, 241 (1990).

[6] Th. Wichert, M. Deicher, G. Grübel, R. Keller, N. Schulz, and H. Skudlik, Appl. Phys. **A48**, 59 (1989).

[7] U. Reislöhner, S. Deubler, P. Dohlus, D. Forkel, G. Pensl, H. Plank, H. Wolf, and W. Witthuhn, Mater. Sci. Eng. B **4**, 83 (1989).

[8] R. Keller, M. Deicher, W. Pfeiffer, H. Skudlik and D. Steiner, Phys. Rev. Lett. **65**, 2023, (1990).

[9] R. Czaputa, Appl. Phys. **A49**, 431 (1989).

[10] A. Chantre and B. Bois, Phys. Rev. B **31**, 7979 (1985).

Materials Science Forum Vol. 83-87 (1992) pp. 173-178

On the motion of iron in silicon at moderate temperatures under high electric fields.

T. HEISER, A. MESLI AND N. AMROUN
Laboratoire PHASE, CNRS, 23 rue du Loess, F67037 Strasbourg Cedex France

ABSTRACT:

Interstitial iron (Fe_i) in silicon is studied by means of capacitance measurements. In particular, the irreversible loss of Fe_i during low temperature annealing in n-type silicon was investigated and the influence of strong electric fields on the defect reaction was studied. From isothermal annealing kinetics and Fe_i depth profile measurements the behaviour of Fe_i is shown to be governed mainly by simple diffusion processes, the driving forces being either the concentration gradient close to the sample surface, or the electric field present in the depletion region.

1. Introduction:

Considerable motion of iron atoms in silicon can occur at moderate temperatures (below 500K) when they are dissolved in the matrix at interstitial sites. The corresponding high diffusion coefficient enables various defect reactions, as for example iron-acceptor pairing, which have been the subject of many interesting works.[1-3] Nevertheless, not all the observed mechanisms involving interstitial iron (Fe_i) are fully understood today.

In a recently published work[4], the irreversible loss of iron in p-type silicon at temperatures around 400K has been investigated by deep level transient spectroscopy. At these temperatures, the well known iron-acceptor pairs are completely dissociated and do not contribute to the loss of Fe_i. First order 'annealing' kinetics were found indicating a diffusion limited precipitation of Fe_i on extended defects. The role of the iron charge state (positive in p-type material and neutral in a depletion region as well as in n-type silicon) in these defect reactions was studied, measuring the precipitation kinetics in p-type silicon under reverse bias. A considerably faster loss of Fe_i was observed following a different temporal evolution. This behaviour was tentatively ascribed to diffusion limited precipitation of neutral Fe_i on growing nucleation centers.

In this work we report on the irreversible loss of Fe_i in n-type silicon. Confirmation of the charge state effect as well as the influence of strong electric fields on the defect reaction were looked for.

2. Experimental details:

For our investigation we used Czochralsky and float-zone grown n-type silicon, with a phosphorus concentration in the range of $0.5 - 5 \cdot 10^{15}$ cm^{-3}. After chemical cleaning, the samples were scratched on both sides with an ultrapure (6N) iron wire. They were then annealed in a sealed argon filled capsule at temperatures between 1200 and 1350 K for 30 to 120 minutes, and quenched in water. Semi-transparent gold dots, were evaporated to form Schottky diodes for electrical characterization.

Fe_i is known to introduce a donor level at 0.39eV from the valence band, which can be easily detected by infrared light induced optical transitions[5]. The resulting photocapacitance signal is proportional to the Fe_i concentration . This technique can also be used in a 'differential' mode in order to analyse the iron depth profile.[5]

Isothermal annealing kinetics have been recorded at temperatures below 470K, the sample being either zero or reverse biased (10V). In the latter case, global photocapacitance measurements have been performed with a reverse bias lower than 10V in order to analyse a region that was situated entirely in the depletion region during annealing.

3. Experimental results:

The results obtained under zero bias condition have already been reported in reference 5. Briefly, the measured kinetics were different from the behaviour observed in p-type silicon for both positively charged (under zero bias) and neutral Fe_i (under reverse bias). The loss of Fe_i was considerably slower following a 'non -first order' evolution. The results could however be successfully adjusted by considering the iron outdiffusion as the responsible mechanism. This was confirmed by the Fe_i depth distribution recorded after isothermal annealing and shown on figure 1. The solid lines are calculated error functions and are seen to fit extremely well the experimental results. The observations indicate that the stability of Fe_i^0 is higher compared to Fe_i^+, the major sink for Fe_i being the sample surface. This is correlated to the absence of long range coulombic forces for Fe_i^0.

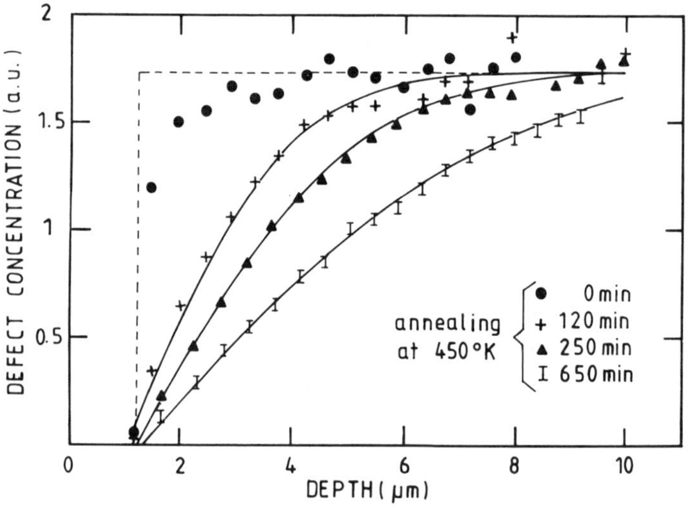

Figure 1: Depth profiles of Fe_i after various annealing times. The dashed curve was taken as boundary condition.

The Fe_i behaviour changes considerably when the samples are reverse biased during annealing. The corresponding isothermal annealing kinetics are shown on figure 2. The concentration of Fe_i decreases faster than without applied voltage and follows an exponential law. The 'loss' rate is found to increase strongly with the phosphorus concentration, as revealed by curve (4) in figure 2.

Figure 3 shows the iron depth profiles obtained after annealing at 400K, with an applied reverse bias (-10V), for various durations. The arrow indicates the edge of the depletion region during annealing. The results clearly show that the Fe_i concentration decreases uniformly over the entire depletion region. Finally, the 'reaction' underlying this loss of Fe_i was irreversible. Indeed, subsequent annealing at zero bias does not yield a uniform increase of the Fe_i concentration, but causes only a pure diffusion process, flattening the initially steep iron depth profile.

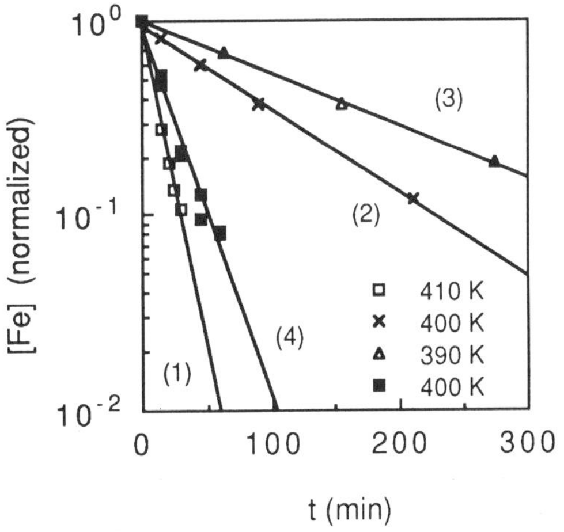

Figure 2: Isothermal kinetics of Fe_i. For curves 1 to 3 the P concentration was $\approx 5{\cdot}10^{14}cm^{-3}$, while for curve 4 it was $\approx 2{\cdot}10^{15}cm^{-3}$.

4. Discussion:

The above results are rather surprising as Fe_i is known to be neutral in both depletion and neutral regions, according to the position of its associated deep level. One could be tempted to explain the first-order kinetic by a simple pairing reaction of the type X + Fe -> XFe, involving Fe_i diffusion over microscopic distances. The dopant dependence of the annealing rate could then be accounted for by identifying the X component as Phosphorus. However, in this frame it seems difficult to understand the role of the electric field and / or the absence of free carriers. Moreover, reversibility of such a reaction should be possible, as the binding energy involved in a complex between a neutral and a donor impurity is expected to be low.

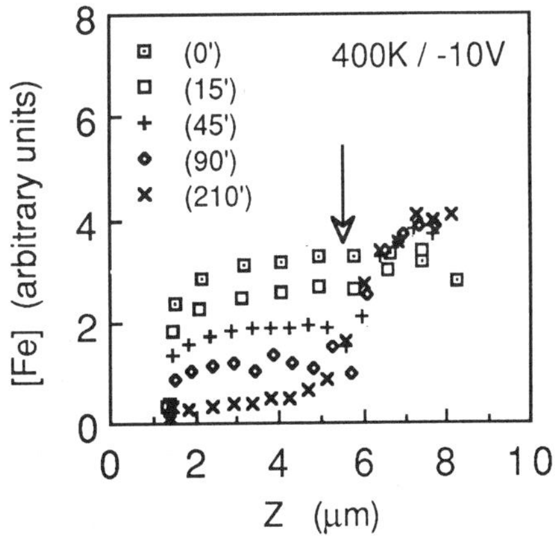

Figure 3: $[Fe_i]$ depth profiles after isothermal annealing under reverse bias

To understand the underlying mechanism, one has to remind, that in the depletion region not all of the iron atoms are neutral. Indeed, from the Shockley-Read-Hall statistics we know that in a high field region, where no carrier capture occurs, the charge state of a defect is completely determined by the electron respectively hole emission processes. For instance, the fraction f^+ of positively charged Fe_i in the depletion region is given by:

$$f^+ = Fe_i^+ / Fe_{itot} = e_n \cdot / (e_n + e_p) \qquad (1)$$

where e_p and e_n are the hole and electron emission rates, and Fe_{itot} the total Fe_i concentration.

For a donor level situated at ≈ 0.4eV from the valence band, having similar cross sections for electron (σ_n) and hole capture (σ_p), the ionized fraction at 400K would be around 0.01% which is usually considered as negligeable. However in the case of Fe_i, it is known that $\sigma_n >> \sigma_p$,[6,7] which rises the ionized fraction to about a few percent. This means that during the annealing under high electric fields, a non negligeable amount of iron is charged and may drift out of the depletion region. In the case of n-type silicon for example, Fe_i would drift towards the sample surface.

Considering this mechanism in a depletion region of a Schottky diode, it is easily shown that the decrease of iron in this region will follow an exponential law. As a matter of fact, the iron concentration is a solution of the diffusion equation:

$$\frac{\partial [Fe_i]}{\partial t} = D \frac{\partial^2 [Fe_i]}{\partial x^2} - \mu_{eff} \cdot \frac{\partial}{\partial x} \cdot \{[Fe_i]\, E(x)\} \qquad (2)$$

where D is the Fe_i^+ diffusion coefficient and E(x) the electric field.
μ_{eff} is equal to $\mu \cdot f^+$, with μ being the iron mobility. It is related to D through Einstein's relation: $D=\mu kT/q$. According to the temperature range investigated, equation (2) does not take into account any loss of iron in the bulk, due for example to precipitation.[5]

The numerical solution of this equation (not given here) shows the iron concentration to decrease uniformely in the depletion region, in agreement with the experimental data. The same result can however be obtained with the following argument. The initial Fe_i concentration is constant across the whole sample. Thus, at least at the beginning of the annealing, the Fe_i concentration gradient ($\partial[Fe_i]/\partial x$) in equation (2) can be neglected, yielding:

$$\frac{\partial [Fe_i]}{\partial t} = - \mu_{eff} \cdot [Fe_i] \cdot \frac{\partial E(x)}{\partial x} \qquad (3)$$

Using the expression of the electric field gradient in a depletion region for uniform doping N_d, equation (3) can be rewritten as:

$$\frac{\partial [Fe_i]}{\partial t} = - \mu_{eff} \cdot [Fe_i] \cdot \frac{qN_d}{\varepsilon} \qquad (4)$$

where ε is the dielectric constant of silicon. This yields an exponential decrease of the Fe_i concentration with the rate :

$$\tau^{-1} = \frac{q^2 N_d D e_n}{\varepsilon kT (e_n+e_p)} \qquad (5)$$

Finally as the rate is independent of depth, the Fe_i distribution will remain flat and equation (4) is valid even for larger times, provided the Fe_i concentration is evaluated far enough from the edge of the depletion region.

Equation (4) explains in a simple way all of the experimental results: - the first order kinetics, - the doping level dependence of the loss rate as described by expression (5), and - the irreversibility of the defect reaction. We tried to fit the experimental temperature dependence of the loss rate, using published data for the parameters in expression (5). The following expressions were used for the capture cross sections:

$$\sigma_p = 5.6\cdot 10^{-16} \exp(-0.048/kT) \quad [cm^2] \qquad (6a)$$

published by Brotherton et.al.[6] and

$$\sigma_n = 10^{-10} \cdot T^{-3/2} \quad [cm^2] \qquad (6b)$$

as published by Lemke[7].

In expression (6a), ΔE_h= 0.048eV is the activation energy for hole capture. The positively charged Fe_i diffusion coefficient was taken from reference (1) as:

$$D = 1.4\cdot 10^{-3} \exp(-0.69\ eV/kT) \quad [cm^2s^{-1}] \qquad (7)$$

According to expressions (5) and (6), and using $e_n << e_p$, the plot of $\ln(\tau^{-1}\cdot T^{5/2})$ over T^{-1}, for constant N_d, yields a linear curve. Its slope should be given by $\Delta E=\Delta E_h+\Delta E_D$, where ΔE_D is the energy barrier for Fe_i^+ diffusion. The experimental data shown on figure (4), was 'normalized' to a doping level of 10^{15} cm^{-3} using:

$$\tau_{norm}^{-1} = \tau_{exp}^{-1} \cdot 10^{15}/N_d$$

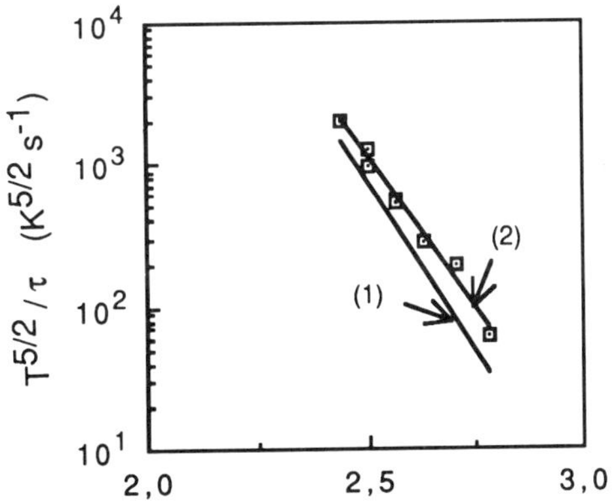

Figure 4: plot of the annealing rate τ^{-1}times $T^{5/2}$ as a function of $10^3/T$.

The expected linear dependence is clearly observed. Curve (1) in figure 4 was obtained by a least square fit of the data. The calculated values are represented by curve (2). Both agree within an acceptable limit, when taking into account that the data used for σ_n and σ_p may be slightly inappropriate. Indeed, they have been measured at a lower temperature range compared to our annealing experiment and vary significantly from one author to another.

In the light of our results, we can expect the same mechanism to happen in p-type material, as the positive fraction of Fe_i in the depletion region is independent of the Fermi level.

The difference, compared to n-type silicon, will be the direction of the electric field. Fe_i should thus drift towards the bulk, yielding a non uniform depth profile and a different rate of Fe_i loss. This is easily verified by profile measurements in p-type silicon.[8] Thus we believe that the in reference 4 reported results on iron precipitation under reverse bias, can be understood within this framework.

Conclusion:

To conclude, we can state that in n-type silicon, the behaviour of Fe_i at moderate temperatures is mainly governed by simple diffusion processes, the driving forces being either the concentration gradient close to the sample surface or the electric field present in the depletion region. The interaction with the electric field is limited by Shockley-Read-Hall statistics of carrier emission processes. Finally, we would like to point out that in operating power devices, strong electric fields and moderate temperatures are easily achieved, exposing these devices to similar defect motion and perhaps to related instabilities. As an example, interstitial iron should accumulate in the p-type region of a PN junction, according to the direction of the electric field in the depletion region.

The authors would like to thank Dr.W.Zulehner from Wacker Chemitronic company for providing the silicon samples used in this work. This work has been supported by the GCIS foundation (Groupement des Circuits Intégrés sur Silicium)

References:

1 L.C. Kimerling and J.L.Benson, Physica **116B**, 297 (1983)
2 A.Chantre and D.Bois, Phys.Rev **B31**, 7979 (1985)
3 J.J.van Kooten, G.A.Weller, and C.A.J.Ammerlaan, Phys.Rev.**B30**,4564 (1984)
4 A.Mesli, T.Heiser, N.Amroun and P.Siffert, Appl.Phys.Lett., **57**,1898 (1990)
5 T.Heiser and A.Mesli, Appl.Phys.Lett., **58**, 2240 (1991)
6 S.D.Brotherton, P.Bradley and A.Gill, J.Appl.Phys, **57**, 1941, (1985)
7 H.Lemke, Phys.Stat.Sol.(a), **64**, 215 (1981)
8 T.Heiser , N. Amroun and A.Mesli, to be published.

Materials Science Forum Vol. 83-87 (1992) pp. 179-184

INTERACTION BETWEEN COPPER AND IRRADIATION-INDUCED DEFECTS IN CRYSTALLINE SILICON

B.G. SVENSSON[1] AND M.O. ABOELFOTOH[2]
[1]Royal Institute of Technology, Solid State Electronics
P.O. Box 1298, S-164 28 Kista-Stockholm, Sweden
[2]IBM Research Divison, T.J. Watson Research Center
P.O. Box 218, Yorktown Heights, N.Y. 10598, USA

ABSTRACT

Schottky barrier structures of the type Cu/Si(100) have been formed by the deposition of thin copper films on electron-irradiated p- and n-type Si(100) substrates. Deep level transient spectroscopy measurements show substantial interaction between Cu and the irradiation-induced defects. A prominent level associated with Cu appears 0.53 eV above the valence band edge (E_v) immediately after Cu depositon at room temperature. It is suggested that this level originates from a complex formed by reaction between Cu and the carbon-oxygen centers which give rise to the 'ordinary' irradiation-induced level at E_v+0.35 eV. This latter level anneals rapidly already at 100 °C in the Cu/Si(100) samples suggesting that passivation by fast diffusing Cu species occurs at elevated temperatures. Finally, no evidence is found for injection of point defects (Si self-interstitials and vacancies) into the silicon substrate during formation of the metal-rich silicide η'-Cu_3Si at 200 °C.

1. Introduction

Three levels in the lower part of the bandgap located ~0.53, ~0.37 and ~0.24 eV above the valence band edge (E_v), respectively, have been associated with Cu[1]. These deep levels can act as traps and recombination centers for free charge carriers and thus strongly affect the electrical properties of silicon devices. On the other hand, it has recently been demonstrated that Schottky barrier (SB) junctions formed by deposition of thin Cu films on both p- and n-type substrates display good current-voltage characteristics[2]. Furthermore, in Si n^+p and p^+n solar cells copper with concentrations of 10^{14} to 10^{15} cm^{-3} was reported to have very small effect on cell efficiency and minority carrier life time[3]. In view of these contradictory results further studies of the electrical behavior of Cu in Si are highly motivated for its potential application in future silicon devices.

In this contribution we have investigated Cu/Si(100) SB structures formed by deposition of Cu on electron-irradiated p- and n-type substrates. Deep level transient spectroscopy (DLTS) reveals a considerable interaction between Cu and the irradiation-induced defects, e.g., a prominent Cu-associated level appears at E_v+0.53 eV immediately after deposition and anneals out below 200 °C while simultaneously the 'ordinary' irradiation-induced levels at E_v+0.35 and E_v+0.25 eV decrease in concentration. We have also studied the injection of vacancies and Si self-interstitials into the silicon during formation of the metal-rich η'-Cu_3Si silicide at 200 °C (here Cu is the dominant moving species). However, in contrast to recent experiments reported in the literature[4], no evidence is found for injection of Si self-interstitials during the growth of η'-Cu_3Si.

2. Sample preparation and analysis

Samples were prepared using Czochralski (Cz) silicon wafers grown in the <100> direction and doped with boron or phosphorus to concentrations of $1x10^{15}$ and $1x10^{16}$ cm^{-3}, respectively. The concentrations of interstitial oxygen and substitutional carbon were $5.5x10^{17}$ at/cm^3 and below $1x10^{16}$ at/cm^3, respectively, according to infrared absorption measurements. The samples were irradiated by 2.0 MeV electrons to doses between $1x10^{15}$ and $1x10^{17}$ cm^{-2}. A low current density was used (≤ 1 μA/cm^2) in order to avoid beam heating effects, and the temperature was kept below 30 °C during the irradiation. The sample surfaces were then prepared using a standard chemical cleaning procedure which included a final dip in diluted hydrofluoric acid[5]. The samples were inserted into the deposition chamber immediately after chemical cleaning, and Cu was deposited in a pressure of $1x10^{-7}$ Torr using electron beam evaporation. The deposition rate was ~10 Å/s, and the total thickness of the Cu films was 1000-1500 Å. The deposition was performed through a metal mask defining dots with diameters of 0.6, 0.8 and 1.0 mm. Annealing of the samples was carried out in an ambient of pure nitrogen for 30 minutes at temperatures between 100 and 200 °C.

In the DLTS measurements eight traditional spectra with rate windows in the range of (20 ms)$^{-1}$ to (2560 ms)$^{-1}$ were recorded during a single temperature scan between 80 and 290 K[6]. The samples were always kept under reverse or zero bias voltage, and no forward injection was applied.

X-ray photoelectron spectroscopy (XPS) and secondary ion mass spectrometry (SIMS) were used to monitor the interfacial reaction between Cu and Si upon annealing, and the silicide phase was identified by electron diffraction.

3. Results

DLTS spectra from two p-type samples (12 Ωcm) irradiated by $1x10^{15}$ e^-/cm^2 and formed with Cu and Ti contacts, respectively, are compared in Fig.1. In the Ti sample two levels appear ~0.25 and ~0.35 eV above E_v where the former is stable up to 150-200 °C and is attributed to the transition between the neutral and singly positive charge state of the divacancy center[7-9].

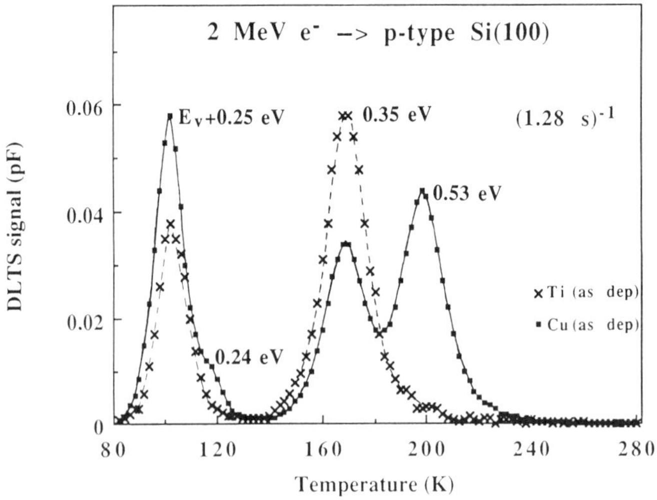

Fig.1 DLTS spectra of two p-type samples irradiated by $1x10^{15}$ e^-/cm^2 and formed with Cu and Ti contacts. Rate window=(1.28 s)$^{-1}$.

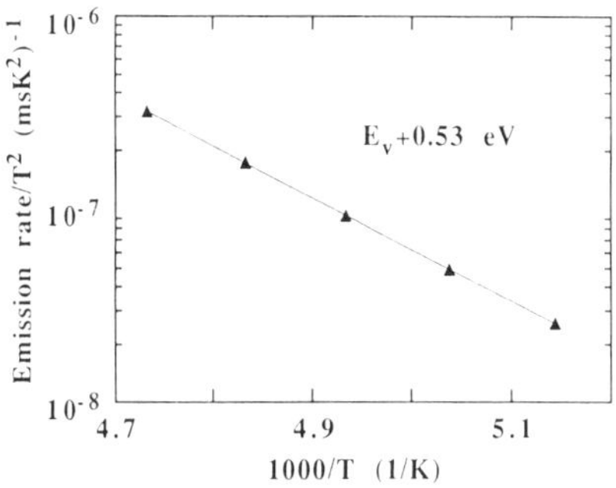

Fig.2 Arrhenius plot for the Cu associated level appearing after Cu deposition at room temperature on irradiated p-type samples.

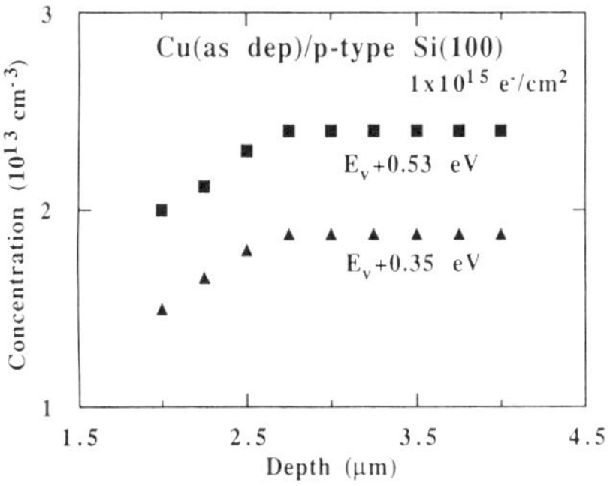

Fig.3 Depth concentration profiles of the levels at E_V+0.35 and E_V+0.53 eV in a Cu/Si(100) sample.

The latter level has been studied extensively by several authors[8-11] and found to consist of two closely spaced energy states, both of them presumably originating from carbon oxygen centers. In the Cu sample, a third major level exists at E_V+0.53 eV with a capture cross section of ~10^{-13} cm^2 according to extrapolation of the Arrhenius plot in Fig.2.

Concentration profiles of the levels at E_V+0.35 and E_V+0.53 eV are constant at depths beyond ~2.8 μm but show a gradual decrease towards the surface, and at 2.0 μm the concentration is ~80 % of that in the bulk, Fig.3.

Cu and Ti samples annealed at 200 °C are compared in Fig.4. Two overlapping Cu-associated levels appear ~0.49 and ~0.58 eV above E_V while the other peaks are observed in both samples. The amplitudes of the peaks are, however, larger in the Ti sample, and in Fig.5, the relative concentration of the E_V+0.35 eV level at a depth of 5.0 μm is plotted versus the annealing temperature for three different specimens, Cu, Ti, and a sample where Cu was first deposited and then chemically removed prior to the deposition of Ti contacts. The as-deposited Cu and Ti-Cu samples have identical DLTS spectra and both show rapid annealing of [E_V+0.35 eV]

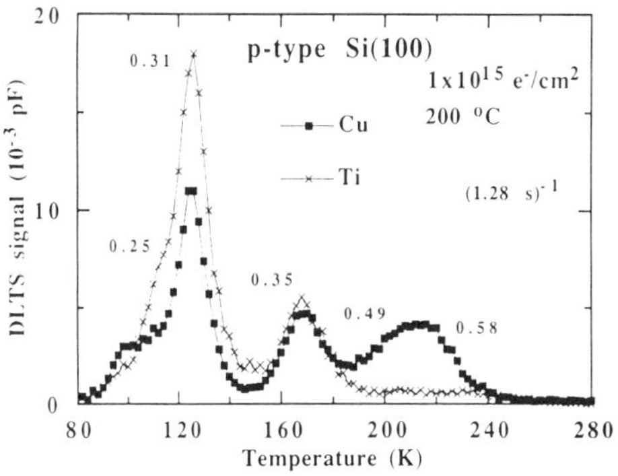

Fig.4 DLTS spectra from two irradiated p-type samples prepared with Cu and Ti contacts and annealed at 200 °C. Rate window=$(1.28\ s)^{-1}$

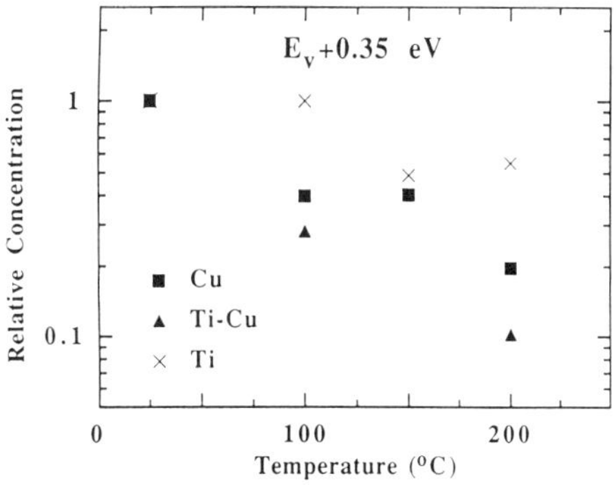

Fig.5 Isochronal annealing of the E_V+0.35 eV level in p-type samples irradiated by $1x10^{15}$ e^-/cm^2 and deposited with Cu, Ti-Cu and Ti contacts.

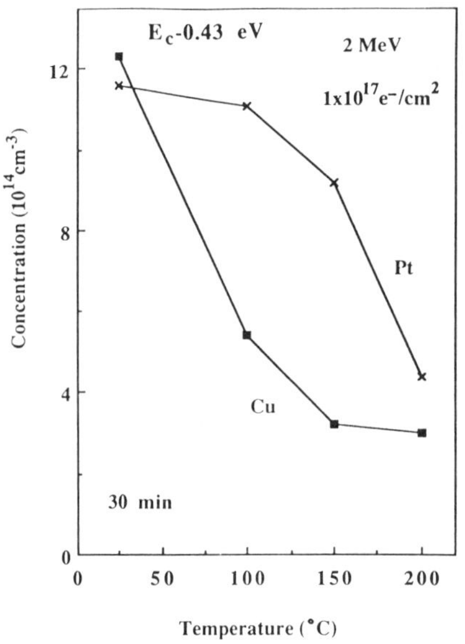

Fig.6 Isochronal annealing of the E_c-0.43 eV level in n-type samples irradiated by $1x10^{17}$ e^-/cm^2 before deposition of Cu and Pt contacts to from SB structures.

(brackets denote concentration values) at 100 °C. In particular, the Ti-Cu specimen exhibits a strong loss, and [E_v+0.35 eV] is reduced by one order of magnitude at 200 °C. The concentration in the Ti sample is more stable and agrees essentially with that previously reported in the literature[8,9].

In Fig.6 we compare the isochronal annealing of the level located 0.43 eV below the conduction band (E_c) in n-type samples irradiated by $1x10^{17}$ e^-/cm^2 and prepared with Cu and Pt contacts. In moderately doped Cz samples (resistivity ≥ 1 Ωcm) this level is mainly due to the transition between the neutral and singly negative charge state of V_2[8,12,13], and on the basis of a comparison with the level at E_c-0.23 eV originating from the doubly negative charge state of V_2 it is found that the relative contribution of V_2 to [E_c-0.43 eV] is more than 80 % in the samples used in this study. Fig.6 shows a rapid decrease of [E_c-0.43 eV] in the Cu samples below 150 °C but an almost constant concentration between 150 and 200 °C. At 200 °C where η'-Cu_3Si is formed [E_c-0.43 eV] is only 30 % lower than in the Pt samples, and no abrupt decrease in [V_2] takes place during the copper silicidation. A similar annealing behaviour of [V_2] is also found in the p-type samples deposited with Cu.

4. Discussion and conclusions

Interstitial Cu in silicon is a fast diffuser[14], and a Cu-associated complex with a level at E_v+0.53 eV forms already at room temperature during deposition of Cu films on irradiated silicon. The position is identical with that previously reported for an acceptor level attributed to Cu in p-type material[1]. This latter level was observed in Cu-doped samples diffused at ~1050-1200 °C and then rapidly cooled to room temperature. However, the present level anneals out already at ~150 °C and is, therefore, not believed to originate from the same center as the level observed in the diffused samples. Normally, DLTS spectra of electron-irradiated p-type Cz samples are dominated by the CO-related level at ~E_v+0.35 eV[8-10] while in the present Cu samples the V_2-related level at ~E_v+0.25 eV is the major peak. If the two spectra in Fig.1 are normalized with respect to the amplitude of the V_2 level it is interesting to note that [E_v+0.35 eV] in the Ti sample and the sum of [E_v+0.35 eV] and [E_v+0.53 eV] in the Cu sample are identical within 10 %. Thus a possible identification of the level at E_v+0.53 eV is a complex

resulting from interaction between CO and Cu, and its growth is accompanied by a corresponding reduction in [E_v+0.35 eV]. This identification is also consistent with the identical shape of the depth profiles for [E_v+0.35 eV] and [E_v+0.53 eV] suggesting that highly mobile Cu species diffuse and react with the stable CO complexes (Fig.3).
Two overlapping Cu-related levels at 0.49 and 0.58 eV above E_v start to emerge at 150 °C while simultaneously [E_v+0.25 eV], [E_v+0.35 eV] and [E_v+0.53 eV] decrease. However, the two growing levels are relatively weak, and there is no one-to-one proportionality with the loss in strength of the three decreasing levels. This indicates a complex generation process, and at this stage, the growing levels are tentatively attributed to higher order complexes involving Cu. Work is in progress to provide further detailed identification[15].
The level at E_v+0.35 eV is normally stable up to ~350 °C but shows a rapid anneal already at 100 °C in the Cu and Ti-Cu samples (Fig.5). Copper has recently been shown both theoretically and experimentally to be a viable candidate for passivation of shallow acceptors in p-type silicon[16-19], and in analogy with that for hydrogen, Cu is also anticipated to passivate vacancy type defects involving broken bonds[20]. The mechanisms for hydrogen and copper passivation are, however, different. For example, while H passivates the boron acceptor by bridging a B-Si bond and forms a stronger bond with Si than with B, in the case of Cu, passivation results from the formation of a Cu-B covalent bond[16], with Cu at the antibonding site to the boron atom. Mooney et al.[9] have estimated that ~80 % of the defects giving rise to the E_v+0.35 eV level originate from the K center which can be viewed as a carbon-oxygen complex bonded within a divacancy[21]. This defect contains dangling bonds, and in analogy with previous results for the V_2 center[20], passivation of the K center by mobile Cu species is expected. In particular, the loss of [E_v+0.35 eV] is more pronounced in the Ti-Cu sample compared with the Cu sample where silicide formation reaction, i.e., formation of Cu(d)-Si(p) bonding, occurs at 150-200 °C. It should, however, also be pointed out that Trombetta et al.[22] have ascribed the E_v+0.35 eV level to an interstitial carbon-interstitial oxygen pair where the electrical activity arises from the unpaired electron predominantly residing on the carbon atom in a p orbital. If this is the case then passivation may occur through formation of a neutral CuCO complex.
In this context injection of point defects during copper silicidation must also be considered. The concentration of V_2 centers was monitored during the formation of η'-Cu_3Si; these centers are normally stable up to temperatures above 200 °C, and a well established annihilation reaction is the recombination with Si self-interstitials (V_2+I --> V). A rapid decrease of [V_2] is observed below 150 °C and is ascribed to Cu passivation of the V_2 centers (Fig.6)[20]. However, in the temperature range 150 to 200 °C where η'-Cu_3Si forms [V_2] stays almost constant, and no abrupt annihilation takes place. This provides strong evidence that Si self-interstitials are not injected into the silicon during the growth of η'-Cu_3Si. If it is assumed that every Cu atom generates one Si interstitial during the silicidation process a uniform interstitial concentration of the order of 10^{19} cm^{-3} is obtained in an underlaying substrate with neglible concentration of traps and recombination centers. This is ~4 orders of magnitude higher than [V_2] in the present samples, and abrupt annihilation of V_2 through the reactions V_2+I --> V and V+I --> Ø is anticipated.
In summary, Cu interacts with irradiation-induced defects in both p- and n-type samples. A major Cu-associated complex with a level at E_v+0.53 eV forms already at room temperature during deposition of copper films on irradiated p-type samples. The absolute peak amplitude and the depth concentration profile suggest that this level is due to a complex formed by reaction between the CO centers giving rise to the E_v+0.35 eV level and mobile Cu species. Furthermore, this latter level shows a rapid anneal already at 100 °C in the Cu and Ti-Cu samples indicating that Cu passivation takes place at elevated temperatures. Finally, no evidence for injection point defects into the silicon substrate during formation of the metal-rich silicide η'-Cu_3Si is found.

Acknowledgments

We greatly appreciate the assistance by Dr J.L. Lindström, NDRE, Linköping, Sweden and the Central Scientific Materials Laboratory at IBM Yorktown Heights, N.Y., in preparing the samples. Financial support was received from the Swedish Board for Technical Development and the Swedish Natural Science Research Council.

References

1. J.W. Chen and A.G. Milnes, Annu. Rev. Mater. Sci. **10**, 157 (1980).
2. M.O. Aboelfotoh, A. Cros, B.G. Svensson and K.N. Tu, Phys. Rev. B**41**, 9819 (1990).
3. A. Rohatgi, J.R. Davis, R.H. Hopkins, P. Rai-Choudbury, P.G. McMullin and J.R. McCormick, Solid State Electron. **23**, 415 (1980).
4. M. Ronay and R.G. Schad, Phys. Rev. Lett. **64**, 2042 (1990).
5. M.A. Taubenblatt, D. Thomson and C.R. Helms, Appl. Phys. Lett. **44**, 895 (1984).
6. B.G. Svensson, K.-H. Rydén and B.M.S. Lewerentz, J. Appl. Phys. **66**, 1699 (1989).
7. G.D. Watkins and J.W. Corbett, Phys. Rev. **138**, A543 (1965).
8. L.C. Kimerling, in *Radiation Effects in Semiconductors 1976*, Inst. Phys. Conf. Ser. No. 31, edited by N.B. Urli and J.W. Corbett (IOP, Bristol, 1977), p. 221.
9. P.M. Mooney, L.J. Cheng, M. Süli, J.D. Gerson and J.W. Corbett, Phys. Rev. B**15**, 3836 (1977).
10. G. Ferenczi, C.A. Londos, T. Pavelka, M. Somogyi and A. Mertens, J. Appl. Phys. **63**, 183 (1987).
11. C.A. Londos, Semicond. Sci. Technol. **5**, 645 (1990).
12. A.O. Evwaraye and E. Sun, J. Appl. Phys. **47**, 3776 (1976).
13. B.G. Svensson and M. Willander, J. Appl. Phys. **62**, 2758 (1987).
14. E.R. Weber, Appl. Phys. A **30**, 1 (1983).
15. M. O. Aboelfotoh and B.G. Svensson, to be published.
16. S.K. Estreicher, Phys. Rev. B**41**, 5447 (1990).
17. T.H. Prescha, T. Zundel, J. Weber, H. Progge and P. Gerlach, Mat. Sci. Eng. B**4**, 79 (1989).
18. B.G. Svensson and M.O. Aboelfotoh, Mat. Sci. Forum **65-66**, 253 (1990).
19. R. Keller, M. Deicher, W. Pfeiffer, H. Skudlik, D. Steiner and Th. Wichert, Phys. Rev. Lett. **65**, 2023 (1990).
20. B.G. Svensson, M.O. Aboelfotoh and J.L. Lindström, Phys. Rev. Lett. **66**, 3028 (1991).
21. Y.-H. Lee, J.W. Corbett and K.L. Brower, Phys. Stat. Sol. (a) **41**, 637 (1977).
22. J.M. Trombetta and G.D. Watkins, Appl. Phys. Lett. **51**, 1103 (1987).

Materials Science Forum Vol. 83-87 (1992) pp. 185-190

GETTERING OF COPPER AND IRON TO EXTENDED SURFACE DEFECTS IN SILICON

M. D. DE COTEAU[1], P. R. WILSHAW[1], AND R. FALSTER[2]
[1]Department of Materials, University of Oxford, Parks Road, Oxford OX1 3PH, United Kingdom
[2] MEMC Electronic Materials, SpA, Novara, Italy

ABSTRACT

The gettering of copper and iron to extended surface defect sites, namely surface pits and oxidation induced stacking faults (OISF), has been studied. Cu was found to be gettered primarily into precipitate colony structures which nucleated at surface pits and the bounding partials of the OISF's. For the conditions investigated, gettering of Fe proceeded by generation of b=1/6<411> interstitial dislocation loops decorated with Fe-rich regions; single $FeSi_2$ precipitates were not observed. The generation of the Fe-related defects was associated with the formation of surface holes. Fe precipitation at the bounding partials was not observed in TEM, however EBIC investigations showed some evidence for decoration of the partials on an atomic scale. Silicon self-interstitials played an important role in the precipitation and gettering of both impurities.

1. Introduction

The precipitation of transition metal impurities in silicon is known to cause catastrophic failures and reduced yields in silicon integrated circuits. Often more than one type of metal contamination is present in a given system. In practice gettering procedures are used to control or eliminate transition metal precipitation and have been shown to be effective in improving device performance. However, the actual mechanisms and specific defect structures involved in gettering are not yet clearly understood. In order to develop more efficient gettering processes it is therefore necessary to understand the precipitation mechanisms of different metals, and how they are modified by a given gettering process. In this study, the effect of an "external" gettering system [1] on the precipitation behaviour of two technologically important impurities, copper, a relatively fast diffuser, and iron, a slower diffuser, is investigated.

2. Experimental

The wafers used in this study were p-type, low oxygen ($5.5 \times 10^{17} cm^{-3}$), (001) CZ-Si. The extended surface defects intended as getter sites were produced by a process known as "soft backside damage" (SBD). In this process, a 0.1μm high-velocity silica slurry was directed at the wafer surface which produces damage in the form of surface pits $\sim$0.1-0.5μm in diameter. For this study SBD was used on the front, polished surfaces of the wafers to facilitate analysis. Four sets of wafers were examined: 1) Damage free. 2) SBD treated; containing $\sim 10^8 cm^{-2}$ surface pits. These pits were irregular in shape and weak beam dark field transmission electron microscopy (TEM) revealed dislocations associated with some of the pits. 3) Type 2 wafers given a dry oxidation at 1000°C for 45 minutes. After oxidation, the pits were smoother and more rounded in appearance than the pits in type 2 wafers. The dry oxidation produced a low density $\sim 4.5 \times 10^4$ - $4 \times 10^5 cm^{-2}$ of OISF <1μm long. Each of the OISF's was observed to lie on a pit located near its centre. Only a very small proportion of the pits had served as sites for OISF nucleation; the remaining pits in these wafers had not nucleated OISF's during oxidation. 4) Type 2 wafers steam oxidised at 1100°C for 80 minutes to produce $\sim 6 \times 10^5 cm^{-2}$ OISF's 10-15 μm long. OISF's lie on {111} planes and have traces in the (001) surface along perpendicular <110> directions. In type 4 wafers the pits and their associated dislocations had been completely removed by the oxidation process. Examples of these defect structures can be found in [2]. The oxides grown for wafer types 3 and 4 were stripped by an HF dip prior to transition metal impurity decoration.

Fe or Cu was pulse sputtered onto the unpolished wafer surfaces opposite the damage, and diffused in for 10 minutes in nitrogen. For Fe the drive-in diffusion temperature was 1150°C; for Cu two different temperatures, 1150°C and 950°C, were studied. Approximately a monolayer of each metal was deposited:

[Fe]=7.0 x 10^{15}cm^{-2}, [Cu]=4.1 x 10^{15} cm^{-2}, as determined by Rutherford backscattering spectrometry measurements. After the metal diffusion, the wafers were quickly pulled from the furnace and cooled in air.

For detailed characterisation of the defect structures and precipitates, (001) plan view and (110) cross-section TEM specimens were prepared from the damaged surface regions. TEM analysis was carried out at either 120kV or 200kV on Philips CM12 or CM20 microscopes, respectively, and EDX analysis of precipitates at 100kV on an HB 501 STEM. Specimens from type 4 wafers were examined using the electron beam induced current mode (EBIC) of a Philips 505 SEM. The EBIC technique and methods used are described [3].

3. Results

3.1 Precipitation of Copper

TEM investigation of the surface regions of the Cu decorated specimens showed that, for all of the wafer types (1-4), the majority of the Cu precipitation was within copper-silicide precipitate colonies which had nucleated at or near the surface. These colonies consist of nearly coplanar arrangements of copper-silicide precipitates bounded by an interstitial edge dislocation loop, examples are shown in Figure 1. Copper silicide precipitate colonies with b=a<100> and b=a/2<110> with {100} and {110} habit planes respectively, are commonly observed and have been described in more detail elsewhere [4,5]. In addition to the smaller colony precipitates (found at the surface and in the bulk) which ranged in size from 50-500Å, larger precipitates 1500-5000Å in size were found at the surface only. These were in the shape of square-based pyramids faceted on {111}, examples are shown in Figure 1a,b. Similar precipitates have recently been reported in rapid-thermal processed silicon [6]. STEM examination of the precipitates confirmed that they were a copper silicide. Precipitates and colonies were observed to nucleate at the undamaged surface, at pre-existing surface pits, and at the bounding partials of the OISF's [2,7]. Precipitates were not observed to nucleate on the OISF fault plane.

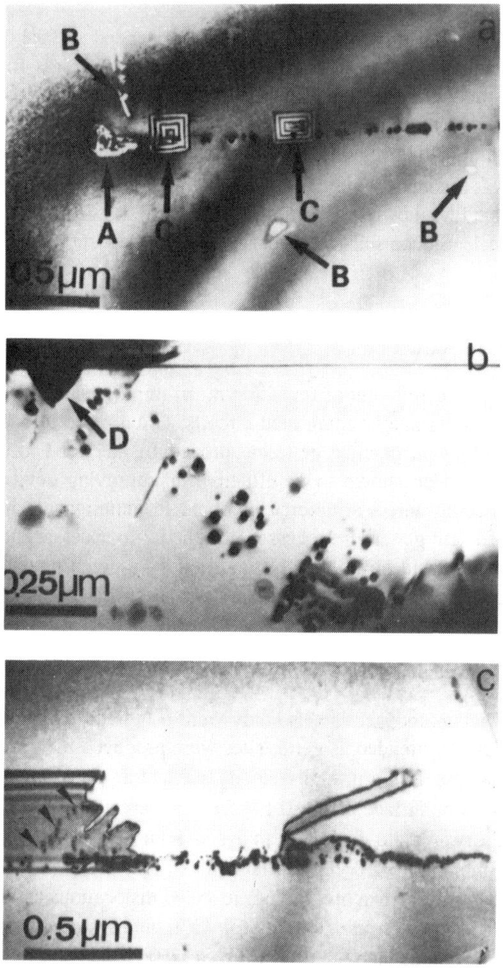

Figure 1. Precipitate colonies in Cu-decorated specimens (a) (001) plan view image of a colony which has nucleated at pre-existing pit (A), undecorated pits which are not nucleation sites (B), extinction contours (C) where pyramidal surface precipitates have been removed during specimen preparation. (b) (110) XTEM image of a colony. Note large pyramidal precipitate (D) at surface. (c) {111} type colony which has nucleated at the 1/3<111> bounding partial of a pre-existing OISF in a type 4 wafer. Arrows indicate the initial position of the bounding partial where a row of precipitates were nucleated.

The surface getter sites in wafer types 2-4 resulted in an increase in the density of colonies intersecting the surface by an order of magnitude or more relative to the undamaged type 1 wafers, from < 5 x 10^4 cm^{-2} (type 1) to 2-7 x 10^5cm^{-2} (types 2-4). The surface getter sites provided suitable sites for the initial nucleation of precipitates. After this nucleation step, precipitation would then proceed in most cases with the growth of a colony structure. The growing colony itself then becomes a centre for further copper gettering. Comparable increases in the density of colonies were achieved for wafer types (2-4) despite the fact that the density of the pre-existing pits in type 2 and 3 wafers was two orders of magnitude higher than the density of

OISF's in type 4 wafers. The majority of the pre-existing getter centres were not observed to be decorated by copper precipitation. We infer that a growing colony, once nucleated, becomes the dominant gettering centre and that increasing the number of nucleation sites above a threshhold level will not alter the gettering efficiency of the system [2].

In type 3 and 4 wafers, the preferential nucleation of colonies at the bounding partials of OISF's was found to alter the distribution of the colonies themselves. In wafer types 1 and 2 both {100} and {110} colonies were observed; however, in wafer types 3 and 4 which contained OISF's, the formation of {100} colonies was suppressed. We believe this is due to preferential nucleation of {111} colonies at the 1/3<111> bounding partials of the OISF's [7]. {111} precipitate colonies in other systems are discussed in detail in [8,9]. Figure 1c is an example of a {111} colony which has nucleated at a bounding partial. The silicon self interstitials (Si_i) which were emitted by the growing colony [4,5,10] have caused the extension of the {111} fault plane during colony growth [7], which is indicated in Figure 1c. The extension of a pre-existing OISF with 1/3<111> bounding Frank partial into a {111} colony would not require the usual colony nucleation steps, i.e. the formation of the interstitial dislocation bounding the colony and the new stacking fault plane [4]. Therefore the nucleation of {111} colonies at the pre-existing OISF's in type 3 and 4 wafers was favoured. The {111} colony configuration was however observed to be unstable and on further growth on {111}, the bounding dislocation would often react with dislocations nucleated at precipitates, eventually converting to a {110} colony.

Figure 2a is an EBIC image of a Cu-decorated type 4 wafer. The image contrast is due to recombination at colonies lying along <110> directions which correspond to colonies observed in TEM.

3.2 Precipitation of Iron

After Fe decoration, similar forms of precipitation were observed at the surface of all four wafer types. Rod-like defects with <100> (Figure 3a) and <110> (Figure 3b) traces were observed at the surface of (001) plan view foils. Those with <110> traces are similar in appearance to the $FeSi_2$ precipitates which have been reported to be associated with device failures [11,12]. However, the defects in the present study were not single $FeSi_2$ precipitates, but instead were Fe-decorated crystallographic defects [13]. TEM contrast analysis similar to that in [14] demonstrated that the rod-like defects with both <100> and <110> traces were interstitial-type, faulted dislocation loops with b=1/6<411> lying on {111} habit planes. These defects were elongated along <110> directions perpendicular to their Burgers vector. STEM examination of these defects found that they were decorated with Fe. Figure 4 is an annular dark field STEM image of a section of a rod-like defect. The bright region labelled "Fe" is an Fe-rich region; only Si signals were detected from the darker regions of the defect. Small (<0.1μm-0.25μm) faceted surface holes were found where these rod defects intersected the wafer surface, these are indicated in Figure 3. The1/6<411> defects were observed individually or as complex defects (Figure 3c) comprised of a central

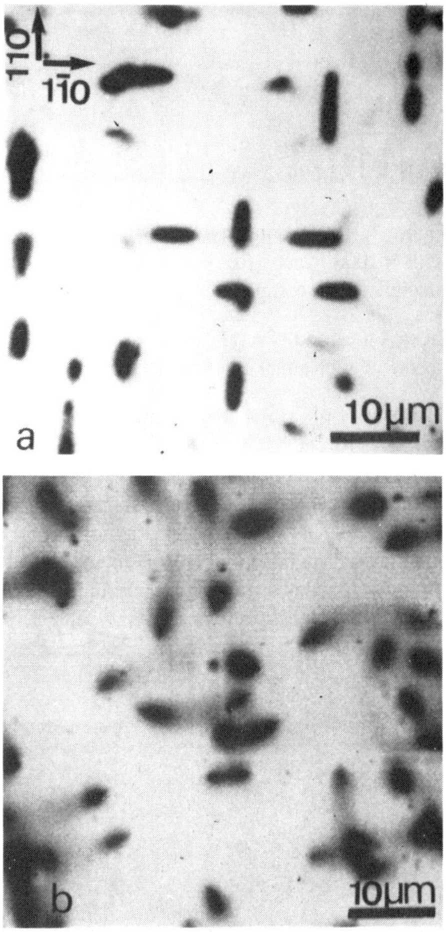

Figure 2. EBIC images of type 4 specimens containing pre-existing OISF's. (a) Cu-decorated, strong recombination observed at colonies along <110> directions. (b) Fe-decorated, showing contrast of 14-18% at the bounding partials.

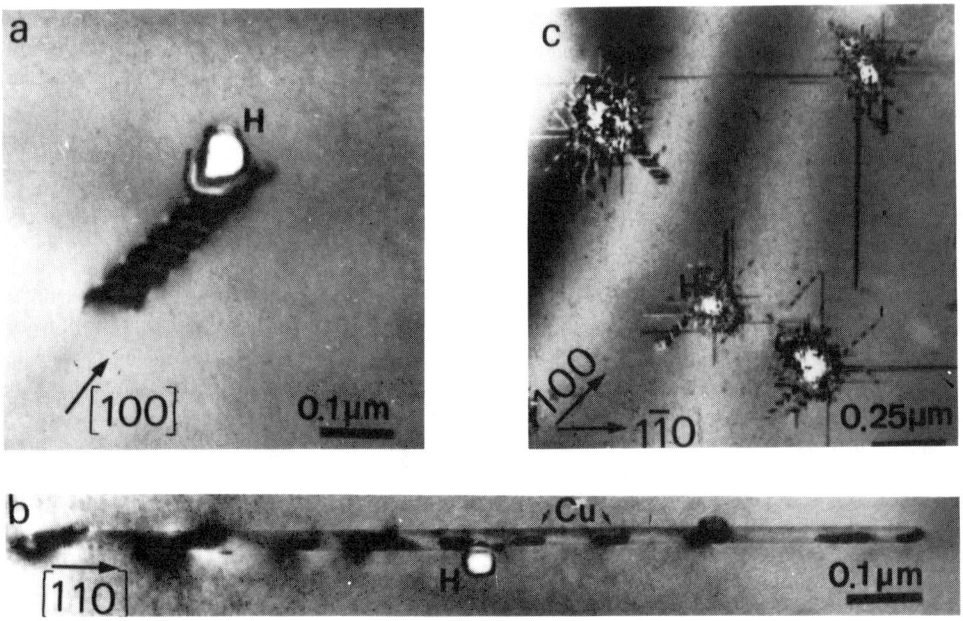

Figure 3. (001) plan view TEM images of rod-like 1/6<411> defects and faceted holes ("H"): (a) defect with <100> trace, (b) defect with <110> trace which is heavily decorated with copper silicide precipitates,(c) complex defects in a type 3 specimen.

faceted hole and several "arms" of rod-like 1/6<411> defects. In some instances, the rod-like defects were decorated with small copper silicide precipitates An extreme example is shown in Figure 3b.

The density of 1/6<411> defects and their associated faceted holes at the undamaged type 1 wafer surface was $\sim 7 \times 10^6 cm^{-2}$. In type 2 and 3 wafers, which contained $\sim 10^8 cm^{-2}$ pre-existing surface damage pits, the 1/6<411> defect density had increased by more than an order of magnitude to $\sim 10^8 cm^{-2}$. The pits (and their associated dislocations in type 2 wafers) were found to be efficient centres for nucleation of 1/6<411> defect complexes. In type 2 wafers, all of the pre-existing pits/dislocations had nucleated 1/6<411> defect complexes. The pits in type 3 wafers which had been smoothed by the dry oxidation step were also efficient for 1/6<411> defect complex nucleation, however "undecorated" pre-existing pits which had not nucleated defects were occasionally observed.

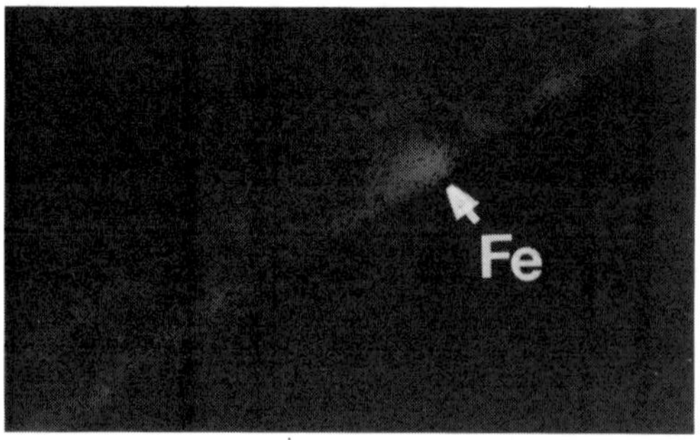

Figure 4. Annular dark field STEM image of a section of a rod-like defeect. Fe= iron-rich region. Defect is ~ 200Å in width.

A high density of pre-existing OISF's were present in type 4 wafers. After Fe decoration, mostly individual, rather than complex 1/6<411> defects were observed, with a density of $\sim 4 \times 10^6 cm^{-2}$. In addition to the 1/6<411> defects, a density of $\sim 5 \times 10^7 cm^{-2}$ of "individual" faceted holes (i.e. not associated with 1/6<411> defects) were observed. The 1/6<411> defects were not observed to nucleate preferentially at the OISF's, and in fact no clear relationship was found between the precipitates and the OISF's.

EBIC investigation of an undecorated type 4 specimen showed no measurable contrast at the bounding partials. After Fe decoration (Figure 2b), the bounding partials were electrically active, with a peak contrast of 14-18%. Because of the lack of observable precipitation in TEM, we infer that the electrical activity of the bounding partials is due to Fe decoration on an atomic scale. This indicates that Fe gettering can take place without necessarily involving precipitation.

4. Discussion

Marked differences have been observed in the precipitation behaviours of Cu and Fe as well as in their interactions with the gettering structures. In this section, the role of point defects in the precipitation processes of the these two metals is considered. The relative efficiencies of the different getter centres are interpreted with respect to these considerations.

Si_i played an important role in the precipitation and gettering processes for Cu. Colonies grow by repeated precipitation on the climbing bounding dislocation [4]. The dislocation provides the interstitial sink needed for precipitate growth, however once the Si_i have caused the dislocation to climb away from the precipitate, it is no longer available as a sink. The colony grows outward, resulting in a high number density of small precipitates at the interior of the colony. The large pyramidal precipitates were observed only at the surface; the free surface provides a continuous interstitial sink and precipitates nucleated at the surface are not constrained in the same way as those nucleated in the bulk. In the present study, the undamaged surface, surface damage pits, and the bounding partials are sites which provide an interstitial sink and are therefore favourable for the growth of colony structures.

Si_i were also found to be important for the precipitation of Fe. The formation of α-$FeSi_2$ precipitates requires consumption of Si_i [10]. Although single $FeSi_2$ precipitates were not observed for the low level of Fe contamination introduced in this study, we have found that Fe-decorated interstitial type dislocation loops were formed which were associated with faceted surface holes. It is reasonable to conclude that the faceted surface holes were a likely source of the Si_i needed for the formation of the interstitial defects. Sites which provide a source of Si_i will be most favourable for gettering Fe by generation of these decorated interstitial defects.The damaged surface resulting from the pre-existing pits in type 2 and 3 wafers may provide many surface steps which would facilitate removal of Si_i, making the pits favourable sites for the growth of defect complexes. It is noted however that the volume of the surface holes produced is larger than the volume of Si estimated to be contained in the rod-like defects; it is possible that some of the Si removed from the surface is used for precipitation in the bulk. The necessity of Si_i sinks and sources for Cu and Fe precipitation, respectively, favours co-precipitation of these two metals, and indeed, the 1/6<411> defects were observed to getter unintentional Cu contamination in the Fe-diffused specimens [13]. Co-precipitation of Cu and Fe has also been reported in [10,11].

The fact that Si_i were emitted in the case of Cu and consumed in the case of Fe can be used as a basis for understanding the efficiency of various gettering centres for each metal. It is important to first consider the undamaged (type 1) wafer surface. In the case of Cu the surface acted as a sink, and for Fe as a source of Si_i, resulting in significant amounts of precipitation of either metal even at an undamaged surface. In particular, in the case of Fe, the precipitation observed resulted in the formation of faceted surface holes and dislocations which would be catastrophic in an active device region.

The pre-existing surface-damage pits in type 2 and 3 specimens were found to be effective gettering centres for both Fe and Cu. In the case of Cu the pits served as sites for the initial nucleation of precipitates and subsequent growth of colonies which then became the dominant gettering centres. Less than 1% of the pits present were used as sites for precipitation and/or colony nucleation [7]. This is in contrast to the Fe case; for type 2 and 3 wafers nearly 100% of the pre-existing pits had nucleated 1/6<411> defect complexes. Although the pre-existing pits were sites for precipitate/colony and 1/6<411> defect nucleation for both Cu and Fe, respectively, for Fe the pits continued to play an active role in the gettering process. During the growth of the 1/6<411> defect complexes, the pre-existing pits were enlarged as they provided Si_i needed for precipitation of the interstitial 1/6<411> defects. Figure 5 is an XTEM image of a 1/6<411> defect complex which has nucleated on a pre-exiting pit/dislocation in a type 2 wafer. As a result of the growth of the 1/6<411> defect complex, the pre-existing pit has become larger and faceted in appearance.

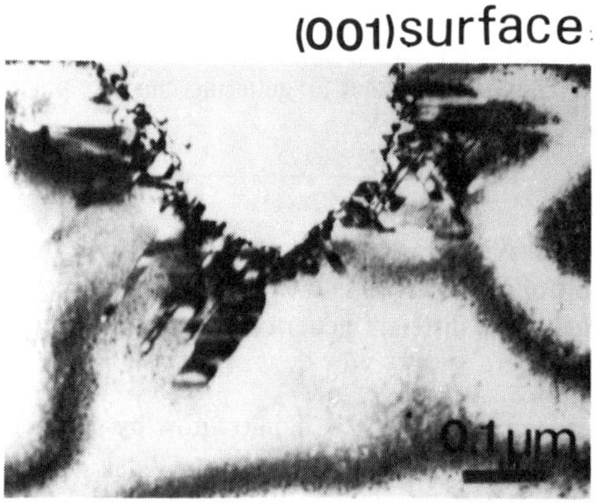

Figure 5. (110) XTEM image of a 1/6<411> defect complex in type 2 specimen. Growth of the interstitial defect complex has enlarged the pre-existing surface damage pit which has become faceted in appearance.

The fact that the bounding partials of the OISF's can act as interstitial sinks has been discussed earlier; the effect of the tensile strain field at the bounding partials is also a factor to be considered for gettering. Although the bounding partials of the OISF'S were observed to be ineffective for nucleation of Fe precipitates, they were observed to be efficient sites for nucleation of copper silicide precipitates and growth of colonies. In Cu-decorated type 4 wafers preferential nucleation of precipitates and {111} colonies at the partials was observed. The efficiency of the partials for Cu precipitate colony growth can be understood as follows. The tensile strain field at the bounding partial dislocation provides a favourable site for the nucleation of a copper silicide precipitate because of the volume increase associated with the formation of such a precipitate [4,10]. The bounding partial then provides a sink for the growing precipitate, and colony growth results. It is the combination of precipitate nucleation at the tensile strain field of the bounding dislocations and the interstitial sink provided by the dislocation that account for the gettering efficiency of the colonies. In addition, the long range strain field associated with the extra colony plane contributes to its continuing effectiveness as a gettering centre during the precipitation process.

Conversely, the presence of Fe on a substitutional site will tend to put the lattice into tension, so that nucleation of Fe related defects is not likely at the tensile strain field of a bounding partial. The EBIC results provide evidence that the Fe is gettered by the bounding partials without observable precipitation taking place. Further EBIC studies are planned in order to characterize the nature of this gettering/recombination mechanism.

Acknowledgement

The authors wish to thank Dr. G. R. Booker for helpful discussions regarding TEM contrast analysis, and Dr. N. J. Long for STEM analysis. MDD gratefully acknowledges support from the Rhodes Trust and AT&T Bell Laboratories (CRFP). PRW gratefully acknowledges support from the Royal Society.

5. References

1. R. Falster, Solid State Phenomena 6&7 13 (1989)
2. M. D. de Coteau, P. R. Wilshaw, and R. Falster, Phys. Stat. Sol. (a) 117 403 (1990)
3. P.R. Wilshaw, Ourmazd, A., and Booker, G.R., J. Physique 44 C4-445 (1983)
4. E. Nes and J. Washburn, J. Appl. Phys. 42 3562 (1971)
5. M. Seibt and K. Graff, J. Appl. Phys.63, 4444 (1988)
6. R.Kola, G.Rozgonyi, J. Li, W. Rogers, T. Tan, K. Bean, and K. Lindberg, Appl. Phys.Lett. 55, 2108 (1990)
7. M. D. de Coteau, P.R. Wilshaw, and R. Falster in *Proc. Microscopy of Semiconducting Materials VII,* (Institute of Physics, Oxford, 1991) in press
8. J. M. Silcock and W. J. Tunstall, Phil. Mag. 10 361 (1964)
9. H. M. Miekk-oja and R. Raty, Phil. Mag. 24 1197 (1971)
10. M. Seibt in *Semiconductor Silicon 1990,* eds. Huff, Barraclough, and Chikawa. (The Electrochemical Society, Pennington, 1990) p. 663
11. A.G. Cullis and L.E. Katz, Phil. Mag. 30, 1419 (1974)
12. P. D. Augustus, J. Knights, and L.W. Kennedy, J. Microscopy 118, 315 (1979)
13. M.D. de Coteau, P.R. Wilshaw, and R. Falster, to be published in *Proc. Gettering and Defect Engineering in Semiconductor Technology, (GADEST'91)*
14. J. A. Lambert and P. S. Dobson, Phil. Mag. A. 37, 441 (1978)

Materials Science Forum Vol. 83-87 (1992) pp. 191-196

SHALLOW EXCITED STATES of the 1014 meV Cu RELATED OPTICAL CENTER in SILICON

M.H.Nazaré, A.J.Duarte and A.G.Steele†, G.Davies†, E.C. Lightowlers†.
Departamento de Física, Universidade de Aveiro, 3800 AVEIRO, PORTUGAL.
†Department of Physics, King's College London, LONDON WC2R 2LS, UK.

ABSTRACT

Uniaxial stress measurements are reported on the shallow excited states, at 1016.4 meV and 1024.1 meV of the 1014.5 meV Cu-related band in Silicon. Data are analysed to show that the transitions occur at a trigonal centre, and have orbital Γ_1 and Γ_3 excited states, thus confirming early results for the 1014 meV line. These states behave under stress in a way that is very similar to the conduction band extrema, and we show that the stress data are consistent with a pseudo-donor model for the copper center. The states at 1014.5 meV, 1016.4 meV and 1024.1 meV arising from the 1s ground state of the donor electron, split by the trigonal, C_{3v}, strain field of the defect.

1. Introduction

Transition metals diffuse very rapidly into crystalline silicon, producing many photoluminescence bands, which present multiple excited states with similar energy separations. One such a case, perhaps the best documented one, is the Cu-related band whose main line occurs at 1014 meV.

This luminescence band, shown in figure 1, is seen in silicon crystals which were Cu diffused and rapidly quenched[1]. At 20 K the band consists of electronic transitions at 1014.2 meV, 1016.2 meV and 1023.8 meV, plus their vibronic sidebands involving a resonance phonon of $\hbar\omega_1 = 7.0$ meV plus other phonons of $\hbar\omega_2 = 16.4$ meV, $\hbar\omega_1 = 25.1$ meV and $\hbar\omega_1 = 57.$ meV[1-3]

Weber et al[1] reported the response of the 1014 meV line to uniaxial stress and from a qualitative analysis proposed a trigonal symmetry for the center. More recently[4] it was shown that the data could be fitted in a trigonal model with a set of parameters similar to others obtained for other TM-related centres in Silicon. However the analysis did not include the excited states for which there were no data available. It is the purpose of the present paper, to investigate the excited states of this optical center in some detail.

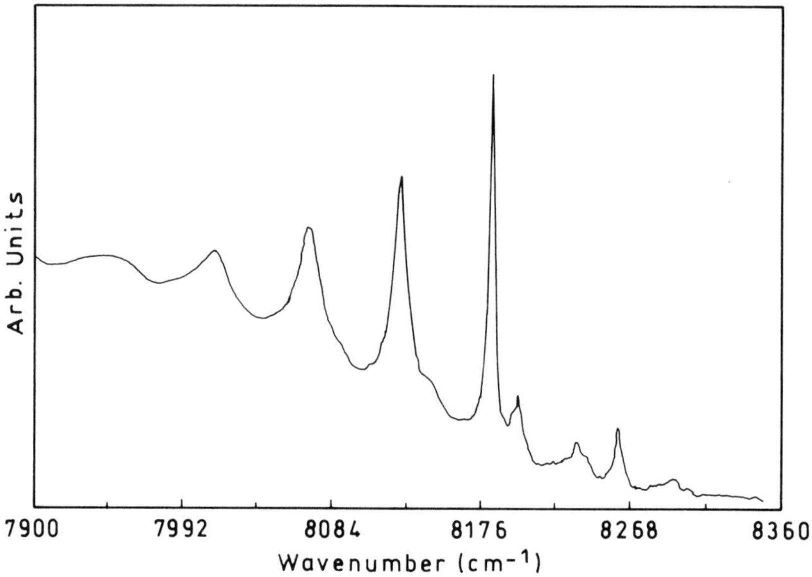

Figure 1: PL spectrum of the copper system recorded at 16 K using a FTIR spectrometer

2. Uniaxial Stress Results and Analysis

Compressive stresses of up to 200 MPa have been applied along the $\langle 001\rangle$, $\langle 111\rangle$ and $\langle 110\rangle$ axes of copper doped silicon samples held at 16 K. The 1014 meV, 1016 meV and 1024 meV lines respond to uniaxial stress as shown in figure 2(a).

We show next that the full set of data is consistent with lines at 1014 meV, 1016 meV and 1024 meV being transitions from orbital Γ_3, Γ_1 and Γ_3 states respectively into a common Γ_1 state at a trigonal center.

In a trigonal environment the stress Hamiltonian is given by

$$H = C_{A_1} s_{A_1} + C'_{A_1} s'_{A_1} + C_{E\theta} s_{E\theta} + C_{E\epsilon} s_{E\epsilon} + C'_{E\theta} s'_{E\theta} + C'_{E\epsilon} s'_{E\epsilon}. \quad (1)$$

Here $C_{A_1} \dots C'_{E\epsilon}$ are electronic operators and their transformation properties, under the operations of the C_{3v} point group, are indicated by the subscripts. $s_{A_1} \dots s'_{E\epsilon}$ are combinations of the stress tensor components that transform in the same way as the related electronic operator. We define relative to the crystal axes:

$$
\begin{aligned}
s_{A_1} &= s_{xx} + s_{yy} + s_{zz} & s'_{A_1} &= s_{yz} + s_{zx} + s_{xy} \\
s_{E\theta} &= s_{xx} + s_{yy} - 2s_{zz} & s'_{E\theta} &= s_{yz} + s_{zx} - 2s_{xy} \\
s_{E\epsilon} &= \sqrt{3}(s_{xx} - s_{yy}) & s'_{E\epsilon} &= \sqrt{3}(s_{yz} - s_{zx})
\end{aligned} \qquad (2)
$$

The matrix describing the stress perturbation, written in the Γ_1, $\Gamma_{3\theta}$, $\Gamma'_{3\theta}$, $\Gamma_{3\epsilon}$ and $\Gamma'_{3\epsilon}$ basis states has the form

$$
\begin{pmatrix}
\delta & \gamma & \gamma' & 0 & 0 \\
\gamma & \alpha - \beta & -\eta & 0 & 0 \\
\gamma' & -\eta & \alpha' - \beta' & 0 & 0 \\
0 & 0 & 0 & \alpha + \beta & \eta \\
0 & 0 & 0 & \eta & \alpha' + \beta'
\end{pmatrix} \qquad (3)
$$

where

$$
\begin{gathered}
\delta = A_1 s_{A_1} + A_2 s'_{A_1} \\
\alpha = A_{1E} s_{A_1} + A_{2E} s'_{A_1} \qquad \beta = B s_\theta + C s'_\theta \\
\alpha' = A'_{1E} s_{A_1} + A'_{2E} s'_{A_1} \qquad \beta' = B' s_\theta + C' s'_\theta \\
\gamma = D s_\theta + E s'_\theta \qquad \gamma' = D' s_\theta + E' s'_\theta \qquad \eta = G s_\theta + F s'_\theta
\end{gathered} \qquad (4)
$$

To simplify the fit of (3) to the data we note that under $\langle 111 \rangle$ stress the shift rates are linear and states Γ_3 (at 1014 meV) and Γ'_3 (at 1024 meV) split only into two components. Consequently stress parameters C and C' can be set to zero as well as the interaction parameters E, E', and F. Further simplifications came from the results under $\langle 001 \rangle$ stress. It is apparent that the only stress induced interaction occurs between line at 1016 meV and one component from line at 1014 meV line. This means that the interaction parameters D and G can also be set to zero.

Lines on figure 2(a) are a fit of this theory to the data and have been calculated with the following set of parameters

$$A_{\Gamma} = -3.94\,\mathrm{meV\,Kbar^{-1}} \qquad A_2 = 3.53\,\mathrm{meV\,Kbar^{-1}}$$

$$A_{1E} = -1.24\,\mathrm{meV\,Kbar^{-1}} \qquad A_{2E} = -1.41\,\mathrm{meV\,Kbar^{-1}}$$

$$B = 1.58\,\mathrm{meV\,Kbar^{-1}} \qquad C = 0\,\mathrm{meV\,Kbar^{-1}}$$

$$A'_{1E} = -1.54\,\mathrm{meV\,Kbar^{-1}} \qquad A'_{2E} = -7.8\,\mathrm{meV\,Kbar^{-1}}$$

$$B' = 1.92\,\mathrm{meV\,Kbar^{-1}} \qquad C' = 0\,\mathrm{meV\,Kbar^{-1}}$$

$$D' = 0. \tag{5}$$

all the other stress coupling parameters have been set to zero.

3. Pseudo-donor model

Weber et al[1] suggested that the no-phonon lines at 1014 meV and 1016 meV were due to radiative recombination, from $J = 2$ and $J = 1$ respectively, to the crystal ground state $J = 0$, of an exciton bound to the copper isoelectronic defect. The nature of the excited state of the 1024 meV line could not be identified.

In this communication we give a different interpretation of the origin of the states involved. A closer examination of the stress data points to two very distinctive facts, one, is that for the Γ_3 like orbital states the C parameter is zero, showing that the states derive from the conduction band minima, which remain degenerate under a $\langle 111 \rangle$ stress[5]. The other is that the only stress induced interaction occur between Γ_1 and Γ'_3 states, as would happen if the 3 states, (Γ_1, Γ_3 and Γ'_3) were derived from valley-orbit states[6] split by the strain field of the defect.

An isolated donor has a EMT 1s ground state that is 6-fold degenerate. In cubic symmetry, T_d point group, the 1s state behaves as $\Gamma_1 + \Gamma_3 + \Gamma_5$, the valley-orbit states. Under stress the only interaction occurs beteen Γ_1 and Γ_5 states. In trigonal symmetry the Γ_5 transforms as $\Gamma_1 + \Gamma_3$; in this model the two states remain degenerate as the lowest states at 1014 meV. We postulate that the copper center plus the bound exciton is a pseudo-donor with a ionization energy of 32 meV[1] and the ionized donor is the hole deeply bound to the isoelectronic defect. The 1s ground state of the donor is split by an axial strain along $\langle 111 \rangle$ giving rise to the 1014 meV ($\Gamma_5 = \Gamma_1 + \Gamma_3$), the 1016 meV ($\Gamma_1$) and 1024 meV ($\Gamma_3$) states.

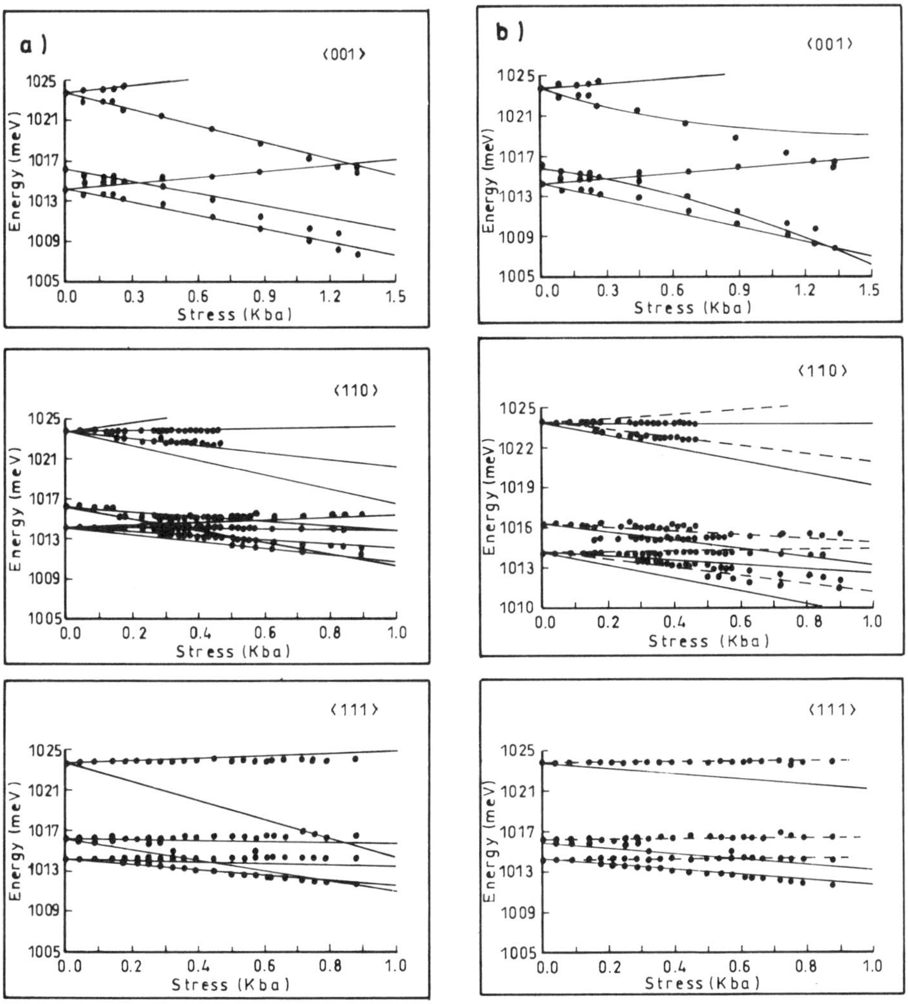

Figure 2: A comparison of theory and experiment for the energies of the 1014–, 1016– and 1024 meV lines under uniaxial compression along ⟨001⟩, ⟨111⟩ and ⟨110⟩. The solid lines show best fit obtained (a) using the theory of section 2 and (b) using the theory of section 3, with coefficients listed in the text. Dashed lines refer to a nonequivalent orientation of the trigonal center under stress.

The effects of the in-built and externally applied uniaxial stress on the, Γ_1, $\Gamma_{3\theta}$,$\Gamma_{3\epsilon}$ and $\Gamma_{5x,y,z}$, valley-orbit states can be obtained from the secular matrix given by Wilson and Feher [6], in terms of two stress parameters, a_1 and b corresponding to hydrostatic and shear deformation respectively. The other parameters involved are, the magnitude of the in-built stress, S_0, Δ and δ. 6Δ and $2\Delta\delta$ give the energy separation of Γ_1 from

Γ_3 and of Γ_3 from Γ_5 respectively.

The best fit of this model to the stress data is shown in figure 2(b). Lines have been calculated with

$$a_1 = -2.67\,\text{meV Kbar}^{-1} \qquad \text{b} = 2.38\,\text{meV Kbar}^{-1} \qquad \text{S}_0 = -1.34\,\text{Kbar}$$

$$\Delta = 1.32\,\text{meV} \qquad \delta = -3.56\,\text{meV}$$

The calculated position of the states deriving from the 1s ground state are, in absence of external stress:

Γ_5 (the lowest), followed by Γ_1 at 1.47 meV and Γ_3 at 9.41 meV.

4. Summary

We have shown that the transitions at 1016 meV and 1024 meV occur at a trigonal centre, and have orbital Γ_1 and Γ_3 excited states, thus confirming early results for the 1014 meV line. These states behave under stress in a way similar to the conduction band extrema, and we showed that the stress data are consistent with a pseudo donor model for the copper center. The states at 1014 meV, 1016 meV and 1024 meV arising from the 1s ground state of the donor electron, split by the trigonal C_{3v} field of the defect. There is a need to investigate the higher effective-mass-like states of the copper center using photoluminescence excitation measurements.

5. Acknowledgements

We thank INIC for support for MHN during a short visit to King's College London.

References

1 Weber J, Bauch H and Sauer R 1982 *Phys. Rev.B* **25** 7688-99

2 Watkins S P, Ziemelis U O and Thewalt M L W 1982 *Solid State Commun.* **43** 687-90

3 Sauer R and Weber J 1984 *Solid State Commun.* **833-36**

4 Davies G and Carmo M C 1988 *International Conference Series* **95**

5 Laude L D, Pollak F H and Cardona M 1971 *Phys. Rev.* **3** 2623-36

6 Wilson D K and Feher F 1961 *Phys. Rev.* **124** 1068-80

Materials Science Forum Vol. 83-87 (1992) pp. 197-202

ELECTRONIC NATURE OF NEUTRAL ZINC IN SILICON: FTIR-ABSORPTION, UNIAXIAL STRESS MEASUREMENTS

B. KAUFMANN[1], A. DÖRNEN[1], M. LANG[2], G. PENSL[2], D. GRÜNEBAUM[3], N. STOLWIJK[3]
[1] 4. Physikalisches Institut, Universität Stuttgart, PO-Box 80 11 40, W-7000 Stuttgart 80, F.R.G.
[2] Institut für Angewandte Physik, Universität Erlangen, Staudtstr. 7, W-8520 Erlangen, F.R.G.
[3] Institut für Metallphysik, Universität Münster, W-4400 Münster, F.R.G.

ABSTRACT

We report optical absorption measurements focusing on the electronic nature of the neutral charge state of double acceptor zinc in silicon (Si:Zn^0). Temperature-dependent measurements indicate that the ground state is probably split up into more than three states, while the excited states are two-fold split. Measurements under uniaxial stress along the [100], [111] and [110] crystal axis were carried out. They evidence that hole-hole interaction is not the dominant interaction forming the ground state in Si:Zn^0, unlike to Ge:Zn^0. The stress response found here is highly anisotropic. We suggest a Jahn-Teller effect to form the ground state.

Introduction

Double acceptors in germanium are widely investigated by the means of optical mid-infrared absorption spectroscopy. For several double acceptors, like Zn, Mg or Hg a hydrogen-like series of absorption lines is found for the neutral charge state and additionally helium-like transitions are observed for Zn^- [1]. Recently, in a single-particle approximation, Fiorentini et al. [2] consider the excitation series of the neutral charge state of double acceptors in germanium. They show that a hydrogen-like effective-mass-(EM) series of excited states is expected for the secondarily bound hole. A split ground state is another feature to be a characteristic of double acceptors. This fact follows from 'hot lines' in absorption measurements. A two-fold splitting of 0.7 meV first was observed for Ge:Hg [3]. Later, a ground state splitting for Mg (1.75 meV) [4] and Zn (2.4 meV) [5] was reported, too. The splitting was explained by j-j-exchange interaction of the two Γ_8 hole states forming a J_0 (Γ_1) and a J_2 (Γ_3 +Γ_5) level. According to Hunds' rule the J_2 state should be the lowest level [3]. For Ge:Hg this 'normal' ordering was identified from the temperature-dependent absorption measurements. The lighter impurities Mg and Zn appeared to have the 'inverted' ordering of states, with a J_0 state as lowest level of the ground state. An inverted ordering of states can principally result from additional local strain fields, which affect the J_0 state stronger than the J_2 state, as discussed by Vouk et al. [6]. In a recent paper the ground state of double acceptors is considered by an LCAO (linear combination of atomic orbitals) approximation [7]. The authors investigate the combined influence of spin-orbit coupling, j-j interaction, and crystal-field splitting. The wave functions are built up by purely p-type orbitals and hence central-cell corrections are neglected, also possible strain effects are not considered.

While in germanium several double acceptors are investigated by absorption spectroscopy, for silicon only few data are available. The neutral charge state of zinc is the only case where spectroscopic data are reported. In recent work, bound-to-bound transitions of a hydrogen-like hole was found [8, 9]. The excited states were found to be well described by effective mass theory [10] while the ground state is rather deep (319.3 meV). The 'hot lines' from temperature-dependent measurements indicate a complicate ground state structure [8, 12]. Following the argumentation for double acceptors in germanium as drawn above, they were identified with Γ_1, Γ_3, and Γ_5 [8, 11]. To get an experimental approach to the couplings within Si:Zn^0, temperature-dependent absorption measurements and absorption studies with applied uniaxial stress were performed.

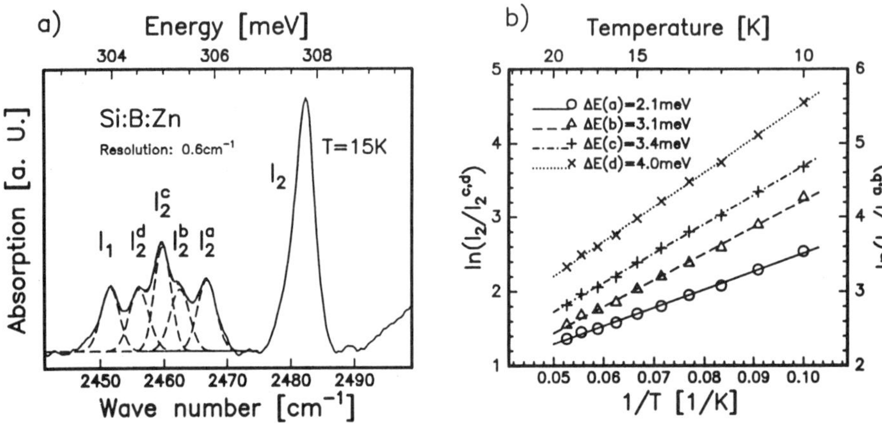

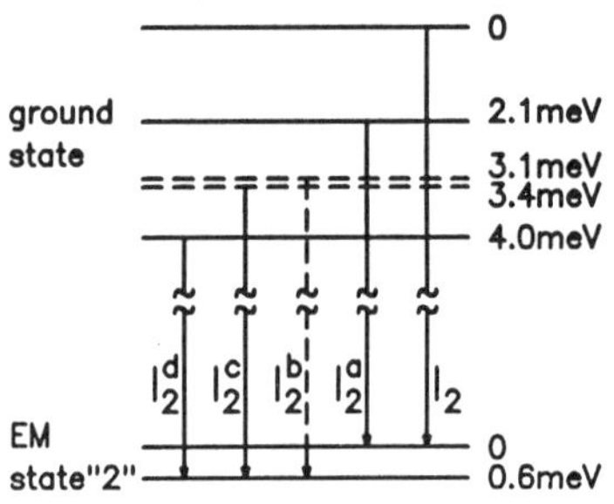

Figure 1: *a) Decomposition of the 'hot lines' $I_2^a \dots I_2^d$. b) Arrhenius diagram and activation energies of the lines $I_2^a \dots I_2^d$ as obtained from the decomposition. Uncertainty: $\delta E = \pm$ 0.3 ($\pm$ 0.6) meV for I_2^a, I_2^c, and I_2^d (I_2^b). c) Level scheme of ground state and EM level '2', the both dashed lines mark states that could be identical.*

Temperature-dependent measurements

In optical absorption spectroscopy 'hot lines', showing up at temperatures above 4 K, have already been reported by Dörnen et al. [8] and Merk et al. [12]. As Merk and coworkers pointed out, the 'hot lines' can be resolved into 5 lines [13]. They find thermal activation energies significantly smaller than the spectroscopic spacings to I_2 and conclude not only the ground state but also the excited state to be split. It has not been stated yet explicitly, that in low temperature spectra, transitions into split excited states are not observed. We have analyzed the temperature-dependent measurements by performing a numerical decomposition of the 'hot lines' and obtain values partly different to those from Merk and coworkers.

In Fig. 1a this decomposition is demonstrated for a spectrum taken at 15 K. The low temperature lines I_1 and I_2 are accompanied by the 'hot lines' I_2^a, I_2^b, I_2^c, and I_2^d (we adopt the nomenclature given in [12]). Each 'hot line' is modeled by a curve of gaussian line shape. While amplitude and spectroscopic position of each line are varied to fit the spectrum, the half widths of the lines are assumed to be identical for a given temperature. Dashed lines represent the components obtained by the decomposition. A superposition formed hereof matches the measured spectrum almost perfectly. From the amplitudes the ratios $I_2/I_2^a \dots I_2/I_2^d$ are formed and plotted in an Arrhenius diagram, see Fig. 1b. The thermal activation energies obtained hereof correspond to the energetic spacings in the ground state.

Combining the activation energies and the spectroscopic data, we can set up a level scheme as shown in Fig. 1c. While the excited state appears to be two-fold split, it cannot be decided whether the ground state is separated into four or five levels. Though the Arrhenius diagram provides a consistent and reasonable fit, the general result has to be considered with some care. For instance, if the hot lines would be superimposed on a broad hump, which could also be thermally active, this could mislead the decomposition. In this hypothetical case the activation energies obtained would be to high. The origin of a complex ground state as followed hereof is unclear and not understood in terms of a simple j-j-coupling scheme.

Evidence of a splitting in the ground state of Si:Zn comes from experiments on phonon scattering [14], by studying excitations below 10 meV. Samples containing Zn^0 less than 10^{15} cm^{-3} were investigated by both optical absorption and phonon scattering. The samples clearly show the optical lines I_1 and I_2 of the Zn^0 excitation series. Phonon scattering experiments performed on these samples show a dominant peak in the phonon scattering 2.0 meV, assigned to an excitation within the ground state manifold of $Si:Zn^0$. Within the experimental error this value fits with the second lowest ground state level found in the optical absorption spectra. Further work is needed to identify the complete phonon scattering spectra with the optical data.

Uniaxial stress measurements

When stress is applied along the [100] direction the lines I_1, I_2, and I_4 split into several components. While some of these stress-induced lines can be followed for stress up to 150 MPa, for others the absorption drops rapidly. All transitions depend linearly on stress. For line I_2 four components ($I_{2.1} \ldots I_{2.4}$) can be resolved. $I_{2.1}$ and $I_{2.2}$, which are shifting to the low-energy region, vanish rapidly as soon as stress exceeds more than 25 MPa. On the other hand the absorption coefficients of $I_{2.3}$ and $I_{2.4}$ remain stable for an applied stress as high as 100 MPa. This fact is is characteristic of a two-fold electronic splitting of both, ground state and excited state. The components $I_{2.1}$ and $I_{2.2}$, origin from the level that shifts to the high-energy region, become depopulated with increasing stress as determined by Boltzmanns' law. The components $I_{2.1}$, and $I_{2.2}$, reflect this thermalization process also by sensitivity on temperature, not shown here. Combining spectroscopic data and thermalization properties, we can evaluate the splitting coefficient of the ground state to be 78 meV/GPa, while the excited EM state

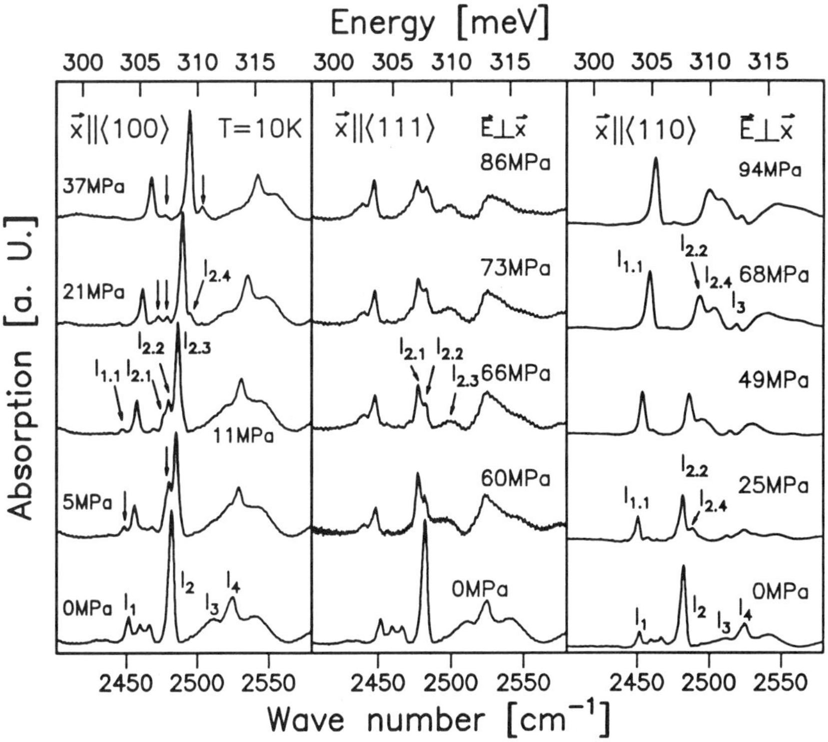

Figure 2: *Series of spectra of the $Si:Zn^0$ excitation series for external stress along [100], [111], and [110] direction.*

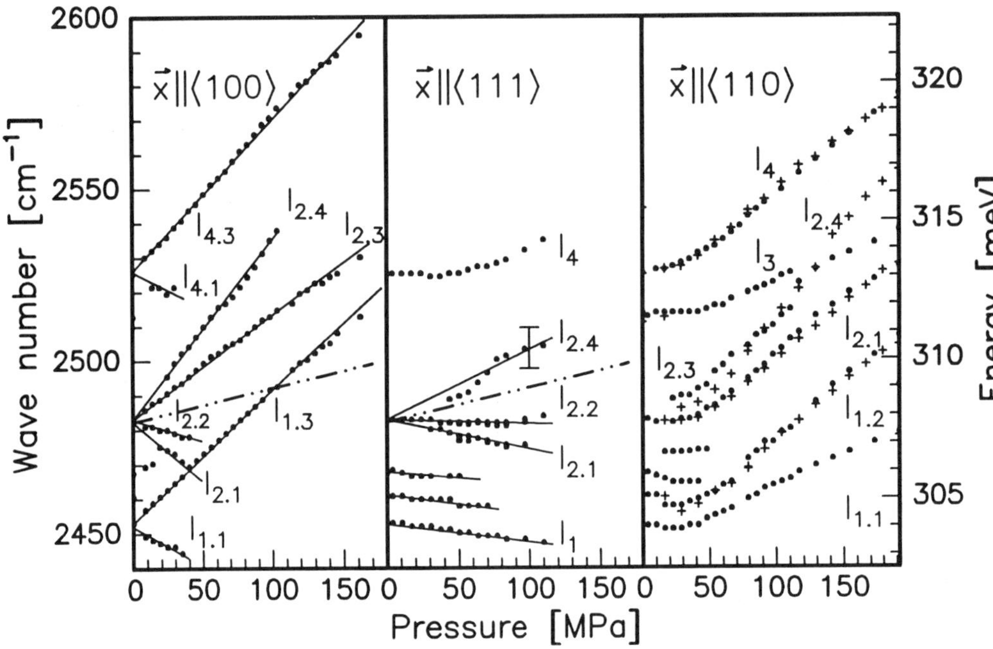

Figure 3: *Splitting of the Si:Zn^0 excitation series for stress along [100], [111], and [110], see text. The 'hot lines', not fully resolved vanish rapidly with increasing [100] stress.*

(labelled as level '2') is found to split by 28 meV/GPa. A splitting into two components is observed for lines I_1 ($I_{1.1}$ and $I_{1.2}$) and I_4 ($I_{4.1}$ and $I_{4.2}$), respectively. Similar to line I_2 the low energy components ($I_{1.1}$, $I_{4.1}$) vanish with increasing stress. Most probably I_1 and I_4 also split into four components like I_2, but because of weaker intensity only the two strongest transitions can be observed. The differences in the piezospectroscopic constants of corresponding lines are identical $((C_{4.1} - C_{4.2}) = (C_{1.1} - C_{1.2}))$ and also agrees with $(C_{2.1} - C_{2.3})$. The relations between the piezospectroscopic constants reflects the fact, that the observed lines I_1, I_2, and I_4 originate from the same ground state. A puzzling fact is, that the 'hot' lines I_2^a, I_2^b, I_2^c, and I_2^d drop even faster with increasing stress than the thermally active components $I_{2.1}$ and $I_{2.2}$.

For stress parallel to the [111] direction a three-fold splitting is observed for line I_2, see Fig. 2 and Fig. 3. The stress-induced components also depend linearly on stress, but the stress response is much smaller as in the case discussed above. The center of gravity, shifting isotropically with stress, indicates two high-energy components to exist. Since $I_{2.3}$ is much broader than $I_{2.1}$ and $I_{2.2}$, this feature may cover two unresolved high-energy components. Most probably for this stress direction I_2 splits into four components. It turns out component $I_{2.1}$ to be thermally active, while $I_{2.2}$ stays constant with increasing stress. So, the ordering of thermally stable and thermally active lines is inverted compared to the previous case. As a consequence hereof the magnitude of the stress response of ground state and excited state is inverted, too. We find the ground state to split by 12 meV/GPa. The excited EM state '2' splits by 26 meV/GPa, comparable to the value found for stress parallel to [100].

For the [110] stress direction considerable nonlinear splitting curves appear. The hamiltonian of the [110] stress splitting contains both, a tetragonal component ([100] stress) and a trigonal component ([111] stress), so nonlinearities observed for the [110] pattern alone is surprising and not understood up to now. Unfortunately, the resolution of the spectra are limited by inhomogeneities of stress, so the complete interaction cannot be revealed. A complete mathematical treatment is impossible. But, as found for the other stress directions two thermally stable stress induced components for line I_2 can be identified.

Discussion

Previously, it was argued from temperature-dependent measurements, the Γ_1 level to be the lowest ground state in Si:Zn^0 [8]. In contrast, the uniaxial stress measurements show that the ground state at least must have a two-fold electronic degeneracy.

Now, we refer to a paper recently published, to consider the stress response of j-j-coupled holes. Mathieu and coworkers [15] investigated this case, to describe the stress-splitting pattern of acceptor bound excitons (A^0X, **ehh**-complex) in InP. The j-j-coupled wave functions are considered to be split into a Γ_1, Γ_3, and a Γ_5 state manifold. The stress response of the single hol eis described by the deformation potential of the valence band. The treatment of the j-j-exchange interaction also applies to the system discussed in this paper. Two important features follow from this theory:

- low stress regime (stress splitting less than spacings of the unperturbed Γ_1, Γ_3 and Γ_5 states): An electronic splitting of the Γ_3 and Γ_5 states of j-j-coupled states only occurs in second order perturbation theory and therefore should result in curves depending quadratic on stress. i) for stress parallel to the [100] direction Γ_3 splits by interaction with Γ_1; but Γ_5 does not split. ii) for stress parallel to the [111] direction Γ_5 splits by interaction with Γ_1; but Γ_3 does not split.
- high stress regime (stress splitting sufficiently exceeds the Γ_1, Γ_3, and Γ_5 spacings): a linear progress is found with an asymptotic slope determined by the deformation potential of the valence band. Anisotropies for high stress should only result from the anisotropy of the deformation potential.

These two results of the model of Mathieu et al. do not fit to the stress response found in our measurements: The splitting of I_2 for stress along the [100] and [111] direction is linear within experimental uncertainty. Furthermore, the stress response found for the ground state appears to be highly anisotropic. The values $\Delta E/\Delta X$ = 78 meV/GPa for stress along [100] and $\Delta E/\Delta X$ = 12 meV/GPa for stress along [111] cannot be caused by the valence band anisotropy. Hence, the ground state does not react like a Γ_3 or Γ_5 state formed by j-j-coupling.

The whole stress splitting pattern of the Zn^0 transitions observed for stress along [100] indicate a ground state consisting of an uncoupled, anisotropic hole. The j-j coupling seems to be diminished or overruled by another interaction. A hint to this interaction is the anisotropic behavior. Anisotropies as found here, are observed in Jahn-Teller (JT) coupled states. In principle, a Γ_8 state can couple to phonons of tetragonal type (ϵ) and trigonal type (τ_2). The static JT effect lifts the electronic degeneracy by splitting the Γ_8 into two Kramers doublets, Γ_7 and Γ_6. If we consider a dynamic Jahn-Teller effect, the ground state would be a Γ_8 level again, but of vibronic nature. When the coupling to ϵ-type phonons is stronger than coupling to τ_2-type phonons a drastic reduction of the [111]-stress response due to the 'Ham' effect results [17]. This reduction may account for the anisotrope stress response observed in our measurements. However it remains unclear why the stress response of the ground state for stress along [100] is almost twice as large than the corresponding deformation potential of the valence band.

The excited ground states could be due to vibrational modes of the Jahn-Teller system, as found for shallow acceptors in silicon [18]. We recall that the 'hot lines' vanish rapidly when stress is applied parallel to [100] but not when stress is applied parallel to [111]. Stress along [100] affects the **X** minimum of the potential surface of the ϵ mode in a different way compared to the **Y** and **Z** minima. This energetic difference disturbs the vibrational modes. In contrast, [111] stress affects the three minima in the same sense and the potential for the nuclear motion is still symmetrical.

The two-fold split exited state (labelled '2') could be due to a possible Jahn-Teller splitting of the primarily bound Γ_8-like hole [18]. It remains unclear why optical transitions into the split excited state '2' are not found in the low temperature spectra.

Recently, Averkiev and coworkers [19] consider the double acceptor system GaAs:Cu^0 by a static Jahn-Teller effect. The authors study the Cu-related electron acceptor recombination band. (A^0,e). From measurements carried out under uniaxial stress, the authors concludes the ground state of GaAs:Cu^0 to be a Γ_8 state statically distorted by a tetragonal-type ϵ phonon.

Summary

By absorption measurements carried out under uniaxial stress we give evidence that in $Si:Zn^0$ the j-j exchange coupling is not the dominant interaction. We suggest in our optical absorption spectra to observe a Γ_8 hole, which is subject to a dynamic Jahn-Teller effect by coupling preferentially to tetragonal ϵ phonons. From the temperature-dependent measurements we obtain at least a four-fold split ground state. We presume that these states are the vibronic levels due to the dynamic Jahn-Teller effect.

Acknowledgement

We thank M.H. Pilkuhn for his steady interest in our work. For many helpful and stimulating discussions we are grateful to K. Laßmann and P. Groß.

References

[1] A.K. Ramdas and S. Rodriguez, Rep. Prog. Phys. **44**, 1297 (1981).

[2] V. Fiorentini and A. Baldereschi, Solid State Commun. **69**, 953 (1989).

[3] R.A. Chapman, W.G. Hutchinson, and T.L. Estle, Phys. Rev. Let. **17**, 132 (1966).

[4] M.L.W. Thewalt, D. Labrie, and B.P. Clayman, Solid State Commun. **56**, 751 (1985).

[5] M.L.W. Thewalt, B.P. Clayman, and D. Labrie, Phys. Rev. B **32**, 2663 (1985).

[6] M.A. Vouk and E.C. Lightowlers in *Proceedings of the Thirteenth International Conference on Physics of Semiconductors*, edited by F.G. Fumi (Tipographia Marves, Rome, 1976), p. 1098.

[7] A. Giesekus and L.M. Falicov, Phys. Rev. B **42**, 8975 (1990).

[8] A. Dörnen, K. Thonke, P. Stolz, G. Pensl, D. Grünebaum, and N.A. Stolwijk, Phys. Rev. B **40**, 12 005 (1989).

[9] E. Merk, J. Heyman, and E.E. Haller, Solid State Commun. **72**, 851 (1989).

[10] A. Baldereschi and N.O. Lipari in *Proceedings of the Thirteenth International Conference on Physics of Semiconductors*, edited by F.G. Fumi (Tipographia Marves, Rome, 1976), p. 595.

[11] A. Dörnen, K. Thonke, P. Stolz, G. Pensl, D. Grünebaum and N.A. Stolwijk in *Impurities, Defects and Diffusion in Semiconductors: Bulk and Layered Structures*, Mat. Res. Soc. Symp. Proc. **163**, edited by D.J. Wolford, J. Bernholc, E.E. Haller (Materials Research Society, 1990), p 21.

[12] E. Merk, J. Heyman, and E.E. Haller in *Impurities, Defects and Diffusion in Semiconductors: Bulk and Layered Structures*, Mat. Res. Soc. Symp. Proc. **163**, edited by D.J. Wolford, J. Bernholc, E.E. Haller (Materials Research Society, 1990), p 15.

[13] By completely resolving their spectra Merk et al. [12] observe 2 more components than [8], in [8] the corresponding lines are shown as shoulders on the three stronger lines

[14] P. Groß and K. Laßmann, private communication

[15] H.Mathieu, J. Camassel, F. Ben Chekroun, Phys. Rev. B **29**, 3438 (1982).

[16] A. Onton, P. Fisher, and A.K. Ramdas, Phys. Rev. B **163**, 686 (1967).

[17] E. Sigmund in *The Dynamical Jahn-Teller effect in Solids*, edited by Yu.E. Perlin and M. Wagner (North-Holland, Amsterdam, 1984), p. 495.

[18] K. Laßmann and H. Schad, Solid State Commun. **18**, 449 (1976).

[19] N.S. Averkiev, A.A. Gutkin, E.B. Osipov, V.E. Sedov, and A.F. Tsatsul'nikov, Sov. Phys. Solid State **30**, 841 (1988) – [Fiz. Tverd. Tela (Leningrad) **30**, 1459 (1988)].

Materials Science Forum Vol. 83-87 (1992) pp. 203-208

DIFFUSION OF GOLD IN AMORPHOUS SILICON

W. FRANK[1], W. GUSTIN[1], S. COFFA[2], J.M. POATE[2], AND D.C. JACOBSON[2]
[1] Max-Planck-Institut für Metallforschung, Heisenbergstraße 1, W-7000 Stuttgart 80, Federal Republic of Germany
[2] AT& T Bell Laboratories, 600 Mountain Avenue, Murray Hill, NJ 07974, USA

ABSTRACT

The diffusion of Au in a-Si has been studied by means of Rutherford-backscattering spectrometry and neutron-activation analysis in combination with serial sectioning. It is concluded that, like the other transition metals, Au undergoes direct interstitial diffusion that is retarded by temporary trapping at the defects intrinsic to the amorphous structure. As a result, Au diffuses faster in a-Si than in dislocation-free c-Si where it diffuses via the kick-out mechanism. While the same is true for Pt and Zn, Cu and Pd diffuse more slowly in a-Si than in c-Si where they are unretarded interstitial diffusers.

1 Introduction

Little was known about the fundamental mechanisms of diffusion in amorphous silicon (a-Si), until recently the present authors [1] have shown that transition metal diffusion in a-Si can be understood in terms of mechanisms which are closely related to those in crystalline silicon (c-Si). This is not surprising, since according to the continuous-random-network model [2] the local order in a-Si is maintained by the directional covalent bonding. On the other hand, it is clear that the high concentration of intrinsic defects in a-Si, which is indicated by the extremely high solubilities of transition metals (e.g., 0.02 at.% Cu at 400° C [3]), must influence diffusion processes.
Both the similarity to c-Si and the "imperfection" of a-Si are distinctly reflected in the diffusion of gold in a-Si, investigations of which will be presented in the present paper. After having introduced the experimental techniques and results (Sec. 2), the latter will be interpreted along the lines described above and compared to the diffusion of other transition metals in c-Si and a-Si (Sec. 3).

2 Experimental techniques and results

2.1 Rutherford-backscattering spectrometry (RBS)

The 2.1 μm thick a-Si layers used in our RBS studies were produced by self-implantation in c-Si at 77 K at different energies (500 keV, 1 MeV, 2 MeV) for a total fluence of 1.5×10^{20} ions/m^2. Subsequently, these layers were implanted with 10^{17} to 10^{20} Au ions/ m^2 at 500 keV. The diffusion anneals were performed in vacuum at temperatures between 350 and 550°C. The as-implanted Gaussian-shaped profiles and the diffusion profiles (Fig. 1) were measured by RBS using 4 MeV He ions and a scattering angle of 170 degree. Fitting of numerical solutions of the diffusion equation to the measured profiles, which was done under the assumption of total reflection at both the surface and the a-Si/c-Si interface, yielded estimates of the diffusion coefficients (full circles in Fig. 2). These show an Arrhenius-type temperature dependence with an activation enthalpy of (1.4 ± 0.1) eV.

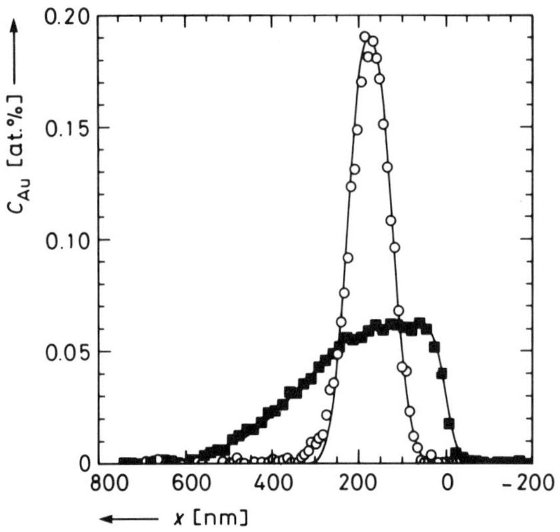

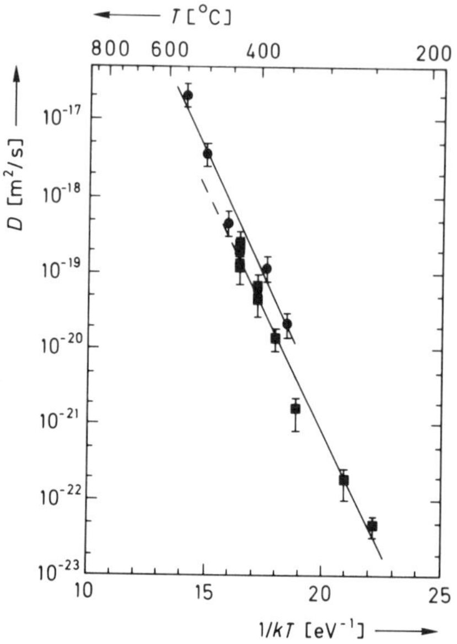

Fig. 1 (left): Gold-concentration profiles $C_{\mathrm{Au}}(x)$ in a 2.1 μm thick a-Si layer measured by RBS, as produced by implantation of 1×10^{19} Au ions/m² with 500 keV (empty circles) and after subsequent diffusion annealing for 1 h at 500°C (full squares). The solid curves represent fits of a Gaussian distribution to the as-implanted profile and a computer-simulated solution of the diffusion equation to the diffusion profile, respectively.

Fig. 2 (right): Arrhenius plot of diffusion coefficients of Au in a-Si measured by RBS (full circles) and NAA (full squares).

2.2 Neutron-activation analysis (NAA)

Our Au-diffusivity measurements were extended to temperatures as low as 250°C by a tracer-profiling technique involving NAA and serial sectioning. The overriding problem in this kind of investigations is that covering of a-Si with a layer of pure Au reduces the crystallization temperature from 650°C to about 160°C. This was avoided by using diffusion specimens in which the Au source contacting the a-Si was an amorphous Si−0.66 at.% Au alloy (Fig. 3).
In the production of the diffusion samples a 150 to 180 nm thick a-Si−0.66 at.%Au layer was deposited on a glass substrate by means of sputter deposition [4] using a 1 keV Au-ion beam. In a subsequent sputtering process this layer was covered by an a-Si layer of about the same thickness. After diffusion annealing and neutron activation the penetration profile of Au in the a-Si layer was determined. This was done by serial sectioning with the aid of an Ar-ion beam [5]. Starting from the top of the a-Si layer (Fig. 3) both the a-Si layer and the a-Si–Au layer were removed in 3 to 4 nm thick sections, and the radioactivity of the Au in the sputtered-off material was measured sectionwise. Fig. 4 shows Au diffusion profiles recorded in this manner. By fitting the appropriate solution of the diffusion equation to such profiles, diffusion

coefficients of Au in a-Si were determined in the temperature regime 250 to 430°C (full squares in Fig. 2). These obey an Arrhenius law with an activation enthalpy of (1.5 ± 0.1) eV.

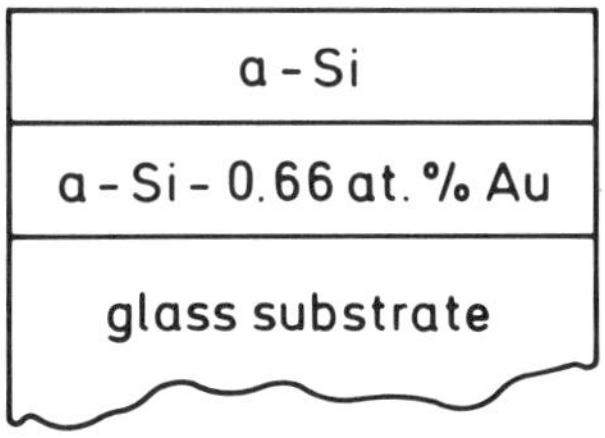

Fig. 3: Architecture of the specimens used in the NAA studies.

3 Discussion

Diffusion in c-Si is well understood [6]. In particular, it is generally accepted that most of the transition metals (e.g., Mn, Fe, Co, Ni, Cu, Pd) behave in c-Si as fast diffusers, which reside in interstices and diffuse by hopping from interstitial site to interstitial site. The coefficient D_{i} for this direct interstitial diffusion obeys an Arrhenius law

$$D_{\mathrm{i}} = D_{\mathrm{i,0}} \exp(-H_{\mathrm{i}}^{\mathrm{M}}/kT), \tag{1}$$

where $H_{\mathrm{i}}^{\mathrm{M}}$ is the migration enthalpy of the interstitials (e.g., 0.2 eV for Pd [6]) and the pre-exponential factor $D_{\mathrm{i,0}}$ and kT have their usual meanings. In a-Si the diffusion of these elements is slowed down considerably, as has been demonstrated for Cu and Pd [1]. A straightforward explanation of this is that transition metals diffusing interstitially in c-Si, do so in a-Si, too. However, in a-Si they become trapped temporarily by the defects intrinsic to the amorphous structure. Thus their effective diffusivity D_{i}^{*} is reduced. For high trap concentrations C_{t} it is given by [7]

$$D_{\mathrm{i}}^{*} = D_{\mathrm{i,0}}^{*} \exp\left[-(H_{\mathrm{i}}^{\mathrm{M}} + H_{\mathrm{it}}^{\mathrm{B}})/kT\right] \tag{2a}$$

with the pre-exponential factor

$$D_{\mathrm{i,0}}^{*} = D_{\mathrm{i,0}}/\alpha C_{\mathrm{t}} \tag{2b}$$

($H_{\mathrm{it}}^{\mathrm{B}}$ = binding enthalpy of the interstitials to the traps (e.g., 0.9 eV in Pd [1]), α = effective number of trapping sites per trap). Further pieces of evidence for this interpretation are the following findings: (i) A reduction of the defect concentration in a-Si by thermal annealing gives rise to an increase of $D_{\mathrm{i,0}}^{*}$ (cf. Eq. (2b)), but leaves the nature of the traps and thus the activation enthalpy in Eq. (2a) unchanged. (ii) At high concentrations of the diffusing elements, at which most of the traps are occupied and thus ineffective, the diffusivity increases.

The observations (i) and (ii) have also been made for the diffusion of Au in a-Si. For instance, the sets A, B, and C of the Au diffusion coefficients presented in Fig. 5 were measured on a-Si produced and/or pretreated in different ways. Obviously, these data sets differ in the pre-exponential factors of the diffusion coefficients, but not with regard to the diffusion enthalpy. This is in accordance with the prediction (2b) of the model of trap-retarded interstitial diffusion for changes in the trap concentration. However, a comparison of the Au diffusivities in a-Si (A

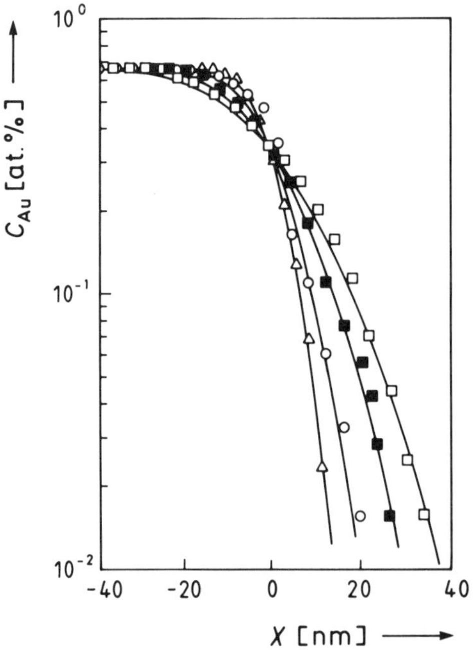

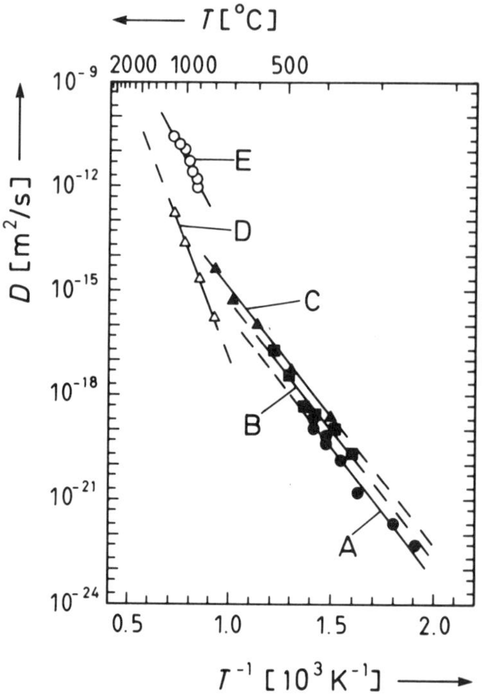

Fig. 4 (left): Gold-concentration profiles $C_{\mathrm{Au}}(x)$ in the a-Si ($x > 0$) and a-Si–Au layers ($x < 0$) of NAA specimens (Fig. 3) in the as-produced state (empty triangle) and after diffusion annealing at 430°C for 0.05 h (empty circles), 0.1 h (full squares), and 0.15 h (empty squares).

Fig. 5 (right): Comparison of the diffusion coefficients of Au in dislocation-free c-Si (D [8]), dislocated c-Si (E [8]), and a-Si (A, NAA data; B, RBS data; C [9]).

to C in Fig. 5) and dislocation-free c-Si (D in Fig. 5) seems to rule out this explanation, since in the regime of low temperatures accessible to studies of Au diffusivities in a-Si these are distinctly higher than the Au diffusivities measured on dislocation-free c-Si (or obtained by low-temperature extrapolation of such data).

A key to the reconciliation of the observations on the diffusion of Au in a-Si and c-Si is the well known fact that Au belongs to the few transition metals (Au, Pt, Zn) which in c-Si do not undergo fast interstitial diffusion, but diffuse via the so-called kick-out mechanism [6]. This mechanism results from the hybrid nature of these solutes, which spend most of their time sitting on substitutional sites (A_s). However, self-interstitials (I), which are present in c-Si under thermal-equilibrium conditions, may kick them out from the A_s-sites according to the reaction

$$A_s + I \rightarrow A_i. \tag{3}$$

In this way the hybrids temporarily become interstitials (A_i) which undergo fast long-range diffusion by the direct interstitial mechanism until reaction (3) is spontaneously reversed. *In c-Si without dislocations* (or other internal sinks for self-interstitials) the outflux to the specimen surface of the self-interstitials produced in this reverse reaction is the rate-controlling step in the in-diffusion of the substitutional fraction of a hybrid solute. This step makes the effective diffusivity of Au_s in dislocation-free c-Si (D in Fig. 5) considerably lower than the diffusivities of fast interstitial diffusers in c-Si such as Cu or Pd. *In dislocated c-Si* the self-interstitials produced in the in-diffusion of gold via the reverse reaction of (3) are instantaneously absorbed at dislocations, so that the rate-controlling process in the penetration of Au_s is the fast diffusion of Au_i atoms (E in Fig. 5). Finally, *in a-Si* the concentration of intrinsic defects is so extremely high that trapping of Au_i precedes and thus suppresses the reversion of reaction (3). As a result, the diffusion of gold in a-Si (A to C in Fig. 5) is controlled by trap-retarded interstitial diffusion. This is faster than the kick-out diffusion of gold in dislocation-free c-Si. As discussed above, in marked contrast to this, for transition metals diffusing purely interstitially in c-Si the diffusivity is lower in a-Si than in c-Si.

References

[1] S. Coffa, J.M. Poate, D.C. Jacobson, W. Frank, and W. Gustin, submitted to Phys. Rev. Lett.

[2] *Amorphous Silicon and Related Materials*, Vol. A and B, edited by H. Fritzsche (World Scientific, New Jersey, 1989).

[3] S. Coffa and J.M. Poate, to be published.

[4] G. Bilger and G. Bauer, Thin Solid Films **119**, 103 (1984).

[5] J. Horváth and H. Mehrer, Cryst. Latt. Def. and Amorph. Mater. **13**, 1 (1986).

[6] W. Frank, Def. and Diff. Forum **75**, 121 (1991).

[7] M. Koiwa, Acta Met. **22**, 1259 (1974).

[8] N.A. Stolwijk, J. Hölzl, W. Frank, E.R. Weber, and H. Mehrer, Appl. Phys. A **39**, 37 (1986).

[9] L. Calcagno, S.U. Campisano, and S. Coffa, J. Appl. Phys. **66**, 1874 (1989).

Materials Science Forum Vol. 83-87 (1992) pp. 209-214

VIBRONIC INTERACTION OF A GOLD-RELATED CENTER IN SILICON

S. GHATNEKAR, M. KLEVERMAN, AND H. G. GRIMMEISS
Department of Solid State Physics, University of Lund
Box 118, S-221 00 Lund, Sweden

ABSTRACT

A new line spectrum with a sharp zero-phonon line at 6125 cm^{-1} (759.5 meV) has been observed in silicon co-doped with iron and gold. Annealing and diffusion experiments strongly indicate the complex to have as constituents two substitutional nearest neighbor gold atoms and an interstitial iron atom along the <111> axis. Up to seven relatively sharp phonon replicas ($\hbar\omega$= 57.4 cm^{-1}) are observed showing interesting fine structures which are shown to be caused by an E⊗E dynamical Jahn-Teller effect.

1. Introduction

The optical properties of Au doped silicon are rather well known. The neutral single substitutional gold center (Au_s^0) gives rise to effective mass like acceptor [1,2] and donor [3] excitation spectra at 611 meV and 793 meV, respectively. In addition to these, several other line series consisting of a sharp zero-phonon line (ZPL) and their phonon replicas have been observed [4,5]. Two different phonon energies were found, depending on the conductivity type of the material: 14.4 meV (p-type and weakly n-type) or 13.1 meV (highly doped n-type), indicating that the center could be found in two different charge states. A trigonal symmetry was established for the 14.4 meV centers from Zeeman and uniaxial stress spectroscopy [5]. An isochronal annealing experiment was performed and the high temperature at which the centers annealed out indicated that the constituents are substitutional. It was therefore suggested that these are caused by excitations at a gold pair center (Au_{2s}) where the Au atoms reside on nearest neighbor substitutional sites (D_{3d} symmetry). Fe doped silicon shows a donor spectrum [6] at about 787 meV. It has been shown from a comparison of results obtained from photo-EPR and photo-conductivity measurements that it is due to excitations at the neutral interstitial iron impurity, Fe_i^0. The spectrum is caused by one-particle transitions from the localized ground state to shallow donor states.
A new spectrum is observed in Au-Fe co-doped silicon, with a ZPL at 6125 cm^{-1} (759.5 meV) followed by several phonon replicas with a phonon energy of about 7.1 meV. The spectrum is presented in Fig.1 and the splitting of the phonon replicas into two, three, and four components for the one, two, and three phonon replicas, respectively, is easily revealed. Also shown is the single substitutional gold donor spectrum.

2. Experimental details

Diffusion experiments were carried out on originally 14 Ωcm p-type (B-doped) FZ silicon.
Gold was rubbed or evaporated on the samples, which then were enclosed into a quartz ampoule. When co-doping with iron, in addition to the sample a small iron piece was enclosed in the ampoule. The samples were heat treated for 2 hours at about 1250°C and then quenched to room temperature in ethylene glycole. The spectra were recorded with a Bomem DA. 03 Fourier transform infrared spectrometer and the temperature was kept at about 9K using a liquid helium continous-flow Leybold cryostat. For the Zeeman experiments, an Oxford Instruments superconducting split-coil spectromagnet was used and the spectra were recorded at 1.9K.

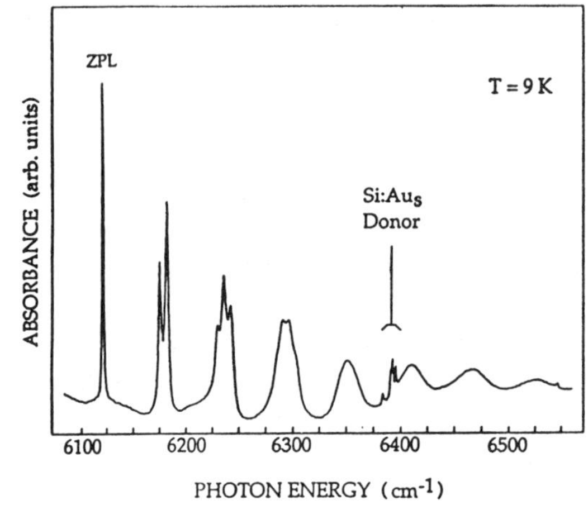

Fig. 1. An absorption spectrum for the Au-Fe center in silicon. The ZPL is located at 6125 cm^{-1}, followed by several phonon replicas with a phonon energy of 57.4 cm^{-1}.

3. Experimental results and discussion

Due to experimental difficulties, the role of iron as a constituent in this 6125 cm^{-1} center is, unfortunately, not completely understood. In some samples, which were only coated with gold prior to the heat treatment, the 6125 cm^{-1} spectrum appeared even though no co-diffusion of iron was performed. It is known from EPR investigations that, during the period these diffusion experiments were carried out, our furnace was contaminated by Fe, a fact that could explain the appearance of the spectrum in these samples. However, later annealing and diffusion experiments all give strong support for iron to be involved in this center.

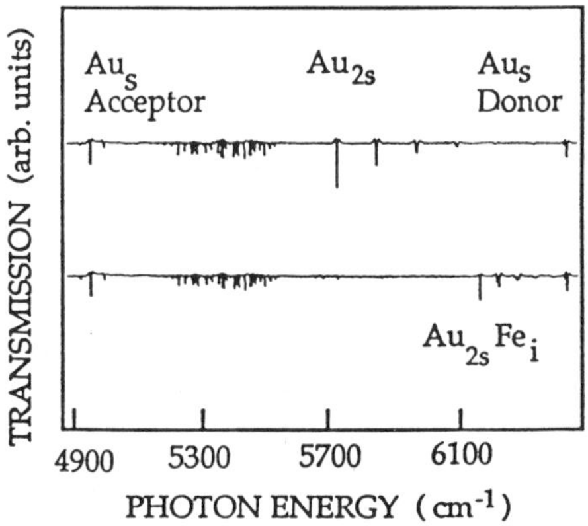

Fig. 2a) A spectrum from a gold doped sample where the Au_S^0 acceptor and donor as well as the Au_{2S} lines are shown and in b), the same sample after an additional Fe diffusion.

It is well known that Au and Fe usually occupies substitutional and interstitial sites in the silicon lattice, respectively. The simplest physical reasonable Au-Fe complexes would therefore consist of one interstitial iron atom and one or two substitutional gold atoms (Au_SFe_i or $Au_{2S}Fe_i$ respectively). A center identified as a Au_sFe_i center has been observed in EPR [7] whereas no EPR studies on a $Au_{2s}Fe_i$ center have sofar been reported.

Fig.2 a) shows a gold doped sample where the Au_s^0 acceptor and donor spectra as well as the Au_{2s} spectra are clearly identified. After a second diffusion procedure, in which the same sample together with a small iron piece was heat treated at 1250 °C for 40 min, the spectrum presented in Fig.2 b) was obtained. The 6125 cm^{-1} spectrum is now clearly identified. It is interesting to note that although the intensity of the Au_s^0 spectrum is almost unaffected by the second heat treatment, the intensity of the Au_{2s} spectrum shows a considerable decrease. It therefore seems plausible that interstitial iron forms a complex with the Au_{2s} center and not with the Au_s center. A second experiment supports this assumption. The Fe_i^0 spectrum is observed in co-diffusion samples using a fast quenching procedure. After annealing at room temperature for a couple of days, the Fe_i^0 spectrum is annealed out whereas the 6125 cm^{-1} spectrum has increased in intensity.

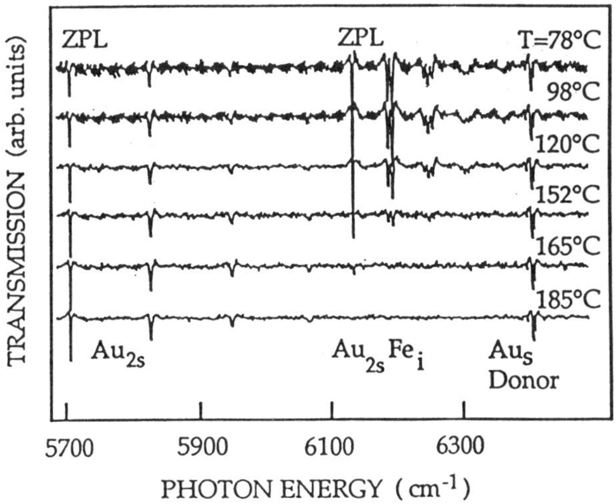

Fig. 3. Annealing characteristics of the new center. When the intensity of the new center decreases, it increases for the Au_{2s} lines, whereas the Au_s donor lines are unaffected.

The result from an isochronal annealing experiment is presented in Fig.3 and give additional proof for the 6125 cm^{-1} center to be a $Au_{2s}Fe_i$ complex. The overall annealing characteristics show that as the intensity of the $Au_{2s}Fe_i$ complex decreases that of the Au_{2s} centers increases whereas the Au_s donor lines are unaffected. Furthermore, the annealing temperature is about 160°C which is similar to that for Fe_i [6].

In EPR, only the previously reported Au_sFe_i spectrum [7] was observed in our samples indicating that the 6125 cm^{-1} center is non-paramagnetic or that the Au_sFe_i EPR center actually is a $Au_{2s}Fe_i$ complex. However, the latter seems unlikely since a hyperfine structure was only observed for one Au atom. All experimental results presented so far thus strongly support the assignment of a $Au_{2s}Fe_i$ complex for the 6125 cm^{-1} center.

The fine structure in the phonon replicas is believed to have a vibronic origin. The number of components for the replicas (n+1 for replica number n) is typical for a linear Jahn-Teller coupling between an E electronic state coupled to E vibrational modes (E⊗E) in the weak coupling limit [8]. This implies that the symmetry of the center must allow for two-dimensional irreducible representations (excluding Kramers doublets). The center could therefore have tetrahedral, tetragonal, or trigonal symmetry. Since the Au_{2s} center has a trigonal symmetry we must assume a trigonal symmetry for a $Au_{2s}Fe_i$ complex and that all three constituents line up in the <111> direction, the two Au atoms preferentially on nearest neighbor substitutional sites and the Fe atom on the nearest interstitial site.

At zero electron-phonon coupling, the electronic and vibrational motions are separable and the hamiltonian is given by $H_0=H_e+H_v$, where H_e is the electronic hamiltonian and H_v is that for a two-dimensional harmonic oscillator. The basis functions could be constructed as simple products between an electronic part and a phonon part, $|E,\mu\rangle|n_\theta\rangle|n_\varepsilon\rangle$, where μ indicates the row index ($\mu=\theta$ or ε) for the electronic E state and n_θ and n_ε are the number of phonons in each of the degenerate phonon states. The eigenvalues of H_0 are then given by $E_e+\hbar\omega(n+1)$, where $n=n_\theta+n_\varepsilon$, and E_e is the electronic energy. The electronic state is two-fold degenerate and the degeneracy of the oscillator is (n+1) thus the degeneracy of the nth electron-phonon state is 2(n+1). The Jahn-Teller effect will reduce this degeneracy but all states remain at least two-fold degenerate.

The coupling term is in trigonal symmetry given by $H_{JT}=V_E(Q_\theta U_\theta+Q_\varepsilon U_\varepsilon)$, where V_E is the coupling coefficient and U_θ and U_ε are the two-by-two matrices defined by Ham [5], which describe the effect on the electronic E state for a Q_θ and Q_ε distortion, respectively. In Fig.4 the result from a simple numerical calculation is presented. Basis functions up to n=7 were included in the calculation. As V_E is increased, the lines split into n+1 components as experimentally observed in the 6125 cm^{-1} spectrum. The eigenvalues depend only on the absolute value of V_E and for $|V_E|\approx 0.25\hbar\omega$, a perfect agreement is obtained between the calculated and experimental energies. Furthermore, it is possible to obtain an analytical eigenvalue solution in the weak coupling regime. However, we instead used a numerical calculation since we later need the eigenfunctions for a relative intensity calculation.

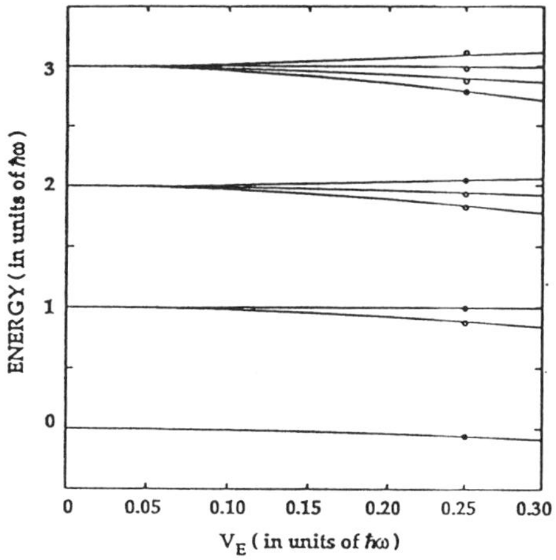

Fig.4. A numerical calculation of the splitting of the phonon replicas as a function of the coupling E⊗E coefficient V_E.

This simple analysis gives in this way a strong support for a dynamical E⊗E Jahn-Teller effect at the 6125 cm^{-1} center. In order to obtain a more detailed understanding, an intensity calculation was carried out. The final states have thus been identified above but now, the symmetry and the electron-phonon coupling in the initial state must also be known in more detail. It is assumed that the center has trigonal symmetry also in its ground state, which by no means is obvious since a static E⊗E Jahn-Teller effect could result in a monoclinic I symmetry. The coupling coefficient V_E for the final states was found to be small and in fact too small for assuming that no lattice relaxation takes place in the initial state. For zero or equal electron-phonon coupling (E-mode) in the ground state and excited states only the ZPL would be visible. Assuming a zero electron-phonon coupling in the ground state and $V_E=0.25\hbar\omega$ for the excited state, the upper component of the first phonon replica would be weakly visible in addition to an intense ZPL. This is easily understood since the lowest final state (ZPL) in the excitations has E symmetry and carries all electric-dipole strength for zero coupling. H_{JT} transforms as A_1 when rotating both the electron and phonon coordinates. Hence, it could only mix states of equal symmetry. The first phonon replica transforms as E+(A_1+A_2), the two singlets being accidentally degenerated for linear coupling. The Jahn-Teller effect in the excited state could therefore only transfer dipole strength to states of E symmetry. This implies that the electron-phonon coupling in the ground state must be considerably larger than that in the excited state. It should be noted that the distortions that have to be considered for the ground state must be of E symmetry. We know from the analysis above, that the excited state equilibrium position in the $Q_\theta Q_\varepsilon$-plane is close to origin. In order to account for the relatively intense phonon replicas, a considerable displacement in the $Q_\theta Q_\varepsilon$-plane must therefore prevail for the ground state. In trigonal symmetry, an E mode couples only to electronic states of E symmetry. The initial state must therefore also have E symmetry and subjected to an E⊗E Jahn-Teller effect.

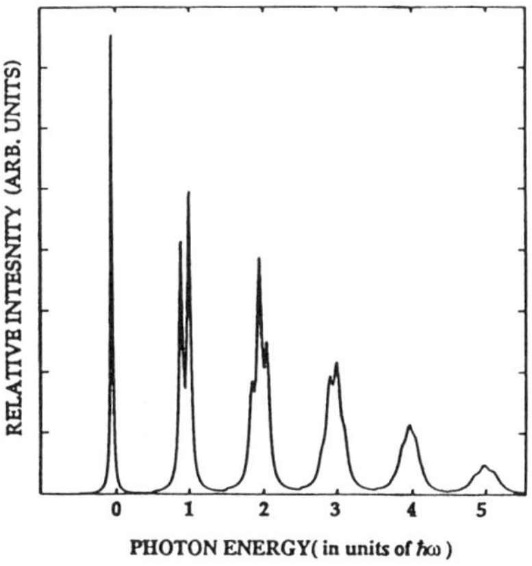

Fig. 5. Simulation of the absorption spectrum. See text for further details.

The relative intensities were calculated in the Condon approximation. The vibrational overlap integral m_{0j} between the phonon ground state and the jth excited state were assumed to obey the Huang-Rhys formula for a two-dimensional oscillator with a Huang - Rhys factor S of about 1. The electric-dipole operator transforms as A_1+E where the z axis is along the C_3 axis and constitutes a basis for the A_1 irreducible representation whereas x and y generate the E representation. The direct product E⊗E=A_1+E shows that the dipole must be considered as a superposition of a π-dipole along the axis and a σ-dipole perpendicular to the C_3 axis, described by two parameters a and e, respectively. It was found in the calculations that a reasonable fit could be obtained for $a/e=-1/\sqrt{2}$. It is interesting to note that this ratio is expected for an E state derived from a T_2 state in T_d symmetry. The calculated spectrum is presented in Fig.5 for $|V_E|=0.25\hbar\omega$, S=1, $a/e=-1/\sqrt{2}$, and using a Lorentzian line shape function and a line width which increases linearly with n. A direct comparison between the calculated spectrum in Fig.5 and the experimentally observed in Fig. 1 gives additional support for our model.

A Zeeman study has been carried out on this spectrum. No splittings were observed indicating that the center is non-paramagnetic. This may explain why only the Au_sFe_i center was observed in EPR.

An uniaxial stress study on the 6125 cm^{-1} ZPL gave conflicting results. The ZPL splits into 2, 4, and 3 components for uniaxial stress parallel to <100>-, <110>-, and <111>-directions, respectively. The number of components equals that expected for a monoclinic I center [10,11] or an A - E transition for an trigonal center [12]. However, neither model could be fitted to the observed polarization pattern and both models are in conflict with the analysis above which lead to an E - E transition at a trigonal center. Another possibility is an E - E transition at a trigonal center [13] but in order to obtain the correct number of components, one has to assume that the E ground state not splits by the stress. Also in this case, however, the observed polarization pattern could not be explained. The stress results are, at the present, therefore not fully understood. One possibility that has to be tested is that the assumption of a trigonal symmetry in the ground state is wrong. A static E⊗E Jahn-Teller could lead to a static distortion which reduces the ground state symmetry to monoclinic I. A different symmetry in the ground state would lead to altered polarization rules and this possibility definitely has to be tested. Such work is in progress.

4. Summary

A new spectrum with a ZPL at 6125 cm^{-1} has been investigated in detail. Annealing and doping experiments give strong support for the center to be a $Au_{2s}Fe_i$ complex. The fine structure of the phonon replicas has been shown to originate from an E⊗E Jahn-Teller effect.

Acknowledgements: We gratefully acknowledge discussions with G. Grossmann and financial support from the Swedish Natural Science Council and the Royal Swedish Academy of Sciences.

REFERENCES

1. G. Armelles, J. Barrau, M. Brosseau, B. Pajot, and C. Naud, Sol. St. Commun. **56,** 303 (1985).
2. M. Kleverman, J. Olajos, and H. G. Grimmeiss, Phys. rev. **B35,** 4093 (1987)
3. D. Thebault, J. Barrau, G. Armelles, N. Laurent, and J. P. Noguier, Phys. Stat. Sol. (b) **125,** 357 (1984).
4. K. Ohta, Sci. Light 22, 12 (1973).
5. M. Kleverman, J. Olajos, and H. G. Grimmeiss, Proc. Int. Conf. on the Physics of Semiconductors, Thessaloniki, 1990.
6. J. Olajos, B. Brian Bech Nielsen, M. Kleverman, P. Omling, P. Emanuelsson and H. G. Grimmeiss, Appl. Phys. Lett. **53,** 2507 (1988).
7. R. L. Kleinhenz, Y. H. Lee, J. W. Corbett, E. G. Sieverts, S. H. Muller, and C. A. J. Am merlaan, Phys. Stat. Sol.(b), **108,** 363 (1981).
8. For a review on the E⊗E Jahn-Teller effect see e.g. M. D. Sturge in Solid State Physics, edited by F. Seitz, D. Turnbull, and H. Ehrenreich (Academic Press Inc., New York, 1967), Vol. 20. F. S. Ham in Electron Paramagnetic Resonance, edited by S. Geschwind (Plenum, New York, 1972). I. B. Bersuker and V. Z. Polinger in Vibronic Interactions in Molecules and Crystals, Springer Series in Chemical Physics Vol. 49, (Springer- Verlag Berlin Heidelberg 1989) and references therein.
9. F. S. Ham, Phys. Rev. 166, 307, (1968).
10. A. A. Kaplyanskii, Optics Spectrosc.,**16,** 329 (1964).
11. G. Davies, E. C. Ligthowlers, M: Stavola, K. Berman, and B. Svensson, Phys. Rev. **B35,** 2755 (1987).
12. A. E. Hughes and W. A. Runciman, Proc. Phys. Soc., **90,** 827 (1967).
13. G. Davies and M. H. Nazare, J. Phys. C: Solid St. Phys., **13,** 4127 (1980).

Materials Science Forum Vol. 83-87 (1992) pp. 215-220

THE GOLD CENTER IN SILICON

A THILDERKVIST[1], M KLEVERMAN[1], G D WATKINS[2], H G GRIMMEISS[1]
[1] **Department of Solid state Physics, University of Lund, Box 118, 221 00 Lund, Sweden**
[2] **Department of Physics, Building 16, Lehigh University, Bethlehem, PA 18015, USA**

ABSTRACT

The electronic structure of the neutral gold center in silicon has been revealed from Zeeman and uniaxial stress studies on the acceptor and donor excitation spectra at 611 meV and 793 meV, respectively. The center is paramagnetic (S=1/2) and has a static <100> tetragonal distortion with $g_{\parallel}\approx2.8$ and $g_{\perp}\approx0$. We observe reorientation effects even at 1.9K, thus establishing the isolated substitutional nature of the defect. The results give for the first time direct evidence for the acceptor and donor levels to be two levels of the same defect. The fact that $g_{\perp}\approx0$ can explain the failure to detect Au_s^{o} by EPR.

1. Introduction

Gold related defects in silicon and particularly the supposingly single substitutional Au center (Au_S) have been of interest for a long time. In spite of all the experimental efforts no details concerning the electronic structure of the neutral Au_S center (Au_S^{o}), nor any confirmation about how it is incorporated in the lattice, have been achieved.
It was established already in 1957 from Hall measurements that the Au_S center gives rise to an acceptor level at E_V+0.62 eV and a donor level at E_V+0.35 eV [1]. These studies were later confirmed by a number of junction space charge experiments [2-5], but also inferred conflicts whether the two levels resulted from the same defect or not.
EPR has proven to give detailed information about the structure of deep centers and several gold-related defects have in fact been observed [6-9]. However, no EPR spectrum has so far been reported that could be associated with the Au_S^{o} impurity. This fact is puzzling, since Au_S^{o} is supposed to behave similarly to the isoelectronic Pt^- center which gives rise to an intense and well studied EPR signal [10-14].
Optical measurements on Au-doped silicon have resulted in spectra with sharp zero phonon structures identified as excitations to effective mass like acceptor (at 611 meV) [15,16], and donor (at 793meV) [17] states. The fine structure of these spectra suggested a lower symmetry core but it was never established whether this was a result of a complex defect or an intrinsic distorsion of the isolated center. Neither has it been established whether the two sets of transitions occur from the same defect or not.
The Zeeman and uniaxial stress results for Au_S^{O} to be presented in this paper establish a simple model that answers some of the important questions on the electronic structure of Au_s^{o}.

2. Experimental details

The samples were prepared by evaporating gold onto n-type, 250Ωcm, floatzone silicon and then introduced into quarz ampoules which were evacuated and sealed off prior to the heat treatment at 1300 °C for 69 hours. The diffusion procedure was followed by a rapid quench in ethylene glycole. The samples were cut in smaller pieces in the main crystallographic directions, 2x2x8 mm^3 for the stress samples and 4x4x10 mm^3 for the Zeeman samples.

All transmission spectra were obtained with a Bomem DA.03 Fourier transform spectrometer equipped with a liquid-nitrogen-cooled InSb photodiode detector.

For the Zeeman experiments, an Oxford Instruments superconducting split-coil spectromagnet in Voigt-configuration was used. The spectra were measured at 1.9K. For the stress measurements a liquid-helium cooled Leybold continuos-flow cryostat was used and the spectra were recorded at approximately 10 K.

3. Experimental results and discussion

In Fig.1, we show the Zeeman spectra for **B**||<100>, for the 793 meV donor. At B=0, four sharp zero-phonon lines are observed, and we label them A, B, B′ and C, according to the convention adopted by Thebault et al [17], who first observed these lines in photoluminescence. Although the Zeeman spectra are complex, showing level anti-crossing effects among the components emanating from the B, B′ and C lines, a simple pattern is apparent as soon as the lines emerge and become linear (Paschen-Back regime). Each of the lines split into four components labeled i=1-4. Each component with a common index i displays the same slope. Components with index 1 and 3 decrease in intensity with increasing magnetic field, while components with index 2 and 4 increase. In addition to this, the A-lines, which are the only lines with linear behavior throughout the range of the magnetic field, show a splitting between the A_1 and A_3 lines which is identical to that between the A_2 and A_4 lines. This splitting would for S=1/2 correspond to g=2.

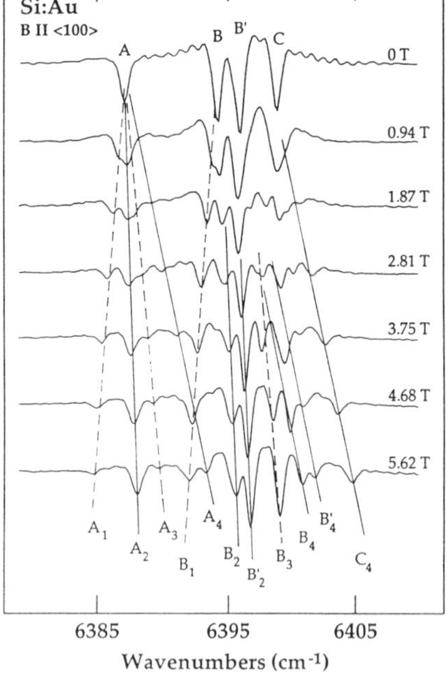

Fig.1 Zeeman spectra for the 793 meV donor recorded at 1.9K. The dashed and full lines indicate transitions originating from centers having their tetragonal axis perpendicular or parallel to the magnetic field, respectively.

These observations lead to the simple model described in Fig.2a for **B**||<100>. Here we consider the ground state to have S=1/2 and a tetragonal distorsion along a <100> direction. Defects distorted parallel to the magnetic field direction split by an amount described by $g_{||}$ ($\Delta E=\pm g_{||}\mu_B B/2$), while perpendicular defects do not split ($g_{\perp}=0$). The paramagnetic electron is excited out of the ground state into effective mass like donor states, where orbital effects are established to be small, and thus the electron is characterized accurately by S=1/2 with a nearly isotropic $g_e=2$.

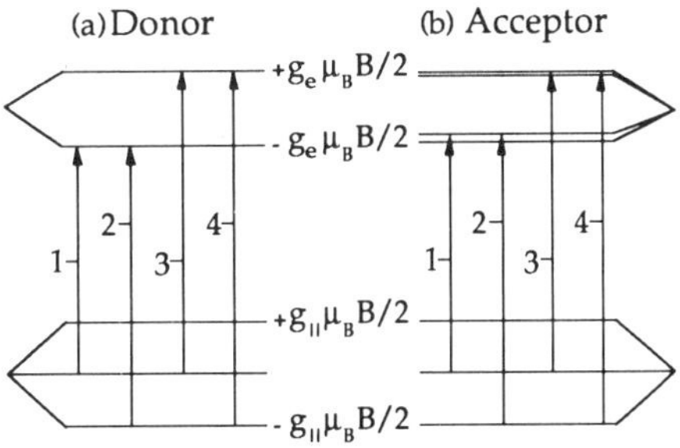

Fig.2 Schematic picture showing the simple electronic model for Au_s^{O} with B||<100> as discussed in the text.

The spectral transitions with index i=1-4, observed in Fig.1 are indicated in the diagram in Fig.2a. Transitions from the upper Zeeman split level are not observed due to thermalization. According to this model, the energy splitting between the 1-3 and the 2-4 components arise purely from the excited effective mass states and we can estimate $g_{||}=2.8$ from the A_2 energy shift $(g_{||}-g_e)\mu_B B/2$ and the A_4 shift $(g_{||}+g_e)\mu_B B/2$. The fact that line 1 and 3 which arise from defects perpendicular to the magnetic field, decrease in intensity, while 2 and 4 (parallel defects) increase, can now be explained as a reorientation of perpendicular to parallel defects. Zeeman spectra for **B**||<110> and **B**||<111> confirm this interpretation. The excited states still split according to Fig.2a but the ground state splitting is altered, now with $g\approx 2.8\cos\theta$ where θ is the angle between the magnetic field and each of the differently oriented defects. Thus for **B**||<111> no reorientation effects are observed since all defects have the same angle to the field, while for **B**||<110> a reorientation from defects perpendicular to the field, to defects oriented 45° to the field is seen.

The acceptor excitation spectra for **B**||<100> presented in Fig.3 show similar effects. The strongest I2 line, identified as excitation to the $2P_{5/2}(\Gamma_8)$ effective mass hole state, splits into four components, qualitatively illustrated in Fig.2b. The excited state is assumed to split with $g_e\approx 1.7$ according to results determined by Merlet et al [18] for the boron $2P_{5/2}(\Gamma_8)$ state. From this we estimate $g_{||}=2.9$ for the ground state in close agreement with the result for the donor spectra. Again we observe an increase in intensity for the lines arising from parallel defects ($I2_2$ and $I2_4$) while the components deriving from the perpendicular defects decrease in intensity. The spectra for **B**||<110> and **B**||<111> confirm our model.

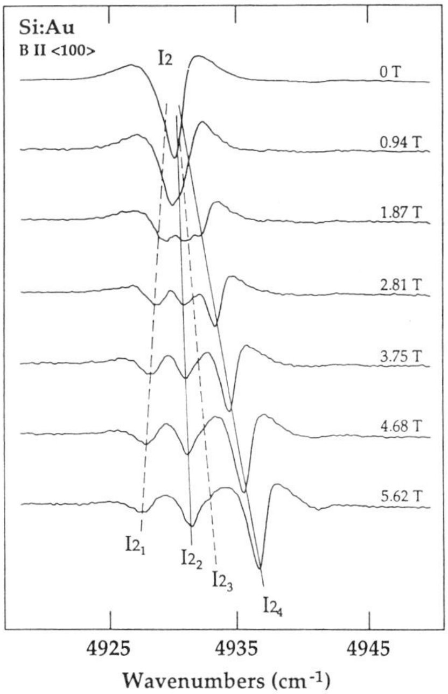

Fig.3 Zeeman spectra for the $2P_{5/2}(\Gamma_8)$ line of the 611 meV acceptor recorded at 1.9K.

In order to establish that the gold center really reorients freely at such low temperatures as 1.9K, we have in Fig.4. plotted the intensity changes versus magnetic field for both the decreasing (perpendicular defects) and increasing (parallel defects) components of the donor and acceptor. We have determined the intensities from components which are clearly resolved and well separated from level anti-crossings where states could be mixed and thus change the intensity. The solid lines are predicted intensities for a Boltzmann distribution between the ground state Zeeman energy levels of Fig.2.

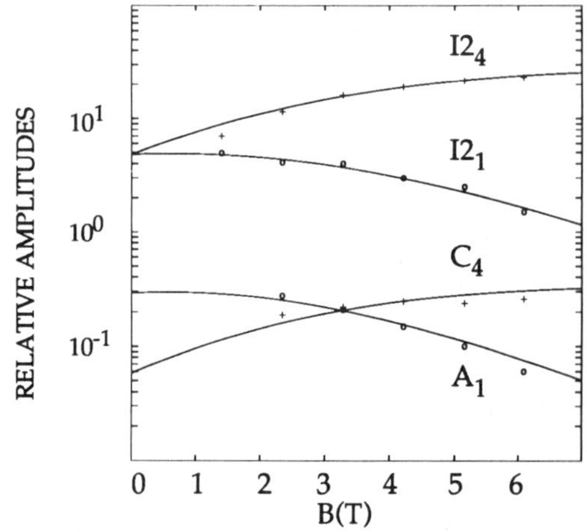

Fig.4 The intensities of donor and acceptor lines originating from perpendicular and parallel defects with B||<100> as a function of magnetic field. The solid lines are the predicted intensities for a Boltzmann distribution in the ground state.

The Zeeman results now allow us to draw some very important conlusions about the Au_s center. The observation of the same ground state g value and same reorientation effect for the donor and the acceptor provides the first direct confirmation that they originate from the same center. The reorientation also establish that Au_s indeed is an isolated center and that the tetragonal distortion must be caused by an intrinsic effect and not by an impurity in the vicinity of the gold center.

In Fig.5 the splittings of the donor lines under uniaxial stress are presented. For stress parallel to <111> only a common shift of the lines is observed whereas the splitting behavior at low stress parallel to <100> and <110> is very complex and the components are difficult to resolve. At higher stress, however, the splitting is typical for shallow donor states in silicon and the components follow different conduction-band valleys. The excited states are therefore believed to be constructed from the 1s manifold which in T_d are split into three states: 1s(A_1), 1s(E), and 1s(T_2) being one-, two-, and three-fold degenerate (excluding spin), respectively. In the case of Au_s^o, a lower symmetry core and spin-orbit interaction may reduce the degeneracies as well as mix the states in the manifold.

The fact that no splitting is observed for <111> is in accordance with our tetragonal model since all defects have the same angle to the stress axis. The complex splitting pattern for the other two stress directions is believed to be caused by reorientation effects which for

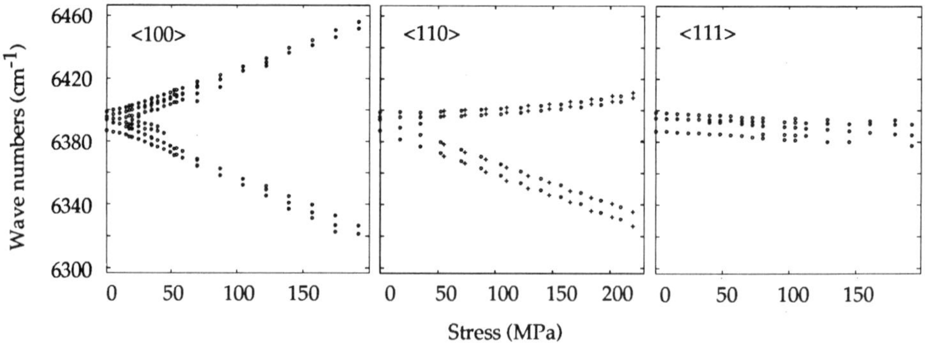

Fig.5 The uniaxial stress results for the donor lines.

higher stress result in a preferential alignment.
At zero stress four lines are observed. The $1s(T_2)$ state can at most only account for three and the fourth must therefore be derived from $1s(A_1)$ and/or 1s(E). The T_2 state splits as the valleys whereas the A_1 and E_θ states must be mixed by the stress in order to account for the observed splitting. It is therefore believed that all six 1s states are involved in the spectrum and that some residual degeneracies account for the fact that only four lines are observed.
The energy level structure of the excited states has not yet been unambiguously identified. However, the non-linear response of the lowest line for stress parallel to <100>, its Zeeman splitting, and its polarization properties enable us to identify it as mainly derived from the $1s(A_1)$ state.
The Zeeman and stress data give a coherent picture of the Au_s^o ground state structure and that it is similar to that for Pt^- which also is an S=1/2 center with a static <100> tetragonal distortion although the g-anisotropy is smaller ($g_{||}$=2.1, $g_\perp$=1.4). The fact that $g_\perp \approx 0$ for Au_s^o explains why EPR investigations have failed [19].
The EPR data for Pt^- has succesfully been explained by two different models, a vacancy model [20,21] and a d^9 model where the Pt atom preferentially bonds to only two silicon neighbors [22]. The clear evidence for reorientation between different inequivalent tetragonal distortions in the Au_s^o Zeeman experiments strongly favors the vacancy model. Calculations show that Au_S introduces an anti-bonding t_2 level in the gap [23] which for Au_s^o is filled with three electrons. In the vacancy model, the observed tetragonal symmetry is caused by a Jahn-Teller E_θ distortion. However, such a distortion does not lead to an energy minimum in the adiabatic potential unless many-particle effects are included [24]. Nevertheless, an E_θ distortion is experimentally observed which splits a t_2 level into a singlet B_2 and a doublet E state. Treating the t_2 state as an effective L=1 state we associate the B_2 state with M_L=0 and the E state with $M_L=\pm1$. The spin-orbit interaction splits the E state into two spin doublets conveniently characterized by $M_J=\pm3/2$ and $M_J=\pm1/2$ where $M_J=M_S+M_L$ as well as mixes the $M_J=\pm1/2$ state with the M_L=0 spin doublet. Using these states as a starting point we note that in order to obtain a g value of about 2.8 the initial state must be one of the spin doublets derived from the $M_L=\pm1$ states. Transitions from a z oriented center to both spin up and down excited states are clearly observed in the A line $(1s(A_1))$ <100>-Zeeman spectrum. According to the vacancy model, this is only possible if the spin-orbit and Jahn-Teller energies are about the same order which then allows for a mixing of the $M_J=\pm1/2$ and the M_L=0 states. The g value for the $M_J=\pm1/2$ states is given by $2g_L$-g_S which shows that g_L must be negative i.e., it derives from the d-like part of the t_2 state.
Both the donor and acceptor spectra show strong polarization effects. A straightforward group theoretical analysis on the polarization properties of the $2P_{5/2}(\Gamma_8)$ acceptor line for **B**||<100> shows, assuming that the results in Ref. 19 are correct, that the initial state must have Γ_7 or Γ_5 symmetry (in S_4). The Γ_5 state has to be rejected since it is a pure spin up state (M_J=3/2). Both the M_J=1/2 and the M_L=0 spin up state belong to Γ_7 and the most physical choice for the initial state is the M_J=1/2 state in agreement with the findings from the donor spectrum.

4. Summary

The Au_s center in Si is a tetragonally distorted isolated center with S=1/2 and $g_{||}$=2.8 and $g_{\perp}\approx 0$ and can reorient freely at cryogenic temperatures. The acceptor and donor level arise from the same defect and the observation that $g_{\perp}\approx 0$ explains why no EPR signal has been detected for Au_s^0.

Acknowledgements: We gratefully acknowledge financial support from the Swedish Natural Science Foundation.

References

1. C. B. Collins, R. O. Carlson, and C. J. Gallagher, Phys. Rev. **105**, 1168 (1957)
2. A. F. Tasch, and C. T. Sah, Phys. Rev. **B1**, 800 (1970)
3. D. V. Lang, H. G. Grimmeiss, E. Meijer, and M. Jaros, Phys. Rev. **B22**, 3917 (1980)
4. L-Å. Ledebo and Zhan-Guo Wang, Appl. Phys. Lett. **42**, 680 (1983)
5. J. Utzig and W. Shröter, Appl. Phys. Lett. **45**, 761 (1984)
6. G. W. Ludwig and H. H. Woodbury in Solid St. Physics ed. by F. Seitz and D. Turnbull (Academic Press, New York) **13**, 223 (1962)
7. M. Höhne, Phys. Stat. Sol. (b) **99**, 651 (1980)
8. R. L. Kleinhenz, Y. H. Lee, J. W. Corbett, E. G. Sieverts, S. H. Müller, and C. A. J. Am merlaan, Phys. Stat. Sol. (b) **108**, 363 (1981)
9. M. Höhne, Phys. Stat. Sol. (b) **109**, 525 (1982)
10. H. H.Woodbury and G. W. Ludwig, Phys. Rev. **126**, 466 (1962)
11. J. C. M. Henning and E. C. J. Egelmeers, Phys. Rev. **B27**, 4002 (1983)
12. R. F. Milligan, F. G. Anderson, and G. D. Watkins, Phys. Rev. **B29**, 2819 (1984)
13. P. Omling, M. Kleverman, P. Emanuelsson, J. Olajos, and H. G. Grimmeiss, Sol. St. Commun. **65**, 557 (1988)
14. F. G. Anderson, R. F. Milligan, and G. D. Watkins, to be published.
15. G. Armelles, J. Barrau, M. Brosseau, B. Pajot, and C. Naud, Sol. St. Commun. **56**, 303 (1985)
16. M. Kleverman, J. Olajos, and H. G. Grimmeiss, Phys. Rev. **B35**, 4093 (1987)
17. D. Thebault, J. Barrau, G. Armelles, N. Laurent, and J. P. Noguier, Phys. Stat. Sol.(b) **125**, 357 (1984)
18. F. Merlet, B. Pajot, Ph. Arcas, and A. M. Jean-Louis, Phys. Rev. **B12**, 3297 (1975)
19. F. G. Anderson, Ball. Am. Phys. Soc. **36**, 862 (1991)
20. G. D. Watkins, Physica **117&118**, 9 (1983)
21. F. G. Anderson, F. S. Ham, and G. D. Watkins, to be published
22. C. A. J. Ammerlaan and A. B. van Osten, Physica Scripta **T25**, 342 (1989)
23. A. Fazzio, M. J. Caldas, and A. Zunger, Phys. Rev. **B32**, 934 (1985)
24. F. G. Anderson, F. S. Ham, and G. Grossmann, at this meeting

Materials Science Forum Vol. 83-87 (1992) pp. 221-226

ELECTRONIC STATES OF Fe_4 AND Mn_4 CLUSTERS IN SILICON

JOSÉ R. LEITE[1], LUCY V.C. ASSALI[1] AND ANTONIO T. LINO[2]
[1]Instituto de Física, Universidade de São Paulo, CP 20516, 01498 São Paulo, SP, Brazil
[2]Departamento de Ciências Físicas, Universidade Federal de Uberlândia, CP 593, 38400 Uberlândia, MG, Brazil

ABSTRACT

We have carried out SCF spin-polarized one-electron state calculations for the Fe_4 and Mn_4 complexes in silicon. The results were obtained by using a molecular cluster model and the multiple-scattering $X\alpha$ theory. It has been found that Fe_4 is a high spin complex ($S = 4$) in agreement with EPR experiments. The ground state of the Mn_4 complex has a closed shell configuration, therefore is a non paramagnetic state. It is suggested that the controversial results $S = 2$ and $S = 6$ found by EPR for the spin of the complex are related to excited electronic states originated from the several 3d-derived impurity levels found in the band gap. Excited spin states of the Mn_4 complex up to $S = 6$ have been calculated and compared with the EPR measurements.

1. Introduction

It is known from pioneering electron paramagnetic resonance (EPR) experiments that when Si samples containing only Mn impurities are cooled slowly from elevated temperatures a dominant paramagnetic center composed of four Mn atoms is formed.[1,2] The complex gives rise to a spectrum displaying 21 hyperfine lines with a fine structure characteristic of $S = 2$ and $L = 0$ values. The symmetry of the spectrum shows that the four Mn atoms are equivalent and arranged in a tetrahedron, probably at the nearest interstitial sites. Recently the Mn_4 cluster was also identified by EPR in high resistivity p-type and n-type Si doped with Mn.[3] The analysis of the observed spectrum leads to the value $S = 6$ for the ground state of the complex. Although these two EPR experiments consistently ascribe an orbital singlet state ($L = 0$) for the complex, they lead to contradictory values for the total spin used to describe the observed spectra.

Besides the discrepancies concerned to the EPR data, the reported positions of the energy levels induced by the complex in the Si band gap are also quite controversial. The photo-EPR results suggest a donor level (0/+) at 0.5 eV below the bottom of the conduction band. The values 0.28 eV and 0.54 eV have been obtained from Hall-EPR and deep level transient spectroscopy-EPR measurements, respectively.[3,4]

It has been also shown from recent EPR measurements that the annealing of Fe-doped high-resistivity Si samples yields a $S = 4$ spectrum which can be ascribed to a complex consisting of four interstitial Fe atoms.[5] The Fe_4 and Mn_4 clusters are very good prototypes to study the interaction mechanisms between transition metal impurities in Si. Nevertheless, very little is known about their microscopic and electronic structures.

In this work we have carried out electronic structure calculations for the Fe_4 and Mn_4 complexes in Si. The starting point of our investigations is the microscopic model which has been proposed for the Mn_4 complex in Si where the four Mn impurities, assumed in the neutral charge state, are placed at nearest-neighbor tetrahedral interstitial sites.[1-3]

2. Theoretical Model

The molecular cluster model and the multiple-scattering $X\alpha$ theory have been used to perform the calculations reported here.[6-9] The results were obtained by adopting initially a 17Si-atom cluster in order to define the band edges of the perfect Si crystal. The simulation of the TM_4 complex in Si (TM = Fe or Mn) was performed by adopting a cluster with 21 atoms comprizing a central Si atom surrounded by its 4 nearest neighbors and 12 next-nearest neighbors and 4 TM atoms placed at the interstitial sites nearest to the Si central atom. The "defect" cluster is schematically shown in Fig. 1.

Touching muffin-tin atomic spheres having the same radii were chosen, consistent with the Si lattice parameter. The exchange parameters adopted were the corresponding atomic values for the Si, Fe and Mn spheres.[10] The value 0.72751 was used for the outer region. For the interatomic region the values 0.72446 and 0.72741 were used for the Fe_4 and Mn_4 cluster, respectively. The inner atomic shells up to 2p in Si were kept frozen as in the free atom. All the electrons of the TM atoms were included in the self-consistent calculations. The basis set for the expansion of the wave functions included $\ell = 0, 1, 2$ for the outer region, TM and central Si atoms. For the remaining spheres the values $\ell = 0, 1$ have been used.

3. Results

In Fig. 2 the Fe 3d-derived impurity levels induced by the Fe_4 complex in the Si bands are shown. By carrying out the calculations to the spin-polarized limit, we verify from Fig. 2 that the Fe_4 complex has a total spin $S = 4$. Twenty electrons fill all the levels up to $2a_1(\uparrow)$ and twelve electrons fill the remaining levels up to $1a_1(\downarrow)$. The total number of electrons filling the 3d-derived impurity levels is 32 as expected from the $3d^6 4s^2$ electronic configurations of the Fe atom. We have calculated the hyperfine contact interaction at the iron nucleus by taking into account contributions from core, valence and impurity states. The obtained result $A = 3.5 \times 10^{-4}\ cm^{-1}$ is about four times smaller than the observed values for the Mn_4 complex.[1-3] No resolved hyperfine interaction with Fe was observed in the EPR spectrum of the Fe_4 complex.[5] Although it has been observed that the Fe_4 complex undergoes a trigonal distortion, our results show that it is not of Jahn-Teller origin.[5] The calculated ground state of the Fe_4 complex corresponds to an orbital singlet.

The resonances and gap levels induced by the Mn_4 complex in the band structure of Si are shown in Fig. 3. All the levels depicted in Fig. 3 have strong contributions from the 3d atomic orbital of Mn and a large charge concentration on the four impurities. The first interesting feature which emerges from the results shown in Fig. 3 is that the ground state of the complex has a total spin $S = 0$. The 28 electrons originated from the $3d^5 4s^2$ electronic configuration of the Mn impurity fill all the levels from the $1t_2$ to the 2e which is occupied by four electrons. Therefore, no EPR sig-

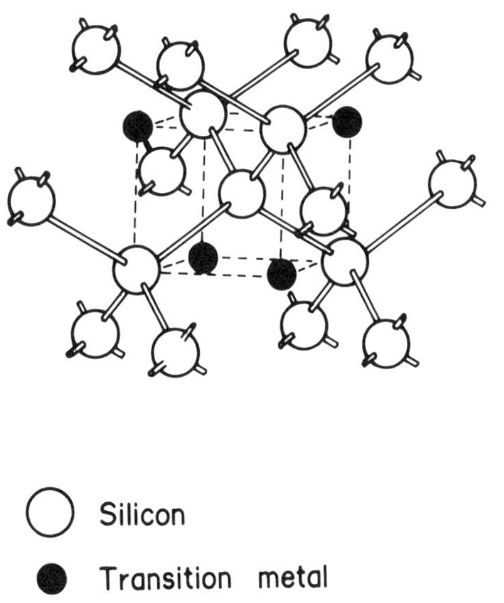

Figure 1. Molecular cluster model used to simulate the microscopic structure of the Fe_4 and Mn_4 complexes in Si .

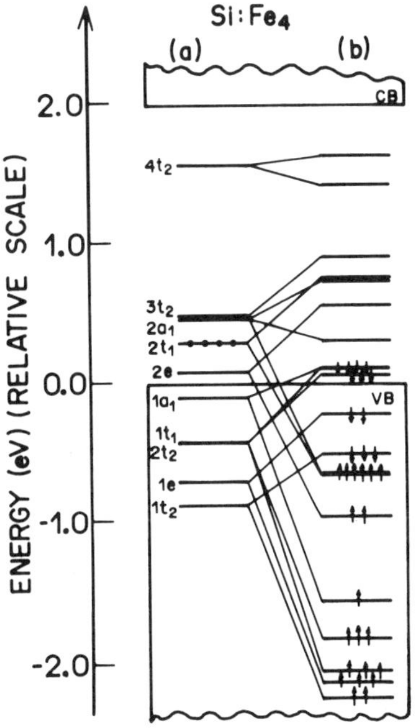

Figure 2. (a) unpolarized and (b) spin-polarized 3d-derived impurity levels induced by the Fe_4 complex in Si .

nal is expected from this closed shell system. However, the calculations show that both, the Fe_4 and Mn_4 complexes induce a sequence of unoccupied states in the gap which are close enough to favor excited states with different values for the total spin. This finding may solve the controversy related to the fact that the Mn_4 complex have been observed in two spin configurations. In order to exploit this idea we have simulated several spin configurations for the complex, by populating the lowest unoccupied impurity levels and by carrying out the calculations to the spin-polarized limit. The excited states of the Mn_4 complex for the total spin values $S = 2$, 4 and 6 are shown in Fig. 3. By taking into account the contributions from the core, valence and impurity states to the hyperfine contact field at the Mn nucleus we calculated the parameter A . The value -31.9×10^{-4} cm^{-1} was found for the spin state $S = 6$ which is comparable with the measured value, $-13.7 \times 10^{-4} cm^{-1}$.[3] For the spin state $S = 2$ the calculations lead to the value 5.0×10^{-4} cm^{-1} , which is also comparable with the result obtained from the first EPR reported data, 12.7×10^{-4} cm^{-1} .

4. Conclusions

The calculations reported here allow us to conclude that the four interstitial TM atoms in the complexes are examples of strong interacting impurities. Let us consider the Mn_4 complex for our analysis once the Fe_4 complex shows similar behavior.

Fig. 3 shows, for the case $S = 0$, the spread of the Mn_4 3d states

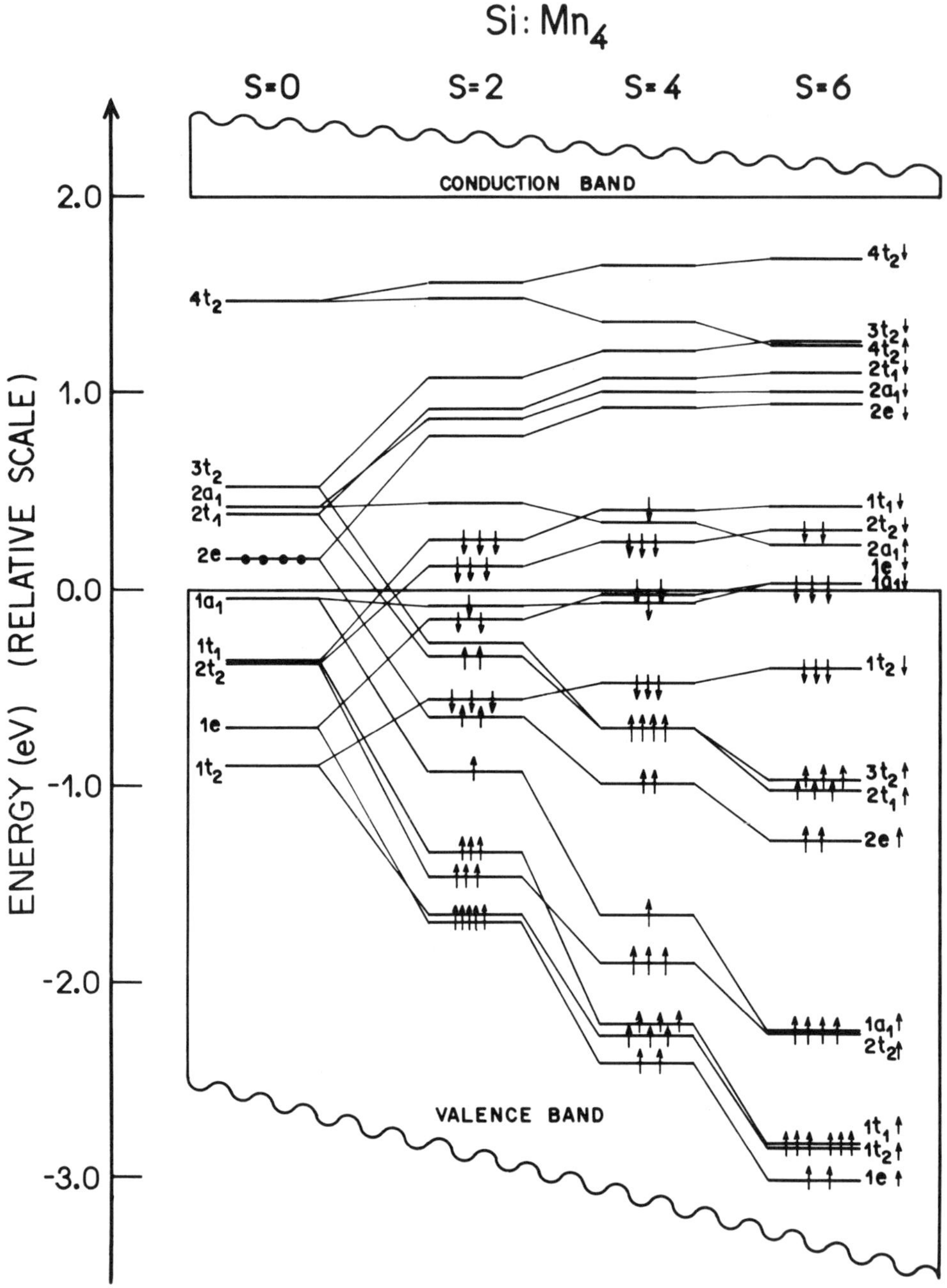

Figure 3. Mn 3d-derived impurity levels of the Mn_4 complex in Si corresponding to the values 0, 2, 4 and 6 of the total spin S. The arrows indicate occupancies of the up(↑) and down(↓) spin states.

which occurs when we bring the four Mn atoms together to form the complex. The band width of the bonding states, which runs from $1t_2$ to $3t_2$ in Fig. 3, span a region of 1.43 eV. This spread, which is mostly due to the Coulomb Interactions between the d states of different Mn impurities, is about five times larger than the crystal field effect on the d states of the interstitial Mn impurity. The crystal field effect on the d states of the isolated impurity is about 0.3 eV according to previous cluster model calculations.[11]

If the spin-polarization effects are operative, as in the case S = 6 for example, the spread of the Mn 3d states, which runs from $1e\uparrow$ to $3t_2\downarrow$ states shown in Fig. 3, increases to 4.1 eV. This spread is even larger, 4.7 eV, if we take into account the fact that the $4t_2\downarrow$ state has 45% of charge inside the Mn shell. In other words, about 16 times larger than the crystal field effect on the d states of the isolated interstitial impurity.

In order to interpret the EPR experiments it has been assumed that the magnetic properties of the complexes can be inferred directly from the magnetic properties of the isolated TM impurities.[1-3,5] This very simple model is based on the assumption that the interactions between the TM impurities in the complexes are small. The strong localization of the 3d-derived orbitals on each TM impurity is a fundamental requirement to preserve their magnetic properties. Our calculations do not support such hypotheses, showing that the different spin configurations of the complexes do not arise from the coupling among the four equivalent independent spins but originates from the different electronic populations of impurity states induced by the complexes which are spread over the entire cluster.

REFERENCES

1. G.W. Ludwig, H.H. Woodbury, and R.O. Carlson, J. Phys. Chem. Solids **8**, 490 (1959).
2. G.W. Ludwig and H.H. Woodbury, Solid State Phys. **13**, 223 (1962).
3. J. Kreissl and W. Gehlhoff, Phys. Status Solidi B **145**, 609 (1988), and references therein.
4. H. Conzelmann, Appl. Phys. A **42**, 1 (1987).
5. S.H. Muller, G.M. Tuynman, E.G. Sieverts, and C.A.J. Ammerlaan, Phys. Rev. B **25**, 25 (1982).
6. L.V.C. Assali, J.R. Leite, and A. Fazzio, Phys. Rev. B **32**, 8085 (1985).
7. J.R. Leite, L.V.C. Assali, and V.M.S. Gomes, Lecture Notes Phys. **301**, 75 (1988).
8. L.V.C. Assali and J.R. Leite, Phys. Rev. B **36**, 1296 (1987).
9. L.V.C. Assali and J.R. Leite, Solid State Commun. **58**, 577 (1986).
10. K. Schwarz, Phys. Rev. B **5**, 2466 (1972).
11. G.G. DeLeo, G.D. Watkins, and W.B. Fowler, Phys. Rev. B **23**, 1851 (1981).

Materials Science Forum Vol. 83-87 (1992) pp. 227-232

DIFFUSIVITIES OF 3D TRANSITION-METAL IMPURITIES IN SILICON

H. Nakashima[1] and K. Hashimoto[2]

[1] **Department of Electrical Engineering, Kyushu University, 6-10-1 Hakozaki Higashi-ku, Fukuoka 812, Japan**
[2] **Department of Physics, Kurume Institute of Technology, 2228-66 Kamitu-machi Kurume, Fukuoka 830, Japan**

ABSTRACT

Diffusivities of Ti, V, Mn and Fe in Si have been investigated in several different temperature ranges. The diffusivities of Ti, V and Fe in a high temperature range are determined from an in-depth profile measurement by deep level transient spectroscopy (DLTS), and those of Ti and V in a lower range are obtained from an annealing experiment which employs a technique for profiling the concentration of a deep level within a depletion region. The obtained profiles are well described by the solution of Fick's equation, and the fit yields the diffusivity of each element. In a room temperature range those of Mn and Fe are studied by monitoring pair reactions of these impurities with substitutional B. From these measurements based on DLTS, the diffusivity of each element is obtained in a wide scale of reciprocal temperature.

1. Introduction

Recently, properties of 3d transition-metal (TM) impurities in Si have received growing interest.[1,2] All 3d-TM are believed to migrate via a simple interstitial diffusion mechanism at a high temperature, and metals from Ti to Fe can be quenched in on interstitial sites. In general, interstitial 3d-TM impurities have four different charge states. Consequently, three levels, i.e., a double donor, a donor and an acceptor, can be measured by deep level transient spectroscopy (DLTS) with two exceptions of Cr and Fe which exhibit only one donor state. Among these elements, Cr, Mn and Fe migrate even at around room temperature because of rather high diffusivity, and positively charged donors of the impurities are captured by negatively charged substitutional B. Hence, the pairs of the donors with B also can be detected by DLTS. Figure 1 shows our DLTS results for deep levels of respective elements,[3-6] where A represents the acceptor, D the donor and DD the double donor: subscripts I and P represent the interstitial and the pair of each donor with B, respectively. These results are in good agreement with published data.[1,2]

In this study, we focus our attention on DLTS signals corresponding to following impurity levels: E_C-0.27 eV (D_I) for Ti, E_C-0.20 eV (A_I) for V, E_V+0.32 eV (DD_I) and E_C-0.50 eV (D_P) for Mn, and E_V+0.40 eV (D_I) and E_V+0.10 eV (D_P) for Fe. By monitoring the concentrations deduced from these DLTS signals, diffusivities of Ti, V, Mn and Fe have been investigated in several temperature ranges. From these experimental results, we show the diffusivity of each element in a wide 1/T scale (T: absolute temperature), and diffusivities of 3d-TM impurities in Si are discussed by comparing with published data.

2. Experimental procedure

Substrates used were P-doped n-type and B-doped p-type Si wafers with shallow impurity concentrations ranging from 2×10^{13} to 4×10^{15} cm^{-3}.

Wafers about 1200 μm thick were used to obtain the diffusivities of Ti, V and Fe in a high temperature range. The wafers were cut into several slices. After standard cleaning procedure of a sliced sample, each metal was evaporated in

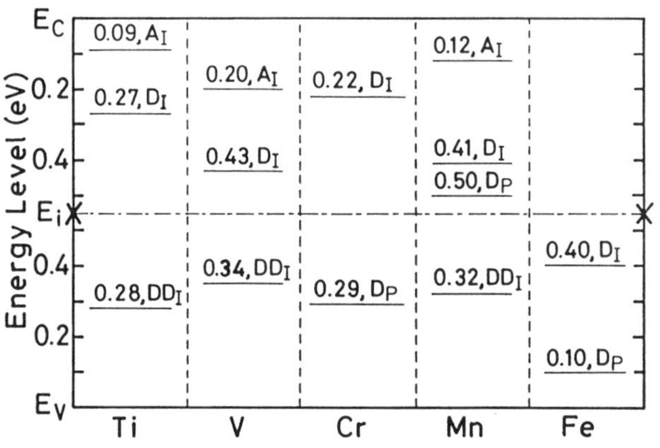

FIG.1. Impurity levels for respective elements from Ti to Fe in Si. A: acceptor, D: donor, DD: double donor, Subscript I: interstitial, Subscript P: pair of each donor with B.

vacuum onto both surfaces of a slice by direct heating of a high-purity wire in which a large electric current was flowing. The slice deposited with each metal was set in an evacuated quartz tube and heat-treated at a temperature between 800 and 1200℃ for a fixed duration, and then sequentially quenched to liq. N_2 temperature. An in-depth profile measurement was performed by DLTS after etching the surface layer little by little and then forming the Schottky barrier with evaporated Au for the n-type sample and with Al evaporation for the p-type sample each time.

Starting materials used for the study on diffusivities of Ti and V in a lower temperature range were n-type samples (thickness: about 500 μm) diffused with Ti at 1200℃ for 15 h and with V at 1170℃ for 24 h, which insured uniform distributions. After fabrication of an evaporated Au Schottky diode, the concentration profile of each element within a depletion region was measured with the profiling technique[7] of DLTS. The measurement for the case of Ti was performed under the condition of reverse bias V_R=20 V and the filling pulse heights V_P ranging from 0 to 20 V in 0.5 V increments, and that for V was V_R=40 V and V_P from 2 to 39 V in 1 V increments. The barrier metal(Au) was then removed with aqua regia, and the sample was annealed at a temperature between 600 and 800℃ for a fixed period; and the Schottky diode fabrication and the spatial distribution measurement were performed again.

B-doped p-type wafers (thickness: 500 μm; B concentration: 1.2×10^{15} cm^{-3}) were used for obtaining diffusivities of Mn and Fe in a room temperature range. For the case of Mn, an n^+p junction wafer fabricated using P diffusion was employed. After a surface of the sliced n^+p junction sample was coated with $MnCl_2$, the coated sample was heat-treated at 1000℃ for 90 min in a quartz ampoule filled with Ar, and the heat treatment was terminated by quickly dropping the ampoule in water. For Fe diffusion, a slice deposited by the direct heating of an Fe wire was heat-treated at 1025℃ for 90 min in vacuum and sequentially quenched. These samples were stored at liq. N_2 temperature prior to the fabrication of mesa-type n^+p diode (Mn) or Schottky diode (Fe) and the subsequent measurement procedure of DLTS.

3. Results and discussion

Profiles of the concentration N(x, t) at depth x for several diffusion temperatures are shown in Fig. 2, where ℓ is the sample thickness and t the diffusion time. It is assumed for the case of Fe of Fig. 2(c) that the sum N(x, t) of concentrations of interstitial Fe and Fe-B pair is interstitial Fe concentration immediately after quenching. All profiles are well described by the solution[6] of Fick's equation assuming a constant surface concentration (solubility) N_S at x=0 and x=ℓ. The fits of the experimental data to the solution are shown by respective solid lines in Fig. 2. The fit yields the solubility N_S and the diffusivity D for each element. The obtained N_S for Ti and

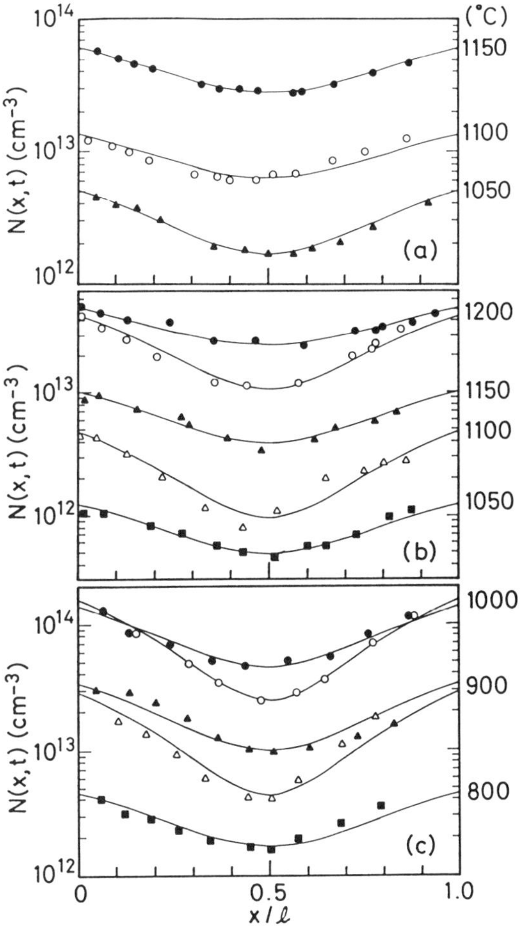

FIG. 2. In-depth profiles of the concentrations of (a) Ti, (b) V, and (c) Fe.

(a) ● : 1150℃, 48h, ℓ=0.116cm,
○ : 1100℃, 72h, ℓ=0.117cm,
▲ : 1050℃, 108h, ℓ=0.103cm.

(b) ● : 1200℃, 8h, ℓ=0.117cm,
○ : 1200℃, 4h, ℓ=0.117cm,
▲ : 1150℃, 12h, ℓ=0.121cm,
△ : 1100℃, 8h, ℓ=0.109cm,
■ : 1050℃, 24h, ℓ=0.118cm.

(c) ● : 1000℃, 8min, ℓ=0.140cm,
○ : 1000℃, 5min, ℓ=0.136cm,
▲ : 900℃, 7min, ℓ=0.111cm,
△ : 900℃, 5min, ℓ=0.114cm,
■ : 800℃, 20min, ℓ=0.114cm.

Fe are in fairly good agreement with the published solubility data,[1,8] and N_S for V is about two orders of magnitude lower than that of Fe. The diffusivity D_{Ti} of Ti in the range 1050-1150℃, D_V in the range 950-1200℃ and D_{Fe} in the range 800-1070℃ will be shown later.

The concentration profiles obtained from annealing experiments for Ti donor and V acceptor are shown in Figs. 3(a) and (b), respectively, where annealing temperature is 600℃. The samples without annealing have uniform distributions of $1.4x10^{14}$ cm^{-3} (Ti) and $2.1x10^{13}$cm^{-3} (V), and the concentrations are in good agreement with the solubilities. By contrast, the concentration profiles for annealed samples show gradual decrease with decreasing distance x. The annealing behavior implies that each element only out-diffuses from the bulk toward the surface without precipitation. Thus, it can be assumed that the diffusion equation for the annealing process, as well as in-diffusion, is given by Fick's equation with the initial condition $N(x, 0)=N_{in}=1.4x10^{14}$ cm^{-3} for Ti ($N_{in}=2.1x10^{13}$ cm^{-3} for V) and the boundary condition $N(0, t)=N_{an}$, where N_{an} is the solubility of each element at the annealing temperature. The solid lines in Figs. 3 (a) and (b) represent the profiles fitted to the solution of Fick's equation with just one parameter D. We could thus determine D_{Ti} at 600, 700, 750 and 800℃, and D_V of at 600, 700 and 800℃.

The pairing reactions of interstitial Mn double donor (Mn_i^{++}) with substitutional B acceptor (B_s^-) and of interstitial Fe donor (Fe_i^+) with B_s^- were examined by monitoring the DLTS spectra after storage for a fixed period at a storage temperature t_s. The concentrations of Mn_i^{++} and $(Mn_iB_s)^+$ are shown in Fig. 4 (a) as a function of storage time. In the same way, the results for Fe_i^+ and $(Fe_iB_s)^0$ are shown in Fig. 4 (b). With the storage time, the concentrations of Mn_i^{++} and Fe_i^+ decrease, those of $(Mn_iB_s)^+$ and $(Fe_iB_s)^0$ increase, and each sum

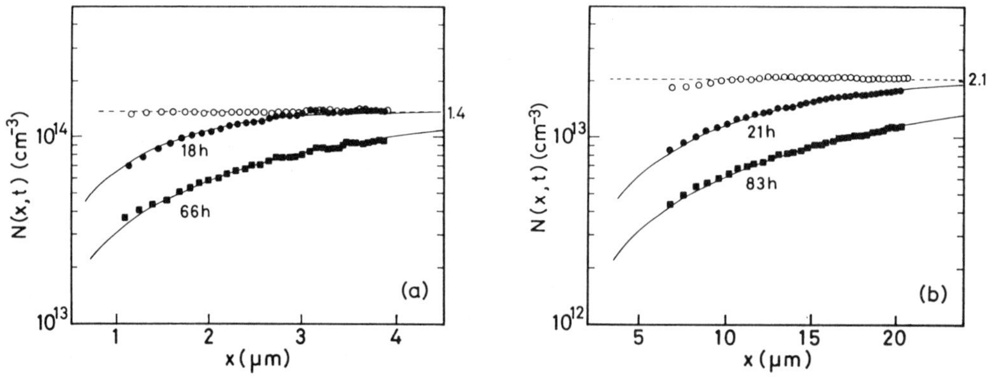

FIG. 3. Concentration profiles of (a) Ti and (b) V. Annealing temperature is 600℃. Annealing times for the case of Ti are 18h (●) and 66h (■), and those for V are 21h (●) and 83h (■). The starting materials are n-type samples in-diffused with Ti at 1200℃ for 15h and with V at 1170℃ for 24h.

remains constant. However, there is a clear distinction between the storage time dependence of Mn_i^{++} and that of Fe_i^{+}. The concentration of Mn_i^{++} continues monotonically decreasing to the detectable limit, whereas that of Fe_i^{+} approaches to a constant value with increasing the storage time. According to a two-point charge Coulomb interaction model,[9] it is expected that a pair binding energy of $(Mn_iB_s)^{+}$ is two times larger than that of $(Fe_iB_s)^{0}$, which has been also verified experimentally.[4] Thus, the difference of the behaviors is explained that $(Mn_iB_s)^{+}$ is hard to dissociate in comparison with $(Fe_iB_s)^{0}$.

Let us now assume that the pairing reactions of Mn_i^{++} with B_s^{-} and of Fe_i^{+} with B_s^{-} are given by

$$Mn_i^{++} + B_s^{-} \underset{k_d}{\overset{k_a}{\rightleftarrows}} (Mn_iB_s)^{+}, \qquad (1)$$

$$Fe_i^{+} + B_s^{-} \underset{k_d}{\overset{k_a}{\rightleftarrows}} (Fe_iB_s)^{0}, \qquad (2)$$

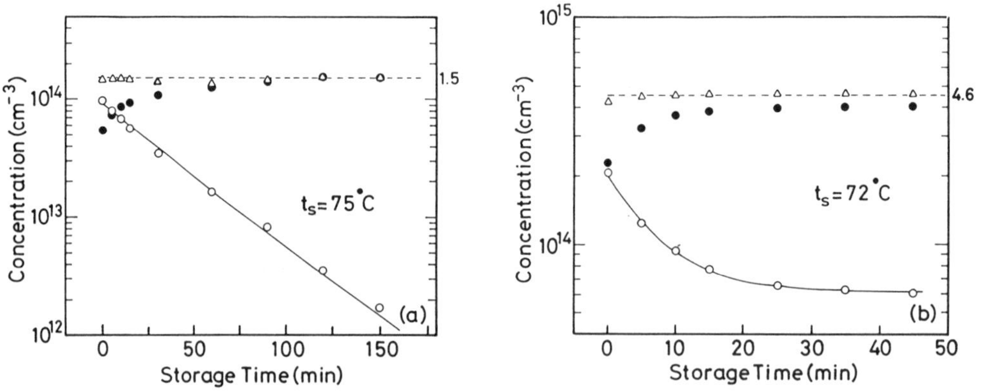

FIG. 4. Storage time dependence of (a) Mn_i^{++} concentration at storage temperature t_s=75℃ and (b) Fe_i^{+} concentration at t_s=72℃. (a) ○: Mn_i^{++}, ●: $(Mn_iB_s)^{+}$, △: sum of the concentrations of Mn_i^{++} and $(Mn_iB_s)^{+}$. (b) ○: Fe_i^{+}, ●: $(Fe_iB_s)^{0}$, △: sum of the concentrations of Fe_i^{+} and $(Fe_iB_s)^{0}$.

where k_a and k_d are the association and dissociation constants for each pair, respectively. When the sum of the concentrations of interstitial donor and the pair remains constant, k_a can be obtained from fitting of the experimental data to the solution[4,5] of the rate equation for the above reactions. The solid lines in Figs. 4 (a) and (b) represent the storage time dependence fitted to the solution. The diffusivity D of interstitial species as a mobile donor can be deduced from the value of k_a according to the equation,[4,10] $D = k_a \varepsilon kT/q$ (ε: dielectric constant of Si, k: Boltzmann constant (eV/K), T: absolute temperature(K), q: elementary electric charge). Hence, we can determine the diffusivity D_{Mn} in the temperature range 14-90℃ and D_{Fe} in the range 0-72℃.

The diffusivity D of each element obtained in the present study is shown in Fig. 5 as a function of absolute temperature T. Each solid line represents diffusivity obtained from a least square fit of all data for each element, and the diffusivity D can be represented as follows:

$$D_{Ti} = 1.2\text{x}10^{-1} \exp(-2.05/kT)\ \text{cm}^2\text{s}^{-1} \quad 873 \leq T \leq 1423\text{K}, \quad (3)$$
$$D_{V} = 9.0\text{x}10^{-3} \exp(-1.55/kT)\ \text{cm}^2\text{s}^{-1} \quad 873 \leq T \leq 1473\text{K}, \quad (4)$$
$$D_{Mn} = 2.4\text{x}10^{-3} \exp(-0.72/kT)\ \text{cm}^2\text{s}^{-1} \quad 287 \leq T \leq 363\text{K}, \quad (5)$$
$$D_{Fe} = 1.1\text{x}10^{-3} \exp(-0.66/kT)\ \text{cm}^2\text{s}^{-1} \quad 273 \leq T \leq 1343\text{K}. \quad (6)$$

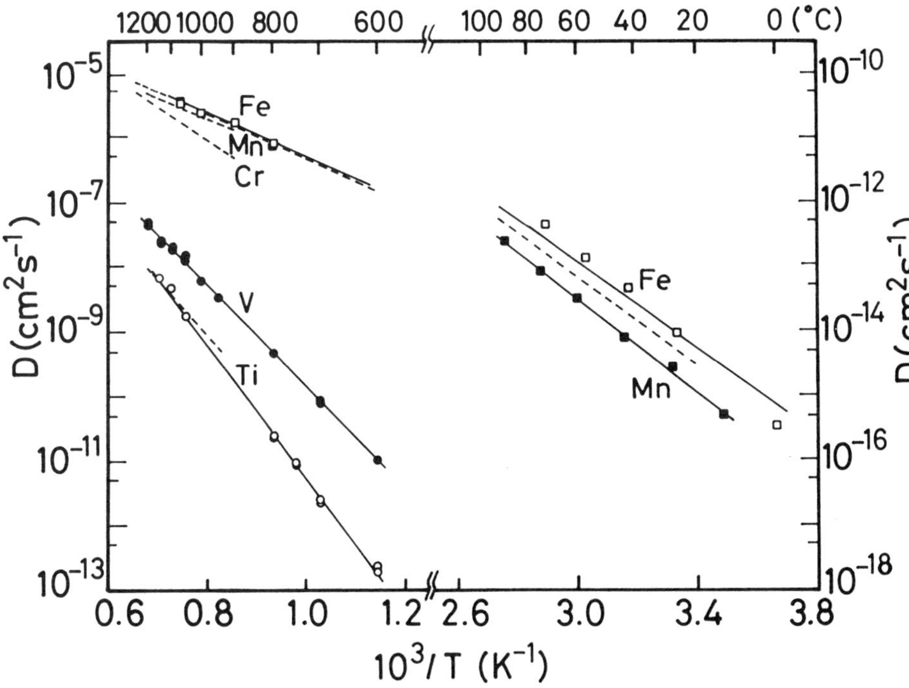

FIG.5. Temperature dependence of diffusivities of Ti(○), V(●), Mn(■) and Fe (□) in Si. The solid lines represent Eq.(3)-(6) and the broken lines represent the published data.[1,8,12,13]

The reliable diffusivity data of 3d-TM elements from Ti to Fe in Si except V have been reported by several researchers in the last few years, as summarized in a recent paper of Utzig.[11] According to their studies, D of each element is as follows: $D_{Ti}=1.45\text{x}10^{-2}\exp(-1.79/kT)$ cm^{2}s^{-1}, obtained in the range 950-1200℃ using DLTS method by Hocine et al.[8]; $D_{Cr}=1.0\text{x}10^{-2}\exp(-1.0/kT)$ cm^{2}s^{-1}, in the range 900-1250℃ using pn-junction method by Bendik et al.[12]; $D_{Mn}=6.9\text{x}10^{-4}\exp(-0.63/kT)$ cm^{2}s^{-1}, in the range 900-1200℃ using DLTS and tracer methods by Gilles et al.[13]; $D_{Fe}=1.3\text{x}10^{-3}\exp(-0.68/kT)$ cm^{2}s^{-1}, indicated in the wide range 20-

1250℃ by Weber.[1] These results are shown by the broken lines in Fig. 5. The present result for Ti in the high temperature range is in full agreement with that by Hocine et al., but the result in the lower range is about three times smaller than the extrapolated value. We believe that D_{Ti} given by Eq. (3) is probably more reliable result, because it is determined from data spanning about five orders of magnitude in D. By the same reason, we are convinced of the reasonability of D_V given by Eq. (4). The value of D_{Mn} at 1000℃ obtained by Gilles et al. is $2.2x10^{-6}$ cm^2s^{-1}, while the value at 1000℃ extrapolated using Eq. (5) is $3.4x10^{-6}$ cm^2s^{-1}. Considering the large span of this extrapolation, this is in excellent agreement. Furthermore, the present results for Fe in the high and room temperature ranges are in good agreement with D_{Fe} by Weber. From the present results and the published data mentioned above, we can clearly see an important chemical trend of 3d-TM impurity, i.e., diffusivity D shows an increase of about four orders of magnitude (at 1000℃) with increasing atomic numbers from Ti to Fe. The increase is not a gradual change and the significant increase occurs from V to Cr. Thus, pair reaction of Ti or V with B can not be expected owing to the very low diffusivity in the room temperature region. Recently, an attempt to calculate diffusivities of 3d-TM has been performed by Utzig.[11] The activation energies obtained in this study are consistent with the the calculated migration enthalpies. In addition, the preexponential factor D_0 of diffusivity of each element is in fairly good agreement with the value calculated from the D_0 theory[11] for interstitial diffusion with the exception of Ti. A further consideration on the disagreement of D_0 for Ti may be needed, and for this purpose the experimental investigation for the diffusivity of Sc will be needed.

4. Summary

The diffusivities of Ti, V, Mn and Fe in Si have been determined in a wide 1/T scale. The diffusivity of each element can be represented by Eq. (3)-(6). The diffusivity shows an increase of about four orders of magnitude with increasing atomic numbers from Ti to Fe. The tendency of the diffusivities is rather consistent with the trend of the calculated migration enthalpies.

Acknowledgements

The authors would like to thank Prof. H. Kitagawa for his encouragement in this study, and Mr. S. Kuge, T. Sadoh and T. Isobe for their assistance in this experiment.

REFERENCES

1. E. R. Weber, Appl. Phys. A **30**, 1 (1983).
2. K. Graff, Mater. Sci. Eng. B **4**, 63 (1989).
3. T. Sadoh and H. Nakashima, Appl. Phys. Lett. **58**, 1653 (1991).
4. H. Nakashima and K. Hashimoto, J. Appl. Phys. **69**, 1440 (1991).
5. H. Nakashima, T. Isobe, Y. Yamamoto and K. Hashimoto, Jpn. J. Appl. Phys. **27**, 1542 (1988).
6. T. Isobe, H. Nakashima and K. Hashimoto, Jpn. J. Appl. Phys. **28**, 1282 (1989).
7. J. H. Zhao, J. C. Lee, Z. Q. Fang, T. E. Schlesinger and A. G. Milnes, J. Appl. Phys. **61**, 1063 (1987).
8. S. Hocine and D. Mathiot, Appl. Phys. Lett. **53**, 1269 (1988).
9. H. Conzelmann, K. Graff and E. R. Weber, Appl. Phys. A **30**, 169 (1983).
10. L. C. Kimerling, J. L. Benton and J. J. Rubin, Defects and Radiation Effects in semiconductors 1980 (Institute of Physics, London, 1981), IOP Conf. Seiries 59, p. 217.
11. J. Utzig, J. Appl. Phys. **65**, 3868 (1989).
12. N. T. Bendik, V. S. Garnyk, L. S. Milevskii, Sov. Phys. Solid State **12**, 150 (1970).
13. D. Gilles, W. Bergholz and W. Schröter, J. Appl. Phys. **59**, 3590 (1986).

Materials Science Forum Vol. 83-87 (1992) pp. 233-238

UNIAXIAL STRESS ALIGNMENT OF Pd^- AND Ni^- IN SILICON

P.M. WILLIAMS, F.S. HAM, F.G. ANDERSON, and G.D. WATKINS
Physics Department and Sherman Fairchild Laboratory,
Lehigh University, Bethlehem, PA 18015 USA

ABSTRACT

We report the results of uniaxial stress alignment studies on the isolated substitutional Pd^- and Ni^- defects in silicon. From these results we may infer the ordering of the Jahn-Teller split t_2 states in the band gap. For both defects, we find the ordering of these t_2 states from lowest to highest energy to be b_2, b_1, a_1, the same as that for Pt^- and the reverse of that for the negative vacancy, V^-.

1. Introduction

The electronic structures of substitutional Pt^-, Pd^- and Ni^- in silicon can be characterized by a triply occupied t_2 manifold in the band gap.[1] In this model, the orbital degeneracy is completely lifted by two spontaneous Jahn-Teller distortions, resulting in a defect of C_{2v} symmetry. Knowledge of the ordering of the resulting states is essential in modeling the electronic structure of the defects. For the special case of a triply occupied t_2 manifold, the final ordering cannot be predicted from first-order considerations, even if the signs of the Jahn-Teller couplings are known. In this work, we report the results of uniaxial stress alignment studies on the isolated Pd^- and Ni^- defects in silicon. The alignment under stress is measured using electron paramagnetic resonance (EPR) and enables us to determine the ordering of the t_2 gap states.

2. Experimental Details

Our samples were prepared from float-zone silicon with P doping of 2.6×10^{16} cm^{-3} and 7.4×10^{14} cm^{-3} for the Pd- and Ni-doped samples, respectively. Either Pd or Ni powder was rubbed into the roughened surface of the sample. The sample was then heated to 1250 °C under a small flow of He gas (Pd) or in a sealed quartz ampoule containing He gas (Ni). The diffusion process lasted 15-20 hours. Afterwards, the samples were quenched, and were then polished to insure that their dimensions were square.

The EPR was performed at 4.2 K using 14 GHz microwaves. The spectra were digitally recorded, and the intensities obtained by numerical integration of the spectra. The uniaxial stress was applied via a plunger inserted into the middle of our TE_{011} cavity and supported from below by the sample. The stress was applied to the plunger via a pneumatic mechanism situated outside the cryostat.

3. Experimental results

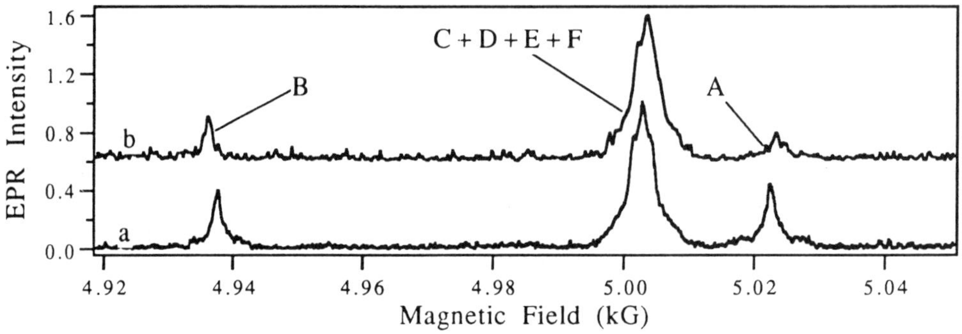

Fig. 1. EPR spectra of Si:Ni^- at 4.2 K and 14 GHz with **B**||[110]. Spectrum (a) is observed without stress, and (b) is observed with compressional stress of 10.6 MPa applied in the [1-10] direction.

Fig. 1 shows a typical spectrum: (a) without stress, and (b) under uniaxial <110> stress. The full angular dependence in the absence of stress reveals that both the Pd^- and Ni^- defects possess C_{2v} symmetry.[2,3] In the absence of stress, there are six equivalent defects. (See Fig. 2.) The principal axis system for a particular defect is defined as follows: (1) the <100> C_2 symmetry axis is the z axis (Fig. 2 shows the z axes for the six possible defect orientations), and (2) the x and y axes are in perpendicular <110> directions. We distinguish the x and y directions by the g values in these directions, as given in Table I.

Also shown in Fig. 2 is the direction of the applied <110> uniaxial stress. This stress removes the equivalence of the six defect orientations: (1) 'A' and 'B' are no longer equivalent, and (2) neither 'A' nor 'B' is equivalent to the remaining four orientations (C-F), which do remain equivalent to one another.

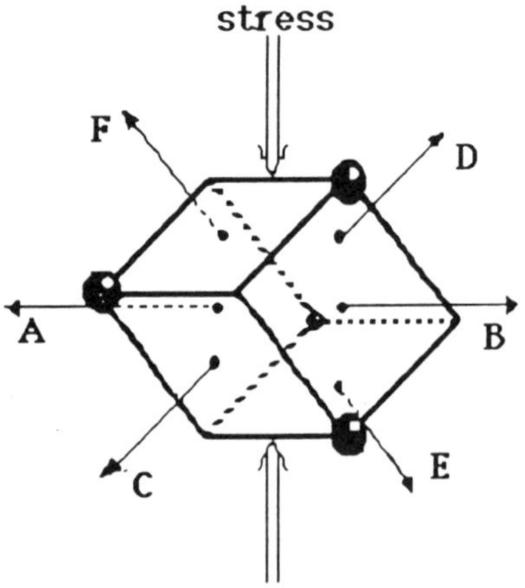

FIG. 2. Schematic of the six possible symmetry axis orientations for a C_{2v} symmetry defect. Also shown is the direction of the uniaxial stress.

Our work focuses on the changes of the relative intensities of the EPR resonances arising from the application of uniaxial stress. This change is easily seen in Fig. 1. We attribute the changes in the relative intensities to a reorientation of the defect, the preferential alignment reflecting the energetically favored defect orientation under stress. This reorientation occurs at 4.2

K with a response time of less than a few seconds, revealing that the barrier for reorientation is very low for all three defects. In Figs. 3 and 4, we plot the intensity ratios I_B/I_A and $I_B/(I_C+I_D+I_E+I_F)$ (obtained by numerically integrating spectra like that in Fig. 1) for the Pd^- and Ni^- defects, respectively, as a function of uniaxial stress.

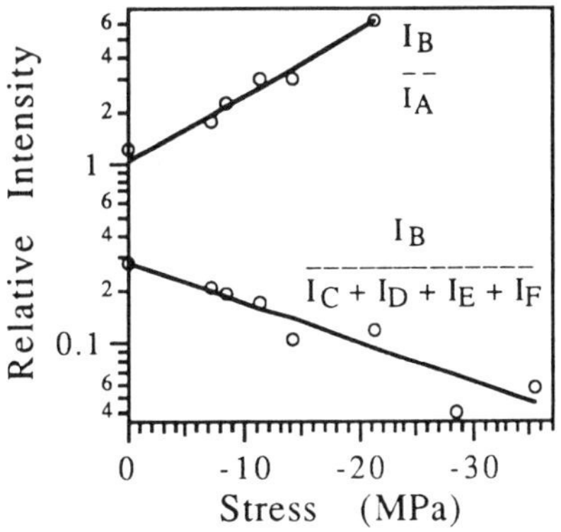

Fig. 3. Ratios of intensities at 4.2 K for the variously oriented defects (See Fig. 2.) versus <110> uniaxial stress for the Pd^- defect.

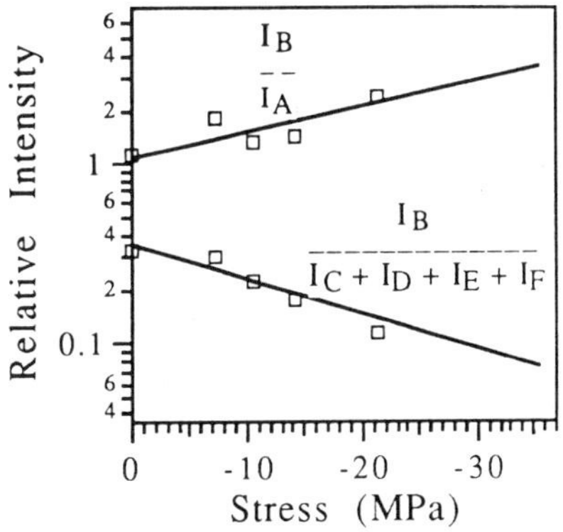

Fig. 4. Ratios of intensities at 4.2 K for the variously oriented defects (See Fig. 2.) versus <110> uniaxial stress for the Ni^- defect.

4. Analysis

The shift in energy (ΔE) of a low symmetry defect in a cubic crystal under uniaxial stress has been given by Kaplyanskii.[5] Following this work, but using the defect principal axis system, we can write the shift in energy for the i-th defect orientation as

$$\Delta E_i = A_{xx}\sigma^i_{xx} + A_{yy}\sigma^i_{yy} + A_{zz}\sigma^i_{zz}\,, \qquad (1)$$

where the A_{kk} are the components of the piezospectroscopic tensor that describe the coupling to the applied uniaxial stress, and σ^i_{kk} are the components of the stress tensor in the principal axis system of the i-th defect. Using Eq. (1) and Fig. 2, we find

$$\begin{aligned} \Delta E_A &= A_{yy}\sigma, \\ \Delta E_B &= A_{xx}\sigma, \\ \Delta E_{C\text{-}F} &= (A_{xx} + A_{yy} + 2A_{zz})\sigma/4, \end{aligned} \qquad (2)$$

Defect	g_{xx}	g_{yy}	g_{zz}
Pt^- (b)	1.3867	1.4266	2.0789
Pd^- (b)	1.9715	1.9190	2.0544
Ni^- (a)	2.018	2.054	2.016
V^- (c)	2.0028	2.0151	2.0038

TABLE I. Principal g values for various defects. (a) Ref. 3. (b) Ref. 2. (c) Ref. 4.

where σ is the applied stress.

Assuming a Boltzman distribution, the ratio of the populations of two different orientations (n_i/n_j) in the presence of applied stress is given by

$$n_i/n_j = \exp\left[-(\Delta E_i - \Delta E_j)/k_B T\right]. \tag{3}$$

From the slope of an Arrhenius plot of the intensity ratios versus σ (See Figs. 3 and 4.), we determine $\Delta E_i - \Delta E_j$. With Eq. (2), this is sufficient to determine the A_{kk} with the exception of the trace. (Since our analysis compares only the relative energy shifts, not absolute shifts, we determine only the traceless part of the A_{kk}.)

The coupling of Eq. (1) can also be expressed in terms of the applied strain. In this case the energy shift for the i-th defect orientation is

$$\Delta E_i = B_{xx}\varepsilon^i_{xx} + B_{yy}\varepsilon^i_{yy} + B_{zz}\varepsilon^i_{zz}, \tag{4}$$

where the B_{kk} give the coupling to the strains (ε_{kk}). The strains are determined from the stresses using the elastic compliance constants, c.f., Ref. 6, which relate the B_{kk} to the A_{kk} In Table II, we give the B_{kk} determined from the data in Figs. 3 and 4. Also given in this table are the B_{kk} for Pt^- and for the negatively charged vacancy in silicon.[6,7]

Defect	B_{xx}	B_{yy}	B_{zz}
Pt^- (a)	0.61	-4.75	4.11
Pd^-	0.13	-4.71	4.58
Ni^-	-0.53	-2.42	2.95
V^- (b)	-2.25	12.15	-9.9

Table II. Values of B_{kk} for various defects, in units of eV/unit strain, expressed in the defect principal axis system. (a) Ref. 6. (b) Ref. 7.

5. Discussion

As described in the introduction, the electronic structure of Ni^-, Pd^-, and Pt^- can be described as that of three electrons occupying a manifold of t_2 states in the band gap. This has come to be labelled the "vacancy" model[1] because simple chemical arguments (confirmed by detailed electronic structure calculation[3,8-10]) predict that the t_2 orbitals are only weakly localized on the central impurity and are mostly on the silicon near-neighbors (the t_2 orbitals of the vacancy into which the central impurity has been inserted). In this model, therefore, the lowered C_{2v} symmetry results, as for V^- (the t_2^3 configuration for the negatively charged vacancy) from two spontaneous Jahn-Teller (JT) distortions, a tetragonal and a trigonal one, that entirely lift the degeneracy of the t_2 gap manifold.

While the vacancy model predicts the occurrence of these two JT distortions for the case of the triply occupied t_2 manifold, the model does not predict the sense of either distortion. That is, the model does not predict the final ordering of the three t_2 states. For example (See Fig. 5), the tetragonal distortion splits the t_2 manifold into a singlet and a doublet. As demonstrated in this figure, for either sense of distortion, the sum of the three one-electron energies gives -3δ. The trigonal distortion has the effect of splitting the doublet (See Fig. 6.); but again, the sum of the three one-electron energies is the same for either sense of the tetragonal or trigonal distortion. This result is peculiar to the case of three electrons.[11-13] Hence, the vacancy model fails to predict the final ordering of the t_2 states.

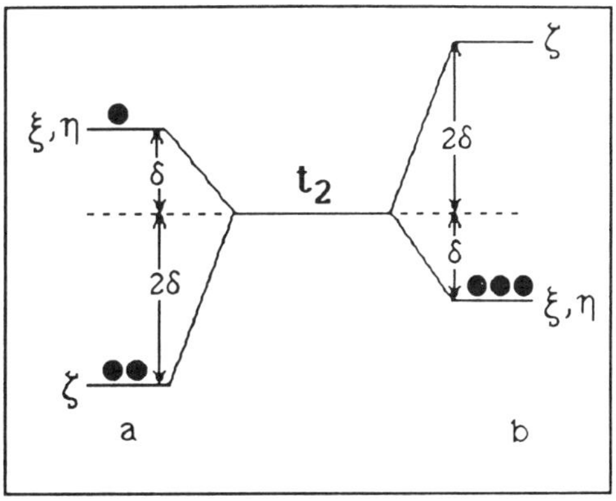

FIG. 5. The Splitting of a t_2 manifold resulting from a tetragonal distortion. Either (a) the singlet or (b) the doublet becomes the ground state. 3δ is the total splitting.

Our experiments are able to determine the ordering of these t_2 gap states, if we make the reasonable assumption that the Jahn-Teller coupling is determined by the vacancy component of the t_2 states, and hence of the same sign as that for the isolated vacancy. The sign of B_{zz} gives the sense of the tetragonal distortion; and the sign of $B_{xx} - B_{yy}$ gives the sense of the trigonal distortion. Examination of the B_{kk} in Table II shows that the senses of the tetragonal and trigonal distortions for the Ni^- and Pd^- defects (and the Pt^- defect) are opposite to those of V^-. Hence, we conclude that the ordering of the t_2 gap states for these transition-metal defects is inverted from that for V^-. (See Fig. 6.) That the ordering of these states is the same for each of these transition-metal defects might be surprising in light of the fact that the g values, in going from Pt^- to Ni^-, become similar in form to the g values for the negative vacancy.

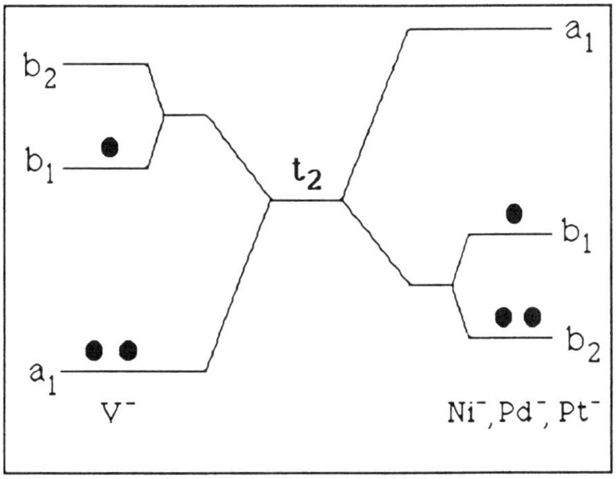

FIG. 6. The ordering of the t_2 gap states for V^- and for Ni^-, Pd^-, and Pt^- defects. The second splitting results from the trigonal distortion.

We note that for either case in Fig. 6, the unpaired (paramagnetic) electron is in the b_1 state (where we have used the irreducible representations of the C_{2v} point group to label the states[14]). This is established by the fact that for all four defects the resolved superhyperfine interaction is observed only for the two ^{29}Si nearest neighbors in the xz plane[1-3].

In addition to determining the state occupied by the unpaired electron, our experiments also allow us to determine which of the two remaining t_2 states is fully occupied. We conclude that in contrast to V^- (but in analogy to Pt^-), the b_2 state is filled and the a_1 state is empty for the Ni^- and Pd^- defects.

7. Acknowledgments

The authors would like to thank R.F. Milligan for his preparation of the Pd-doped samples. This research has been supported by the U.S. Navy Office of Naval Research, Contract No. N00014-90-J-1264.

8. References

1. G.D. Watkins, Physica **117B & 118B**, 9 (1983).

2. H.H Woodbury and G.W. Ludwig, Phys. Rev. **126**, 466 (1962).

3. L.S. Vlasenko, N.T. Son, A.B. van Oosten, C.A.J. Ammerlaan, A.A. Lebedev, E.S. Taptygov, and V.A. Khramtov, Solid State Comm. **73**, 393 (1990).

4. G.D. Watkins, in Radiation Damage in Semiconductors, (Dunod, Paris, 1965), p. 97.

5. A.A. Kaplyanskii, Optics and Spectroscopy **16**, 329 (1964).

6. F.G. Anderson, R.F. Milligan, and G.D. Watkins, submitted to Phys. Rev. B.

7. G.D. Watkins, in Deep Centers in Semiconductors, edited by S.T. Pantelides (Grodon and Breach, New York, 1986), p. 147.

8. L.A. Hemstreet, Phys. Rev. B **15**, 834 (1977).

9. A. Zunger and V. Lindefelt, Phys. Rev. B, **27**, 1191 (1983).

10. J.A.L. Alves and J.R. Leite, Phys. Rev. B, **30**, 7284 (1984).

11. F.G. Anderson, F.S. Ham, and G.D. Watkins, submitted to Phys. Rev. B.

12. F.G. Anderson, C. Delerue, M. Lannoo, and G. Allan, submitted to Phys. Rev. B.

13. M. Lannoo, Phys. Rev. B **36**, 9355 (1987).

14. M. Tinkham, in Group Theory in Quantum Mechanics (McGraw-Hill Book Company, New York, 1964), p. 325.

Materials Science Forum Vol. 83-87 (1992) pp. 239-244

ELECTRICAL AND OPTICAL PROPERTIES OF TITANIUM, MOLYBDENUM AND TUNGSTEN RELATED DEFECTS IN SILICON

K. SCHMALZ[1], H. G. GRIMMEISS[2], H. PETTERSSON[2], L. TILLY[2] AND K. TITTELBACH[1]
[1]Institute of Semiconductor Physics, POB 409, O-1200 Frankfurt(Oder), Federal Republic of Germany
[2]Department of Solid State Physics, University of Lund, Box 118, S-221 00 Lund, Sweden

ABSTRACT

Titanium-, molybdenum- and tungsten doped silicon has been investigated using junction space charge techniques. For the Ti-doped samples three energy levels were observed at E_C -0.065 eV (A level), E_C -0.295 eV (B level) and E_V +0.255 eV (C level) at 80 K. The energy levels were located at E_V +0.298 eV and E_V +0.379 eV for Mo and W, respectively, at 80 K. In the Ti-, Mo- and W-doped samples a midgap level, probably process induced, was observed. Optical threshold energies of the Ti-, Mo- and W-related centers are in good agreement with calculated changes in Gibb's free energy, indicating a negligible lattice relaxation.

1. INTRODUCTION

The recombination properties of titanium-related centres in silicon have previously been characterized electrically [1,2,3]. In most of these studies, two Ti-related centers have been observed at E_C -0.28 eV (B level) and at E_V +0.25 eV (C level). Recently a third Ti-related center at E_C -0.08 eV (A level) has been reported. The three levels have been suggested to be different charge states of the same center with the B and C levels as donorlike levels. In addition to these levels several papers report on a fourth level in the range from E_C -0.51 to E_C -0.55 eV (D level) [4]. A Mo-related center was found at E_V +0.30 eV [5,6]. Somewhat different results with E_V +0.36 eV and E_V +0.53 eV were reported later [7]. The W doping generates an energy level at E_V +0.40 eV [6].
The purpose of this paper is to present for the first time optical studies of the Ti-, Mo- and W-related centers in silicon. The optical data are compared with a comprehensive investigation of thermal emission and capture rates, which enable us to present the thermodynamic quantities such as changes in Gibb's free energy ΔG, enthalpy ΔH and entropy ΔS of these transition metals.
The results presented in this paper suggest that lattice relaxation effects connected with the energy levels studied are small if not neglible. This is in agreement with other transition-metal impurities in silicon, such as Au, Ag and Pt, which all seem to have very small Frank-Condon shifts [8,9,10].

2. EXPERIMENTAL DETAILS

The samples used were either p^+n or n^+p diodes implanted with Ti, Mo, or W. The p-type- and n-type-doped silicon wafers , respectively, had a free carrier concentration of about $10^{15}cm^{-3}$. The metals were implanted onto the front side of the silicon wafer with an energy of 300 keV and a dose of $10^{13}cm^{-2}$, and diffused into the n- and p-type regions by heating the samples to 1100 °C for 30 min. The Ti, Mo, and W, respectively, was finally electrically activated by rapid thermal annealing at about 1200 °C. The thermal emission rates were determined using DLTS and single-shot measurements. By combining the different methods we were able to determine the corresponding emission rates and capture cross sections of the Ti-, Mo-, and W-related levels over a large temperature range. Some of the optical measurements were performed using a 0.5 m vacuum grating monochromator (Acton Research Corporation) to avoid disturbances from atmospheric absorption lines. For the same reason the globar light source and the collimating system were also evacuated. The resolution of the monochromator used was about 1 meV. The other optical measurements were performed using a double prism monochromator.

3. RESULTS AND DISCUSSION

Assuming that the temperature dependence of the capture cross section of electrons σ_n can be expressed as $\sigma_n = \sigma_o * \exp(-\Delta H_{cn}/kT)$, from the activation energy $E_{DLTS} = \Delta H_n + \Delta H_{cn}$, obtained from the T^2 corrected Arrhenius plot of the thermal emission rate of electrons, ΔH_n was calculated. The change of Gibb's free energy ΔG_n as a function of temperature was calculated by inserting the absolute values of the emission rate e_n and the capture cross section σ_n in the detailed balance relationship [11].
From $\Delta G_n = \Delta H_n - T * \Delta S_n$ the values for ΔH_n and ΔS_n were obtained, where ΔH_n is the change in enthalpy and ΔS_n the total change in entropy due to ionization [12].
In the absence of lattice relaxation effects it can be expected that ΔG_n is close to the change of the optical Gibb's energy ΔG_n^o, at least at lower temperature [13].
Similar considerations are valid for hole transitions.

Our Ti-doped samples showed all four levels reported previously [1-4]. In case of the A level the electron capture cross section σ_n was found to be temperature independent with $\sigma_n = 1.6 * 10^{-15} cm^2$ and the calculated activation energy $E_{DLTS} = 0.077$ eV is therefore close to ΔH_n. The change in Gibb's energy was calculated as $\Delta G_n = 0.077 - 1.47 * 10^{-4} T$.
For the C level $E_{DLTS} = 0.267$ eV was obtained. The hole-capture cross section σ_p showed a slight increase with increasing temperature (see figure 1).
The capture barrier was calculated to be about $\Delta H_{cp} = 10$ meV, thus giving an enthalpy change of $\Delta H_p = 0.257$ eV.
The change in Gibb's energy was calculated as $\Delta G_p = 0.258 - 3.86 * 10^{-5} T$. The slope for ΔG_p versus T gives a value of about 0.5k for the entropy change.
One would expect an optical threshold energy of about 0.255 eV at 77 K for the photoionization cross section of holes. However, the threshold energy deduced from the spectral distribution of σ_{po} reveals a slightly larger value of 0.27 eV.
The best fit to the spectral distribution was obtained by using an expression suggested for doubly forbidden optical transitions not involving excited states [14] (see figure 2, dashed curve).

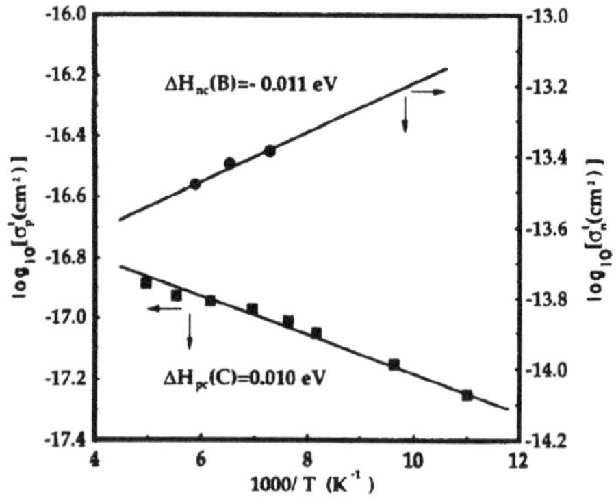

FIG.1. Arrhenius plot of the capture cross section for electrons (B level) and holes (C level)

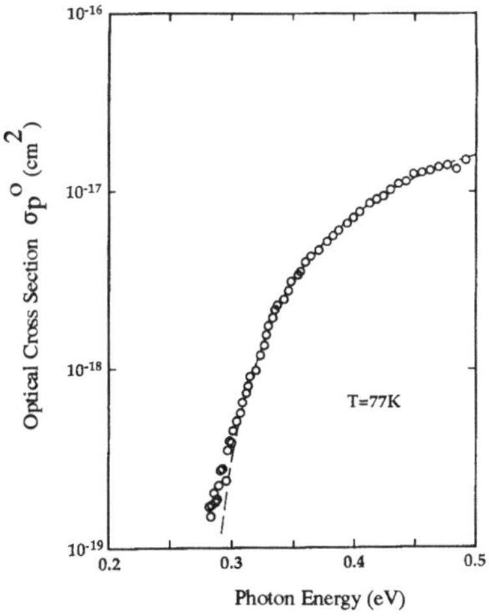

FIG.2. Spectral distribution of photo-ionization cross section for holes (C level) at 77 K

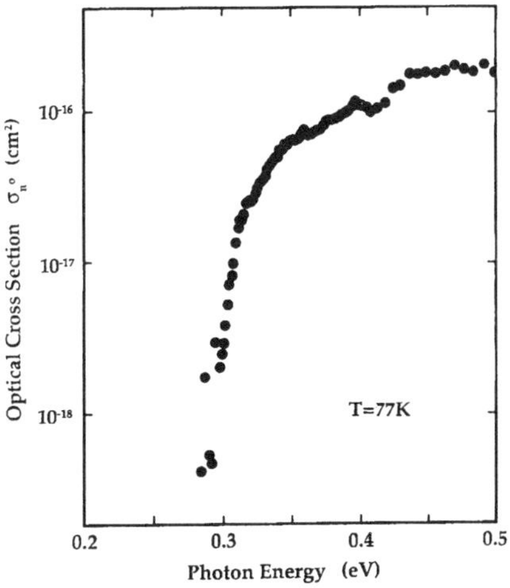

FIG.3. Spectral distribution of photo-ionization cross section for electrons (B level) at 77 K

In the case of the B level E_{DLTS}= 0.255 eV was obtained. The electron-capture cross section σ_n decreases slightly with increasing temperature (see figure 1). The enthalpy of the capture process ΔH_{cn} was found to be about -11 meV, resulting in ΔH_n= 0.266 eV for the thermal emission of the B level. The ΔG_n was calculated as ΔG_n= 0.268 + 3.42 $*10^{-4}$T.
The slope corresponds to ΔS_n of about -4k.
The dependence ΔG_n versus T suggests that the photoionization cross section spectrum of electrons should have a threshold of about 0.295 eV at 77 K if no lattice relaxation occurs. Figure 3 shows that a threshold at about 0.3 eV is indeed observed. For comparison, the spectral distribution of $\sigma_p^{\,o}$ was also measured in n^+p-diodes using the photocurrent storage effect [15] (see figure 4). The data obtained from this measurement method are in good agreement with the photocapacitance measurements resulting in a threshold energy of 0.270 eV. A second threshold was observed at about 0.6 eV. Extrapolating the $\sigma_p^{\,o}$ value, obtained at lower energies, to higher energies and substracting these values from the measured data, the dashed line (open squares) was obtained with a threshold energy of 0.51 eV, which can probably be attributed to the energy distance of the level from the valence band. The energy level at about 0.6 eV is also observed in the

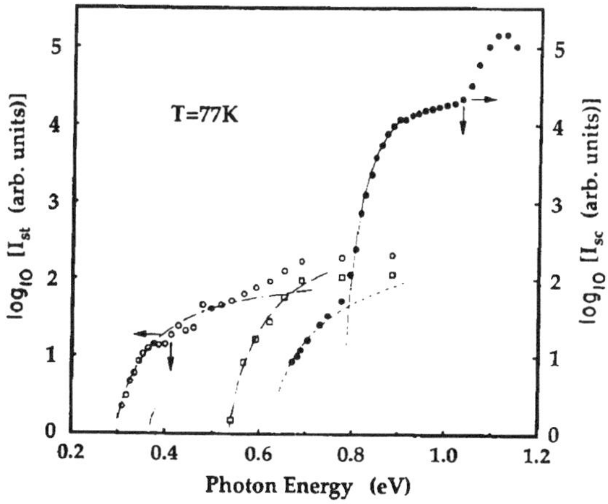

FIG.4. Spectral distribution of the short-circuit current I_{sc} (solid circels) and the photocurrent storage effect I_{st} (open circels and squares) measured for holes in n^+p Si:Ti diodes at 77 K

spectral distribution of the photoionization cross section obtained from measurements of the short circuit current [15] in n^+p-diodes giving a threshold of 0.585 eV (see figure 4). This energy probably corresponds to the energy distance of the D level from the conduction band. A second threshold at about 0.8 eV is difficult to assign since the energy distances of the B and C levels from the nearest energy band are very similar.
The midgap level at about 0.6 eV was observed by us not only in the Ti-doped samples but als in samples doped with Mo and W, respectively. There are good reasons to believe that the midgap level is a kind of process induced built-in defect.

Several of the properties observed in titanium-doped silicon support the suggestion that the levels A, B and C are likely due to the same defect. In the framework of this tentative model it is also evident that the hole capture at the C level is not supposed to involve excited states whereas the electron capture of the B level may well employ excited states. If this is true one would expect that the electron capture cross section decreases with increasing temperature, whereas the hole capture cross section may increase with increasing temperature if multiphonon-emission processes are involved. Figure 1 shows that this is indeed the case.Further support of the tentative model is obtained from the photoionization cross-section data. It has previously been shown for chalcogenes in silicon that photoionization cross sections due to 0/+ or +/++ transitions exhibit much sharper thresholds than +/0 or ++/+ transitions [16]. Comparing the spectra presented in figures 2 and 3 it is quite evident that the threshold of $\sigma_n{}^o$ is considerably sharper than the one for $\sigma_p{}^o$.

For our Mo-doped n^+p-diode the Arrhenius plot of the thermal emission rate e_p of holes gives an activation energy E_{DLTS}=0.292 eV in fair agreement with results already published [5,6]. The capture cross section of holes σ_p was in the order of $10^{-15}cm^2$ with ΔH_{cp}= -4 meV (see figure 5). Although this value is within the experimental error, we calculate a change of enthalpy ΔH_p= 0.296 eV.

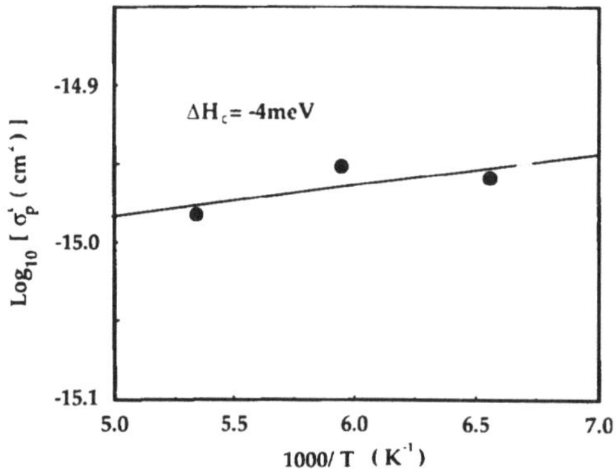

FIG.5. Arrhenius plot of the capture cross section for holes for the Mo-related level

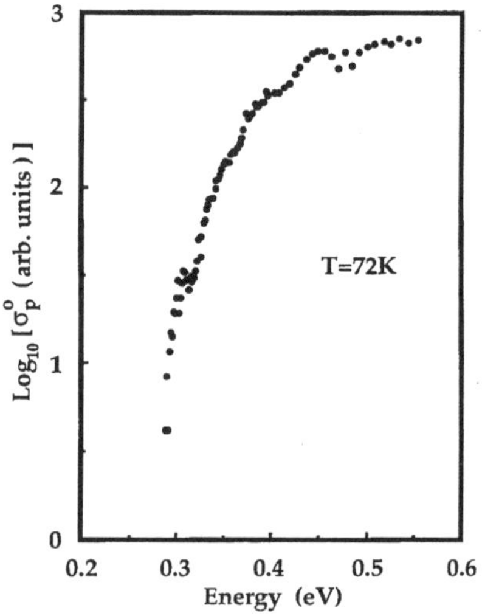

FIG.6. Spectral distribution of the photoionization cross section for holes in Si:Mo n^+p diodes at 72 K

The change in Gibb's free energy was obtained to be ΔG_p= 0.296 + 2.70 $*10^{-5}$T, giving a negative value ΔS_p= -0.3k. In the absence of larger lattice relaxation effects these data suggest an optical threshold energy of about 0.298 eV at 72 K.
This value is in reasonable agreement with the spectral distribution of the photoionization cross section for holes measured at 72 K using the photocapacitance technique in n^+p-diodes (see figure 6). A threshold energy around 0.29 eV is determined.
When measuring the short circuit current in p^+n-diodes at 45 K two thresholds were observed. One of the thresholds clearly revealed the presence of the midgap level while the other indicated a threshold energy of about 0.95 eV.

DLTS measurements of our W-doped n^+p-diodes exhibited only one peak in the temperature range between 70 and 300 K. The calculated activation energy E_{DLTS}= 0.4 eV is in fair agreement with previously published results [6]. The temperature dependence of the capture cross section of holes σ_p was quite considerable with ΔH_{cp}= 18 meV (see figure 7), resulting in a value of ΔH_p= 0.382 eV.

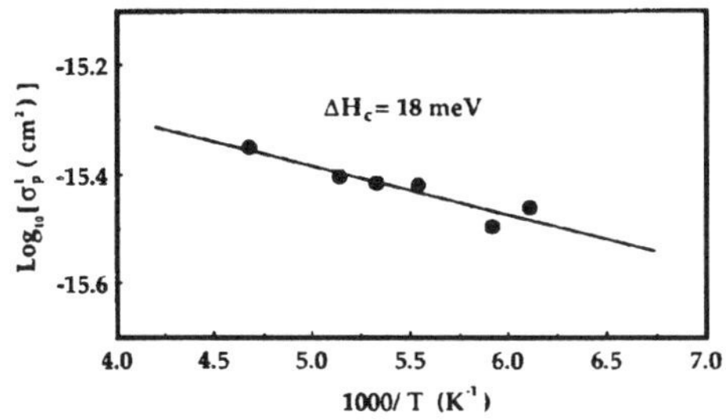

FIG.7. Arrhenius plot of the capture cross section for holes for the W-related level

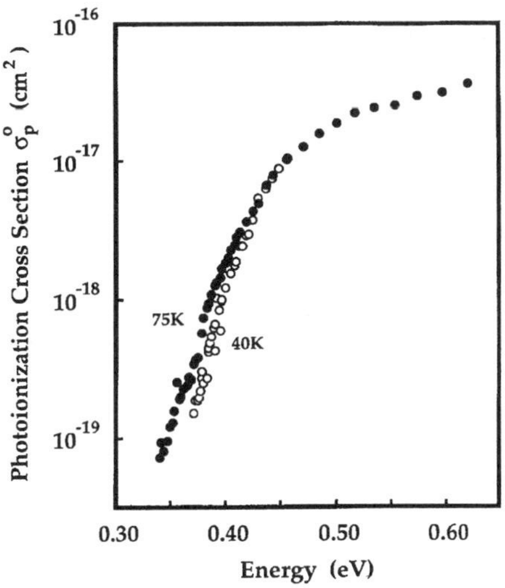

FIG.8. Spectral distribution of the photoionization cross section for holes in Si:W n^+p diodes at 75 and 40 K

The ΔG_p as a function of temperature was calculated to be ΔG_p= 0.383 -5.2 $*10^{-5}$T, giving ΔS_p= 0.6k. These data suggest an optical threshold energy of about 0.38 eV at 40 K.
A somewhat smaller value of about 0.37 eV is obtained from the spectral distribution of the photoionization cross section of holes at 40 K (see figure 8). Similar measurements performed at 75 K show that the threshold shifts to about 0.34 eV, suggesting that a value close to 0.38 eV may be expected for the optical threshold at 0 K.
Short circuit measurements of the W-doped p^+n-diodes resulted in very similar spectra to those obtained for the Mo-doped p^+n-diodes.

REFERENCES

1. K. Graff and H. Pieper, in Semiconductor Silicon 1981, edited by H. R. Huff, R. J. Kriegler, and Y. Takeishi (The Electrochemical Society, Pennington, NJ, 1981), p.331
2. D. Mathiot and S. Hocine, J. Appl. Phys. 66, 5862 (1989), and references therein
3. Alex C. Wang and C. T. Sah, J. Appl. Phys. 56, 1021 (1984)
4. L. C. Kimerling, J. L. Benton, and J. J. Rubin, Inst. Conf. Ser. 59, 217 (1980)
5. A. Rohatgi, J. R. Davis, R. H. Hopkins, and P. G. McMullin Solid-State Electron. 26, 1039 (1983)
6. H. Lemke, phys. stat. sol. (a) 76, K193 (1983)
7. Jie Zhou, Xiujing Ji, Shuying Li, Jian Wu, Jilın Gao and Zhiyoung Han Mater. Sci. For. 38-41, 457 (1989)
8. M. Kleverman, J. Olajos, and H. G. Grimmeiss, Phys. Rev. B 37, 2613 (1988); J. Olajos, M. Kleverman, and H.G. Grimmeiss, ibid. 40, 6196 (1989), and references therein
9. J. Olajos, M. Kleverman,and H. G. Grimmeiss, Phys. Rev. B 38, 10633 (1988), and references therein
10. M. Kleverman, J. Olajos, and H. G. Grimmeiss,Phys. Rev. B 35, 4093 (1987), and references therein
11. O. Engström and A. Alm, Solid State Electron. 21, 1571 (1978)
12. L. Tilly, H. G. Grimmeiss, H. Pettersson, K. Schmalz, K.Tittelbach, and H. Kerkow, Phys. Rev. B 43, 9171 (1991)
13. C-O. Almbladh and G. J. Rees, J. Phys. C 14, 4575 (1981)
14. H. G. Grimmeiss, L. Å. Ledebo, C. Ovrén, and T. N. Morgan, in Proceedings of the Twelfth International Conference on the Physics of Semiconductors, edited by M. H. Pilkuhn (Teubner, Stuttgart, 1974)
15. H. G. Grimmeiss and C. Ovrén, J. Phys. E 14, 1032 (1981)
16. H. G. Grimmeiss and B. Skarstam, Phys. Rev. B 23, 1947 (1981)

Materials Science Forum Vol. 83-87 (1992) pp. 245-250

NICKEL RELATED DEEP LEVELS IN GERMANIUM

F.X. ZACH[1], H. GRIMMEISS[2], AND E.E. HALLER[1,3]
[1]Mat. Science Div., Lawrence Berkeley Laboratory, 1 Cyclotron Rd., Berkeley CA 94720 USA
[2]Department of Solid State Physics, University of Lund, Box 118, S-22100 Lund, Sweden
[3]Mat. Science and Mineral Engineering Dept., University of California, Berkeley CA 94720 USA

ABSTRACT

By means of deep level transient spectroscopy we have studied nickel related deep levels in germanium. The crystals used in this work were grown in a nitrogen or a hydrogen atmosphere which allowed us to study complex formation between nickel and hydrogen. Not only the emission rates but also the capture cross sections were determined as a function of temperature. The activation energy of each emission rate was corrected for the temperature dependence of the capture cross section. In p-type, high purity material we observe three majority traps with activation energies of 0.215, 0.154 and 0.416 eV, respectivly. We tentatively assign the 0.215 eV level to the first ionization stage of substitutional nickel, whereas the levels with 0.154 and 0.416 eV activation energy are most probably due to Ni-H complexes. In addition we present the complete signature for several copper and copper-hydrogen related levels, including the temperature dependence of the capture cross section.

1. Introduction

Studies on nickel related centers date back to as early as 1955, when Newman, Tyler and Woodbury [1] reported that nickel introduces two acceptor levels at 0.23 eV above the valence band and 0.3 eV below the conduction band. Later Pearton [2] presented DLTS-data on nickel doped germanium crystals in a general survey of transition metals in germanium. He found a total of 4 different, nickel related levels, three acceptor levels with activation energies for hole emission of 0.14, 0.21 and 0.40 eV and a donor level with 0.18 eV activation energy for electron emission. Two of them (0.21 eV and 0.40 eV) were assigned to lithium-nickel complexes. As observed for the trivalent acceptor copper, the formation of nickel hydrogen complexes could give rise to additional levels seen in DLTS [4]. We therefore started a systematic study of nickel diffused germanium samples, especially adressing the question of complexes with hydrogen and lithium.

2. Sample preparation

High purity, p-type germanium crystals with a net acceptor concentration of 1 to 6×10^{11} cm^{-3} shallow acceptors were used. The crystals were grown in a nitrogen or a hydrogen atmosphere. After etching the samples in a mixture of HNO_3-HF (4:1), they were rinsed in methanol, immersed in KCN solution and then coated with the respective metal. Nickel was deposited on the germanium samples by e-gun evaporation or electroplating out of a $NiCl_2$- H_2BO_3 containing aqueous solution. Copper was deposited on the samples by electroplating from a $CuSO_4$ solution. The diffusion was performed at 400°C in an evacuated quartz ampoule for several hours. In order to study the influence of lithium on the defect levels observed, pn junctions were formed in three different ways : (i) Li- diffusion, (ii) low temperature phosphorus implantation with subsequent annealing, and (iii) sputtering amorphous germanium on the samples and evaporating a thin layer of gold. In all cases, the backside ohmic contact was prepared by boron implantation and evaporating palladium or simply pressing indium foil on the contact.

3. Experimental

A standard DLTS- setup with a 1 MHz capacitance meter, pulse generator, constant bias supply and a Miller correlator for signal averaging were used to record the DLTS-spectra of the samples. For an analysis of the temperature dependence of the emission rate, the full transients were recorded with a digital oscilloscope and transferred to a computer to analyze the time constants. Single pulse measurements allowed us to extend the temperature range for the determination of the emission rate : The transient, following a single bias pulse, is recorded with a x-t recorder. The capture constant was measured by recording the height of the capacitance transient as a function of the filling pulse length (0.1 to several hundred microseconds). Defect concentrations were determined from the slope of a depth profile, i.e., the change in the magnitude of the capacitance transient as a function of pulse height. The necessary information on the shallow background doping was determined from C-V curves measured at the peak of the DLTS-signal.

4. Results

Copper in germanium is one of the most commonly observed traps which is introduced, often unintentionally, during heat treatment of germanium even at moderately high temperatures. In order to distinguish these levels from nickel related levels, we first studied the DLTS signals arising from copper.

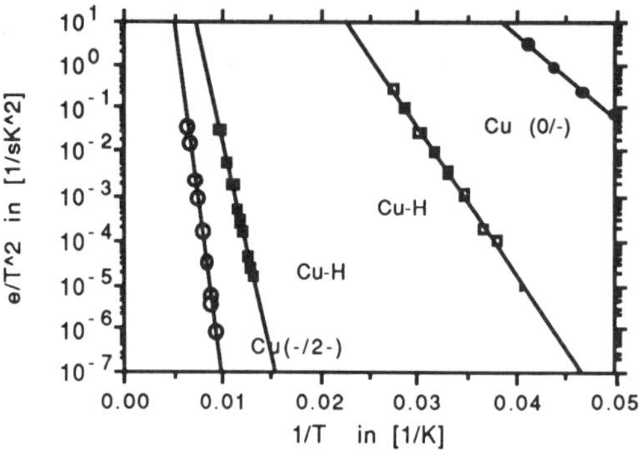

Figure 1 : Emission rate e, divided by the square of the temperature T^2 versus 1/T for copper related levels. The symbols represent experimental data, the lines have been fitted using $e/T^2 = K\exp(-E_{act}/kT)$.

filled circles :	Cu(0/-)	E_{act} = 38 meV	$K = 2.16\cdot10^8\ s^{-1}K^{-2}$
open circles :	Cu(-/2-)	Eact = 318 mev	$K = 8.36\cdot10^8\ s^{-1}K^{-2}$
open squares :	Cu-H	E_{act} = 66 meV	$K = 3.84\cdot10^8\ s^{-1}K^{-2}$
filled squares :	Cu(-/2-)	Eact = 193 mev	$K = 9.97\cdot10^8\ s^{-1}K^{-2}$

In the copper doped crystals, which were grown in a nitrogen atmosphere and thus do not contain hydrogen, we observe two main peaks. In figure 1 we plot the emission rate e, divided by the square of the temperature T^2 for these levels. e/T^2 can be expressed in the form $K\exp(-E_{act}/kT)$ with the activation energy of the emission rate E_{act}, the Boltzman factor k, the absolute temperature T and a prefactor K. The quantities E_{act} and the prefactor K were obtained from the fit to the curves and are given in the figure caption. The signature of these levels is quite similar to what has been reported for the first - Cu(0/-) - and second - Cu(-/2) - ionization stage of copper with activation energies of 38 and 318 meV [4,5]. Even though the prefactors differ somehow from what can be

Table 1 : Summary of the parameters obtained from a fit of the capture cross section as a function of temperature assuming a temperature dependence of the form $s_0\exp(-E/kT)$. Literature data at single temperatures are compared to the values obtained here.

Level	S_0 [cm^2]	E [meV]	This work [cm^2]	References
Cu(0/-)	$3.3\cdot10^{-13}$	-3.3	$2.2\cdot10^{-12}$ (20K) $1.9\cdot10^{-12}$ (22K) $1.4\cdot10^{-12}$ (27K)	$2.5\cdot10^{-12}$ Ref. [6] $2.4\cdot10^{-12}$ Ref. [5] $9.5\cdot10^{-13}$ Ref. [7]
Cu(-/2-)	$2.86\cdot10^{-14}$	-15	$9.5\cdot10^{-14}$ (145K) $8.1\cdot10^{-14}$ (167K)	$2.0\cdot10^{-13}$ Ref. [5] $2.0\cdot10^{-14}$ Ref. [6]
Cu-H (66 meV)	$4.29\cdot10^{-13}$	-0.8	$5.6\cdot10^{-13}$ (35K)	$1.6\cdot10^{-12}$ Ref. [5]
Cu-H (193 meV)	$2.12\cdot10^{-13}$	-2.0	$2.7\cdot10^{-13}$ (100K) $2.6\cdot10^{-13}$ (119K)	$7.6\cdot10^{-13}$ Ref. [5] $2.0\cdot10^{-14}$ Ref. [6]

Table 2: Summary of the enthalpy and entropy changes H and S for the copper related levels. The entropy S is given in units the Boltzman constant k.

Level	Cu(0/-)	Cu(-/2-)	Cu-H (66 meV)	Cu-H (193 meV)
H [meV]	41.3	33	67	195
S	-0.8	3.2	-0.2	-0.8

found in literature, most probably due to small differences in the activation energies, the actual emission rates at the peak temperatures are quite similar.
In crystals containing hydrogen two additional levels can be observed. The emission rate for these levels is also included in figure 1. We obtain activation energies of 66 and 193 meV respectivly in agreement with reported values for Cu-H complexes [4,5].
In figure 2 we plot the capture cross section as a function of temperature for the level corresponding to the second ionization stage of copper. The data have been analysed assuming a functional dependence of the form $s = s_o \exp(-E/kT)$. The parameters obtained from these fits are summarized in table 1. Apart from the case of Cu(-/2-), the temperature dependence is small with activation energies of less than a few meV. For the second ionization stage of copper, we observe a somewhat higher energy of 15 meV. For all these levels we find a decrease of the capture cross section with increasing temperature, indicating that the first step in the capture to the deep level involves shallow excited states of the defect. A comparison of the capture cross section obtained in this work with previous papers is given in table 1.

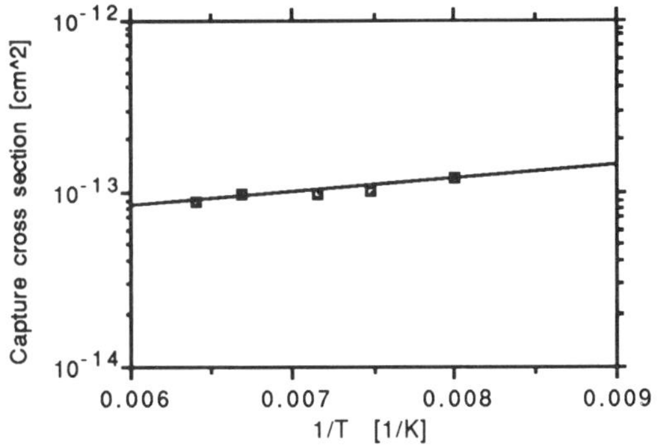

Figure 2 Capture cross section as a function of temperature for the second ionization stage of copper. The symbols represent experimental data, the line corresponds to a fit of the form $s_o \exp(-E/kT)$.

By further analysis of our data we can calculate the change in free energy G for the carrier emission process as a function of temperature : The emission rate e as a function of temperature is given by the expression : $e = s\, v_{th}\, N_V \exp(-G/kT)$
with the capture cross section s, the thermal velocity of the holes in the valence band v_{th} and the effective density of states in the valence band N_V. Due to the temperature dependence of the capture cross section, the activation energy as determined from a plot of the emission rate versus the inverse of temperature includes a contribution from the capture cross section. Dividing the emission rate by the capture cross section for the same temperature allows one to calculate the change in free energy G. G as a function of temperature for the Cu(-/2-) level is plotted in figure 3. The curve has been calculated from fits of the emission rate and capture cross section versus temperature. The change in free energy is related to the enthalpy change H and the change in entropy S by : $G = H - TS$. From the slope of the G versus T curve and the extrapolated value of G at T= 0 we can therefore determine the values of H and S. The activation energy for the emission rate E_{act}, H and the activation energy for the capture cross section E are related to each other : $E_{act} = H + E$. The parameters we found for the copper related levels are summarized in table 2.
We now turn to the results obtained for the nickel doped crystals. The DLTS spectrum for the crystal grown in a nitrogen atmosphere is shown in figure 4. Besides the peaks already observed at 20.5 and 136 K, corresponding to the first and second ionization stage of copper, we reproducibly find an additional peak at about 100K. Our diffusion conditions lead to concentrations of the order of $1\text{-}4 \times 10^{10}$ cm^{-3} . The peak observed at 45 K does not seem to be related to the presence of nickel. The standard Arrhenius plot of the emission rate divided by the square of temperature T^2 versus 1/T gives an activation energy of 215 meV for the level observed at about 100K (See figure

5). As for the case of copper, we fitted our experimental data on the capture cross section to the form $s_0\exp(-E/kT)$. The capture cross section for this level is thermally activated with an activation energy of about 7.6 meV and we obtained a prefactor of $8.9\times10^{-13}cm^2$ as summarized in table 3. In analogy to what has been presented for the copper related levels, the resulting change in free energy as a function of temperature has been calculated and analyzed in terms of the change in enthalpy H and entropy S. We obtain an enthalpy of 207 meV and an entropy change of -1.9 in units of the Boltzman constant k (for a summary see table 4).

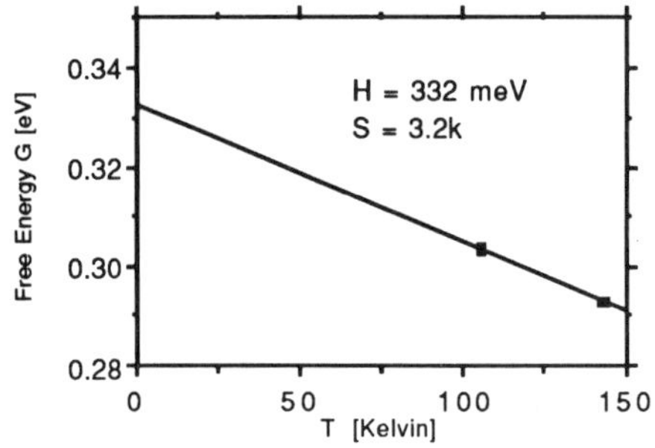

Figure 3 Change in Gibbs free energy as a function of temperature for the second ionization stage of Cu, Cu(-/2-). The parameters are obtained from a fit of G = H -TS.

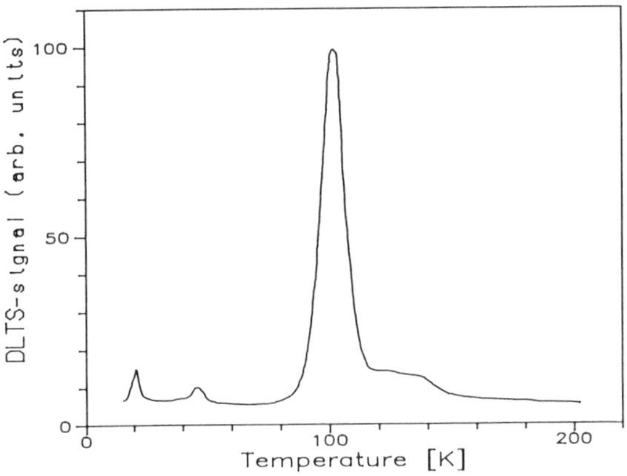

Figure 4
DLTS spectrum of a Ni-doped germanium diode not containing hydrogen. The experimental conditions are : Bias voltage 6V, Pulse height 3V, Pulse length 100 usec, the miller correlator is set to an emission rate of 37 $msec^{-1}$.

Table 3: Summary of the parameters obtained from a fit of the capture cross section as a function of temperature assuming a temperature dependence of the form $s_0\exp(-E/kT)$.

Level	Ni(0/-)	Ni-H(154 meV)	Ni-H (416 meV)
S_0 [cm^2]	$8.9\cdot 10^{-13}$	$3.4\cdot 10^{-15}$	$1.82\cdot 10^{-11}$
E [meV]	7.6	-3.2	117

Table 4 : Summary of the enthalpy and entropy changes H and S for the nickel related levels

Level	Ni(0/-)	Ni-H (154 meV)	Ni-H (416 meV)
H [meV]	207	159	299
S	-1.9	3.3	-4.3

In samples containing hydrogen, we observe two different peaks at about 77K and 183K, which were not observed in the crystals free of hydrogen. On the other hand the level with a peak temperature at about 100K is not found in these crystals. The concentration of these levels is of the order of $1\text{-}4\times10^{10}$ cm^{-3} and $1\text{-}3\times10^{9}$ cm^{-3} respectivly. As shown in figure 5, we obtain activation energies of 154 and 416 meV from the Arrhenius plot of the emission rate for these levels. The results for the capture cross section are summarized in table 3. The capture cross section for the 0.15 eV center decreases with increasing temperature which seems to indicate the presence of shallow excited states for these level. On the other hand, the level at .416 eV shows a thermally

activated capture cross section with a quite significant activation energy of about 120 meV (see figure 7). Taking into account this strong temperature dependence of the capture cross section, we calculated, as above, the change in free energy as a function of temperature. The results are presented in table 4.
In addition, several other levels with lower concentrations have been observed in the range from 30 to 50K. Up to now it is not clear, if one (or more) of these levels is related to nickel. ·

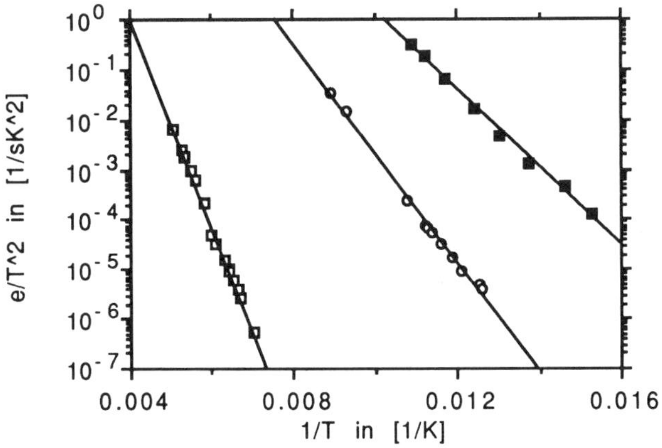

Figure 5
Emission rate e, divided by the square of the temperature T^2 versus the inverse of temperature for nickel related levels. The symbols represent experimental data, the lines have been fitted using a functional dependence of the form $e/T^2 = K \exp(-E_{act}/kT)$. We use the tentative assignments which have been made in this paper for the different levels.

open circles :	Ni(0/-)	Eact = 215 meV	$K = 1.5\times10^{8}\ s^{-1}K^{-2}$
filled squares :	Ni-H	E_{act} = 154 meV	$K = 8.85\times10^{7}\ s^{-1}K^{-2}$
open squares :	Ni-H	Eact = 416 meV	$K = 2.69\times10^{8}\ s^{-1}K^{-2}$

5. Discussion

We would like to stress, that the tentative assignment of the levels mentioned sofar is based on a comparison of copper and nickel diffused samples. As already pointed out earlier, copper is one of the most pervasive contaminants in germanium. Because it is practically impossible to avoid copper contamination completely, especially at the low concentrations encountered in this work, we decided to minimize the introduction of Cu in our samples and to determine with Cu-doped reference samples all the Cu-related peaks. We further assume, that every contamination during the diffusion process from the furnace or the quartz used will affect both type of samples in the same way. These precautions seemed necessary, as space charge techniques alone cannot unambigously identify a defect. The most consistent interpretation of our data is as follows :

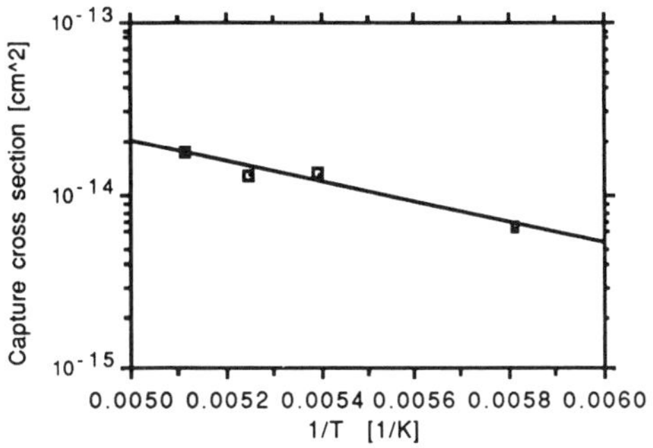

Figure 7
Capture cross section as a function of temperature for Ni-H (0.416 eV) complex. The symbols represent experimental data, the line corresponds to a fit of the form $s_0\exp(-E/kT)$.

The defect level observed at 100K is most probably due to the first acceptor level of substitutional nickel. Variable temperature Hall measurements indicate, that nickel is a double acceptor in germanium with two energy levels at 0.24 eV above the valence band and 0.3 eV below the conduction band respectivly. Our results on the crystal without hydrogen show a nickel related level at about this energy : The change in free energy G is about 0.23 eV in the range where the Hall measurements of ref.8 have been performed. In our measurements, only the first ionization stage of the acceptor could be observed and no attempt was made to identify the second ionization stage. The two levels observed in samples containing hydrogen with activation energies of 0.15 and 0.41 eV are most probably due to nickel-hydrogen complexes. Taking into account the results obtained in the samples with phosphorus implanted or amorphous germanium blocking contacts, none of the levels assigned to nickel here, including the level with 0.22 activation energy seems to include lithium as a part of the defect. We would like to note that the two levels assigned to Ni-H could also be observed in crystals which were not grown in hydrogen when the nickel was deposited by electroplating out of a $NiCl_2$-H_2BO_3 solution. The concentrations however were significantly lower than in the hydrogen doped samples. We propose that these levels can be seen because some of the hydrogen from the boric acid could migrate into the crystal during nickel diffusion.
DLTS measurements on nickel doped germanium have been performed by Pearton [2]. He observes a total of three different levels in nickel doped p-type crystals with activation energies obtained from the emission rate of 0.14, 0.21 and 0.40 eV. The energies of these levels are in good agreement with the energies obtained in this work, however the prefactors for the emission rates seem to differ significantly from what is observed in Pearton's work. Furthermore, we disagree on the assignments which have been made for these levels. Whereas Pearton assigns the 0.14 eV level to a nickel acceptor level, our work seems to indicate that this level is most probably due to a Ni-H complex. We would like to point out that the crystals used in [2] were grown in hydrogen atmosphere and no attempt has been made to identify possible metal- hydrogen complexes. In addition Pearton identifies the levels at 0.21 eV and 0.4 eV with a nickel - lithium complex, based on a comparison of the depth profiles for the three observed levels. The levels at 0.21 and 0.4 eV increase significantly relative to the other peak when measuring closer to the Li-contact. Based on our results for diodes prepared by phosphorus implantation or by sputtering amorphous germanium we disagree with respect to this identification.
Comparing our DLTS- results with published data, we find that the signature for the 0.15 eV level is similar to the one which has been ascribed to a Cu-Li-H complex [4]. However, as already mentioned above, we exclude that Li is involved in any of the levels described here. A DLTS signal with an activation energy of about 0.41 eV has been reported for the third ionization stage of copper. As we do not observe the 0.41 level reported here in copper doped samples, but only in nickel doped samples, we conclude that the level observed here is not related to copper.
Finally the DLTS- signal with a activation energy of 0.215 eV is close to what has been reported for Pb [9]. In view of the fact, that the diffusion depth for Pb for our experimental conditions is neglegible Pb can be excluded as the origin of this peak.

6. Summary

In summary we report three different DLTS signals which have been observed in nickel doped p-type germanium crystals. They show activation energies of 0.154, 0.215 and 0.416 eV respectivly. Our results indicate, that the 0.22 eV level is related to the first ionization stage of the nickel double acceptor, whereas the other two levels are most likely related to Ni-H complexes.

References

1. W.W. Tyler, R. Newman, H.H. Woodbury, Phys. Rev. 98 (1955) 461
2. S.J. Pearton, Solid State Electr. 25 (1982) 305
3. H.H. Woodbury, W.W. Tyler, Phys. Rev. 105 (1957) 84
4. E.E. Haller, P.P. Li, G.S. Hubbard, W.L. Hansen, IEEE Trans. Nucl. Sci. NS-26 (1979) 265
5. E. Simoen, P. Clauws, M. Lamon, J. Vennik, Semicond. Sci. Technol. 1 (1986) 53
6. A.O. Evwaraye, R.N. Hall, T.J. Soltys, IEEE Trans. Nucl. Sci. NS-26 (1979) 271
7. P. Norton, H. Levinstein, Phys. Rev. B 6 (1972) 489
8. R. Newman, W.W. Tyler in Solid State Physics, Vol.9 ; editors F. Seitz, D. Turnbull, Academic Press, New York 1959, p. 73
9. S.J. Pearton, Aust. J. Phys. 35 (1982) 53

Materials Science Forum Vol. 83-87 (1992) pp. 251-256

MAGNETIC RESONANCE FROM A METASTABLE SULFUR-PAIR-RELATED COMPLEX DEFECT IN SILICON

W.M. CHEN[1], M. SINGH[1,*], A. HENRY[1], E. JANZÉN[1], B. MONEMAR[1], A.M. FRENS[2], M.T. BENNEBROEK[2], J. SCHMIDT[2], K.J. REESON[3] and R.M. GWILLIAM[3]

1 Department of Physics and Measurement Technology, Linköping University, S-581 83 Linköping, Sweden
2 Department of Physics, Huygens Laboratory, University of Leiden, P.O. Box 9504, 2300 RA Leiden, The Netherlands
3 Department of Electrical and Electronic Engineering, University of Surrey, Guildford GU2 5XH, United Kingdom
* Permanent address : Physics Department, King's College London, Strand, London WC2R 2LS, United Kingdom.

ABSTRACT

We present a study of a recently reported metastable sulfur-related complex defect in silicon by magnetic resonance techniques, with the emphasis on the electronic structure and the chemical identity of the defect. The lowest bound exciton (BE) state of the neutral charge state for both configurations is confirmed to be a spin triplet. Data obtained from optical detection of magnetic resonance (ODMR) without magnetic field show that the hyperfine interaction is responsible for the unusually broad linewidth observed in conventional ODMR. Comprehensive information gained from the ODMR investigations with and without the magnetic field provides evidence for the presence of a copper atom in the complex, which localizes the BE wavefunction. Electron spin resonance results predict a (-/0) level of the defect in the band gap, at least in one configuration.

1. Introduction

A sulfur-related complex defect in silicon [1-4], produced in S-doped silicon after a few seconds annealing at about 800°C followed by a rapid quenching to room temperature, has recently attracted much attention due to its interesting configurational metastability, with two photoluminescence (PL) spectra S_A and S_B characteristic of configuraton 1 and 2, respectively. This complex gives rise to deep bound exciton (BE) PL emissions which at low temperature have zero-phonon lines at 0.968 eV (S_A^0) and 0.812 eV (S_B^0), respectively [2], and exhibit PL fingerprints of the conversion between the two corresponding configurations either by light ($S_A \rightarrow S_B$ at T<40K) or by thermal annealing ($S_B \rightarrow S_A$ at T>40K) [3,4]. Preliminary optically detected magnetic resonance (ODMR) results showed that the S_A^0 and S_B^0 lines both correspond to excited spin-triplets [5]. Due to the weak and broad ODMR signal, however, no detailed ODMR data which could lead to a better understanding of the complex was obtained in earlier work.

Recently it was suggested that the defect was related to a sulfur pair since the energy separation between S_A^0 and S_B^0 is nearly exactly the same as between the $1s(A_1^+)$ and the $1s(E^-)$ states of the neutral sulfur pair [4].

In this work, we shall present our recent results from a more detailed and more accurate X-band (≈ 9.23 GHz) ODMR investigation. In addition new experimental data from zero-field ODMR and electron spin resonance (ESR) will be discussed.

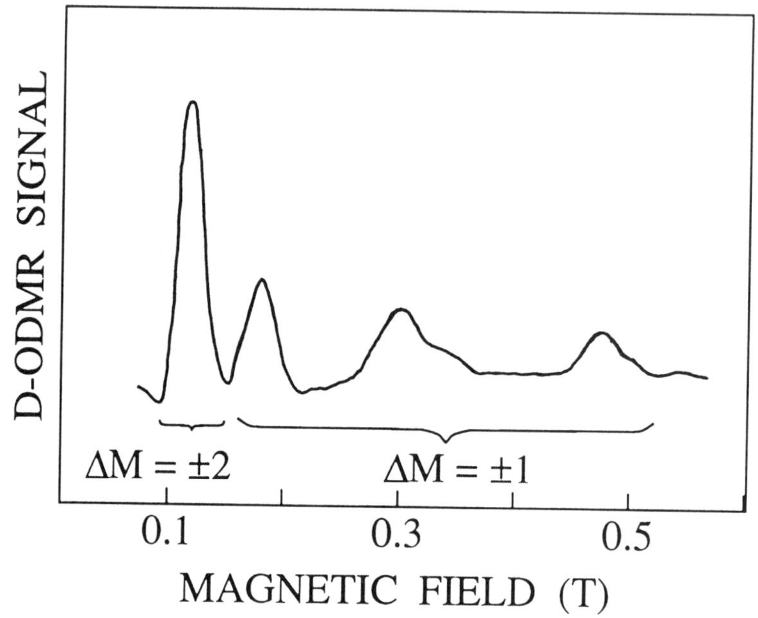

Fig.1 : D-ODMR spectrum from the S_A BE triplet taken at 4K with the magnetic field direction close to the <111> axis.

These data altogether lead to a deeper physical insight of the electronic structure and the possible defect constituents. We shall concentrate mainly on results obtained from the S_A configuration, since the relatively stronger ODMR signal in this case enables us to achieve more reliable information.

2. Experimental

The samples used in this work were from single crystalline silicon wafers grown either by the Czochralski (Cz) or the float-zone (Fz) method. The S doping was done either by high temperature diffusion or by high temperature annealing of S-implanted samples. An annealing procedure at about 800°C was done as a second step, followed by a rapid quenching. More details about the sample preparation can be found elsewhere [2,3].

The X-band ESR and ODMR experiments were performed with the aid of a modified Bruker ER-200D ESR spectrometer, equipped with a TE_{011} microwave (MW) cavity with optical access in all directions. Either the MW amplitude or the magnetic field could be modulated in the experiments. In many cases, a delayed ODMR (D-ODMR) technique was employed to avoid a strong background signal of the optically detected cyclotron resonance (ODCR) related to free carriers. Details about the experimental procedure and setup were described previously [6]. In the zero-field ODMR experiments, a set of tunable re-entrant MW cavities was used to cover MW frequencies in the range of 1-10 GHz [7].

3. Results and discussion

3.1. X-band ODMR results

A strong ODCR background signal was observed in all of the samples studied,

Table 1 : Spin Hamiltonian parameters for the S_A BE from X-band data

	D-tensor (MHz)	g-tensor	A-tensor (MHz)
x	-2888	≈2	≈200
y	-585	≈2	≈200
z	+3473	≈2	≈200

where x is 17 degrees off the [111] axis towards the [001] axis in the $(1\bar{1}0)$ plane, z=$[1\bar{1}0]$, y is perpendicular to both x and z

which obscured a detection of any defect-related ODMR signal, in a standard ODMR experiment. Instead, the D-ODMR technique was successfully applied. In Fig.1, we show a D-ODMR spectrum from the S_A BE, when the magnetic field direction is close to the crystallographic <111> axis.

A full angular dependence of the ODMR spectrum for the S_A could be obtained, which was fitted by a spin Hamiltonian

$$\mathbf{H} = \mathbf{SDS} + \mu_B \mathbf{SgB} + \mathbf{SAI} \quad (1)$$

Here the first term describes the zero-field splittings of the spin-triplet, which is introduced by the spin-orbit interaction and the magnetic dipole-dipole interaction [8]. The second term is the electronic Zeeman term, the third the central hyperfine (HF) interaction between the electronic and nuclear spins of the defect atoms present in the complex. **S** represents the BE electronic spin and **I** the nuclear spin of the defect atom. μ_B is the Bohr magneton, **B** the external magnetic field. **D**, **g** and **A** are the tensors describing the zero-field interaction, magnetic Zeeman and central hyperfine interaction, respectively. The nuclear Zeeman term and the nuclear quadrapole interaction were not listed in Eq.(1), for simplicity. The so-deduced spin-Hamiltonian parameters for the best fit to the experimental data are listed in Table 1. The symmetry was deduced to be monoclinic I, in agreement with the previous conclusion by analysing the preliminary ODMR data [5]. One of the axis perpendicular to the primary axis is rotated 17° in the $(1\bar{1}0)$ plane from the [111] axis.

An unusually broad ODMR linewidth was again observed, where no large anisotropy could be detected. No difference in the ODMR linewidth could be seen in samples implanted by ^{33}S (I=3/2) or in ^{13}C (I=1/2)-rich materials. This observation clearly indicates that the line broadening is not induced by either S or C atoms. In other words, the BE wavefunction is localized at a defect atom other than a S or a C atom.

3.2. Zero-field ODMR results

For a spin-triplet of monoclinic I symmetry, the spin degeneracy is completely lifted already at zero external field. Therefore three different ODMR lines should be observed. This is actually the case for S_A. In Fig.2 we show all three MW-induced transitions. These transitions are consistent with the energy scheme deduced from the X-band ODMR results, and confirm the spin-triplet nature of S_A^0.

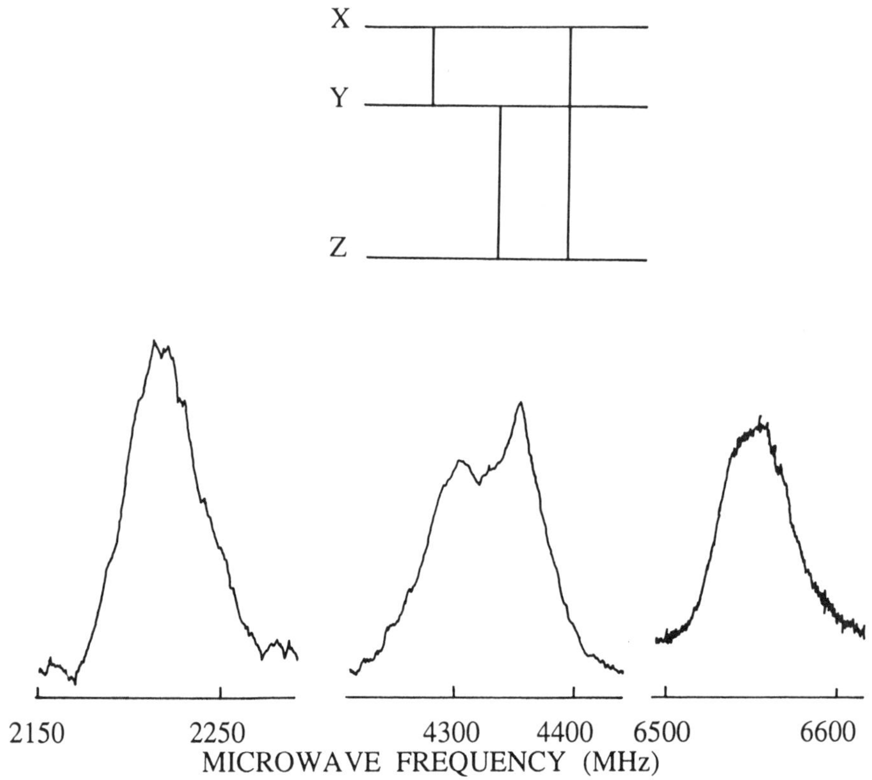

Fig.2 : Zero-field ODMR spectra from the S_A BE triplet taken at 4K.

It is interesting to note that the ODMR linewidth at zero-field is observed to be about 50 MHz, whereas the observed X-band linewidth corresponds to about 200 MHz. This is a fingerprint for the fact that the X-band ODMR line broadening is due to a strong HF interaction since the HF interaction is quenched in the first order at zero field within the framework of perturbation theory, since :

$$\langle u \mid \mathbf{S}_u \mid u \rangle = 0, \qquad \text{where } u = x, y, z \qquad (2)$$

A more detailed study by a double resonance technique shows a multiplet nature for each zero-field ODMR line, due to the HF interaction involving a defect atom with a nuclear spin I=3/2 of a dominating abundance. As an example, we show in Fig.3 experimental double resonance result for the S_A BE triplet, by pumping the Y-Z resonant transition while detecting at the X-Y transition. In this way, we are able to resolve the doublet nature for each electronic level. In addition the corresponding nuclear sublevels are linked due to the selection rule $\Delta m_{\pm} = 0$ in the ODMR experiments. By fitting to the energy shift induced by the second-order effect, the HF interaction can be estimated to be about 200 MHz. This is in turn consistent with the X-band ODMR linewidth.

Among the common residual impurities in the samples used are carbon, oxygen and some transition metals such as copper and iron. The only candidate which possesses a nuclear spin I=3/2 of a dominating abundance is thus Cu, with ^{63}Cu and ^{65}Cu both having I=3/2, differing in their nuclear g values by only a few percents.

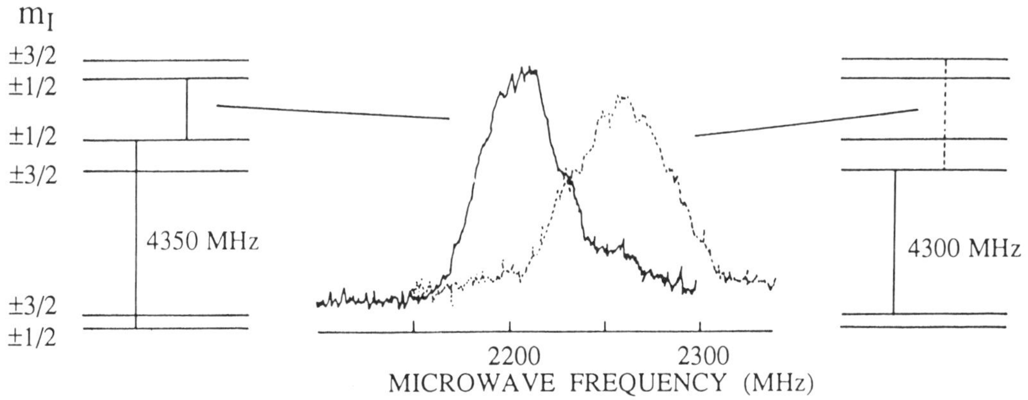

Fig.3 : Double resonance spectra from the S_A BE triplet obtained while scanning over the ODMR region of the X-Y transition by pumping the Y-Z resonant transition.

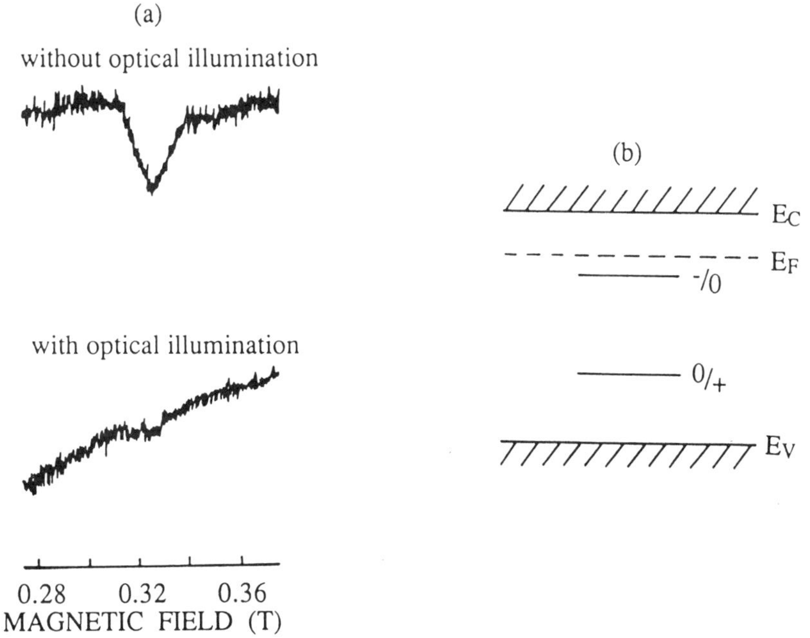

Fig.4 : (a) ESR spectra obtained with or without optical illumination. The sample was cooled down in darkness.(b) Energy level scheme for the configuration 1 of the complex which gives rise to the S_A BE.

3.3. ESR results

From PL experiments it is known that by cooling the sample in dark the defect is frozen in the configuration 1 (giving S_A PL), whereas by a subsequent laser excitation the defect converts to the configuration 2 (giving S_B PL) [3,4]. A similar behaviour was also observed in the ESR experiments. In Fig.4(a) ESR spectra taken just after cooling down the sample in dark and taken after light excitation, respectively are shown. The broad ESR signal around g=2 was only seen when the defect under study is present. We believe that this ESR signal is related to the same S-related defect, in a paramagnetic charge state of configuration 1, presumably the negative charge state since the Fermi level after S doping is usually high (in the upper half of the forbidden gap). The ESR linewidth is similar to that of the ODMR signal, which further supports the assignment. The (0/+) ground state level of the defect is known from the knowledge of the recombination energy of the S_A BE and the binding energy of the loosely bound electron of the BE [2]. In Fig.4(b) we schematically illustrate the energy levels for configuration 1 of the defect.

4. Summary

We have investigated the electronic structure and chemical identity of a S-related complex defect, by a combination of various magnetic resonance techniques, including ESR, D-ODMR and zero-field ODMR. We have presented here detailed investigations for one configuration of the metastable defect, via the S_A emission. We have confirmed the spin-triplet nature of the lowest BE states in the neutral charge state of the defect. The defect has a distorted monoclinic I symmetry with an axis perpendicular to the primary axis rotated 17° from the [111] direction in the $(1\bar{1}0)$ plane. We have provided evidence for the presence of a Cu atom in the complex where the BE wavefunction is believed to be localized. The presence of a (-/0) level in the upper part of the band gap is predicted for this configuration. A corresponding study for the S_B BE is currently in progress.

Acknowledgement

One of us (M.S) would like to thank the British Council Sweden and Linköping University for financial assistance.

References

1. T.G. Brown and D.G. Hall, Appl. Phys. Lett. 49, 245 (1986).
2. D.J.S. Beckett, M.K. Nissen and M.L.W. Thewalt, Phys. Rev. B40, 9618 (1989).
3. M. Singh, E.C. Lightowlers, G. Davies, C. Jeynes and K.J. Reeson, Mat. Sc. and Eng. B4, 303 (1989).
4. A. Henry, W.M. Chen, E. Janzén and B. Monemar, *Proc. 20th Int. Conf. on the Phys. of Semiconductors*, Greece, Aug. 6-10, 1990, eds. E.M. Anastassakis and J.D. Joannopoulos (World Scientific, Singapore, 1990) p.545.
5. W.M. Chen, A. Henry, E. Janzén, B. Monemar and M.L.W. Thewalt, Mat. Res. Soc. Symp. Proc. Vol. 163, 303 (1990).
6. W.M. Chen and B. Monemar, J. Appl. Phys. 68, 2506 (1990).
7. M.C.J.M. Donckers, J. Schmidt, W.M. Chen and B. Monemar, Phys. Rev. B42, 1684 (1990).
8. B. Monemar, U. Lindefelt and W.M. Chen, Physica 146B, 256 (1987).

Materials Science Forum Vol. 83-87 (1992) pp. 257-262

PIEZOSPECTROSCOPY OF TWO BERYLLIUM RELATED DOUBLE ACCEPTORS IN SILICON

J. N. HEYMAN[1], A. GIESEKUS[2] AND E. E. HALLER[3]
[1]Physics Dept., University of California at Berkeley, and Lawrence Berkeley Laboratory, 1 Cyclotron Rd., Berkeley CA 94720 USA
[2]Institut für Physik, University of Dortmund, Postfach 500-500, W 4600 Dortmund 50, Germany
[3]Materials Science and Mineral Eng. Dept., University of California at Berkeley, and Lawrence Berkeley Laboratory, 1 Cyclotron Rd., Berkeley CA 94720 USA

ABSTRACT

We present a detailed study of two beryllium related defects in silicon using infrared spectroscopy combined with uniaxial stress. We have investigated the electrically active centers with ionization energies E_V+191.9 meV (Be_1) and E_V+145.8 meV (Be_2). We find that the Be_1 center possesses tetrahedral symmetry. The ground state and excited states of this system are split at zero stress. Uniaxial stress measurements indicate that the ground state of the system transforms according to the Γ_5 representation of T_d. Our data agree qualitatively with recent calculations[1] of the electronic structure of the double acceptor ground state manifold in silicon. We conclude that the Be_1 double acceptor is the isolated substitutional beryllium impurity.

The excitation spectrum of the Be_2 center consists of two acceptor effective mass line series, displaced by 2 meV. Sets of Fano resonances corresponding to each of these series are seen in the Be_2 continuum, shifted from the line series by 63.4 meV which is the energy of the zone center optical phonon in silicon. The relative intensity of the two series is constant in all samples measured. In addition, stress induced line splittings are equal for corresponding lines of the two series. These data indicate that both effective mass series arise from the same center. Stress data indicate that the center is trigonal. Our experimental results are consistent with a distorted double acceptor model in which hole-hole correlation effects are important in the ground state but may be neglected in the description of the excited states. We suggest that the Be_2 center consists of two beryllium atoms on nearest neighbor substitutional sites.

1. Introduction

The characterization of double acceptors (DA) defects in germanium[2,3] by infrared spectroscopy has established the importance of hole-hole coupling in these centers. The effect of hole-hole coupling in neutral double acceptors has been described by several authors[1,4,5]. The $(1s)^2$ envelope state is 6 fold (15 fold if the split-off band states are included). In a first approximation, the Coulomb interaction splits the $(1s)^2$ envelope states into J=0 (Γ_1) and J=2 (Γ_3+Γ_5) levels. A more sophisticated treatment[1] takes into account the split-off band states. This introduces additional levels in the $(1s)^2$ manifold with much higher energies, and leads to a finite splitting between the Γ_5 and Γ_3 levels. Splitting of the DA excited states is analogous to the singlet-triplet splitting of neutral helium. However, a more complex splitting pattern is expected. For example, hole-hole coupling may split the $1s(\Gamma_8)$-$2p(\Gamma_8)$ excited state into seven levels. Giesekus and Falicov[1,4] have calculated values of splittings in the DA $(1s)^2$ states and the (1s-2p) manifold states. They find a strong dependence on the localization of the defect. In particular, for reasonable choices of material parameters they obtain a splitting between the Γ_1 and the Γ_3+Γ_5 $(1s)^2$ states of order 1 meV in Ge DA's, and 40 meV in Si DA's. In Si and Ge they find a splitting of about 1 meV between the Γ_3 and Γ_5 states. They obtain a ground state level ordering of $\Gamma_3<\Gamma_5<\Gamma_1$.

Any other level ordering requires an attractive interaction between holes on neighboring atomic sites. This may arise from many-body effects. Thewalt et. al[2]. have found the ground state of Ge DA's to be Γ_1.

Valence arguments suggest that a substitutional group II Be impurity will form a double acceptor. However, doping silicon with beryllium introduces several electrically active and inactive centers which have been studied by Hall effect[6], DLTS[7], photoluminecence[8,9] as well as infrared spectroscopy. In particular, IR line spectra of four acceptor centers have been reported[6,10-12]. Crouch and Robertson[6] have proposed that the centers with ionization energies 191.9 meV and 145.8 meV are the substitutional center and a Be related complex, which may be the pair. This conclusion was based on annealing studies which show that the concentration of the 145.8 meV center is increased by slow cooling, but is reduced by a rapid quench form T>600°C. Recent first principles calculations[13] have investigated the total energy of Be_s, Be_i and a number of multi-atom Be complexes in Si. The Be_s double acceptor is predicted to be stable at the high temperature doping conditions. In addition a Be_s-Be_s pair may form under certain conditions, and the lowest energy configuration of this center is predicted to be a double acceptor.

It is possible to identify neutral double acceptors with infrared spectroscopy. The IR spectrum of a DA is very similar to the acceptor effective mass line series. However, the hole-hole interaction may cause observable splittings of the ground and excited states. In addition, uniaxial stress measurements give information about the symmetry properties of the two-particle wavefunctions, which differ from the single particle states. These techniques have been used to study double acceptors in Ge. IR spectroscopy has also identified[14,15] splitting of the ground and excited states of Si:Zn.

2. Experimental Procedure

Bulk Be doped silicon was obtained by high temperature (1300°C) diffusion of Be metal into high purity Fz Si in sealed quartz ampules. SIMS measurements on selected samples gave total Be concentrations of 1 - 5 $\cdot 10^{16} cm^{-3}$. Room temperature carrier concentrations in the samples used in this study were also in this range. Transmission measurements were obtained with a Digilab 80E-V Fourier transform spectrometer. Samples were mounted in an exchange gas optical cryostat. Uniaxial stress measurements were made with a stress cell which may be mounted in our cryostat. In this system <111>, <100> or <110> oriented samples are placed between two pistons. Force is applied to the upper piston through a pushrod from outside the cryostat. A known pressure is obtained from a gas piston at room temperature. Stress induced dichroism was investigated with a wire grid polarizer.

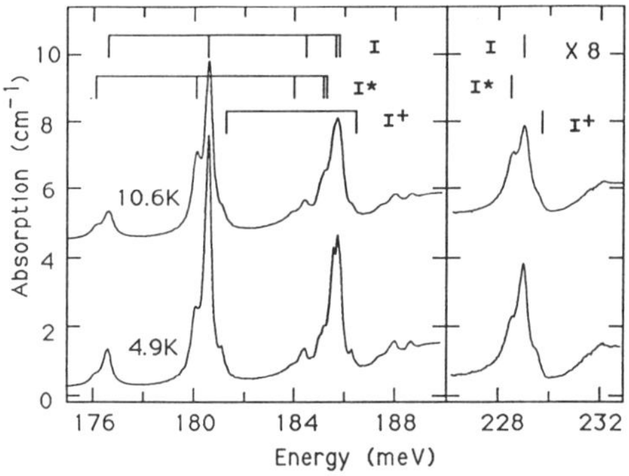

Figure 1. Sharp line spectrum of the Be_1 center. A thermally activated series (I*) accompanies the main series (I). Features due to an excited state splitting (I^+) are seen for lines I_2, I_{4A}, and $I_{2p'}$. The unapodized resolution is 0.03 meV.

3. Be_1 Experimental results

The infrared absorption spectrum of the Be_1 center is shown in figure 1. The main features are similar to the effective mass line series of simple acceptors. Transitions to bound states of the j=3/2 valence band (the $p_{3/2}$ series) and transitions to bound states of the j=1/2 split off band (the $p_{1/2}$ series) are observed. The lines show a fine structure. A second series (I*) displaced by -0.51(5) meV from the main (I) series grows with increasing temperature. We find

$$\frac{I^*}{I} = 0.6 \exp\left(\frac{-E_a}{kT}\right) \qquad E_a = 0.35(10)\text{ meV}$$

The (I-I*) splitting and the thermal activation energy compares well with the energy of a FIR transition in Si:Be of 0.50(5) meV which has been observed by Peale et. al.[16], who have also observed the temperature dependent (I*) series in Si:Be. The temperature dependence arises because the initial state of (I*) must be thermally populated. This indicates that the (I-I*) splitting is due to a split ground state. In addition three lines clearly show a weak component 0.53(5) meV above the main feature (I+). The relative intensity of the (I+) features to the (I) series is almost temperature independent, showing that this structure arises from a splitting of the final state.

The splitting of line I_{2p} of the $p_{1/2}$ series under uniaxial stress is shown in figure 2. The stress dependences of the lines of the $p_{3/2}$ series are complicated by the splitting of the j=3/2 valence band, and will not be discussed in detail. The strong feature I_{2p} splits under all orientations of stress. As above, we can distinguish splitting of the initial state from splitting of the final state by the temperature dependence of the line intensities. We observe that the final state of I_{2p} does not split under <111> stress. The splitting under <100> and <110> stress are linear and are $\Delta_{<100>} = 7.9 \cdot 10^{-3}$ meV/MPa and $\Delta_{<110>} = 1.6 \cdot 10^{-2}$ meV/MPa. Thermalizing components of I_{2p} appear under <111> and <100> stress, indicating a splitting of the initial state.

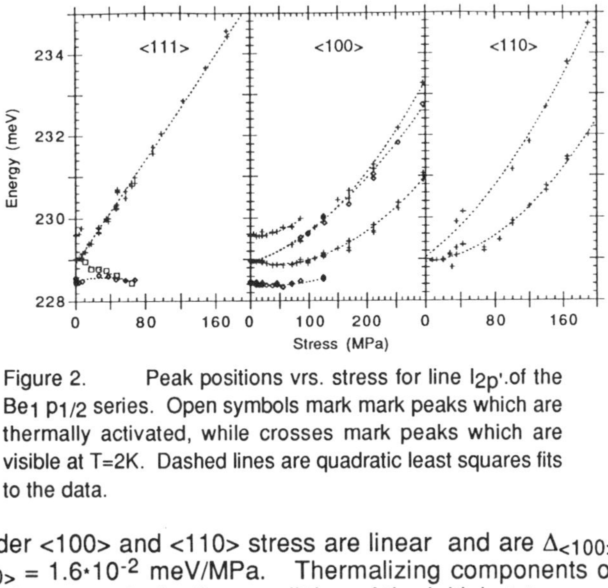

Figure 2. Peak positions vrs. stress for line $I_{2p'}$.of the Be_1 $p_{1/2}$ series. Open symbols mark mark peaks which are thermally activated, while crosses mark peaks which are visible at T=2K. Dashed lines are quadratic least squares fits to the data.

4. Be_1 Discussion

Our data determine the site symmetry of the Be_1 center. Transitions at non-cubic centers may split under stress due to a lowering of the orientational degeneracy. The lack of such splitting under <111> stress indicates that the center is tetragonal or tetrahedral. The splitting of lines at a tetragonal center under <100> and <110> is given by $\Delta_{<100>} = 2\Delta_{<110>}$, which does not fit our data. We conclude that the center is tetrahedral.

We anticipate that the ground state wavefunction will transform like either the Γ_1, Γ_3 or Γ_5 representations of T_d. Of these, only Γ_5 splits under all orientations of stress. The final state of I_{2p}, which does not split under <111> stress, transforms like Γ_3. We are not able to resolve splittings of the weak components I^*_{2p} and I^+_{2p}. Stress data from the $p_{3/2}$ series are consistent with the observations made above. Significantly, they also show that the initial state of the I* series splits under <100> stress, but not under <111> stress. We conclude that the wavefunctions of the two lowest levels of the $(1s)^2$

manifold in Be_1 transform like Γ_5 and Γ_3. The splitting between the Γ_5 and Γ_3 states is 0.5 meV.

5. Be_2 Experimental results

Infrared spectra of samples containing $>10^{16}cm^{-3}$ electrically active Be which have been slowly cooled from T>600°C show absorption lines in the region 130-150 meV (fig. 4). The spacing and relative intensity of 5 strong lines match the $p_{3/2}$ series of an acceptor with ionization energy of 145.8 meV. The remaining lines comprise a second series shifted 2 meV to higher energy. We label the series α and β respectively. The relative intensity of the two series is approximately 5:1 and is both temperature independent, and constant in all samples measured. Optical phonon Fano resonances of both series are observed[11], and are approximately equal in intensity. The $p_{1/2}$ series lines are not seen, being obscured by much stronger Be_1 $p_{3/2}$ lines.

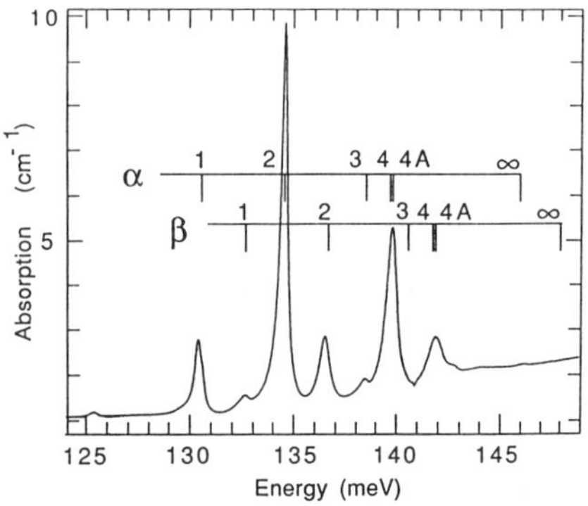

Figure 3. Sharp line spectrum of the Be_2 center. Two overlapping effective mass series separated by 2 meV are seen. A weak feature at 124.4 meV is not related to the Be_2 center.

Line splittings due to <100> and <111> uniaxial stress are summarized in figure 5. The relative intensities of all components are temperature independent, indicating that the initial state does not split under stress. All splittings of the α and β series under <100> stress agree with well known models of the final states of acceptor $p_{3/2}$ lines. The magnitude of the splitting of lines $I_{1\alpha}$ and $I_{2\alpha}$ are comparable to results obtained from other acceptors[17] in Si, and from theory[18]. Within experimental error, effects in the α and β series are identical.

Under <111> stress we also observe splittings of the α series lines predicted for acceptor excited states. However, there is an additional linear splitting of 0.027 meV/MPa of all lines. For example, line 2α splits into 4 components. The splitting between the the first and second component (third and fourth) are comparable to the splitting of the $2p_{5/2}(\Gamma_8)$ line in Si:B. The splitting between the first and third (second and fourth) is 0.027 meV/MPa. Splittings of the β series are only poorly resolved, however a splitting of about 0.027 meV/MPa is also visible in lines 2β and 4β.

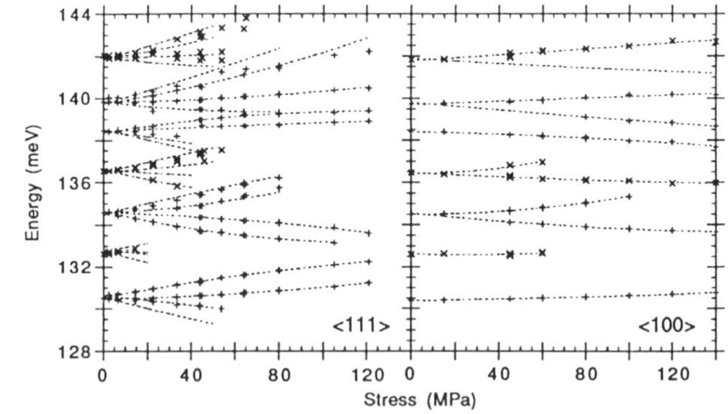

Figure 4. Peak positions of the Be_2 $p_{3/2}$ series vrs. stress. Dashed lines are obtained from our model. This data shows that the Be_2 center is trigonal.

The splitting pattern observed in our data show that the Be_2 center is trigonal. Under <100> stress all orientations of trigonal centers are equivalent. Under <111> stress the magnitude of the splitting depends on the Be_2 center orientation with respect to the stress direction. The spectra show an additional splitting due to the simultanious measurement of nonequivalent centers. A linear theory[19] gives the shift of transition energies in a trigonal center:

$$\Delta = \sum_i A_1 \sigma_{i,i} + \sum_{i \neq j} A_2 \sigma_{i,j}$$

where $\sigma_{i,j}$ are elements of the stress tensor in a cubic coordinate system relative to the orientation of the center. From our data we obtain $A_1 = 2(2) \cdot 10^{-3}$ meV/MPa, $A_2 = 1.02(5) \cdot 10^{-2}$ meV/MPa. The relative intensity and polarization dependence of line components of 2α are consistent with the trigonal model. Finally, we note that as the constants A_1 and A_2 measured for the β series are identical to the α series, the microscopic structure of the centers which give rise to these lines must be identical. In addition, we have found that the ratio of intensities of the two series is constant in all samples. We conclude that both series arise from the same center.

6. Be_2 Discussion

Any model for the Be_2 center must explain the origin of the β series. The β series is of electronic origin and is not a phonon sideband, because the typical temperature dependence of phonon sidebands[20] is absent. We interpret our data with a trigonal double acceptor model. We assume that the 1s-2p states may be treated in an independent hole approximation: Splittings due to the Coulomb interaction are not observed (The unresolved high energy shoulder of line 1α may indicate the breakdown of this assumption). Within the framework of this simplification, the two-hole 1s-2p excited states transform like a direct product of the single-hole symmetries: The 1s-state transforms like Γ_8, and the 2p-states split into two Γ_8 components (lines 1 and 2), a Γ_6 and a Γ_7 component (lines 4 and 4A).

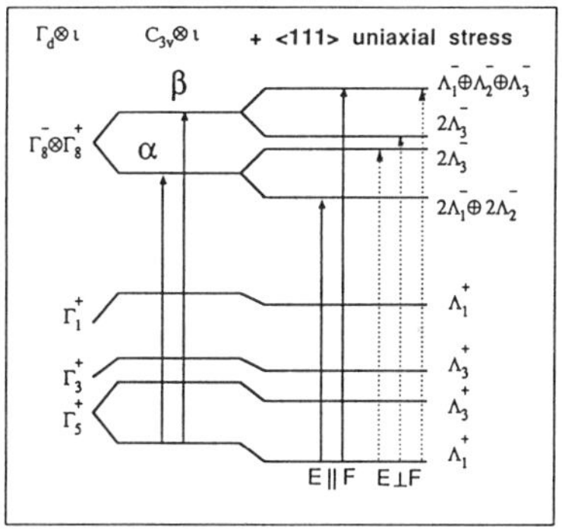

Figure 5. Proposed level diagram for Be_2 showing the $(1s)^2$ state and the 1s-2p excited state for line 2. The left side corresponds to a tetrahedral double acceptor. The center shows the effect of the trigonal central cell. The right hand side shows the effect of an additional <111> uniaxial stress. This stress splitting is only appropriate for centers oriented along the stress axis, e.g. lines $2\alpha a$, $2\alpha b$, $2\beta a$, and $2\beta b$.

The trigonal symmetry of the central cell will split the 1s-states into two Kramers doublets. In principle, the 2p Γ_8 states will split as well, however, an estimate of the splitting due to the quadrupole term of the central cell potential yields 0.04 meV, which is unresolvable. As a consequence, the excited states are split because of the 1s-state splitting only, leading to the additional line series (β series).

The ground-state manifold of a double acceptor in a trigonal potential transforms according to $2\Lambda_1 + 2\Lambda_3$ of the group C_{3v}. If the effect of the trigonal potential is large, the lowest state will have Λ_1 symmetry and the others may not be thermally accesible. Electric dipole transitions will only connect the Λ_1 ground state to both the α and the β series if the Coulomb interaction strongly mixes the two Λ_1 components. This explains why the β series is much weaker than the α series.

7. Summary

We have presented data from IR studies which indicate that the Be related defect at Ev+191.8 meV is a fully tetrahedral center, while the 145.8 meV center is trigonal. In the trigonal center, we observe two effective mass line series. We show that the two series may arise from the splitting of the 1s- wavefunction in the 1s-2p configuration of a trigonal double acceptor. Hole-hole coupling is important in the description of the ground state of the system. Our data support the Be_s-Be_s model for this center originally suggested by Crouch and Robertson[6]. IR spectra of the tetrahedral center show splittings of the initial and final states of the $p_{3/2}$ and $p_{1/2}$ series. Uniaxial stress studies show that the lowest levels of the groundstate manifold transform like Γ_5 and Γ_3, and the Γ_5-Γ_3 splitting is 0.5 mev. These data indicate the 191.8 meV center is a neutral DA. Recent theories[1] are consistent with our data, and require an attractive intersite Coulomb interaction in this system. We conclude that this center is isolated substitutional Be.

8. Acknowledgments

The authors would like to recognize the invaluable assistance of Professor L. M. Falicov. In addition we thank W. Walukiewicz, E. Merk and W. Hansen for their suggestions and guidance. Finally, we wish to thank J. Beeman for fabricating our Ge:Cu IR detectors, and J. Walton for the use of his facilities. This work was supported by USNSF grant DMR-88-06756. One of us (A. G.) acknowledges the support of the Deutsche Forschungsgemeinschaft (D.F.G).

References

1. A. Giesekus and L. M. Falicov, Phys. Rev. B **42**, 8975 (1990).
2. M. L. W. Thewalt, D. Labrie, I. J. Booth, B. P. Clayman, E. C. Lightowlers, and E. E. Haller, Physica B**146**, 47 (1987).
3. D. Labire, I. J. Booth, M. L. Thewalt, and E. E. Haller, Phys. Rev. B **38**, 5504 (1988).
4. A. Giesekus and L. M. Falicov, Phys. Rev. B **42**, 11725 (1990).
5. E. P. Kartheuser, S. Robriguez, and P. Fisher, Phys. Stat. Sol. (b) **64**, 11 (1974).
6. R.K. Crouch, J. B. Robertson, and T. E. Gilmer, Jr., Phys. Rev. B **5**, 3111 (1972).
7. H. Tomokage, M. Hagiwata, K. Hashimoto, Jpn. J. Appl. Phys. **21**, 402, (1982).
8. M. O. Henry, K. G. McGuigan, M. C. do Carmo, M. H. Nazare, and E. C. Lightowlers, J. Phys.: Cond. Matter **2**, 9697 (1990).
9. D. Labrie, T. Timusk, and M. L. Thewalt, Phys. Rev. Lett. **52**, 81 (1984).
10. J. N. Heyman, A. Giesekus, L. Falicov, and E. E. Haller, Bull. Am. Phys. Soc. **35**, 279 (1990).
11. J. N. Heyman and E. E. Haller, Bull. Am. Phys. Soc. **36**, 862 (1991).
12. M. Kleverman and H. G. Grimmeiss, Semicond. Sci. Technol. **1**, 45 (1986).
13. Eugen Tarnow, S. B. Zhang, K. J. Chang, and D. J. Chadi, Phys. Rev. B **42**, 11252 (1990).
14. E. Merk, J. Heyman, and E. E. Haller, Solid State Commun. **72**, 851 (1989).
15. A. Dörnen, R. Kienle, K. Thonke, P. Stoltz, G. Pensl, D. Grünebaum, and N. A. Stolwijk, Phys. Rev. B **40**, 12005 (1989).
16. R. E. Peale, K. Muro, and A. J. Sievers, Phys. Rev. B. **41**, 5881 (1990).
17. H. R. Chandrasekhar, P. Fisher, A. K. Ramdas, and S. Rodriguez, Phys. Rev. B **8**, 3836 (1973).
18. R. Buczko, Il Nuovo Cimento **9**, 669 (1987).
19. A. A. Kaplyanskii, Opt. Spectrosc. **16**, 329 (1964).
20. W. Hayes, A. M. Stoneham in *Defects and Defect Processes in Nonmetallic Solids*, (John Wiley & Sons Inc., New York, USA., 1985) p. 185.

Materials Science Forum Vol. 83-87 (1992) pp. 263-268

ON THE DETERMINATION OF NITROGEN IN CZOCHRALSKI SILICON

M.W.Qi[1], T.S.Shi[1], S.S.Tan[1], B.Zhu[1], P.X.Cai[1], L.Q.Liu[1], D.L.Que[2] and L.B.Li[2]

[1] **Shanghai Institute of Metallurgy, Academia Sinica, Shanghai 200050, China.**
[2] **Semiconductor Materials Research Institute, Zhejiang University, Hangzhou 310027, China.**

ABSTRACT

A modified calibration curve for determining the nitrogen content in Czochralski silicon Fourier Transform infrared spectroscopic techniques is presented. This modified calibration curve is based upon the fact that part of the nitrogen in Czochralski silicon which is complexed with oxygen impurity contributes to the 801 cm^{-1} band, instead of the wellknown 963 cm^{-1} band and should be included in the calibration curve.

1. INTRODUCTION

Nitrogen is receiving increased attention as an impurity in silicon after oxygen, carbon and hydrogen due its stronger strengthening effect than oxygen and its ability to diminish the swirl and D-defects in floating-zone (FZ) melted silicon[1,2]. Little is known about N-doped Czochralski (CZ) silicon. Much of the understanding about the behaviour of nitrogen in silicon has come from studies of ion-implanted Si. Nitrogen atoms exist in silicon as N-N pairs, it was concluded from two separate infrared absorption experiments by simultaneously implanting[3] or doping[4] two different N-isotopes, ^{14}N and ^{15}N, where "mixed"pairs were observed. Furthermore, the interaction of nitrogen and oxygen has been first pointed out by Stein[3]. The other characteristic of N-doped CZ-Si is the series of shallow effective mass-like single donors which are related to the presence of N-O complexes in annealed Si[5].

The strength of an IR absorption band associated with the N-N pair centre at 963 cm^{-1} has been calibrated against the nitrogen content, determined by charged particle activation analysis with the $^{14}N(p,\alpha)^{11}C$ reaction[6]. The calibration curve for as-grown FZ-Si is expressed as $[N] = (1.83 \pm 0.24) \times 10^{17} \times \alpha_{963cm^{-1}} at/cm^3$. For heat-treated FZ-Si and CZ-Si, this relation was found to give an underestimated value. In this paper, we show that some of these N-N-O complexes already exist in the as-grown crystal, thus reducing the number of N-N pairs, and the intensity of the 963 cm^{-1} band in CZ-Si. We obtain the proportionality constant for 801 cm^{-1} to 963 cm^{-1}, one of the strong bands in the N-N-O complexes, so we are able to estimate the nitrogen concentration in CZ-Si by modifying the calibration curve suggested by Itoh et al[6].

2. EXPERIMENTAL

$< 111 >$ oriented n-type low carbon CZ-Si crystal containing nitrogen with nominal resistivity of 6-9 Ω cm were grown under N_2 atmosphere at reduced pressure[7]. 5mm thickness samples were originally cut from $< 111 >$ orientation N-doped CZ-Si, and chemically polished to an optical surface on both sides and diced into 10 ×10 mm^2 pieces for IR absorption measurement at room temperature. Isochronal or isothermal annealing of samples was performed under flowing N_2 ambient. IR absorption measurements were carried on a NIC-7199C Fourier transform infrared spectrometer (FTIR). Typical measurement conditions for FTIR were a resolution of 4 cm^{-1} and accumulation of 100 scans. An undoped high purity vacuum FZ-Si was used as reference. It is difficult to make both the sample containing nitrogen and the reference have the same thickness, surface condition and free carrier concentration, so the subtraction method was used. The sample spectrum and the reference spectrum were recorded respectively with the air spectrum as the background, then the latter was subtracted from the former by varying the scale coefficient to minimise the influence of the difference in samples thickness, surface condition and free carrier concentration on the resultant spectrum. Because the above mentioned effect varied with wave number, the subtraction was made region by region.

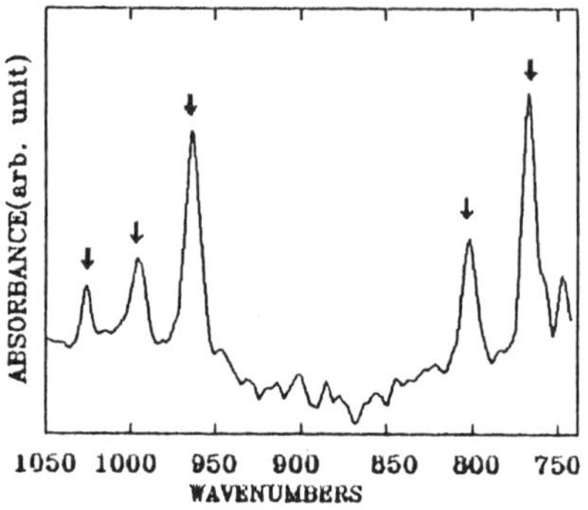

Fig.1 Infrared absorption spectrum of as-grown N-doped CZ-Si measured at 300K.

The nitrogen content in the crystals was determined by measuring the 963 cm^{-1} absorption band with the calibration factor[6] of 1.83 $\times 10^{17}$. The characteristics of tested crystals are shown in table I.

Table I The characteristics of tested crystals

Sample No.	Resistivity Ω cm	Oxygen at./cm^{-3}	Carbon at./cm^{-3}	Nitrogen at./cm^{-3}
6	6	4.7×10^{17}	$< 2 \times 10^{16}$	1.8×10^{16}
4	6.2	5.3×10^{17}	$< 2 \times 10^{16}$	9.2×10^{15}

3. RESULTS AND DISCUSSION

It is well known[3,4] that nitrogen pairing is the dominant mode for nitrogen incorporation into crystalline silicon producing two infrared absorption band 963 and 766 cm^{-1}. Three additional absorption bands at 1026, 996 and 801 cm^{-1} are observed in as-grown low carbon N-doped CZ-Si as shown in figure 1, probably an N-O complex as was suggested first by Stein[3] for Si co-implanted with nitrogen and oxygen. Recently, five additional absorption bands related to oxygen and nitrogen concentration were reported by Wagner[8] in N-doped CZ-Si at low temperature ($< 15K$). It is difficult, in general, to identify these two extra additional absorption bands at 1018 and 810 cm^{-1} in a room temperature measurement, because they are submerged in the noise and it is impossible to distinguish them. So there are 5 absorption bands, 963, 766, 1026, 996 and 801 cm^{-1}, which are related to nitrogen in the low carbon N-doped CZ-Si at room temperature.

The annealing behaviour for low carbon N-doped CZ-Si is quite different from that of N-doped FZ-Si[9]. As for N-doped FZ-Si, the intensities for the 963 and 766 cm^{-1} bands keep their original value almost unchanged until 800°C, then fall monotonically and disappear completely at 1100°C. Fig. 2 shows the intensity variation of these five absorption bands during one hour isochronal annealing for the low carbon N-doped CZ-Si, where the absorption coefficient of these five absorption bands are plotted against isochronal annealing temperature. It can be seen from Fig. 2 that the intensities of the 963 and 766 cm^{-1} bands have their minimum and maximum values at $600°C$ and $800°C$ respectively, then decrease after $800°C$ and anneal out at $1100°C$. The annealing behaviour for 963 and 766 cm^{-1} bands is very similar, in agreement with the suggestion that they originate from the same N-N pair centre[3].

The annealing behaviour of these three absorption bands in low carbon N-doped CZ-Si is most interesting. As can be seen clearly from Fig. 2, they not only have a similar behaviour but also change in an opposite way as compared to absorption bands related to the N-N pair, i.e. those at 963 cm^{-1} and 766 cm^{-1}.

It may be concluded from the above results that there are two sets of N-related absorption bands in the low carbon N-doped CZ-Si. The first set of absorption bands is the well known N-N pair. The second set of absorption bands belong to N-N-O complexes as has been suggested by Wagner[8].

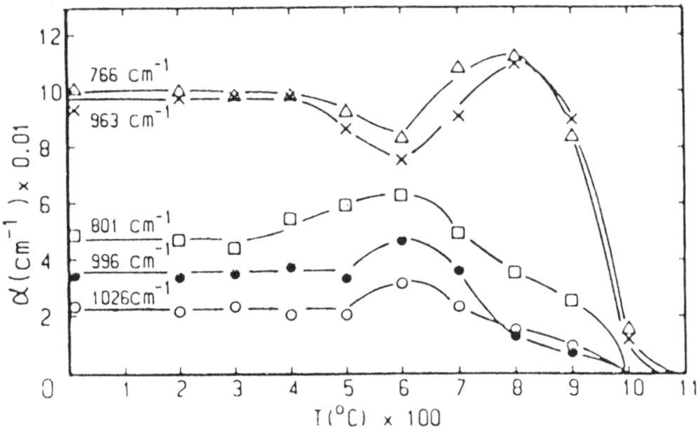

Fig.2 One hour isochronal annealing behaviour for two sets of IR bands corresponding to the N-N pairs and N-N-O complexes in N-doped CZ-Si measured at 300K.

We have shown[10] that a dynamical equilibrium exists between N-N pairs and N-N-O complexes below 800°C, such that (N-N)+O$\rightleftharpoons$ (N-N-O). When the temperature is above 400°C, the interstitial oxygen atoms in Si start to move and as they reach the nearest neighbour sites to the N-N pairs they produce new N-N-O complexes, with vibrational frequencies different from the N-N pairs. Correspondingly the intensity of N-N-O complexes begins to increase and the intensity of N-N pairs decreases up to 600°C. So at lower temperatures, the reaction for dynamical equilibrium shifts towards the right the equilibrium concentration for N-N-O complexes increases. As the temperature is lifted and the equilibrium shifts to the left, part of N-N-O complexes being decomposed to N-N pairs and interstitial oxygen atoms up to 800°C. The dynamical equilibrium was concluded from two experimental results[9] as following: 1) The intensity variation of N-N pairs with temperature is reversible during one hour of repeated heating and cooling in the temperature range of 600 $\sim$ 800°C. In contrast the intensity of N-N-O complexes changed with the same tendency in the opposite direction, but the data were more scattered because of the weakness of these bands. Some indication of these phenomena was noticed by Wagner[8] in his preliminary results on the stability of the N-N pairs and N-N-O complexes. 2) The complementary character between N-N pairs and N-N-O complexes is more clearly shown when the sum of the intensities of all five absorption bands are plotted against isochronal annealing temperature. A straight line parallel to the axis is obtained for temperatures up to 800°C. The idea of interaction of nitrogen pairs and oxygen in CZ-Si is quite conclusive considering the results shown above.

The strongest band related to the N-N-O complexes is that at 801 cm^{-1}. We found that all three IR bands of the N-N-O complex, at 801, 996 and 1026 cm^{-1} in the as-grown state and after various heat treatment up to 800°C have a good correlation coefficient as shown in Fig.3 amd Fig.4. The average ratio of between 801 and 996 or 1026 cm^{-1} are 1.47 and 2.34, respectively, by regression analysis. So the 1026 cm^{-1} band is the weakest in the N-N-O complexes. It is further confirmed that these N-N-O complexes originate from the same vibration centre. Some of these N-N-O complexes already exist in the as-grown crystal thus reducing the number of the N-N pairs and the intensity of the 963 cm^{-1} band in CZ-Si.

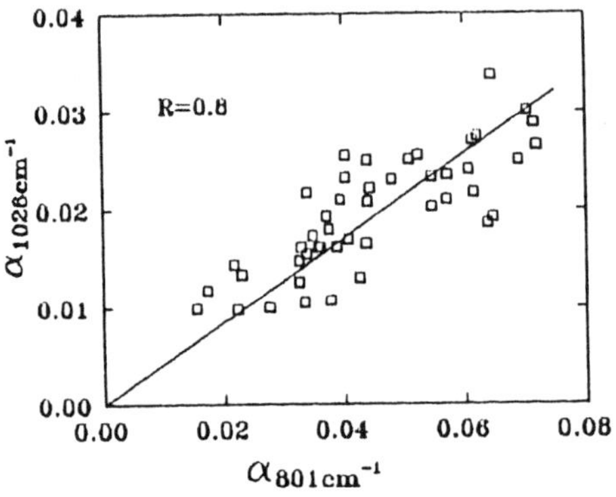

Fig.3 Correlation between the absorption coefficient of the absorption bands at 801 and 996 cm^{-1}.

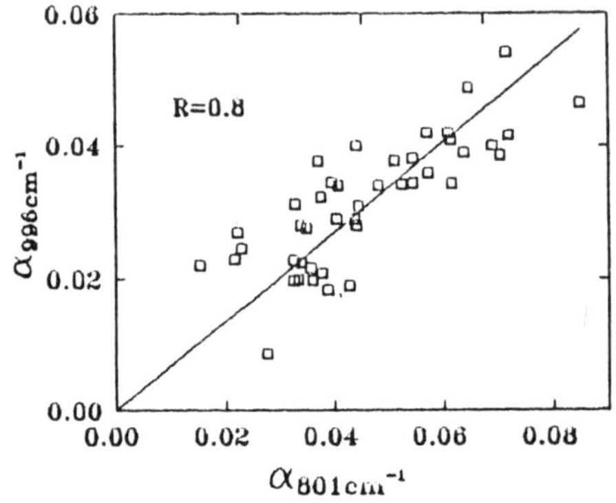

Fig.4 Correlation between the absorption coefficient of the absorption bands at 801 and 1026 cm^{-1}.

A calibration curve for the determination of nitrogen in Si by measuring the 963 cm^{-1} band was obtained by Itoh[6] et al using charged particle activation analysis with the ^{14}N $(p,\alpha)^{11}C$ reaction. The absorptivity of nitrogen at 963 cm^{-1} band in CZ-Si was found to be often considerably less than in SZ-Si. This value is also different from sample to sample. It is clear from the above discussion that part of the nitrogen in CZ-Si which is complexed with oxygen atoms will not contribute to the intensity of the 963 cm^{-1} and 766 cm^{-1} bands. Thus a refined method for determination of the nitrogen content in CZ-Si is required. We assumed that the tested samples did not contain a significant concentration of precipitated nitrogen. All these five bands are shifted to higher frequency at low temperature ($< 15k$)[8], presumably due to the local mode vibration. The charged particle activation analysis gives the absolute nitrogen content. Our results show that there are two sets of absorption bands, related to two kinds of N-containing complexes, i.e. N-N and N-N-O in the low carbon N-doped CZ-Si. Therefore, the contribution of N-N-O complexes should be taken into account for determining the conversion factor for IR measurement. But the increase of intensity for the 963 cm^{-1} band is different from that for 801 cm^{-1}. The variants of isolated $\Delta\alpha_{963cm^{-1}}$ plotted against the variants of $\Delta\alpha_{801cm^{-1}}$ are shown in Fig.5. Individual aberrant data points were excluded. Regression analysis for these 31 data points was used to calculate a slope for a straight line constrained to pass though the origin, Y=bX, and the resulting slope was taken as the best estimate for the proportionality constant between 963 and 801 cm^{-1} bands. The average value calculated gives Y=1.4X. A check made for significance of the sample correlation coefficient gives a value of 0.81. The systematic error of our results may arise from the possible crystal inhomogeneity and the calculation of the α due to noise. Room temperature measurement for the determination of the nitrogen content in silicon has some limitation. That is, all bands related to nitrogen are very weak due to its low concentration in melt-doped silcon. Low temperature measurement for determination of nitrogen content in silicon is needed.

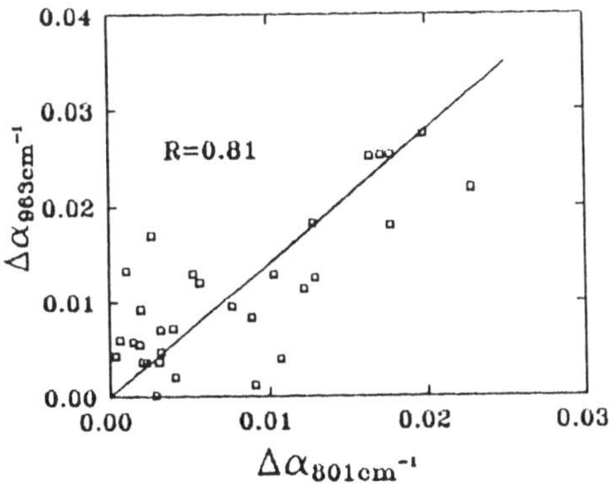

Fig.5 The variance of isolated $\Delta\alpha_{963cm^{-1}}$ plotted against the variance of $\Delta\alpha_{801cm^{-1}}$.

4. CONCLUSION

The infrared measurement of nitrogen content in low crabon N-doped CZ-Si detailed in this paper serves as the modified method for determination of nitrogen after Itoh et al. The calibration curve for as-grown or heat treated CZ-Si is expressed as $[N]=(1.83\pm0.24)\times10^{17}\times(\alpha_{963cm^{-1}}+1.4\alpha_{801cm^{-1}})at./cm^3$. The second term in the relation accounts for the nitrogen in the N-N-O complexes, the proposed calibration curve has a large uncertainty, however, it is considered to give a better reasonable value.

5. REFERENCES

1. T.Abe, H.Harada and J.Chikawa, Mat. Res. Soc. Symp. Proc., 14, 1,1983.
2. K.Sumino, I.Yonenaga, M.Imai and T.Abe, J.Appl.Phys., 54, 5016, 1983.
3. H.J.Stein, Mat. Res. Soc. Symp. Proc., 59, 523, 1986.
4. T.Abe, H.Harada, N.Ozawa and K.Adomi, Mat. Res. Soc. Symp. Proc., 59, 537, 1986.
5. M.Suezawa, K.Sumino, H.Harada and T.Abe, Jpn. J.Appl.Phys., 27, 62, 1988.
6. Y.Itoh, T.Nozaki, T.Masui and T.Abe, Appl.Phys.Lett., 47, 488, 1985.
7. D.L.Que, L.B.Li, X.Z.Chen, Y.P.Lin, J.X.Zhang, X.Zhou and J.S.Yang, to be published in Science in China, Series A, 1991.
8. P.Wagner, R.Oeder and W.Zulehner, Appl.Phys.A, 46, 73, 1988.
9. M.W.Qi, S.S.Tan, B.Zhu, P.X.Cai, W.F.Gu, X.M.Xu, T.S.Shi, D.L.Que and L.B.Li, J.Appl.Phys., 69, 3775, 1991.
10. M.W.Qi, S.S.Tan, B.Zhu, P.X.Cai, W.F.Gu, X.M.Xu, T.S.Shi, D.L.Que and L.B.Li, Chinese.J.Semiconductors, 12, 214, 1991. (in Chinese)

Materials Science Forum Vol. 83-87 (1992) pp. 269-272

β NMR OF NONSUBSTITUTIONAL ^{12}B AFTER IMPLANTATION INTO P–TYPE SILICON

B. FISCHER, W. SEELINGER, E. DIEHL, K.–H. ERGEZINGER, H.–P. FRANK, B. ITTERMANN, F. MAI, G. WELKER, H. ACKERMANN, AND H.–J. STÖCKMANN
Fachbereich Physik und Wissenschaftliches Zentrum für Materialwissenschaften der Philipps-Universität, W–3550 Marburg, Germany

ABSTRACT

β–radiation detected nuclear magnetic resonance of nonsubstitutional ^{12}B ($\tau = 29$ ms) implanted into p–type silicon has been observed at $T = 320$ K. From NMR transition frequencies an electric field gradient parallel to a <111> direction was deduced. From the effects of simultaneous irradiation of several radio frequency fields it could be shown that the nonsubstitutional ^{12}B ions perform diffusive motions without reaching substitutional sites during their lifetime τ.

1. Introduction

In two recent studies the lattice locations of ^{12}B (β lifetime $\tau = 29$ ms, $I = 1$) implanted into variously doped Si samples were investigated [1,2]. It turned out that for all samples irrespective of conduction type and dopant concentration the fraction f_S of ^{12}B stopped on substitutional sites increased from about zero to nearly 100 % if the temperature T was raised from $\lesssim 50$ K to about 1000 K. At room temperatue values of $f_S = 20...30$ % were obtained in most samples. Fig. 1 shows as an example a $f_S(T)$ curve for a moderately boron doped p–type sample. It is the purpose of the present paper to report on the search for ^{12}B–NMR signals of the missing nonsubstitutional species.

2. Experimental

The applied technique was β–radiation detected nuclear magnetic resonance (β NMR) on implanted ^{12}B [1–3]. Spin polarized ^{12}B nuclei, created by the nuclear reaction $^{11}B(d,p)^{12}B$ in a thin boron target-foil, were implanted into single–crystal Si samples with recoil energies distributed between 0 and 450 keV. The β–radiation distribution of the ^{12}B ensemble is asymmetric relative to the direction of the nuclear polarization. Thus NMR transitions can be detected as a reduction of the β–decay asymmetry. The β particles were registered by a pair of scintillation counter telescopes mounted on both sides of the sample at the pole faces of a magnet. The typical flux of polarized ^{12}B ions onto the crystal surface ($14\times14\times0.5 mm^3$) was about 10^5/s.

3. Measurements and Results

On normal substitutional sites the energy levels of diamagnetic ^{12}B are split by the Zeeman effect only and are separated by $h\nu_0$ where ν_0 is the Larmor frequency. As was discussed in [1,2] boron does not occupy the tetrahedral interstitial site in Si and therefore the NMR–signal intensity at ν_0 reflects the fraction f_S. At any other locations the electric field gradient (efg) does not vanish and two quadrupole transitions $m = 1 \leftrightarrow 0$ and $0 \leftrightarrow -1$ shifted from ν_0 occur. Under our experimental conditions the relative sublevel populations created by the nuclear reaction are 4:3:3 for $m = -1, 0, +1$, respectively. Thus only the transition $-1 \leftrightarrow 0$ produces an β–NMR signal representing the complete spectrum for a given quadrupole interaction. Its resonance frequency in a strong magnetic field B is given in second-order perturbation theory for the case of an axial symmetric efg [4]

$$\nu_i = \nu_0 + \frac{3}{8h} e^2qQ(3\cos^2\vartheta_i - 1) + \frac{3}{16\nu_0 h^2}(e^2qQ)^2 \left[\frac{3}{2}\cos^2\vartheta_i(1-\cos^2\vartheta_i) + \frac{3}{8}(1-\cos^2\vartheta_i)^2\right]. \quad (1)$$

ϑ_i is the angle between B and the direction of the axis of the largest component of the efg.

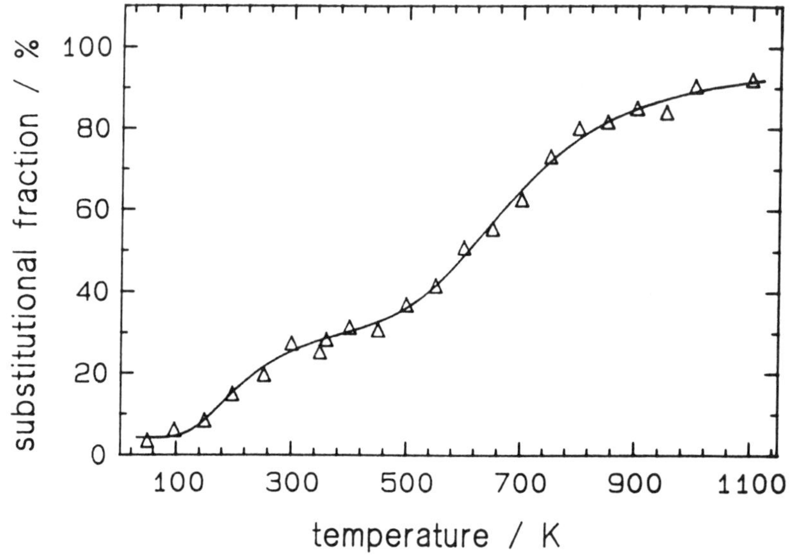

Fig. 1. Temperature dependence of the substitutional fraction $f_S(T)$ of ^{12}B implanted into p–type silicon ($2 \cdot 10^{15}$ B/cm^3).

The insert in Fig. 2 shows a β–NMR scan. Five resonances at ν_0 and $\nu_1,...,\nu_4$ were found. The angular dependence of the quadrupole-shifted ones, shown in Fig. 2 too, could be well fitted by (1) yielding the results: (i) The four detected efg's are axially symmetric, (ii) the resonances at $\nu_1,...,\nu_4$ collapse for $B\|$<100> and therefore the axes of the efg's are parallel to the four <111> directions, (iii) the coupling constant has the value $e^2qQ/h = 360(15)$ kHz.

Each of the five resonance frequencies $\nu_0,\nu_1,...\nu_4$ corresponds to ^{12}B in one definite lattice location, namely ν_0 to the substitutional site and the other ν_i to a nonsubstitutional site with its efg parallel to one of the four <111> axes. This fact can be used to search for diffusive motions of ^{12}B between these five locations during τ. The technique is based on the depolarizing effects produced by the simultaneous irradiation of several rf fields at some of the frequencies $\nu_0,\nu_1,...,\nu_4$, which we designate by rf(ν_i) [5]. Fig. 3 shows two measurements performed at 320 K in the p–type Si sample already specified above. In measurement (A) exclusively the quadrupole shifted resonances $\nu_1,...,\nu_4$ are involved. The signal (a) represents the collapsed resonances at ν_1 and ν_2 for $\alpha = 45°$ (cf. Fig. 2) obtained by scanning rf(ν) over the resonance region. If additionally to the scanning field rf(ν) the two fields rf(ν_3) and rf(ν_4) (with fixed frequencies at the resonance positions ν_3 and ν_4, respectively) were irradiated, the signal (a) was blown out as is shown in (b). The difference signal (a) – (b) demonstrates the significance of the blow-out effect which we interpret in the following way: The nonsubstitutional ^{12}B probes resonating at $\nu_1 = \nu_2$ must have become depolarized during temporal residences in positions corresponding to resonance frequencies ν_3 and ν_4, i.e. in positions with different angles ϑ_i of the efg relative to B. This can be caused by diffusive motions between positions with differently oriented efg's or by any other reorientation mechanism of the efg's between the four <111> directions. In any case the corresponding jump rate $1/\tau_c$ has to be larger than $1/\tau$ in order that a ^{12}B ion is able to exchange its location prior to its decay. On the other hand $1/\tau_c$ has to be smaller than the quadrupole interaction frequency otherwise motional averaging of the quadrupole interactions at different locations would occur. From these conditions the jumprate in case (A) can be estimated to $10^2 s^{-1} \lesssim 1/\tau_c(300\ K) \lesssim 10^5 s^{-1}$.

The same measurements were repeated at $T = 280$ K. This time no blow–out effect showed up which means that the jumps of nonsubstitutional ^{12}B became too slow to occur several times during τ, i.e. $1/\tau_c(280\ K) < 10^2 s^{-1}$.

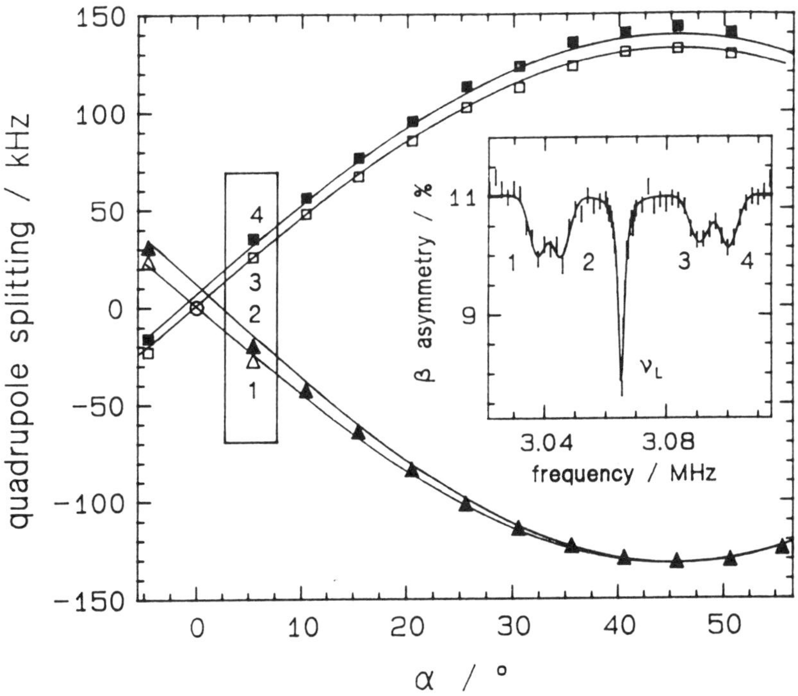

Fig. 2. Angular dependence of the quadrupole shifted resonance frequencies ν_i of ^{12}B in p–type Si ($2 \cdot 10^{15}$ B/cm^3). α is the angle between $\boldsymbol{B}$ and the [001] direction and the sample was rotated about the [100] axis. The doubling of the top and bottom curves is caused by a small deviation of the rotation axis from [100] below 1°. The curves are fits by eq. (1). Insert: Example of β–NMR signals in the same sample for the orientation $\alpha = 5°$. The central line at ν_0 corresponds to substitutional ^{12}B. Each of the four quadrupole shifted lines represents a full spectrum of nonsubstitutional ^{12}B with the efg parallel to one of the four <111> directions. $T = 320$ K, $B = 401$ mT, rf induction $B_{1rot} = 50\ \mu$T.

These results can be compared with EPR measurements on electron irradiated Si [6]. In that study the B_i defects in p–type Si were aligned along a specific [111] direction by illumination with polarized light. Afterwards the recovery time τ_r of the alignment was monitored yielding the relation $1/\tau_r = 5.0 \cdot 10^{12} \exp(-0.60(5)\mathrm{eV}/kT)$ s^{-1}. The reorientation rates calculated from this equation at our T values 320 K and 280 K are 1.8×10^3 s^{-1} and 80 s^{-1}, respectively. These numbers fit well to the limits for $1/\tau_c$ extracted from our measurements and corroberate the interpretation that the observed quadrupole spectra were caused by the same interstitial boron as has been seen in the EPR experiments.

We return to Fig. 3 and consider the measurements (B), where both ν_0 and $\nu_1,...,\nu_4$ are involved. The signal (c) represents the Larmor resonance of the substitutional ^{12}B species obtained by scanning rf(ν) over the resonance region. Additional irradiation of the four fields rf(ν_1),...,rf(ν_4) had no effect on the Larmor signal as demonstrates curve (d). This means that at 320 K the nonsubstitutional ^{12}B species does not jump into substitutional sites during its lifetime τ.

From the NMR–signal intensities we can roughly estimate that the fraction of the nonsubstitutional ^{12}B species resonating at $\nu_1,...,\nu_4$ amounts to about 30 % in our p–type crystal at 320 K. Since the substitutional fraction at the same Temperature amounts equally to about 30 % there remains an undetected fraction of about 40 %. The search for it is still under way.

The quadrupole shifted signals have also been observed in a heavily doped p–type Si sample ($5 \cdot 10^{18}$ B/cm^3) at $T = 300$ K. Up to now, however, the search for quadrupole split spectra in n–type Si has not been successful.

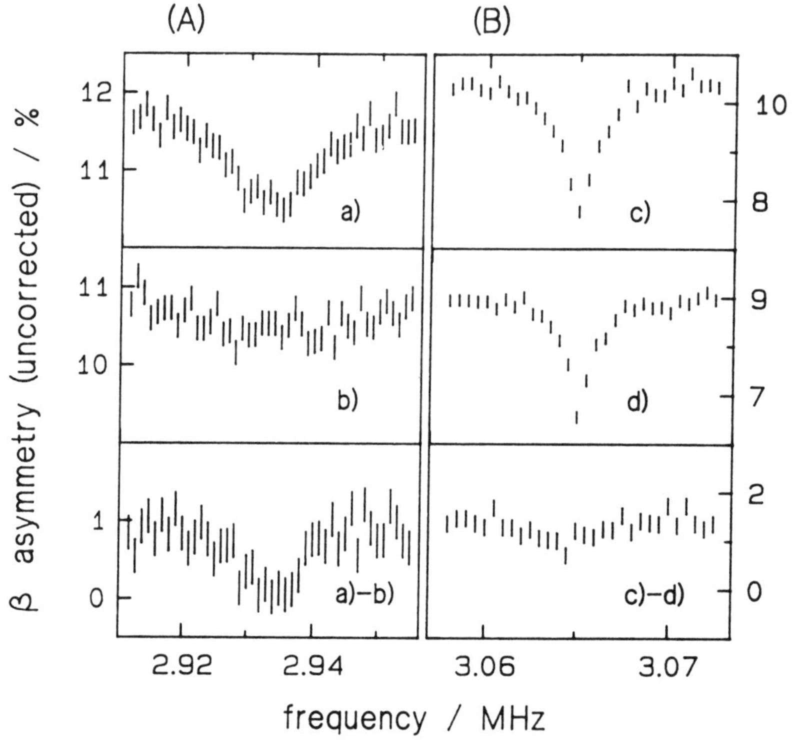

Fig. 3. β NMR of ^{12}B modified by simultaneous irradiation of several rf fields. $B = 401$ mT, $T = 320$ K. Column (A) – (a): collapsed quadrupole lines at $\nu_1 = \nu_2$; (b): blow–out effect by additional irradiation of rf(ν_3) and rf(ν_4); (a) – (b): difference of spectra (a) and (b). Column (B) – (c): Larmor line at ν_0; (d): additional irradiation of rf(ν_3),...,rf(ν_4) has no effect on the Larmor line; (c) – (d): difference of spectra (c) and (d).

Acknowledgements

We gratefully acknowledge the hospitality of the Strahlenzentrum der Justus-Liebig-Universität Gießen and the support of its scientific and technical staff. This work was sponsored by the Bundesministerium für Forschung und Technologie.

REFERENCES

[1] H. Metzner, G. Sulzer, W. Seelinger, B. Ittermann, H.–P. Frank, B. Fischer, K.–H. Ergezinger, R. Dippel, E. Diehl, H.–J. Stöckmann, and H. Ackermann, Phys. Rev. B42, 11419 (1990)
[2] W. Seelinger, B. Fischer, E. Diehl, K.–H. Ergezinger, H.–P. Frank, B. Ittermann, F. Mai, G. Welker, H. Ackermann, and H.–J. Stöckmann, Nucl. Instr. and Meth. B (1991), in press
[3] H. Ackermann, P. Heitjans, and H.–J. Stöckmann in *Hyperfine Interactions of radioactive nuclei*, ed. J. Christiansen (Springer, Berlin-Heidelberg-New York,1983) p. 291
[4] A. Abragam: *The principles of nuclear magnetism* (Oxford University Press, London, 1961)
[5] B. Ittermann, thesis, Universität Marburg, 1989 (unpublished)
[6] G. D. Watkins, Phys. Rev. B12, 5824 (1975)

Materials Science Forum Vol. 83-87 (1992) pp. 273-278

DEFECT IMPURITY COMPLEX FORMATION AT HIGH DONOR CONCENTRATION IN SILICON

A.Nylandsted Larsen and G.Weyer,
Institute of Physics, University of Aarhus,
DK-8000 Aarhus C, Denmark

ABSTRACT

Enhanced diffusion of Sb in silicon at very high donor concentrations is found to correlate with the appearance of a new impurity complex. Based on this observation and on the strong binding between a vacancy and a donor impurity, a new diffusion mechanism is proposed in which the complex Sb-vacancy-donor is the moving entity.

1. Introduction

Recently, non- and quasi-equilibrium techniques altering point defect concentrations in silicon have elucidated impurity diffusion mechanisms [1]. In particular the fractions of interstitial- and vacancy-assisted mechanisms of common donor and acceptor dopants have been determined. Here we report on experiments utilizing Sb donor atoms, which are strongly indicated to diffuse predominantly by a vacancy mechanism. Extremely high diffusivities for Sb as well as for As and Sn impurities have been found under extrinsic conditions in heavily doped n^+-silicon at relatively low temperature ($\lesssim 1100^0$C). Atomic scale information about their interactions with point defects has been obtained from Mössbauer spectroscopy (MS). A non-substitutional Sb or Sn related component in the spectra occurred in correlation with the fast diffusion mechanism. This has been previously interpreted in the framework of a vacancy percolation model [2,3,4]. Here we persue an approach based on impurity complex formation as deduced from the MS data.

2. Experimental

A comprehensive account of the experimental details can be found in reference [5]. Suffice it here to mention that all the silicon crystals were p-type (boron-doped), 150 Ωcm, <100>-oriented, float zone refined monocrystals. Most of the experiments reported in the following are isoconcentration experiments in which a small concentration of Sb (probe atom B) is implanted into a Si crystal pre-doped with a constant concentration of either P or As (donor atom A). Results from a few experiments in which a high concentration ($\sim 5x10^{20} cm^{-3}$) of only Sb is studied will also be reported. In addition to the stable Sb a small amount ($1x10^{13} cm^{-2}$ or less) of radioactive ^{119}Sb was implanted and Mössbauer spectra were measured for the 24-keV γ radiation of the ^{119}Sn daughter. The hyperfine interaction parameters isomer shift δ and quadrupole splitting ΔE_Q as well as the Lamb-Mössbauer factor, from which impurity Debye temperatures were estimated, were deduced from the Mössbauer spectra. All annealings were done with rapid thermal annealing (RTA) in typically 10s. Diffusion coefficients were extracted from chemical profiles measured by Rutherford backscattering spectrometry, and carrier density profiles were measured by Hall effect combined with layer removal.

3. Results and discussion

Figure 1 shows carrier density profiles (P+Sb) and Sb chemical profiles, and the corresponding Mössbauer spectra after RTA at temperatures, where the Sb implantation profiles are practically

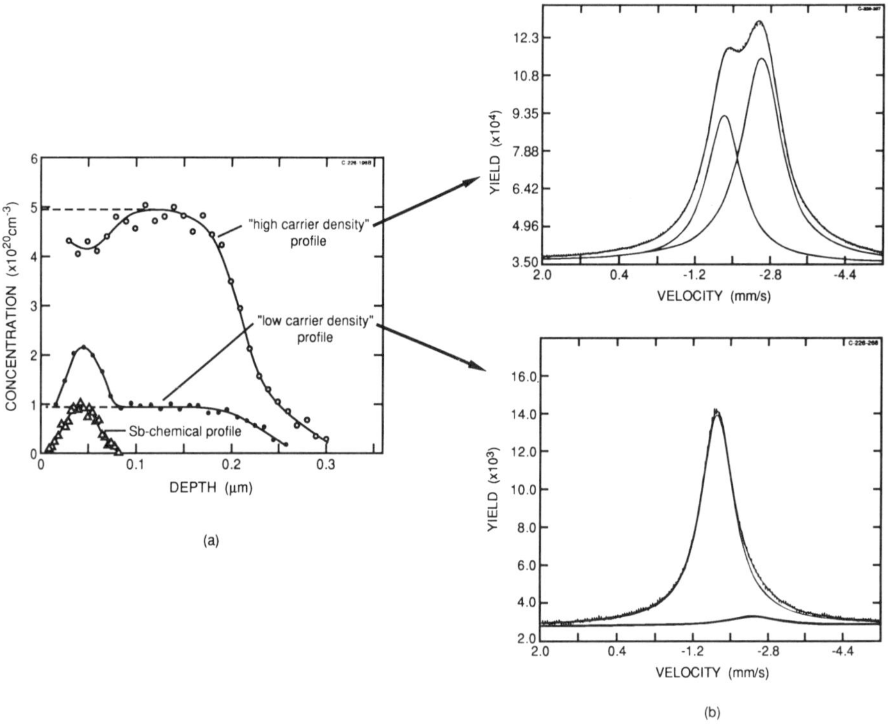

Fig.1. Chemical and electrical carrier profiles after RTA at 850°C (a). The flat carrier profiles before the Sb implantations are indicated by dashed lines. Corresponding Mössbauer spectra measured at 77K (b).

unbroadened by diffusion. The flat P profiles before the Sb implantation as measured for control samples are indicated by dashed lines. Evidently, for the lower P concentration all of the Sb is also electrically active. This is in agreement with the MS, which shows a predominant line stemming from substitutional Sb. For the high P concentration a major Sb fraction is electrically inactive. The MS shows a dominant intensity of an additional line for a Sb fraction, which is in quantitative agreement with the electrically inactive fraction. For similarly As background or purely Sb doped samples the ^{119}Sb Ms decompose into the similar lines, their parameters are listed in Table I. The parameters of the substitutional line are invariant under all conditions, the assignment to electrically active, substitutional Sb is considered unambiguous [6,7]. The electrically inactive fraction from MS is plotted in Fig.2b as a function of n/n_i and likewise the diffusivities in Fig.2a. The temperature dependent intrinsic carrier concentration n_i has been calculated according to ref.8. Very similar n/n_i dependencies are found for slightly different

Table I: MS parameters of lines (δ is given at 77K relative to $CaSnO_3$ at 297K, Θ relative to substitutional Sn)

Line assignment	isomer shift δ (mm/s)	line width Γ (mm/s)	Debye temp. $\Theta/\Theta_{sub.}$	dopant
subst.Sn	1.76(3)	0.87(3)	1	all
Sn-V-Sb	2.34(4)	1.38(8)	0.89(3)	Sb
Sn-V-As	2.48(4)	1.30(8)	0.72(5)	As
Sn-V-P	2.55(3)	1.20(8)	0.82(3)	P

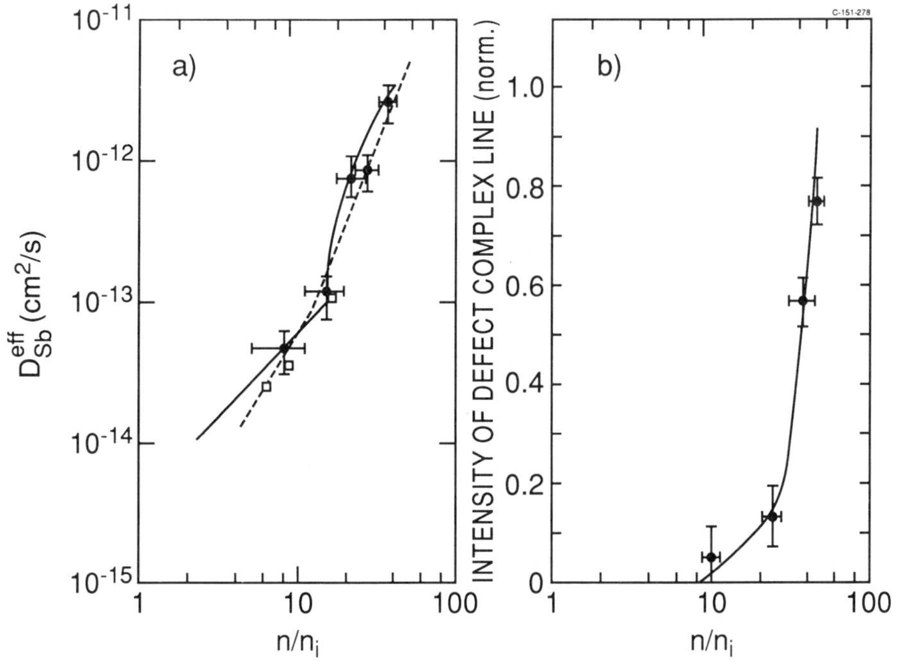

Fig.2. Comparison between effective diffusion coefficients at 1050^{0}C from Ref.5, theoretical predictions from the percolation model (full line) from ref.4 and the predictions of the present model (dashed line) (a). The impurity complex fractions after annealing ar 1050^{0}C as deduced from MS are shown in (b). The curve is the prediction of the present model.1

temperatures as well as for As or Sn probe atoms irrespective of P- or As-background doping [4,9]. Precipitation effects obscure a comparison with purely Sb doped samples, however, much lower electrically inactive fractions are found [6,7].

In the following we consider the data within an impurity complex formation model assuming:
1. All vacancies are bound to impurities.
2. The dependence of the vacancy concentration C_V on n and T is given by

$$C_V = [1 + \delta^- \frac{n}{n_i} + \delta^= (\frac{n}{n_i})^2] C_V{}^0 \quad (1)$$

where the δ values are taken from ref.10. Then the total vacancy concentration is given by C_{PV}, the P-V pair concentration, in the notation of Fahey et al. [1]:

$$C_{PV} = \Theta_{PV} \frac{C_P C_V}{C_S} e^{E^B_{PV}/kT} \quad (2)$$

where E^B_{PV} is the charge state dependent PV binding energy, $C_S = 5 \cdot 10^{22}$ cm^{-3} the number of lattice sites per cm^3, and the coordination number is $\Theta_{PV}=4$.

As

$$\frac{C_V{}^0}{C_s} = e^{S^F_V/kT} e^{-H^F_V/kT}, \quad (3)$$

eq.2 may be rewritten in terms of the P-V formation enthalpy $H^F_{PV} = H^F_V - E^B_{PV}$

$$\frac{C_{PV}}{C_s} = \Theta_{PV} \frac{C_p}{C_s} [1+\delta^- \frac{n}{n_i} + \delta^= (\frac{n}{n_i})^2] e^{S^F_V/k} e^{-H^F_{PV}/kT} \quad (4)$$

This approximation is used even up to concentrations where the impurities interact.

For dilute Sb probes in such heavily P-doped samples their association with P-V pairs with a binding energy ΔE^B may be calculated as

$$C_{SbVP} = \Theta_{SbVP} = \frac{C_{Sb} C_{PV}}{C_s} e^{\Delta E^B/kT} \quad (5)$$

Adopting $\Theta_{SbVP}=8$ (n.n. pairs) and S^F_V = 3K for numerical purpose, H^F_{SbVF} has been deduced from the equation

$$\frac{C_{SbVP}}{C_{Sb}} = 640 \frac{C_P}{C_S} [1+\delta^- \frac{n}{n_i} + \delta^= (\frac{n}{n_i})^2] e^{-H^F_{SbVP}/kT} \quad (6)$$

We note that both numerical choices may introduce an uncertainty of several times 0.1 eV on the H^F_{SbVP} values. Identifying the electrically inactive fraction from MS with the Sb-V-B fractions, the formation enthalpies listed in Table II have been obtained.

Table II: Sample parameters and deduced H^F values

Temp.(^{0}C)	n/n_i	donor atom A	probe atom B	C_{BVA}/C_B	H^F(eV)
1050	10	P	Sb	0.01	$H^F_{Sb^+V^=P^+}=0.8$
1050	40	P	Sb	0.60	$H^F_{Sb^+V^=P^+}=0.8$
900	108	P	Sb	0.73	$H^F_{Sb^+V^=P^+}=0.8$
820	191	P	Sb	0.65	$H^F_{Sb^+V^=P^+}=0.8$
700	529	P	Sb	0.53	$H^F_{Sb^+V^=P^+}=0.8$
880	86	As	Sb	0.55	$H^F_{Sb^+V^=As^+}=0.8$
700	339	Sb	Sb	0.05	$H^F_{Sb^+V^=} = 1.1$

Except for the sample with $n/n_i=10$, these represent values for B^+-$V^=$-A^+ pairs. From the difference $H^F_{Sb^+V^=} - H^F_{Sb^+V^=P^+}$ a value of ΔE^B = 0.3 eV is inferred, which is about the difference in binding energy between neutral P^+V^- and Sb^+V^- pairs [1]. This implies also $H^F_{P^+V^=}$ = 1.1 eV. The low ΔE^B value suggests the complex structure to be $Sb^+(P^+V^=)^-$. The MS parameters for the second line in the spectra are characteristic for a vacancy-associated Sn atom [7,11], thus the assignment Sb^+-$(V^=$-$P^+)^-$ has been made. The systematic trend in particular in the isomer shift for B equal to P, As, or Sb, respectively, strongly supports this proposal. Under the present conditions both SbV and PV pairs are negatively charged ((0/-) level at about E_c-0.4 [12]. Such a complex fulfills the charge neutrality condition locally and a rotation of $V^=$ around either impurity is sufficient to bring the vacancy into a 3.n.n. position, the saddle point position for long-range vacancy diffusion. Thus the diffusivity in this model for $n/n_i \geq 10$ and constant temperature, may be approximated by

$$D_{Sb} \simeq D_{(SbV)^-} [\delta^- (\frac{n}{n_i})^2 + (\frac{n}{n_i})^3 \delta^=].$$

Figure 2a shows that this n/n_i dependence is in agreement with experiment. This applies also for the fraction $C_{Sb}V_P/(C_{Sb})$ as shown in Fig.2b. There might be a weak dependence of the migration enthalpy on n; thus, the n^3 dependence as given above should be considered as a lower limit. The temperature dependence for a given n value of this fraction can be approximated by

$$\delta^- (\frac{n}{n_i}) \simeq e^{0.6 \text{ eV}/kT} \quad , \qquad \delta^=(\frac{n}{n_i})^2 \simeq 0^{0.7 \text{ eV}/kT}.$$

Hence

$$\frac{C_{SbVP}}{C_{Sb}}(T) \simeq e^{-0.1 \text{ eV}/kT},$$

i.e. the ratio is nearly temperature independent in the range of interest. This is believed also to assure that despite slow cooling rates the measured C_{SbVP}/C_{Sb} values are representative also for the nominal RTA temperature. If this holds not true H^F values are upper limits.

It is interesting to compare the present model to previous propositions [2,4,9,13]. Fair et al. found D_{Sb} for $n/n_i \lesssim 10$ to be well represented by

$$D_{Sb} = D_i^0 + D_i^{=}\left(\frac{n}{n_i}\right)^2,$$

where D_i^0 and $D_i^{=}$ are the (intrinsic) diffusivities mediated by V^0 and $V^{=}$, respectively. No contribution of V^- was indicated. However, the present model suggests an $(n/n_i)^2$ for P^+V^- i.e. formed with V^-. The onset of an $(n/n_i)^m$ dependence with $2<m\lesssim 3$ occurs for $(n/n_i)\approx 17$. This value coincides with the onset of the percolation phenomenon as the P atoms are on the average separated by 6 n.n. distances [4]. Above this value on enhancement of the diffusivity by a factor

$$F = e^{\Delta E_A^3/kT}$$

is predicted as shown in Fig.2a. The accuracy of the experimental data does not permit to distinguish between the percolation and the present model.

This work has been supported by the Danish Accelerator Physics and Synchrotron Radiation Council.

References

1) P.M.Fahey, P.B.Griffin, and J.D.Plummer, Rev.Mod.Phys.61, 289 (1989)
2) D.Mathiot and J.C.Pfister, J.Appl.Phys.66, 970 (1989) 970
3) A.F.Burenkov, F.F.Komarov, and S.A.Fedotov, Rad.Eff.and Def.in Solids 115, 45 (1990)
4) K.Kyllesbech Larsen and A.Nylandsted Larsen, Mat.Sci.Forum 65-66, 293 (1991)
5) A.Nylandsted Larsen, P.Tidemand-Petersson, P.E.Andersen and G.Weyer, Inst.Phys,Conf.Ser.95, 499 (1989)
6) A.Nylandsted Larsen, F.T.Pedersen, G.Weyer, R.Galloni, and R.Rizzoli, MRSEUROPE IV, eds. V.T.Nguyen and S.G.Cullis (Editions de Physique, 1985) 319
7) A.Nylandsted Larsen, F.T.Pedersen, G.Weyer, R.Galloni, R.Rizzoli, and A.Armigliato, J.Appl.Phys.59, 1908 (1986)
8) C.D.Thurmond, J.Electrochem.Soc.122, 1135 (1975)
9) R.B.Fair, M.L.Manda, and J.J.Wortman, J.Mater.Res.1, 705 (1986)
10) D.Mathiot and J.C.Pfister, J.Appl.Phys.55, 3518 (1984)
11) S.Damgaard, J.W.Petersen, and G.Weyer, Hyp.Int.10, 751 (1981)
12) A.D.Evwaraye, J.Appl.Phys.48, 1840 (1977)
13) K.Nishi, K.Sakamoto, and J.Meda, J.Appl.Phys.59, 4177 (1986)

Materials Science Forum Vol. 83-87 (1992) pp. 279-284

ELECTRONIC STRUCTURE OF ISOLATED ALUMINIUM POINT DEFECTS AND DEFECT PAIRS IN SILICON

HARALD OVERHOF [1] and GÀBOR CORRADI [2]
[1]**Physics Department, University of Paderborn, D-4760 Paderborn, F. R. G**
[2] **Research Lab. for Crystal Physics, Hung. Acad. Sci., H-1502 Budapest , Hungary**

ABSTRACT

We use the self-consistent LMTO-ASA method for the calculation of the electronic structure, total energies and Hyperfine fields for isolated Al point defects and for trigonal Al defect pairs in silicon. Many-body effects are treated in the Local-Density approximation of the Density-Functional theory and the band gap appropriate for Si is obtained using the scissor operator technique. We discuss in particular the hyperfine interaction (HFI) matrix elements of the $(Al_i\text{–} Al_{Si})^+$ pair in comparison with the Al_i^{++} point defect. For a pair where one Al atom is located on the tetrahedral nearest neighbor interstitial site to the substitutional Al atom the HFI data compare well with the data obtained experimentally for the defect Si-G19. However the assignment of the experimental HFI data to the shells for the $\langle 111 \rangle$ ligands has to be reversed. We discuss results obtained for several trigonal $(Al_i - Al_{Si})^+$ pairs whith different distances between the two Al atoms. None of these pairs shows the large contact HFI interaction characteristic for Si-G20, the other trigonal $(Al_i - Al_{Si})^+$ pair.

1. Introduction

In recent papers [1, 2] we have shown that for isolated substitutional point defects and defect pairs hyperfine interaction (HFI) matrix elements at the impurity nuclei and at the ligands can be calculated which agree with experimental results obtained by EPR and ENDOR techniques. The HFI interaction constants for the impurity nuclei were found to be entirely due to the state in the gap which gives rise to the paramagnetism. In contrast the contact interaction with the ^{29}Si ligand nuclei was largely due to the spin polarization of the valence band states. It is the aim of the present paper to extend the investigations to deep interstitial defects. As examples we have chosen Al_i^{++} isolated point defects and $(Al_i - Al_{Si})^+$ defect pairs first observed by Watkins [3] as Si-G18, Si-G19, and Si-G20. The pairs are very interesting since the same kind of atomic species is built into two nonequivalent lattice positions.

Ligand HFI data have been determined experimentally by EPR and ENDOR techniques [3, 4, 5]. The experimental methods, however, can at most determine the direction but not the length of the distance vector from the impurity center to the ligand nucleus. It was claimed [5] that a comparison of experimental HFI data for Al_i^{++} and $(Al_i - Al_{Si})^+$ allows for a definite assignment of the experimental data to the different shells. We shall show that such an assignment is not unambiguous - in fact our results show that the data for the two inner $\langle 111 \rangle$ shells must be interchanged.

2. Computation

Our calculations have been performed within the general framework of the spin-density functional theory (DFT). Exchange and correlation are treated in the local spin-density approximation (LSDA) of the spin-density functional theory. For the computations we have made use of the self-consistent linear muffin-tin orbital method in the atomic sphere approximation (LMTO-ASA) [6]. In this method the crystal is divided into (overlapping) spheres in which the potential is assumed to have spherical symmetry. The Green function G_0 for this crystal is calculated self-consistently. It is assumed that the presence of the defect changes the potential by ΔV in a small region around the defect only. For this "perturbed region"a Dyson equation is solved self-consistently with the potential ΔV. The use of the LMTO method has the advantage that it allows for the accurate calculation of the wavefunctions in the region near the nuclei which dominates the contributions to the HFI matrix elements. The use of the ASA unfortunately does not allow to include lattice relaxation effects.

Since the computational effort rises drastically with the size of the perturbed region the self-consistent calculation is limited to a rather small perturbed region. For the case of the isolated interstitial Al_i point defect it contains the point defect, the four nearest Si ligands and the six next nearest Si ligands. The perturbed region also consisted of 14 "empty" ASA spheres introduced to fill the open structure of the diamond lattice. The potential ΔV obtained from a self-consistent solution of the Dyson equation is used to solve the Dyson equation non-self-consistently for perturbed regions that also contained Si ligands of more distant shells. In that way it was possible to calculate the HFI matrix elements for eight shells of neighbors in the case of isolated Al_i^{++} point defects and for 11 shells for the pairs. By a shell we denote all the atoms that are transformed into each other by the point group symmetry operations of the perturbed region. Data for only one member of a shell will be given in the following, and the distances to the ligands will be measured from Al_i^{++} in units of $d/4$ where d is the lattice constant.

It is well known that the single-particle band gap of the DFT-LSDA is too small by about 0.5 eV if compared to the experimental band gap. In previous calculations [1, 2] on chalcogen in Si this small gap resulted in a shift of the defect-induced state in the gap into the bottom of the conduction band where it formed a resonance state. In the case of Al_i we use the scissor-operator technique described by Beeler et al [7, 8] in order to obtain the experimental band gap for the crystal. This ensures that the defect-induced state has its single-particle energy in the fundamental band gap and not in the valence band. Alternatively we have calculated the HFI interaction for the paramagnetic defects *not self-consistently* using an extra potential to move the defect state into the small LSDA band gap. The small differences between both calculations (less than 10 % for the contact HFI at the Al_i^{++} nucleus) shows that the use of the scissor operator technique does not critically influence the results.

3. Isolated point defects

For p-type Si the total energy of an isolated interstitial Al point defect is calculated to be by about 4 eV higher than that of a substitutional Al point defect. We thus do not expect to find Al interstitials in thermal equilibrium. In Si crystals bombarded by electrons at lower temperatures there are mobile Si self-interstitials. The total energy on the right hand side of the reaction

$$Si_i^{++} + Al_{Si}^{-} \leftrightarrow Si_{Si} + Al_i^{++} + e^-$$

is by about 2 eV smaller than that on the left hand side in p-doped Si. Si self-interstitials therefore will transform Al_{Si} into Al_i. For this interstitial state, we find two different charge states, Al_i^{++} and Al_i^{+} with a removal energy $E^{++/+} = E_V + 0.58$ eV which compares with the value $E^{++/+} = E_V + 0.17$ obtained experimentally [9], an agreement that is not too convincing. According to our calculations there is no neutral state for Al_i as e. g. postulated by Troxell et al. [9]. However we do find a resonance transforming according to the T_2 irreducible representation at the bottom of the conduction band . This state will be subject to a Jahn-Teller distortion (not included here) in general agreement with the prediction of Troxell et al. that the neutral state should be accompanied by a significant lattice relaxation.

The single-particle wave function of the A_1 state in the gap is quite different from the

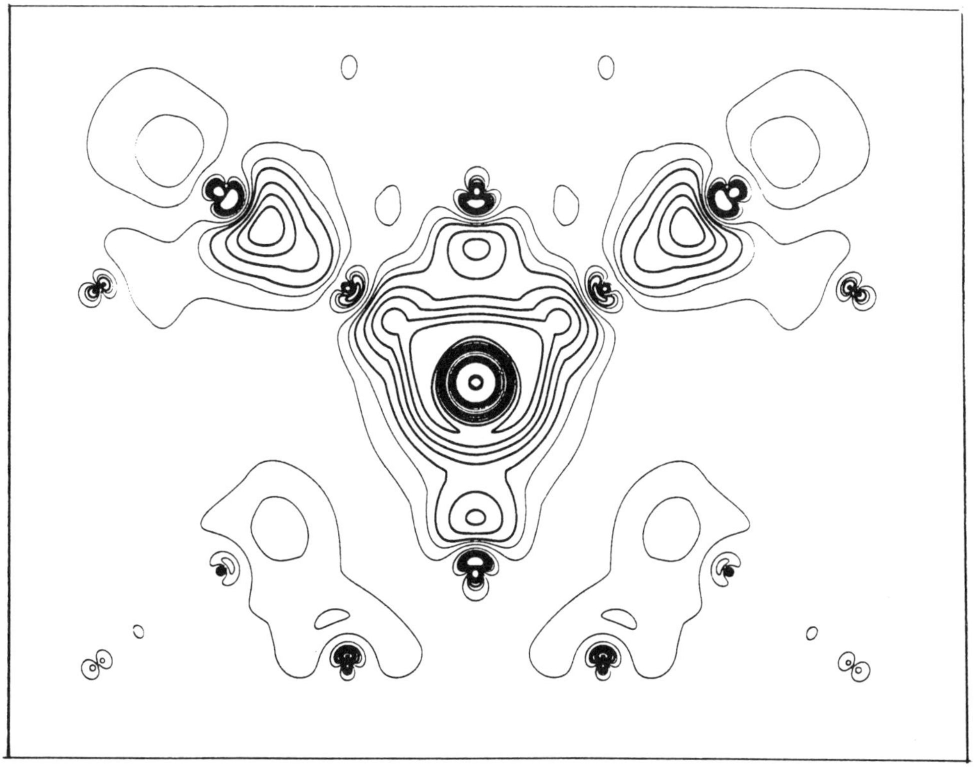

Figure 1: Contour plot of the particle density for the gap state of the Al_i^{++} point defect in the (110) plane.

corresponding wave functions of deep levels of e.g., substitutional chalcogen in Si. About 15 % of the particle density is found in the Al ASA sphere compared to only 4 % in the case of S_{Si}. In figure 1 we show a contour plot of the particle density of the gap state of Al_i^{++} in the (110) plane. We see that the state in the gap is bonding and that there are bond charges extending into the ⟨111⟩ directions. Since the state at the (1,1,1) ligand is predominantly s/d-like this bond extends further whereas the s/p-like state at (2,2,2) terminates the bond charge: there is an order of magnitude less particle density arising from the gap state at the next ligand at (3,3,1). There are also bonds directed toward the next nearest neighbors in the [100] directions but again the particle density due to the gap state does not extend further. We thus have the picture that the interstitial Al_i^{++} ion is bound by the gap state to the Si ligands whereby the bond charges extend as long as there is a straight chain of ligands. Very little particle density extends into directions where the ligands are far apart. Except for the oscillations of the wave function within the core regions (we show in fig. 1 the true LMTO-ASA wave function, not a pseudo-function) the wave function has no nodes within the rectangle spanned by the shell of (2,2,2) Si ligands. However the wave function changes sign outside this rectangle where its magnitude is small since about 75 % of the gap state charge is contained in the region covered by our calculation.

A comparison of the calculated HFI data with EPR and ENDOR data of Brower [4] and with ENDOR data of Niklas et al. [5] confirms these results. Table 1 shows that the HFI interaction at the defect nuclei and at all ligands is predominantly given by a_l which is the

Table 1: Comparison of the calculated HFI data (in MHz) for isolated Al_i^{++} point defects and $(Al_i - Al_{Si})^+$ defect pair a with experimental data taken from Brower [4] for the interaction with the $^{27}Al_i^{++}$ nucleus and from Niklas et al. [5] for the interaction with the ^{29}Si ligand nuclei.

	Al_i^{++}						$(Al_i - Al_{Si})^+$				
ligand	a_l	a_{tot}	a_{exp}	b_{tot}	b_{exp}	ligand	a_l	a_{tot}	a_{exp}	b_{tot}	b_{exp}
$Al_i^{++}(0,0,0)$	1287.	1365.	1320.	0.0	0.0	$Al_i^{++}(0,0,0)$	1225.	1253.	1180.	2.0	3.3
Si(1,1,1)	51.3	54.7	50.09	0.41	-0.05	$Al_{Si}^-(1,1,1)$ Si(1,$\bar{1}$,$\bar{1}$)	33.5 50.7	32.7 48.7	37.75 56.76	0.65 $\simeq$0	1.1 -0.41
Si(2,0,0)	32.6	32.8	22.6	2.2	1.44	Si(2,0,0) Si($\bar{2}$,0,0)	36.4 39.4	35.3 39.8	25.46 24.41	1.82 2.45	1.02 1.42
Si($\bar{3}$,1,1)	3.92	3.38	3.51	1.0	0.89	Si($\bar{3}$,1,1)	0.13	-0.39		0.29	
Si(2,2,2)	51.4	51.6	52.31	5.24	3.91	Si(2,2,2) Si(2,$\bar{2}$,$\bar{2}$)	44.4 51.8	45.7 51.4	31.32 60.76	4.35 5.15	3.98 6.08
Si($\bar{2}$,$\bar{2}$,$\bar{2}$)	13.4	12.9	11.31	0.05	0.16	Si($\bar{2}$,$\bar{2}$,$\bar{2}$)	12.1	12.0	13.53	0.02	0.16
Si(3,3,1)	1.11	1.06	1.24	0.56	0.16	Si(3,3,1) Si($\bar{3}$,$\bar{3}$,1)	0.17 0.98	-0.43 0.96		0.29 0.54	
Si($\bar{3}$,$\bar{3}$,$\bar{3}$)	0.20	0.11		0.02							

contribution of the gap state. Contributions from the valence band polarization caused by the spin of the gap state are of relative significance only at ligands where a_l is very small, in contrast to our results for the substitutional chalcogen defects in Si [1, 2]. Note that for the HFI interaction at the Si ligands all calculated HFI data should be negative due to the negative sign of the nuclear g-factor. We give the incorrect sign because the experimental data are given with a positive sign and a negative value of the dipolar term simply means that the sign is opposite to that of a_{exp}. We have interchanged the experimental data for the Si(1,1,1) and the Si(2,2,2) ligands with respect to Niklas et al.[5]. Both shells have approximately the same contact HFI while the dipolar HFI differ by nearly one order of magnitude since the dipolar HFI for d-electrons is much smaller than for p-electrons. The negative sign of b_{exp} with respect to a_{exp} for the Si(1,1,1) ligand can be taken as further evidence that the wave function has a strong d-admixture (see van Wezep et al. [10]).

4. Trigonal Al – Al pairs

In order to investigate the stability of the trigonal aluminum pairs with respect to dissociation we have calculated total energies for several different trigonal pair configurations involving two Al atoms. The pair formation energies are calculated as a function of the Fermi energy in analogy to ref [2]. For the $(Al_{Si} - Al_{Si})$ and $(Al_i - Al_i)$ pairs the binbding energies are negative for all charge states of the pairs. These pairs therefore are not stable.

Experimentally one finds two pairs which are trigonal and involve two Al atoms, one of which is located at an interstitial position while the other substitutionally replaces one Si atom. The most extensively studied metastable pair is Si-G19, identified as $(Al_i - Al_{Si})^+$ by Watkins [3]. Since there is another paramagnetic trigonal $(Al_i - Al_{Si})^+$ pair reported in the literature (Si-G20, [3]) we have performed total-energy calculations for several trigonal $(Al_i - Al_{Si})$ pair configurations which differ with respect to the distances between the Al atoms. These configurations are schematically shown in Fig 2 together with total energies of the pairs.

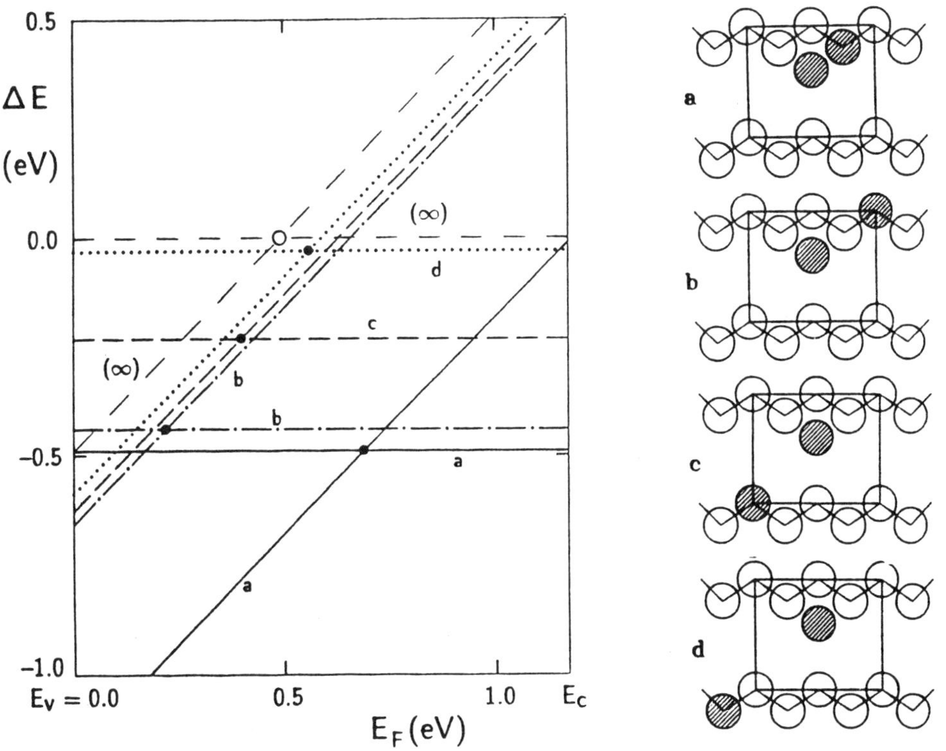

Figure 2: Pair formation energy for the ($Al_i - Al_{Si}$) pairs a-d in the positive and neutral charge state as function of the Fermi energy.

We find that for all positions of the Fermi energy the most stable pair is that for which the two Al atoms are separated by a single nearest neighbor distance. The binding energy about 0.7 eV. The total energies for the other pairs in the positively charged state are not very much different from the total energy of the dissociated pair, although all pairs appear to be stable. The total energies of pairs a and b in the neutral charge state are virtually equal and also pair c appears to be stable. It appears that the larger distance of the Al atoms in pair b with respect to pair a does not alter the binding energy of the pair because a Si atom is situated in-between the defects.

We shall concentrate our discussion of the paramagnetic properties on pair a which will be identified with Si-G 19. For this pair the calculated transition energy is $E^{+/0} = E_V + 0.78$eV. The particle density for the pair is remarkably similar to that of the isolated Al_i^{++} defect with the exception that the bond charge directed toward the Al_{Si} ligand is somewhat stronger than the bonds to the nearest neighbor Si ligands. This insensitivity of the wavefunction on the replacement of a Si atom by Al is also borne out by the HFI matrix elements listed in Table 1. We find a remarkably good agreement of our calculated data with the experimental results of Niklas et al [5] if we again interchange the assignment of the experimental data to the Si(1,$\bar{1}$,$\bar{1}$) and the Si(2,$\bar{2}$,$\bar{2}$) ligands. As in the case of isolated Al_i^{++} point defect the larger b_{exp} value belongs to the more distant ligands. Unfortunately we cannot take into account a lattice relaxation which should be present since the Al atoms are bonded.

The calculated HFI data at the ^{27}Al nuclei for all trigonal pairs are listed in Table II. The

Table II: Comparison of the calculated hyperfine data for the different trigonal $(Al_i - Al_{Si})^+$ defect pairs (in MHz) with experimental HFI interactions at the ^{27}Al nuclei for Si-G19 and Si-G20 taken from Watkins [3].

	Al_i^{++}		Al_{Si}^-	
pair	*a*	*b*	*a*	*b*
pair a	1253.	2.01	32.7	0.65
pair b	1200.	0.66	31.8	4.34
pair c	1300.	1.33	11.0	0.14
pair d	1286.	0.52	1.61	0.017
Si-G 19	1180.	3.31	37.75	1.1
Si-G 20	1161.	3.0	127.	0.5

agreement between the experimental dipolar HFI data for Si-G19 and calculated results in Table II is slightly better for pair a than for pair b. However for pair b the HFI with Si(111) (a_{tot}= 11.6 MHz and b_{tot} = 2.4 MHz not listed in Table II) has no counterpart in the experimental HFI data. None of the trigonal pairs studied by us shows the rather large value a_{exp} = 127 MHz which is characteristic for the HFI contact interaction *at* the Al_{Si} nucleus in the $(Al_i - Al_{Si})^+$ pair Si-G20. We have performed preliminary calculations for more complicated defect structures involving different trigonal $(Al_i - Al_{Si} - Al_{Si})^0$ and different trigonal $(Al_i - Al_{Si} - C_{Si})^+$ aggregates as tentatively suggested by Watkins [11]. None of these pairs had a contact interaction *at* the Al_{Si} nucleus that was larger than about half the value observed experimentally for Si-G20. Thus there is still no valid atomic structure for Si-G20, a trigonal complex for which one interstitial aluminum and one substitutional aluminum have been identified as constituents by EPR [3] some 25 years ago.

ACKNOWLEDGMENTS

The authors are grateful to Holger Weihrich for his computation of the anisotriopic HFI data. Financial support from the Deutsche Forschungsgemeinschaft and from the National Scientific and Research Fund of Hungary (OTKA) is thankfully acknowledged.

References

[1] H. Overhof, M. Scheffler, and C. M. Weinert, Mat. Sci. Forum **38 - 41**, 293 (1989), Mat. Sci.and Eng. **4**, 315 (1989) .

[2] H. Overhof, M. Scheffler, and C. M. Weinert, Phys. Rev. **B 43**, 12494 (1991) .

[3] G. D. Watkins, in: *Radiation Damage in Semiconductors* , P. Baruch, editor, (Academic Press, N. Y. 1965), p. 97.

[4] K. L. Brower , Phys. Rev. **B 1**, 1908 (1970) .

[5] J. R. Niklas, J.-M. Spaeth, G. D. Watkins, Proc. MRS Meeting San Francisco 1985, Vol. **46**, 237.

[6] O. Gunnarsson, O. Jepsen, and O. K. Andersen, Phys. Rev. **B 27**, 7144 (1983).

[7] F. Beeler, M. Scheffler, O. Jepsen, and O. Gunnarsson, Phys. Rev. Lett. **54**, 2525 (1985).

[8] F. Beeler, O. K. Anderson, and M. Scheffler, Phys. Rev. **B 41**, 1603 (1990).

[9] J. R. Troxell, A. P. Chatterjee, G. D. Watkins, and L. C. Kimerling, Phys. Rev. B **19** 5336, (1979) .

[10] D. A. van Wezep, T. Gregorkiewicz, E. G. Sieverts, and C. A. J. Ammerlaan, Phys. Rev. B **34** , 4511 (1986) .

[11] G. D. Watkins, *private communication.*

Materials Science Forum Vol. 83-87 (1992) pp. 285-290

INTERACTION AND DYNAMICS OF HIGH-TEMPERATURE DEFECTS IN CARBON-RICH SILICON

NATKO B. URLI and BRANKO PIVAC
Rudjer Boskovic Institute, P.O.B. 1016, 41001 Zagreb, Croatia, Yugoslavia

ABSTRACT

Various EFG polycrystalline silicon ribbons, with carbon concentration in excess of 10^{18} cm^{-3}, and with large carbon-to-oxygen ratio, were annealed isochronally and isothermally in the temperature range from 450 to 1250°C. IR spectroscopy and DLTS measurements were used to monitor decays and growths of various carbon, oxygen and SiC microprecipitates and agglomerates. A large absorption peak at 800 cm^{-1}, present only in the surface layers, has been associated with carbon-selfinterstitial-SiO_x precipitates, and a broad band from 810-960 cm_{-1} with SiC nanoprecipitates in the bulk. The whole series of bands just below the Raman frequency was attributed to vibrations of different carbon pairs. Several energy levels associated with intragrain defect complexes as well as a distribution of the grain boundary states after the thermal treatment have been detected by DLTS measurements.

1. Introduction

Oxygen and carbon are the two most common impurities unintentionally introduced during the crystal growth of silicon by various methods. Their behavior has extensively been investigated for several decades[1,2] because of their potential influence on the quality of various silicon electronic devices. When distributed homogeneously in the crystalline bulk as point-type defects, they are electrically inactive and harmless; however, during high temperature processing steps, they may migrate, interact, and pair with other defects forming new electrically active complexes or promoting different precipitating processes.

Most of the studies have been done on Czochralski- grown silicon crystals[1,3] with supersaturated oxygen concentration in excess over the carbon concentration. It has been indicated previously[4,5] that the ratio of oxygen-to-carbon concentration may influence significantly the precipitation processes at high temperatures, that is, the high carbon silicon may behave differently. Therefore, in the present study we have used the Edge-defined Film-fed Growth (EFG) silicon samples with different carbon-to-oxygen ratio obtained by deliberate changes in the ambient gases reacting with the silicon melt. The high pull rates of EFG ribbons (about 2 cm/min) in combination with large interface temperature gradients and high cooling rates are the origin of various manifestations of stress in this material with relatively high dislocation densities (10^3 to 10^7 cm^{-2})[6]. In this paper, we present new results on defect interactions after high-temperature heat treatments of EFG samples, obtained by infrared spectroscopy (IR) and DLTS measurements.

2. Experimental

The samples used in this research were cut from EFG ribbons grown in the atmosphere of CO_2 or without it, or from a closed nonagon cylinder, and supplied by the Mobil Solar Energy Co. They were about 0.4 mm thick and boron-doped with resistivities of the order of 1 ohm cm. Isochronal and isothermal treatments up to 1250°C were performed in flowing dry nitrogen atmosphere. Differential infrared spectra were taken at room temperature with a Perkin Elmer 1710 FTIR spectrophotometer at 4 cm^{-1} resolution and with float-zone oxygen- and carbon-free monocrystalline silicon as a reference. In order to check the influence of spatial distribution of defects in the bulk and close to the sample surface, a 10-μm-thick layer of a nonagon sample was etched off initially using a chemical planar etch after the RCA standard cleaning procedure. After the thermal traetments, the samples were dipped in HF solution to remove the oxide layer that may be formed at the surface.

Schottky diodes on p-type samples were formed by evaporating Al-dots far from grain boundaries. Ohmic back contacts were formed by sintering evaporated aluminum in hydrogen atmosphere at 400°C for 15 minutes. Capacitance-voltage (C-V) profiling and DLTS measurements were performed on a Sula Technologies (Palo Alto, CA) computerized Deep Level Spectrometer.

3. Results

Figure 1 shows the IR spectra of the as-received EFG ribbon (CO_2-off) (curve 1) and nonagon (curve 2) samples. They look quite differently. The presence of segregated oxygen in the nonagon sample is manifested by a broad peak from 1000 to 1130 cm^{-1} with a shoulder up to 1250 cm^{-1}. There are no such broad oxygen-related peaks in the sample grown without CO_2 atmosphere, and the other peak at 800 cm^{-1} is less pronounced than that shown in curve 2.

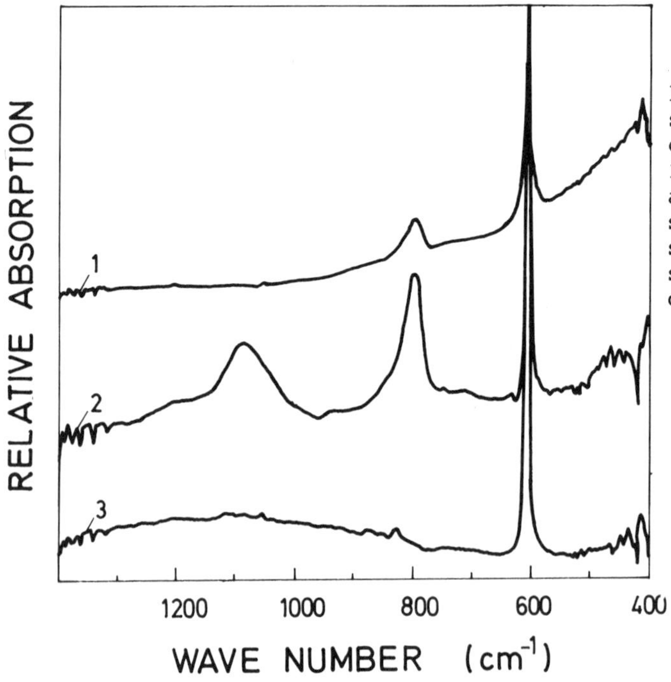

Fig. 1 Absorption IR spectra of the as-received EFG samples: 1- grown without CO_2-atmosphere, 2-nonagon sample, 3- nonagon sample with 10 μm of surface layer etched-off.

Curve 3 of the nonagon sample taken after approximately 10 μm of surface layer etched off has proven that the 800 cm^{-1} peak is associated with defects close to the ribbon surface. The substitutional carbon concentration (C_s) at 605 cm^{-1} was equal to 1.6 x 10^{18} cm^{-3} (using the ASTM conversion factor[7] of 1.1 x 10^{17} atoms/cm^3, and 5.4 x 10^{16} cm^{-3} for the interstitial oxygen at 1107 cm^{-1} (using the conversion factor of 2.45 x 10^{17} atoms/cm^3 [8]). In addition, some small peaks at 828, 854, 882, and 1052 cm^{-1} are visible, too, as well as a whole series in the 400 to 520 cm^{-1} region. The CO_2-off sample has no detectable interstitial oxygen, and C_s = 1.7 x 10^{18} cm^{-3}.

Figure 2 shows the spectra of the same samples after multistep heatings in the range from 450 to 1250°C. The main feature is a broad peak from 800 to 960 cm^{-1} in all samples with well-resolved structure in the case of etched nonagon sample. It has grown substantially only after the last 6 hr-1250°C-step, although some of its components were visible in the as-received samples, too (Fig. 1, curve 3). This broad band is a bulk property of EFG materials[9].

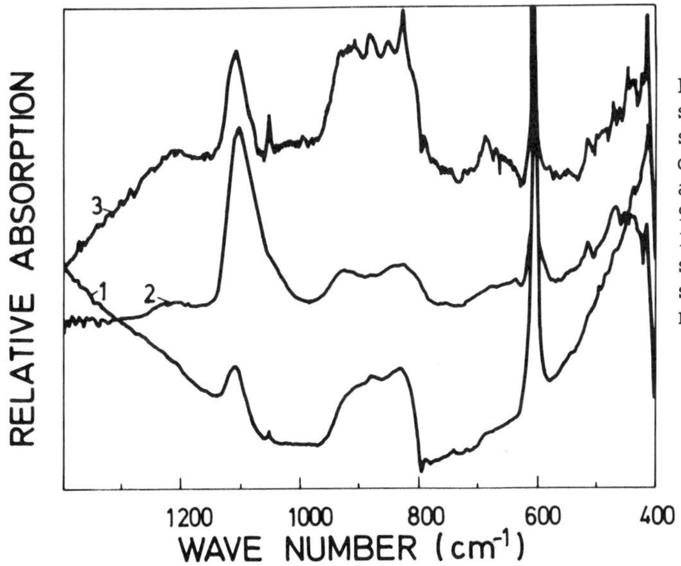

Fig. 2. Absorption IR spectra of the EFG samples heated sequentially for 6 hrs. at 450, 650, 750, 950, 1050, 1150, and 1250°C: 1- CO_2-off sample, 2- nonagon sample, 3- etched nonagon sample.

Curve 2 in Fig.2 also shows a high peak close to 1100 cm^{-1} with an asymmetric shoulder down to 1000 cm^{-1} which also developed during the last heating step at 1250°C. The broad band in that region of the as-received sample vanished completely even after the 650°C heating step[9]. The broad band from 810-960 cm^{-1} is less pronounced, and the substitutional carbon concentration (C_s) decreased by a factor of 2 only after the last heating step.

Curve 3 in Fig.2 reveals an interstitial oxygen concentration of 1.7 x 10^{17} cm^{-3}, a significant increase in the 1052 cm^{-1} peak, a large decrease in C_s from 1.6 x 10^{18} to 2.5 x 10^{17} cm^{-3}, and the appearance of the 800 cm^{-1} band enclosed in the broad peak. The CO_2-off sample (curve 1) has shown the interstitial oxygen in concentration equal to 1.3 x 10^{17} cm^{-3}, otherwise undetectable in the as-received sample. There is also a gradual disappearance of the

800 cm^{-1} peak starting even during the 650°C heating step.

The dynamics of the growth and decay of the high-temperature precipitates is shown in Fig. 3 for a sample grown in the CO_2 atmosphere.

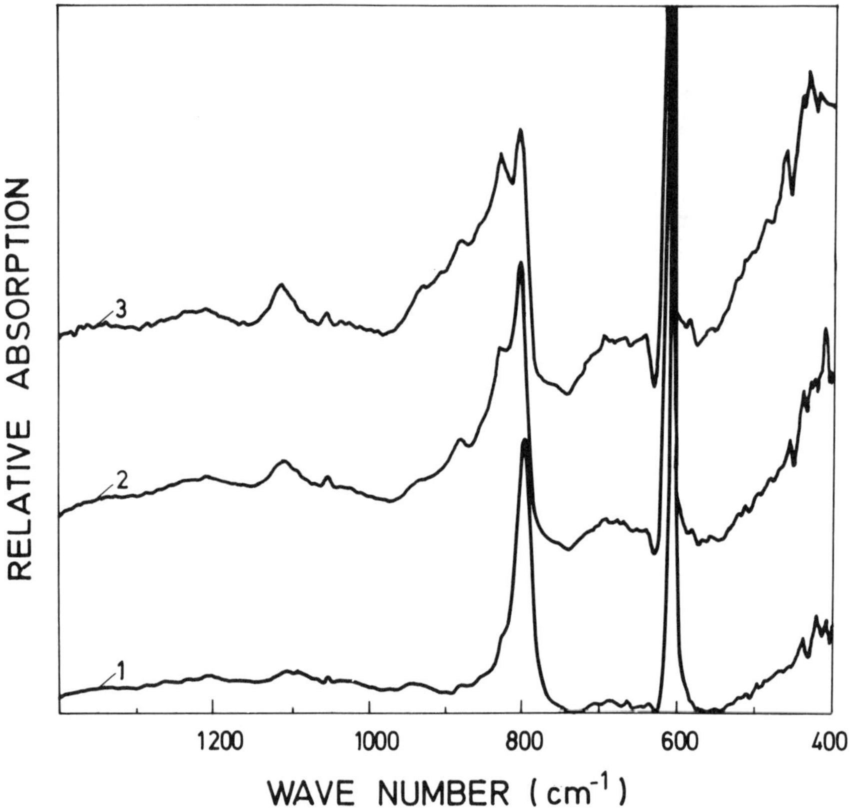

Fig. 3. Absorption IR spectra of the EFG sample grown in CO_2 atmosphere and isothermally annealed at 1250°C: 1- before heating, 2- heated for 30 minutes, 3- heated for 1 hour.

There is a simultaneous decrease in the 800 cm^{-1} peak and the substitutional carbon concentrations, and a gradual increase in concentration of the interstitial oxygen and of the components of the broad 810-960 cm^{-1} band.

DLTS measurements of the as-received EFG samples show an energy level distribution between 0.18 and 0.60 eV above the valence band in agreement with other DLTS studies of polycrystalline silicon[10], with dominant levels around 0.20 eV. Results of the thermal annealing at 1250°C of two nonagon samples are shown in Fig. 4. The DLTS spectra of the initially etched sample (curves 1a to 1d for different rate windows) enabled determination of the ionization energy and the hole capture cross section, i.e. E_a = 0.26 eV and σ_p = 3 x 10^{-15} cm^2, by the following equation

$$\tau_m^{-1} = \sigma_p v_p N_v \exp(-E_a/kT_m) \quad (1)$$

where τ_m is the rate window, T_m temperature of the peak, v_p thermal velocity of holes, and N_V density of states of the valence band.

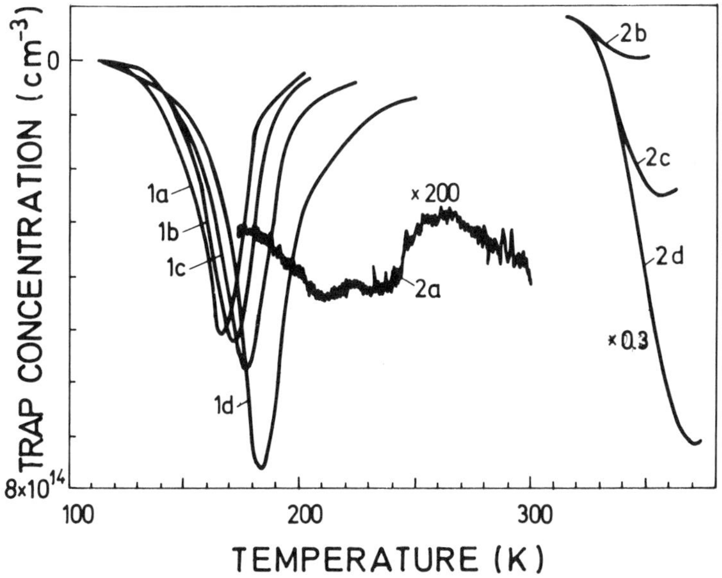

Fig. 4. DLTS spectra of the etched nonagon sample (1a- τ_m=4.3 ms, 1b- τ_m=2.2 ms, 1c- τ_m=0.9 ms, 1d- τ_m= 0.4 ms), and of the unetched nonagon sample (2a to 2d with the same rate windows, respectively) after heating at 1250°C; V(bias)= -3.5V, V(pulse)=3.5V, pulse width=550 μs.

A large shift in the magnitude and position of DLTS peaks at higher temperatures (curves 2b, 2c, and 2d), and the observed dependence on the pulse width indicate a distribution of the density of states associated with the grain boundary states[11], and peaked at E_V + 0.44 eV.

4. Discussion

In an analysis of the nature of the 800 cm^{-1} peak in the IR spectrum it should be mentioned that it was attributed previously to the existence of SiC[12]. However, as it starts to decompose at relatively low temperatures[9], such an assignement should be ruled out. It appears in a subsurface layer of the as-grown EFG crystals, and its magnitude is proportional to the oxygen concentration deliberately added in the form of CO_2 gas during the ribbon growth. SIMS depth profile analysis of EFG samples have shown a strong pile-up of oxygen and carbon close to the surface to a depth of 1 μm[13]. During the fast cooling of crystallized EFG ribbons, at temperatures of about 1200°C, in higher oxygen materials (Fig. 1, curve 2) there is enough oxygen to form clusters in the surface layer consisting of SiO_x and probably C-O compexes (as indicated by the broad peak from 1000 to 1130 cm^{-1}) beside precipitating in the form of SiO_x on carbon-selfinterstitial agglomerates acting as nuclei for precipitation. The latter we associate with the 800 cm^{-1} band. Both complexes are unstable at higher temperatures and start to dissociate even at 650°C. In oxygen-lean EFG materials (Fig.1, curve 1) oxygen atoms, trapped by carbon-selfintersitial nuclei, have been released and dispersed predominantly to the interstitial sites upon a complete dissolution of the 800 cm^{-1} peak after the 1250°C heating step.

The broad band from 810 to 960 cm^{-1} could be attributed to the SiC nanoprecipitates of different radii distributed in the bulk and nucleated originally on the solid-liquid interface[14]. It is visible in the etched nonagon sample (Fig.1, curve 3), and becomes the dominant IR band in the oxygen-lean samples after carbon migration from the substitutional sites in the temperature range from 1150 to 1250°C (Fig.2, curves 1 and 3). Similar broad and asymmetric peak has been found by Bean and Newman in monocrystalline samples with a high-carbon concentration[15].

Heating at 1250°C helps oxygen to get to the interstitial sites and enhances various C-O complexes in the 1000-1100 and 635-750 cm^{-1} spectral regions.

The IR spectra of EFG samples with high carbon concentration and high carbon-to-oxygen ratio show the existence of the whole series of bands in the band mode absorption region just below the Raman frequency of 523 cm^{-1}, expected but not observed previously[16]. Their absorption coefficients are in the range from 0.1 to 0.7 cm^{-1}, and we associate them tentatively with vibrations of carbon pairs with different interatomic distances.

As a conclusion, we may stress again the importance of the relative ratio or the difference in carbon and oxygen concentration on the final configuration and kinds of defect complexes in thermally treated silicon at high temperatures.

REFERENCES

1. J.R.Patel in "Semiconductor Silicon 1977",eds. H.R.Huff and E.Sirtl (Electrochem.Soc.,Princeton,1977),p.521.
2. R.C.Newman, Mat.Res.Soc.Symp.Proc. 1986, Vol. 59, p.403.
3. C.Claeys and J. Vanhellemont, Solid State Phenomena 6&7, 21 (1989).
4. F.Shimura,J.P.Baiardo, and F.Fraundorf, Appl.Phys.Lett. 46, 941 (1985).
5. N.B.Urli and B.Pivac, Rad.Effects and Defects in Solids 111&112, 449 (1989).
6. J.P.Kalejs in "Silicon Processing for Photovoltaics II", ed.by C.P.Khattak and K.V.Ravi (North Holland,Amsterdam,1987),p.185.
7. Annual Book of ASTM Standards (ASTM, Philadelphia, 1981), procedure F 120, p.543.
8. idem, Part 43, p.F-121.
9. B.Pivac and U.V.Desnica, J.Appl.Phys. 65, 4759 (1989).
10. Y.Yang,S.Milstein,J.T.Borenstein, and J.Hanoka, Appl.Phys.Lett. 56, 2222 (1990).
11. L.J.Cheng and C.M.Shyu in "Semiconductor Silicon 1981", ed. by H.R.Huff,R.J.Kriegler,and Y.Takeishi (Electrochem.Soc.Inc. 1981), p.390.
12. J.A.A.Engelbrecht, Infrared Phys. 26, 243 (1986).
13. B.Pivac et al., J.Mat.Science Lett. 9, 397 (1990).
14. J.P.Kalejs and B.Chalmers, J.of Crystal Growth 79, 487 (1986).
15. A.R.Bean and R.C.Newman, J.Phys.Chem.Solids 32, 1211 (1971).
16. R.C.Newman and R.S.Smith, J.Phys.Chem.Solids 30, 1493 (1969).

Materials Science Forum Vol. 83-87 (1992) pp. 291-296

PHOTOLUMINESCENCE OF EDGE-DEFINED FILM-FED GROWTH SILICON

WILLIAM D. SAWYER◊ AND JÜRGEN MICHEL*
◊ Mobil Solar Energy Corporation, 4 Suburban Park Drive, Billerica, Mass. 01821-3980
* Massachusetts Institute of Technology, Department of Materials Science and Engineering
Cambridge, Mass. 02139

ABSTRACT

Low temperature photoluminescence measurements are made on edge-defined film-fed growth polycrystalline silicon. Processed and as-grown samples are investigated. The luminescence spectra of the samples are of an intensity similar to the spectra of high quality electronic grade silicon. Most of the photoluminescence lines detected are well known from previous investigations of crystalline silicon. The as-grown samples occasionally show an unusually intense electron-hole-droplet line. The intensity of the electron-hole-droplet line varies across the surface and as a function of depth in the samples. Lines due to dislocations are consistently observed in grains with a high dislocation content. A deep band at about 1.526 μm is observed in every luminescence spectra. Possible causes of enhanced electron-hole-droplet formation and the cause of the deep band are presented.

Introduction

Various crystal growth methods are currently employed to produce substrates for the production of solar cells. Ribbon growth techniques provide substantial cost savings compared to conventional silicon growth techniques because they do not require the slicing of a boule which results in the loss of large amounts of material, as well as the use of expensive slicing machines. At Mobil Solar Energy Corporation, one of the ribbon growth techniques known as the edge-defined film-fed growth (EFG) technique is used to grow a hollow eight sided polycrystalline silicon ribbon several meters in length. In the EFG technique, the molten silicon liquid is shaped prior to solidification by means of a slotted graphite die. The sides of the "octagons" are cut into 4 inch square blanks which are then processed into solar cells.

As in all ribbon growth techniques, the large surface to volume ratio and fast cooling rate cause macroscopic stresses to be built into the material. Cooling to room temperature in a few minutes, the silicon contains residual strains and stresses which, if not carefully controlled, can result in ribbon buckling and even breakage. They also cause the formation of dislocations during the solidification process. The dislocations are distributed inhomogeneously throughout the EFG Si.

The use of graphite crucibles and shaping dies in the EFG growth method introduces a high concentration of carbon, typically in the 10^{18}atoms/cm^3 range. However, no carbon precipitates have been observed in the as-grown as well as processed material [2]. In addition, there is no experimental evidence as yet for the existence of point defects involving carbon. It is thought that the high carbon concentration in EFG material leads to "atomic" stresses, which are caused by the Si atoms in the lattice having a varying number of carbon atoms in a next neighbor position. There is also no experimental evidence for this phenomenon.

Experimental

The grains of EFG polysilicon are typically oriented in the growth direction, and can be up to several centimeters in length. Assuming that the characteristics of a particular grain do not change over its length, it is possible to cut "sister" samples from a given EFG blank which contain identical grains. One can then compare the results of different processes on samples which contain the same grain. In this manner, sets of 2 x 1 cm samples were cut so that the growth direction runs across the shorter dimension. Samples were prepared from various B doped 2.5 Ωcm EFG material. In addition, an EFG tube was-grown in which the resistivity (B concentration) of the silicon varied across the length of the tube from 1.12 to 200 Ωcm.

One sample out of each pair of sister samples was Syton polished using Buehler Mastermet solution. The polished sample was then given a Schimmel etch [3] for 30 seconds. A Schimmel etch is preferred over other etches because its etch rate is less dependent on the crystallographic direction, which, for EFG Si, is normally [110] on the outside octagon face, but can be several degrees off. The other sample from each pair of sister samples was given a short etch in a 9:1 HNO_3 and HF solution to remove approximately 5 μm of its surface. This eliminates any SiC particles left in the EFG material due the interaction of the graphite die and the molten Si during growth. These samples are considered to be in their "as-grown" state.

Unless otherwise noted, the photoluminescence (PL) spectra of the EFG samples were recorded at liquid-helium temperature (4.2 K) using a Spex 3/4 m spectrometer, a cooled Ge detector, conventional lock-in techniques, and an argon laser (514.5 nm). The laser was operated at 150 mW and illuminated the samples in most measurements without the use of a lens in order to avoid a large excitation intensity which can result in the formation of an electron-hole-droplet (EHD). At the start of each PL measurement session, the spectra of a high purity FZ sample was measured in order to make sure the excitation intensity was not high enough to cause the EHD line to dominate the PL spectrum.

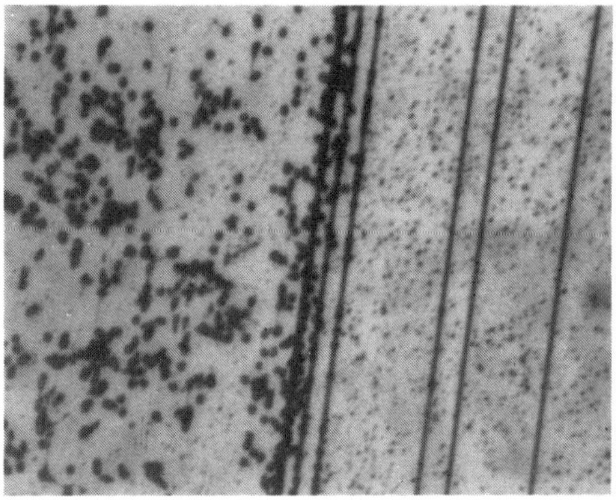

Fig. 1. Optical micrograph of a 2.4 Ωcm EFG polished and etched sample taken at 500x using interference contrast optics.

Figure 1 shows an optical micrograph of a polished and etched EFG sample. Several grain boundaries run vertically across the sample[1]. The larger black dots are the result of the etching of dislocations. They are easily distinguished from the smaller black dots, which are etch pits believed to be caused by the presence of microdefects [4]. One can see in Fig. 1 that the concentration of microdefects is relatively constant across the grains of a sample. The concentration of dislocations, however, can vary greatly between any two grains. Some grains contain no dislocations.

The pairs of sister samples were mounted on the sample holder in the cryostat with the Syton polished sample directly above the as-grown sample. In this manner, it is possible to perform a PL scan on any given grain in the polished sample, and then follow the same grain into the as-grown sample and repeat the PL scan. This allows the correlation of the observed PL lines with the physical characteristics of the EFG Si grains, such as the dislocation concentration.

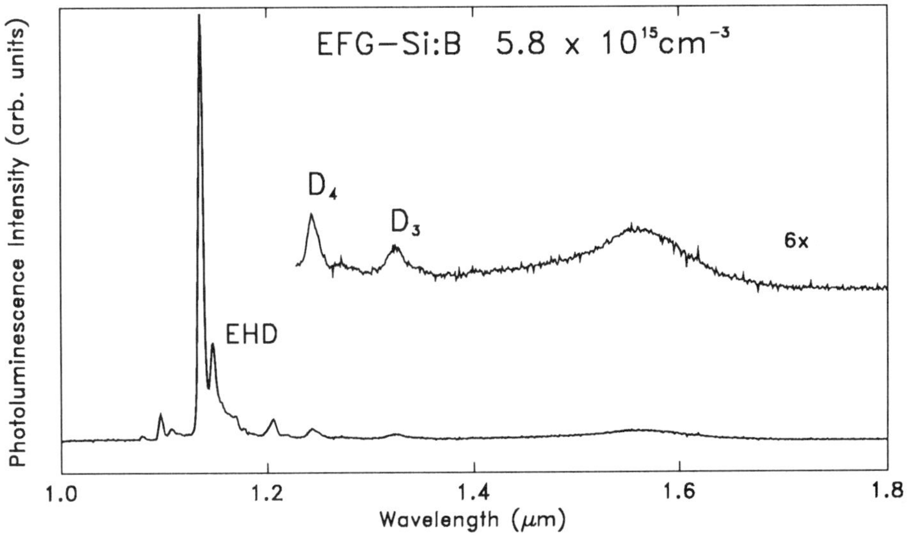

Fig. 2. Photoluminescence spectrum of an as-grown EFG sample.

Results

In general, the PL intensity of the EFG samples was quite strong, whereby the polished samples showed the strongest intensity. In fact, the luminescence intensity of the polished EFG samples was similar to the intensity of PL spectra of electronic grade Si with a comparable doping concentration.

Figure 2 shows a PL spectrum of an as-grown EFG sample. One can see the usual near band gap luminescence due to excitons that are bound to neutral B atoms [5]. In addition to the excitonic luminescence, there is a strong EHD line at about 1147 nm [6].

The intensity of the EHD line varies greatly. In some EFG samples, the luminescence spectrum showed an EHD line as large as or larger than the B^{TO} line, while in other samples the EHD line could not be detected. In order to determine the cause of the EHD line, grains with varying dislocation content were investigated. The results show no correlation of the dislocation concentration to the EHD line intensity.

According to Shiraki and Nakashima [7,8], A- and B-type swirl defects act as nucleation centers for electron hole droplets. In order to investigate the possibility that microdefects in EFG Si can cause EHD formation, several grains which contained a high microdefect and low dislocation concentration were investigated. The results show that the presence of the microdefects has no relation to the measured intensity of the EHD line. By etching off layers of samples in a mixture of 5:3:3 HNO_3, HF and CH_3COOH (CP-4A), the depth profile of the intensity of the EHD line is studied. The depth profile varies significantly and also cannot be correlated to the microdefect or dislocation concentration. One must conclude that the intensity of the EHD line is not related to the dislocations, or the microdefects. It is caused by some other characteristic of EFG Si and is inhomogeneous throughout the material.

The luminescence spectrum in Fig. 2 also contains the D_3 and D_4 lines, which have been attributed to the presence of dislocations [9]. Unlike the EHD line, the intensity of the D_3 and D_4 lines in the

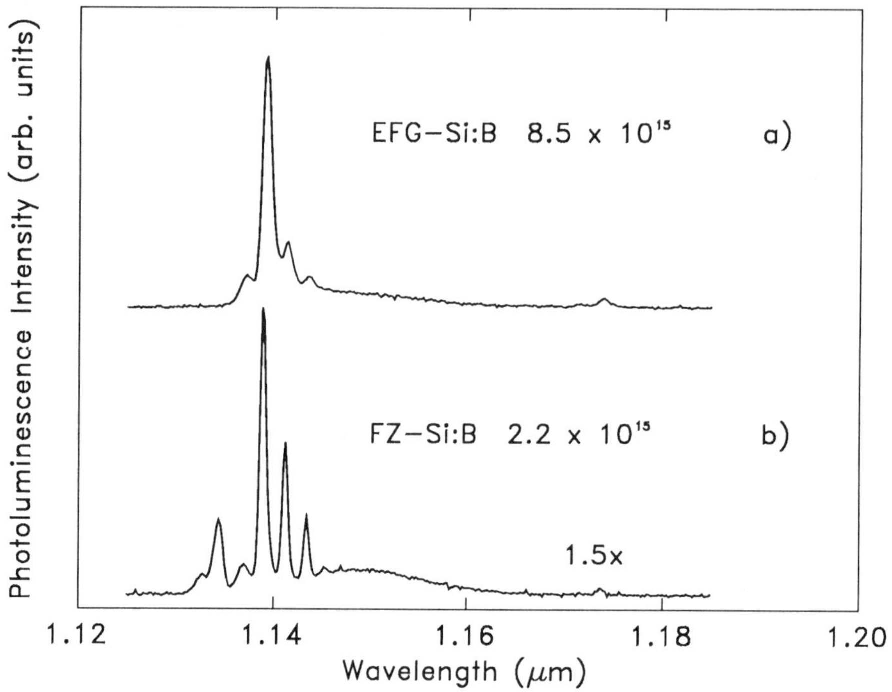

Fig 3. (a) Photoluminescence spectrum of an as-grown EFG sample. (b) Photoluminescence spectrum of a high purity FZ sample. Spectra a) and b) were measured with the same spectral resolution.

as-grown samples increases with an increasing concentration of dislocations seen in the optical micrographs of the grains measured [10]. Large dislocation concentrations cause the bound exciton luminescence to be reduced. All of these results are consistent with previous work on the dislocation lines.

At about 1.56 μm a broad band is present. One is tempted to identify this line with the D_1 line [9], which is also associated with dislocations and should be located at about 1.526 μm. However, this band is observed in every EFG sample or grain investigated, even when there are no dislocations present in the area of the sample which is illuminated by the laser beam. In addition, the D_2 line which should be at about 1.4165 μm in the PL of dislocated samples along with the D_1 line is not observed in the luminescence spectrum of any of the samples. Measurements of the broad band in Syton polished samples show a very large line width, and sometimes a line shape that looks as if the line consists of more than one transition.

Figure 3a shows the near band-gap luminescence of an as-grown EFG sample measured in an area with a high dislocation concentration. In Fig. 3b, a PL spectrum of a high purity FZ sample is shown. Both spectra were measured with the same spectral resolution. The line widths of the B^{TO} and its multi-bound exciton replicas [5] in the spectrum of the EFG sample are larger than in the FZ spectrum. Nevertheless, one can see in Fig. 3a that the most intense luminescence line in the EFG samples is the B^{TO} line, not the FE^{TO} line. The line broadening of the bound exciton lines in EFG material is also seen in areas with a low concentration of dislocations as well as in areas where there are no dislocations,

Discussion

The near band-gap luminescence of EFG Si consists of the usual bound exciton luminescence lines. The energies of the excitonic PL lines are not shifted to lower or higher energies. The energetic line widths of the excitonic lines are "occasionally" broader than they are in high purity FZ Si. The degree to which the excitonic lines are broadened cannot be correlated to the density of dislocations in the area illuminated by the laser beam. Indeed, the excitonic lines are also broadened in areas with no dislocations. There is an additional cause of line broadening, such as a high impurity concentration. The most likely candidate for the impurity is C, because it exists in EFG Si in a large concentration due to the growth method. The results presented in this paper, however, do not allow an unequivocal identification of the cause of the excitonic line broadening.

The EHD line is observed to be present with an unusually strong intensity in the PL spectrum of some samples. In a manner similar to the excitonic line broadening, the intensity of the EHD line cannot be correlated with any observable feature of the grain illuminated during measurement, such as the dislocation or microdefect concentration. It is possible that increased formation of electron-hole-droplets is due to the random stresses and strains which are inhomogeneously distributed in the EFG Si lattice during the cooling phase of the growth process. Previous investigations of Ge have shown that free excitons, carriers and droplets are accelerated in a strain gradient towards points of maximum sheer strain [11]. It is also possible that the high C concentration exerts "microscopic" strains in the lattice, which enhances EHD formation. It is interesting to note that the size of these stresses in EFG Si cannot be very large because the energetic position of the excitonic lines is not shifted [12].

The D_3 and D_4 lines, in contrast to the previously mentioned PL lines, can be directly correlated to a characteristic of EFG Si. Their intensity increases in areas with a high concentration of dislocations. The broad band at 1.56 μm, however, is not the D_1 line. It is not present at the right wavelength, nor is it accompanied by the D_2 line, which is considered to be related to the D_1 line [9,10,13]. In addition, it is present in every PL spectrum of EFG Si (with various intensities), including spectra of areas which contain no dislocations. The energetic width of this band increases significantly after Syton polishing. This could be related to the interaction of the defect that causes the band with Cu, which is known to permeate into Si during Syton polishing [14,15]. The only physical property of EFG Si which could be associated with it is the ubiquitous microdefect concentration.

The observation of the band at 1.56 μm is interesting because of recent results showing that the D_1 and D_2 lines are related to metallic impurities located in the strain field of the dislocations [16]. Edge-defined film-fed Si has a high dislocation concentration, and contains higher impurity concentrations in its as-grown state than FZ Si. Nevertheless, the D_1 and D_2 lines are not observed, even though the D_3 and D_4 lines are. These results suggests that the origin of the D_1 and D_2 lines is more complicated than the present interpretation in the literature.

Conclusion

Low temperature PL measurements of EFG Si reveal the usual bound exciton luminescence with an intensity similar to electronic grade silicon, but a greater line width. The larger line widths of the excitonic lines in EFG silicon are believed to be caused by the high concentration of dislocations and the high carbon concentration in EFG Si. The high concentration of dislocations in some EFG Si grains results in the well known D_3 and D_4 lines. A line which is in a similar position as the D_1 line is also observed. Because this line does not correlate with the dislocation concentration, but is observed in every EFG spectrum, it is not related to dislocations. One possible cause of this line is the existence of microdefects which are observed in every grain of polished and etched EFG samples.

References

1. H. Strunk, B. Cunningham and D. Ast, in *Materials Research Society Symposia Proceedings*, Vol. 2, Defects in Semiconductors, edited by J. Narayan and T.Y. Tan (North-Holland, New York, 1981) p. 297.
2. R. Gleichman, B. Cunningham and D. G. Ast, J. Appl. Phys. **58**, 223 (1985).
3. Designation: F 47 - 87, in *Annual Book of ASTM Standards*, Vol 10.05, (ASTM, Philadelphia, 1988), p. 63.
4. N. W. Marr, M. C. Cretella and B. R. Bathey, in *Microstructural Science*, Vol. 2, edited by R. T. DeHoff, J. D. Braun and J. L. McCall (Elsevier, New York, 1983), p. 319.
5. For a review see M. Thewalt, In *Excitons*, ed. by E. I. Rashba and M. D. Sturge (North-Holland, New York, 1982) p. 393.
6. For a review see Y. A. Pokrovski, Phys. Stat. Sol. (a) **11**, 385 (1972) or A. A. Rogachev in *Progress in Quantum Electronics*, Vol. 6, edited by T. S. Moss and S. Stenholm (Pergamon Press, Oxford, 1980) p. 141.
7. Y. Shiraki and H. Nakashima, Solid State Commun. **27**, 1033 (1978).
8. H. Nakashima and Y. Shiraki, Solid State Commun. **40**, 195 (1981).
9. R. Sauer, J. Weber, J. Stolz, E. R. Weber, K.-H. Küsters, and H. Alexander, Appl. Phys. A **36**, 1 (1985).
10. N. A. Drosdov, A. A. Patrin, and V. D. Tkachev, Zh. Eksp. Theor. Fiz., Pis' ma Red. **23**, 651 (1976) [Soviet. Phys. J. Exper. Theor. Phys. Lett. 23, 597 (1976)].
11. Markiewicz, J. P. Wolfe, and C. D. Jeffries, Phys. Rev. B **15**, 1988 (1977).
12. K. Yasutake, M. Umeno, H. Kawabe, H. Nakayama, T. Nishino and Y. Hamakawa, Jpn. J. Appl. Phys. 21, 1715 (1982).
13. M. Suezawa, K. Sumino and Y. Nishina, Jpn. J. Appl. Phys. **21**, L518 (1982).
14. P. Jones, Y. Zhang, J. Liu, J.-Z. Yuan, C. Ortiz, B. Baufeld, H. Bakhru, J. W. Corbett and S. J. Pearton in *Defect Control in Semiconductors*, edited by K. Sumino, (Elsevier, New York, 1990) p. 317.
15. H. Prigge, P. Gerlach, P. O. Hahn, A. Schnegg and H. Jacob, J. Electrochem. Soc., **138**, 1385 (1991).
16. V. Higgs, E. C. Lightowlers and P. Kightley in *Materials Research Society Symposia Proceedings*, Vol. 163, Impurities, Defects and Diffusion in Semiconductors: Bulk and Layered Structures, edited by D. J. Wolford, J. Bernholc and E. E. Haller (Materials Research Society, Pittsburgh, 1990) p. 57.

Materials Science Forum Vol. 83-87 (1992) pp. 297-302

EPR Studies of NTD-produced As donors in Isotopically Enriched ^{74}Ge Single Crystals

V.S. Weiner
University of Chicago, Chicago, Illinois 60637
James Franck Institute, 5640 Ellis Ave.

ABSTRACT

EPR spectra of the neutron transmutation produced As donors in the 74-isotope enriched Germanium have been investigated. Reduction to 1% 73-isotope content and high level of uniformity produced by the n-irradiation resulted in considerable narrowing of the EPR lines and observation of the well resolved single and exchange coupled donor spectra at concentration of ~$2\cdot10^{16}$ cm^{-3}. Additional to the main quadruplet, hyperfine structure of the exchange donor clusters in Ge have been observed for the first time. No spectral effects caused by the electron hopping motion have been observed.

I. Introduction

The slow neutron transmutation reactions are well known to produce the uniform distribution of ^{75}As donors in the ^{74}Ge isotope enriched Ge. It is less known that reduction of the ^{73}Ge isotope content in the enriched composition results in the appreciable narrowing of the EPR spectrum lines[1]. Till now Ge was considered inappropriate for EPR studies which is the case for the normal isotope Ge.
The present study revises the problem of the EPR spectroscopy of defects and impurities in Ge to provide an affirmative answer. Two main sources of the line broadening should be eliminated to obtain well resolved EPR spectra: excess of the ^{73}Ge isotope and nonuniformity of the defect distribution during the crystal growth. Both technological procedures used at the stage of crystal preparation for this study, reduction of the ^{73}Ge isotope content and the neutron irradiation for the donor production, meet these requirements.
Spectral data presented below have been obtained for the first time. We observed well resolved hyperfine structure of the exchange coupled As donor clusters as well as the single donor quadruplet at the As concentration of ~$2\cdot10^{16}$ cm^{-3}. It corresponds to the hopping conductivity concentration range where the shallow donor spectrum in the normal isotope Ge is presented by a single motion narrowed line[2]. The linewidth was about 4 G in the H‖[100] orientation. The angular variation of the linewidth did not exceed a factor of 1.5 off the [100] direction.

II. Effects of the line broadening

The excessive line broadening of the shallow donor spectra in Ge are determined by the following factors:

1. Unresolved hyperfine interactions with the remote ^{73}Ge nuclei.

2. Random g-factor shift produced by the nearby dislocation strain.

The first of these results in appreciable content (7.8% natural abundance) of the ^{73}Ge isotope having the large nuclear momentum μ =-0.879 n.m. and the spin I = 9/2. Since the central atom wave function density spreads over many lattice sites, many ^{73}Ge isotope nuclei are involved in the hyperfine interactions. Calculations of the interactions require explicit knowledge of the wave function distribution which is unknown. Nevertheless, probability estimation of their contributions in the line broadening can be made as follows.
Probability P_m to find m particular atoms in any coordination sphere around the defect center consisting of n equal sites is

$$P_m = C_n^m x^m (1-x)^{n-m} , \qquad (1)$$

where x the partial content of those atoms in the bulk composition.

The table below summarizes the probability calculations at the different x:

x,%	n	P_0	P_1	P_2
7.8	4	0.789	0.240	0.030
	12	0.387	0.382	0.173
2.0	4	0.922	0.075	0.003
	12	0.785	0.192	0.022
1.0	4	0.960	0.040	-
	12	0.886	0.107	0.007

Obviously, the fractional spectral line intensities corresponding to the central atom interaction with 0,1 or 2 nuclei follow P_m distributions.

Considering the numbers of the bottom row and noticing that P_1 intensity is distributed among 2I+1=10 hyperfine components one concludes that the isotopic broadening is removable by the reduction to the 1% ^{73}Ge isotope content.

The second source of the line broadening originates from the strain-induced bound electron g-factor shift from the average single valley electron g-factor $g_0 = (g_\parallel + 2g_\perp)/3$, where $g_\parallel \approx 0.87$ and $g_\perp \approx 1.92$ (see[1,3]).
In the absence of strains the As donor ground states formed by the four equivalent conduction band valley states are nearly degenerated and separated by the small $4\Delta_c$ valley-orbit splitting of 4.2 meV. A small strain Δs produces a significant valley strain $X = (\Xi/4\Delta_c)\Delta s$, where Ξ= 19 eV is the deformation potential for strain, that lifts the valley degeneracy. As a result of the new averaging the bound electron g-factor deviates from its former value.
Even in moderately dislocated crystals an averaged strain $\Delta s = 10^{-8} \varepsilon^{\frac{1}{2}}$ arising from dislocations, where ε is the dislocation density, is sufficient to cause the g-factor shift1.

The line broadening caused by such a shift is the largest along the longitudinal axis [111] and is proportional to

$$g - g_0 = \frac{2X}{9}(g_{\parallel} - g_{\perp}) \qquad (2)$$

It is very small in the [100] direction and rises remarkably at a small declination from this direction at the angle θ. The broadening was found to be proportional to[1]

$$g - g_0 = \frac{X}{9}(g_{\parallel} - g_{\perp})\sin 2\theta . \qquad (3)$$

More details on the strain dependent shallow donor spectrum parameters can be found in[3].

III. Experimental results

A. Samples

The studied crystal was prepared by pulling from the ^{74}Ge enriched melt by Czochralski method. The isotopic content of the crystal as compared with the normal Ge and studied in[1] was as shown

	^{70}Ge	^{72}Ge	^{73}Ge	^{74}Ge	^{76}Ge
This study	1.70%	2.40%	1.00%	93.90%	1.00%
Wilson's study[1]	0.80%	1.03%	0.86%	96.75%	0.56%
Normal	20.45%	27.41%	7.77%	36.58%	7.79%

The crystal was aligned with a laser beam reflected from the etched cross-section and cut off across (111) plane in discs of 1 mm thickness. The alignment accuracy did not exceed few degrees. The discs were irradiated with a slow neutron flux in the nuclear reactor for the transmutation reaction $n + {}^{74}Ge \rightarrow {}^{75}As$. Subsequently discs were annealed at 450°C for 24 h to remove radiation damages. At the neutron flux of $\sim 4 \cdot 10^{13}$ $cm^{-2}s^{-1}$ and the exposure for ~3000 s the doping level reached $(N_d - N_a) = (1.85 \pm 0.08) \cdot 10^{16}$ cm^{-3} as determined by the Hall effect measurements. For EPR measurements one disc was cut in bars with the long side along the [100] direction and etched.

B. EPR study

All measurement were performed in the basic mode of a conventional 10 GHz EPR spectrometer at a temperature of 4.2 K. No saturation effects at the low microwave power level were observed. The spectra exhibited the angular dependence of the linewidth with ΔH ~4 G in the $H \parallel [100]$ orientation. ΔH did not exceed 6 G in any other orientation. In the $H \parallel [100]$ orientation the spectrum was fairly resolved as shown in the Figure.

The spectrum exhibits a hyperfine satellite structure overlapping the single As donor quadruplet (I=3/2) with the known hyperfine splitting of ~35 G. The satellite spectrum structures are typical for exchange coupled shallow donor clusters and known to observe in the moderately doped Si⟨P⟩ (see [4,5]).

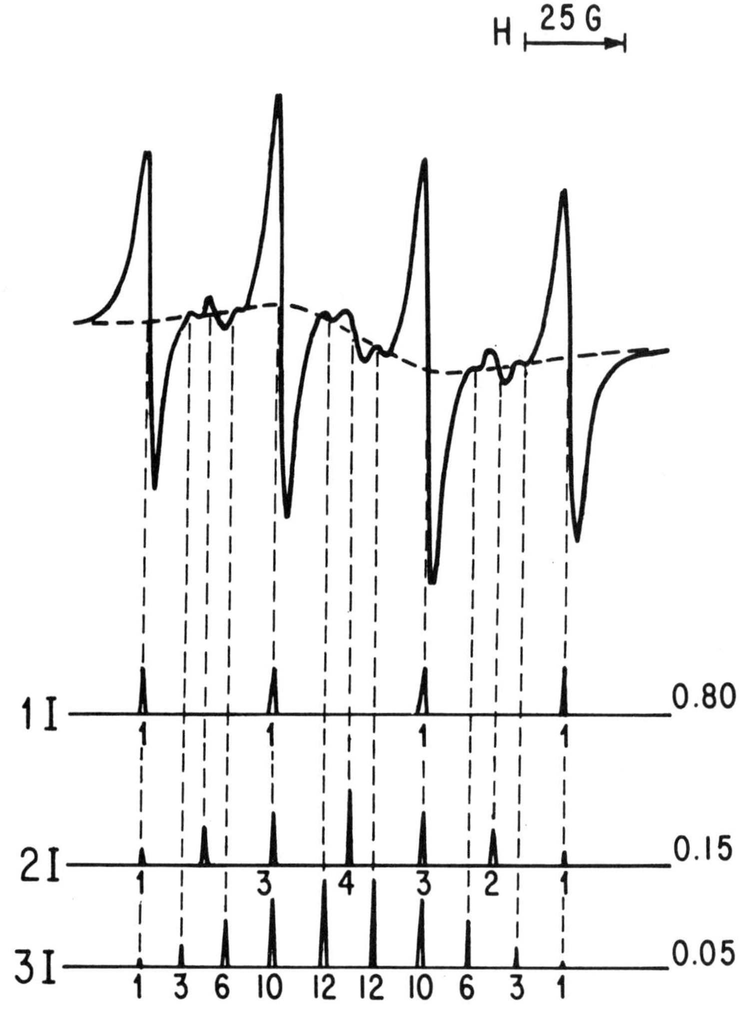

Figure. As donor absorption spectrum in the ^{74}Ge single crystal at $(N_d-N_a)=1.85\cdot10^{16}$ cm^{-3}. H‖[100], T=4.2 K.
Below: Equivalent splitting and intensity distribution structure for single, double and triple donor spectra. Numbers on the right-fraction of each spectrum in the total resolved spectrum.

The spectra are described by the spin Hamiltonian of similar $S_i = 1/2$ spins coupled by the pair exchange interaction $J\mathbf{S}_i\mathbf{S}_j$ and interacting with own nuclear spin over $A\mathbf{S}_i\mathbf{I}_j$. Then the full spin Hamiltonian can be written over the total spin $\mathbf{S} = \sum_i \mathbf{S}$ of the n interacting atoms as follows

$$\mathcal{H} = g\beta HS + \frac{A}{n}\left\{\sum_{i\neq j}^{n} (\mathbf{S}_i-\mathbf{S}_j)(\mathbf{I}_i-\mathbf{I}_j) + S\sum_{i}^{n} I_i\right\} + 1/2J\left\{S(S+1) - \sum_{i}^{n} S_i(S_i+1)\right\} . \quad (4)$$

The exchange term in (4) splits $(2S_i+1)^n$ spin levels into multiplets still retaining the Zeeman degeneracy except a singlet. The cross-multiplet products in the hyperfine term mix different multiplet states. When A~J this mixing results in multiple resonance transitions and the appearance of the wide unresolved background line as seen in the Figure. When $J \gg A$, the multiplets are separated enough to neglect mixing and (4) becomes simply

$$\mathcal{H} = g\beta HS + \frac{A}{n} S\sum_{i}^{n} I_i + 1/2J\left\{S(S+1) - \sum_{i}^{n} S_i(S_i+1)\right\} . \quad (5)$$

The corresponding EPR spectrum consists of $2nI_i+1$ hyperfine lines spaced by A/n and has the characteristic intensity distribution as shown in the Figure for two- and three- atom clusters. Since the exchange interaction is defined to be isotropic, its value effects the multiplet energy separations and hence, the multiplet populations. The hyperfine spectra remain regardless of the repopulations, i.e. the exchange interaction values.
Utilizing the exchange cluster models which have been proved one can proceed to the analysis of the donor bulk distribution. The intensity ratio between the resolved and background spectra is close to 1:1, and the single atom: double: triple atom spectrum intensity ratio is 0.80:0.15:0.05. From the first ratio one can deduce very unexpected result of the donor exchange coupling. Weakly interacting donors (J~A) form the giant exchange cluster in the whole crystal volume. The average separation between atoms is very close to $R_m = (2\pi N)^{-1/3}$ in the maximum of the random distribution curve:

$$\frac{dP}{dR} = 4\pi R^2 N \exp[(-4/3)\pi R^3 N] \quad (6)$$

where P is probability of the nearest-neighbor atoms to be separated by R. At the concentration $N = 2\cdot10^{16}$ cm^{-3} R_m = ~210 A that exceeds considerably the As Bohr radius of ~60 A.
For the close pairs ($J \gg A$) the probability to find them separated by the R or less is given by integrating (6)

$$P = 1 - \exp[(-4/3)\pi R^3 N] \quad (7)$$

and the pair spacing is $R \lesssim 100$ A with P =0.075 derived from the above ratios.
Another evidence of the high donor bulk uniformity in the studied sample arises from the absence of the line motional narrowing effects in the hopping conductivity concentration range. It was shown for Ge⟨P⟩[2] the motional narrowing of the single donor line to occur in the concentration range of $(1.2-10)\cdot10^{16}$ cm^{-3}. The linewidth in the [100] orientation narrowed exponentially from 25 G to 5 G. As a consequence of the close Bohr radii the same concentration

narrowing effect should be expected for the donor spectrum in Ge⟨As⟩. The premature line narrowing in the normal Ge should apparently be interpreted as a result of the nonuniform donor distribution.

Conclusion

Highly resolved EPR spectra of the double and triple exchange donor clusters as well as unresolved wide spectrum of the weakly coupled volume donor cluster have been observed in the moderately doped isotopically enriched Ge. The spectra apparently reflects the high level of the NTD-produced donor bulk uniformity.

I would like to thank Prof. Fritzsche for the interest in this study.

References

1. D.K.Wilson, Phys. Rev. **134A**, A265 (1964)
2. K.Morigaki and M.Mitsuma, J.Phys. Soc. Japan. **20**,62 (1965)
3. M.Nakayama and H.Hasegava,J.Phys. Soc. Japan. **18**,229(1963)
4. G.Feher, R.C.Fletcher and E.A.Gere, Phys. Rev. **100**, 1784 (1955)
5. S.Maekawa and N.Kinoshita, J.Phys. Soc. Japan. **20**, 1447 (1965)

Materials Science Forum Vol. 83-87 (1992) pp. 303-308

IN-SITU HVEM STUDY OF DOPANT DEPENDENT DEFECT GENERATION IN SILICON DURING 1 MeV ELECTRON IRRADIATION

Albert Romano-Rodríguez[1,2] and Jan Vanhellemont[1]
[1]Interuniversity Micro-Electronics Centre (IMEC), Kapeldreef 75, B-3001 Leuven, Belgium
[2]LCMM, Departament de Física Aplicada i Electrònica, Universitat de Barcelona, Diagonal 647, E-08028 Barcelona, Spain

ABSTRACT

Results are presented of an in-situ study of electron irradiation induced extended defect generation in silicon wafers with B or P doped surface layers formed by ion implantation. During 1 MeV electron irradiation of cross-section specimens in a high voltage transmission electron microscope, extended defects are preferentially generated in areas with a well defined dopant concentration range. An attempt is made to explain the observations on the basis of point defect reactions.

1. Introduction

Accelerated electrons with energies above 140 keV are known to introduce lattice damage in silicon at room temperature. For the creation of extended lattice defects, which can be imaged using transmission electron microscopy, electron fluxes ϕ larger than 10^{18} $cm^{-2}s^{-1}$ and higher electron energies and/or irradiation temperatures are required. These experimental conditions can easily be obtained by in-situ electron irradiation in a high voltage transmission electron microscope (HVEM) [e.g. 1-10]. The observed extended defects lie predominantly in {113} habit planes and are identified as extrinsic defects, indicating a local supersaturation of self-interstitials [1,2].

In the present paper, results are presented of an in-situ study of dopant dependent extended defect generation in silicon, using 1 MeV electron irradiation of cross-section specimens in a HVEM. The defect nucleation and growth is studied as a function of the irradiation conditions and of the dopant type and concentration profile.

2. Experimental

Cross-section TEM samples are prepared from n- or p-type CZ silicon wafers with heavily doped surface layers obtained by B or P ion implantation. Some implantations are performed through a patterned silicon oxide or polycrystalline silicon mask to study the influence of two dimensional dopant profiles on the electron irradiation induced defect generation. Cross-section TEM specimens are prepared following a technique described elsewhere [11]. The 1 MeV electron irradiations are performed in the JEOL1250 microscope of the University of Antwerpen (RUCA). The temperature of irradiation is varied between 300 and 640K and the electron flux ranges from $2x10^{18}$ to $2x10^{19}$ $cm^{-2}s^{-1}$. The defect nucleation and growth kinetics are studied by irradiating for up to 30 minutes. Some of the irradiated samples, received a second irradiation at another temperature for various times and electron fluxes. Specimens were re-irradiated in the same orientation or after turning the specimen upside down in the microscope holder.

3. Observations

Boron implanted wafers

A typical time sequence of electron irradiation induced defect nucleation and growth at 643K is shown in figure 1 for a boron implanted n-type CZ silicon substrate. It is observed that during the irradiation a band with a very high density of defects develops in an area corresponding to dopant concentration between $5x10^{17}$ and $5x10^{18}$ cm^{-3}. Irradiation at lower temperatures (300K or 493K) does not change the position of the irradiation induced defect band and only the nucleation and growth kinetics of the defects are influenced. In figure 2 a typical result of a re-irradiation experiment is shown. When re-irradiating at room temperature a specimen which was irradiated before at 643K, shrinkage of the irradiation induced defects is always and only observed when the specimen is turned upside down in the microscope. No shrinkage of defects is observed if the specimen is not turned. This observation suggests that the extended defect nucleation during the first irradiation occurs preferentially at one side of the specimen, i.e. close to the specimen surface oriented towards the incident electron beam [6, 8].

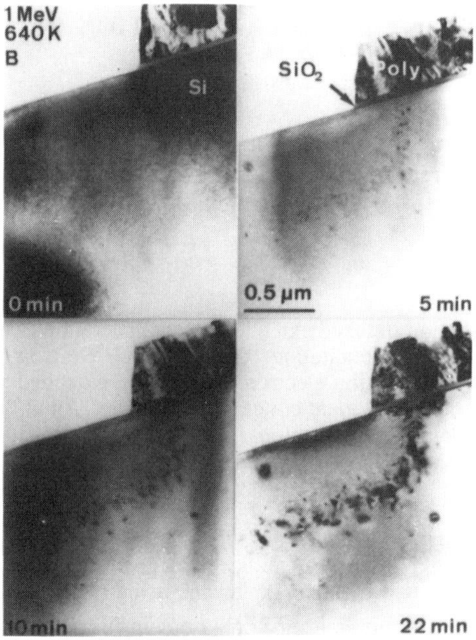

Figure 1: *Time sequence of defect formation in B implanted silicon ($\phi = 2x10^{18}\ cm^{-2}s^{-1}$).*

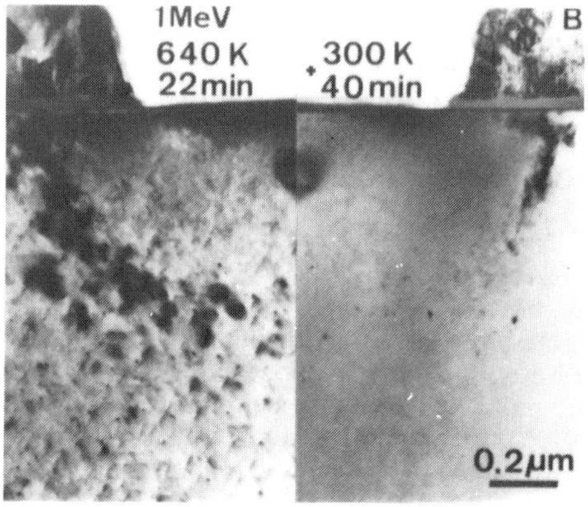

Figure 2: *Typical re-irradiation experiment in boron implanted silicon: a) after the first high temperature irradiation ($\phi = 2x10^{18}\ cm^{-2}s^{-1}$); b) after turning the specimen and irradiating at room temperature ($\phi = 2x10^{18}\ cm^{-2}s^{-1}$) shrinkage of the defects is observed.*

Phosphorous implanted wafers

In phosphorous implanted silicon, the defect nucleation is strongly depending on the irradiation temperature, as illustrated in figure 3. Between 300 and 410 K, defects nucleate preferentially in the highly doped areas (> 10^{19} cm^{-3}). At higher irradiation temperatures, defects are observed in areas with a P concentration below $5x10^{17}$ cm^{-3} (figures 3 and 4). Re-irradiation at 300K, as shown schematically in figure 5, leads to a dissolution of the defects as was the case for the boron doped substrates. One of the samples which was irradiated at 300K for 22 min, received a second irradiation at 520K for 22 min using the same electron flux and energy. The defects formed in the highly doped areas during the first irradiation, disappear during the second high temperature irradiation while new defects are formed in the lowly doped areas.

4. Discussion

The nucleation rate and the growth velocity of the extended defects increase with irradiation temperature. Annihilation of the defects created during high temperature irradiation is observed when the specimen is irradiated again at room temperature from the backside. A similar effect was reported by Hasebe et al [6] who observed an initial enhanced extended defect formation at the exit surface of the specimen which was intentionally doped with a high concentration of carbon, followed by a dissolution of these defects and the usual nucleation of defects at the entrance surface. They attributed this behaviour to an accumulation of the slower moving point defect, i.e. the vacancy, leading to the observed dissolution. In the next paragraphs different aspects of electron irradiation induced point and extended defect generation are discussed in an attempt to explain the experimental observations.

4.1 Point defect generation during irradiation

During irradiation with high energetic particles, silicon lattice atoms are displaced, resulting in the formation of Frenkel pairs. In a fraction of these Frenkel pairs the self-interstitials and vacancies will be separated far enough to be considered as independent from each other. In silicon, self-interstitials are much more mobile than vacancies so that it can be assumed that changes in the self-interstitial concentration are immediate compared to changes in the vacancy concentration. As the damaged silicon lattice is far from equilibrium, the generated intrinsic point defects will try to lower the total free energy by interaction with available sinks in the material.

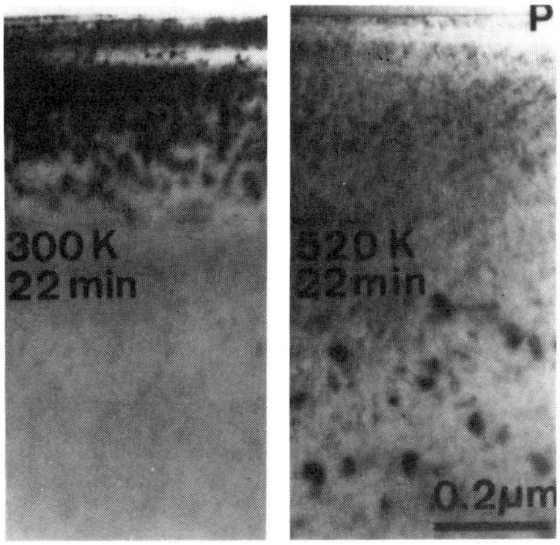

Figure 3: *Irradiation at 300 K ($\phi = 1.5x10^{19}$ $cm^{-2}s^{-1}$) (left) and at 520 K ($\phi = 2x10^{18}$ $cm^{-2}s^{-1}$) (right) of a phosphorous implanted sample. At 300 K, defects are formed in areas with a P concentration above 10^{19} cm^{-3}, while at 520 K, they are formed only for concentrations below $5x10^{17}$ cm^{-3}.*

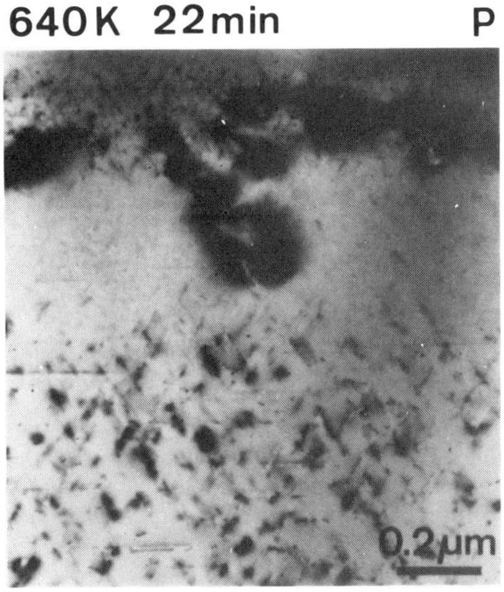

Figure 4: *Result of irradiation ($\phi = 2x10^{18}$ $cm^{-2}s^{-1}$) of a sample with a buried P profile . At 640 K, a buried defect free zone is formed corresponding with a phosphorous concentration above $5x10^{17}$ cm^{-3}.*

4.2 Extended defect nucleation

Different hypotheses to explain the observed localised extended defect nucleation are explored. The most important parameters which have to be taken into account to describe the behaviour of the generated intrinsic point defects are:

- the position of the Fermi level with respect to the position of the deep levels associated with the created point defect complexes,
- the electrical fields due to the dopant concentration profile and to the presence of a junction,
- the electrical field which can develop over the XTEM specimen thickness during electron irradiation,
- the interaction of extrinsic point defects with the created intrinsic point defects,
- the temperature of irradiation, and
- the presence of localized strain fields.

<u>Fermi level effects</u>

In the diluted solution approximation, the concentration of charged (C_X^c or C_X^{cc} with c = - or +) and uncharged (c = 0) point defects X (X = V, I) is given by [14]:

$$\frac{C_X^c}{C_X^0} = \frac{\theta_X^c}{\theta_X^0} e^{c\frac{E_X^c - E_F}{kT}} , \quad \frac{C_X^{cc}}{C_X^c} = \frac{\theta_X^{cc}}{\theta_X^c} e^{c\frac{E_X^{cc} - E_F}{kT}} . \tag{1}$$

θ_X^c is the number of degrees of internal freedom of the defect on a lattice site and E_F the Fermi level.

The position of the Fermi level E_F itself depends on the dopant concentration. Assuming Boltzman statistics, one can write

$$E_F = - c\, kT\ln \frac{d}{n_i} + E_F^i \ , \text{ and } \ \frac{C_X^c}{(C_X^c)^i} = \frac{d}{n_i} \text{ and } \frac{C_X^{cc}}{(C_X^{cc})^i} = [\frac{d}{n_i}]^2 . \tag{2}$$

d = n and d = p for c = - and c = +, respectively. n_i is the intrinsic carrier concentration and E_F^i the intrinsic Fermi level. $(C_X^c)^i$ and $(C_X^{cc})^i$ are the concentrations of charged point defects under intrinsic conditions. The concentration of negatively charged point defects will thus be higher in n-type silicon than in p-type and vice versa for positively charged point defects in p-type silicon. In highly doped areas and at the beginning of the electron irradiation, the Fermi level will be close to the valence band for p-type and close to the conduction band for n-type material, so that most of the created point defects will be charged. The continuous formation of radiation induced, ionized defects, shifts the Fermi level towards the corresponding energy level in the band gap. As a large number of interstitial related levels is lying around 0.4 eV above the valance band, the Fermi level will move towards this level during prolonged irradiation of p-type material. Increasing further the irradiation dose will make the material intrinsic and can even cause a transformation from p- to n-type [15].

Due to athermal diffusion processes, the self-interstitials are mobile even at the lowest irradiation temperatures. Compared to the slower vacancies they diffuse nearly instantaneously to available sinks making it difficult to observe isolated interstitials. Most of the self-interstitial related defects occur under the form of self-interstitialcy defects (= di-interstitials).

Influence of electrical fields

The presence of large electrical fields will enhance the diffusion of charged point defects. Berman et al [17] studied the influence of an externally applied electrical field on the profile of radiation defects. They observed a strong defect reduction in the space charge region and explained this effect by the influence of the electric field on the diffusion of the charged intrinsic point defects. The preferential generation of extended defects at the entrance surface of the specimen during irradiation in the HVEM, might be explained as follows. During the irradiation of a transparent silicon foil with 1 MeV electrons, an electrical field develops over the thickness of the sample, between the entrance and exit surface of the electron beam. At the start of the electron irradiation, all generated defects will be positively charged for a substitutional boron concentration above 10^{17} cm^{-3}. The generation of charged defects leads to a gradual increase of the Fermi level towards the centre of the bandgap. Once the Fermi level is above the level for positively charged vacancies, but still below E_V + 0.4 eV, only positively charged self-interstitials will be generated, as illustrated schematically in figure 6. The electrical field developing over the specimen thickness will cause a build-up of charged self-interstitials at the front surface of the specimen. Preferential nucleation of self-interstitial clusters will thus occur at the entrance surface, while a vacancy-rich layer will be formed at the exit surface of the specimen. A similar influence is expected from the build-in electrical field related to the dopant concentration profile and from the presence of a junction. Charged point defects will indeed diffuse rapidly by electrical field enhanced athermal diffusion towards the bulk leaving a relatively defect lean area. In case of a buried dopant concentration maximum, this diffusion will also occur towards the surface of the wafer, leading to a double defect band (figure 4).

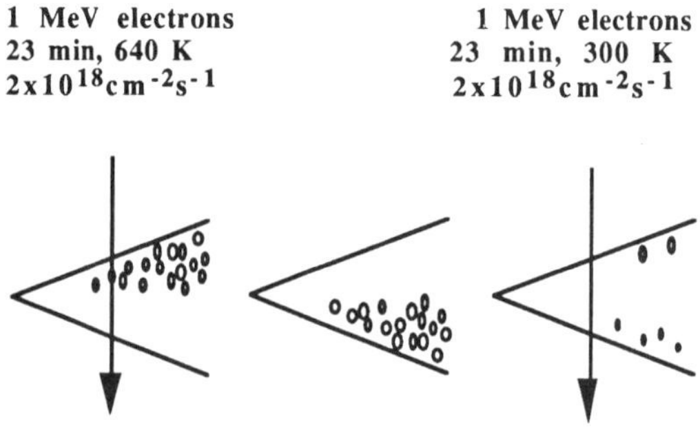

Figure 5: *Schematic view of the re-irradiation experiment. Left: during the first high temperature irradiation defects are preferentially nucleating at the entrance surface. Right: after turning the specimen, re-irradiation at room temperature leads to a dissolution of the defects.*

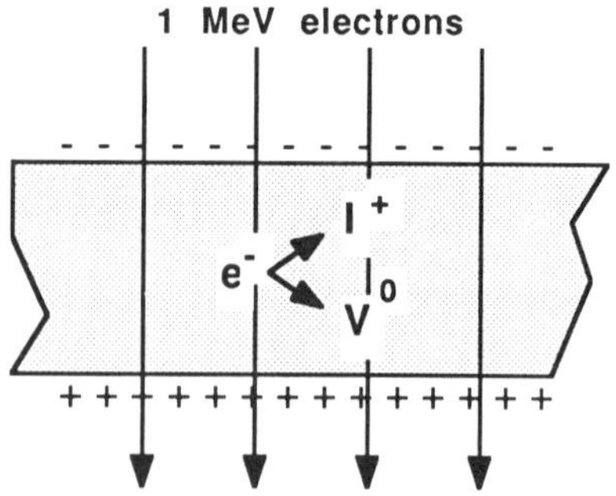

Figure 6: *Generation of positively charged self-interstitials and neutral vacancies, leading to an accumulation of positively charged self-interstitials near the entrance surface for a p-type sample.*

Interaction with extrinsic atoms

From extensive deep level transient spectroscopy (DLTS) studies, it is well known that both types of intrinsic point defects can form complexes with extrinsic point defects [e.g. 12,13]. In CZ silicon, which always contains a high concentration of interstitial oxygen (10^{17}-10^{18} cm^{-3}), dominant sinks for vacancies are the formation of double vacancies (VV) and vacancy-interstitial oxygen pairs or "A-centres" (VO_i). In CZ substrates one thus expects mainly the self-interstitial related reactions to be influenced by the dopant concentration. For low dose irradiations, the concentration of both types of intrinsic point defects will be depleted efficiently by complex formation with extrinsic point defects.

For high dose irradiations however, the number of available sinks for self-interstitials will saturate leading to an increase of the self-interstitial concentration above the critical value for extended defect nucleation. Small interstitial clusters will then nucleate, forming preferential sinks for further generated self-interstitials. As locally the number of self-interstitials will depend strongly on the concentration of extrinsic point defects, also the extended defect nucleation will be influenced by the dopant distribution. The lack of extended defects in highly boron doped areas (figure 1) can be explained by the creation of interstitial boron/substitutional boron (B_iB_s) pairs [12]

$$B_s + I \leftrightarrows B_i \quad \text{and } B_s + B_i \leftrightarrows (B_sB_i) \; . \qquad (3)$$

The probability of the formation of (B_sB_i) pairs is proportional to ϕ $[B_s]^2$. These (B_sB_i) pairs are stable up to relatively high anneal temperatures (> 700K). For B_s concentrations above > 10^{19} cm^{-3}, (B_iB_i) pairs might also be formed. The formation with high efficiency of interstitial boron atoms B_i or B_i related clusters, represents a strong sink for the irradiation induced self-interstitials, explaining why in the highly boron doped areas no extended defects are observed.

Defect nucleation and growth kinetics

When the self-interstitial concentration reaches a critical value, small interstitial clusters are formed which are the nucleus for {113} defect formation. The build-up of the intrinsic point defect concentration is a dynamic process which depends on the local material parameters. In general one can write for the concentration of self-interstitials or vacancies [7]:

$$C_X = G - R, \qquad (4)$$

$$G = (1 - C_V)(1 - k_V C_V) zJ = k_a z \sigma J, \qquad (5)$$

$$R = k_{IV} C_I C_V D_I + k_{I,VD} C_D C_{I,V} D_{I,V} + k_{I,VS} C_S C_{I,V} D_{I,V}. \qquad (6)$$

G is the generation and R the annihilation rate of intrinsic point defects. k_V is the annihilation coefficient of the first order reaction, z is the number of Frenkel pairs generated per electron, σ is the displacement cross-section for the formation of Frenkel pairs, k_a is the probability that the Frenkel pair separates and J is the electron current density. The first term on the right side of (6) describes the mutual annihilation of I and V, taking into account that I is the moving species. The second term describes the absorption of point defects by extended defects (D) while the last term describes the absorption of point defects by other sinks (S).

Important sinks for intrinsic point defects are extrinsic point defects Y (dopant atoms, O, C...), surfaces, interfaces and extended lattice defects such as stacking faults and dislocations. Important for the irradiations at different temperatures is the stability of the formed point defect pairs. For each interaction one can write reactions as schematically shown in table I. All these possible reactions have different temperature dependent reaction constants k. In many cases however it is possible to determine which of them are the most important ones (e.g. O and C in low doped CZ silicon).

Table I: Basic point defect reactions in silicon.

Intrinsic reactions	Extrinsic reactions	Secondary reactions
$V + V \leftrightarrows VV$	$Y + V \leftrightarrows YV$	$Y_s + Y_i \leftrightarrows Y_sY_i$
$VV + I \leftrightarrows V$	$Y + I \leftrightarrows YI$	$Y_i + Y_i \leftrightarrows Y_iY_i$
$I + I \leftrightarrows II$	$Y_s + I \leftrightarrows Y_i$	$Y^1 + Y^2 + V \leftrightarrows Y^1Y^2V$
$II + V \leftrightarrows I$	$Y_i + V \leftrightarrows Y_s$	$Y^1 + Y^2 + I \leftrightarrows Y^1Y^2I$
	$YV + I \leftrightarrows Y$	$VV + VV \leftrightarrows VVVV$
	$YI + V \leftrightarrows Y$	

Influence of temperature of irradiation and re-irradiation

As mentioned above, the stability of the formed point defect complexes plays an important role in the extended defect nucleation and growth. Important to notice are the high stability of the A-centre and of the divacancy in CZ silicon, which both can survive temperatures up to 700K.
At temperatures above 400 K, both B_i and the B_iC_s pairs are unstable and no longer form an efficient sink for self-interstitials. During the low temperature re-irradiation of B doped silicon, a large number of B_s atoms are thus available as sinks for self-interstitials, leading to the dissolution of defects. The E-centre (VP) is unstable above 400K so that phosphorous atoms no longer form an efficient sink for vacancies leading to an increased divacancy formation. During re-irradiation at low temperature, these divacancies are sinks for self-interstitials leading to dissolution of extended defects. When the first irradiation is performed at 300K, the highly P doped areas form a strong sink for vacancies, leading to a supersaturation of self-interstitials and the creation of extended defects. During a second irradiation at 640K, the VP pairs are unstable and large numbers of vacancies are released leading to a dissolution of the defects which were formed in the highly doped areas during the first irradiation.

Influence of strain

Coulombic and elastic interaction are the two main types of interaction between dopants and intrinsic point defects. Calculations show that the Coulombic interaction is at least one order of magnitude smaller than the observed one which is thus dominated by elastic strain compensation. Italyantsev et al [16] observed a strong decrease of electron irradiation induced A-centre and divacancy formation in silicon doped with a high concentration of germanium ($3x10^{18}$ to $5x10^{20}$ cm^{-3}). As Ge is not electrically active, this behaviour can only be attributed to the strain resulting from the difference between the covalent radius of germanium and that of silicon.
Compressive strained areas will be sinks for self-interstitials while tensile stressed areas will be a sink for vacancies leading to a local supersaturation of self-interstitials and an enhanced extended defect nucleation. Close to the silicon oxide/silicon interface a reduction of the self-interstitial content will occur due to vacancy concentration build up leading to a shallow zone free of extended defects at the silicon surface as can be observed in figures 1-4 [7,9].

Acknowledgements

A. De Keersgieter kindly supplied the ion implanted silicon wafers and the dopant concentration profiles. A. R.-R. is indebted to the Spanish Ministry of Science and Education for his fellowship.

REFERENCES

1. I.G. Salisbury and M.H. Loretto, Phil.Mag.A 39, 317 (1979).
2. H. Asahi, R. Oshima and F.E. Fujita, Inst.Phys.Conf.Ser. 59, 454 (1981).
3. L.M. Brown and D. Fathy, Phil. Mag. B 43, 715 (1981).
4. F. Reynaud and B. Legros-de Mauduit, J.Microsc.Spectrosc.Electron. 9, 199 (1984).
5. L.I. Fedina and A.L. Aseev, phys.stat.sol. (a) 95, 517 (1986).
6. M. Hasebe, R. Oshima and F.E. Fujita, Jpn.J.Appl.Phys. 25, 159 (1986).
7. A.L. Aseev, L.I. Fedina, D. Hoehl and H. Bartsch, Solid State Phenomena 6&7, 235 (1989).
8. G.-C. Hua, R. Oshima and F.E. Fujita, in "Defect Control in Semiconductors", ed. K. Sumino, Elsevier Science Publishers B.V. (North-Holland), 535 (1990).
9. A. Romano, J. Vanhellemont, J. Morante and W. Vandervorst, Micron and Microscopica Acta 20, 151 (1990).
10. A. Romano-Rodríguez, J. Vanhellemont, A. De Keersgieter, W. Vandervorst and J. Morante, proc. GADEST 91, to be published in Solid State Phenomena, Sci-Tech Publications, (1991).
11. A. Romano, J. Vanhellemont and H. Bender, Ultramicroscopy 31, 183 (1989).
12. L.C. Kimerling, M.T. Asom, J.L. Benton, P.J. Drevinsky and C.E. Caefer, Materials Science Forum Vols. 38-41, 141 (1989).
13. P.J. Drevinsky, C.E. Caefer, L.C. Kimerling and J.L. Benton, in "Defect Control in Semiconductors", ed. K. Sumino, Elsevier Science Publishers B.V. (North-Holland), 341 (1990).
14. P.M. Fahey, P.B. Griffin and J.D. Plummer, Reviews of Modern Physics 61, 289 (1989).
15. V.I. Kuznetsov, T.A. Logvinenko, P.F. Lugakov and V.D. Tkachev, Sov.Phys.Semicond. 9, 491 (1975).
16. A. Italyantsev, L. Khirunenko, V. Mordkovich, E. Rubinova, E. Temper, V. Trunov, V. Shakhovtsov, L. Shpinar and V. Yashnik, Materials Science Forum Vols. 38-41, 1163 (1989).
17. L.S. Berman, N.A. Vitovskii, V.B. Voronkov, V.N. Lomasov and V.N. Tkachenko, Sov.Phys.Semicond. 24, 765 (1990).

Materials Science Forum Vol. 83-87 (1992) pp. 309-314

ATOMIC STRUCTURE OF THE INTERSTITIAL DEFECTS IN ELECTRON-IRRADIATED Si AND Ge

S. TAKEDA, S. MUTO and M. HIRATA
Department of Physics, College of General Education, Osaka University, Toyonaka, Osaka 560

ABSTRACT

Assuming that a self-interstitial chain is nucleated in the hexagon in the <110> projection of the diamond structure, the atomic model of the {113} planar interstitial defects , which has recently been obtained from the experiments [S. Takeda, Jpn. J. Appl. Phys. **30**, L639 (1991)], is described in detail. The steps of the planar defects, which are observed in cross-sectional electron microscopy in the present study, are described by the fault of the nucleation sites of the chains. Such faults are presumably due to the stress of other neighboring defects, the distribution of impurities and the thermal stress at elevated temperatures. Similar quasi-planar defects in many Si-based materials have a strong tendency to be on {113}, but do not exhibit the well-defined structure. In the present context, they are of self-interstitial type and have the disturbed structure of the strictly planar {113} defect due to the incoherent nucleation sites or to accommodation of the local strain.

1. Introduction

The planar {113} defect or rod-like defect is a well-known lattice imperfection in Si and Ge[1-2]. It was discovered *in situ* in a high energy electron microscope[1]. After some incubation period to form nuclei, the defects become visible in the electron-irradiated area. Under the succeeding irradiation, the defect occasionally changes to a perfect dislocation loop of interstitial-type on the same plane, and leaves no visible precipitates. It was, therefore, concluded that the nature of the defect was of self-interstitial type[1,2]. Since no aggregate of point defects on {113} has been reported in any elemental material except Si and Ge, the defect has been attractive for many researchers to find out an unknown configuration of self-interstitials of lower energy . Recent transmission electron diffraction[4, 5] and high-resolution transmission electron microscopy experiments[3, 5] have offered an advanced structural data concerning the planar defects, and an atomic model of the defects has been deduced from the experiment[3]. The model contains self-interstitials only and shows the unexpected arrangement of Si atoms without any dangling bond and chemical impurity. The interstitials constitute the hexagonal structure locally on the {113} plane.

Similar defects appear in various Si-based materials such as an oxygen-implanted and annealed Si wafer[6, 7]. The defects in such materials are not always strictly on {113}, but show a complicated shape. A part of the defects was at first analyzed to be coesite (SiO_2), and later re-analyzed to be hexagonal packing of the {111} planes of the diamond structure[7]. Recent careful analysis of the hexagonal diamond structure introduced by indentation has leaded the conclusion[8] that the hexagonal diamond structure is generated by the martensitic transformation and is not a thermodynamically equilibrium state. There has been no experimental evidence of the martensitic transformation in irradiated materials, so that there is another mechanism to induce the hexagonal diamond structure locally. The purpose of the present paper is to describe the structure of the {113} interstitial defect with assuming the nucleation of interstitial chains inside the hexagon in a <110> projection, and to show some experimental evidences of the atomic steps in the planar defect. The steps may be induced by the same origin to form the defect in complicated shape which tends to be on {113} and has the hexagonal structure.

2. Experimental

Experimental procedures were described precisely in the previous papers[3-5]. The defects were introduced at 250℃ in Ge and 400℃ in Si in a high-voltage electron microscope, accelerating voltage of which was 2MV. The irradiated specimens were observed in a 200kV-type conventional electron microscope (Jeol JEM2000EX II). Since the threshold electron energy for the displacement damage in Si is around 170kV, we operated the microscope down at 160kV in order to avoid the damage during the HRTEM observation. All micrographs in the paper were taken in a <110> incidence.

3. Structure of the {113} interstitial defects

A cross-sectional image of a {113} defect in electron-irradiated germanium is shown in Fig. 1. The image exhibits a similar image feature to that in Si[3], indicating clearly the defect structure is identical to that in Si. The locations of the self-interstitial chains are pointed by the arrows. The images of the defect in Ge which have been previously reported[5] show the reverse contrast, and hence the analysis of them has not been straight.

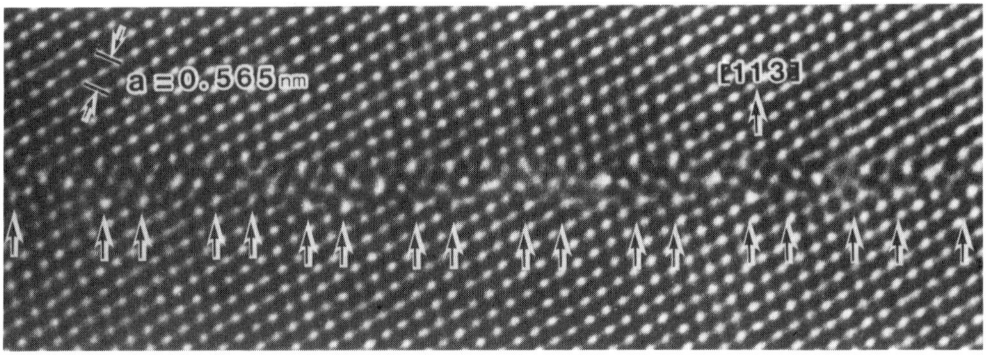

Fig. 1 A cross-section of a {113} defect in Ge observed by high resolution electron microscopy.

The model for the defect is described by the arrangement of the <110> zigzag chains of self-interstitials, which are sandwiched by the atomic model for the reconstructed {113} surface. The analogy of the surface structure to the {113} defect in bulk suggests the reason for self-interstitials to aggregate on {113}[3]. This is shown rather pictorially in Fig. 2. Figure 2(a) shows the <110> projection of the 1x1 reconstructed model for the surface with 'rebonding at steps'[9]. Even though a recent scanning tunneling microscopic study[10] has revealed that the clean surface exhibits an extended 3x2 reconstruction, in some part of which the 'rebonding at steps' atoms are seen, we assume here the reconstructed 1 x 1 surface. Suppose the two surfaces facing each other adhere as in Fig. 2(b). An atom with one dangling bond on the surface forms an inter-surface bond with an equivalent atom on the other surface. This hypothetical procedure results in the periodic arrangement of the 8-membered atomic rings as seen in Fig.2(b). A zigzag chain of self-interstitials in the <110> direction is then inserted in some of the 8-membered rings. After the rebonding and the relaxation of atomic positions, the final model is deduced and shown in Fig. 2(c). It should be stressed that the structure of the {113} defect is strictly planar and well defined. An interstitial atom belongs to a 6-membered atomic ring which constitutes the hexagonal arrangement.

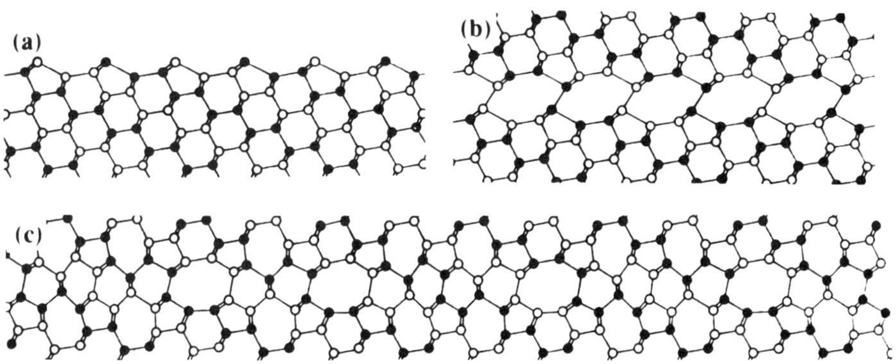

Fig. 2. A pictorial representation of the analogy of the {113} defect structure to the surface reconstructed structure. (a) The model for the 1x1 reconstructed {113} surface by Gibson et al.[9]. (b) The two reconstructed surfaces are facing each other so as to form the periodic arrangement of the 8-membered atomic rings. (c) The <110> zigzag chains of the self-interstitials are inserted in some of the 8-membered rings and the model of the defects[3] is reached.

4. Description of the atomic model on the assumption of the nucleation of the interstitial chains

Let us consider the nucleation of the interstitial defects. Even though impurities such as carbon contribute to the nucleation of the defects, there have been no firm experimental data of the nuclei of the defects. We assume here that the nucleation site is inside the hexagon in the <110> projection and denoted by the asterisk in Fig.3(a). In the figure, the bond between the atoms 1 and 7 and that

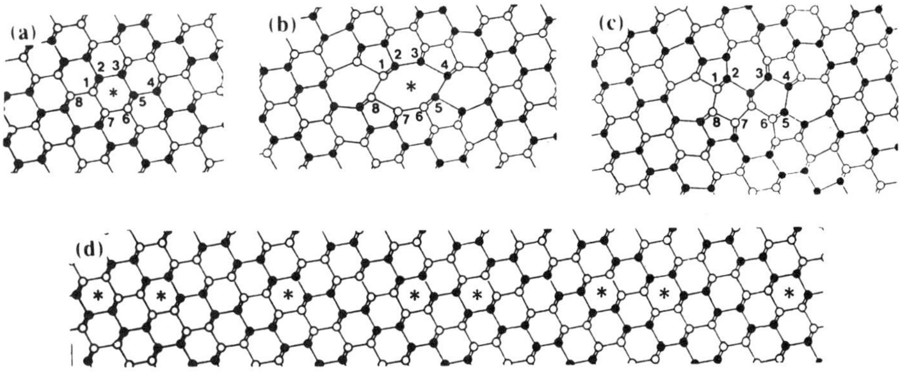

Fig. 3. Description of the model on the assumption of the nucleation of the interstitial chains. (a) The interstitial chain is assumed to nucleate in the hexagon denoted by the asterisk. (b) The intermediate defect with the 5-, 6-, 7 and 8-membered ring. (c) The minimum unit of the {113} defect. (d) The coherent nucleation sites on {113} denoted by the asterisks. The resultant defect structure is shown in Fig. 2(c).

between the atoms 3 and 5 are cut, and then new bonds form between the atoms 3 and 4 and between 7 and 8, resulting in the intermediate defect structure shown in Fig.3(b). The edge of the defect is obtained in a similar way. The formation of the intermediate defect is conservative; no vacancy or interstitial is required in the defect. The interstitial chain is inserted in the 8-membered atomic ring. The model shows the minimum unit of the {113} defect and is presumably for the line-shaped defects observed at higher temperatures. The nucleation of the chain at the coherent sites on a {113} plane indicated by the asterisks in Fig.3(d) results in the strictly planar {113} defect, which is shown in Fig.2(c).

The self-interstitials in the model are located at the lattice sites on the two extra {113} net planes which are inserted between the two existing planes, but the extra net planes are not fully occupied. During the unfaulting of the defect, the shrink of the defects has not yet been reported[1, 2]. Hence, additional self-interstitials may take part in the defects and induce the unfaulting[3]. One may see, after unfaulting, that the model of the edge consisting of 5-, 6- and 7-membered rings shown in Fig. 3(c) changes to the model of the edge dislocation without any dangling bond[11].

5. Atomic steps of the planar defects

The nucleation of the defects are probably affected by the impurity distribution, the stress due to the other neighboring defects and thermal stress at elevated temperatures. Under the circumstances, the nucleation sites of the chains may not be exactly on the {113} plane, and these nucleation sites of incoherence give rise to the defect structure with faults. Examples of such nucleation sites are shown in Fig. 4(a) and (b), and the corresponding atomic models are depicted in Fig.4 (c) and (d). The faulted structures are characterized by the atomic steps of the planar fault. The steps are 2a/11[113] and a/11[113] in height, respectively, in Fig. 4(c) and (d). The edge of the steps is assumed in the models along the <110> direction. Such faulted structures have been observed in the present study, and are shown in Fig. 5(a) and (b). The images fit well the model depicted in Fig. 4(c) and (d).

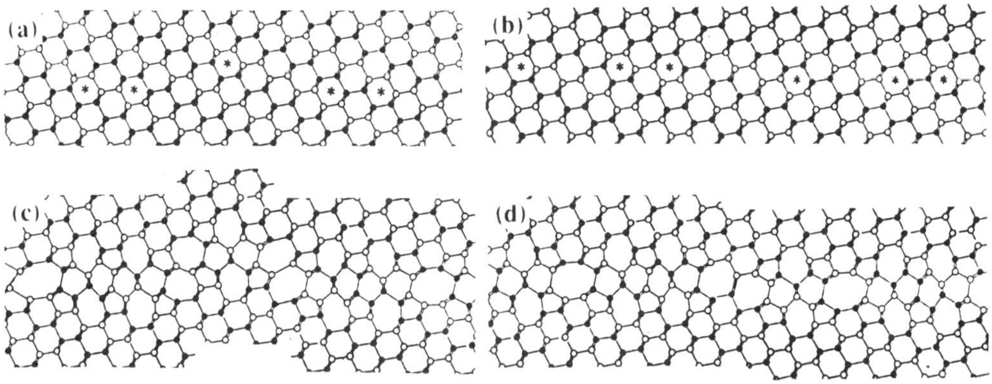

Fig. 4. The possible nucleation sites of incoherence are shown in (a) and (b). The atomic models corresponding to (a) and (b) are obtained in a similar ways as in Fig. 3 and depicted in (c) and (d), respectively.

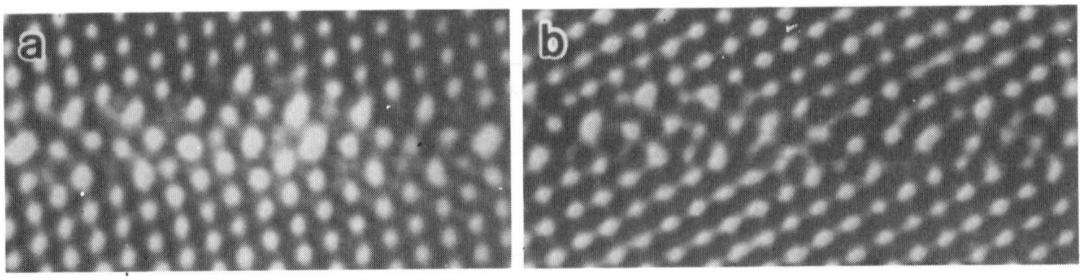

Fig. 5. Observed atomic steps in the {113} defects (a) in Si and (b) in Ge. The images (a) and (b) fit well the models shown in Fig. 4(c) and (d), respectively.

6. Quasi-planar defects in electron-irradiated materials

The quasi-planar defects in various shapes have been observed in thermally annealed, ion-implanted and electron-irradiated Si and Ge[7]. Figure 6(a) shows an image in electron-irradiated Ge. The defect A, enlarged image of which is shown in Fig.6(b), is a planar {113} defect with

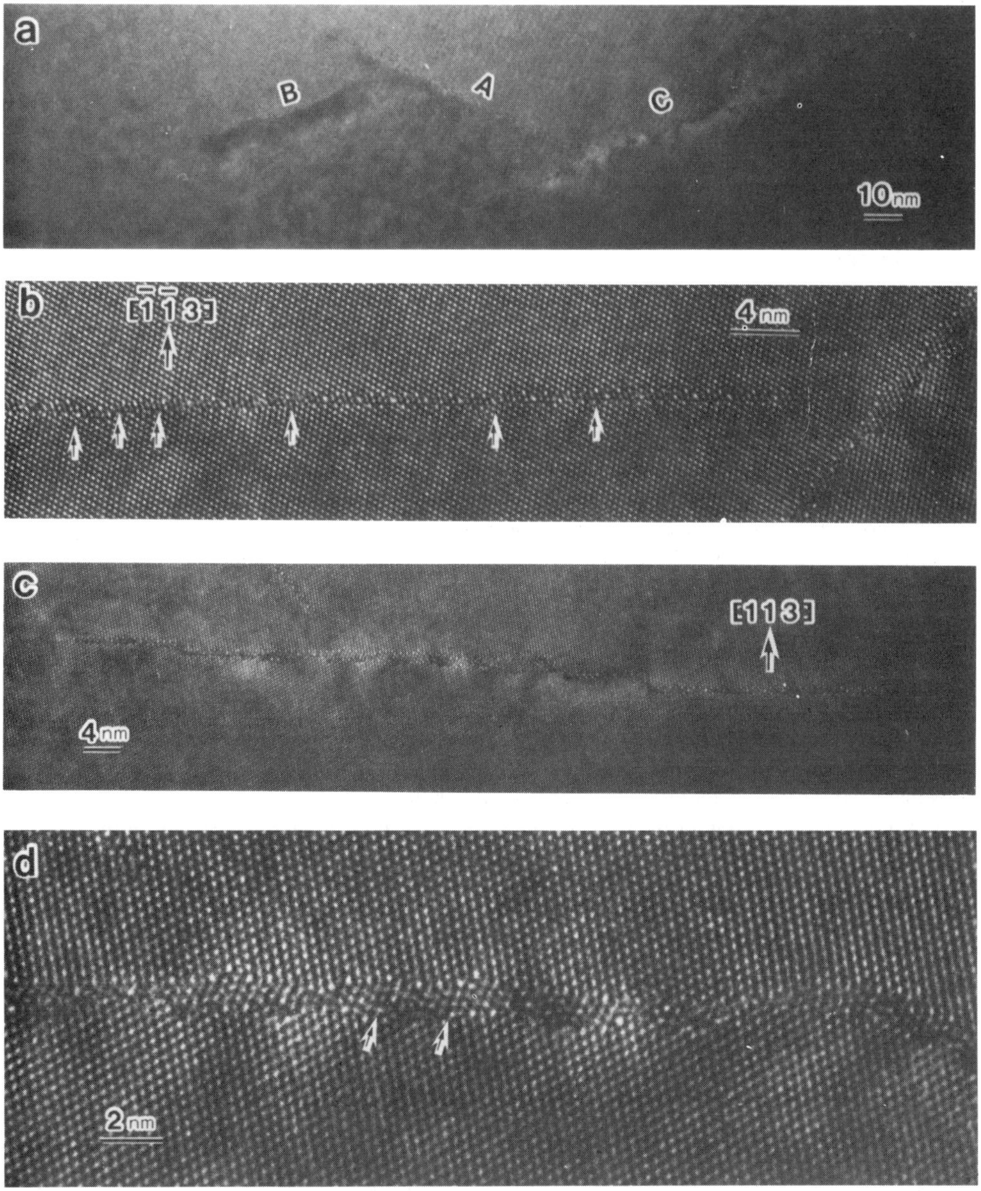

Fig. 6 Planar and quasi-planar defects coexist in irradiated Ge. (a) A lower magnified image. (b) An enlarged image of the planar defect. (c) A medium magnified image of the quasi-planar defect which has the planar part near the edge and exhibits rather complicated shape in the central part. (d) A higher magnified image of the central part of the quasi-planar defect shown in (c). A hexagonal arrangement and the stacking faults are seen. Note no clear habit plane is seen.

several steps of the defect plane. The steps are denoted by the arrows. The defects B and C in Fig. 6(a) near the defect A are not strictly planar and exhibit complicated shape. As seen in the medium magnified image (Fig. 6(c)), the defect C has the planar part near the edges and exhibits rather complicated shape in the central part. A higher magnified image of the central part of the quasi-planar defect is shown in Fig. 6 (d). One can trace the zigzag stacking of the {111} plane. A part of the complicated defects is certainly hexagonal structure, as suggested by Tan et al.[12] and Bourret[7]. The orientation relationship is described as <0001>hex.//<110>cubic and <0110>hex.//<110>cubic. The arrows in (d) indicate stacking faults in the hexagonal structure. The structure of the defect is, therefore, not well defined. The habit plane between the defect and the matrix is not on $\{115\}_{cubic}$, which is observed in an indented materials by the martensitic transformation[8]. The stress near the defect may induce the rebonding and the shuffling of the atomic position, and may be an origin of the hexagonal structure as well as the nucleation of the interstitial chains in incoherence.

7. Conclusions

The structure of the self-interstitial aggregate on {113} in Si and Ge is described in detail. The structure is well defined and strictly planar. The well defined structure may be disturbed to form complicated shape. The atomic steps of the planar defects are observed in the present study and they are explained in terms of the nucleation sites of incoherence. The part of the defects in complicated shape has the hexagonal structure. The orientation relationship between the hexagonal and cubic structures is the same as in the indented Si, but the habit planes are not well defined in irradiated materials. Therefore, as suggested by Pirouz et al.[6], the hexagonal structure in irradiated materials is probably not induced by the martensitic transformation. The hexagonal structure is, in our opinion, the disturbed structure of the well defined {113} defect due to the incoherent nucleation sites or to accommodation of the local strain.

ACKNOWLEDGMENTS

Electron irradiation was performed at the High-Voltage Electron Microscope Center of Osaka University. The authors are grateful to Prof. H. Mori of the Center for his useful comments. The authors thank Mr. K. Ibe of Jeol Co. for his invaluable help in electron microscopy. Part of this work was supported by the Grant-in-Aid from the Ministry of Education, Science and Culture.

REFERENCES

1. C. A. Ferreira-Lima and A. Howie, Philos. Mag. **34**, 1057 (1976).
2. I. G. Salisbury and M. H. Loretto, Philos. Mag. **39**, 317 (1979).
3. S. Takeda, Jpn. J. Appl. Phys. **30**, L639 (1991).
4. S. Takeda, S. Muto and M. Hirata, Jpn. J. Appl. Phys. **29**, L1698 (1990).
5. S. Takeda, M. Hirata, S. Muto, G.C. Hua, K. Hiraga and M. Kiritani, Ultramicroscopy (1991).
6. P. Pirouz, U. Dahmen, K. H. Westmacott and R. Chaim, Acta metall. mater. **38** (1990) 329.
7. A. Bourret, Inst. Conf. Ser. **87**, 39 (1987).
8. U. Dahmen, K. H. Westmacott, P. Pirouz and R. Chaim, Acta metall. mater. **38** (1990) 323.
9. J. M. Gibson, M. L. McDonald and F. C. Unterwald, Phys. Rev. Lett. **55**, 1765 (1985).
10. J. Knall, J. B. Pethica, J. D. Todd and J. H. Wilson, Phys. Rev. Lett. **66**, 1733 (1991).
11. J. Hornstra, J. Phys. Chem. Solids. **5**, 129 (1958).
12. T. Y. Tan, H. Foll and S. M. Hu, Philos. Mag. **A44**, 127 (1981).

Materials Science Forum Vol. 83-87 (1992) pp. 315-320

INTERACTIONS BETWEEN DEFECTS DURING THE ANNEALING OF CRYSTALLINE SILICON

Gordon Davies, Kwok Tat Kun and Tim Reade

Physics Department, King's College London, Strand, London WC2R 2LS, UK

ABSTRACT

We show that even when annealing at one fixed temperature, the decay times of the di-carbon 'G' center in different samples span an order of magnitude. This effect is caused by interactions between the G centers and substitutional carbon atoms (which increase the decay time) and between the G centers and radiation damage complexes (which decrease the decay time). The variation is described quantitatively over five orders of magnitude in the decay time.

1. INTRODUCTION

The di-carbon G center is one of the most thoroughly investigated of "point defects" in silicon[1]. It may be observed by optical absorption, photo- or cathodo-luminescence, photoconductivity, optically detected magnetic resonance or paramagnetic resonance. It is known[2] to be formed via a metastable state when a substitutional carbon atom C_s traps an interstitial carbon atom C_i. Despite the wealth of information available on the G center, its destruction by thermal annealing is uncertain. The first measurement[3] of the annealing, using neutron-irradiated samples, gave an activation energy of 1250 meV. The second study[4] used Czochralski silicon which had been irradiated with 10^{18} cm^{-2} 3-MeV electrons. The strength of the luminescence in the G band decreased exponentially as

$$[G] = [G_o]\exp(-t/\tau), \quad \text{with} \quad \tau = \tau_o \exp(E/kT). \tag{1}$$

The time constant factor τ_o and its activation energy E, measured at annealing temperatures of 195 °C to 225 °C, were found to be

$$\tau_o = 8 \times 10^{-11}\text{s}, \quad E = 1340 \text{ meV}. \tag{2}$$

The third measurement of the annealing rate[5] used four float-zone samples with an initial concentration $[C_s]$ of carbon atoms of 5×10^{16} cm^{-3}, which were irradiated with 5×10^{16} cm^{-2} 2-MeV electrons and annealed at between 100 and 200 °C. Decay constants were

$$\tau_o = (4 \mp 3) \times 10^{-14} \text{ s}, \quad E = 1700 \pm 50 \text{ meV}. \tag{3}$$

Each of these studies was made on a set of samples which was simply annealed at different temperatures to give the activation energy. We show here that even when annealing at one fixed temperature, the decay times from different samples span an order of magnitude. This effect is caused by interactions between the G centers and the carbon atoms (which increase the decay time) and between the G centers and radiation damage complexes (which decrease the decay time).

2. EXPERIMENTAL ASPECTS AND SAMPLES.

Samples 2 mm thick were mounted in gallium/indium eutectic mixture on a water-cooled copper block and were irradiated with 2 MeV electrons at a typical beam current of 20 μA cm^{-2}. Holes of 1 mm diameter drilled in trial samples showed that the bulk temperature during the irradiation was 100 ± 5 °C. At these temperatures the C_i are highly mobile[6], allowing formation of the G centers, while the G centers themselves are very stable. The G absorption line was measured using a dispersive spectrometer with the sample mounted in the full white light of a tungsten strip lamp and immersed in liquid helium pumped below the λ point. Checks on the accuracy of the spectroscopy could be made by ensuring that the silicon isotope structure in the G line was being observed in the correct relative proportions[7]. For temperatures up to 230 °C, annealing was carried out with the samples wrapped in aluminum foil and immersed in a temperature-stabilised silicone oil bath, and at 260 °C with each sample secured by Al foil to the tip of a temperature probe which was plunged into a furnace. All the float-zone silicon slices used here had oxygen concentrations less than half the carbon concentrations as measured by optical absorption in the local vibrational modes of the impurities. Similarly the concentrations of electrically active shallow donors and acceptors were in all cases less than 5×10^{14} cm^{-3}.

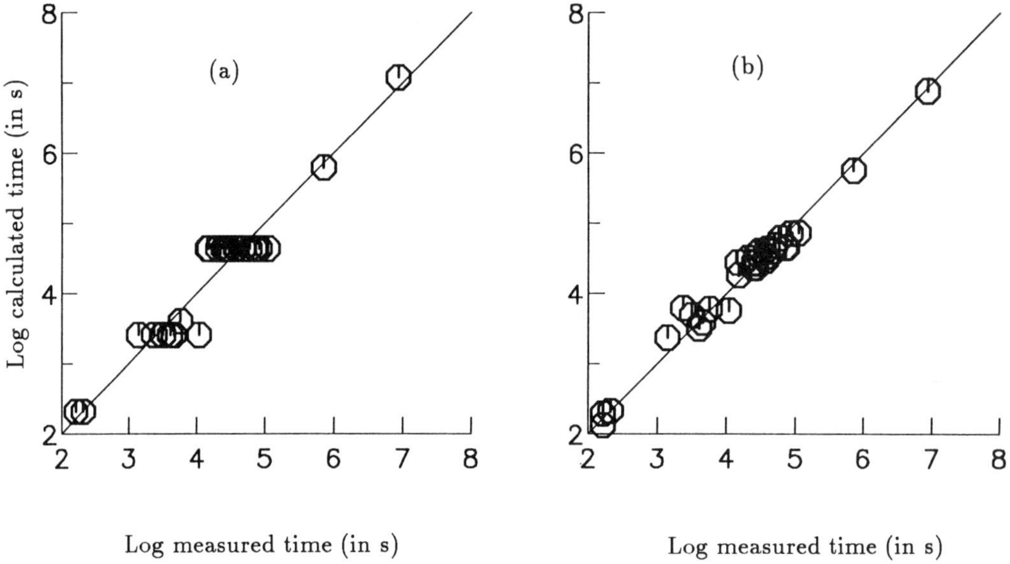

Fig. 1 Points compare the measured decay times with the calculated decay times for 31 float-zone samples. Ideally the points should lie on the diagonal lines. In (a) the calculated decay times use only the breakup mechanism, $\tau = \tau_b \exp(E_b/kT)$. In (b) the decay times are calculated using Eqs. (11) and (12).

3. ANNEALING STATISTICS

The absiscae of figure 1a show the decay times measured for 31 samples at temperatures between 150 and 260 °C. The ordinate of each point has been calculated allowing only for the temperature, as in equation 1, with

$$\tau_o = 1.1 \times 10^{-16} \text{ s}, \quad E = 1934 \text{ meV}. \tag{4}$$

The reason for these values is given later; the important point is that variations by an order of magnitude can be observed at each temperature for different samples.

To investigate this variation we examine the dose-dependence of the G center's annealing kinetics in three sets of samples, cut respectively from single slices of float-zone silicon with initial concentrations of substitutional carbon of 1.1, 3.3 and 16 $\times 10^{16}$ cm^{-3}. The samples were irradiated with between 5×10^{16} and 1×10^{18} cm^{-3} 2 MeV electrons and annealed at the same temperature of 200 °C. The decay *rates* increase linearly with radiation dose R for the three groups of samples annealed at the same temperature, (figure 2a) giving

$$1/\tau = 1/\tau_e + cR, \tag{5}$$

where c is evidently the same for the three values of $[C_s]$. Measurements at different temperatures show c to be thermally activated. Assuming that

$$c = c_o \exp(-E_r/kT), \tag{6}$$

$c_o \approx 2.5 \times 10^{-11} \text{cm}^2\text{s}^{-1}$ and $E_r \approx 1120$ meV. These values are refined later.

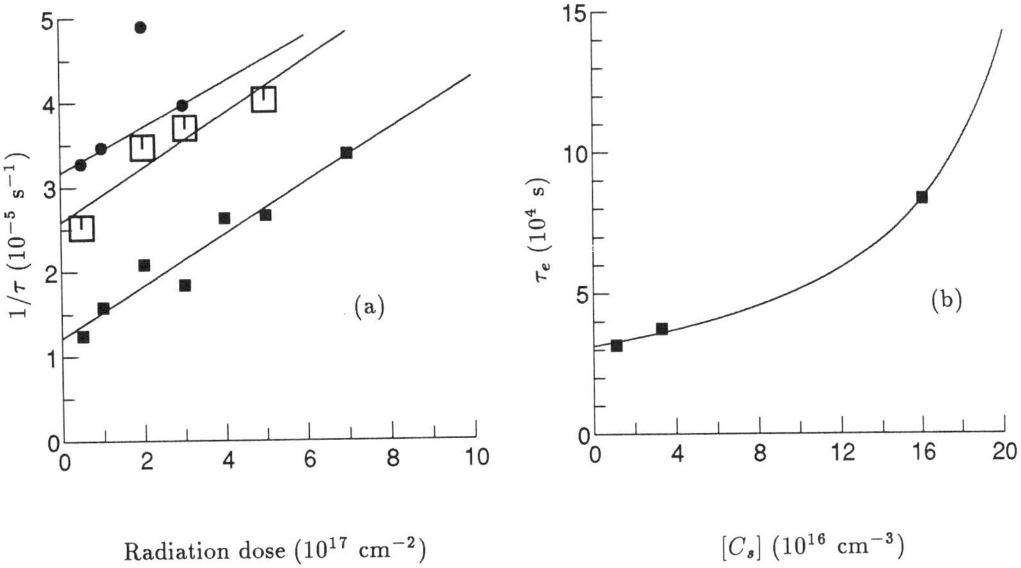

Fig. 2 (a) The decay rates $1/\tau$ for three sets of samples with, from top to bottom, $[C_s] = 1.1,\ 3.3,\ 16 \times 10^{16}$ cm^{-3} as functions of radiation dose R for annealing at 200 °C. (b) The decay times τ_e in the limit of $R \to 0$ for the three sample sets of (a). The line is Eq. (10).

The same data show that for a given radiation dose the rate of annealing slows down with increasing carbon concentration $[C_s]$. The values τ_e of τ extrapolated to zero dose are shown on Fig. 2b; the line will be justified in Sec. 4. It follows from the dependence of the decay rate on $[C_s]$ that inhomogeneous distributions of carbon will produce complications in the annealing kinetics. In some of our samples we find that the G centers decreased as the sum of two exponentials (Fig. 3a) :

$$[G] = [G_{oa}] \exp(-t/\tau_a) + [G_{ob}] \exp(-t/\tau_b). \tag{7}$$

This form of decay can occur if $[G_{oa}]$ is the initial concentration of G centers in one environment which decay with a time constant τ_a and $[G_{ob}]$ is the concentration of G centers in a second environment which decay with a time τ_b. Confirmation that two independent environments exist in these samples is given by the behavior of the width (*i.e.* the full width at half height) of the G line. The width of a zero-phonon line in silicon is determined by the various random strains experienced by the optical centers. The widths Γ in these inhomogeneous samples increase during the annealing (Fig. 3b), from an initial value Γ_o to a final value (near the end of the annealing) of Γ_f, and can be simply described by a saturating exponential

$$\Gamma = \Gamma_o + (\Gamma_f - \Gamma_o)[1 - \exp(-t/\tau)] \tag{8}$$

with a time constant τ equal to the fast decay time of the G absorption. The width data establish that the G centers in regions of low $[C_s]$, and hence low strain fields, anneal more rapidly than those in regions of high $[C_s]$. It can be shown that the increase in width is quantitatively consistent with this interpretation.

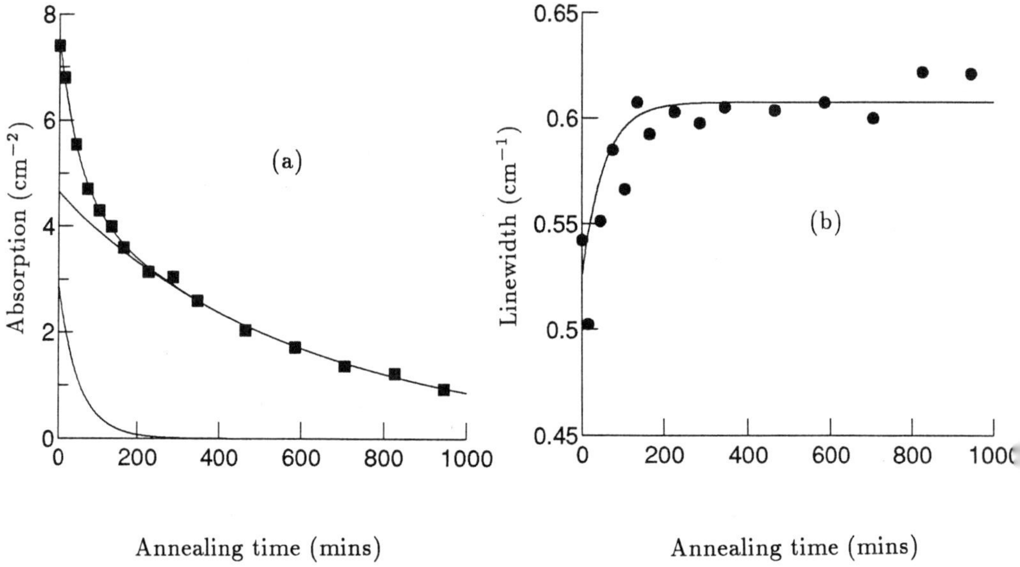

Fig. 3 (a) shows that the decay of G absorption in a sample annealed at 230 °C can be described by the sum of two exponential decays with time constants $\tau_a = 0.9$ hrs and $\tau_b = 10$ hrs. The width of the G line in the same sample increases in the form of Eq. (8) with the 0.9 hr time constant.

During the thermal destruction of the G center, a small increase is observed near the beginning of each isothermal anneal in about half of all our samples. The peak produces about a 10% increase in the G line relative to a smoothly varying background, and occurs especially in samples irradiated with smaller doses. The time τ_p at which the peak occurs varies by more than three orders of magnitude in our samples (figure 4a) and obeys

$$\tau_p = \tau_{p0} \exp(E_p/kT), \ \tau_{p0} = 2.5 \times 10^{-6} \text{ s}, \ E_p = 890 \text{ meV}. \tag{9}$$

4. DISCUSSION

The pre-factor τ_o (Eq. 4) has a magnitude consistent with very few atomic jumps taking place during this mode of destroying the G center — *i.e.* the G center breaks up during this mode of annealing. Since the core of the G center consists of two carbon atoms and one silicon atom occupying two silicon lattice sites, breakup may result in the emission of either a carbon interstitial atom or a silicon interstitial atom. We have measured an increase in the concentration of substitutional carbon indicating that carbon is ejected from the G center during breakup.

Equation 5 shows that with increasing radiation dose two independent decay mechanisms are active — the G centers may break up at a rate $1/\tau_e$, or they may be destroyed by a process whose rate is proportional to the concentration of radiation damage centers in the neighborhood. The term c is not strongly dependent on the carbon concentration (Fig. 2a) but varies with temperature (Eq. 6).

In the limit $R \to 0$, when ideally the only crystalline imperfections are the C_s atoms and a very small number of G centers and vacancy-related centers, the breakup of a G center will liberate a C_i atom which has a finite probability p of being trapped at another C_s atom to form a new G center. The process of breaking up and reforming does not change the total G absorption. Since the transit time of the C_i atom between the C_s atoms is small, the retrapping lengthens the observed mean decay time of the G centers to

$$\tau_e = \tau_b/(1-p). \tag{10}$$

The retrapping process cannot occur when $[C_s] \to 0$, identifying τ_b as the intrinsic breakup time-constant at the relevant temperature. It is reasonable to assume that the probability p of retrapping is proportional to $[C_s]$. The line on Fig. 2b is the least squares fit of Eq. (10) to the three points, giving preliminary estimates of $\tau_b = 3.14 \times 10^4$s and $p = c'[C_s]$ with $c' = 3.9 \times 10^{-18}$cm^3. As long as the capture radii and the species of the competing traps are independent of temperature, or vary in the same way, c' is independent of temperature.

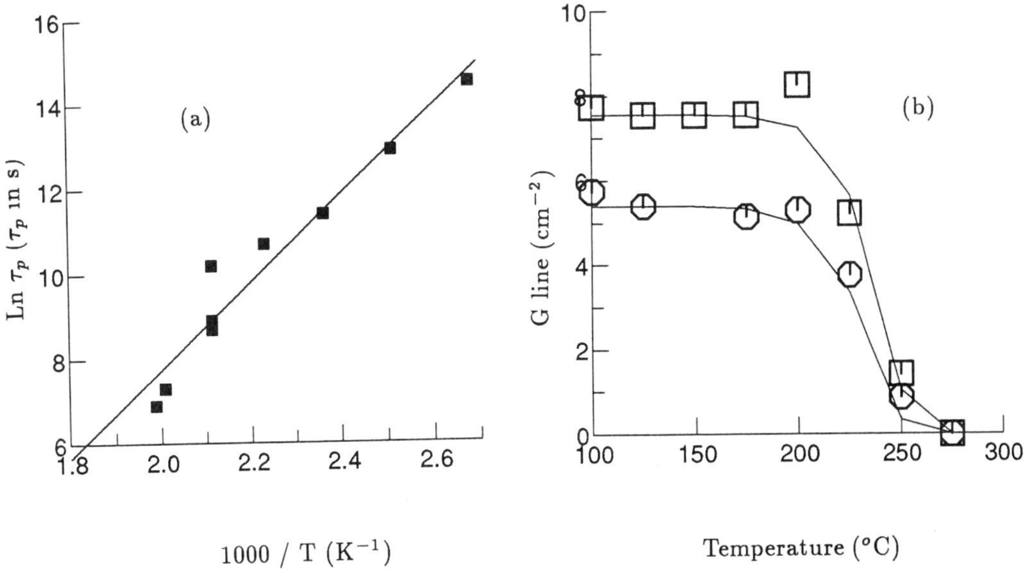

Fig. 4 (a) Arhennius plot of the time τ_p at which a small growth occurs in the G absorption; the line is Eq. (9). (b) Measured datum points for the isochronal annealing (in 20 min stages) for two samples with initially $[C_s] = 1.6 \times 10^{17}$ cm^{-3} and irradiated with 4×10^{17} (upper curve) and 1×10^{18} cm^{-2} 2-MeV electrons. The lines join points calculated at 25 °C intervals using τ_{pred} of Eq. (11).

We now have three elements in the destruction of the G center. The thermally activated breakup has a temperature dependence $\tau_b = \tau_{bo} \exp(E_b/kT)$ in the limit of low carbon concentration $[C_s]$ and small radiation dose R. This time is lengthened to $\tau_b/(1 - c'[C_s])$ by multiple retrapping among the C_s atoms, Eq. (10). Finally, the decay rate is enhanced by cR, with $c = c_o \exp(-E_r/kT)$, Eqs. (5) and (6). Combining these we predict a decay time τ_{pred} of

$$\frac{1}{\tau_{pred}} = \frac{(1 - c'[C_s])}{\tau_{bo} \exp(E_b/kT)} + c_o R \exp(-E_r/kT), \tag{11}$$

where preliminary estimates have been made for the five parameters τ_{bo}, E_b, c', c_o, and E_r. A least squares fit to the 31 decay times of figure 1 gives

$$\tau_{bo} = 1.1 \times 10^{-16}\ \mathrm{s}^{-1},\ E_{bo} = 1934\ \mathrm{meV},\ c' = (3.1 \pm 0.5) \times 10^{-18}\ \mathrm{cm}^3,$$
$$c_o = (1 \pm 0.5) \times 10^{-9}\ \mathrm{cm}^2\mathrm{s}^{-1},\ E_r = 1254 \mp 30\ \mathrm{meV}. \qquad (12)$$

Figure 1b demonstrates the fit by plotting the measured decay time against the predicted decay time for each of the 31 samples. The data lie close to the diagonal line, where they would be for perfect agreement; the correlation coefficient between the calculated and measured decay times is 0.99.

E_r is the diffusion energy of a radiation-damage species which destroys a G center by migrating to it. The value $E_r = 1250$ meV agrees with the first measurement of the destruction rate of the G center[3], suggesting that in that work the dominant mode of destruction was this extrinsic process. E_r is close to the diffusion energy (of 1280 meV) which has been recently reported for the divacancy[8].

In general terms the growth process, Eq. (9), is possibly caused by a Si_i being liberated into the lattice by the annealing, creating a C_i atom and hence a new G center. The growth of the G centers is then diffusion limited with, from Eq. (9), an activation energy E_p equal to the migration energy of the C_i atoms[6] and a pre-factor τ_{p0} consistent with long-range migration of a C_i atom to a surviving C_s atom.

Using Eqs. (11) and (12) it is now possible to predict explicitly the behavior of the G centers during isochronal annealing. On figure 4b the intensity remaining after each anneal is shown by the apex on the line at that anneal temperature. During isochronal annealing an increase (by of the order of 10 %) frequently occurs at 200 °C. This increase requires a growth process with a time constant which is presumably comparable with the 20 mins anneal time. The peaks observed in isothermal annealing have similar magnitudes but at 200 °C occur after 120 ± 60 mins (Eq. 9). The increase observed during isochronal annealing is more rapid than we expect from the isothermal data.

This study illustrates how the annealing kinetics of this very simple defect are determined by interactions between the defect and impurities in the lattice and by defect/radiation-damage interactions. To reach a quantitative understanding of the annealing kinetics of a defect it is clearly not sufficient to examine only a few samples.

This work was supported by the Science and Engineering Research Council.

References

1. L.W. Song, X.D. Zhan, B.W. Benson and G.D. Watkins, Phys. Rev. **B42**, 5765 (1990), and references there-in.

2. G. Davies, E.C. Lightowlers, M.F. Thomaz and J.G. Wilkes, Semicond. Sci. Technol. **3**, 608 (1988).

3. V.D. Tkachev and A.V. Mudryi in *Radiation effects in semiconductors 1976*, edited by N.B. Urli and J.W. Corbett (Institute of Physics, London, 1977) Inst. Phys. Conf. Ser. **31**, 231 (1977).

4. E.H. Wong and B.G. Streetman J. Appl. Phys. **42**, 5882 (1971).

5. G. Davies and Kwok T.K. Semicond. Sci. Technol. **4**, 327 (1989).

6. G.D. Watkins and K.L. Brower, Phys. Rev. Lett. **36**, 1329 (1976) and S.P. Chappell, G. Davies, E.C. Lightowlers and R.C. Newman, Mat. Sci. Forum **10-12**, 109 (1987); however a smaller migration energy of 720 to 750 meV has recently been reported by L.W. Song and G.D. Watkins, Phys. Rev. **B42**, 5759 (1990).

7. K. Thonke, H. Klemisch, J. Weber and R. Sauer, Phys. Rev. **B24**, 5874 (1981).

8. B.G. Svensson, K. Johnsson, D.-X. Xu, J.H. Svensson and J.L. Lindström, Radiat. Effects and Defects in Solids **111**, 439 (1989).

Materials Science Forum Vol. 83-87 (1992) pp. 321-326

AN EFFECT OF THE BEAM CURRENT AND ENERGY OF FAST ELECTRONS ON THE PRODUCTION RATES OF A-CENTRES AND DIVACANCIES IN n-Si

V.V.EMTSEV, P.M.KLINGER AND T.V.MASHOVETS
A.F.Ioffe Physico-Technical Institute, 194021 Leningrad, USSR

ABSTRACT

The production rates of A-centres and divacancies in Cz-Si were studied as a function of the beam current and energy of fast electrons. It has been found that all the parameters of pulsed electron irradiation including the duty cycle of accelerators should be taken into account while studying the production rates of secondary defects as well as monitoring the production rate of primary defects.

1. Introduction

So far we still do not have clear evidence showing the true role of the irradiation intensity in the defect production in Si. Irradiations of Si with fast electrons of 1-2 /1-4/ and 10 MeV /5/ gave only fragmentary information. Moreover, the data seem to be at variance indicating either a decrease /1-3/ or an increase /4/ in the production rate of secondary defects (η) with increasing intensity (J) or even nonmonotonic behaviour of η(J) /5/. According to /5/, an explanation of such a peculiar dependence of η on J may be offered on the belief that the irradiation intensity affects essentially the annihilation rate of Frenkel pairs produced as primary defects. Another way of argueing is presented in /1,4/ giving attention to the secondary defect production.
In the present paper we made a detailed examination of the production rates of A-centres (η_A) and divacancies (η_{VV}) in Cz-Si as a function of the energy E_{el} and intensity J_{el} of pulsed electron irradiation. Two aspects of this question generate a profound interest. Firstly, such irradiations whose parameters can be varied over a wide range are intensively exploited in the defect study. Secondly, the formation of A-centres follows a simple quasichemical reaction where one component (oxygen) available in a high content is electrically inactive, the latter allows to prevent possible complications associated with the charge states of interacting defects. Because of this, η_A is commonly used as a monitor for the production of primary defects, i.e. free vacancies V and self-interstitials I (after the dissociation of Frenkel pairs introduced during irradiation), see for instance /6-8/.

2. Experimental

A commercial Cz-Si crystal doped with P in a content of $(4\div 7)\cdot 10^{14}$ cm^{-3} was used in this study.
Irradiations were perfomed with pulsed electron beams from three accelerators whose parameters (the pulse length τ and frequency f) are as follows: τ =237 μs and f=468 pps for E_{el}=0.9 MeV (an effective resolution of $\lesssim$0.1 MeV), τ=3.7 μs and f=468 pps for E_{el}=3.0 MeV as well as τ =4.5 μs and f=200 pps for E_{el}=4.4, 5.5

and 7.0 MeV (an effective resolution of $\lesssim 0.2$ MeV). The integrated flux varied from 10^{14} to a few 10^{15} cm^{-2} for the most part. The samples were cooled by an air blast during irradiation and the sample temperature was in the range of $50° \div 70°$C.
The temperature dependences of the electron concentration n(T) over the range of 20-300 K were taken for the samples prior to and after irradiation. A computerized analysis of n(T) with the aid of the electroneutrality equation /9/ made it possible to determine the total concentration of donors at E_c-0.044 eV (phosphorus) and the concentrations of A-centres (the acceptor level at $\approx E_c$-0.16 eV) and divacancies (the acceptor level at $\approx E_c$-0.23 eV), separately. For all samples studied in this work A-centres and divacancies dominate over other radiation-induced acceptors absolutely; prior to irradiation all samples were weakly compensated, the compensation ratio being less than 0.1.

3. Experimental results and discussion

Below we outline the main experimental findings.
i) As would be expected, the concentration of shallow donors at E_c-0.044 eV remains practically unaltered for the E-centre formation is strongly suppressed in Cz-Si, in a sharp contrast to n-Si(FZ) where a substantial loss of the shallow donor states was observed following electron irradiation /10/. The electron concentration decreases with increasing flux due to the formation of A-centres and divacancies. The fraction of divacancies in compensation of the electron conductivity does not exceed 25% of that of A-centres.
ii) In general, the production rate of A-centres decreases as the irradiation intensity J is increased (Fig. 1). However, there are the intensity regions where η_A (J) is practically constant.
iii) As can be seen from Fig.1, the region of J where η_A decreases is sensitive to the pulse length and frequency.
iv) In general, the production rate at lesser J increases with increasing energy of fast electrons (Fig. 2).
v) The production rate of divacancies in n-Si as well as p-Si (η_{VV}) turned out to be independent of the irradiation intensity and increases with increasing E_{el} (Fig. 1 and 2); in p-Si(Cz) with a boron content of $4 \cdot 10^{14}$ cm^{-3} the defects are identified by their donor levels at $\approx E_V$+0.21 eV. It should be noted that η_{VV} observed was found to be appreciably less than the calculated one using a threshold energy 80 eV, i.e. twice as large when compared to the threshold energy for displacement of the lattice atom ($E_d \approx 40$ eV /12,13/, see also below).
For divacancies η_{VV} being independent of J means that they are produced in primary events for the most part, i.e. through displacement of two neighbouring host atoms provided the energy transfer is enough for such kick-out. Thinking in this way, one should take into consideration a possible intermediate configuration of the divacancy anchored at its parent self-interstitial(s) (so-called non-reorientable divacancies /14,15/). These complex defects are unstable at room temperature and disappear via annihilation /15/, only a small fraction of divacancies of the normal type (i.e. being reorientable) are produced during irradiation.
As the energy of fast electrons approaches 1 MeV, the direct production of divacancies becomes minor (Fig. 2) and another mechanism of their production - via association of two migrating

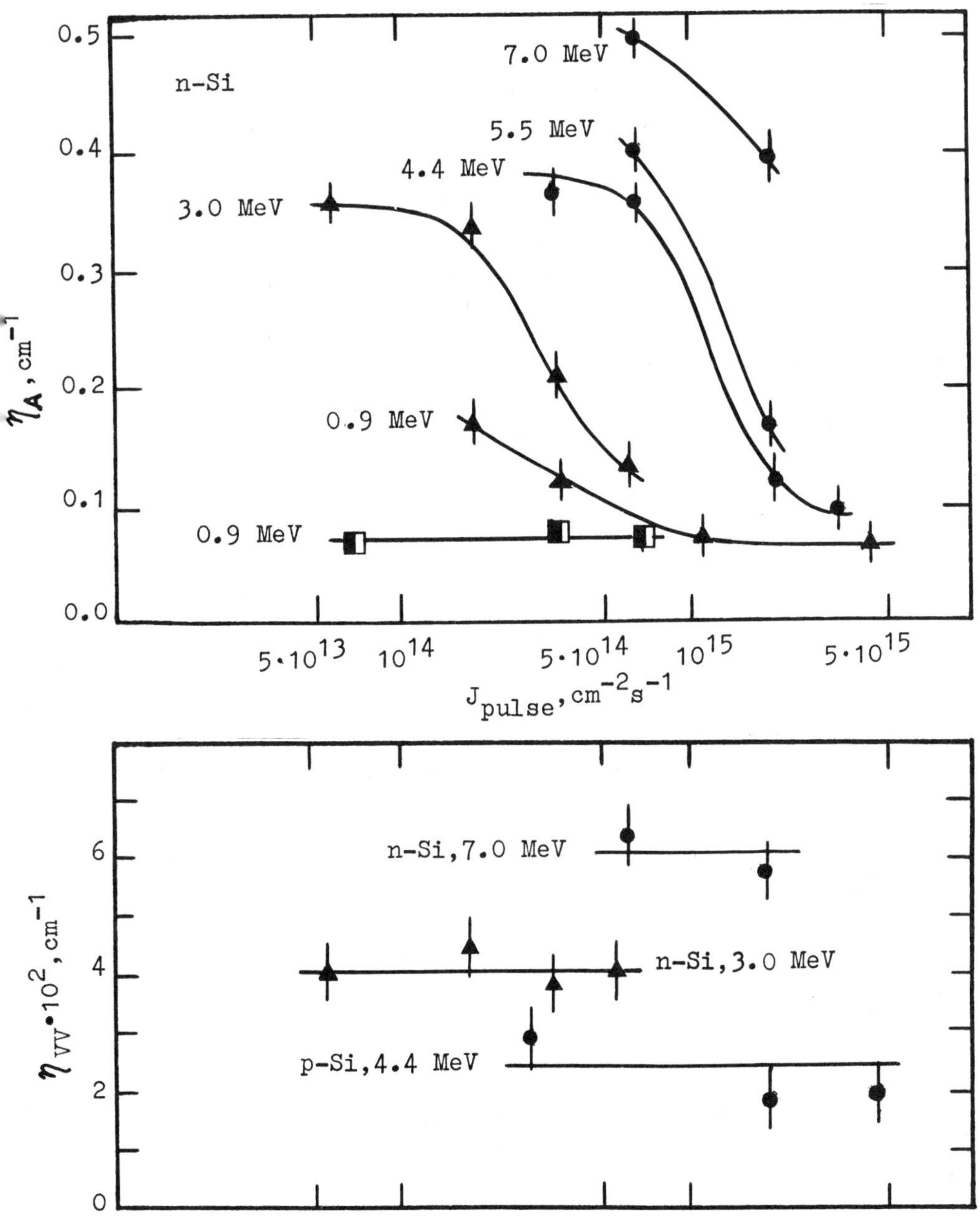

Fig.1 The production rates of A-centres and divacancies in Cz-Si as a function of the irradiation intensity. η_A at 0.9 MeV is taken from /1/; η_{VV} in p-Si(Cz) at 4.4 MeV is taken from /11/. The pulse length (τ) and duty cycle (θ) of the accelerators used: τ =237μs and θ=1.1·10^{-1} (squares), τ =3.7 μs and θ = 1.7·10^{-3} (triangles), τ =4.5μs and θ=9·10^{-4} (circles).

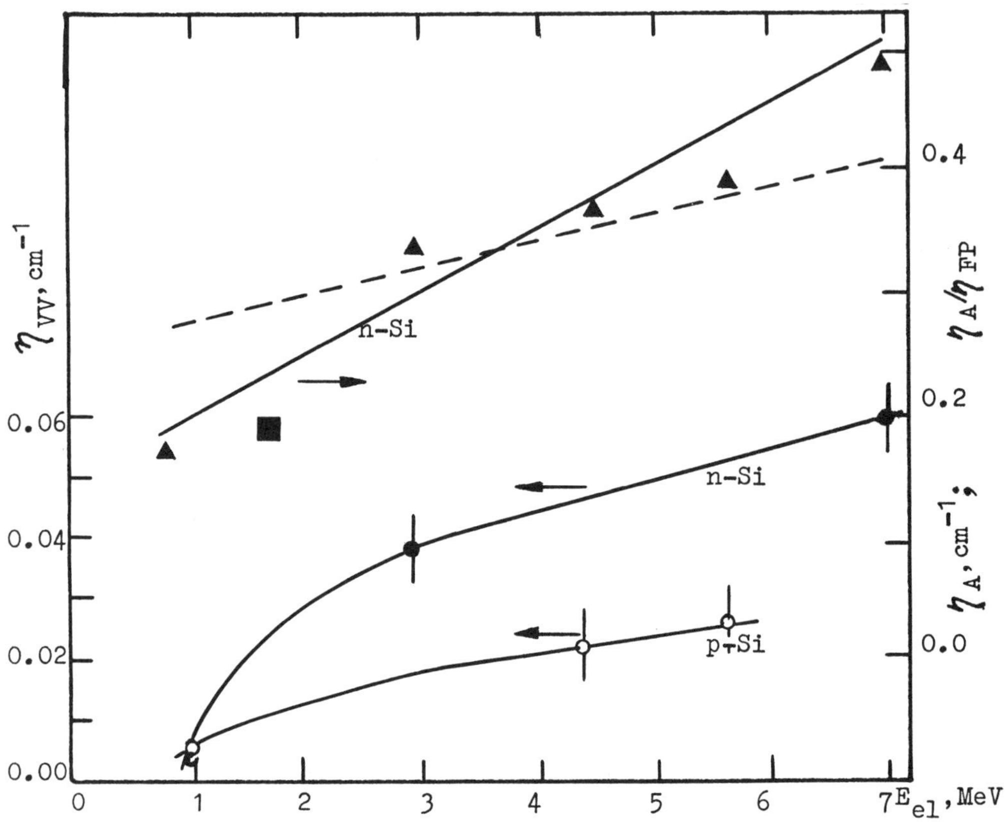

Fig.2 The production rates η_A (triangles), η_A/η_{FP} (dashed line) and η_{VV} (circles) in Cz-Si as a function of the energy of fast electrons. η_A in n-Si(Cz) at 100 K (square) is taken from /7/; η_{VV} in Cz-Si at 0.9 MeV is taken from /19/.

vacancies - is operating. As a consequence, the production rate of divacancies for $E_{el} \lesssim 1$ MeV is claimed to be dependent on the irradiation intensity /19/. In much the same way the indirect production of divacancies may be expected for high-resistivity FZ-Si irradiated with fast electrons of $E_{el} > 1$ MeV.

Now we turn to the production of A-centres.

Rough estimate will suffice to show what scale should be used while considering the lifetime of a Frenkel pair whose components are two or three lattice spacings apart. Primary defects V and I being characterized by a very low activation energy for motion (a few tenths eV) will walk that distance in $\tau_{FP} \ll 10^{-8}$s at $T \approx 70°C$. This time is too short for primary defect recharging even though high excitation conditions can be realized during the pulse (for all judgements we have taken the parameters commonly used for diffusion and recombination processes). This implies that variations in the beam current should have a strong impact on interactions of V and I with impurities. Of course, the conclusion we arrived at is reasonable under appropriate conditions. A certain change in conditions (say, by stepping down the temperature) may have also a pronounced effect on the fate of Frenkel pairs and, thus, on the production of secondary defects, see /16/.

In a similar way, one can show that all the vacancies coming from dissociated Frenkel pairs have enough time to be captured by oxy-

gen during the pulse. This is why we have plotted the production rate η_A versus the irradiation intensity J_{pulse} relating to the duty cycle of pulsed electron beam (Fig.1).
Now we are going on to the problem of radiation-induced annealing of A-centres. Thermal annealing of A-centres during the pulse may be ruled out. Actually, examination of Fig.1 discloses an interesting fact: for fast electrons of 3.0 and 4.4 MeV (with nearly the same energy loss) the dissipated power during the pulse (and, therefore, the peak temperature) is expected to be practically the same at $J \approx 5 \cdot 10^{14}$ $cm^{-2}s^{-1}$; however, we observed a significant difference in η_A. To adduce an argument, we refer to the study of A-centres produced with fast electrons of 1.7 MeV in n-Si(Cz) at $T \lesssim 100$ K /7/: under such conditions thermal annealing of A-centres is eliminated and the production rate η_A measured correlates well with our data (Fig. 2).
Another possible way to reduce η_A is radiation-induced annealing and the most probable candidate should be carbon. These substitutional impurity atoms C_s interact with self-interstitials I giving rise to the formation of interstitial carbon C_i (C_s+I→C_i). Similar to the production of A-centres the latter reaction is thought to be also produced during the pulse because of high mobility of self-interstitials. Interstitial carbon atoms diffusing through the lattice can, in turn, be captured by other impurities and defects (O_i, C_s, A-centres and so on). However, it is well known that thermally activated motion of C_i is relatively slow /17/ and a rough guess answers that radiation-induced annealing of A-centres couldn't be due to such a process only.
A hint that ionization-enhanced migration of C_i may be involved (ionization is reported to produce an effect on motion of interstitial carbon /18/ is given by further inspection of Fig.1 revealing a general trend: the longer the duty cycle is the lesser the irradiation intensity ought to be used to reach the maximal η_A. No doubt, the nature of the phenomena is complex; however, a simplified model permits realizing some problems. Taking into consideration the production and radiation-induced annealing of A-centres the kinetic equation for η_A may be written in the form (at steady-state conditions)

$$\eta_A = \eta_V \tau_V r_1 N_0 - r_2 N_A N_{C_i}/J. \qquad (1)$$

Here η_V is the production rate of free vacancies; τ_V is their lifetime; N_O, N_A and N_{C_i} are the concentrations of oxygen, A-centres and interstitial carbon, respectively; r_1 and r_2 are the reaction rate constants for the formation and annealing of A-centres, respectively. Generally speaking, from Equation (1) it follows that η_A should increase with increasing J. However, η_A may be also dependent on J through some parameters entering in the equation. By way of example, the lifetime of free vacancies takes the form: $\tau_V^{-1} = r_1 N_0 + r_3 N_{C_i}$ if we consider another possible interaction of vacancies with C_i (in this case r_3 refers to annihilation V + C_i → C_s). In doing so, we have two coefficients r_2 and r_3 which, in turn, may be dependent on J because of ionization-enhanced migration of C_i. At low excitation conditions the contribution of ionization-enhanced migration can be ignored and η_A reaches its peak. At higher excitation conditions this contribution can be saturated and η_A levels off at a minimum. This implies the sensitivity of annealing processes to a duty cycle, too.
And the last point we would like to touch on. A set of the maximal η_A for all energies of fast electrons permits to monitor a fraction of free vacancies and self-interstitials produced after dissociation of Frenkel pairs. Recent detailed studies of the

production rates of A-centres and carbon-related defects in Si at near-threshold energies of electrons (300 - 500 keV) /13/ point to the fact that the probability of displacement is strongly dependent on the energy transferred to an atom in the region of 30-40 eV and it is close to unity at $E_d \approx 40$ eV. Fig.2 displays $\eta_V (= \eta_A/\eta_{FP})$ of free vacancies versus E_{el}. As one would expect, an increase in the energy of fast electrons leads to an increase in the fraction of dissociated Frenkel pairs (because of expanding separation between genetic V and I).

REFERENCES

1. A.G. Abdusattarov, V.V. Emtsev, V.N. Lomasov and T.V. Mashovets, Fiz. i Techn. Poluprov. (Sov. Phys. Semicond.) 20, 164 (1986); 23, 2221 (1989).
2. A.A. Zolotukhin, A.K. Kovalenko, T.M. Meshcheryakov, L.S. Milevskii and T.A. Pagava, Fiz. i Techn. Poluprov. (Sov. Phys. Semicond.) 9, 1201 (1975).
3. A.N. Kraichinskii, L.V. Mizrukhin and V.I. Shachovtsov, Fiz. i Techn. Poluprov. (Sov. Phys. Semicond.) 17, 437 (1983).
4. N.N. Gerasimenko, A.V. Dvurechenskii, V.I. Panov and L.S. Smirnov, Fiz. i Techn. Poluprov. (Sov. Phys. Semicond.) 5, 1644 (1971).
5. P.F. Lugakov and V.V. Lukyanitsa, Fiz. i Techn. Poluprov. (Sov. Phys. Semicond.) 18, 345 (1984).
6. J.W. Corbett and G.D. Watkins, Phys. Rev. 138, A555 (1965).
7. H.J. Stein and F.L. Vook, Phys. Rev. 163, 790 (1967).
8. T. Wada, K. Yasuda, S. Ikuta, M. Takeda and H. Masuda, J.Appl. Phys. 48, 2145 (1977).
9. J.S. Blakemore, Semiconductor Statistics (Pergamon Press, Oxford - London - New York - Paris, 1962).
10. V.V. Emtsev, T.V. Mashovets and A.G. Abdusattarov, Fiz. i Techn. Poluprov. (Sov. Phys. Semicond.) 21, 2106 (1987).
11. V.V. Emtsev, P.M. Klinger, T.V. Mashovets and K.M. Mirazizyan, Fiz. i Techn. Poluprov. (Sov. Phys. Semicond.) 25, 561 (1991)
12. N.A. Vitovskii, D. Mustafakulov and A.P. Chekmareva, Fiz. i Techn. Poluprov. (Sov. Phys. Semicond.) 11, 1747 (1977).
13. L.S. Berman, N.A. Vitovskii, V.N. Lomasov and V.N. Tkachenko, Fiz. i Techn. Poluprov. (Sov. Phys. Semicond.) 24, 1816 (1990).
14. G.D. Watkins in Radiation Effects in Semiconductors, ed. by F.L. Vook (Plenum Press, New York, 1968) p. 67.
15. R.D. Harris and G.D. Watkins in Proc. of the 13th Int. Conf. on Def. in Semicond., eds. L.C. Kimerling and J.M. Parsey,Jr. (The Metallurgical Society of AIME, Warrendale, Penn., USA, 1985) p. 799.
16. V.V. Emtsev, T.V. Mashovets, V.V. Mikhnovich and N.A. Vitovskii, Rad. Eff. 111-112, 99 (1989).
17. G.D. Watkins and K.L. Brower, Phys. Rev. Letts. 36, 1329 (1976).
18. L.W. Song, B.W. Benson and G.D. Watkins, Phys. Rev. B33, 1452 (1986).
19. N.V. Kolesnikov, V.N. Lomasov and S.E. Malkhanov, Fiz. i Techn. Poluprov. (Sov. Phys. Semicond.) 23, 1921 (1989).

Materials Science Forum Vol. 83-87 (1992) pp. 327-332

THE PSEUDODONOR ELECTRONIC STATES OF A METASTABLE DEFECT IN SILICON STUDIED BY UNIAXIAL STRESS SPECTROSCOPY

J. H. SVENSSON*, E. JANZÉN, O. KORDINA, AND B. MONEMAR
Department of Physics and Measurement Technology, Linköping University
S-581 83 Linköping, Sweden
*Present address: Max-Planck Institut für Festkörperforschung
Heisenbergstraße 1, Postfach 800665
D-7000 Stuttgart 80, BRD

ABSTRACT

The absorption spectrum of the metastable configuration of a complex defect in electron-irradiated silicon is investigated with uniaxial stress perturbation spectroscopy. The zero-phonon line with the lowest energy (615.0 meV) is found to split according to what is expected from the energy splitting of an electronic transition at a defect with monoclinic I symmetry. One of the principal axes of the defect perpendicular to the primary axis is found to be rotated in a {110} plane a few degrees from the <111> direction lying in that plane. The dipole that belongs to the zero-phonon transition is also perpendicular to the primary axis and close to the <111> direction.

1. Introduction

Recently, an absorption spectrum related to a metastable configuration of a defect in silicon was reported [1]. The absorption spectrum was interpreted as reflecting electronic transitions from a deep ground state of a neutral defect to the excited pseudodonor states consisting of an electron bound in the Coulomb field of the positively charged defect core. This excited electron could be described by effective-mass-like donor s-states perturbed by a low symmetry central-cell potential [2].

From the results of uniaxial stress spectroscopy on the lowest zero-phonon line at 615.0 meV, we show in this paper that the symmetry of the metastable defect configuration is monoclinic I [3-5]. We also show that both one of the principal axes of the defect and the dipole that belongs to the 615.0 meV line are close to the <111> direction.

2. Sample preparation

In this paper we report absorption measurement results from the metastable defect configuration obtained by applying uniaxial stress to samples along the <100>, <111> and <110> directions. The samples were all cut from the same crucible-grown ingot with an initial resistivity of ≈ 40 Ωcm (phosphorus doping) prior to electron irradiation. The samples were irradiated at room temperature with 2 MeV electrons and then isothermally annealed at 312 °C for 0.5 h. To populate the optically active metastable configuration of the defect, the samples were excited with the 1.06 μm line of a YAG-laser for several hours at temperatures below 65 K.

3. Uniaxial stress : Orientational degeneracy

When uniaxial stress is applied to a crystal, the splitting of an electronic transition at a defect can be caused by the lifting of the degeneracy of the electronic states or by the defect having different inequivalent orientations in the crystal lattice with

Table I. Number of components observed for unpolarized light incident normal to the stress axis for defects with orientational degeneracy in cubic crystals (see Refs. 3,4.). It is assumed that the transitions are between non-degenerate states.

Type of defect	Stress direction <001>	<111>	<110>
1. Tetragonal	2	1	2
2. Trigonal	1	2	2
3. Orthorombic I	2	2	3
4. Orthorombic II	3	1	3
5. Monoclinic I	2	3	4
6. Monoclinic II	3	2	4
7. Triclinic	3	4	6

respect to the stress axis. The effect on the transitions between non-degenerate electronic states of crystal defects caused by the application of uniaxial stress has been analyzed by Kaplyanskii [3,4] (see Table I).

A defect with monoclinic I symmetry has 12 equivalent orientations in the silicon crystal lattice. One of these possible orientations can be defined by a primary unique axis, Z, along the [110] crystallographic direction. Perpendicular to [110] two orthogonal axes can be defined: X parallel to [001] and Y parallel to [$1\bar{1}0$]. Note that there are no symmetry restrictions on the orientation of the X- and Y-axes but that they can be rotated together in the (110) plane. If uniaxial stress is applied in the [100] direction, the 12 equivalent orientations can be disintegrated into two groups. The four orientations in the (100) plane will have their Z-axis orthogonal to the stress direction, whereas the Z-axis of the eight orientations in the (010) and (001) planes will form an angle of 45 degrees with the stress direction. There is thus a possibility for the electronic states of the defect to have different energy positions depending on the orientation of the defect when stress is applied. The energy of the transition between two electronic states can thus split into two possible values, if the stress sensitivity for the initial and final states differs. In this way the behavior under uniaxial stress in different directions and for different defect symmetries can be analyzed and Table I constructed.

For a defect with monoclinic I symmetry all electronic states are orbitally non-degenerate. Since there are no interactions of the states involved in the transition with any other electronic states, they can only be affected to first order by perturbations that transform according to the totally symmetric representation of the defect point group [6]. The stress perturbation of the potential, ΔV, in the first order can be expressed as [5]

$$\Delta V = P_{XX}S_{XX} + P_{YY}S_{YY} + P_{ZZ}S_{ZZ} + P_{XY}S_{XY}$$

where P_{ij} are electronic operators and S_{ij} are stress tensor components with X, Y and Z defined as above.

The energy shift of the zero-phonon transition due to ΔV is conventionally expressed by four stress parameters, A_1, A_2, A_3 and A_4, [3-5] (see Table II), which are the components of a symmetric tensor. The definitions of the four A_i and the form of the tensor are shown below for monoclinic I symmetry:

Table II. Uniaxial stress perturbations of the 615.0 meV zero phonon absorption line The notations follow Ref. 5 assuming a monoclinic I center.

			Shift rates	
Stress axis	Component	Theoretical expression	Experimental (meV/GPa)	Bestfit (meV/GPa)
<001>	a	A_1	15.6	16.3
	b	A_2	-8.3	-8.6
<111>	c_1	$(A_1 + 2A_2 - 2A_3 - 4A_4)/3$	-16.2	-16.7
	c_2	$(A_1 + 2A_2 - 2A_3 + 4A_4)/3$	21.6	21.9
	d	$(A_1 + 2A_2 + 2A_3)/3$	-2.9	-3.2
<110>	e_1	$(A_1 + A_2 - 2A_4)/2$	-9.6	-10.6
	e_2	$(A_1 + A_2 + 2A_4)/2$	18.2	18.3
	f	$A_2 - A_3$	-5.2	-4.3
	g	$A_2 + A_3$	-13.5	-12.9

$$A_1 = \langle f|P_{XX}|f\rangle - \langle i|P_{XX}|i\rangle$$
$$A_2 = \{\langle f|P_{YY} + P_{ZZ}|f\rangle - \langle i|P_{YY} + P_{ZZ}|i\rangle\}/2$$
$$A_3 = \{\langle f|P_{ZZ} - P_{YY}|f\rangle - \langle i|P_{ZZ} - P_{YY}|i\rangle\}/2$$
$$A_4 = \{\langle f|P_{XY}|f\rangle - \langle i|P_{XY}|i\rangle\}/2^{3/2}$$

$$\begin{pmatrix} A_2 & A_3 & -A_4 \\ A_3 & A_2 & A_4 \\ -A_4 & A_4 & A_1 \end{pmatrix}$$

$\langle f|$ denotes the final and $\langle i|$ the initial state of the transition. The tensor is expressed in a cubic coordinate system, x,y,z. The principal axes of this tensor coincide with the fundamental symmetry axes of the defect, which for monoclinic I centers are the <110> axis and two axes in the perpendicular plane.

An A → E transition at a trigonal defect gives the same splitting pattern under uniaxial stress as a transition between non-degenerate states at a monoclinic I defect [7]. In trigonal symmetry the intensities of the lines are unique, whereas the intensities in monoclinic I symmetry depend on the orientation of the dipole of the transition. For the 615.0 meV line the trigonal case can be ruled out, since it is impossible to extinguish any of the two lines obtained under <001> uniaxial stress when using light polarized parallel to the stress axis (see Fig. 1b.).

In monoclinic I symmetry the dipole can be either parallel or perpendicular to the primary axis [110]. The parallel dipole is treated in Ref. 4. and the perpendicular dipole in Ref. 5. When the dipole is parallel it is possible to extinguish one of the two lines obtained under <001> uniaxial stress for light polarized parallel to the stress axis [4]. Since this extinguishment does not occur for the 615.0 meV line (see Fig. 1b.), the dipole must be perpendicular to the primary axis. In table III calculated intensities of the stress-split lines for polarized light are shown for this case [5]. θ is the angle between the dipole and the [001] direction. In Fig. 2. intensities calculated from table III for unpolarized light are presented. The results for the parallel dipole [4] and for the trigonal case [7] are also shown.

4. Results and discussion

For uniaxial stress applied in the <001>, <111> and the <110> direction the 615.0 meV line splits into two, three and four components, respectively (Fig. 1.). The experimental shift rates are given in Table II. The number of stress-split

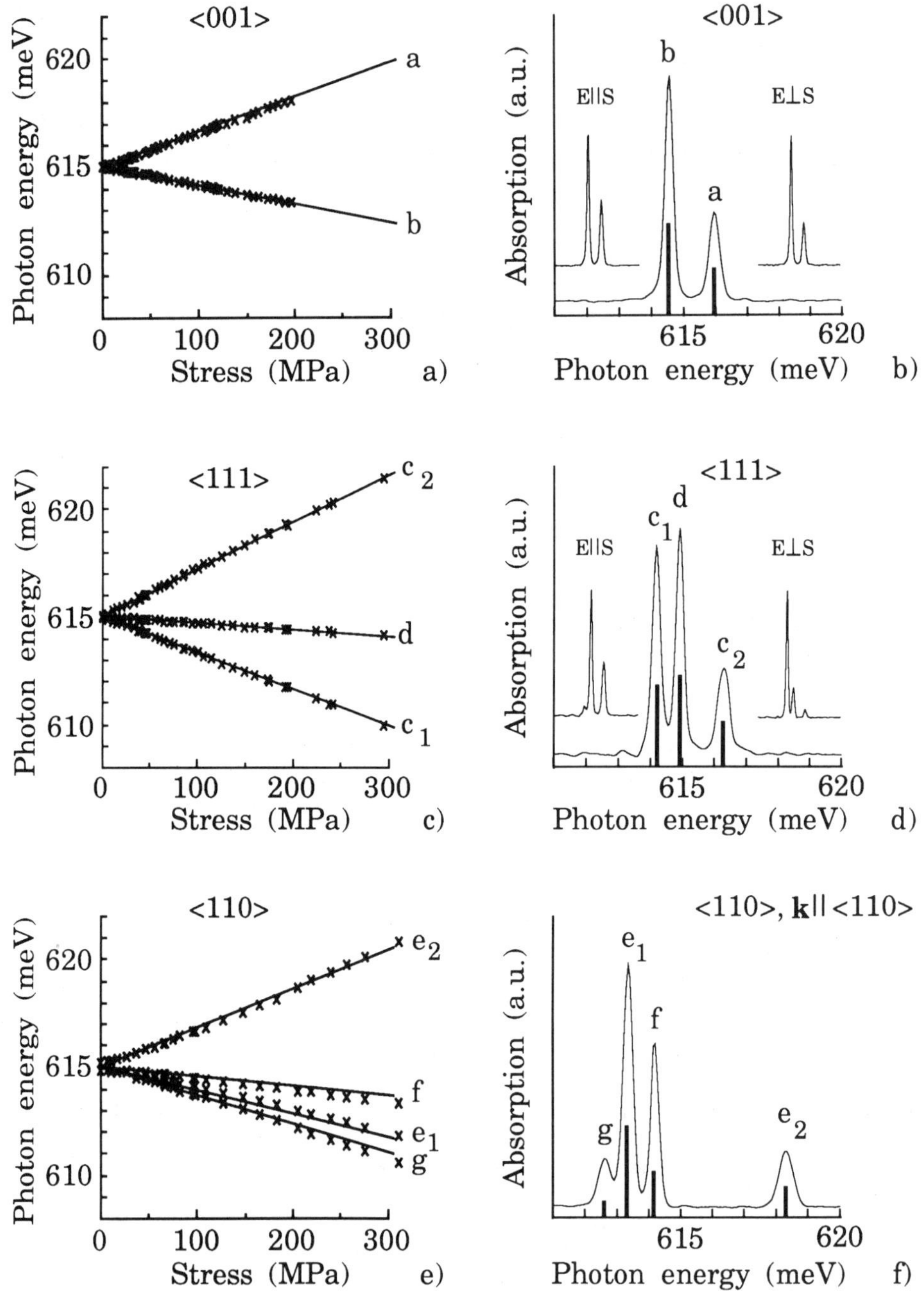

Fig. 1. Uniaxial stress perturbation data of the 615.0 meV line obtained at 20 K. In a, c and e the energy shifts for the different components are shown for the <001>, <111> and <110> directions. X are experimental points and the solid lines are obtained from the best fit for all directions. In b (56 MPa), d (55 MPA) and f (180 MPa) the corresponding absorption spectra are shown. Vertical lines in the spectra represent calculated intensities with $\theta = 126°$. In the insets of b and d polarized measurements are shown.

Table III. Intensities of the different components of an absorption line obtained by uniaxial stress perturbations following Ref. 5. The transition is between non-degenerate states and the defect symmetry is monoclinic I. For a defect with the primary axis in the [110] direction the dipole is in the (110) plane θ degrees from the [001] direction towards the [110] direction. $\theta_1 = \arccos(1/\sqrt{3}) \approx 55$ degrees.

Stress axis	Components: E ∥ S	E ⊥ S: E∥[001] for <110> stress	E ⊥ S: E∥[110] for <110> stress
<001>	$a_{\parallel} = 4\cos^2(\theta)$	$a_{\perp} = 2\sin^2(\theta)$	
	$b_{\parallel} = 4\sin^2(\theta)$	$b_{\perp} = 4 - 2\sin^2(\theta)$	
<111>	$c_{1\parallel} = 3\cos^2(\theta + \theta_1)$	$c_{1\perp} = 1.5\sin^2(\theta + \theta_1)$	
	$c_{2\parallel} = 3\cos^2(\theta - \theta_1)$	$c_{2\perp} = 1.5\sin^2(\theta - \theta_1)$	
	$d_{\parallel} = 2\cos^2(\theta)$	$d_{\perp} = 2 + \sin^2(\theta)$	
<110>	$e_{1\parallel} = 3\sin^2(\theta - \theta_1)$	$e_{1,001} = 2\sin^2(\theta)$	$e_{1,110} = 3\sin^2(\theta + \theta_1)$
	$e_{2\parallel} = 3\sin^2(\theta + \theta_1)$	$e_{2,001} = 2\sin^2(\theta)$	$e_{2,110} = 3\sin^2(\theta - \theta_1)$
	$f_{\parallel} = 2\sin^2(\theta)$	$f_{001} = 2\cos^2(\theta)$	$f_{110} = 0$
	$g_{\parallel} = 0$	$g_{001} = 2\cos^2(\theta)$	$g_{110} = 2\sin^2(\theta)$

components is as expected for splittings induced by the inequivalence of the defect orientations of a monoclinic I defect under uniaxial stress applied in the above three crystal directions [3-5] (see Table I). Raising the temperature to ≈ 40 K did not change the relative intensities of the stress-split components for any of the three stress directions, implying that the ground state is non-degenerate and supporting the interpretation that the splitting is induced by the orientational degeneracy of the defect.

The best fit between experiment and theory for shift rates was given for the following stress parameters: A_1 = 16.3 meV/GPa, A_2 = -8.6 meV/GPa, A_3 = -4.3 meV/GPa and A_4 = 14.5 meV/GPa. Using these values the principal values were found to be λ_1 = -12.9 meV/GPa, λ_2 = -16.9 meV/GPa and λ_3 = 28.9 meV/GPa and the principal axes $X_1 = 1/\sqrt{2}\,[110]$, $X_2 \approx 1/\sqrt{3}\,[1\bar{1}1]$ and $X_3 \approx 1/\sqrt{6}\,[1\bar{1}\bar{2}]$. X_2 and X_3 are rotated together in the (110) plane with X_2 3.6° from the $[1\bar{1}1]$ direction towards the $[1\bar{1}0]$ direction.

By comparing the ratio of the integrated intensities of the two lines in Fig. 1b. with the results of Fig. 2. it is clear that the dipole angle (θ or 180 - θ) is somewhere between 50° and 60° Since $e_{2\parallel}$ (not shown in Fig. 1.) and $c_{1\perp}$ (see Fig. 1d) are close to zero, θ must be ≈ 180 - θ_1 (see Table III). Using the results from the <111> direction (see Fig. 1d.) the angle θ was fitted to be 126° i.e. close to the $[1\bar{1}\bar{1}]$ direction. The calculated unpolarized intensities for θ = 126° are shown in Figs. 1b., 1d. and 1f.. and the agreement between the calculations and the experimental results is good.

As pointed out in Ref. 5., it is not possible to distinguish between θ , 180 + θ and 180 - θ. There is also a complete ambiguity in the sign of A_4. Changing θ to 180° - θ and A_4 to $-A_4$ simply interchanges the stress-split components e_1 with e_2 and c_1 with c_2.

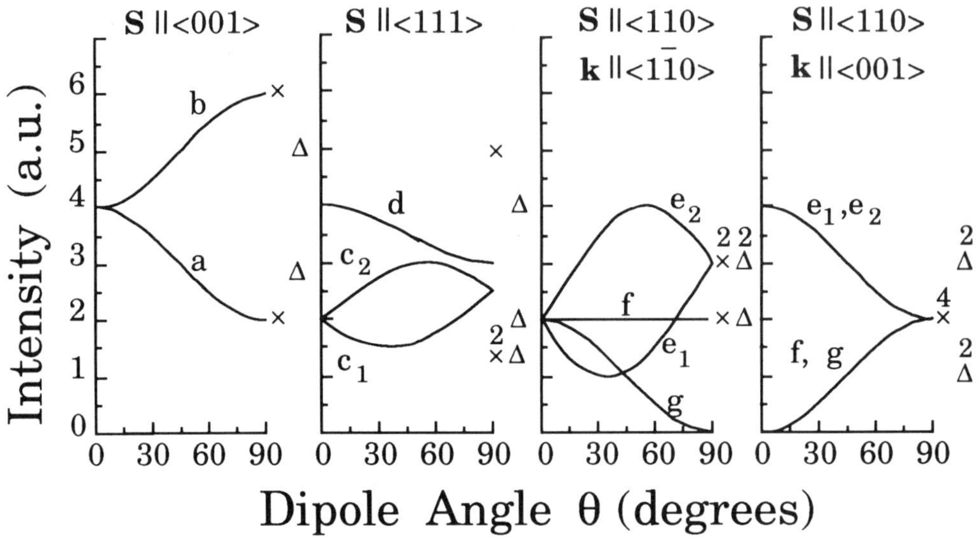

Fig. 2. Calculated intensities of stress-split lines in monoclinic I symmetry for unpolarized light with the stress axis in the <001>, <111> and <110> directions. The dipole is in the plane perpendicular to the primary axis [110] θ degrees from the [001] direction towards the [1$\bar{1}$0] direction. The results for parallel dipole (X) and for the trigonal symmetry (A → E) (Δ) are also shown. The numbers above the symbols (X or Δ) gives the number of lines with equal intensities.

5. Summary

By means of uniaxial stress perturbation spectroscopy we have shown that the stress-induced energetical splitting of the lowest zero phonon line at 615.0 meV is in agreement with a monoclinic I symmetry of the metastable configuration of the corresponding defect. Both one of the principal axes of the defect and the dipole of the transition are close to the <111> direction.

6. Acknowledgement

The authors want to thank M. Kleverman and M. Singh for useful discussions and A. Henry for help with the preparation of the figures.

REFERENCES

1. J. H. Svensson and B. Monemar, Phys. Rev. B**40**, 1410 (1989).
2. J. H. Svensson, B. Monemar, and E. Janzén, Phys. Rev. Lett. **65**, 1796 (1990).
3. A. A. Kaplyanskii, Opt. Spectrosc. **16**, 329 (1964).
4. A. A. Kaplyanskii, Opt. Spectrosc. **16**, 557 (1964).
5. G. Davies, E. C. Lightowlers, M. Stavola, K. Bergman, and B. G. Svensson, Phys. Rev. B**35**, 2755 (1987).
6. M. Tinkham, *Group Theory and Quantum Mechanics* (McGraw-Hill Book Company, New York, 1964)
7. A. E. Hughes and W. A. Runciman, Proc. Phys. Soc. **90**, 827 (1967)

Materials Science Forum Vol. 83-87 (1992) pp. 333--338

A NEW DEFECT OBSERVED IN ANNEALED PHOSPHORUS-DOPED ELECTRON-IRRADIATED SILICON

J.L. LINDSTRÖM[1],B.G. SVENSSON[2] AND W.M. CHEN[3]

[1]Swedish Defence Research Establishment, P.O. Box 1165, S-581 11 Linköping, Sweden

[2]Royal Institute of Technology, Solid State Electronics, P.O. Box 1298, S-164 28 Kista-Stockholm, Sweden

[3]Department of Physics and Measurement Technology, University of Linköping, S-581 83 Linköping, Sweden

ABSTRACT

In this contribution we report on results from infrared absorption (IR), deep level transient spectroscopy (DLTS) and optical detection of magnetic resonance (ODMR) measurements in electron-irradiated (2 MeV) phosphorus-doped silicon. A new defect is observed directly after irradiation at nominal room temperature (< 30 °C), and it increases in concentration upon annealing above 200 °C. The defect is an efficient electron trap and gives rise to two local vibrational IR modes at 864 and 916 cm^{-1} (sample temperature 10 K). Correlation of the IR results with ODMR studies suggests that the IR bands originate from an A center like defect involving two adjacent phosphorus atoms. We favour a model where the oxygen atom occupies a nearly substitutional site and bonds to the two phosphorus atoms (P_2VO center). DLTS spectra of annealed samples show a major level located ~0.27 eV below the conduction band which may originate from the transition between the two charge states corresponding to the 864 and 916 cm^{-1} bands.

1. Introduction

One of the most important and extensively studied defects in electron-irradiated Czochralski (Cz) grown silicon is the vacancy oxygen pair (VO or the A center). The first experimental observations of VO were reported already in the late 1950's[1,2], and it has subsequently been established by different experimental techniques that the oxygen atom in the VO pair occupies a central tetrahedral substitutional site but is displaced in a <100> direction bonding to two silicon atoms[2,3].

The VO center exists in two charge states, neutral (VO^0) and singly negative (VO^-), and the corresponding infrared (IR) bands arising from local vibrational modes of the oxygen atom appear at 835 cm^{-1} and 885 cm^{-1}, respectively, at low temperature measurements (10K). The transition between the neutral and singly negative charge states takes place at ~0.18 eV below the conduction band (E_c)[2,3,4].

The VO pair is normally stable up to temperatures above 300 °C where a new dominant defect appears with a vibrational IR band at 895 cm^{-1}. This defect was found to be a vacancy dioxygen center (VO_2) and is formed by a reaction between diffusing VO centers and interstitial oxygen atoms ($VO+O_i \rightarrow VO_2$)[5,6]. Typically, about 50% of the VO centers are transformed to VO_2 in moderately doped Cz silicon. However, in phosphorus-doped material of ~0.1 Ωcm an anomalously rapid annealing of VO is observed, and the formation rate of VO_2 is very low[7]. A new IR band emerges at 916 cm^{-1} but the maximum strength of this band is only about one tenth of that of the initial 835 cm^{-1} band indicating that several reactions take place during the annealing process.

In this investigation experimental results are reported on a novel defect observed in phosphorus-doped samples using IR absorption, deep level transient spectroscopy (DLTS) and optical detection of magnetic resonance (ODMR). We have studied both Cz and float-zone (Fz) samples subjected to different doses of 2 MeV electrons and heat treatments between 150 and 425 °C. The new defect is shown to involve two phosphorus atoms and one oxygen atom, and it exhibits several similarities with VO concerning the electrical and optical properties. We present evidence for a model where the oxygen atom bonds to the two phosphorus atoms and occupies a nearly substitutional site in the silicon lattice (P_2VO center).

2. Experimental

Cz and Fz samples doped with phosphorus to a nominal resistivity of 0.1, 1.0 and 10 Ωcm were used in the present study. The samples were polished to an optical surface on two sides and then irradiated at nominal room temperature (≤ 30 °C) to doses between 5×10^{17} and 5×10^{18} cm^{-2} using 2.0 MeV electrons. Subsequently, the concentration of interstitial oxygen was determined by standard IR measurements and found to be between 4×10^{17} and 6×10^{17} at/cm^3 in the Cz samples and below 1×10^{16} at/cm^3 in the Fz samples. The substitutional carbon concentration was below 1×10^{16} at/cm^3 in both types of samples (DIN calibration constants were assumed). Annealing of the samples was undertaken in an inert atmosphere at temperatures between 150-425 °C.

IR absorption analysis was carried out using a Fourier Transform Infrared Spectrometer (FTIR) with a spectral resolution of 0.5 cm^{-1}, and the sample temperature was varied in the range of 10 to 300 K. The experimental setups for the DLTS and ODMR measurements performed in this study have been described in detail elsewhere[8,9].

3. Experimental results

Fig. 1(a) shows an IR spectrum obtained at 10 K for a 0.1 Ωcm Cz sample irradiated to 3×10^{18}

e⁻/cm². In addition to the well-known bands at 835 and 885 cm^{-1} originating from VO^o and VO^-, two new bands are clearly resolved at 864 and 916 cm^{-1}. The relative strength of these two new bands depends strongly on the Fermi level position, and in samples irradiated to a dose of $1x10^{18}$ e^-/cm^2 and annealed at temperatures above 200 °C the 916 cm^{-1} band dominates ,Fig. 1(b). As the sample temperature is varied a reversible one-to-one correlation holds between the absorption coefficients a_{864} and a_{916}, Fig. 2. A similar type of correlation has previously been reported for the VO bands observed at 835 and 885 cm^{-1} and is the hall-mark of a defect existing in two charge states[10].

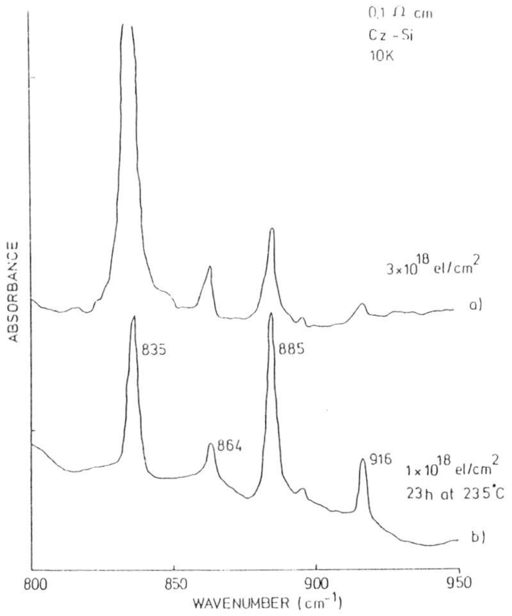

1. Infrared spectra at 10K for electron-irradiated (1 and $3x10^{18}$ e^-/cm^2) and subsequently annealed 0.1 ohmcm Cz sample (23 h at 235 ^OC).

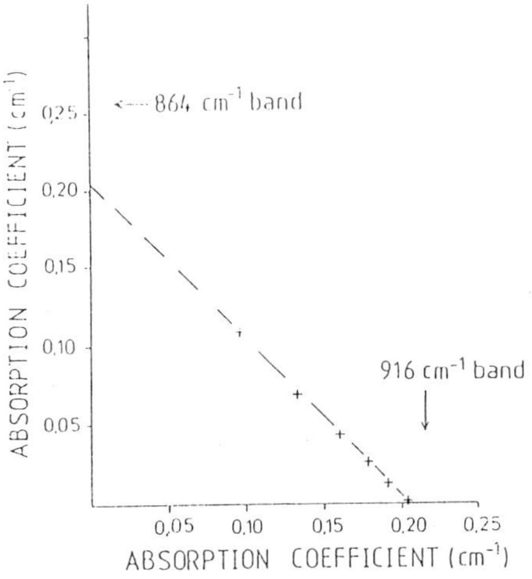

2. The reversible correlation between the strength of the 864 and 916 cm^{-1} IR-bands at different temperatures.

The 864 and 916 cm^{-1} bands were not observed in the Fz samples (0.1, 1 and 10 Ωcm) nor in the moderately doped Cz samples (1 and 10 Ωcm) after irradiation or annealing.

DLTS spectra of the annealed 0.1 Ωcm Cz samples show three peaks at 0.27, 0.51, and 0.56 eV below E_c, Fig. 3. Concentration profile measurements revealed that the latter level is located close to the surface and is not due to a true bulk defect. For the levels at E_c-0.27 and E_c-0.51 eV the concentration is low close to the surface and increases with depth until a constant value is reached beyond 2 µm; the profile shape is very similar to that reported for the VO and V_2 centers[8,11].

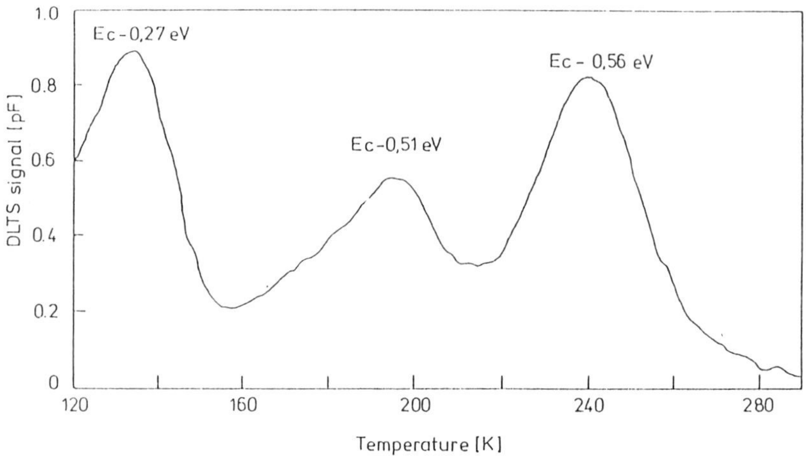

3. DLTS-spectrum for a sample where the VO-center has been annealed out and the 916 cm^{-1} band has reached a maximum.

The samples used in the IR and DLTS measurements were also analysed by ODMR, and a novel signal appeared in the 0.1 Ωcm Cz samples, Fig. 4. The strength of this signal increases after annealing at 250-300 °C, and the spectrum exhibits several similarities with that for the VO center. The presence of two phosphorus atoms is revealed by a resolved hyperfine structure from the two ^{31}P atoms of nuclear spin I=1/2 (100% naturally abundant). The defect is found to be a strong electron trap in the annealed 0.1 Ωcm Cz samples judging from its role in reducing the radiative carrier recombination via other defects, which are detected in the ODMR experiment. The concentration of this new defect is below the detection limit in the Fz samples and the moderately doped Cz samples.

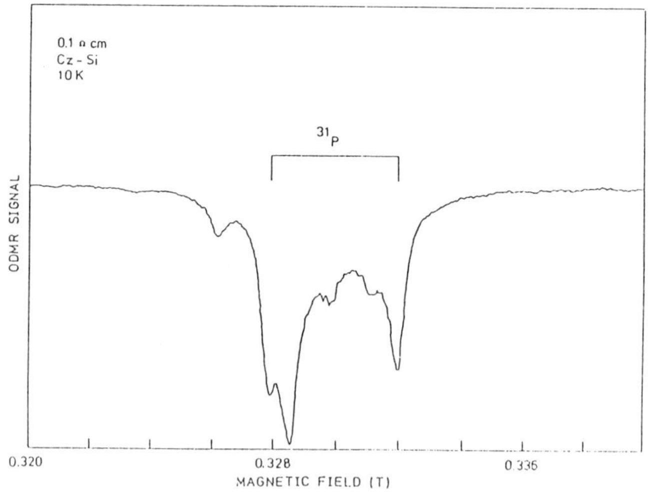

4. ODMR-spectrum obtained at 10K. The central doublet is related to the P-donors.

4. Discussion and conclusions

We conclude that the 864 and 916 cm^{-1} absorption bands are local vibrational modes of an oxygen atom existing in a configuration involving two phosphorus atoms, and the two bands originate from the different charge states of the defect. They show several striking similarities with the two VO related vibrational modes at 835 and 885 cm^{-1}, e.g., the separation is almost identical (52 versus 50 cm^{-1}) and the positions shifts by 5 to 7 cm^{-1} for all the four bands as the sample temperature changes from 10 to 300 K. Furthermore, both pairs of the bands exhibit a reversible one-to-one correlation when varying the sample temperature. In the ODMR spectra, strong hyperfine interactions from two Si ligand atoms can be resolved, indicating strong localization of two un-paired electron wavefunctions. Such a strong localization has often been observed in irradiated silicon, and is known to be a fingerprint of vacancy-related defects when the strong p character of the wavefunction is taken into account [12]. In fact the observed hyperfine interaction from the two Si ligand atoms resemble that for the VO, and can be attributed to the molecule-orbit bonding of the dangling bonds adjacent to the vacancy[3]. The presence of two P atoms in the complex is evident from a characteristic 1-2-1 ODMR intensity pattern in the hyperfine structure, corresponding to two ^{31}P atoms of $I=1/2$. It is interesting to note that the hyperfine splitting due to the ^{31}P atoms involved in this complex is smaller than that for the shallow P substitutional donors, indicating a weaker overlap of the unpaired electron wavefunction at the P sites. These experimental facts provide evidence for a VO like model where the oxygen atom occupies a nearly subsitutional site and bonds to the two phosphorus atoms. The difference in vibrational frequency compared with VO is attributed to a stronger force constant experienced by the oxygen atom in the 'P-O-P molecule' versus that in the Si-O-Si configuration.

Possible positions for the charge state transition of the P_2VO center are E_c-0.27 and E_c-0.51 eV, Fig.3. The latter level can, however, be ruled out since its concentration is found to be non-proportional to the strength of the 916 cm^{-1} band. The former level is a more promising candidate but further correlation studies are required before a definite assignment can be made.

The formation mechanism for the P_2VO center is not clear. During annealing above 200 ^{o}C several radiation induced defects (or dissociated constituents from the defects) can interact. A study of the reaction kinetics during annealing of the VO center in 0.1 ohmcm phosphorus doped material indicated a second-order process[7] and a reaction between two defects of comparable concentration was suggested. These two defects could be the PV center and the VO center in our study. As the PV center is known to anneal out at about 150 ^{o}C and the VO center at about 300 ^{o}C one formation mechanism could be the trapping of vacancy-assisted diffusing phosphorus atoms at VO centers.

Finally, it should be pointed out that the position of the 864 cm^{-1} band is close to that of the well-known 865 cm^{-1} band originating from the C(3) center (carbon oxygen complex). The 865 cm^{-1} band

and its accompanying band at 1116 cm^{-1} are, however, not observed because of the low carbon concentration in the present samples.

5. Acknowledgements

The authors are indebted to Professor J.W.Corbett and Dr. B.Liedberg for helpful discussions. Experimental assistance by M.Sc. J.Önnegren is gratefully aknowledged. Financial support was partly received from the Swedish National Board for Technical Development.

REFERENCES

1. G.Bemski, J.Appl.Phys. 30,1195 (1959)
2. G.D.Watkins, J.W.Corbett and R.M.Walker, J.Appl.Phys.30,1198 (1959)
3. G.D,Watkins and J.W.Corbett, Phys.Rev.121,1001(1961);J.W.Corbett, G.D.Watkins, R.M. Chrenko and R.S.McDonald, ibid 121,1015 (1961)
4. See e.g. L.C.Kimerling , *Radiation Effects in Semiconductors 1976*, eds. N.B.Urli and J.W.Corbett, (Conference Series No.31, The Institute of Physics, Bristol,1977), p.221
5. J.W.Corbett,G.D.Watkins and R.S.McDonald, Phys.Rev.135, A1381 (1964)
6. J.L.Lindström and B.G.Svensson, Mat.Res.Soc.Symp.Proc. 59, 45 (1986)
7. B.G.Svensson and J.L.Lindström, Mat.Sc.Forum,Vol.10-12, 1087 (1986)
8. B.G.Svensson, K.-H.Ryden and B.M.S.Lewerentz, J.Appl.Phys.60, 1699 (1989)
9. W.M.Chen and B.Monemar, Phys.Rev.,B40, 1365 (1989)
10. A.R.Bean and R.C.Newman,Sol.State Com., 9, 271 (1971)
11. B.G.Svensson and M.Willander, J.Appl.Phys. 62, 2758 (1987)
12. Y.-H.Lee and J.W.Corbett, Phys.Rev. B13, 2653 (1976)

Materials Science Forum Vol. 83-87 (1992) pp. 339-344

CONFIGURATIONAL METASTABILITY OF CARBON-PHOSPHORUS PAIR DEFECTS IN SILICON

Emir Gürer[1], Brent W. Benson[2], George D. Watkins[2]
[1]Zettlemoyer Center for Surface Studies, Sinclair Lab. # 7, Lehigh Univ., Bethlehem, PA 18015 USA
[2]Physics Department, # 16, Lehigh University, Bethlehem PA 18015 USA

ABSTRACT

We report new results on previously observed multistable carbon phosphorus pairs, C_i-P_s in electron-irradiated silicon[1-4]. Conversion kinetics studies suggest that C_i moves in the lattice during the conversions among four configurations of C_i-P_s pairs, while a new fifth configuration is more likely to be a result of a bond rearrangement. Experimentally determined energy level diagrams and configuration coordinate (CC) energy surfaces are consistent with the simple ionic model which takes into account the Coulomb interactions between C_i and P_s atoms at different lattice separations in a uniform dielectric medium of silicon. The CC energy surfaces suggest that the C_i-P_s pairs may exhibit a charge state effect and a recombination-enhancement mechanism under electronic excitations.

INTRODUCTION

C_i is the precursor defect for the C_i-P_s pairs. The single donor (o/+) state of C_i is observed in p-type material and it introduces an energy level at E_v+0.28 eV[5]. The single acceptor (-/o) state of C_i has an energy level at E_c-0.10 eV as observed in n-type material[6]. EPR studies[7,8] conclude that the microscopic structures for the C_i^+ and C_i^- are very similar. They consist of a <100> oriented carbon-silicon molecule occupying a single substitutional site, consistent with the observed C_{2v} symmetry. So far only a single configuration of C_i has been observed experimentally. However, the carbon-carbon pair (C_i-C_s) which is observed after annealing of C_i, is a bistable defect. Recent experimental results led to the microscopic structures, CC energy surfaces, and a possible mechanism for the bistability[9,10]. Configuration A is stable in both the positive and the negative charge states. It involves a C-Si molecule (each atom is three fold coordinated) on a single lattice site, like C_i, but slightly perturbed by the presence of the C_s atom. Configuration B is stable in the neutral charge state and involves two C_s atoms (each four fold coordinated) with a two fold coordinated Si atom in between. It was concluded that the mechanism for the C_i-C_s bistability is molecular bond-switching between two configurations of similar energy.

The first evidence of C_i trapped by a group-V donor atom came in P doped Si where a defect was identified as an C_i-P_s pair[1-3]. The C_i-P_s pair displays a remarkable configurational multistability. At that time four distinct configurations were observed using electrical measurements. However, lack of a reliable charge state assignment for the observed energy levels prevented the development of a model to explain the unusual multiplicity of C_i-P_s pairs. This paper describes our recent research on the C_i-P_s pairs which resulted with the discovery of yet another, fifth configuration. One of the main emphases will be on the charge state assignment. Detailed conversion kinetics results will be presented under various bias conditions. The CC energy surfaces will be described for all five configurations in three charge states. The possible mechanisms that drive the C_i-P_s multistability will also be discussed. We recently discovered two similar multistable defects in As[4] and Sb[4,11] doped Si that we identified as C_i-As_s and C_i-Sb_s pairs. Although these results will be discussed in future publications, we will refer to these defects throughout the paper. Finally, under the light of the general properties of carbon-group V pairs, C_i-D_s and the other C_i-related defects, we will propose a possible microscopic structure for the stable configuration of C_i-P_s.

DLTS RESULTS

Electron irradiation of carbon rich FZ silicon at 300K starts a series of defect reactions which involves C_i-P_s as well. The DLTS spectrum in Fig. 1a was recorded immediately after irradiation. The peak at E_c-0.1 eV corresponds to the acceptor (-/o) level of C_i. The C_i peak disappears after annealing the sample at 350K and a peak corresponding to the stable configuration of C_i-P_s is observed at E_c-0.38 eV along with the C_i-C_s peak, as shown in Fig. 1b. The E_c-0.38 eV peak disappears after injecting holes electrically at 240K and a new peak is observed at E_c-0.23 eV as shown in Fig. 1c.

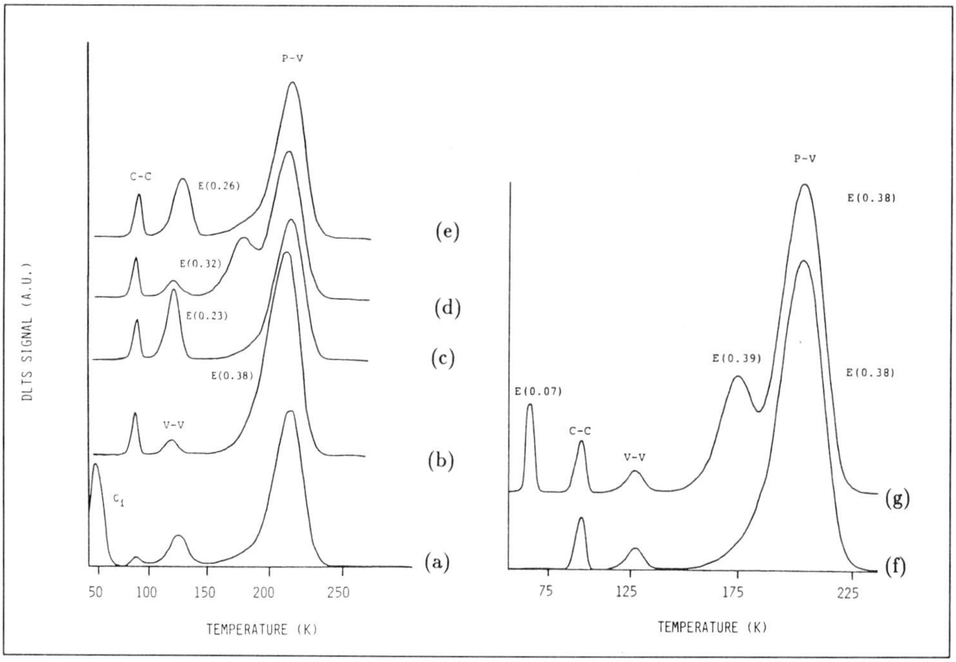

Fig. 1 DLTS spectra showing the five configurations of C_i-P_s pairs. a) After e^- irrad. at 295K; b) After 350K, 1800 sec zero bias (zb) annealing of C_i; c) After injection at 240K, 300 sec (0.38 →0.23 eV); d) After (zb) annealing at 260K, 300 sec (0.23→0.32 eV); e) After zb annealing at 290K, 300 sec (0.32→0.26 eV); f) same as b); g) After injection of holes at 140K, 300 sec (0.38→0.07/0.39 eV).

A new DLTS peak is observed at E_c-0.32 eV after the E_c-0.23 eV peak disappears as a result of a 260K anneal as shown in Fig. 1d. Further annealing of the sample at 290K under zero bias gives another peak at E_c-0.26 eV and the E_c-0.32 eV peak disappears as shown in Fig. 1e. The configurations with 0.32 eV and 0.26 eV are bistable in that anneal of the sample at 250K under reverse bias regenerates the E_c-0.32 eV peak. The cycle is completed when the sample is annealed at 350K at which conversion back to the stable configuration take place. This cycle can be repeated indefinitely as long as the injection temperature is below 260K. Above this the pairs start to dissociate and C_i is regenerated.

Figures 1f and 1g show two DLTS spectra associated with the fifth configuration. Fig. 1f is the same as Fig. 1b. Electrical injection of holes at 140K introduces two new energy levels at E_c-0.07 eV and E_c-0.39 eV as shown in Fig. 1g. These peaks always appear and disappear (185K) together. They have the same peak height independent of the sample cooldown conditions. During the DLTS scan shown in Fig. 1g, the new configuration would have converted back to the stable configuration before the E_c-0.38 eV peak was observed at 215K. The DLTS peaks at E_c-0.07 eV and E_c-0.39 eV are not observed if the sample is injected at 140K before C_i is annealed or after the C_i-P_s is dissociated. Neither are they observed if one of the three metastable configurations of C_i-P_s is populated and then the sample is injected at 140K.

TSCAP RESULTS

The TSCAP spectra shown in Fig. 2a establish the connection between the fifth and the stable configurations. Spectrum A was recorded after annealing of C_i and zero bias cooldown. Spectrum B is the reference run recorded after reverse bias cooldown with no charging step. Spectrum C was recorded after injection of holes at 140K followed by a zero bias cooldown to 77K. The 0.07 eV emission is too shallow to be observed in a TSCAP experiment but its presence can be inferred from its replica[12]. By comparing capacitance changes we conclude that the number of electrons trapped in the E_c-0.39 eV configuration is equal to the number of holes emitted during the 0.07/0.39 eV → 0.38

eV conversion. This is in turn the same as the number of electrons in the E_c-0.38 eV emission. We therefore conclude that the 0.07/0.39 eV configuration is a fifth configuration of C_i-P_s and it converts to the stable configuration after injection of holes at 140K. Reverse conversion takes place at 185K.

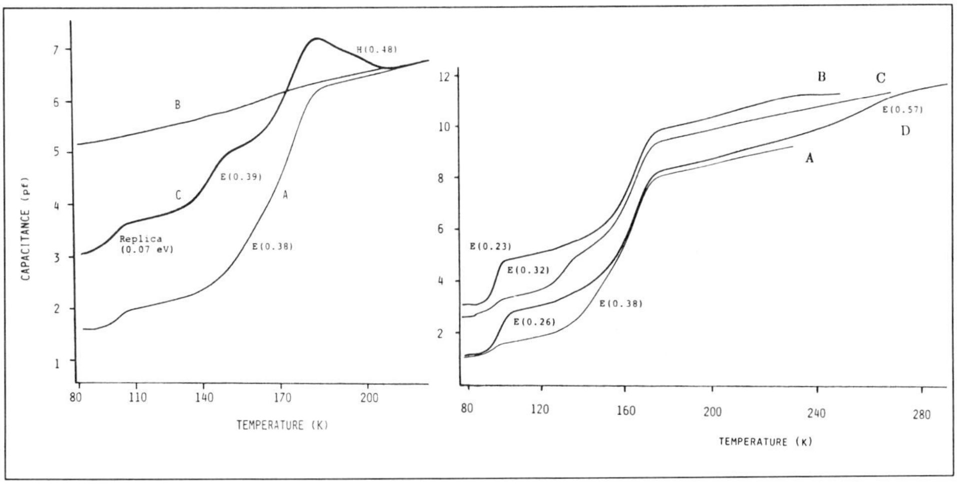

(a) (b)

Fig. 2 TSCAP spectra showing the multistability of the ***a) fifth configuration*** : *A), After annealing of* C_i *(zb cooldown) ;B) reverse bias cooldown ; C) After injection at* $140K$ *(zb cooldown);* ***b) original four configurations of*** C_i-P_s ***pairs***: *A) after anneal of* C_i; *B) After injection of holes at* $240K$ *(zb cooldown); C) After zb anneal at* $260K$ *(zb cooldown); D) After zb anneal at* $290K$ *(zb cooldown).*

Fig. 2b shows a set of TSCAP spectra illustrating junction capacitance changes upon the various conversions among four original configurations. Spectrum A was recorded after annealing of C_i and zero bias cooldown to 77K. The overall junction capacitance increases as a result of injection of holes at 240K as shown in spectrum B. After the 0.23 eV → 0.32 eV conversion only a small portion of capacitance is recovered as shown in spectrum C. But after the 0.32 eV → 0.26 eV conversion, capacitance almost completely returns to its value before 240K injection of holes. This is shown in spectrum D. Also in spectrum D, we observed a new electron emission around 250K. If the sample is cooled down after this run the result is the same as spectrum C. Therefore the new electron emission which we associate with an energy level at E_c-0.57 eV is accompanied by the 0.26 eV → 0.32 eV conversion during the TSCAP run. These results suggest that E_c-0.38 eV and E_c-0.26 eV levels are of the same charge state. Similarly, E_c-0.23 eV and E_c-0.32 eV levels have the same charge state. Additionally, the E_c-0.57 eV emission has the same step height as the other four emissions of the C_i-P_s (after an edge effect correction[12]). We therefore conclude that these two groups differ by one charge and the first group is more negative. To assign the charge states we further investigated junction capacitance changes during the formation and dissociation of the pair. During the formation, junction capacitance decreases; a further evidence of the involvement of P_s^+ in this defect. After the dissociation of the pair around 400K, the capacitance increases and stays very close to the level observed after 240K injection which populates the E_c-0.23 eV configuration. This means the E_c-0.23 eV level has the same charge state as P_s^+, namely it is a donor (o/+) level. Similar results were obtained using 1 MHz and 12 MHz capacitance meters.

We propose the energy level diagram for the four original configurations in Fig. 3a. It consists of a pair of an acceptor (-/o) and a donor (o/+) level which are separated by U ~ 0.3 eV for each configuration. The donor (o/+) levels of IA and IIA were observed for the first time . According to Fig. 3a, energy levels get shallower as the conversions IA → IIA → IIB → III are induced. Fig. 3b shows the charge state assignment for the two energy levels of the fifth configuration. It consists of a single acceptor (-/o) level at E_c-0.07 eV and a single donor (o/+) level at E_c-0.39 eV. After 140K injection, junction capacitance increases ; consistent with the introduction of a donor level at E_c-0.39 eV in the upper half of the Si band gap. This assignment is also consistent with the observed hole emission $(C_i\text{-}P_s)^+ \rightarrow (C_i\text{-}P_s)^o + h^+$ during the conversion back to the stable configuration.

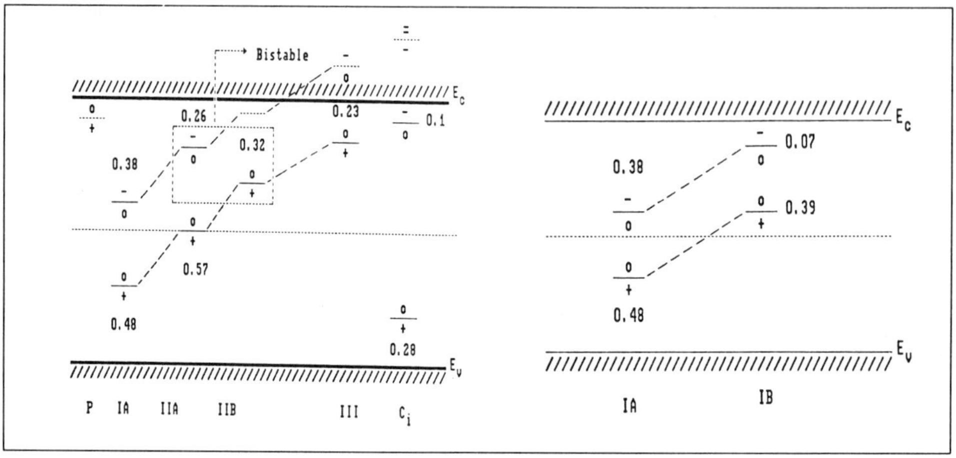

(a) (b)

Fig. 3 Charge state assignment for the a) four original configurations; b) fifth configuration.

KINETICS STUDIES

All of the conversions studied are thermally activated and they can be described by first order kinetics equations. Table 1 shows the resultant kinetics equations obtained for various conversions under the indicated bias conditions. Current density dependence of the 0.38 eV → 0.23 eV conversion rate clearly shows two distinct regions. In the hole capture limited region ($J< 0.1$ A/cm^2), rate $\propto J^2$ suggesting that the conversion involves a capture of two holes. In the electron capture limited region ($J \geq 0.1$ A/cm^2), the conversion rate saturates. This implies a greater hole capture cross section than that for electrons for the E_c-0.38 eV level. This conclusion is consistent with the (-/o) charge state assignment of the same level. As shown in Table 1, the energy barrier for the same conversion is independent of the current density (hole concentration) whereas the preexponential factor decreases with decreasing current density. These results are suggestive of the charge state effect as the conversion mechanism where 0.54 eV is the barrier for the motion of C_i with a nearby P_s. The preexponential factor simply represents the probability of the defect being in the positive charge state. But our analysis of the preexponential factor for the 0.38 eV → 0.07 eV conversion suggest that an energy-release mechanism[13] may be responsible for this injection-induced conversion. The rest of the conversions take place under thermal annealing conditions. The energy barriers are consistent with C_i motion during the conversions and the preexponential factors suggest single jump atomic motion.

$\nu=9.8 \times 10^8\ exp(-0.54\ eV/kT)$	0.38 → 0.23 eV	J=2.5 A/cm^2
$\nu=1.9 \times 10^8\ exp(-0.54\ eV/kT)$	0.38 → 0.23 eV	J=0.08 A/cm^2
$\nu=3.0 \times 10^6\ exp(-0.22\ eV/kT)$	0.38 → 0.07 eV	J=2.5 A/cm
$\nu=2.0 \times 10^{12}\ exp(-0.49\ eV/kT)$	0.07 → 0.38 eV	zero-reverse
$\nu=1.7 \times 10^{13}\ exp(-0.77\ eV/kT)$	0.23 → 0.32 eV	zero bias
$\nu=9.2 \times 10^{12}\ exp(-0.86\ eV/kT)$	0.32 → 0.26 eV	zero bias
$\nu=1.5 \times 10^{12}\ exp(-0.70\ eV/kT)$	0.26 → 0.32 eV	reverse bias
$\nu=2.8 \times 10^{16}\ exp(-1.25\ eV/kT)$	0.26 → 0.38 eV	zero bias

Table 1. Experimental conversion rate equations for C_i-P_s pairs.

STRUCTURAL MODELS

A Coulomb interaction exists between the oppositely charged $C_i^{=,-}$ and P_s^+. In the simple ionic model of point charges in a uniform medium of dielectric constant ϵ, the total energy is lowered upon the pair formation by $2e^2/\epsilon r$ for $(C_i^{=}\text{-}P_s^+)^-$ and $e^2/\epsilon r$ for $(C_i^-\text{-}P_s^+)^o$ from that of $C_i^=$ and C_i^-.

In this model the level positions of the various configurations are predicted from the total energies as

$$(C_i\text{-}P_s)\ (\text{-}/o) = E_{c_i}\ (=/\text{-}) - \frac{e^2}{\epsilon r}$$

$$(C_i\text{-}P_s)\ (o/+) = E_{c_i}\ (\text{-}/o) - \frac{e^2}{\epsilon r}$$

These level positions are consistent with Fig. 3. Donor (o/+) and acceptor (-/o) levels get shallower as the defect moves through its configurations. Also shown are the energy levels of C_i for $r \rightarrow \infty$. The pair separation for the stable configuration is estimated, using the above equations, as 1.9 A° which is very close to a typical nearest-neighbor C-P bond length. Similar results were obtained for the C_i-As_s and C_i-Sb_s pairs where the stable configuration corresponds to the closest pair separation of the typical carbon-donor bond length[4]. Under the light of these results we propose the structure shown in Fig. 4 for the stable (close pair) configuration of C_i-P_s pairs. It is a <100> oriented C-P molecule (each atom is three fold coordinated) which occupies a single lattice site. This structure is very similar to that of C_i^+ and C_i^- and has a C_{2v} symmetry consistent with the recent EPR results[14].

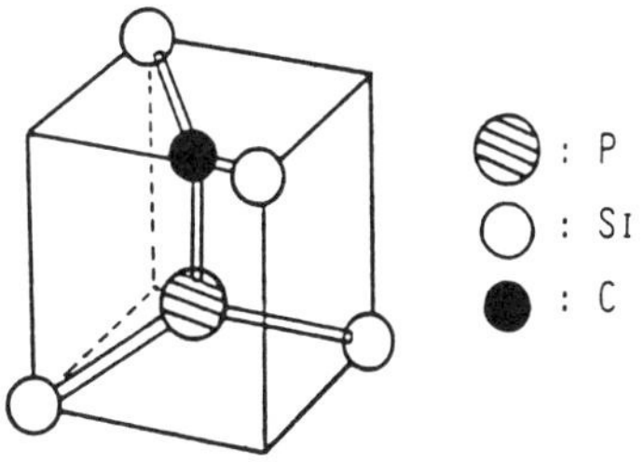

Fig. 4 Proposed structure for the stable configuration of the C_i-P_s pair defects.

In the case of C_i-As_s and C_i-Sb_s pairs[4], we observed a C_i-C_s like bistability for which a molecular bond-switching mechanism was proposed[9,10]. The fifth configuration of C_i-P_s appears to have a different bonding arrangement than other configurations. This is supported by the smaller barrier height (0.22 eV) for the 0.38 eV → 0.07 eV conversion as compared to that of the 0.38 eV → 0.23 eV conversion (0.54 eV). Therefore a bond-switching mechanism may be responsible for the conversion to the fifth configuration. By comparing all three carbon-donor pairs we see that energy levels get deeper and the energy barriers for the conversions become larger with increasing donor atom size. This suggests that, in addition to the ionic model (similar to that was proposed for the Fe_i-Al_s pairs[15]) and the bond-switching mechanism, there is a short range, attractive interaction between C_i and D_s whose magnitude is proportional to the size of the donor atom.

CONFIGURATION COORDINATE ENERGY SURFACES

We have constructed the CC energy surfaces for the four original configurations of C_i-P_s pairs depicted in Fig. 5a. Solid lines represent the actual experimental measurements whereas the dotted lines represent estimates that have larger uncertainty. Kinetics results under injection were not included in Fig. 5 due to the complications of relating the measured activation energy to a barrier height. However, these results were used extensively to investigate conversion mechanisms. A small lattice relaxation model was used in Fig. 5. Conversion due to a charge state effect involves the capture of two holes during injection which puts the defect in the positive charge state. This switches off the Coulomb interaction between C_i and P_s and C_i can move a few lattice sites away. If the sample temperature is kept around 240K during injection, C_i locks on the farthest (and the least stable) configuration with a level at E_c-0.23 eV. Since the pair does not dissociate there must be an additional lattice mediated attractive interaction between C_i and P_s. The rest of the conversions are energetically downhill and they take place under thermal annealing conditions. Energy barriers get bigger as the configurations become more stable which is consistent with smaller pair separation. The CC energy surfaces shown in Fig. 5a also show the 0.26 eV ↔ 0.32 eV bistability. After the electron from the donor (o/+) level at E_c-0.57 eV is emitted, conversion to the 0.32 eV configuration takes place in the positive charge state over the 0.7 eV barrier.

The fifth configuration only interconverts with the stable configuration and it may constitute a separate branch. Fig. 5b shows the proposed CC energy surfaces for these two configurations. The 0.38 eV→0.07 eV conversion is uphill in energy for all three charge states and may be due to a recombination-enhancement mechanism[13]. The 0.07 eV → 0.38 eV conversion first takes place in the positive charge state and it is followed by the hole emission from the donor (o/+) level at E_v+0.48 eV.

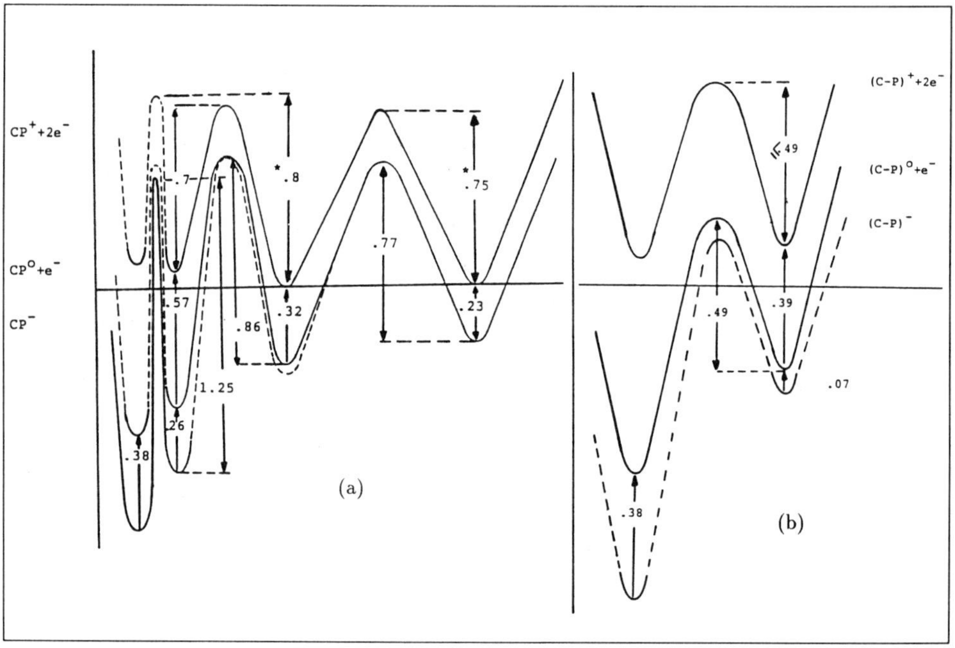

*Fig. 5 CC energy surfaces for the a) four original configurations; b) fifth configuration; *estimates.*

This work was made possible by support from the Office of Naval Research Contract No. N00014-84-K-0025.

REFERENCE

1) L. W. Song, B. W. Benson and G. D. Watkins, Phys. Rev. B 33, 1452 (1986).
2) A. Chantre, J. L. Benton, M. T. Asom and L. C. Kimerling, Mat. Sci. For. Vol. 10-12 (1986) p. 1111.
3) L. C. Kimerling, M. T. Asom, J. L. Benton, P. J. Previnsky and C. E. Caefer, Mat. Sci. For. Vol 38-41 (1989) p.141.
4) E. Gürer and B. W. Benson, in Impurities, Defects and Diffusion in Semiconductors, Mat. Res. Soc. Symp. Proc. Vol. 163, ed. by D. J. Wolford, J. Bernholc and E. E. Haller (MRS Pittsburgh, 1990) p.295.
5) P.M. Mooney, L. J. Cheng, M. Suli, J. D. Gerson and J. W. Corbett, Phys. Rev. B 15, 3836 (1977).
6) L. C. Kimerling, P. Blood and W. M. Gibson in Defects and Radiation Effects in Semiconductors 1978, (Inst. Phys. Conf. Ser. 46, London, 1979) p. 273.
7) G. D. Watkins and K. L. Brower, Phys. Rev. Lett. 36, 1329 (1976).
8) L. W. Song and G. D. Watkins, Phys. Rev. B 42, 5759 (1990).
9) L. W. Song, X. D. Zhan, B. W. Benson and G. D. Watkins, Phys. Rev. Lett. 60, 460 (1988).
10) L. W. Song, X.D. Zhan, B.W. Benson, and G. D. Watkins, Phys. Rev. B 42, 5765 (1990).
11) B. W. Benson and E. Gürer, Mat. Sci. For. Vol. 38-41 (1989) p. 391.
12) C. T. Sah and J. W. Walker, Appl. Phys. Lett. 22, 384 (1973).
13) J. D. Weeks, J. C. Tully and L. C. Kimerling, Phys. Rev. B 12, 3286 (1975).
14) X. D. Zhan and G. D. Watkins, Bull. Am. Phys. Soc. 35, 279 (1990).
15) A. Chantre and D. Bois, Phys. Rev. B 31, 7979 (1985).

Materials Science Forum Vol. 83-87 (1992) pp. 345-350

EPR STUDY OF MULTISTABLE C_i-P_s IN SILICON

X.D.ZHAN and G.D.WATKINS
Physics Department and Sherman Fairchild Laboratory
Lehigh University, Bethlehem, PA 18015 USA

ABSTRACT

Three new electron paramagnetic resonance (EPR) centers are reported, Si-L8, Si-L9 and Si-L10, which we associate with the stable and two of the four metastable configurations, respectively, of a multistable defect recently discovered by deep level transient capacitance spectroscopy (DLTS) in electron-irradiated phosphorus-doped silicon. ^{31}P and ^{13}C hyperfine interactions are observed in the spectra which serve to confirm the identificaton as a C_i-P_s pair. Models for the atomic configurations of the defect will be presented.

1 Introduction

A multistable defect has recently been discovered by DLTS in electron-irradiated phosphorus doped silicon.[1–6] As reported, the defect has five distinct lattice configurations which produce various energy levels in the forbidden gap of the material. The stable configuration, formed after electron irradiation and 30 min. 350K anneal, introduces a single acceptor level $(-/0)$ at $E_c - 0.38$ eV or E(0.38).[7] It converts to other metastable configurations under various injection and thermal conditions. In particular, the stable configuration converts to one metastable configuration after electrical minority carrier injection at T$\sim$ 140K, which produces a single acceptor level $(-/0)$ at E(0.07) and a single donor level $(0/+)$ at E(0.39). It converts back to the stable configuration at T$\sim$ 180K. Electrical injection at higher temperatures (260K$\sim$300K) brings the defect into another metastable configuration with an acceptor level at E(0.26).

Concentration dependence studies have shown that the DLTS signal of the defect increases linearly with the concentration of the carbon[4] and phosphorus[3] in the material. In addition, the emergence of the defect signal was found to accompany in one-to-one correspondence the disappearance of an acceptor level at E(0.10), previously identified with interstitial carbon.[8] Therefore the defect was tentatively identified as an interstitial-carbon-substitutional-phosphorus C_i-P_s pair. Direct confirmation, however, requires correlated EPR work.

In this paper, we report our EPR experiments on the defect, which we believe provide, for the first time, a clean, unambiguous identification of the defect. We confirm that the multistable defect observed in the phosphorus-doped material arises from the C_i-P_s pair.

2 Experimental Procedure

The samples used in this study fell essentially into two categories: normal samples and ^{13}C enriched samples. They were all phosphorus-doped floating zone silicon. The normal samples were obtained from Merck Chemical Co. and the ^{13}C-enriched samples were specially grown

Table 1: EPR Samples

Sample	[P] (cm^{-3})	[^{12}C] (cm^{-3})	[^{13}C] (cm^{-3})	[O] (cm^{-3})
C85-15B	3.5×10^{16}	2.0×10^{16}	—	—
WAC-2	2.3×10^{16}	$< 5.0\times10^{15}$	1.1×10^{17}	$< 1.0\times10^{16}$

for us by Wacker-Chemitronic of Germany. Table 1 gives the chemical information for the samples used in the EPR study.

The EPR experiments were carried out on a 14 GHz spectrometer operating in dispersion at T = 20 ~ 40K. Optical injection was accomplished by shining ~2W of 1.16 eV Nd:YAG laser light through a long quartz pipe onto the sample.

3 Results and Discussions

Normal samples

In the C85-15B series samples, a new EPR spectrum has been detected at 30K in dispersion after room temperature electron irradiation (electron fluence ~ 6.0 to $8\text{x}10^{16}e^-/cm^2$) and 30 min. 350K anneal. The new EPR spectrum, labeled Si-L8, is a simple S=1/2 spectrum with C_{2v} symmetry, and with no resolved hyperfine structure. Upon 1.16eV Nd:YAG laser excitation at ~140K, Si-L8 disappears and another new EPR signal, Si-L9, grows in. The Si-L9 spectrum is also a simple S=1/2 spectrum but with lower symmetry of C_{1h}.

Fig.1(a) and (b) shows the EPR spectra of Si-L8 and Si-L9 with the magnetic field $\vec{B} \parallel <100>$. As seen in the figure, the Si-L8 spectrum has two resonance lines with intensity ratio 1:2, which is characteristic for a center of C_{2v} and C_{1h} symmetry. In the Si-L9 spectrum, however, each line is now split into two equally intense lines, revealing the hyperfine interaction with a nucleus of spin 1/2 (100% abundant). In our sample, phosphorus is the only possibility. The resolved hyperfine structure therefore reveals the incorporation of phosphorus in the defect center.

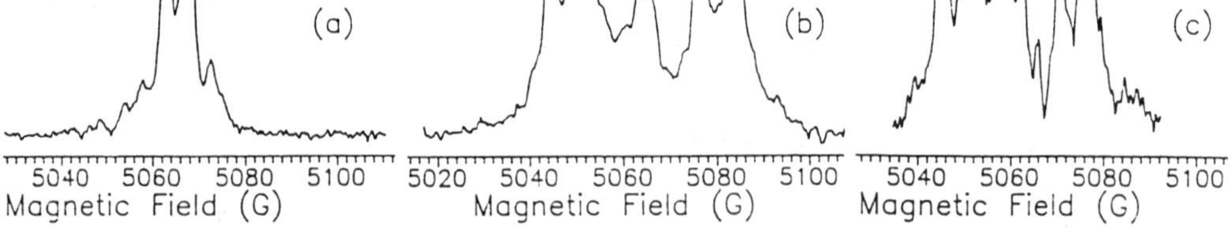

Figure 1: EPR spectra in n-type irradiated silicon at T~30K, ν = 14GHz, $\vec{B} \parallel <100>$. (a) Si-L8; (b) Si-L9; (c) Si-L10.

A third spectrum, Si-L10, was seen immediately after electron irradiation. The spectrum is again an S=1/2 spectrum, but with even lower symmetry, C_1. ^{31}P hyperfine lines were also resolved in Si-L10, as seen in Fig.1(c), confirming again the invovlement of a phosphorus atom.

All of the three spectra can be described by a simple Hamiltonian $\beta\vec{B}\cdot\tilde{g}\cdot\vec{S}+\vec{S}\cdot\tilde{A}\cdot\vec{I}$. We find very similar principal g-values for the three centers, all close to 2. The hyperfine interactions with ^{31}P in Si-L9 and Si-L10 are nearly isotropic and of the same order of magnitude as that found for phosphorus-vacancy pair,[9] where only 1% of the spin wavefunction was accounted for on the phosphorus.

Si-L8 and Si-L9 are found to be interconvertible. As mentioned above, 140K optical excitation converts Si-L8 into Si-L9, consistent with the corresponding DLTS measurements.[10] The recovery of Si-L8 from Si-L9 is a thermally activated process. Isochronal and isothermal anneals were performed in the temperatures range 160-180K on the sample after Si-L8 had been converted to Si-L9. A monotonic decrease was observed in the intensity of Si-L9 with a corresponding increase in Si-L8. The kinetics of the recovery process was studied and the time constant of recovery was found to be

$$\tau \sim 10^{13}\exp(-0.5\mathrm{eV/kT}) \quad \mathrm{sec.} \tag{1}$$

This is also in good agreement with the corresponding results from the DLTS studies.[10] From the generation and anneal of Si-L9, combined with the energy level estimate in the EPR studies for Si-L8 and Si-L9,[11] we conclude that we are observing the same defect by EPR as that discovered in DLTS. In particular, Si-L8 corresponds to the E(0.38) and Si-L9 to the E(0.07/0.39) configuration.

The correlation of Si-L10 with DLTS emission E(0.26) is based on the observation that they both appear and disappear under the same conditions. In other words, they are both seen immediately after electron irradiation and they both anneal out at around 340K. Therefore we tentatively associate Si-L10 to the E(0.26) configuration.

Application of uniaxial stress along a $< 110 >$ direction at T~330K produces a preferential alignment of Si-L8, which can be studied by cooling the sample to room temperature with the stress on, removing the stress, and monitoring the EPR intensities. The change in the intensities of the spectral lines associated with different defect orientations, as shown in Fig. 2(a), reflects the defect response to the uniaxial stress, and therefore provides additional clues as to the structure of the defect.

Writing the energy of the defect in an applied strain, as $E = \sum B_{ij}\epsilon_{ij}$, analysis of the observed alignment gives for the traceless components of the piezospectroscopic tensor $B_{11} \simeq 8.3$eV, $B_{22} \simeq -0.5$eV, and $B_{33} \simeq -7.8$eV. All other components are zero by symmetry. The axes are the 1,2,3 principal axes of the defect as shown in Fig. 3(a). These components are remarkably similar to those previously determined for the isolated C_i^- Si-L6 center[8] and C_i^+ Si-G12 center.[12] They are sizeable coupling coefficients and may be interpreted as reflecting substantial bonding character for the carbon atom.

Recovery from the alignment was studied by a series of isothermal anneals. The results are shown in Fig. 2(b). The activation energy required for the reorientation was estimated to

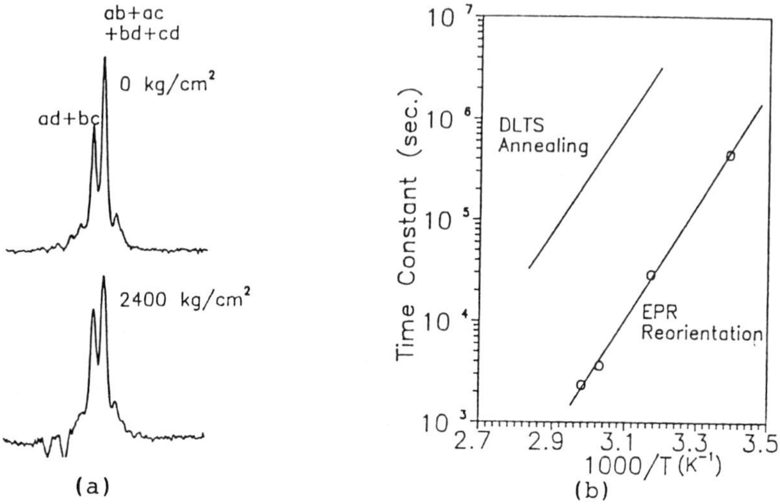

Figure 2: (a) The Si-L8 spectrum before and after 330K compressional stress $\sigma \parallel < 0\bar{1}1 >, \vec{B} \parallel < 100 >$; (b) the kinetics for the reorientation of Si-L8 after releasing the stress.

be 1.1eV, essentially the same as that determined previously for the loss of the defect upon annealing,[13] also shown in the figure. This suggests the same type of motion involved in the reorientation and the annealing, the C_i interstitial separating from the phosphorus atom but either returning in a different orientation or, less often, migrating away, leading to the loss of the defect. The activation energy is 0.3 eV higher than that for the migration of isolated C_i,[8] reflecting perhaps the component of binding energy between C_i^- and P^+ which must be overcome in the separation.

^{13}C-enriched samples

In ^{13}C-enriched samples, the three EPR centers were also seen under the same conditions. In particular, in spectrum Si-L8, well resolved hyperfine interaction with a single ^{13}C nucleus was observed, confirming unambiguously the presence of the carbon in the defect complex. Angular dependence of the spectrum reveals that the hyperfine parameters ($^{13}A_1 = 4.9$, $^{13}A_2 = 49.8$, $^{13}A_3 = 5.2 \times 10^{-4}$ cm^{-1}) are very close to those previous determined for isolated carbon C_i^+.[12] The hyperfine interaction is therefore very anisotropic and has almost axial symmetry around the 2-axis (See Fig.3). An LCAO analysis indicates about 50% of the defect wavefunction localized on the carbon, with mostly 2p-like character (95%) pointing along the 2-axis.

No ^{13}C hyperfine is resolved for either Si-L9 and Si-L10, indicating redistribution of the spin wavefunction away from the carbon atom in these configurations. We note the similarity to C_i^- and $(Ci\text{-}C_s)^-$, where no or only weak ^{13}C hyperfine was observed.[8,14] .

4 Models and Conclusions

Our EPR results appear to confirm the model for the stable configuration of the defect C_i-P_s pair proposed by E.Gürer et al.[5,6] The atomic structure is shown in Fig.3 for isolated interstitial carbon C_i^+ when X=Si and for C_i-P_s when X=P. The carbon and phosphorus atoms form a < 100 >-oriented interstitialcy, consistent with the C_{2v} symmetry of the Si-L8 spectrum and the almost identical piezospectroscopic tensor $\tilde{B}$ for Si-L8 and isolated interstitial carbon.[8,12]

A simple molecular orbital model, also shown in Fig.3, can explain the strong ^{13}C hyperfine interaction observed in Si-L8, as it does for C_i^+.[8] In the case of C_i^+, the unpaired electron is localized in the nonbonding *p* orbital of the carbon atom due to its larger electronegativity than silicon. The ^{13}C hyperfine interaction is therefore strong, as observed. In the case of $(C_i$-$P_s)^0$, which has two additional electrons, the electrons tend to go first into the phosphorus orbital since it has one extra nuclear charge. After filling the non-bonding *p* orbital of the phosphorus with two electrons, the remaining unpaired electron again ends up in the *p* orbital of the carbon, as shown in the figure. This explains the same type of ^{13}C hyperfine interaction detected for $(C_i$-$P_s)^0$ as that for C_i^+.

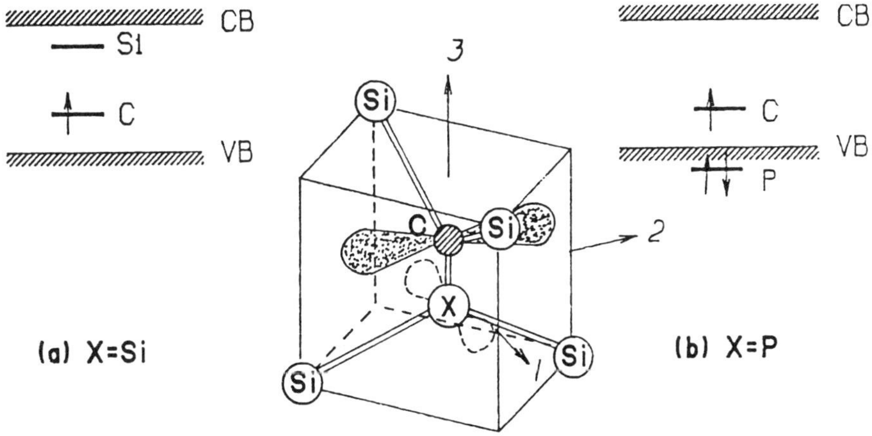

Figure 3: Microscopic models proposed for (a) interstitial carbon C_i^+ (X=Si) and (b) the stable configuration of the (C_i-P_s $)^0$ pair (X=P).

As soon as the carbon moves away, however, our results suggest that the phosphorus becomes substitutional and the neutral defect pair becomes essentially $C_i^- - P_s^+$. In this case the EPR properties reflect therefore those of C_i^- perturbed by a nearby P_s^+. (The weak carbon hyperfine interaction for C_i^- follows naturally from Fig.3.(a), where with two additional electrons to form C_i^-, the unpaired electron is no longer in the C_i-orbital[8]). Based on this simple argument and considering that the Si-L8 to Si-L9 conversion is easy to achieve (at 140K), we propose a possible structure for the Si-L9 center. The carbon moves away from phosphorus and forms a split C-Si interstatialcy with adjacent silicon atom. (In the model of Fig.3, this would correspond to X=Si to form the C-Si interstitial but with one of the upper corner Si atoms replaced by P). This structure is consistent with the EPR observations: (a) The Si-L9 spectrum has C_{1h} symmetry. (b) The ^{31}P hyperfine structure appears in the spectrum since the phosphorus is no longer at the node of the spin wavefunction. (c) ^{13}C hyperfine interaction is absent because now the unpaired electron is not in the non-bonding *p* orbital of the carbon atom, as argued for the C_i^-

Si-L6 center.[8]

In conclusion, we have observed three configurations of the multistable defect C_i-P_s pair by EPR. With the power of magnetic resonance, we are able to identify the chemical entities involved in the defect and provide confirmative evidence to the microscopic structure of the atomic configurations of the defect.

This work was supported by the United States Office of Naval Research under contract No. N00014-84K-0025 and grant No. N00014-90J-1264.

References

1. L.W.Song, B.W.Benson and G.D.Watkins, Phys Rev. **B33**, 1452 (1986)
2. A.Chantre and L.C.Kimerling, Appl. Phys. Lett. **48**, 1000 (1986)
3. A.Chantre, J.L.Benton, M.T.Asom and L.C.Kimerling, Mat. Sci. Forum, **10-12**, 1111(1986
4. L.W.Song, B.W.Benson and G.D.Watkins, Appl. Phys. Lett. **51**, 1155 (1987)
5. E. Gürer and B.W.Benson, in *Impurities, Defects, and Diffusion in Semiconductors: Bulk and layered Structures*, ed. by D.J.Wolford, J.Bernholc, and E.E.Haller (MRS Symposium Proceedings Vol.163, Pittsburgh, 1990), p.295.
6. E. Gürer, B.W.Benson and G.D.Watkins, this conference
7. The level position for the stable configuration is given as $E_c - 0.40$ eV in Ref. 1, $E_c - 0.30$ eV in Ref.2-4, and $E_c - 0.38$ eV in Ref.5. All refer to the same DLTS peak, the uncertainty in the level positon coming from the difficulty in estimating the temperature dependence of the capture cross section. In this paper, we will use the more recent estimate of Ref.5
8. L.W.Song and G.D.Watkins, Phys. Rev. **B42**, 5759 (1990)
9. G.D.Watkins and J.W. Corbett, Phys. Rev. **134**, A1359 (1964)
10. E. Gürer, Ph.D. dissertation, Lehigh University, 1990 (unpublished)
11. X.D.Zhan and G.D.Watkins, Appl. Phys. Lett., **58**, 2144 (1991)
12. G.D.Watkins and K.L.Brower, Phys. Rev. Lett., **36**, 1329 (1976)
13. L.W.Song, Ph.D. dissertation, Lehigh University, 1988 (unpublished)
14. L.W.Song, X.D.Zhan, B.W.Benson, and G.D.Watkins, Phys. Rev. **B42**, 5765 (1990)

Materials Science Forum Vol. 83-87 (1992) pp. 351-356

OBSERVATION OF A CONFIGURATIONALLY UNSTABLE DEFECT IN Si

C.A. Londos
University of Athens, Physics Department (Solid St. Section)
Panepistimiopolis - Zografos, GR 157 71 GREECE

ABSTRACT

Boron-doped, Czochralski Si Schottky diodes were irradiated *in-situ* at 80K with 1.5 MeV electrons. DLTS studies were concentrated on a defect state at E_V+0.34eV which exhibits charge-dependent characteristics. It is stronger when quenching the samples under zero bias than under reverse bias prior to begining the DLTS pulsing. No complementary peaks were detected in these spectra when scanning upwards in the latter case. Surprisingly enough, when the temperature is ramped downwards from T> 180K a peak H appears irrespective of a previous annealing with or without the application of a reverse bias. When this peak arises the height of the E_V+0.34eV peak is decreased. A model is suggested invoking metastability to account for our observations. A microscopic picture of the defect is also suggested based on the variation of oxygen clustering around the defect core.

INTRODUCTION

The DLTS technique has been proven very valuable in studying alternate structures of metastable defects. In general, each structure has its own electronic states which give rise to corresponding DLTS peaks. The appearance of a specific defect configuration is achieved by preparing the defect in the proper charge state by electrical or optical excitation following by (a fast) cooling to low temperature[1].

A number of studies of various metastable defects in Si has been reported in the recent years[2]. In this letter we report the detection of a configurationally multistable center in boron-doped Si. It is suggested that entropy changes cause changes in the energy ordering of local minima of Gibbs free energy, leading to spontanious thansformations between the defect configurations at a critical temperature. A microscopic modeling of the phenomenon is attempted.

EXPERIMENTAL DETAILS

Schottky diodes were prepared by evaporation of Aluminum on carefully cleaned, boron-doped, Czochralski-grown prepolished Si wafers. This was done in high - vacuum evaporator at a presure of 10^{-6} Torr. Irradiation was performed *in-situ* at 80 K with 1.5MeV electrons to a constant dose of a about $2x10^{16}$ e^{-}/cm^{2} . Measurements were taken by using a standard DLTS experimental set - up[3], with increasing /decreasing temperature at a constant heating / cooling rate of 0.15K/ s, in the range 80-220K.

EXPERIMENTAL RESULTS AND DISCUSSION

Fig.1 presents the DLTS spectrum in the temperature range below 220K. This spectrum was received after annealing out of the vacancies by heating up the specimens at 220K for 30 mins.It contains four traps specified as H(0.13), H(0.29) , H(0.34) and H(0.38). In this notation H stands for holes and the number in brackets denotes the energy depths of the traps from the valence band edge respectively. Preliminary discussion on the obtained spectra appeared in reference 4. The solid curve shows the observed peaks after a 5 min anneal at 220K with bias off (V_R= 0) quench to 80K and then scanning upwards. The dashed curve is the corresponding of the solid one but with the application of a reverse bias (V_R=-5V) during the 5 min anneal. The cooling down procedure in both the above cases was done by keeping the same voltage conditions as during the corresponding anneal. Such a procedure allows us to prepare specific charge states of our samples before applying the DLTS. The variation of the peak amplitude of H(0.34) as seen in figure 1 (solid and dashed curve) indicates that the corresponding center may alternate between two different configurations. The changes in the amplitudes of the H(0.34) peak are observed only for slow enough rate windows which allow the detection of the corresponding peak below 195K. For faster rate windows H(0.34) is always low.

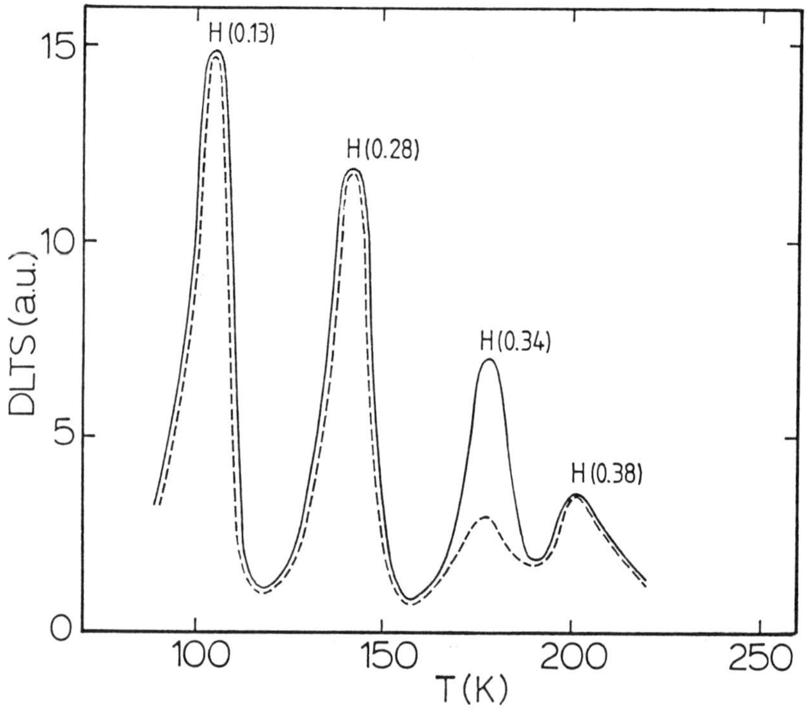

Fig.1 The DLTS spectrum of Si:B irradiated with electrons at 80K. The solid curve was received after cooling down from 220K with bias off and the dashed curve with bias on. The specta were recorded with the temperature ramped upwards.

Surprisingly enough, when the temperature is ramped downwards from values above T-180K an additional feature H appears in the spectra (Fig. 2) irrespective of the previous annealing with or without the application of a reverse bias. Peak H exhibits the following systematic behaviour. Initially it appears on the left side of peak

H(0.13). After a couple of days, while measurements were always restricted below T-220K, H suddenly emerges on the right side of H(0.13) where it finally remains. In this position it appears always in the temperature range just below 120K no matter what rate window is operated. It appears always in the downwards scanning mode when begining from T>180 K and it has the appearance of a peak which is abruptly interrupted during its tracing. For temperatures lower than 120K,peak H disappears but peak H(0.34) is still low. These changes indicate that a specific relationship exists between the appearance/ disappearence of the H peak and theH(0.34) one. When H emerges the amplitude of H (0.34) is low and remains low having an intermediate value between that of the solid and that of the dashed curve in Fig. 1. It is worth noticing that detailed studies and analyses of the data have shown that the two peaks emerge and anneal out simultaneously from the spectra. These facts strongly indicate that the two peaks should be considered as alternative manifestations of the same metastable defect. Finally it is noted that the appearance of the additional DLTS peak in the downward scanning as compared to the upward scanning, was not observed using float zone Si instead the pulled Si.

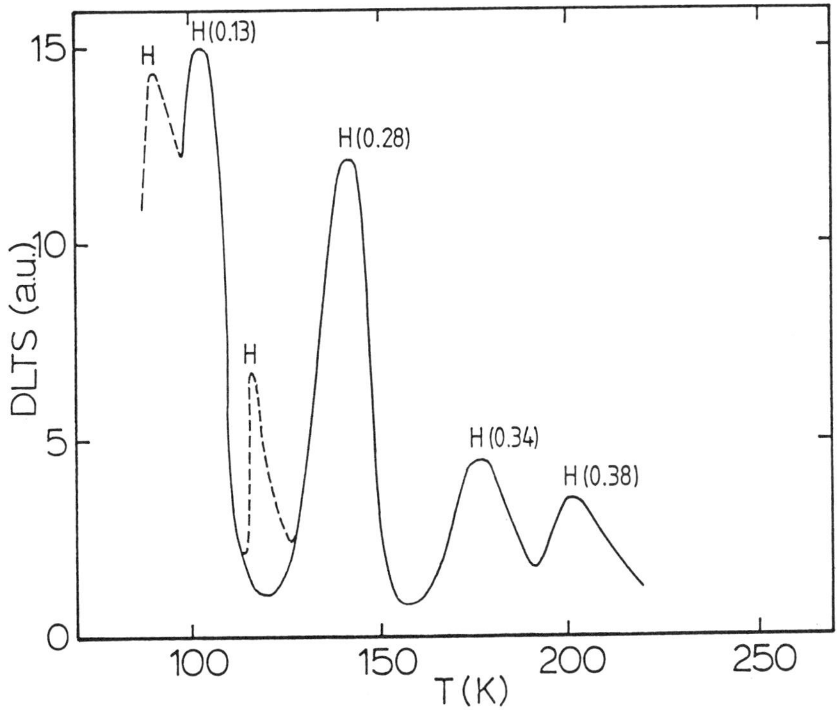

Fig. 2 The DLTS spectrum obtained with the downwards scanning mode from 220K. The additional feature H initially emerges on the left side of H (0.13) peak and finally on its right side.

Recently, Hamilton, Peaker and Pantelides [5] (HPP) reported defect metastabilities detected in electron-irradiated float-zone Si. As the metastabilities they observed did not indicate any changes in their charge-state, they proposed according to earlier theoretical developments[6] that these phenomena manifest entropy variations in the configuration space and they were able to give a macroscopic explanation of their DLTS spectra.

In this letter, we will show that our DLTS spectra can be equally well explained according to the arguments of the HPP theory. Based on the macroscopic explanation of the HPP theory we will discuss a possible microscopic description based on short range order and randomness.

The HPP theory anticipates that stable defect configurations imply a minimum in the Gibbs free energy of formation, G, of the defect. The variation of G with a configuration parameter, as shown in fig.3, indicates regions where stable or metastable defect configurations may exist. This implies that it is not only the formation enthalpies, H, (formation energies) of the defects which will determine the kind of the defect stability but it is also the formation entropies, S, which play a dominant role. Thus according to the HPP theory, critical temperatures Tc exist at which the changes in the formation entropy and the formation enthalpy determine transition regions between various metastable configurations of the defect .

In the present case, like in the HPP theory, we can associate our experimental findings with variations of the Gibbs free energy of formation along a configuration parameter. Thus, we can argue that, annealing the sample with bias-off, it results in free hole capturing into the level leaving the defect in its more positive charge state (C^+). In this charge state the defect can exist in three configurations :A stable, B1 and B2 metastable (curve C^+in Fig.3). Cooling down the sample, the defects freeze in configuration A which gives rise to a strong peak H(0.34) in the spectra. Annealing the sample with bias on leaves the level emply o f holes setting the defect in the less

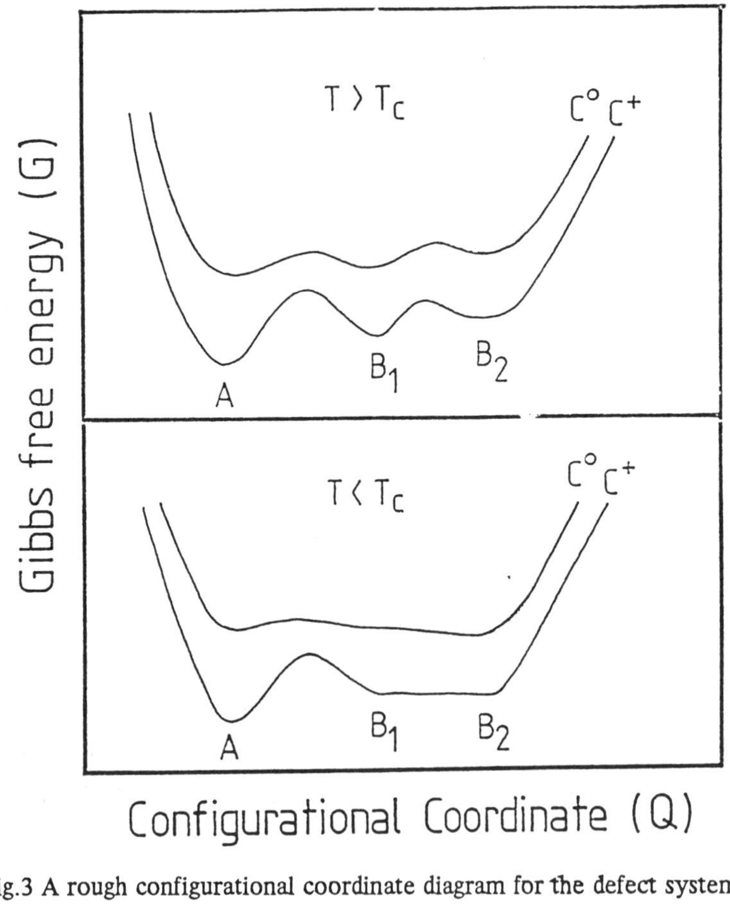

Fig.3 A rough configurational coordinate diagram for the defect system.

positive charge state (C^0),(curve C^0in Fig.3). In this charge state, configurations A and B1 have similar values of Gibbs free energy and therefore both of them could be populated. This explains why H(0.34) still remains in the spectra but with low amplitudes. B_2 is assigned to the defect configuration which is not observed in DLTS, and to which the defect is transformed when the temperature becomes less that T=120K. Its presence is justified from the fact that H(0.34) peak remains low after the disappearance of the H peak. Thus annealing with bias-on sets the defect in B1 but quenching to 80K transforms it to B_2.

Cooling down from 220K while the DLTS system is operating, the defect is partly in configuration B_1 and partly in configuration A. The defect has the tendency to be in A during the applicaton of the filling pulse and in B_1 when the filling pulse is removed[7]. The conversion from A to B_1 during the cool down scanning does not take place in the same extension as in the case leading to the dashed curve of fig.1. This explains the intermediate heightof the H(0.34) peak of Fig.2. It is evident from the proposed configurational coordinate diagram of Fig.3 that the defect could convert to the B_2 configuration only in the C^0 charge state for $T<T_c$.

In the followingwe will atempt a microscopic explanation of our results based on the macroscopic concepts discussed above.The microscopic picture we prepose is based on the additional observation that float-zone Si does not exhibit the above described behaviour in the cool down scanning mode. This allows us to assume that the H peak and the associated changes in H(0.34) are due to the excess oxygen in pulled Si[4].

Irrespective of the charge state of the defect, temperature changes allow either oxygen concentration to vary around the defect or the number of atoms/ions participating in the defect to undergo substantial structural changes. These introduce local electrostatic and strain field changes at the defect as well as changes in the configurational entropy, σ, of the defect. An estimation of the configurational entropy of the defect may be found by considering the defect as a cluster of a Bethe lattice whose central atom has a(temperature dependent in general) coordination number Z. If we assume that all species except oxygen participating in the defect act as A-type and oxygen acts as B-type constituent of the defect, we can find the configuration entropy, σ, of such a defect from the equation[8,9]

$$\sigma=kZ\left(\sum_{i=A,B} x_i \ln x_i - \frac{1}{2}\sum_{i,j=A,B} P_{i/j}\ln\left(\frac{1}{2}P_{i/j}\right)\right) \qquad (1)$$

where k is Boltzman's constant, x_A is the concentration of the A-species, $x_B=1-x_A$ and $P_{i/j}$ denotes the probality a lattice site to be occupied by species i when the neighbouring one is of j-type. The entropy σ, represents the number of possible ways one can populate the sites of a cluster (with the conditions $\sum_{i=A,B} x_i=1$ $\sum_{i,j=A,B} P_{i/j}=1$)[8]. To the entropy σ one can add entropy changes due to any other structural changes of the defect. The various $P_{i/j}$, i,j,=A or B are explicitly known by giving $P_{A/B}$ and x_A (see for example reference 9). The variation of entropy σ for various values of x_A, $P_{A/B}$ and given Z are not so large compared to variations due to changes in Z. For Z values between 6 and 12 and for varous x_A and $P_{A/B}$ the values of σ predicted by eqn.(1) may vary in the range of 2k - 9k justifying the assumptions of the HPP theory.

According to the above description the change in the entropy which results from the variation of the oxygen concentration of the defect is assumed to be much greater than that due only to structural changes of a defect whose oxygen concentration is not changing with the variation of temperature. On the other hand the observed relationship between the H(0.34) and the H peak supports our assumption of oxygen induced changes in the entropy of the defects because it reflects the creation of one type of a defect configuration at the expense of some other one (of the same or different charge state), a condition that can be justified by oxygen movement.

It is therefore the short range order correlation ($P_{A/B}$) and the coordination number Z (for certain concentration value x_A) that specify the conditions for a defect configuration. The temperature may specify the local changes in the coordination number Z. Such changes may be the result of structural changes of/and the capturing of (unbound) oxygen atoms[10].

According to the above discussion the variation of G is due to the changes of both the formation entropy of the defect as predicted by eqn.(1), and the changes in the formation enthalpy. It is understood that at this stage we cannot go beyond the present model assumptions. We cannot, for example, separate the effect of the electronic changes which accompany the structrural changes of the defect from the effect of the entropy changes. Also we cannot exclude possibilities where Si self-interstitials may play a dominant role in capturing oxygen and promote or not its diffusion parameters[10]. Nevertheless the arguments of the present microscopic description indicate that entropy changes should not be underestimated in the consideration of the stability of the defects in Si. At the same time we propose a possible microscopic process which, when engaged in extensive numerical work, may justify the present and the HPP description.

Acknowledgements: I wish to thank Dr. P.C. Bandury of the Physics Department of Reading University, England for the provision of the research facilities and Dr A. N. Andriotis of the Institute of Electronic Stucture and Laser, Heraklion Crete for very helpful discussions

References

1. J.L. Benton and M. Levinson , in Defects in Seminconductors, edited by S. Mahazan and J. W. Corbett (North_Holland, New York, 1983) p.95
2. A. Chantre, in Defects in Electronic Materials, Edited by N. Stavola. S.J. Pearton and G. Davies, MRS Synposia Proccedings (North -Holland, New York 1988) p. 37
3. D. V. Lang, J. Appl. Phys. 45, 3023 (1974)
4. C.A. Londos, Solid State Communications 62, 719 (1987)
5. B. Hamilton, A.R. Peaker and S.T. Pantelides, Phys. Rev. Letters 61 1672 (1988)
6. S.T. Pantelides, Phys. Rev. B36, 3462 (1987), and references therein.
7. M. Levinson, J. Appl. Phys. 58, 2678 (1985).
8. R. Kikuchi, Phys. Rev. 81, 988 (1951)
9. see for example A.N. Andriotis and J.E. Lewther, J. Phys. F Met Physics 16, 1189 (1986) and references therein.
10. V. Gosele and H. Strunk, Appl. Phys. 20, 265 (1979)

Materials Science Forum Vol. 83-87 (1992) pp. 357-362

ODMR AND ELECTRON SPIN ECHO STUDIES ON THE NON-RADIATIVE TRIPLET STATE OF THE $(V\text{-}O)^0$ DEFECT IN SILICON.

A.M. Frens[1], J. Schmidt[1], W.M. Chen[2] and B. Monemar[2]

[1] Department of Physics, Huygens Laboratory, University of Leiden, P.O. Box 9504, 2300 RA Leiden, The Netherlands.

[2] Department of Physics and Measurement Technology, Linköping University, S-581 83 Linköping, Sweden.

ABSTRACT

The excited triplet state of the $(V\text{-}O)^0$ defect in silicon has been studied with Optical Detection of Magnetic Resonance (ODMR) and Electron Spin Echo (ESE) spectroscopy in zero-magnetic field. The relative populating rates of the triplet sublevels behave remarkably different from the decay rates, suggesting that these processes are determined by different mechanisms. The ESE technique in combination with pulsed tunable laser light allows us to study the transfer process between the 0.79 eV (C) defect and the $(V\text{-}O)^0$ complex.

Introduction

Defects in irradiated silicon are among the most widely studied defects in semiconductors. The electronic structure of such defects is in most cases not known, or only partly known. Well known examples that exhibit a characteristic luminescence are the carbon-related 0.97 eV (G-line) and the 0.79 eV (C-line) complexes [1,2]. These centres have been studied extensively via optical spectroscopy [3]. In contrast, the non-radiative (V-O) centre has been investigated mainly by EPR techniques [4,5]. The atomic constituents of these defects are understood quite well but their dynamical properties and the communication amongst them are still mostly unknown.

In this paper we report the results of a study, using ODMR and ESE techniques, on the excited triplet state of the $(V\text{-}O)^0$ complex. A remarkable aspect of these ODMR experiments is that the absorption of resonant microwaves in this non-radiative defect is detected via the luminescence of other centres in the sample. This is a consequence of the fact that the (V-O) defect is an important recombination channel for charge carriers in silicon [6]. The ODMR study, that is carried out in zero-magnetic field, allows us to determine the exact zero-field parameters of the excited triplet state.

The dynamical properties of the triplet sublevels have been investigated with ESE spectroscopy. To our knowledge this is the first application of this technique to an excited state of a defect in a semiconductor. By detecting the ESE signal as a function of the wavelength of a tunable pulsed laser with an output energy below the bandgap, we find that a very efficient transfer of excitation takes place from the luminescent 0.79 eV defect to the $(V\text{-}O)^0$ complex.

Experimental

In the ODMR and ESE experiments we used n-type (74 Ωcm) Czochralski silicon irradiated with 2 MeV electrons ($2x10^{17}$ e^-/cm^2). The sample was immersed in liquid helium pumped down to a pressure corresponding to 1.2 K. The photoluminescence at this temperature showed strong emissions characteristic for the 0.79 eV (C) and the 0.97 eV (G) defects.

In the ODMR experiments, the sample was continuously excited above the bandgap with a Ar-ion laser (Spectra Physics 2016, 514 nm, 50 mW). The luminescence was passed through a 830 nm optical high pass filter and detected with a liquid nitrogen cooled germanium detector (North Coast EO 817S). The sample was placed in a short circuited helix and the microwaves were generated by a frequency synthesizer (HP 8660B).

In the ESE experiments band to band excitation was provided by a pulsed Nd:Yag laser (Spectra Physics GCR-3) at 1064 nm. For selective excitation we shifted the output of a Nd:Yag pumped dye laser system to the infrared region with a Raman shifter filled with 20 bar H_2 gas. With this method it is possible to produce pulses of 5 ns with an energy of several millijoules that are continuously tunable between 0.8 and 3 microns. In the case of the ESE experiments the sample was placed in a tunable (0.9 - 2 GHz) re-entrant cavity [7,8]. The advantage of this system is that a higher B_1 field with a well-defined polarization can be obtained. The echo signal is measured with a heterodyne detection system.

Results and discussion

The optically detected zero-field transitions of the excited triplet state of the $(V\text{-}O)^0$ complex are presented in figure 1. They are observed by monitoring the total intensity of the luminescence of all radiative centres present in the sample (mainly the G and C defect). All three transitions in figure 1 give rise to a decrease of the total luminescence, just as the ODMR experiments in the presence of a magnetic field [9]. The zero-field parameters are given in table 1 and correspond very well with the values derived by Brower from conventional K-band EPR experiments [5]. From this reference we have also taken the labels for

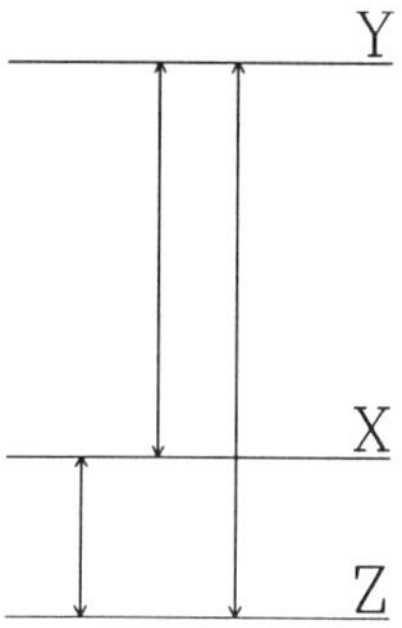

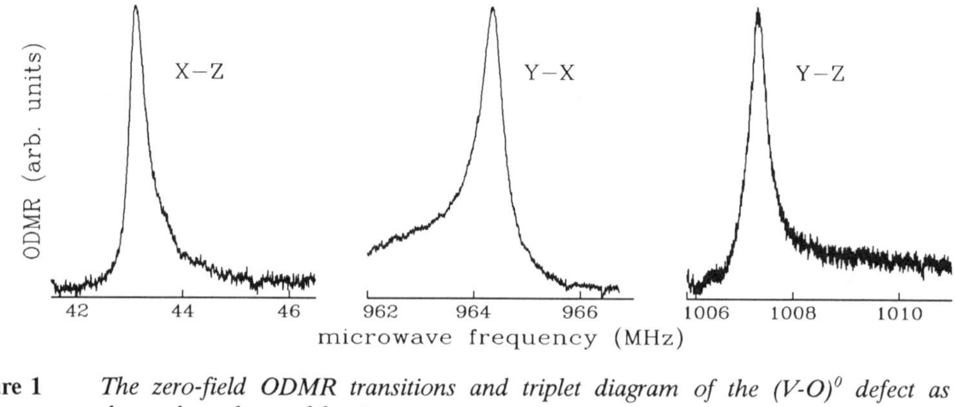

Figure 1 *The zero-field ODMR transitions and triplet diagram of the $(V\text{-}O)^0$ defect as detected on the total luminescence.*

	zero-field energies (MHz)	$k_u(s^{-1})$	P_u^{rel}
T_x	-307.1	4000	1.00
T_y	657.2	769	1.00
T_z	-350.1	1220	1.00

Table 1. *Zero-field energies, decay rates and relative population rates of the triplet sublevels*

the triplet spin sublevels.
The lineshapes of the zero-field transitions are clearly asymmetric. The shoulders on the high energy side of the 1007.3 MHz and the 43 MHz transitions and on the low energy side of the 964.3 MHz transition can be accounted for by second order hyperfine interaction with the ^{29}Si nuclei (I=½, 4.7% abundance) [10]. The asymmetries are dominated by the repulsion of the two closest lying levels T_x and T_z.
The microwave pulse sequence for the ESE experiments in zero-field is shown in figure 2. Three types of experiments were performed. In the first we varied the interval τ between the microwave pulses while keeping the delay t_d between the laser flash and the first $\pi/2$ microwave pulse constant. From the decay of the echo as a function of 2τ we find the dephasing time T_2 to be about 70 microseconds. This corresponds with a homogeneous linewidth of:

$$\Delta\nu_{hom} = \frac{1}{\pi T_2} \simeq 5\,kHz$$

This value has to be compared with the total linewidth of the zero-field transitions of 0.5

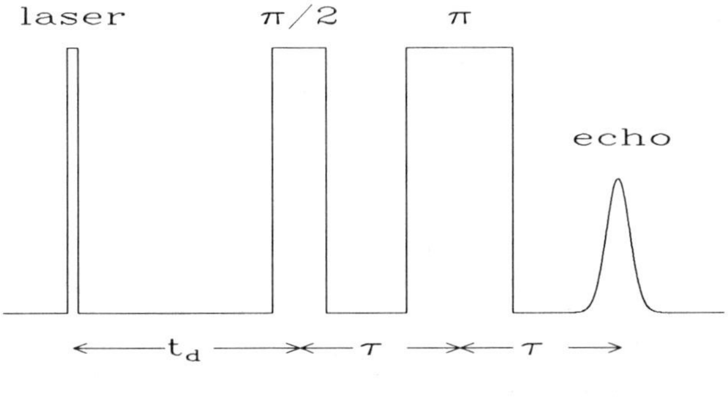

Figure 2. *The Electron Spin Echo pulse sequence.*

MHz. This finding confirms the contention that the zero-field lines are inhomogeneously broadened by second order hyperfine interactions.

In the second type of experiment the delay time t_d is varied while keeping τ fixed. The time dependence of the echo signal then reveals the evolution of the population difference of two triplet sublevels that are connected by the microwaves. The results of such an experiment on the excited triplet state of the $(V\text{-}O)^0$ defect are shown in figure 3 for the 964 and 1007 MHz transitions. The curves show a typical bi-exponential behaviour with time constants and pre-exponential factors that are obtained from a fitting procedure (solid line). The two time constants represent the lifetimes of the two substates involved in the transition and the preexponential factor gives the population difference between these sublevels at $t_d=0$. From the fitting procedure we obtain the decay and relative populating rates as given in table 1.

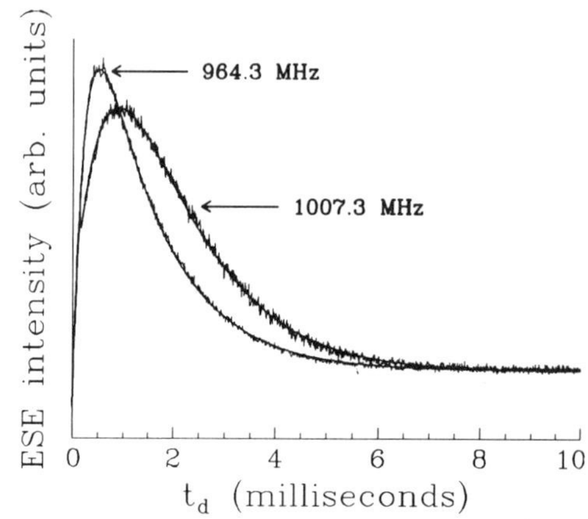

Figure 3. *Echo intensity versus t_d for the 964.3 MHz and 1007.3 MHz transition. The solid line is a fitted curve.*

The remarkable finding from these ESE experiments is that the populating rates of the triplet sublevels are exactly the same within the experimental accuracy whereas the decay rates differ strongly. It is expected that the decay rates are determined by the amount of singlet character mixed into the triplet sublevels via spin-orbit interaction in second order. Since the spin functions correspond with different irreducible representations of the symmetry group C_{2v} of the $(V\text{-}O)^0$ defect we expect this singlet admixture to be very specific and consequently that the decay rates of the sublevels differ strongly. The fact that the populating rates are equal demonstrates that the mechanism governing the creation of the triplet state is not determined by symmetry properties of the defect.

In the third type of ESE experiment we vary the wavelength of the excitation below bandgap with a Raman shifter while keeping t_d and τ fixed. By setting t_d to an optimum value of 500 μs and monitoring the echo intensity as a function of the excitation wavelength we obtain the spectrum of the states in the bandgap connected with the triplet of the $(V\text{-}O)^0$ complex. The result presented in figure 4 shows that the spectrum so obtained is essentially identical to the excitation spectrum of the 0.79 eV defect as reported by Wagner et al. [11]. The spectrum shows the zero-phonon bound exciton (BE) line at 0.79 eV, together with a range of discrete excited donor states, all related to the same C-line BE, superimposed on a slowly varying background. From this we conclude that there is a very efficient transfer of excitation from the excited states of the 0.79 eV defect to the triplet state of $(V\text{-}O)^0$. Comparison of the echo intensities shows that the efficiency of the transfer for selective excitation is similar to that of band to band excitation. We also measured the echo intensity as a function of t_d with this selective excitation. This experiment gives exactly the same results for the relative populating and decay rates as with above bandgap excitation, showing that there is no spin selectivity in the excitation transfer process.

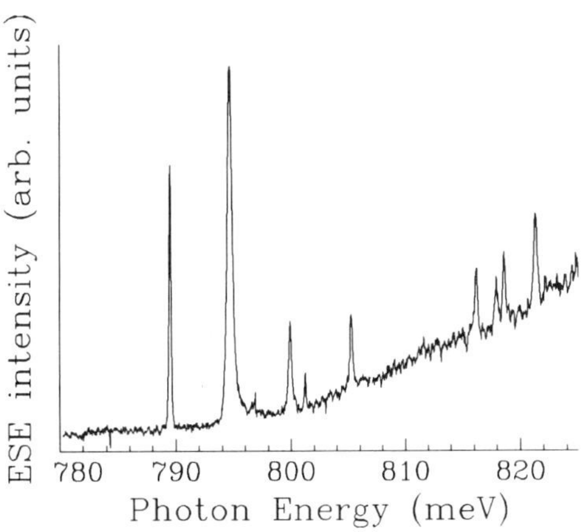

Figure 4. *ESE-detected excitation spectrum of the $(V\text{-}O)^0$ triplet state*

Conclusion

With the help of ESE experiments on the excited triplet state of the $(V\text{-}O)^0$ defect, we have determined the relative populating and decay rates of the triplet sublevels. The remarkable finding is that the decay process is strongly spin selective but that in contrast the relative populating rates are equal. Qualitatively similar results were recently obtained for a Cu-related triplet BE state in GaP [12]. Apparently the population mechanism is not governed by the symmetry properties of the defect in its triplet state.
By using tunable laser light with an energy below the bandgap, in combination with the ESE technique, we find a very efficient transfer of excitation from the 0.79 eV (C) radiative defect to the $(V\text{-}O)^0$ complex. The mechanism responsible for this transfer is not understood yet, and experiments to elucidate this are in progress.

Acknowledgement

We are grateful to J.L. Lindström for the kind assistance in electron irradiation of the samples used in this work. This research is supported by the Commission of the European Communities Science Plan via grant SCI-0363-M.

References

1. K.P. O'Donnell, K.M. Lee and G.D. Watkins, Physica 116B, 258 (1983).
2. K. Thonke, A. Hangleiter, J. Wagner and R. Sauer, J. Phys. C 18, L795 (1985).
3. G. Davies, Physics Reports 176, 82 (1989).
4. G.D. Watkins and J.W. Corbett, Phys. Rev. 121, 1001 (1961).
5. K.L. Brower, Phys. Rev. B 4, 1968 (1971).
6. W.M. Chen, B. Monemar, E. Janzén and J.L. Lindström, to be published.
7. L.E. Erickson, Phys. Rev. 143, 295 (1966)
8. T. Moreno, Microwave Transmission Design Data (McGraw-Hill, New York, 1948)
9. W.M. Chen, O.O. Awadelkarim, J.H. Svensson, B. Monemar and F. Wang, Proc. 20th Int. Conf. on the physics of Semiconductors, eds. E.M. Anastassakis and J.D. Jounnopoulos (World Scientific, Singapore, 1990) p. 601.
10. J. Schmidt and J.H. van der Waals, Chem. Phys. Lett. 3, 546 (1969)
11. J.Wagner, K. Thonke, R. Sauer, Phys. Rev. B 29, 7051 (1984).
12. M.C.J.M. Donckers, J. Schmidt, W.M. Chen and B. Monemar, Phys Rev B 42, 1684 (1990).

Materials Science Forum Vol. 83-87 (1992) pp. 363-368

SPIN-MARKING METHOD IN SILICON: INTERSTITIAL AND PAIRING DEFECTS

S.M.YAKUBENYA

I.V.Kurchatov Institute of Atomic Energy, 123182 Moscow, USSR

Abstract.
A method of charge state identification of a pairing defects and interstitial defects with unfulling core shell in silicon is considered. The role of local effects in the change of a "marking level" in a gap is discussed.

Introduction

The identification of a defect type and charge state is one of the main tasks in doped semiconductor physics. A new method, named by us "spin-marking method " of charge state identification of nonmagnetic defects was proposed [1]. It is based on the fact, that a ground state of the defect, induced by 3d impurity ion , is high-spin. Therefore, different charge states of such defects may be identified very easily by their EPR-spectra .At the same time a deep level position in the gap connected with recharge of such defect will be defined mainly by the properties of the original semiconductor and the influence of the impurity atom will be very small if some requirements on the wave function structure of impurity ion are met .We have used this effect as a basis for a spin-marking method of charge state identification of interstitial centers in silicon [1].Manganese impurity ions located in interstitial positions in a crystal lattice have been used as spin-markers.

The feasibility of these conditions for pairing defects in silicon will be discussed in the paper. Besides, we shall consider the situation when interstitial defects in silicon have an incompletely fulled core sub shell.

Method and its discussion

We define a pairing defect as a defect consisting of one defect located in an interstitial position in a crystal lattice and another defect located in the nearest or the next nearest neighbor substitutional position. A particular case of a substitutional defect is a silicon vacancy. A pairing defect [2] is a particular case of a double defect with a variable distance Δl between the partners. As Δl→∞ we have one limiting case of a double defect (interstitial defect). The other limiting case takes place as Δl→0 and corresponds to a substitutional centers.The intermediate case, Δl~ d, where d is the lattice constant refers to the case of a pairing defect. In a general case, we would have an infinite quantity of such paring defects with different distances between partners. But since the degree of localization of the wavefunction $\eta(r)$ decreases very rapidly with the increasing distance between partners (following an exponential law $\eta_0^2*\exp(-r/r_o)$), the series becomes truncated at Δl ~ d. Increase or decrease of the distance between partners would result in one or the other limiting case of a double defect (either the interstitial or the substitutional defect would realize respectively).

For a example , η_0^2 and r_0 for V_{si}^- - silicon vacancy is equal to 0.27 and 3.5 A respectively [3]. Hereinafter the notation t^{DBH} is used for an electron state formed of band states [4].

In a particular case, state $(t_2)^{DBH}$ of transition-metal impurity ions can be represented as a superposition of band and impurity states [4].

$$| t_2)^{DBH} = \alpha | t_2)^{d} - \beta | t_2)^{v} \qquad (1)$$

For all other cases, $(t_2)^d$ in this formula should be substituted for $(t_2)^X$ where $(t_2)^X$ is the p-state of an impurity ion. In other words, $(t_2)^{DBH}$ is an antibonding component of a t_2-state. For a bonding component $(t_2)^{VBH}$ similar to $(t_2)^{DBH}$, the following expression can be written:

$$| t_2)^{VBH} = \beta | t_2)^{d} + \alpha | t_2)^{v} \qquad (2)$$

where α , β are the constant and relative magnitudes of their are determined by normalization condition.

The problem of determining the deep level position in the gap connected with the change in the occupation of t_2^{DBH} orbital types (a scattering potential includes both the resonance (d-component) and the potential term , (p-states)) was solved by

Flerov, Kikoin [5].

In this case, the eigenstates of the energy, ε^{DBH}, are found from the solution of a secular equation:

$$(\varepsilon - \varepsilon^{d}) - \frac{V_0^2 \, M(E)}{1 - \varepsilon^{v} M(E)} = 0 \qquad (3)$$

where $M(E) = 1/(\sum_{ka}(E - E_{ka}))$ is the average of the Gilbert image of an electron density state, ε^{d} and ε^{v} are the initial level positions before hybridization, V_0 is the Anderson hybridization parameter. A graphical solution of this equation is shown in Fig.1. Thus, we find that a pinning-effect takes place if the initial state ε^{d} is more than ε^{v}.

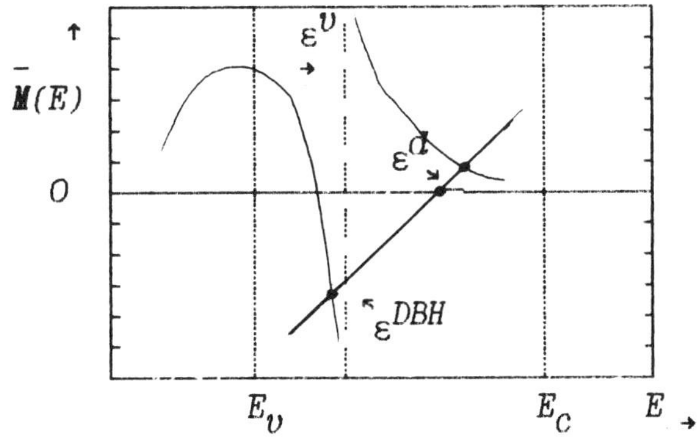

Fig.1.
Graphic solution of secular equation.
$-\Delta = \varepsilon^{DBH} - \varepsilon^{V}$

In this case the determination of the charged state of X-defect located in an interstitial position becomes divided into two separate tasks:

(1) the determination of the degree of occupation of a $(t_2)^{DBH}$ orbital (n^{DBH}) at a given Fermi level position. (2). The determination of the degree of occupation of core orbitals (n^{core}) located below the valence band gravity center in the energy scale; due to this location the additional level in a gap is not pushed out. The total number of electrons located on a defect with a given Fermi level position $n_X^{\Sigma}(E_{FO})$ can be defined by :

$$n_X^{\Sigma}(E_{FO}) = n^{DBH}(E_F = E_{FO}) + n^{core} \qquad (4)$$

As it follows from the said above, to pin a deep level the following conditions must be met:

(a) (t_2^d) should be located above $(t_2)^{DBH}$;

(b) - band component of wave function $(t_2)^{DBH}$ should dominate.

As it was shown in [1] the pinning is sufficiently strong: Δ is of the order of 0,05 eV(see Fig.1.).

At the same time, some local effects can most important effects result from lattice distortions of various types.

Therefore, there arises one more condition on a spin-marker: (c) a spin marker and a defect under consideration should have a similar structure of core states.

Taking into account the fact that the total number of electrons in p-orbitals cannot be more than 6 and the ground state of the majority of defects are of low-spin type we deduce, that it is necessary to distinguish between next defect types with respect to the degree of occupation of the core orbitals n^{core}:

n^{core}=0,2,6 (completely filled shells)

n^{core}=1 (one more electron above the completely filled shell)

n^{core}=2 (two more electron above the completely filled shell)

n^{core}=3 (three more electron above the completely filled shell).

For defects with n^{core}= 4,5,6, the combinations of initial and final multi - electron term occuring for defects with n^{core}=2,1,0 are repeated.

It should be noted that the difference maximum in marking level position will be take place for the defects with n^{core} = 0 and n^{core} = 1

Proceeding from above we can conclude that the use of manganese impurity ions (which have a half-filled 3d-shell) as spin markers imposes the following requirements on X-defects:

$$I_X^k > I_{Mn}^2 \quad \text{or} \quad I_X^k < I_{Mn}^3 \qquad (5)$$

where k is the integer, and I_{Mn}^2, I_{Mn}^3, I_X^k are the second, third and k^{th} ionization potentials of manganese and X ion respectively For manganese $I_{Mn}^3 = 33.67$ eV; $I_{Mn}^2 = 15.64$ eV with respect to vacuum level [6].

In this case we deal with defects of the first type, n^{core}=0 (a completely filled core shell).

Now come down directly to considering piaring defects. As we have already mentioned a pairing defect consist of an interstitial atom and a vacancy located in a nearest or next nearest site. The formation of such a defect from two isolated defects (the interstitial and the vacancy) is accompanied by a redistribution of electron density between the t_2^{DBH} and t_2^{VBH} orbitals. In other words, an additional distortion of C_{3v} symmetry is imposed on a defect. As a result, the binding energy E_b, increases.

This process can be described by the following expression:

$$(t_2^0)^{VBH} + (t_2^n)^{DBH} \Leftrightarrow (t_2^1)^{VBH} + (t_2^{n-1})^{DBH} \quad (6)$$

for each of the partners entering in such a pairing defect. The limiting value of the binding energy increase, ΔE_b, can be easily estimated as follows: Indeed, the following process takes place when a silicon vacancy, undergoes a change in charge state, $V_{Si}^{++/+}$.

$$(t_2^0)^{DBH} \Leftrightarrow (t_2^1)^{DBH} + h_v^+ \quad (7)$$

The thermal activation energy of such a process is well known [7] and is equal to 0.13 eV. This value is the limit of a binding energy variation, ΔE_b, since it is in the case of a vacancy that the maximum possible increase energy of a system due to local lattice relaxations for a particular crystal matrix (Jahn-Teller effect) is realized. Thus, the marking level positions for pairing defects are shifted downward in energy by not more than 0.13 eV.
(E_c - 0.26 eV; E_c - 0.55 eV ; E_v + 0.17 eV)
Fig.2 is a good illustration of this situation.

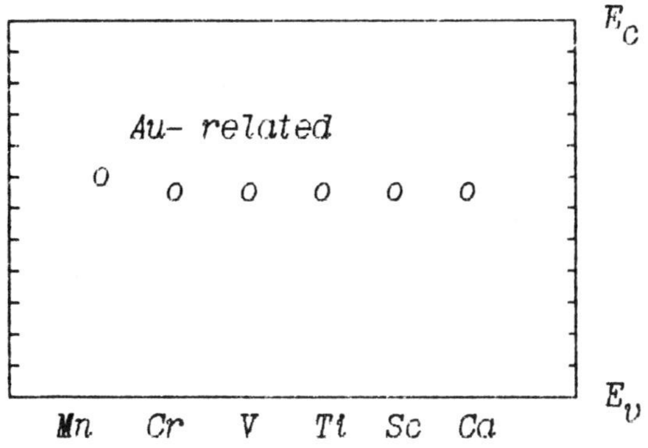

Fig.2.

Electron traps connected with transition metal impurities in silicon . [8].

The degree of occupation is defined by the following form:

$$n_x^{\Sigma}(E_{FO}) = n^{DBH}(E_F = E_{FO}) + n^{core} + n^{VBH} \qquad (8)$$

where $n^{VBH} = 1$.

Note also that the expression given is valid for an interstitial defect containing one more electron in addition to a filled shell, i.e. it corresponds to $n^{core} = 1$.

At the same time, these defects (pure interstitial ones and those incorporated in paring defects) have different activation energies to diffusion. Moreover, a pairing defect will be stable if the charge states of the defects are different $[X^- - Y^+]$ but will be unstable if the defects are of the same sign $[X^- - Y^-]$ or $[X^+ - Y^+]$. . The latter type of defects will be dissociated into two isolated defects: $[X$ and $[Y]$.

Conclusion

We have discussed the possibility of the spin-marking method construction for pairing and interstitial defects in silicon. Moreover, we have the classified different defects into the four classes and given conditions of dividing defects into the separate classes. The problem of substitutional defects will be discussed in future.

REFERENCES

1.P.A.Alexandrov,S.M.Yakubenya, Proceedings ICSTDS-16 (Yokohama 1989 edited by K.Sumino (Elsevier Science Publishers B.V.,Amsterdam,1990)
2.S.M.Yakunenya, Sov. Sol. St. Phys.,22,.1462, (1991)
3. M.Sprenger, R. van Kemp, E.G.Sieverts, C.A.G.Ammerlaan, Phys.Rev.,1987, 35B,1582 , (1987).
4. A.Zunger , U.Lindefelt , Phys.Rev., 27B, 1991, (1983)
5. V.Flerov, K.Kikoin, J.Phys. C, 19, 887, (1986)
6. A.Radshig, B.Smirnov, Handbook "Atom's and Ion's parameters", published by Atomic Energy Agency (1986).
7.J.L.Newton, A.P.Chatterjee,R.D.Harris,G.D.Watkins, Physica B + C , 116B ,219, (1983)
8.L.C.Kimerling,J.L.Benton,J.J.Rubin Proceedings of 11th Conference on defect and Radiation Effect in Semiconductors edited by R.R.Hasiguti (IOP, London, 1981)

Materials Science Forum Vol. 83-87 (1992) pp. 369-376

ELECTRICALLY ACTIVE OXYGEN IN GALLIUM ARSENIDE

H. CH. ALT*
Siemens Research Laboratories for Materials Science and Electronics,
Otto-Hahn-Ring 6, 8000 München 83, Germany

ABSTRACT

The mystery around the role of oxygen in GaAs has been solved very recently after thirty years of research. The isolated oxygen impurity occurs in GaAs in an off-center substitutional and in an interstitial position. Off-center substitutional oxygen, the subject of this paper, exhibits extremely fascinating infrared-spectroscopic as well as electrical properties - both being due to a negative-U ordering of the two gap levels. Oxygen is a residual impurity in as-grown semi-insulating GaAs with maximum concentrations of the electrically active oxygen defect in the 10^{15} cm^{-3} range. Much higher concentrations of about $1x10^{18}$ cm^{-3} are produced by oxygen implantation and annealing. Such oxygen-implanted layers are useful for the electrical isolation of devices.

1. Introduction

Recent experimental observations have drawn again the scientific interest to oxygen as an impurity in GaAs. Oxygen is among the earliest impurities which have been studied in GaAs.[1] When semi-insulating (SI) GaAs came up in the beginning of the 1960's there were several reports claiming that increasing oxygen partial pressure during horizontal Bridgman growth or adding Ga_2O_3 to the melt leads to higher resistivities of the bulk material. Therefore it was believed that oxygen introduces a deep level into GaAs. This assignment seemed quite reasonable as substitutional oxygen on the arsenic lattice site (O_{As}) should behave as a donor and, moreover, in the related III-V compound GaP the electrical activity of O_P as a deep donor was well established.[2]

---

*present address: Fachhochschule München, Fachbereich 06, Lothstraße 34,
8000 München 2, Germany

It was not before 1979 that it became clear from a careful comparison of quantitative deep level transient spectroscopy (DLTS) data with secondary ion mass spectrometry that not oxygen is the dominating deep trap in SI GaAs but an intrinsic defect, at that time already known as EL2.[3] The role of oxygen in this context apparently is the reduction of the silicon contamination of the growing crystal due to a chemical reaction with Ga_2O gas.[4] The reason for the long period of misinterpretation is in part caused by the lack of any spectroscopic signature unambiguously due to oxygen. Also, different from the situation in Czochralski grown Si or GaP, the total chemical O content in GaAs turned out to be rather low, in most cases below 1×10^{16} cm^{-3}.

The first observation of oxygen by an optical technique was reported by Akkerman et al. who assigned a local vibrational mode (LVM) in the infrared absorption spectrum of SI GaAs:O at 840 cm^{-1} to interstitial oxygen.[5] However, as no electrical activity seemed to be correlated with this defect, not much attention was paid to this observation. The situation concerning oxygen changed dramatically when Song et al. found in chromium-doped SI material two LVM bands at 730 and 714 cm^{-1} which were photosensitive at low temperatures.[6] Originally again related with the metastability of EL2, it became quite rapidly clear that these bands are due to off-center substitutional oxygen bonding with two gallium atoms.[7,8]

The origin of the photosensitivity and the related properties of this defect are the main subject of this paper. However, at this point it must be mentioned that another phenomenon was in the past the source of continuous interest in oxygen. This is the observation that implantation of O and subsequent annealing leads to effects in the implanted layer which can only be explained by an electrical activity of O or O-related defects.[9,10] It is shown at the end of the paper that this technological interesting effect is at least in part also due to the off-center substitutional O defect.

2. Experiment and structural model of the defect

If not stated differently, the experiments described in the following were done with samples coming from not intentionally doped SI GaAs crystals which contained oxygen as a residual impurity. The crystals were grown after the liquid-encapsulated Czochralski technique using a pyrolithic boron nitride crucible and a boric oxide encapsulant. The concentration of electrically active oxygen was $1-2\times10^{15}$ cm^{-3} as estimated from the strength of the LVM absorption bands (see below). Optical absorption studies were carried out with a Fourier transform spectrometer with a resolution limit of 0.03 cm^{-1} (Bruker IFS 113v).

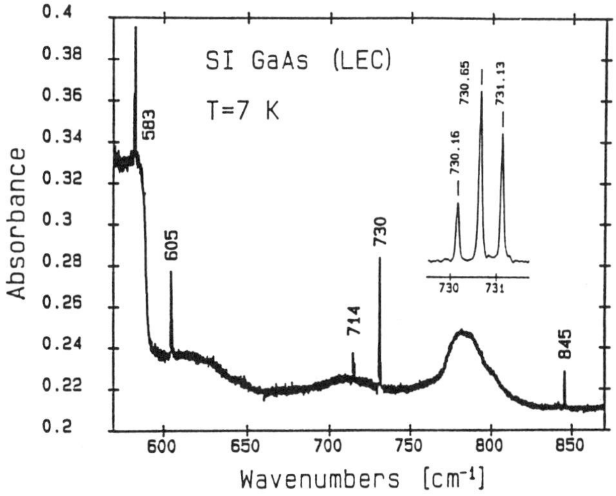

Figure 1. Low-temperature infrared absorption spectrum of oxygen-rich SI GaAs. Inset: fine structure of band B.

A typical low-temperature infrared absorption spectrum of an oxygen-rich sample in the spectral region of interest is shown in Fig. 1. The bands at 730 and 714 cm^{-1} (731 and 715 cm^{-1} at T=10 K), denoted as A and B,[6] are due to off-center substitutional oxygen. The other bands are due to the LVM of interstitial oxygen (845 cm^{-1}), the carbon acceptor impurity (582 cm^{-1}), and a so far unidentified defect (605 cm^{-1}).

When the bands A and B are investigated with higher spectral resolution, a remarkable triplet fine structure becomes visible (inset in Fig. 1). The separation between the individual lines is 0.48 cm^{-1} and the relative intensities are ~3:4:1 from high to low energy. From this it is inferred that the splitting is due to the Ga isotopes 69 and 71, occurring in relative natural abundances of 60 % and 40 %, respectively. Three lines and the experimentally observed intensities are expected for a defect bonding with two Ga atoms in the GaAs matrix. In conjunction with the relatively high vibrational frequency it was speculated[7] that the defect involved might have the molecular structure Ga-O-Ga in analogy to the oxygen vacancy center in Si.[11] The crucial experiment was done by doping a crystal with ^{18}O.[8] This crystal exhibited band B at 715 cm^{-1} and, in addition, a band at 679 cm^{-1}. The ratio of these two LVM frequencies is close to the value expected from a simple harmonic oscillator model $\{M(^{18}O)/M(^{16}O)\}^{1/2}=1.06$. The vibrating entity, therefore, contains as its essential constituent one oxygen atom. It is, from all of this, well established that the defect giving rise to the LVM bands at 731 and 715 cm^{-1} is the analogue of the oxygen vacancy centre in Si. In the following this center will be denoted as Ga-O-Ga defect. Piezospectroscopic studies on the band A are compatible with the assumed defect symmetry C_{2v}.[12]

3. Photosensitivity, charge states and negative-U behavior

The most fascinating property of the LVM bands A and B is their photosensitivity. The origin of the photosensitivity could be resolved in systematic studies using optical excitation and thermal cycling. The results are published in detail in a series of papers.[13-16] In SI samples, where the Fermi level is located near midgap, a conversion of the band at 731 cm^{-1}, A, into the band at 715 cm^{-1}, B, is observed during illumination with below-bandgap light ($h\nu \geq 0.8$ eV, $T \leq 77$ K). Also, the reverse process occurs under certain circumstances.

Experimentally it is also found that after cooling a SI sample in the dark only band A appears. The conversion rate from A to B is proportional to the photoionization cross section σ_n^0 of the omnipresent deep donor EL2. From this it is concluded that the conversion is caused by a charge-transfer process from the partially occupied EL2 level to the unoccupied Ga-O-Ga defect via the conduction band. It is worth mentioning that the only other example for a charge-state-induced shift of the LVM frequency is again the oxygen-vacancy center in Si.[17]

More detailed studies revealed that a third band B' exists, shifted from B by 0.66 cm^{-1} to lower energy. This band emerges during the optically induced conversion from A to B at an intermediate stage (Fig. 2). The kinetics is quantitatively described by a model

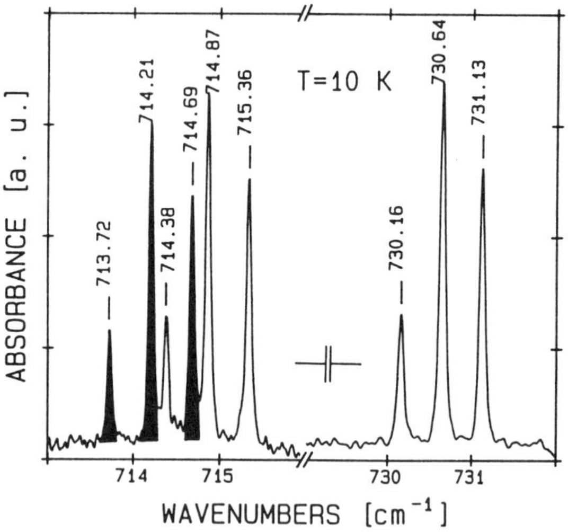

Figure 2. The LVM bands of the Ga-O-Ga defect: band A (731 cm^{-1}), band B' (714 cm^{-1}), and band B (715 cm^{-1}). Band B' is only observed during conversion from A to B and from B to A, respectively.

considering successive capture of conduction-band electrons, which have been photoexcited from the EL2 defect level. The corresponding depopulation of the EL2 level ($EL2^0 \rightarrow EL2^+$) has been observed experimentally.[18] Thus, there is conclusive experimental evidence that the band B' corresponds to a third charge state, where A, B', and B now must be assigned to the zero-, one-, and two-electron states, respectively. From a comparison of samples with different positions of the Fermi potential, it is concluded that the corresponding two energy levels are located in the upper half of the bandgap.

However, only the states with zero or two electrons (bands A and B) are detected at thermal equilibrium. This observation led to the speculation that a new negative-U center has been found, that is a system where the second electron is bound more strongly than the first. Experimentally, this is proved as follows. A SI sample is cooled in the dark to 10 K. The band A, present at this stage, is converted in part to band B' by illumination with hv=1.35 eV. Now, the sample is warmed up in the dark to 95 K and, after a while, cooled to 10 K again. The result is that the band B' disappears completely, band A increases, and, as a new feature, band B emerges. This is the so-called disproportionation which is indicative for a negative-U center.[19] The reason is that the binding energy for the two-electron system (band B) is larger than twice the binding energy for the one-electron system (band B').[20]

The next question is where the gap levels are. The binding energy of the first and the second electron was derived from the thermally induced decay of the local mode bands B' and B, respectively. Arrhenius plots of the time constants of the decay give activation energies of 0.14±0.02 eV and 0.58±0.03 eV, respectively. As a further independent experiment for the assignment of the gap levels, the photoexcitation properties of the bands A, B', and B were investigated. A threshold energy of 1.37 eV is observed for the optically induced transfer of electrons from the valence band to the unoccupied Ga-O-Ga center (A --> B'). This gives again a position of E_c-0.15 eV for the first level. Similarly, the depopulation of B starts at a threshold energy of 0.60-0.65 eV, in excellent agreement with the thermal decay data of band B.

In some samples, where in the dark both bands, A and B, are observed, temperature-dependent Hall-effect measurements give an activation energy of 0.43 eV for the conduction-band electrons. If the Fermi level is pinned by a negative-U center, theory says that the activation energy in the low-temperature limit should be half the binding energy of the two-electron state, that is $(0.14+0.58)/2$ eV $\approx$ 0.4 eV. Thus, the Hall data are consistent with the level positions found optically and, furthermore, provide additional support for the negative-U characteristics of this defect. Recent comparative studies of n-type conductive samples containing the Ga-O-Ga defect with DLTS brought convincing experimental evidence that the deeper gap state is identical with the well-known, so far unidentified EL3 level.[21,22]

4. Oxygen-implanted layers

In a variety of compound semiconductors, ion implantation is used to obtain highly resistive layers for the electrical isolation of devices. Normally, the major role for the isolation behavior is played by damage-induced deep levels.[10] These layers are thermally stable up to temperatures of typically 450 - 550 ^{0}C. However, it is known that oxygen-implanted layers in GaAs sometimes show thermal stability up to much higher temperatures.[9] This has been the origin of speculations that oxygen itself may act as a deep electron trap in such layers.

We have investigated the dose and temperature dependence of the sheet resistance of oxygen and boron isolation implants in n-type modulation-doped AlGaAs/GaAs heterostructures. Boron atoms are considered to be chemically and electrically inactive in the implanted layer. Using different implantation energies, the concen-tration of the implanted atoms was kept roughly constant between 50 and 250 nm depth. The concentration was varied between $3x10^{16}$ and $3x10^{18}$ cm^{-3} in both cases.

Considerable differences between these two implant species are observed in the annealing behavior of the layer resistances. Provided a proper annealing step is carried out, in the case of boron the maximum sheet resistance is constant irrespective of the implanted dose. This is consistent with the model that an optimum concentration of irradiation-induced deep levels exists, given by the number of free carriers to be removed and the increasing hopping conductivity at high concentrations.[10] In contrast to this, for oxygen the maximum sheet resistance is increasing with increasing dose. In this case, this behavior is a clear hint that not only irradiation-induced deep levels are present in the O-implanted layer.

It is tempting to associate this observation with the newly found electrical properties of the Ga-O-Ga defect. As the structures used for the sheet resistance measurements are not suitable for infrared studies, the behavior of O was simulated by an equivalent isolation process in Si-implanted GaAs layers. Unannealed samples were compared with samples annealed between 450 and 550 ^{0}C. All samples subjected to annealing showed two additional local modes at 721 and 734 cm^{-1} (Fig. 3). These bands stem from the implanted layer as they disappear after a short etch removing 3 μm of the surface. We attribute these bands to the Ga-O-Ga defect, corresponding to the two bands at 715 and 731 cm^{-1} in bulk GaAs:O. There are no other absorption features which could lead to an ambiguity in this assignment. The slight shift to higher frequencies of 6 and 3 cm^{-1}, respectively, is probably due to the distorted environment of the Ga-O-Ga center caused by the large number of damage-induced defects.

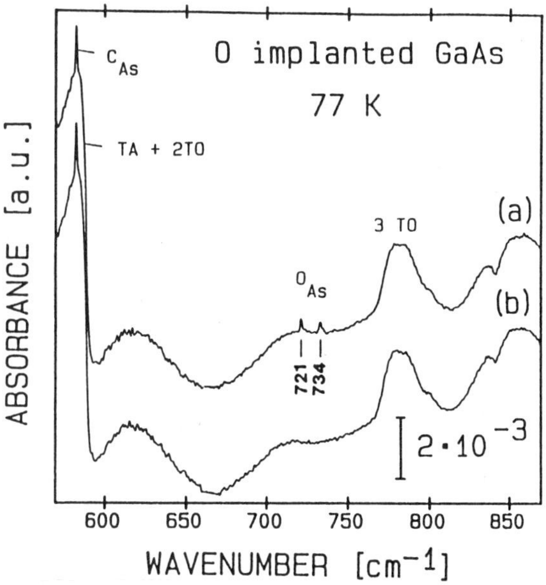

Figure 3. Infrared absorption spectra of oxygen-implanted GaAs: (a) after annealing at 500 ^{0}C, (b) after removing the implanted layer by a 3-μm etch.

From the studies in bulk GaAs it is possible to estimate the sheet concentration of O atoms involved in the two LVM bands to about $5x10^{13}$ cm^{-2}. This value amounts to 50 % of the implanted dose. The average density of electrically active oxygen is of the order of $1x10^{18}$ cm^{-3} in the implanted layer and thus more than two orders of magnitude higher than in bulk crystals. The most important result is that oxygen implantation is capable to introduce a high density of electrically active oxygen defects into GaAs and is a promising technique for the electrical isolation of devices.

REFERENCES

1. see for reference: G.M. Martin and S. Makram-Ebeid in *Deep Centers in Semiconductors*, ed. S.T. Pantelides (Gordon and Breach, New York, 1986), p. 399.
2. see for reference: P. Dean ibid., p. 185.
3. A.M. Huber, N.T. Linh, M. Valladon, J.L. Debrun, G.M. Martin, A. Mitonneau and A. Mircea, J. Appl. Phys. 50, 4022 (1979).
4. G.M. Martin, G. Jacob, J.P. Hallais, F. Grainger, J.A. Roberts, J.B. Clegg, B. Blood, G. Poiblaud, J. Phys. C: Solid State Phys. 15, 1841 (1982).
5. Z.L. Akkerman, L.A. Borisova and A.F. Kravchenko, Sov. Phys.-Semicond. 10, 590 (1976).
6. C. Song, W. Ge, D. Jiang and C. Hsu, Appl. Phys. Lett. 50, 1666 (1987).
7. X. Zhong, D. Jiang, W. Ge and C. Song, Appl. Phys. Lett. 52, 628 (1988).
8. J. Schneider, B. Dischler, H. Seelewind, P. M. Mooney, J. Lagowski, M. Matsui, D.R. Beard and R.C. Newman, Appl. Phys. Lett. 54, 1442 (1989).
9. P.N. Favennec, J. Appl. Phys. 47, 2532 (1976).
10. S.J. Pearton, Mat. Sci. Rep. 4, 313 (1990).
11. G.D. Watkins and J.W. Corbett, Phys. Rev. 121, 1001 (1961).
12. C. Song, B. Pajot and C. Porte, Phys. Rev. B 41, 12330 (1990).

13. H.Ch. Alt, Appl. Phys. Lett. 54, 1445 (1989).
14. H.Ch. Alt, Appl. Phys. Lett. 55, 2736 (1989).
15. H.Ch. Alt, in *The Physics of Semiconductors*, eds. E.M. Anastassakis and J.D.Joannopoulos (World Scientific, Singapore, 1990), p. 734.
16. H.Ch. Alt, Phys. Rev. Lett. 65, 3421 (1990).
17. R.C. Newman, *Infra-red Studies of Crystal Defects* (Taylor&Francis, London, 1973), pp. 121-124.
18. The oscillator strength of the bands A and B was derived from the well-known optical cross sections for the interaction of EL2 with the bands.
19. G.D. Watkins in *Festkörperprobleme: Advances in Solid State Physics*, Vol. XXIV, ed. by J. Treusch (Vieweg, Braunschweig, 1984), p. 163.
20. The same conclusion was drawn independently by M. Skowronski, S.T. Neild and R.E. Kremer, Appl. Phys. Lett. 57, 902 (1990).
21. S.T. Neild, M. Skowronski and J. Lagowski, Appl. Phys. Lett. 58, 859 (1991).
22. U. Kaufmann, E. Klausmann, J. Schneider and H. Ch. Alt, Phys. Rev. B 43, 12106 (1991).

Materials Science Forum Vol. 83-87 (1992) pp. 377-388

OXYGEN RELATED POINT DEFECTS IN GaAs

MAREK SKOWRONSKI
Department of Metallurgical Engineering and Materials Science, Carnegie Mellon University, Pittsburgh, PA 15213, USA

ABSTRACT

Oxygen in GaAs crystals produces an abundance of different point defects, including isolated oxygen interstitials and complexes of oxygen with native defects and impurities. This paper will review types of oxygen related defects found in GaAs and their identification based on Localized Vibrational Mode absorption measurements. Of particular interest are electrical properties of oxygen related centers, their microscopic models and thermal annealing behavior.

1. Introduction

Interest in oxygen in gallium arsenide dates from the early 1960s when semi-insulating GaAs crystals were successfully grown by adding either gallium oxide to the melt or by high temperature annealing in oxygen atmosphere.[1,2] Initially it had been suggested that oxygen introduced deep level in GaAs bandgap which compensated shallow contaminants and pinned the Fermi level close to the middle of the bandgap. However, shortly thereafter it was realized that primary role of oxygen is to prevent dissolution of silica crucibles used for GaAs growth and the incorporation of silicon donors into the crystal.[2-4] This "silicon gettering" mechanism has been independently verified by several groups.[5-7] Similar results have been obtained for Vapor Phase Epitaxy (VPE); the addition of oxygen to the gas stream resulted in lower levels of silicon contamination, and therefore reduced layer conductivity.[8,9]

The problem of oxygen incorporation into the growing crystal, however, remained unresolved as well as the origin of the deep donor center required by the electrical compensation model.[10] This center, named EL2 by Martin *et al.*[11], and located at 0.75 eV below the conduction band, has been suspected for many years of being due to oxygen. By comparison with GaP it has been commonly assumed that oxygen substitutes for arsenic, and forms a deep donor in GaAs. This assignment has been suggested by theoretical calculations of O_{As} level position within the GaAs [12] and by experimental data on shift of EL2 ionization energy with phosphorus content in $GaAs_xP_{1-x}$ alloys.[13] The linear extrapolation of EL2 position to GaP agreed well with the substitutional oxygen level. Careful oxygen content measurements using Secondary Ion Mass Spectroscopy (SIMS) proved, however, that oxygen cannot be a constituent of the EL2 center. The typical concentration of EL2 is $1x10^{16}$ cm^{-3} while that of oxygen is about an order of magnitude lower.[14,15] The same

SIMS experiments also showed that oxygen is a frequent contaminant in GaAs, typically present in the concentration range 1-3x10^{15} cm^{-3}. No known DLTS peaks could be correlated with oxygen content. This observation gave rise to speculation that oxygen induced DLTS peak overlaps that of the EL2 center forming level named EL2O or ELO.[16,17] However, further careful DLTS spectra analysis could not corroborate such suggestions.[18]

A fundamental problem in the early studies of oxygen related centers in GaAs was that of being able to detect relatively low oxygen concentrations. The detection limit of chemical analysis techniques such as SIMS or GDMS is on the order of 10^{15}-10^{16} cm^{-3}. This limit is in part due to high oxygen background levels, and is close to the highest oxygen concentrations reported thus far in the literature. Additionally, because oxygen forms nonradiative centers in GaAs, photoluminescence, which facilitated oxygen identification in GaP, cannot be used. The breakthrough in our ability to detect and identify oxygen related centers came from infrared absorption measurements. Similarly, as in the case of silicon, germanium, and gallium phosphide, light oxygen atoms produce in GaAs infrared absorption bands due to localized vibrational modes (LVM). One of the bands located at 845 cm^{-1} at 77 K (836 cm^{-1} at 300 K).[19] shifts upon replacing ^{16}O with the ^{18}O isotope. It was a conclusive proof of oxygen involvement . Several other oxygen related absorption features has been reported recently and are discussed below.

2. Off-center substitutional O_{As}

The oxygen related center, which is of primary interest, is an analog of the substitutional O_P defect in GaP. In n-type GaAs crystals, the center's LVM signature consists of three equidistant, narrow lines centered at 715 cm-1 with intensity ratio of 1.3:4:3.[20] This fine structure is due to the presence of two gallium isotopes, ^{69}Ga and ^{71}Ga ,with relative abundances of 60 and 40%, respectively. Three lines correspond to three possible combinations of oxygen bonding with two gallium neighbors: ^{69}Ga-O-^{69}Ga, ^{69}Ga-O-^{71}Ga, and ^{71}Ga-O-^{71}Ga. By analogy the defect was suggested to resemble the A center in silicon.[21,22] Oxygen, rather than retaining its symmetric substitutional position, bonds with only two gallium ligands, and forms a Ga_2O molecule with C_{2v} symmetry, while the other two gallium neighbors form a molecular Ga-Ga bond. This center will be referred to as distorted substitutional oxygen on the arsenic side in order to stress similarity of its electronic structure to that of O_P. The LVM band observed at 715 cm^{-1} is due to asymmetric ν_3 bond stretching vibration.

Similarly as substitutional oxygen in GaP this defect is electrically active and in semi-insulating crystals there are three sets of LVM triplets which can be transformed one into another by near infrared illumination.[23,24] The three LVM triplets correspond to the three charge states of the defect; their most intense components are located at 730.7, 714.2 and 714.9 cm^{-1} which correspond to two, one and zero electrons on the center.[24] It should be pointed out that the number of electrons (0, 1, or 2) is intended to indicate the change of number of electrons only and it does not correspond to the total number of electrons on the center. In other words we do not know if O_{As} levels are of donor or acceptor character.

The next step in the characterization process of the O_{As} center was the determination of the location of its two levels in the GaAs bandgap. This has been accomplished by thermal de-excitation of photo-populated charge states in high resistivity material. The Ahrenius plot for emission from the one electron state yielded an activation energy of 0.14 eV and that for the two electron state, 0.57 eV (Fig. 1).[25,26] Apparently, it takes more energy to ionize electron from the two electron state than from the one electron state. The O_{As} center, therefore, has a negative Hubbard correlation energy[27], exhibiting a negative-U property.[28] One consequence of this peculiar sequence of energy levels is that the center at thermal equilibrium cannot be occupied by one electron. The system can lower its energy if two singly occupied centers disproportionate into one with two electrons and one with zero electrons. Thus, although there are two energy levels due to O_{As} in the upper part of the band gap, one should observe only one peak in DLTS measurements. During the filling pulse O_{As} captures two electrons. An energy of 0.57 eV is required to emit the first electron to the conduction band, and as determined from LVM measurements, this emission occurs at approximately 250 K. When one of the two electrons is emitted the other follows immediately because of the much smaller activation energy. Therefore, the DLTS peak height reflects emission of two electrons from each center and its height should correspond to twice the concentration of O_{As} defects.

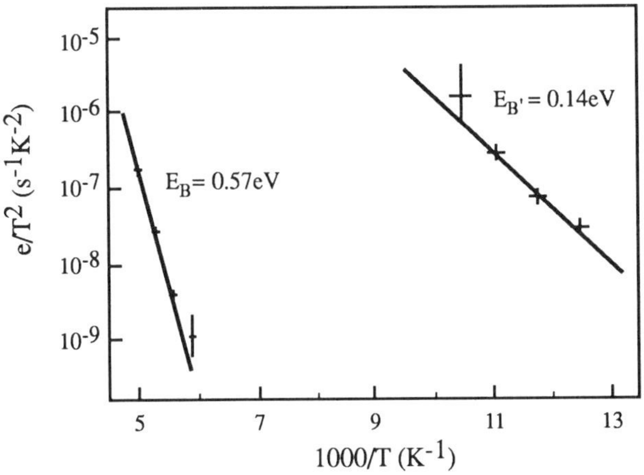

Fig. 1. Thermal activation plot of the two LVM bands of the O_{As} center.

The DLTS spectrum of an intentionally oxygen doped crystal is shown in Fig. 2.[29] Compared to typical DLTS spectra of Bridgmann grown crystals, it is readily apparent that a new major peak is present. It is located at about 250 K and has a concentration equal $9x10^{15}$ cm^{-3}. Its activation energy is 0.55 eV, which agrees well with the results of infrared absorption measurements. There is no peak with an activation energy of 0.14 eV and comparable amplitude in agreement with the negative-U property of the center suggested above. The final argument supporting identification of oxygen's DLTS signature is the positive correlation between DLTS peak height and intensity of LVM absorption.[29] The increase of the DLTS peak located at 250 K upon oxygen doping has been reported in early investigations of electron traps in GaAs[7,30] but has never been fully ex-

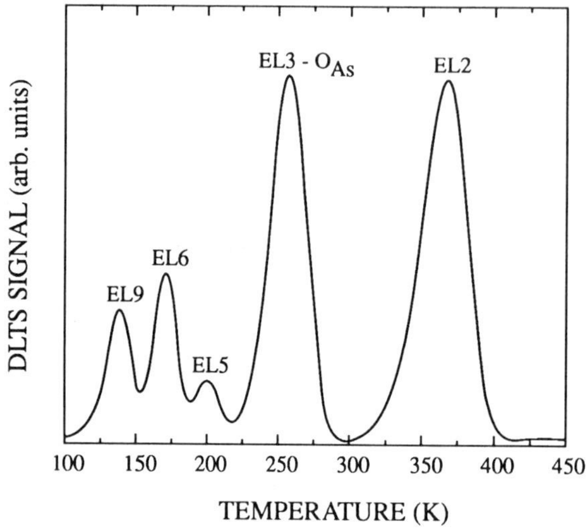

Fig. 2. DLTS spectrum of horizontal Bridgman grown GaAs doped with oxygen.

plained. Comparison of oxygen induced DLTS peak parameters with a catalogue of electron traps in GaAs, compiled by Martin *et al.*[7], identifies the O_{As} signature as the EL3 electron trap, frequently observed in VPE and melt grown GaAs crystals. Typical concentration of EL3 traps is in the $5x10^{14}$-$3x10^{15}$ cm^{-3} range in agreement with SIMS data. Moreover, the annealing behavior of O_{As} LVM lines discussed below and that of EL3 trap reported in literature agree quite well.[31,32] It appears EL3 traps in undoped GaAs are due to oxygen.

In order to understand the effects of oxygen on compensation mechanism in GaAs, one has to determine if the two electron state acts as acceptor or donor. Since the concentration of O_{As} centers is low and usually lower than the density of frequently observed unidentified traps, EL5 and EL6, the technological data are not reliable and cannot be used. One should also try to explain the nature of the large lattice relaxation necessary to overcome the electron-electron repulsion, which is responsible for negative-U property. A simple molecular orbital model suggested by D.J. Chadi[33], and described below, provides a tentative explanation of these properties.

The starting point in the analysis is an undistorted, substitutional oxygen atom located on arsenic site, reminiscent of oxygen in GaP. The bonding of O_P has been analyzed by Baraff *et al.*[34,35] who concluded that bonds of positively charged center are weak compared to gallium-phosphorus bonds. The major reason for this weakness is the relative depth of oxygen orbitals (especially s-type). The resulting small spatial extent of molecular orbitals reduces the overlap of oxygen sp3 hybrids with that of its gallium neighbors. Moreover, the bonds grow weaker still upon capture of two subsequent electrons which in turn must be placed on antibonding a_1 orbitals. As a consequence the negatively charged substitutional O_P becomes unstable and relaxes to a new equilibrium position.

The lattice constant of GaAs is larger than that of GaP (5.65 Å vs. 5.45 Å) which further reduces the strength of covalent sp^3 bonding. In response oxygen does not occupy substitutional arsenic site for any of three possible charge states. It moves from the arsenic substitutional site along the <001> direction toward a pair of gallium neighbors in order to take advantage of increased overlap. The neutral oxygen center has nine electrons: three contributed by gallium neighbors and six valence electrons of the oxygen atom. Four of these electrons can be accommodated on oxygen-gallium bonds (two electrons per bond) and additional four on two oxygen's lone pair orbitals. Each of these orbitals is located within the valence band while corresponding antibonding states are in the conduction band. The Ga-O-Ga bridge is not electrically active and any additional electrons must populate states of the other half of the oxygen center.

Because oxygen moves toward one pair of gallium neighbors, the other pair is left with two dangling bonds, which are probably responsible for the appearance of 0.14 eV and 0.57 eV levels in the GaAs bandgap. Lets consider first the bonding in a positively charged center, i.e., with no electrons on dangling gallium hybrids. An analogous situation occurs on the Ga(111) surface of GaAs crystals where both experimental data and theoretical calculations explain well the electronic structure and lattice relaxation.[36,37] The Ga(111) unreconstructed surface consists of layer of gallium atoms, each of which bonding with three underlying arsenic atoms, and a partially filled dangling bond perpendicular to the surface. Electrons from high energy gallium dangling bonds can be transferred to low energy As-dangling bonds by creation of gallium vacancies on the surface, which also acts to neutralize surface polarity. Once the gallium hybrid is empty, gallium atoms can relax almost to the arsenic plane forming planar sp^2 bonds. Because the bonding arrangement is virtually the same, one could reasonably expect that the same lattice relaxation will occur in the center under discussion. This situation is illustrated in Fig. 3a, where two gallium

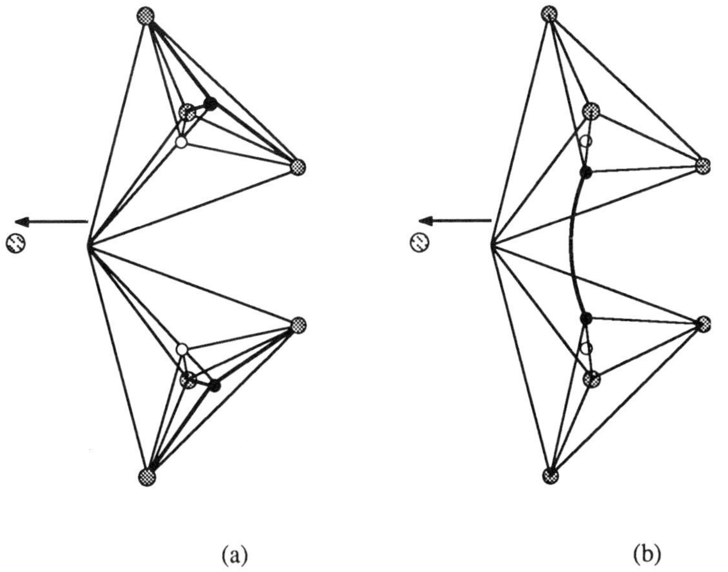

Fig. 3. Lattice relaxation in off-center substitutional oxygen for (a) positively charged center and (b) either neutral or negatively charged O_{As}. Empty circles denote original position of gallium atoms, filled circles mark relaxed gallium configuration.

atoms have relaxed from their lattice positions toward the base of a tetrahedrons with apexes on a former arsenic site now left vacant by oxygen. The sp^2 Ga-As bonds are expected to be stronger than sp^3 bonds and therefore should be located within the valence band.

In the neutral center the ninth electron should occupy the state originating from dangling gallium hybrids. The unreconstructed Ga sp^3 orbitals are located within the conduction band but their energy is expected to be lowered by the relaxation of two gallium atoms toward each other, and the formation of Ga-Ga bond, similar to that observed on gallium stabilized (001) GaAs surface. The energy gain, however, cannot be large because the overlap of gallium hybrids can be accomplished only at the expense of straining the Ga-As backbonds (Fig. 3b). The +/0 level of O_{As} therefore should be located either within or close to the edge of the conduction band. The situation is different for the capture of the second electron. In order to accommodate the ninth electron the center relaxed and formed bonding orbital which can hold two electrons. The capture of the second electron into an already prepared bonding state could produce larger energy gain than for the first one. There is no large penalty in the form of strain energy, only minor one due to electrostatic repulsion between electrons. This argument could explain the fact that the ionization energy of the second electron is larger than that of the first. It also predicts that the 0.57 eV level is of acceptor character.

It appears that it will be very difficult to experimentally verify the above model's validity, since the expected type of distortion would not change the center's symmetry in agreement with LVM signature of all three charge states. The best possible argument to corroborate or invalidate the above model could come from first principles calculation of the center energy. However, it is quite clear from experimental spectra that the major change in the center's configuration occurs upon capture of the first electron. The frequency difference between LVM triplets of zero and one electron states is 16.5 cm^{-1} while the difference between one and two electron states is only 0.7 cm^{-1}. These values agree well with the above model.

3. Oxygen interstitial

Oxygen in GaAs can also occupy an interstitial position as it does in silicon, germanium and GaP. In this configuration it bonds to arsenic and gallium atoms on the two nearest lattice sites. As a result of bonding with only one Ga neighbor the LVM lines splits into two components with an intensity ratio corresponding to the relative abundance of ^{69}Ga and ^{71}Ga isotopes. Such a doublet has been observed in the infrared spectrum of oxygen doped crystals; its association with oxygen unambiguously proved by the presence of isotope shift when ^{16}O has been replaced by ^{18}O.[20] The doublet is not affected by illumination and/or doping, indicating that O_i is not electrically active. It should not affect the electrical and optical properties of GaAs crystals but can provide an easy, low temperature diffusion route.

The existence of isolated oxygen interstitial defects provides an alternative way of O_{As} center perception. It can be considered as a complex of the oxygen interstitial and the arsenic vacancy. This is a natural approach in silicon where an analogous center is created by capture of irradiation induced vacancies by isolated oxygen interstitials. The bonding scheme is virtually identical in both cases although possible charge states appear to be different. If one was to extend the analogy further, one could expect that the constituents of a complex could be separated by a short term, high temperature annealing. Rapid thermal annealing experiments support this idea. At temperatures around 700° C, the concentration of O_{As} rapidly decreases with concomitant increase of isolated oxygen interstitial LVM intensity. Thermal activation energy of the dissociation process was determined to be 3.7 eV.[38,39] The transformation of $O_{As} \rightarrow O_i$ could also be explained as the annihilation of an interstitial. In arsenic-rich GaAs excess arsenic is present in form of small precipitates which can release arsenic interstitials during thermal annealing followed by the reaction $O_{As} + As_i \rightarrow O_i$.

3. Oxygen - impurity complexes

Oxygen's small size, high electronegativity and high diffusivity at moderate temperatures should facilitate complex formation with other impurities and/or native defects during crystal post-growth cooldown. Two centers discussed above, O_{As} and O_i, are therefore not the only oxygen containing species expected to be present in as-grown crystals. Two experimental observations support this suggestion. The first is the presence of additional high energy LVM absorption bands in GaAs crystals containing oxygen.[39,40] Oxygen atoms, even if present in the form of complexes, should still lead to the appearance of localized vibration bands due to their light weight and ability to form strong bonds. Among LVMs reported so far in GaAs:O, and suggested to be associated with oxygen, are: a characteristic triplet located at 605 cm^{-1}, a doublet at 794 cm^{-1}, and two groups of singlets observed between 960-980 cm^{-1} and 1100-1200 cm^{-1}. Although at this time there is no reliable experimental evidence as to their nature, the correlation of high energy LVMs and the presence of

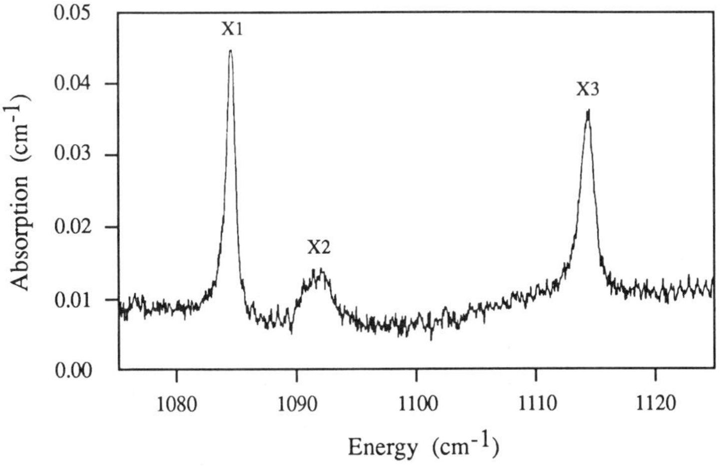

Fig. 4. Localized vibrational mode X1, X2, and X3 absorbtion lines.

oxygen indicates common origin. The most intense LVM lines in our samples are triplet at 605 cm^{-1} and three singlets (X1, X2, and X3) located at 1084.2, 1092.0 and 1114.3 cm^{-1}, respectively (Fig. 4).

The second set of experimental evidence has been obtained from thermal annealing experiments. Two different annealing regimes have been used to study the effects of thermal processing on oxygen related centers. The short term annealing (5 min.) followed by quenching to room temperature has been used to break up any complexes which could have formed during crystal cooldown, while long term (14 hrs.) ampoule annealing was intended to allow sufficient time for impurities to diffuse and form complex centers. Both regimes are frequently employed in GaAs processing and therefore the data presented below are relevant to semiconductor technology. The long term annealing is used in whole ingot annealing to reduce internal stresses in GaAs ingots and to homogenize defect distribution, whereas the short term is used in rapid thermal processing of

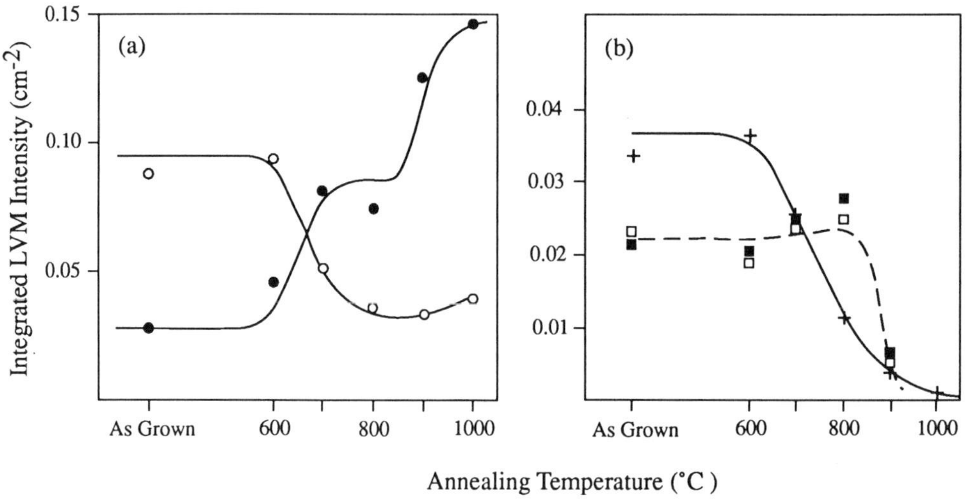

Fig. 5. Changes of integrated intensities of LVM bands upon 5 min. annealing followed by quenching. Open circles - O_{As}, filled circles - O_i, crosses - 605 cm^{-1} triplet, filled squares - X3 line.

wafers. The short term annealing experiment results are presented in Fig. 5. Two distinctly different behavior types can be noticed. The intensity of all the lines attributed to complex centers dramatically decreased with lines X1, X2, X3, and the 605 cm^{-1} triplet disappearing altogether above 900 °C, while the LVM intensity of isolated interstitial oxygen steadily increased after annealing. The total integrated LVM intensity of all lines remained approximately constant during short term annealing experiments. This behavior is fully consistent with the interpretation of high energy LVM lines as resulting from impurity-oxygen complexes.

The results of long term annealing experiments are shown in Fig. 6. At temperatures above 600

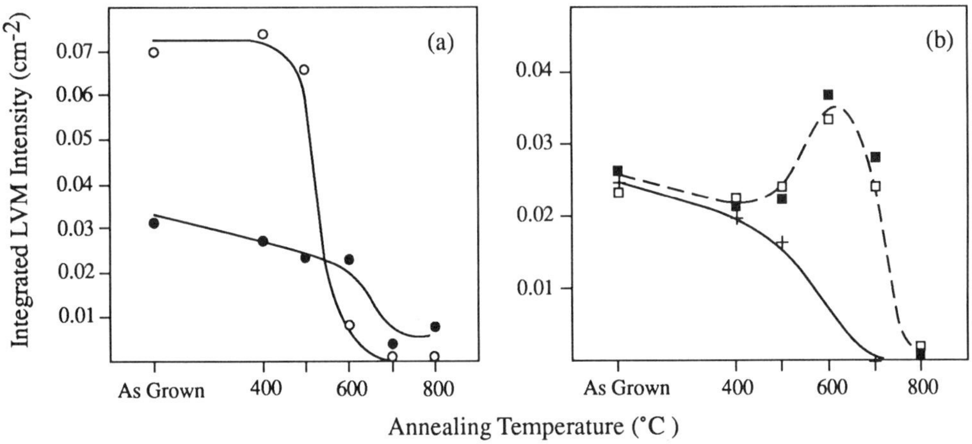

Fig. 6. Effects of long term (14 hrs.) annealing on integrated LVM line intensities. Open circles - O_{As}, filled circles - O_i, crosses - 605 cm^{-1} triplet, filled squares - X1 line, open squares - X3 line.

°C the O_{As} LVM disappeared, followed by a decrease of O_i LVM at temperatures higher than 700 °C. The decrease at temperatures below that of the dissociation temperature, indicated the onset of long range diffusion of both centers, and the formation of complexes and/or precipitates. In addition to the disappearance of O_{As} LVM, there was a significant increase of high energy LVMs, again in full accord with the suggested origin of X lines. The remaining unresolved issue is the nature of the impurity involved in the complex and microscopic structure of the complex. Chemical analysis of our samples shows relatively high concentrations of aluminum $N_{Al}=2x10^{15}$ cm^{-3} and silicon $N_{Si}=5x10^{15}$ cm^{-3}. The well known affinity of oxygen for both of these elements and large Al-O and Si-O bond strength makes these impurities the most likely candidates. Further experiments are in progress.

4. Oxygen - divacancy complex

Careful examination of Fig. 5 shows an interesting trend. The O_{As} center LVM intensity decreases after short term annealing above 600 °C most likely due to dissociation of the center. It does not, however disappear completely even at much higher temperatures. In fact the intensity increases slightly with annealing temperature above 900 °C. This effects can be explained as the capture of thermally generated vacancies during quenching. With increasing temperature the vacancy concentration increases and for quenching from 1200 °C the ratio of LVM intensities $I(O_{As})/I(O_i)$ exceeds 1. In other words most of the oxygen interstitials can capture an arsenic vacancy during cooldown. In addition to the increase of O_{As} LVM, a new triplet appears in the semi-insulating crystal spectrum with the most intense component at 733.4 cm-1 (Fig. 7a). The characteristic ratio of component intensities 1.3:4:3 indicates bonding with two gallium atoms while the proximity to the O_{As} LVM triplet (the new one is shifted by only 2.6 cm^{-1} with respect to O_{As}) suggests in-

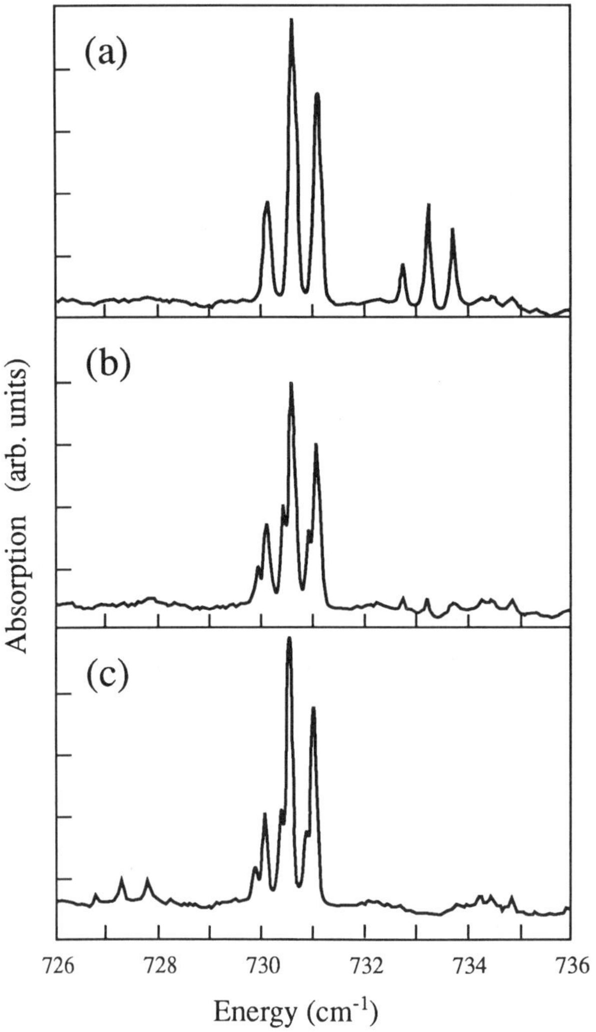

Fig. 7. (a) Spectrum of O_{As} - V_{Ga} - V_{As} center in the dark, (b) after 30 seconds wide band illumination,and (c) after 30 minutes illumination.

volvement of oxygen. Since the total quenching time form 1200 °C to room temperature was approximately 30 seconds it is unlikely that the new center involves any impurity, yet possible that some kind of native defect is captured. According to the model of O_{As} center, the 730 cm^{-1} triplet corresponds to the O_{As}^{+} charge state with gallium atoms relaxed as shown in Fig. 3a. The position of the 733.4 cm^{-1} triplet would therefore indicate the presence of a similar relaxation. Based on the above arguments and the analogy with the center observed in silicon the new triplet is tentatively interpreted to be due to an oxygen interstitial - divacancy complex. Similarly as in silicon, the center is electrically active and its LVM spectrum is photosensitive (Fig. 7). Without illumination only the 733.4 cm^{-1} triplet is present in semi-insulating crystal. Upon wide band infrared illumination its intensity decreases and a new triplet appears on the low frequency side of the O_{As}^{+} signature. With prolonged illumination the center is transferred to yet another charge state (or metastable state) with the corresponding triplet appearing at 727.3 cm^{-3}. Relatively small shift of O_i-V_{As}-V_{Ga} LVM lines upon illumination and normal ordering of energy levels in the bandgap indicate a lack of large lattice relaxation which in turn suggests that three sets of LVM lines are due to the change of the charge state. Detailed discussion of the properties of this center will be published elsewhere.

5. Conclusions

Infrared absorption measurements of localized vibrational modes of oxygen related centers in bulk GaAs crystals allowed the identification of the microscopic structure of several centers, and the determination of their properties. In particular, the dominant electrically active oxygen center in as-grown crystals, the off-center O_{As}, is a negative-U center with an acceptor level located at 0.57 eV below the conduction band. Its DLTS signature is identical to that of the commonly observed trap in VPE and melt grown GaAs, namely EL3. Thermal annealing experiments indicate that oxygen can also form complexes with impurities and native defects. One of these has been tentatively identified as an oxygen interstitial and V_{As}-V_{Ga} divacancy complex.

References:

[1]C.H. Gooch, C. Hilsum and B.R. Holeman, J. Appl. Phys., **32**, 2069 (1961).
[2]N.G. Ainslie, S.E. Blum and J.F. Woods, J. Appl. Phys., **33**, 2391 (1962).
[3]C.N. Cochran and L.M. Foster, J. Electrochem. Soc., **109**, 149 (1962).
[4]J.F. Woods and N.G. Ainslie, J. Appl. Lett., **34**, 1469 (1963).
[5]J.M. Woodall, Trans. Metall. Soc. AIME, **239**, 378 (1967).
[6]M. Kaminska, J. Lagowski, J. Parsey, K. Wada and H.C. Gatos, Inst. Phys. Conf. Ser., **63**, 197 (1982).
[7]G.M. Martin, G. Jacob, J.P. Hallais, F. Grainger, J.A. Roberts, B. Clegg, P. Blood and G. Poiblaud, J. Phys. C, **15**, 1841 (1982).
[8]D.S. Ruby, K. Arai and G.E. Stillman, J. Appl. Phys., **58**, 825 (1985).
[9]B. Lee, K. Arai, B.J. Skromme, S.S. Bose, T.J. Roth, J.A. Aguilar, T.R. Lepkowski, N.C. Tien and G.E. Stillman, J. Appl. Phys., **66**, 3772 (1989).
[10]J. Blanc and L.R. Weisberg, Nature, **192**, 155 (1961).
[11]G.M. Martin, A. Mitonneau and A. Mircea, Electronics Lett., **13**, 191 (1977).
[12]M. Jaros, J. Phys. C, **8**, 2455 (1975).
[13]L. Samuelson, P. Omling, H. Titze and H.G. Grimmeiss, J. Cryst. Growth, **55**, 164 (1981).
[14]A.M. Huber, N.T. Linh, M. Valladon, J.L. Debrun, G.M. Martin, A. Mitonneau and A. Mircea, J. Appl. Phys., **50**, 4022 (1979).
[15]K. Shikano, K. Kobayashi and S. Miyazawa, Appl. Phys. Lett., **46**, 391 (1985).
[16]M. Taniguchi and T. Ikoma, Semi-Insulating III-V Materials, 283 (1982).
[17]J. Lagowski, D.G. Lin, T. Aoyama and H.C. Gatos, Appl. Phys. Lett., **44**, 336 (1984).
[18]M. Skowronski, J. Lagowski and H.C. Gatos, J. Appl. Phys., **59**, 2451 (1986).
[19]Z.L. Akkerman, L.A. Borisowa and A.F. Kravchenko, Sov. Phys. Semicond., **10**, 590 (1976).
[20]J. Schneider, B. Dischler, H. Seelewind, P.M. Mooney, J. Lagowski, M. Matsui, D.R. Beard and R.C. Newman, Appl. Phys. Lett., **54**, 1442 (1989).
[21]G.D. Watkins and J.W. Corbett, Phys. Rev., **121**, 1001 (1961).

[22]J.W. Corbett, G.D. Watkins, R.M. Chrenko and R.S. McDonald, Phys, Rev., **121**, 1015 (1961).
[23]C. Song, W. Ge, D. Jiang and C. Hsu, Appl. Phys. Lett., **50**, 1666 (1987).
[24]H.C. Alt, Appl. Phys. Lett., **55**, 2736 (1989).
[25]M. Skowronski, S.T. Neild and R.E. Kremer, Appl. Phys. Lett., **57**, 902 (1990).
[26]H.C. Alt, Phys. Rev. Lett., **65**, 3421 (1990).
[27]J. Hubbard, Proc. Roy. Soc. London, **A 276**, 238 (1963).
[28]P.W. Anderson, Phys. Rev. Lett., **34**, 953 (1975).
[29]S.T. Neild, M. Skowronski and J. Lagowski, Appl. Phys. Lett., **58**, 859 (1991).
[30]A. Mircea and A. Mitonneau, Appl. Phys., **8**, 15 (1975).
[31]S. Chichibu, N. Ohkubo and S. Matsumoto, J. Appl. Phys., **64**, 3987 (1988).
[32]D. Wong, H.K. Kim, Z.Q. Fang, T.E. Schlesinger and A.G. Milnes, J. Appl. Phys., **66**, 2002 (1989).
[33]D.J. Chadi, private information.
[34]G.A. Baraff, E.O. Kane and M. Schluter, Phys. Rev. Lett., **47**, 601 (1981).
[35]G.A. Baraff, E.O. Kane and M. Schluter, Phys. Rev. B, **25**, 548 (1982).
[36]S.Y. Tong, G. Xu and W.N. Mei, Phys. Rev. Lett., **52**, 1693 (1984).
[37]D.J. Chadi, Phys. Rev. Lett., **52**, 1911 (1984).
[38]M. Skowronski, S.T. Neild and R.E. Kremer, Appl. Phys. Lett., **58**, 1545 (1991).
[39]M. Skowronski and R.E. Kremer, J. Appl. Phys., **69**, 7825 (1991).
[40]C. Song, B. Pajot and F. Gendron, J. Appl. Phys., **67**, 7307 (1990).

Materials Science Forum Vol. 83-87 (1992) pp. 389-394

NEW EVIDENCES FOR THE DEFECT MODEL OF PHOTOCONVERSION BY OXYGEN IN SEMI-INSULATING GaAs

Xue-Fu Zhong*

Institute of Semiconductors, Chinese Academy of Sciences

P.O.Box 912, Beijing,100083, People's Republic of China

*Present address: Department of Physics, University of Missouri-Kansas City, 1110, E. 48th St. Kansas City, MO64110, U.S.A.

ABSTRACT

Two experiments: piezospectroscopic and gap level measurements for the oxygen defect of photoconversion in GaAs are analyzed on the basis of the defect model suggested by author and coworkers. It is shown that the model works well for these new progresses. The piezospectroscopic study demonstrates that the dipole moment of Ga-O bond is along the (110) direction. The model calculation with a large lattice relaxation provides all the bond lengths around the oxygen atom and leads to the same geometry. The measurements on electronic states indicate that the gap level from $Ga\text{-}O^{-2}$ bond is lower than that to $Ga\text{-}O^{-1}$ bond, so the defect behaves like a negative U center. Our model analysis indicates that the more stable $Ga\text{-}O^{-2}$ bond is due to the opposite direction of charge transfer in comparison with $Ga\text{-}O^{-1}$ bond. A simple quantitative estimate gives a reasonable energy difference between two gap levels. It suggests a new mechanism for negative affinity which, in a combined action with exchange interaction, interprets the defect behavior.

The local vibrational modes (LVM) by oxygen defect in semi-insulating (SI) GaAs attract a great attention due to their photo-conversion and possible relationship with EL2 center.[1,2] Recently, some new experiments: the high resolution spectroscopic measurements by Alt,[3] the piezospectroscopic study by Song et al.[4] and the measurements on electronic energy level by Alt[5] and Skowronski et al.[6], were carried out so that more information is available in understanding their behaviors.

For the high resolution spectroscopic measurements a charge state analysis has been given,[7] based on the defect model which was suggested in an earlier paper[2] and was later confirmed by Schneider et al. in an isotope doping experiment.[8] Because the defect model has taken various basic interactions, within a reasonable approximation, into account, we are able to relate the high resolution spectroscopic measurements to the properties of electronic states involved in photoconversion process. Our purpose is therefore the physical aspect of the problem, not the numerical accuracy, although the theoretical method -- self-consistent bond orbital model (SCBOM) is of independent significance.

In this paper the defect model will be applied to the other two experiments: the piezospectroscopic study and the electronic level measurements. It is quite encouraging to see that these experiments actually provide new evidences for the defect model. The SCBOM can be easily extended to the new circumstances and the physics involved in the charge state analysis provides a profound basis for the model so that we sometimes need to borrow concepts from that paper.

I. *PIEZOSPECTROSCOPIC STUDY*

Song et al. reported[4] that the dipole moments of Ga-O bonds are along the (110) direction or equivalent, this determines the atomic configuration of the defect center. But, from our model calculation, this local structural rearrangement also can be derived. In fact, it has been already implied in the data provided in our earlier paper[2,7]. For example, the calculated bond lengths for Ga-O and corresponding Ga-As bonds are 1.83 Å and 2.72 Å respectively.[2] By the impurity lattice relaxation assumed in our model, the Ga atoms move along the (111) direction or equivalent and the oxygen atom along the (100) direction, it is easy to check that the distance between two Ga atoms after relaxation is almost twice as large as the Ga-O bond length. This means that the three atoms should meet each other on a plane which is parallel to the x-z plane. In this case the only direction for Ga-O bonds is the (110) direction or equivalent and the three atoms must lie on a straight line as the core of the defect. This geometry of defect center is shown in Fig.1.

The same analysis can be also applied to the other two charge states, because the bond lengths are not much different in these cases (they were calculated to be 1.917 Å and 1.921 Å for $Ga\text{-}O^{-1}$ and $Ga\text{-}O^{-2}$ bonds, respectively[7]) and the dipole moments of $Ga\text{-}O^{-1}$ or $Ga\text{-}O^{-2}$ bonds are also reported to be along the (110) direction or equivalent. So theoretically we should have almost the same atomic configurations for all three charge states. Note that the bond lengths of Ga-O (or Ga-O^{-1}, Ga-O^{-2}) and Ga-As are compensated with each other in the cluster to keep joint with the host crystal, so the longer the Ga-As bonds, the shorter the Ga-O (Ga-O^{-1}, Ga-O^{-2}) bonds. But a longer Ga-As bond means two closer Ga atoms, therefore, their distance can be always about twice as long as the corresponding Ga-O (O^{-1}, O^{-2}) bond length and the straight line geometry of defect center is thus kept in all cases.

Note that this structural rearrangement is definitely not trivial. There are some three atom molecules, such as H_2O, which are not arranged in a straight line. Especially in our case, the environmental symmetry of the defect center is rather complicated. So the straight line geometry can only be interpreted as a result of the internal interactions. Therefore the coincidence of theoretical prediction with observation should be a strong support to the model calculation. In particular, it supports the choice of overlap parameter *C* of Ga-O bond in SCBOM which has a great effect on the bond length and is subject to a very weak limitation. (No problem with *C* parameter of Ga-As bond, it has been determined by Harrison to provide the correct bond length of GaAs crystal[9]). Since the piezospectroscopic study represents an independent data source of structure analysis, so the model calculation is not designed only for the purpose of correct LVM frequencies.

2. *MEASUREMENTS ON GAP ENERGY LEVELS*

The motivation of the experiment is to confirm the negative U behavior. Since the high resolution spectroscopic measurement has shown that oxygen atom with two extra electrons is more stable and a kinetic analysis is favorite to this conclusion. For this purpose, Alt[5] and Skowronski et al.[6] had carried out, independently, the measurement of electronic levels, trying to find out some gap levels for different charge states. If the gap level corresponding to O^{-1} state is really higher than that to O^{-2} state, the conjecture of negative U center is thought to be true. They both succeeded. The energy level for O^{-1} state locates about 0.45 eV higher than O^{-2} state in the gap.

The interpretations to the experiment depend on the stability of charge states. Our recent paper[7] has given the answer to this question, but no estimate is made for the gap levels. The essential idea is the dependence of orbital energy on atomic charge state. We illustrated this for oxygen atom in Fig.2. All data are from an exact self-consistent calculation by Clementi and Roetti[10] except for O^{-2} which we create from an extrapolation scheme.

It is realized that the gap levels are from unbonded electrons in the oxygen atom, because their energies are, in general, higher than those of bonded ones which should enter into the valence bands. The orbital energies of unbonded electrons need not to be degenerate due to the environment, but, as a rough estimate, we neglect the details. The unboned electrons may be either from unhybrized atomic orbit p_z or from the third sp^2 hybrid orbit pointing to the opposite direction of the two Ga-O bonds. But they do not make much difference, because we are only interested in the relative energy of two gap states and it is easy to see from Fig.2 that the slopes of energy curves between O^{-1} and O^{-2} states are almost same for p_z and for sp^2 orbits.

We need the effective atomic charge to determine the "real" orbital energies of p electron in O^{-1} and O^{-2} states. This is obtained by calculating the charge transfer which is given by the polarity from SCBOM. The results are also represented on Fig.2. For O^{-1} state, the gain of electrons on oxygen atom is 0.606 (the polarity of $Ga\text{-}O^{-1}$ bond is 0.303, there are two bonds related to the oxygen), so the effective atomic charge of oxygen is 7.606. On the other hand, for O^{-2} state the loss of electrons on the oxygen atom is 0.648 (the polarity of $Ga\text{-}O^{-2}$ bond is 0.324), its effective atomic charge is only 7.376, or the actual number of electrons on oxygen atom is less in O^{-2}, and more in O^{-1} states. Therefore the orbital energy of unbonded electrons is higher in O^{-1} state and lower in O^{-2} state. We then have a clear interpretation of the experiment qualitatively.

As for quantitative estimate, we note that the orbital energies in Fig.2. are for free atom or ions. In crystal we should have dielectric screening. The actual difference of atomic potential in these two cases can be evaluated by the difference of orbital energies divided by the static dielectric constant (12 for GaAs). This gives 0.3 eV in comparison with 0.45 eV from observation.

Apparently, the result is not totally satisfactory. But, remember that the exchange energy will provide another negative U tendency. For O^{-1} state the last electron is spin-unpaired. When adding one more electron to become O^{-2} state, the last electron should be spin-paired.

In this case, the orbital energy will be further reduced. Although here we do not go so much in detail as to estimate this contribution from exchange energy, the physical aspect of our model is quite clear.

The negative U center we discussed here has never been reported in literatures up to now and represents a new mechanism. It is essentially from the dependence of orbital energy of unbonded electron on charge state. In our case it is due to the fact that the orbital energy of Ga sp^3 hybrid happens to fall into the separation between the sp^2 for O^{-1} and O^{-2} states and thus the charge transfer takes an opposite direction for Ga-O^{-1} and Ga-O^{-2} bonds. Although the mechanism seems rather particular and may play part only in special cases, it is also of fundamental importance, because the mechanism stands for an independent factor in bond forming, not deducible from any other physical tendency.

As we know, there are other mechanisms for negative U center in an impurity or defect. Some of them are related to the lattice distortion. For a strong coupling between the lattice and electronic state an extra electron might cause a large distortion which sometimes can reduce the total energy of the system. This mechanism makes the state with one more extra electron more stable and thus may create a negative U center . Another possibility is due to the spin-pairing, i.e., one spin unpaired electron calls for another opposite spin-orientated electron. One may always think of a tendency of negative affinity if there is a spin-unpaired electron in impurity or defect. Of course, only when exchange interaction is strong enough, in comparison with Coulombic interaction, the spin-pairing can really make an impurity to be a negative U center. In most cases, we have a combined action of these mechanisms, a negative U center is rarely a simple object.

In our case, we see there is not much difference in bond lengths for different charge states, so the lattice distortion can not have a appreciable or even a detectable effect for the negative U tendency. The dominate mechanism should be the opposite directions of charge transfer. But there is a spin unpaired electron in O^{-1} state, so the exchange interaction also play a part. We accept a combined action of two independent physical origins in this case as we said above.

It is not surprising that oxygen atom is involved in such a complicated process as light-induced interconversion. Since the element is belong to the first period, it has much deeper orbital energy in comparison with Ga or As atoms of third period element, so there is a large space in energy scale to provide atomic bounded states to participate in the process of bond forming. This is one of the reasons why the investigation on oxygen defect is so rich by its contents.

* Work supported by the National Foundation of Natural Sciences.

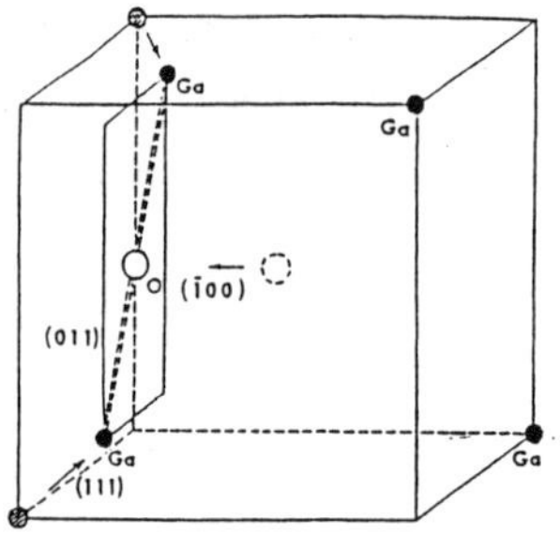

Fig. 1.

The local structural rearrangement from model calculation. The length of Ga-O bond is one half of the distance between the two Ga atoms after relaxation, so the oxygen and two gallium atoms are embedded in a plane which is parallel to the y-z plane and the dipole moments of Ga-O bonds are along the (110) direction or equivalent. The agreement with experiment is excellent.

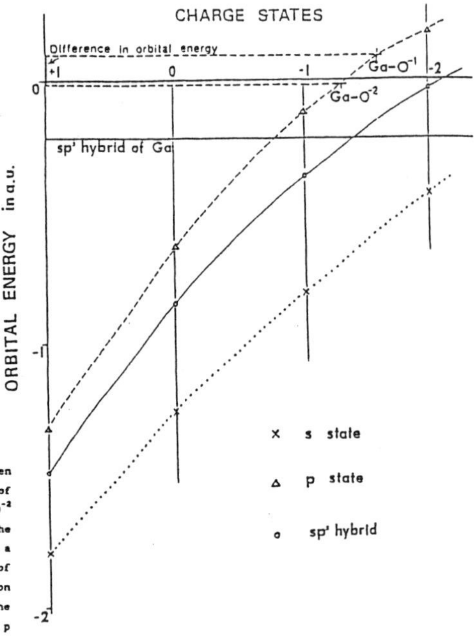

Fig. 2. The dependence of orbital energy on charge state for oxygen element. Data are from the self-consistent calculation of atomic structure by Clement and Roetti[12], except for O^{-2} which is from extrapolation scheme. For reference the orbital energy of sp^3 hybrid of Ga atom is also shown by a horizontal line. It locates between the orbital energies of O^{-1} and O^{-2} sp^2 hybrids. The effective atomic charges on oxygen in $Ga-O^{-1}$ and $Ga-O^{-2}$ bonds are indicated. From the data one can deduce the difference in orbital energy of p states for $Ga-O^{-1}$ and $Ga-O^{-2}$.

REFERENCES

1. C.Song, W.Ge, D.Jiang, and C.Hsu, Appl.Phys.Lett.50,1666(1987).
2. X.Zhong, D.Jiang, W.Ge, and C.Song, Appl.Phys.Lett.52,628(1988).
3. H.Ch.Alt, Appl.Phys.Lett.55,2736(1989).
4. C.Song, B.Pajot, and C.Porte, Phys.Rev.B41,12330(1990).
5. M.Skowronski, S.T.Neild, and R.E.Kremer, Appl.Phys.Lett.57,902(1990).
6. H.Ch.Alt, Proc. 20th Conf.Intern.Phys.Semicond. Tu2-A5,Greece(1990).
7. X.Zhong, D.Jiang, and C.Song, J.Appl.Phys.69,1124(1991).
8. J.Schneider, B.Dischler, H.Seelewind, P.M.Mooney, J.Lagowski, M.Matsui, D.R.Beard, and R.C.Newman, Appl.Phys.Lett.54,1442(1989)
9. W.Harrison, Phys.Rev.B27,3592(1982).
10. E.Clementi and C.Roetti, Atomic Data and Nuclear Data Tables, 14, 179(1974).

Materials Science Forum Vol. 83-87 (1992) pp. 395-400

THEORETICAL STUDIES ON THE STRUCTURE FOR THE CORE OF OXYGEN THERMAL DONORS IN SILICON

P. DEÁK,[1] M. HEINRICH,[2] L. C. SNYDER[3] and J. W. CORBETT[3]
[1]Max-Planck-Institut für Festkörperforschung, 7000 Stuttgart 80, Heisenberg Str. 1, Germany (perm. address: Technical University Budapest, H-1521, Hungary)
[2]Technische Universität Wien, 1040 Wien, Gusshausstr. 25-27., Austria
[3]State University of New York, 1400 Washington Ave., Albany, NY.12222, USA

ABSTRACT

The results of semi-empirical molecular and cyclic cluster calculations on the interaction of oxygen atoms with the silicon lattice are summarized. A new model for the core of the oxygen related 450 °C thermal donors (TD) involving (at most) two oxygens and a self-interstitial is presented. It is shown that this model emerges as the only one being in accord with available experimental information about TD's.

1. Introduction

After almost four decades of intensive research following their discovery,[1] the atomic structures of the double donor centers arising after heat treatment around 450 °C in oxygen rich silicon are still not known. The properties of these "thermal donors" (TD's) have been recently summarized[2] as follows. There is a series of TD_n complexes: usually up to n=9 are mentioned but further species up to 16 have recently been identified.[2,3] The initial rate of formation is proportional to the fourth power of the original interstitial concentration $[O_i]$. The maximum concentrations of TD_1 and TD_2 are proportional roughly with $[O_i]^1$ and $[O_i]^2$, respectively, indicating that very few oxygens are actually present in these first species. The activation energy for TD formation, about 1.8 eV,[4] is significantly smaller than the 2.5 eV barrier for O_i diffusion. Carbon inhibits the formation of TD's while hydrogen plasma treatment enhances it. The growth of TD's is not correlated with vacancy-oxygen complexes. No local vibration mode characteristic to TD's is known. The symmetry of the TD's is C_{2v} - with possible small deviations[5] - and they exert compressive stress along [001] (the C_2 axis). It is generally assumed that all TD's have a common core to which an increasing number of oxygen atoms are added in a (110) plane. Recently it has been proven that TD_1 and TD_2 are bistable.

TD's are double donors with the donor orbital constructed from the pair of conduction band minima along the C_2 axis. The (0/+) and (+/++) occupancy levels are around E_c - 0.07 and 0.15 eV, respectively, getting gradually shallower for higher TD's. The excited states of each TD_n are in agreement with predictions of effective mass theory. The EPR center NL8 (studied mainly in B-doped Si) correlates with the characteristic electronic transitions while NL10 (studied mainly in Al-doped Si) correlate with the change in carrier concentration related to TD's. The wavefunctions deduced by ENDOR are similar but that of NL10 is considerably less localized.[6,7] ENDOR experiments resolve 2 to 4 oxygens in early TD's, but none on the C_2 axis. Direct evidence exists for the involvement of only 2 oxygens in the core.[5] No other impurity has been found in the core of NL8, while NL10 shows direct incorporation of Al in aluminium-doped samples.

A variety of structural models for the core of TD's have been put forward throughout the years. They have to be tested against new experimental findings but also against quantum mechanical calculations, i. e., have to be checked whether they are really in accordance with what is assumed about them. Although, calculations on compressed VO and VO_2 complexes,[8,9,10,11] on the OBS[12] model[13,14] and on the YLID[15] model[14,16] all indicate possible donor activity (the C_{2v} symmetry has been *a priori* assumed), these models can now all be discarded because of the oxygen on their C_2 axis. On the other hand, calculations have proven the models based on NL8 and NL10 ENDOR data[6,7] to be instable.[17,18] As it is, none of the models proposed so far seem to be sufficiently well-founded. In a situation like this, a systematic theoretical study is desirable in order to find new possibilities. Such calculations at a level of high accuracy are largely hampered by the immense difficulty inherent with a complex like this: it involves a number of non-substitutional atoms, induces long range relaxation, gives rise to a delocalized defect orbital and contains oxygen (an atom with an incompletely screened core). Due to these reasons it was only very recently that even the energetic feasibility of oxygen aggregation could be reproduced computationally.[19] Systematic studies (compromising on calculational accuracy to various degrees) on TD models remained as yet inconclusive but have established the importance of trivalent (threefold coordinated) oxygen[11,17,20] and the self-interstitial.[10,21]

We have performed extended calculations on small oxygen complexes with and without the presence of a vacancy and/or a self-interstitial, using molecular and cyclic cluster models (MCM and CCM) to simulate the host, and semi-empirical hamiltonians[22] to solve the Schrödinger equation. The same computational framework has been used to study hydrogen in silicon[23] and is described in detail elsewhere.[24,25] We have found[26] that the complex shown in Figure 1 is stable in C_{2v} symmetry and exhibits shallow double donor character. The aim of this paper is to show that complexes built around such a core are up to now the exclusive candidates capable of explaining the whole scale of TD properties. A more detailed account of our calculations will soon be published.[25]

2. Oxygen complexes without a self-interstitial

Oxygen complexes without other impurities can be compiled into four groups. The first involves complexes of oxygen interstitials (O_i's) without the participation of intrinsic lattice defects, i. e., vacancies (V) or self-interstitials (I). We have studied the aggregation of O_i in the (110) plane in a large molecular cluster ($Si_{47}H_{60}$) with elongated shape along [110]. We find the bond-center (BC) site as the equilibrium position of isolated O_i. The Si-O distances are 1.61 Å with a Si-O-Si angle of 180°. The latter is at variance with the 162° deduced from experiment[27] and is a basis set effect of little consequence considering the flexibility of the bond angle.[28] There are no levels in the gap. The energy barrier for diffusion from one BC site to the next is calculated to be 2.5 eV at the YLID configuration. In case of two O_i's in the cluster, we have found earlier[29] two configurations with about equal bonding energies of -0.1 eV . Needels et al.[19] reported -1.0 eV for one of them (two O_i's with one common Si neighbor), establishing the stability of this complex beyond doubt. Electrical activity can only be expected[29] from the ring-like configuration (two trivalent O_i's with two common Si neighbors) due to its trivalent oxygens but its symmetry and the nature of the donor wavefunction excludes it as a core for TD's.[20] Adding further oxygens to these complexes we find that the aggregation of O_i's in [110] chains is energetically favorable, but the most stable complexes up to 5 O_i's always contain a ringlike structure. Neither a YLID[12] nor a di-YLID[15,20] with adjoining O_i's on both sides are stable with respect to symmetry lowering. Since TD_1 does not contain more oxygens than that, we conclude that "pure" [110] chains of O_i's cannot be the TD's. In studies using symmetric molecular ($Si_{35}H_{36}$) and cyclic clusters (Si_{32}) we have found that neither $(O_i)_4$ (four O_i's with a common Si neighbor), nor two O_i's in the center of parallel Si-Si bonds are electrically active. In principle, if oxygens in the latter complex formed a bond and became trivalent, they would be forced to emit two electrons into the bottom of the conduction band (CB).[11] According to our calculations, however, the necessary ionization energy is comparable with the gap, and the neutral system is not bonded anyway. In agreement with others[10,11,19] we also find that an O_2 molecule is not stable in the silicon lattice. In summary, we may exclude "oxygen-only" complexes as TD's.

Another group of oxygen complexes is the one involving a vacancy. Our calculations with up to 4 oxygens in the vicinity of a single vacancy show that the most stable structures are the A-center (VO) and the VO_2 complex (the latter with no bond between the oxygens). Further oxygens align in a (110) plane on both sides of VO_2 as O_i's. No donor activity can be found (the oxygens are still unbonded). Since they also contain oxygens on the C_2 axis, these "oxygen+vacancy" complexes can be excluded as TD's as well.

A special group of TD models[6,7,12] contain a semi-dissociated Frenkel-pair in addition to oxygens. The self-interstitial is on the C_2 axis somewhere between the substitutional and the tetrahedral interstitial sites. The "semi-vacancy" is filled with 4 oxygens in the model of Michel et al.,[6] by 3 in the OBS model,[12] while in the model of Gregorkiewicz et al.[7] there are only 2 O_i neighbors beside it in the (110) plane. According to our calculations all of these models are highly unstable with respect to V-I recombination. Therefore, we may exclude "oxygen + semi-vacancy" complexes from further consideration.

3. The "oxygen + self-interstitial" model of the TD core

The only class of oxygen complexes left is the one involving self-interstitials. Even though the calculated energy of Frenkel-pair formation, 7.4 eV, is high, there is experimental evidence that self-interstitials are generated previous to or during TD formation.[30,31,32] Their incorporation into TD's would readily explain the known formation kinetics.[4] In CCM-MINDO/3 calculation we have found[26] that the structure shown in Figure 1 is stable in C_{2v} symmetry and has an extremely low ionization energy. In a subsequent check of the donor wavefunction by CNDO/S we have found that its characteristics correspond to a combination of two conduction band states along [001].[26] This orbital with an energy level above the cluster VB edge by 1.83 eV (cf. the CNDO/S cluster gap of 2.14 eV) is rather delocalized: 0.3 % is on the IO_2 unit, 12.0 % in all on the three silicon atoms to which the IO_2 unit is bonded and altogether 31.0 % on first neighbors of those silicon atoms while the rest, 56.7 %, is spread out on the remainder of the cluster. Figure 1 also shows a comparison between the squares of CNDO/S coefficients and the spin localizations deduced by ENDOR for NL8.[33] The calculated values are generally higher by a factor of 10

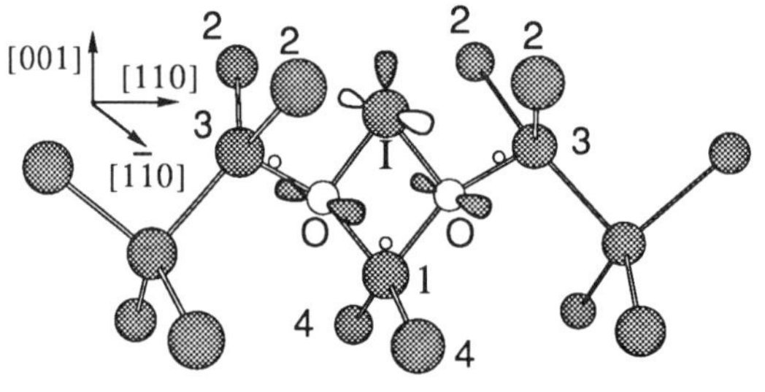

	CNDO/S		ENDOR	
	s	p_z	s	p_z
1	1.42	1.18	0.09	0.06
2	1.94	0.84	0.16	0.05
3	0.52	1.34	0.13	0.34
4	1.40	1.42	0.11	0.12
I	0.02	0.02	0.07	0.03
O	0.04	0.02	<	0.02

Figure 1

due to the confinement in a cluster too small for an effective mass-like orbital. The trend in the s/p ratio is, however, well reproduced. It should be emphasized that all hyperfine interactions with atoms of the core are of the same magnitude. The overall qualitative agreement is by far the best we have found for any of the complexes investigated. The IO_2 complex conforms also with what is known about the structure of the TD core.[34] It has C_{2v} symmetry and there is no silicon atom in the very center. The oxygens are (within 1°) on <111> axes and none on [001]. The relaxation of the neighbors of the IO_2 unit indicate that the complex exerts a compressive stress along [001]. The double donor activity is the result of two trivalent oxygens with one excess electron each. The self-interstitial is in a divalent configuration, with two of its electrons in a sp^2 hybrid pointing toward [001] (the corresponding level is in the VB). It also has an empty perpendicular p-orbital above the CB edge. Due to the defect, the degeneracy of the six equivalent CB valleys is locally lifted, and the symmetric combination of the two valleys along [001] is the lowest energy state available for the excess electrons. In short, the electrical activity comes from the trivalent oxygens, but the donor electrons are in an effective mass like orbital induced by the presence of the complex. The role of the self-interstitial is to transform the divalent oxygens into trivalent ones. In the ++ charge state the net charge is -0.4 on each of the oxygens, +0.8 on I, +0.5 on atom 1 and +0.3 on atoms 3 (each). The binding energy of the I atom to the $(O_i)_2$ complex is about -1 eV, i.e., at elevated temperatures there is a dynamic equilibrium between IO_2 and free I + $(O_i)_2$.

All the above properties make the IO_2 complex an excellent candidate for the core of TD's. We have studied an even smaller IO unit by removing one of the oxygens from IO_2. The resulting structure[25] is basically the same with slight deviation from C_{2v} symmetry. The complex is still a double donor: in the dipositive charge state it can be thought of as an O_i trapping an I^{++}. The CNDO/S calculation at this geometry but for the neutral charge state puts the donor electrons on an orbital similar to that in IO_2. In contrast, the calculation for the neutral charge state with MINDO/3 ends up with the two electrons in a localized antibonding state between the two silicons around the removed oxygen. Unlike in the case of IO_2, there is a considerable rearrangement of the atoms relative to the dipositive charge state, with the main effect of increasing the distance between the silicons which host the occupied antibonding orbital. It is hard to say whether this rearrangement is an artifact due to the incorrect description of the CB by MINDO/3 resulting in the occupation of a localized antibonding orbital, or the donor orbital predicted by CNDO/S is an artifact of its inability to take lattice relaxation into account. More accurate calculations are necessary to decide, whether IO itself is TD_1. This assumption is supported by the finding[25] that IO is bistable with a negative U behavior as seen in Figure 2. A metastable state of IO_2 has also been found but more accurate calculations are necessary to prove bistability.

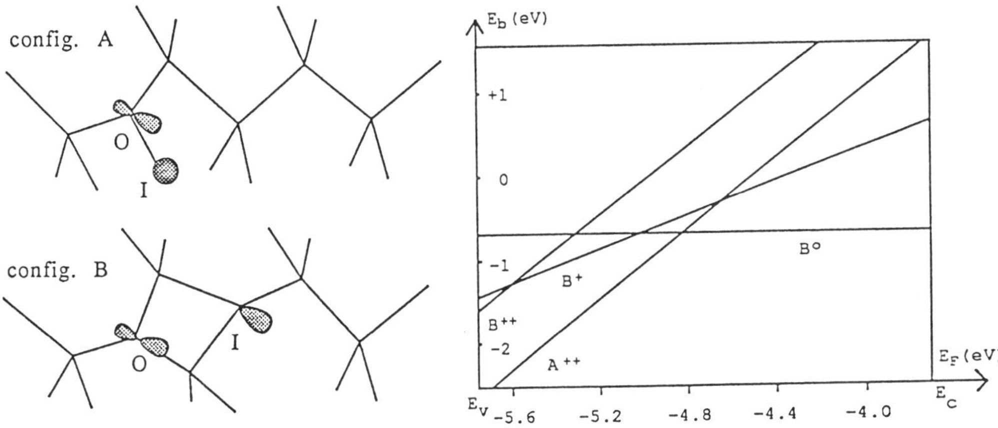

Figure 2

As a summary, we may say that the complex of a self-interstitial with at most two oxygens at the core of TD's is capable of explaining most of the TD properties. The calculated binding energy (-0.8 eV) and diffusion barrier (~1.6 eV) of the IO complex is relatively low. An I atom may pick up and carry an O_i for a short time with a virtual enhancement of the oxygen diffusion rate. Assuming that TD's grow as this I-mediated O-diffusion mechanism deposits further O_i's in [110] sidechains around the IO_2 complex, the kinetics of TD formation can be explained (see ref. 4). Since I atoms are produced at the cost of two O_i's,[4] the ~ $[O_i]^4$ dependence of initial formation rate can be understood. Finally, our calculations predict the vibration frequency associated with the trivalent oxygens in TD's around 900 (±50) cm^{-1}.

Acknowledgement: Support by the Humboldt Foundation and the National Scientific Research Fund of Hungary is appreciated.

References

[1]C. S. Fuller, J. A. Ditzenberger, N. B. Hannay and E. Buehler, Phys.Rev. **96**, 833 (1954).

[2]P. Wagner and J. Hage, J. Appl. Phys. A **49**, 123 (1989).

[3]W. Götz and G. Pensl, to be published

[4]R. C. Newman, in *"Proc. of the 20th Int. Conf. on the Physics of Semicond."'* ed. by E. M. Anastassakis and J. D. Joannopoulos (World Scientific, Singapore 1990) p. 525.

[5]T. Gregorkiewicz, H. H. P. Th. Bekman and C. A. J. Ammerlaan, Phys. Rev. B **41**, 12628 (1990).

[6]J. Michel, J. R. Niklas and J.-M. Spaeth, Phys. Rev. B **40**, 1732 (1989).

[7]T. Gregorkiewicz, H. H. P. Th. Bekman and C. A. J. Ammerlaan, Phys. Rev. B **38**, 3998 (1988) and **39**, 1648 (1989).

[8]G G. DeLeo, W. B. Fowler and G. D. Watkins, Phys. Rev. B **29**, 3193 (1984).

[9]G. G. DeLeo, C. S. Milstead, Jr. and J. C. Kralik, Phys. Rev. B **31**, 3588 (1985).

[10]P. J. Kelly, in*"Proceedings of the 15th Int. Conf. on Defects in Semicond."*, ed. by G. Ferenczi, Mater. Sci. Forum Vols. **38-41**, (Trans Tech Publications, Switzerland, 1989) p. 269.

[11]D. J. Chadi, Phys. Rev. B **41**, 10595 (1990).

[12]A. Ourmazd, A. Bourret and W. Schröter, J. Appl. Phys. **56**, 1670 (1984).

[13]J. Robertson and A. Ourmazd, Appl. Phys. Lett., **46**, 559 (1985).

[14]M. Saito and A. Oshiyama, Phys. Rev. B **38**, 10711 (1988).

[15]M. Stavola and L. C. Snyder, in *"Defects in Silicon"*, ed. by L. C. Kimerling and M. W. Bullis (The Elechtrochemical Society, Inc. Pennington, N.J. 1983) p. 61.

[16]L. C. Snyder and J. W. Corbett, in *"Proc. of the 13th Int. Conf. on Defects in Semicond."*, ed. by L. C. Kimerling and J. M. Parsley, Jr. (The Metallurgical Society of AIME, Warrendale, Pa. 1985) p. 693.

[17]R. Jones, Semicond. Sci. Technol., **5**, 255 (1990).

[18]L. C. Snyder, P. Deák, R. Z. Wu and J. W. Corbett, in *"Proceedings of the 3rd Int. Conf. on Shallow Impurities in Semicond."*, ed. by B. Monemar, IOP Conf.Ser. **95** 221 (1989).

[19]M. Needels, J. D. Joannopoulos, Y. Bar-Yam and S. T. Pantelides, Phys. Rev. B **43**, 4208 (1991).

[20]L. C. Snyder, P. Deák, R. Z. Wu and J. W. Corbett, in*"Proceedings of the 15th Int. Conf. on Defects in Semicond."*, ed. by G. Ferenczi, Mater. Sci. Forum Vols. **38-41**, (Trans Tech Publications, Switzerland, 1989) p. 329.

[21]P. Deák, L. C. Snyder, J. W. Corbett, R. Z. Wu and A. Sólyom, *ibid.* p. 281.

[22]the MINDO/3 approximation has been used for geometry optimizations while the CNDO/S method was applied to calculate the electronic structure of donor complexes at the geometry supplied by MINDO/3 (for references see ref. 24)..

[23]P. Deák, L. C. Snyder and J. W. Corbett, Phys. Rev. B **37**, 6887 (1988) and **43**, 4545 (1991).

[24]P. Deák and L. C. Snyder, Phys. Rev. B **36**, 9619 (1987).

[25]P. Deák, L. C. Snyder and J. W. Corbett, to be published

[26]P. Deák, L. C. Snyder and J. W. Corbett, Phys. Rev. Lett. **66**, 747 (1991).

[27]D. R. Bosomworth, W. Hayes, A. R. L. Spray and G. D. Watkins, Proc. Roy. Soc. London Ser. A **317**, 133 (1970).

[28]A. H. Edwards and W. B. Fowler, J. Phys. Chem. Solids **46**, 841 (1985).

[29]L. C. Snyder, J. W. Corbett, P. Deák and R. Z. Wu, in *Defects in Electronic Materials*, ed. by M. Stavola, S. J. Pearton and G. Davies, Symp. Proc. Vol. **104** (MRS, Pittsburgh, Pa. 1988) p. 179.

[30]R. C. Newman, A. S. Oates and E. F. Livingstone, J. Phys. C **16**, L667 (1983).

[31]M. Reiche, J. Reichel and W. Nitzsche, Phys. Stat. Sol. (a) **107**, 851 (1988).

[32]W. Kürner, R. Sauer, A. Dörnen, and K. Thonke, Phys. Rev. B **39**, 13327 (1989).

[33]Calculated using the hyperfine parameters of ref. 6. in the way described in ref. 26.

[34]We were not able to find a stable configuration of C_{2v} symmetry with two oxygen pairs in perpendicular planes. It was shown, however, in ref. 5 that such an assumption in ref. 6 is not warranted by the ENDOR results themselves.

Materials Science Forum Vol. 83-87 (1992) pp. 401-406

REORIENTATION OF STRESS INDUCED ALIGNMENT OF THERMAL DONORS IN SILICON

P. WAGNER[1], J. HAGE[1], J.M. TROMBETTA[2,3] and G.D. WATKINS[2]
[1]Wacker-Chemitronic GmbH, P.O. Box 1140, D- 8263 Burghausen, Germany
[2]Sherman Fairchild Lab., Lehigh University, Bethlehem, Pa. 18015, USA
[3]Present address: Texas Instruments, P.O. Box 655012, Dallas, Tx. 75265, USA

ABSTRACT

The reorientation of preferentially aligned thermal double donors follows above 400°C an exponential decay law with activation energies between 1.9 and 2.7 eV. These values are very close to the activation energy for single jumps of interstitial oxygen. The reorientation rates decrease with increasing cluster size and are much larger than the growth rate of the thermal double donors but smaller than the rate of jumps of interstitial oxygen. These findings indicate a close relation between the thermal double donors and clusters of interstitial oxygen. Below 400°C an additional mechanism enhances the reorientation rates.

1 Introduction

A series of double donors can be generated in silicon containing a sufficient amount of oxygen by annealing such material at temperatures around 450°C /1, 2, 3, 4/. These double donors are therefore called thermal double donors (TDD) and they can be observed by low temperature infrared absorption in their two charge states. The ground state binding energies of these TDDs differ only slightly and the donors are enumerated in the order of increasing ground state binding energies. So far eleven members of the TDD-series have been identified (TDD_n, n=1,...,11) /4/.

Two centers with C_{2v}-symmetry called NL8 and NL10 also can be observed with Electron Spin Resonance (ESR) after such heat treatments /5/. Both centers have been investigated with ENDOR (Electron Nuclear Double Resonance) by which the presence of oxygen in these centers was revealed /6, 7/. The center NL8 could be related to the TDD-series by using piezospectroscopy at low temperatures /8, 9/. The ground state of the TDDs is constructed essentially from the wavefunctions of only a single pair of conduction band minima according to these studies.

Applying uniaxial stress during the formation of the TDDs induces a preferential alignment of these C_{2v}-symmetric centers /10/. The alignment is revealed by the dichroism of the electronic transitions at the TDDs. It is found that the TDDs are preferentially aligned with their C_{2v}-axis perpendicular to the stress direction, indicating that the TDDs exert a compressive stress on the lattice along this axis and a net tensile stress along one of the C_{2v} mirror planes. These findings are paralleled by an identical behaviour of NL8 as observed with ESR /10, 11/ thus confirming the tight relation of this center with the TDDs. The compressive stress along the [100]-axis induces a downward shift of the corresponding two conduction bands with respect to the other four ones. This shift is the origin of the above mentioned, very special nature of the electronic structure of the TDDs.

Annealing of samples containing preferentially aligned TDDs without applied stress at temperatures around 450°C results in a relaxation of the alignment. Results of such investigations are reported in the present paper. A comprehensive and more detailed study about the reorientation of NL8 and the TDDs as observed with ESR and infrared absorption spectroscopy will be reported elsewhere /11/.

2 Experimental Details

For the present investigations samples of Czochralski-Si containing 10^{18} cm^{-3} O_i and $4 * 10^{15}$ cm^{-3} B were used. The samples were cut in the shape of small bars with dimensions of 3*3*15 mm^3. The long axis of the bars was oriented parallel to the [100]-direction. A thermal donor quench treatment of 10 min at 770°C was used to remove the TDDs in the as-grown material. These samples were annealed at 453–454°C for 90 min (samples A– C) or 8 h (samples D–F, H), respectively, with applied stress ranging from 232 to 305 MPa parallel to the [100]-direction. The stress was kept applied until the samples were cooled below 300°C except sample H where the stress was removed accidentally at about 400°C.

The isothermal recovery annealing was performed in a tube furnace. The samples were enclosed in Al-foil together with a thermocouple in order to record the temperature profile during annealing. The annealing was performed in a temperature range in between 375 and 490°C. Two samples were annealed successively at two different temperatures (410 and 470°C, 430 and 490°C).

The infrared absorption spectra were obtained at 8 K with a Fourier transform spectrometer equipped with a closed cycle cryogenerator using a KRS- 5 polarizer between sample and detector. A small polarization of the incident beam was corrected for by using an appropriate background spectrum. Each sample was measured twice after each annealing step, once with polarization parallel and perpendicular, respectively, to its long axis.

3 Results

3.1 Preferential Orientation

The TDDs or NL8, respectively, are centers with C_{2v}-symmetry. The three different components A, B and C of the ESR spectrum of NL8 result from the three inequivalent orientations of the center's g-tensor (components g_1, g_2 and g_3) with respect to magnetic field direction where the g_3-direction corresponds to the C_2-axis of the center /10/. The ratio of the line intensities can then be expressed in terms of the normalized probabilities a, b and c as

$$A : B : C = a : 4b : c \tag{1}$$

with

$$a + 4b + c = 1. \tag{2}$$

The ratios of the intensities are a direct measure of the defect alignment. In thermal equilibrium the different probabilities are equal and they have the value

$$a = b = c = 0.167. \tag{3}$$

In a similar way the optical absorption coefficients α_p and α_s for light polarized parallel and perpendicular to the stress direction during stress alignment (SA) heat treatment can be expressed in terms of the probabilities a, b and c /10/. The dipole moment for the n_{p0}-transitions is parallel to the C_2[001]'-axis (= g_3-axis) and the $np_{\pm A}$- and $np_{\pm l}$-dipoles are parallel to the g_1- and g_2- axes /10/, respectively. (The prime denotes the defect's axis system.) These expression are reproduced here only partly for stress ‖[001]:

$$\begin{aligned} 2p_0 - lines: \quad & \alpha_p \sim a + c \\ & \alpha_s \sim 2b \\ 2p_{\pm l} - lines: \quad & \alpha_p \sim 4b \\ & \alpha_s \sim a + c + 2b \end{aligned} \tag{4}$$

Table 1: Activation energies ΔE and preexponential factors t_0 of the decay constants $t_{1/2}$ of the TDD_n- relaxation evaluated from the $2p_0$- und $2p_\pm$-lines.

	T < 400°C		T > 400°C	
Line	ΔE/eV	t_0/sec	ΔE/eV	t_0/sec
$1(2p_\pm)$	1.0±0.1	$10^{-5\pm1}$	-	-
$2(2p_\pm)$	0.6±0.2	0.05	1.9±0.1	$10^{-11\pm1}$
$3(2p_\pm)$	0.4±0.15	1	-	-
$3(2p_0)$			2.6±0.1	$10^{-16\pm1}$
$4(2p_0)$			2.1±0.2	$10^{-12\pm1}$
$5(2p_0)$			2.7	10^{-15}
$6(2p_0)$			≈2.5	
NL8			2.7±0.2	$2*10^{-16\pm1}$
O_i a)	2.54	$1.8*10^{-13}$		
TDD-formation			1.7 b)	

a) /12/, different evaluation
b) /13, 14/

A polarization P was defined for the evaluation of the alignment decay as follows:

$$P = (\alpha_p - \alpha_s)/(\alpha_p + 2\alpha_s) \tag{5}$$

P is a measure for the relative difference of centers oriented with their g_3- component parallel and perpendicular to the stress direction. In a similar way the remaining fraction p_b of centers with g_3 ‖ stress direction is expressed for the ESR-results as

$$p_b = (b - b_0)/(b_i - b_0) \tag{6}$$

where the suffixes i and 0 denote the initial state and the thermal equilibrium (final) state of the samples.

The degree of SA as expressed by P_i ranges from about - 0.15 to -0.250 for the $2p_0$-lines with the exception of TDD1 and in TDD2 in samples D, E and F where interference with other lines prohibited an exact determination of P. These ratios indicate deviations of the probability b (g_3 perpendicular to stress) from the equilibrium distribution of about +20%.

3.2 Relaxation of Preferential Alignment

The polarization P/P_i decays in general according to an exponential law for the different samples and TDDs during isothermal relaxation annealing (RA). Exceptions from this behaviour are observed in the samples A, B and C which were annealed at temperatures between 375 and 395°C. Incubation times before the beginning of the exponential decay of the order of 15 min occurred in these samples. This effect decreases with increasing temperature. The total TDD concentration as measured by $2\alpha_s + \alpha_p$ flucuates in general around an mean value by up to ±25–30% during RA. These fluctuation are most pronounced in samples A, B and C. A slight decrease of the total TDD concentration of about 25% for samples A, B and C could be weakly concluded from the fluctuating values.

The times $t_{1/2}$ for which the P/P_i= 0.5 were derived from the decay curves neglecting some non-exponential behaviour of the samples A, B and C. The activation energy for the relaxation

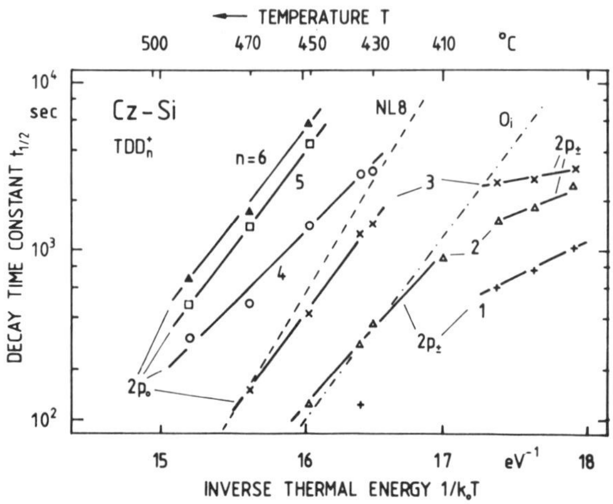

Figure 1: Arrhenius plot of the reorientation times $t_{1/2}$ of different TDDs (as measured by the polarization of the absorption lines $2p_0$ or $2p_\pm$, respectively), NL8 and O_i.

of the alignment or the decay of the polarization P/P_i, respectively, is derived from an Arrhenius plot of the times $t_{1/2}$ (Fig.1). Pre-exponential factors and activation energies are collected in Tab.1.

Two temperature regions with a border at about 400°C can be distinguished in Fig.1. Above 400°C the decay constants $t_{1/2}$ are thermally activated with energies of about 1.9–2.7 eV whereas below 400°C much smaller energies are observed. The decay constants for the relaxation of O_i /12/ and NL8 /13, 14/ are also displayed in Fig.1 as well as in Tab.1.

The reorientation of NL8 as described by equ.6 follows a nonexponential behaviour which can be explained by the superposition of the contributions of several different TDDs reorienting with an exponential law /15/. The reorientation of NL8 clearly is thermally activated, too. Results for the reorientation of TDDs or NL8, respectively, aligned with stress ‖[110] will be reported elsewhere /11, 15/.

4 Discussion

4.1 Temperature > 400°C

Most models for TDDs agree that the small differences in binding energies are due to small perturbations introduced by the successive addition of a clustering species to a common core /e.g. 16/. This is confirmed by corresponding model calculations /17, 18/ in a way that larger clusters have the smaller binding energy. The main features of the electronic structure of the TDDs are determined by the— up to now unknown —structure of the TDDs' core. An increase of reorientation time constants with increasing TDD- number is found in the present investigation whereas the activation energies for reorientation of the different TDDs are very similar. These findings support the assumption that the basic mechanism for reorientation is the same for each TDD with modifications due to the different cluster sizes. The close agreement of the reorientation activation energies of the TDDs and of O_i indicates the participation of oxygen in the reorientation process if not its incorporation in the TDD-core. The similarity of the reorientation time constants of TDD1 and TDD2 on one hand and O_i on the other hand might then be an argument for the incorproration of only a small number of $O_i$atoms in the core of the TDDs'.

The reorientation time of NL8 and its activation energy is very close to the corresponding parameters of the TDDs, especially to TDD3 and TDD4. The idea that NL8 consists of the unresolved contributions of different TDDs /11, 15/ is therefore supported again.

The reorientation of the TDDs occurs on a much faster time scale than their formation /14, 15, 19/. Therefore, the TDDs can reorient several times before another clustering species is added. The activation energies for reorientation and clustering differ, maybe with the exception of TDD1 or TDD2, as can be seen from Tab.1. The polarization and the alignment after the SA treatment as well as the components of the piezospectroscopic tensor derived from them therefore correspond to a thermal equilibrium situation /10/. One of these components, the one describing the coupling to stress along the TDDs' C_2- axis ($\parallel g_3$[001]' axis), is very similar to the corresponding component of the A- center /10/ indicating the incorporation of a similar structure as the A- center in the core of the TDDs. Such models have been suggested already previously /20, 21, 22/.

The growth of the TDDs cannot be explained by diffusion coefficients for $O_i$derived from high temperature experiments /16, 23/ or O_i- reorientation /12/. Therefore, models assuming enhanced diffusion /e.g. 16/ or the diffusion of O_2 - molecules at temperatures < 600°C /23/ have been suggested. However, the addition of an O_i-atom or an O_2- molecule should change the C_{2v}-symmetry of the centers under consideration. The observed fast reorientation of the TDDs would help to resolve this conflict.

4.2 Temperature < 400°C

The results obtained for the temperature region < 400°C are much less reliable due to a larger scatter of the data and the observed incubation time at the beginning of the reorientation. In general, this indicates that at temperatures below 400°C probably an additional mechanism acts to reduce the alignment.

5 Summary

The reorientation of preferentially aligned TDDs was found to be thermally activated above 400°C with activation energies of the different TDDs similar to each other and to the reorientation of O_i. The reorientation time increases with increasing size of the clusters ascribed to the TDDs. The reorientation of the different TDDs above 400°C therefore is governed by only one mechanism with a highly probable participation of O_i. These findings again confirm the incorporation of $O_i$in the TDDs. Below 400°C an additional reorientation mechanism contributes to the reorientation. The stress induced alignment of TDDs and its reorientation provide a stringent test for the different TDD-models.

6 References

1. D. Wruck, P. Gaworzewski, phys. stat. sol. (a) 56, 557 (1979)
2. B. Pajot, H. Compain, J. Lerouille, B. Clerjaud in *16th Int. Conf. on the Physics of Semiconductors*, Montpellier, 1982

3. R. Oeder, P. Wagner in *Defects in Semiconductors II*, Materials Research Society Symposia Proceedings Vol. 14, North-Holland, New York, 1983, S. Mahajan, J.W. Corbett, Eds., p. 171
4. P. Wagner, J. Hage, Appl. Phys. A49, 123 (1989)
5. S. Muller, Thesis, University of Amsterdam, Amsterdam, 1981
6. J. Michel, J.R. Niklas, J.-M. Spaeth, Phys. Rev. B49, 1732 (1989)
7. D.A. van Wezep, T. Gregorkiewicz, H.H.P.Th. Bekman, C.A.J. Ammerlaan in *Defects in Semiconductors*, Materials Science Forum Vol. 10-12, Trans Tech Publications, Schweiz, 1986, H.J. von Bardeleben, Ed., p. 961
8. K.M. Lee, J.M. Trombetta, G.D. Watkins in *Microscopic Identification of Electronic Defects in Semiconductors*, Materials Research Society Symposium Proceedings Vol. 46, Materials Research Society, Pittsburgh, Pa., 1985, N.M. Johnson, S.G. Bishop, G.D. Watkins, Eds., p. 263
9. M. Stavola, K.M. Lee, J.C. Nabity, P.E. Freeland, L.C. Kimerling, Phys. Rev. Letters 54, 2639 (1985)
10. P. Wagner, H. Gottschalk, J.M. Trombetta, G.D. Watkins, J. Appl. Physics 61, 346 (1987)
11. J.M. Trombetta, G.D. Watkins, J. Hage, P. Wagner, to be published
12. M. Stavola, J.R. Patel, L.C. Kimerling, P.E. Freeland, Appl. Phys. Letters 42, 73 (1983)
13. P. Wagner in *Oxygen, Carbon, Hydrogen, and Nitrogen in Silicon*, Symposia Proceedings Vol. 59, Materials Research Society, Pittsburgh, Pa., 1986, J.C. Mikkelsen, S.J. Pearton, J.W. Corbett, S.J. Pennycook, Eds., p. 139
14. M. Claybourn, R.C. Newman, Appl. Phys. Letters 51, 2197 (1987)
15. J.M. Trombetta in *EPR and Optical Studies of Oxygen Complexes in Silicon*, Thesis, Lehigh University, Bethlehem, Pa., 1988
16. A. Ourmazd, W. Schröter, A. Bourret, J. Appl. Phys. 56, 1670 (1984)
17. C.S. Chen, D.K. Schroder, J. Appl. Phys. 63, 5761 (1988)
18. J.T. Borenstein, J.W. Corbett in *Oxygen, Carbon, Hydrogen, and Nitrogen in Crystalline Silicon*, Materials Research Society Sympo- sia Proceedings Vol. 59, Materials Research Society, Pittsburgh, Pa., 1986, J.C. Mikkelsen, Jr., S.J. Pearton, J.W. Corbett, S.J. Pennycook, Eds., p. 159
19. V.P. Markevich, L.F. Makarenko, L.I. Murin in *15th Int. Conf. on Defects in Semiconductors*, Materials Science Forum Vol. 38-41, Trans Tech Publication Ltd., Schweiz, 1989, p. 589
20. D. Helmreich, E. Sirtl in *Semiconductor Silicon 77*, The Electrochemical Society, Inc., Princeton, NJ, 1977, H.R. Huff, E. Sirtl, Eds.
21. G.G. DeLeo, B. Fowler, G.D. Watkins, Phys. Rev. B29, 3193 (1984)
22. G.G. DeLeo, C.S. Milsted, Jr., J.C. Kralik, Phys. Rev. B31, 3588 (1985)
23. U. Goesele, T. Tan, Appl. Phys. A28, 79 (1982)

Materials Science Forum Vol. 83-87 (1992) pp. 407-412

SILICON THERMAL DONORS: PHOTOLUMINESCENCE AND MAGNETIC RESONANCE STUDY OF BORON- AND ALUMINUM-DOPED SILICON

B.J. HEIJMINK LIESERT, T. GREGORKIEWICZ, and C.A.J. AMMERLAAN
Van der Waals – Zeeman Laboratorium, Universiteit van Amsterdam
Valckenierstraat 65-67, NL-1018 XE Amsterdam, The Netherlands

ABSTRACT

The generation of thermal donor centers (TD's) in boron- and aluminum-doped Czochralski-grown silicon as monitored by photoluminescence and magnetic resonance has been compared. In photoluminescence the TD-bound excitons were investigated; the results were examined against the magnetic resonance data on the development of Si-NL8 and Si-NL10 TD-related EPR spectra. The results of the photoluminescence study provide evidence that the generation of the TD centers is independent of the original (acceptor) doping of the material. At the same time the rate of the acceptor loss is clearly different with the aluminum atoms being much faster removed from their substitutional positions. Taken together, the results of the study indicate that the different TD generation kinetics in Al- and B-doped Czochralski silicon as evidenced by EPR measurements should be attributed to different behavior of the Fermi level position in these two materials rather than to real differences in the generation of TD centers.

1. Introduction

Silicon thermal donors (TD's) continue to be one of the most puzzling issues of the defect physics of silicon. In spite of concerted effort by both experimental and theoretical physics no consensus could be reached neither on the core structure of this evidently extended defect nor on the growth-development mechanism responsible for its multispecies character. Nevertheless a vast amount of evidence has been gathered and several aspects of TD centers and related phenomena could be clarified [1]. The fingerprints of TD centers were identified by most of the experimental techniques applied in semiconductor materials science. Among those the most direct information was provided by the infrared absorption spectroscopy where two TD-related ionization series were identified [2]. More information on the multispecies character of TD's was obtained from the EPR/ENDOR studies where two spectra namely Si-NL8 and Si-NL10 were found to be TD related [3]. More recently further confirmation of the multiple nature of the centers came from photoluminescence (PL) spectroscopy as the luminescence lines originating from TD-bound excitons (BE's) have been identified and studied [4].

In the present study the TD generation kinetics, as visualized by TD-bound exciton luminescence, has been followed for boron- and aluminum-doped Czochralski-grown silicon. The results were compared with data for the same material obtained from magnetic resonance studies. The objective of the study was to gain new insight into the TD issue by intercorrelation of the information provided by two independent techniques sensitive to different charge states of the same defect. In this case the magnetic resonance spectroscopy delivers information on paramagnetic states of thermal donors: the Si-NL8 spectrum on TD^+ [5] and, following a somewhat speculative identification, the Si-NL10 spectrum on the overcharged TD^- thermal donor state [6]. On the other hand the information obtained from optical measurements would come from the photoluminescence of excitons bound to the neutral TD^0 state of thermal donors.

2. Experimental

In the present study two kinds of samples were used with the following starting parameters:

1. boron-doped Czochralski-grown silicon: $[B_s] = 1.3 \times 10^{16}cm^{-3}$, $[O_i] = 1.3 \times 10^{18}cm^{-3}$, $[C] \leq 10^{15}cm^{-3}$
2. aluminum-doped Czochralski-grown silicon: $[Al_s] = 4 \times 10^{15}cm^{-3}$, $[O_i] = 1.3 \times 10^{18}cm^{-3}$, $[C] \leq 10^{15}cm^{-3}$

Prior to the experiment the samples were given the standard initial heat treatment of 1/2 h, 1350°C in ambient gas atmosphere in order to disperse the oxygen and ensure the same, well-defined starting conditions. This heat treatment had no effect on the level of acceptor concentration. Following the initial heat treatment, thermal donor centers were subsequently generated in the samples by annealing them for various time intervals at a temperature of 470°C. After each annealing stage photoluminescence spectra have been measured.

The photoluminescence measurements were performed at 4.2 K. Luminescence was excited with a cw Ar^+-ion laser operating at 514.5 nm. To avoid spurious plasma lines, a 514.5-nm interference filter was used. The emerging luminescence was collected from the laser-irradiated side. It was dispersed by a high-resolution 1.5-m F/12 monochromator (Jobin-Yvon THR-1500) with a 600-grooves/mm grating blazed at 1.5 μm and detected by a nitrogen-cooled Ge detector (North Coast EO-817). The detector output was amplified using conventional lock-in techniques. The signals were then digitized and fed into a computer for further data acquisition.

3. Results and discussion

As has already been mentioned in the introduction the magnetic resonance investigations revealed two TD-related EPR spectra labelled Si-NL8 and Si-NL10. Following this conclusion the generation kinetics of these centers has been investigated in different materials [3, 7]. The studies have shown that both centers could be generated in (originally) p-type material regardless of the actual type of acceptor while in n-type only Si-NL10 centers could be observed. The most important conclusion was that in aluminum-doped material the generation of Si-NL10 centers was considerably enhanced in comparison to more commonly used boron-doped silicon. Since at the same time the actual microscopic identification of the Si-NL10 center is a matter of considerable controversy this particular finding promoted the idea of relating this center to some form of aluminum-based thermal (single?) donor or Al-TD complex. Such idea was further supported by the fact that the hyperfine interactions with the ^{27}Al nuclei have indeed been observed for, at least some, Si-NL10 centers [8, 9]. Simultaneously the possibility that in the aluminum-doped material some distinctly different from *normal* TD's form of thermal donor centers is generated has also been investigated by infrared absorption [10].

In addition to the above mentioned information on the generation kinetics of TD centers the EPR studies revealed also that, while the individual TD species could not be directly resolved, the multispecies character of thermal donors manifested itself in nonhomogeneous character of both spectra and the so called *g-shifting effect* as the relative concentration of the individual species varied. The g-shifting effect is experimentally observed as the lowering of anisotropy of both spectra for longer annealing times and can be conveniently expressed as a change of the splitting of two resonance lines as observed in the [111] crystallographic direction. Fig.1 summarizes briefly the EPR results; Fig.1(a) compares the generation kinetics of Si-NL8 and Si-NL10 centers in two types of material used in the current study; in Fig.1(b) the gradual decrease of the anisotropy of the Si-NL8 (TD^+) spectrum for these materials is depicted.

The generation kinetics as deduced from EPR measurements has to be treated with certain care. One should bear in mind that the EPR measurement is sensitive to a particular charge state of the defect; in case of thermal donors in total three different charge states are available: TD^0, TD^+, and TD^{++} (if the possibility of an overcharged TD^- paramagnetic state is not considered). Out of these only one, TD^+, is paramagnetic and as such will be monitored by EPR. Therefore the EPR-based TD kinetics not only reflects the generation process of thermal donors but also the change of the Fermi level position. Since the process of the loss of acceptors has been found to occur in parallel to TD generation [11], the actual position of the Fermi level for a given heat-treatment time will be influenced by two different processes: creation of donors and annihilation of acceptors. The EPR experiment is not capable of separating the two contributions.

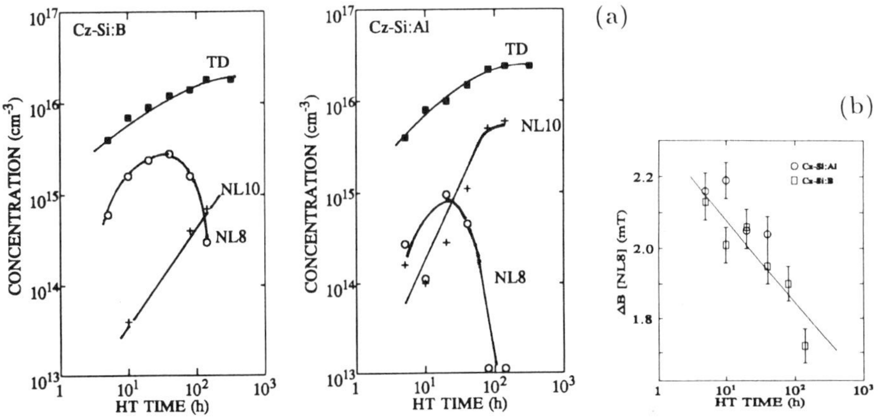

Figure 1:(a) *Generation kinetics of TD's (on the basis of resistivity measurements) and TD-related EPR centers, and* **(b)** *the annealing time dependence of g-shifting for the Si-NL8 EPR spectrum in Czochralski-grown silicon.*

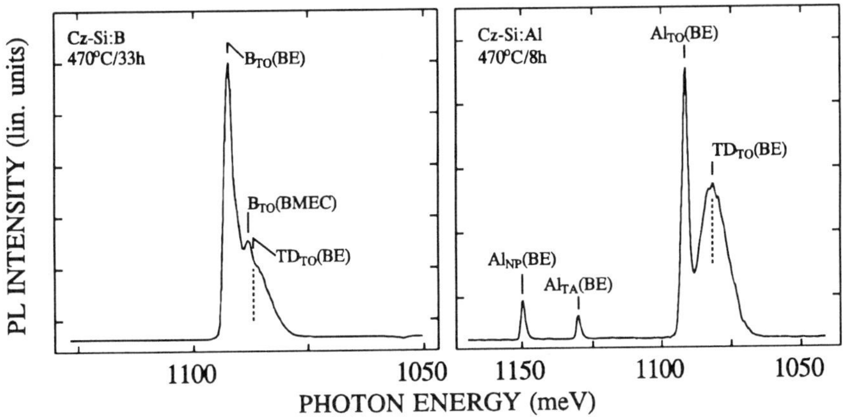

Figure 2: *Photoluminescence spectra of boron- and aluminum-doped Czochralski silicon under Ar^+-excitation.*

A very similar kind of argument can also be made for the TD generation kinetics as derived from infrared absorption studies [2]. Also in this case the level at which the absorption is to be studied has to be populated, i.e., the Fermi level has to be sufficiently high. In case of p-type material this effect may be a source of severe deformation of the results.

The photoluminescence experiment, performed under nonequilibrium conditions of laser excitation, is free from the above outlined limitation. One can therefore expect that TD generation kinetics derived from PL measurements should be far more reliable than those based on resistivity, EPR or IR absorption data.

Fig.2 presents the PL spectra as recorded for boron- and aluminum-doped material. The acceptor and TD-related luminescence lines are marked. In both cases the TO replica of the TD-bound exciton line may clearly be recognized in the vicinity of the strong acceptor lines. The spectra are in full analogy with earlier data on TD luminescence [4, 11] which were able to separate the contribution from the individual thermal donor species. A similar goal has also been attempted here and is illustrated in Fig.3. Fig.3(a) presents details of the no-phonon (NP) TD-bound exciton luminescence for boron-doped silicon. As can be seen in the present experiment some of the earlier identified lines [4] could be resolved and are marked in the figure. For the sake of completeness also the positions of boron-related exciton lines are depicted.

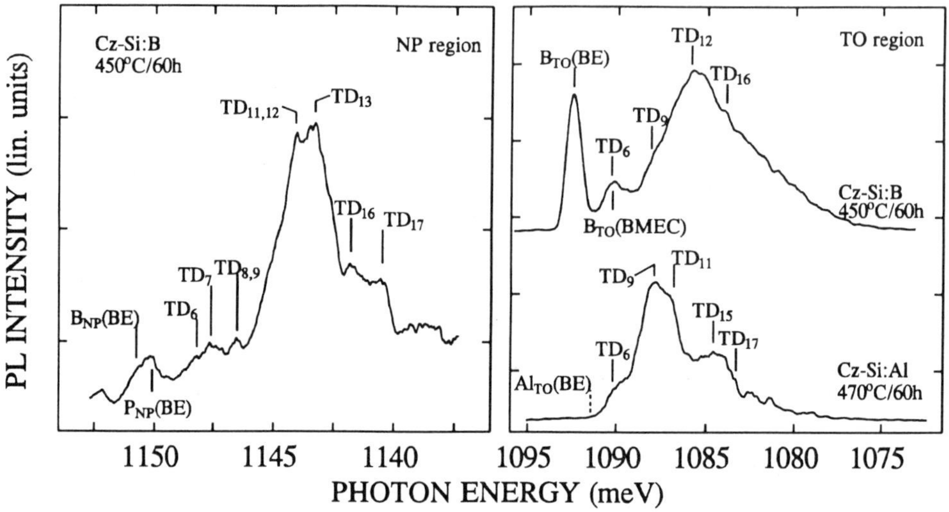

Figure 3: *TD-bound exciton photoluminescence spectra of Czochralski (Cz) silicon:* (a) *NP range, Cz-Si:B,* (b) *TO replica range, Cz-Si:B and Cz-Si:Al. Individual TD species at the energies identified earlier by Steele and Thewalt [4] are marked.*

Fig.3(b) shows detailed spectra as recorded for B- and Al-doped silicon in the TO replica region of the TD-bound exciton luminescence. For the B-doped sample the same TD species (as in the NP region) can be accounted for, thus supporting the identification. Additionally, a strong signal of boron-bound excitons, reflecting the concentration of substitutional boron acceptors, is present. Also for the Al-doped material, some of the TD species identified earlier may be resolved. At the same time the aluminum-bound exciton line is no longer visible.

Upon annealing the relative concentration of individual TD species changes as does the total number of thermal donor centers. In the less detailed luminescence spectra this process may be visualized by the shift of the TD-related emission band towards higher photon energies and by a change of its total intensity. Both effects are illustrated in Fig.4. Fig.4(a) presents the shift of the center of the TO replica of the TD-related luminescence for two kinds of samples used in the experiment. As can be seen both materials behave in a similar manner and the shift of the luminescence towards higher energies resembles very much the shift of g-values of the Si-NL8 and Si-NL10 spectra towards lower anisotropy. In Fig.4(b) the annealing time dependence of the intensity of the TD band is compared for two kinds of acceptor dopants. It coincides well with the behavior as expected for TD centers on the basis of IR absorption studies. It should be noted that the kinetics appears to be a little slower in the aluminum-doped material. However, the effect is small and as such can probably be attributed to a somewhat lower oxygen contents (TD formation rate is known to depend on the fourth power of the initial oxygen concentration).

The results of the current study as depicted in Figs.2,3, and 4 indicate that the generation of thermal donors is not dependent on the actual acceptor dopant. In more detail the study shows that the aluminum doping has no accelerating effect for TD kinetics which is (predominantly) governed by the initial oxygen concentration. For both boron and aluminum doping the generation kinetics of TD's as monitored by luminescence proceeds in a very similar manner. This conclusion is illustrated in Fig.3. As shown in this figure, identical TD species are formed in both materials, thus indicating identical TD generating processes. It should be mentioned here, that such conclusion does not rule out the possibility that in aluminum-doped silicon in addition to *normal TD's* also some other aluminum-based form of TD-like centers is formed. This however seems to be rather unlikely in view of very similar generation kinetics of *normal*

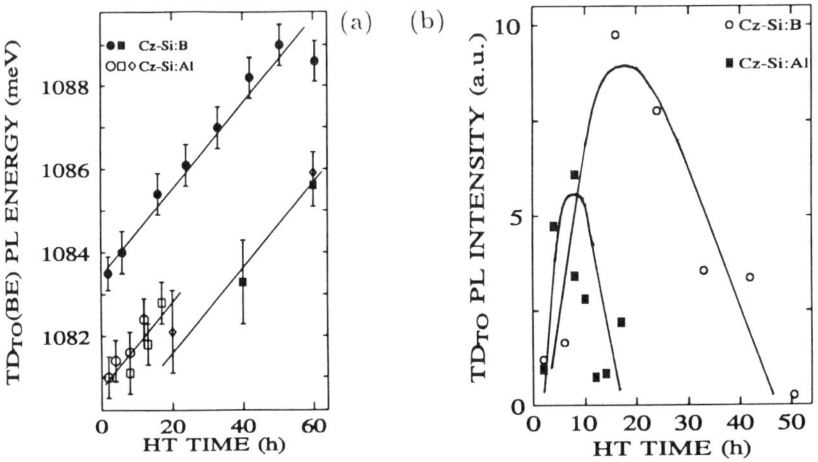

Figure 4: *The annealing time dependence of the TD-bound exciton emission band in the TO region:* (a) *the shift of its center, and* (b) *the change of its total intensity.*

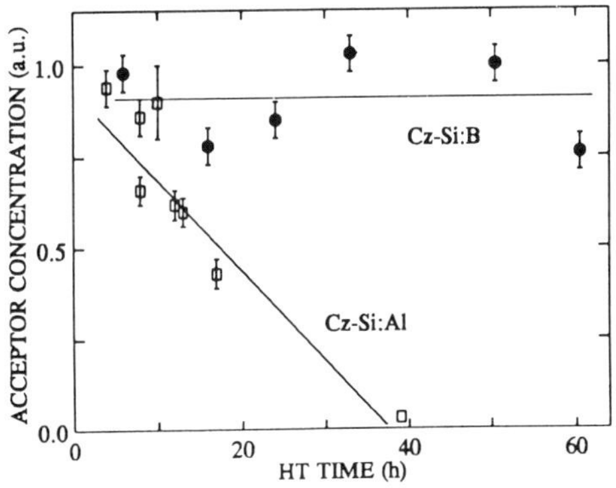

Figure 5: *Decay of acceptor-bound exciton luminescence for the Czochralski silicon as used in this study.*

TD's for both materials.

The above conclusion appears in contradiction to the results of IR absorption studies [10] as well as to the resistivity and magnetic resonance findings as summarized by Fig.1. This apparent contradiction may be understood in view of the remarks presented at the beginning of this section. In contrast to the luminescence of bound excitons, all the above mentioned techniques are sensitive not only to the generation of TD centers but also to the actual position of the Fermi level. Different results obtained from these techniques could then be explained if in the aluminum-doped material the Fermi level position would be influenced also by yet another mechanism, independent from the generation of TD's. Such mechanism could be provided by gradual removal of acceptors from the substitutional positions and has indeed been concluded to take place in parallel to the generation of TD's [11]. In the present study it is further confirmed as can be concluded from Fig.5, where the intensity of acceptor-related luminescence is depicted as a function of the annealing time. It can clearly be seen that, while the concentration of boron remains almost constant all through the annealing procedure, the concentration of aluminum is quite soon diminished below the detection limit. In this situation the behavior of the Fermi level is very much different in both materials (especially since interstitial aluminum has donor character) and thus could provide a very natural explanation of the seemingly catalytic role of aluminum in the TD generation process.

Taken together the results of the current study argue very much against the idea that in aluminum-doped silicon a different kind of TD centers is being created.

Acknowledgment

This project has been partially sponsored by the Netherlands Foundation for Fundamental Research on Matter (Stichting voor Fundamenteel Onderzoek der Materie, FOM).

References

[1] A. Bourret, in *Proceedings of the Thirteenth International Conference on Defects in Semiconductors,* edited by L. C. Kimerling and J. M. Parsey, Jr. (TMS-AIME, New York, 1985), p.129.

[2] P. Wagner and J. Hage, Appl. Phys. A **49**, 123 (1989).

[3] T. Gregorkiewicz, D. A. van Wezep, H. H. P. Th. Bekman, and C. A. J. Ammerlaan, Phys. Rev. B **35**, 3810 (1987).

[4] A. G. Steele and M. L. W. Thewalt, Can. J. Phys. **67**, 268 (1989).

[5] K. M. Lee, J. M. Trombetta, and G. D. Watkins, in *Proceedings of the International Conference on Microscopic Identification of Electronic Defects in Semiconductors, San Francisco, 1985,* edited by N. M. Johnson, S. G. Bishop, and G. D. Watkins (MRS, Pittsburgh, 1985), p.263.

[6] H. H. P. Th. Bekman, T. Gregorkiewicz, and C. A. J. Ammerlaan, Phys. Rev. Lett. **61**, 227 (1988).

[7] H. H. P. Th. Bekman, T. Gregorkiewicz, D.A. van Wezep, and C. A. J. Ammerlaan, J. Appl. Phys. **62**, 4404 (1987)

[8] T. Gregorkiewicz, H. H. P. Th. Bekman, and C. A. J. Ammerlaan, Phys. Rev. B **38**, 3998 (1988).

[9] J. Michel, N. Meilwes, J. R. Niklas, and J.-M. Spaeth, in *Shallow Impurities in Semiconductors 1988*, (Institute of Physics, Bristol, 1989), p.201.

[10] M. Claybourn and R. C. Newman, Mater. Sci. Forum **38-41**, 613 (1989).

[11] A. C. T. Drakeford and E. C. Lightowlers, Mat. Res. Soc. Symp. Proc. **104**, 209 (1988).

Materials Science Forum Vol. 83-87 (1992) pp. 413-418

DEFECT DISTRIBUTION IN LARGE CZ-SILICON WAFERS INVESTIGATED BY POSITRON ANNIHILATION SPECTROSCOPY

P. Mascher[1], W. Puff[2], S. Hahn[3], K.H. Cho[4] and B.Y. Lee[4]

1 Centre for Electrophotonic Materials and Devices,
Department of Engineering Physics, McMaster University,
Hamilton, Ontario, Canada
2 Institut für Kernphysik, Technical University of Graz, Austria
3 Department of Materials Science and Engineering,
Stanford University, Stanford, CA 94305
4 Lucky Advanced Materials Inc., #283 Imsoodong,
Gumi Kyungsangbukdo, Korea

ABSTRACT

Results of the first stage of a positron annihilation study of defect characteristics across six-inch CZ-Si wafers are reported. The positron response is found to essentially stem from small oxygen clusters and "perfect" bulk silicon only. Possible contributions from vacancy-type defects are at or just below the detection limit. Wafers from material grown at high pull speeds generally show less of the oxygen cluster component but overall the wafers are quite homogeneous in terms of their across-the-wafer characteristics.

1. Introduction

Silicon crystals grown by the Czochralski (CZ) method are saturated with oxygen at the melting temperature, but are supersaturated at typical device processing (i.e. much lower) temperatures [1]. As a consequence, oxygen precipitation occurs, both homogeneously and inhomogeneously. It is generally accepted that point defects play an important role in the precipitation process as evidenced by a multitude of observations of oxygen-defect complexes after heat treatment [2].

In this contribution, we report results of the first stage of a systematic investigation by both positron lifetime and Doppler-broadening spectroscopy in an effort to monitor the involvement of vacancies in the oxygen precipitation process. During the recent years this technique has been demonstrated to be a valuable tool for the characterization of defects in crystalline silicon [3]. The work was triggered by two previous investigations where it was found that significant differences in the defect characteristics can be detected across a given wafer [4] and that heat treatment can give rise to interesting impurity-defect complexes, at least at a transient level [5].

2. Experimental

The wafers investigated here were p-type material with resistivities ranging from 6 to 10 Ωcm. Two sets of wafers were investigated, one cut from a boule grown under normal pull speed (0.85 mm/min), the other from a boule grown under fast pull speed (1.26 mm/min). Each set consisted of wafers taken from the seed-, middle- and tail-sections of the boules, respectively. Interstitial oxygen concentrations, $[O_I]$, were determined using a BIORAD QS-100 Fourier transform infrared spectrometer. These measurements were performed at 14 equidistant locations across the wafers with the extreme positions located 6 mm from the respective edges.

The positron lifetime spectra were obtained using state of the art spectrometers with time resolutions (FWHM) of between 205 and 230 ps. Each spectrum contained at least 6×10^6 counts. The positron sources were rather weak, typically 3.5 µCi of $^{22}NaCl$, deposited on thin (1–3 µm) Al-foil. Numerical analysis of the lifetime data was performed using the programs PATFIT88 [6] or PFPOSFIT [7] and agreement of results on test spectra was always found to be within statistical uncertainty. After correction for annihilations in the Al-foil and the NaCl itself, all the spectra contained three lifetime components, the longest of which however, was an essentially constant term of about 1.5 ns with no more than 0.3% intensity. This term could easily be separated from the actual spectra leaving two significant terms from the samples themselves.

Doppler-broadened energy spectra of the annihilation quanta were obtained using an intrinsic Ge detector with a resolution (FWHM) of 1.18 keV at 497 keV (close to the 511 keV annihilation γ-energy). A total of 2.5×10^6 counts was recorded for each spectrum. In this first investigation the numerical analysis of the Doppler spectra was restricted to the determination of the shape parameter, S, defined as the ratio between the total amount of counts in a central portion of the spectrum and the total amount of counts [5].

All the positron measurements were performed at the above mentioned 14 locations across the wafers. The size of the source (~2 mm diameter) determined the spatial resolution. The experiments were conducted at room temperature.

3. Results and Discussion

Unconstrained fits to the lifetime spectra yielded two significant components in all wafers. The value of the first (shorter) components, τ_1, varied strongly but not systematically from position to position, ranging from 95 ps to about 140 ps. Corresponding intensities, I_1, were low, ranging from 6 to 15%. The second component was much better defined (due to the much higher intensity) and had values between 220 and 230 ps. If we now make the assumption, as is usually done, that the short-lived τ_1 component is actually the bulk lifetime modified by trapping into defects which give rise to the longer lived τ_2 component, we can use the simple trapping model [8] to calculate what the bulk lifetime should have been according to:

$$1/\tau_B^{TM} = I_1/\tau_1 + I_2/\tau_2 \qquad (1)$$

where $\tau_B{}^{TM}$ is the bulk lifetime resulting from the application of the trapping model (TM). This approach is important because the bulk lifetime is a materials parameter which should not be affected by the presence of positron traps. The established value for silicon is 218-220 ps [9]. The results for $\tau_B{}^{TM}$ in the present study lie in a range from 207 to 212 ps, which is significantly shorter, and thus causes doubt as to the applicability of the trapping model. It rather seems as if τ_1 represented a distinct positron state as opposed to the modified bulk lifetime. Results of this type have been obtained earlier for CZ-Si [4, 5, 10] and in all these cases the short lifetime component had values of about 100 ps and was associated with positrons trapped at interstitial-type clusters. Taking the average of all the short components in the present spectra yields $\tau_1 = 113\pm15$ ps which is in excellent agreement with these earlier data. This component was then fixed in the subsequent analyses. In the first two panels of Fig. 1 are shown the lifetime data from this analysis for a typical wafer (IH1-193, see Table 1) as a function of the measurement position. Also shown in the figure are the interstitial oxygen concentrations, $[O_I]$, determined at the very same locations on the wafers.

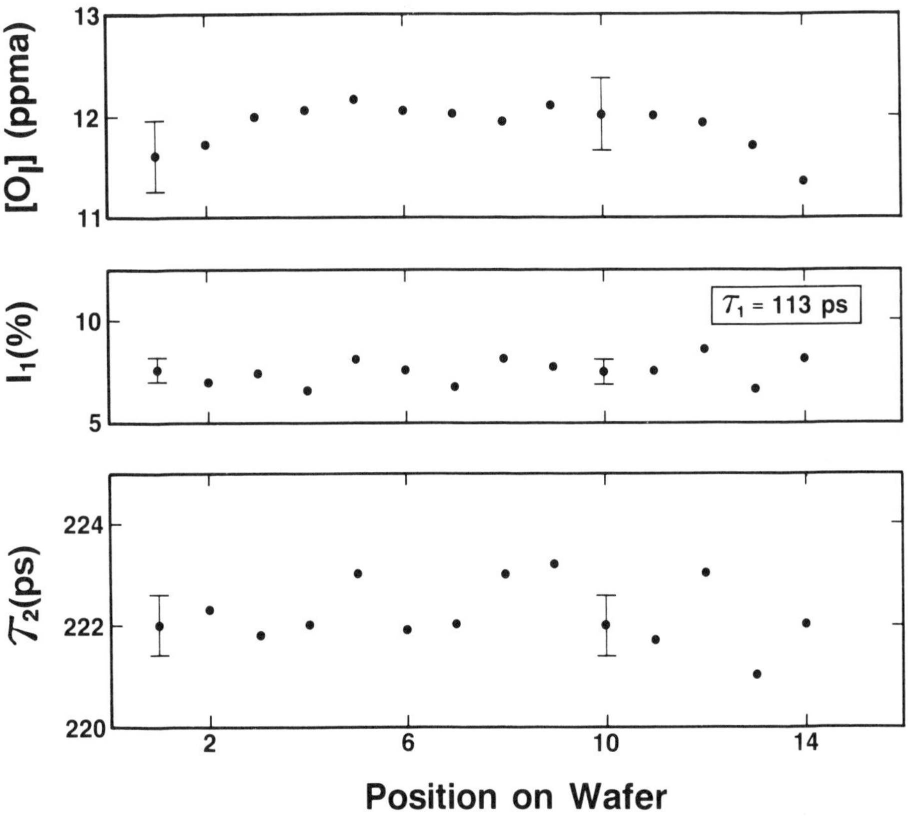

Figure 1. Main lifetime component, τ_2, intensity, I_1, of the short-lived component (τ_1 = 113 ps) and interstitial oxygen concentration, $[O_I]$, as a function of measurement position on wafer IH1-193 (see Table 1). Positions 1 and 14 correspond to locations 6 mm from the respective edges.

Sample	Pull Speed (mm/min)	Position in Boule	$[O_I]$ (ppma)		I_1 (%) (τ_1 = 113 ps)	
NL1–028	0.85	seed	16.5	(0.4)	11.1	(1.3)
NL1–290		middle	14.1	(0.3)	11.3	(1.0)
NL1-592		tail	14.0	(0.2)	11.4	(1.2)
IH1–008	1.26	seed	16.8	(0.4)	9.1	(0.5)
IH1–193		middle	11.9	(0.3)	7.4	(0.7)
IH1–260		tail	11.3	(0.2)	8.5	(0.6)

Table 1. Interstitial oxygen concentration, $[O_I]$, and intensity, I_1, of the short-lived component (τ_1 = 113 ps) in two groups of wafers grown at different pull speeds. The values in brackets indicate maximum overall deviations across the wafer rather than experimental uncertainties.

There are several interesting observations to be made. First, the main lifetime component, τ_2 (with intensity $I_2 = 100–I_1$), is now very close to the expected value for the bulk lifetime, i.e. 218-220 ps. This further corroborates our assessment of τ_1 as being due to a distinct positron state, namely one of positrons trapped at oxygen clusters where the local electron density is about twice that for silicon. A detailed analysis of the trapping process involved is given in [4]. The fact that τ_2 is actually slightly longer than the bulk lifetime suggests a contribution of vacancy-type defects (with $\tau_D > \tau_B$) but attempts to resolve such a component were inconclusive, most likely to due to too small a contribution to the spectrum.

Second, there seems to be no obvious change in the lifetime characteristics as one moves from one edge of the wafer (position 1) to the other (position 14). This is in contrast to the earlier study [4] – on different wafers – which showed significant changes in the spectrum parameters across a given wafer, especially close to the edges. This region was clearly dominated by a vacancy response, a feature which is completely absent in all the wafers investigated in the present study. It seems as if the wafer material simply is much more homogeneous (and of better quality) despite the fact that the wafers are six inches in diameter as compared to four-inch wafers in [4]. This finding is further emphasized by the generally random distribution of the S-parameter values across the wafers (Fig. 2).

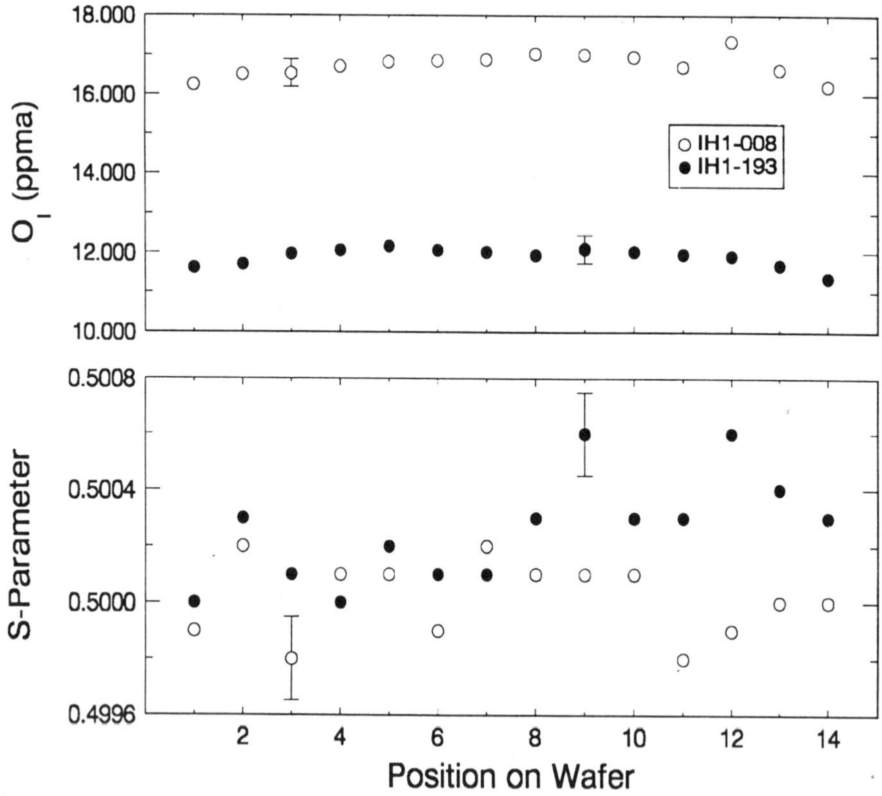

Figure 2. Doppler-broadening S-parameter and interstitial oxygen concentration, $[O_I]$, as a function of measurement position on wafers IH1-008 and IH1-193 (see Table 1). Positions 1 and 14 correspond to locations 6 mm from the respective edges.

Third, there seems to be no correlation between [O_I] and the intensity of the oxygen-cluster related lifetime component. The last two columns in Table 1 further illustrate that point. As expected, this is also reflected by the Doppler-broadening data. Fig. 2 shows nearly identical S-parameter values for two wafers with substantially differential interstitial oxygen concentrations. Both these observations are not really surprising since, as was argued in [4], interstitial oxygen by itself is not a likely candidate for the positron trap in question.

Another interesting observation from Table 1 is that there seems to be some influence of the pull speed on the concentration of oxygen clusters, giving rise to the short lifetime component, with high pull speeds resulting in lower cluster concentrations. Also, in this material, there seems to be more scatter in the intensity of this component as one moves from seed to tail despite the fact that the deviations across the wafers are quite comparable to the standard pull speed material.

At the present time the wafers are being subjected to a first step in a commercial heat treatment process. After that, the very same across-the-wafer measurements will be performed to investigate the involvement of defects in oxygen precipitation phenomena.

Acknowledgements

This work was supported by the Natural Sciences and Engineering Research Council of Canada (NSERC) and the Fonds zur Förderung der Wissenschaftlichen Forschung of Austria.

References

1. J.C. Mikkelsen in: Oxygen, Carbon, Hydrogen and Nitrogen in Crystalline Silicon, eds. J.C. Mikkelsen, Jr., S.J. Pearton, J.W. Corbett and S.J. Pennycook (Materials Res. Soc., Pittsburgh), 1986, p. 19.
2. J.W. Corbett, P. Deak, J.L. Lindström, L.M. Roth and L.C. Snyder in: Defects in Semiconductors 15, ed. G. Ferenczi, Materials Sci. Forum 38-41, 579 (1989) and many references therein.
3. S. Dannefaer in: Defect Control in Semiconductors, ed. K. Sumino, Elsevier Science Publishers B.V. (North-Holland), 1990, p. 1561.
4. S. Dannefaer and D. Kerr, J. Appl. Phys. 60, 1313 (1986).
5. S. Dannefaer, W. Puff, P. Mascher and D. Kerr, J. Appl. Phys. 66, 3526 (1989).
6. P. Kirkegaard, N.J. Pedersen and M. Eldrup, RISO-M-2740, Riso National Laboratory 1989, DK-4000 Roskilde, Denmark.
7. W. Puff, Comput. Phys. Commun. 30, 359 (1983).
8. R.N. West, Adv. Phys. 22, 263 (1977).
9. S. Dannefaer, Phys. Status Solidi A 102, 481 (1987).
10. W. Puff, S. Dannefaer, P. Mascher and D. Kerr, Phys. Status Solidi A 102, 527 (1987).

Materials Science Forum Vol. 83-87 (1992) pp. 419-424

INFRARED ABSORPTION BY INTERSTITIAL OXYGEN IN GERMANIUM-DOPED SILICON CRYSTAL

HIROSHI YAMADA-KANETA[1], CHIOKO KANETA[2] and TSUTOMU OGAWA[1]

[1] Basic Process Development Division, Fujitsu Limited, 1015 Kamikodanaka, Nakahara-ku, Kawasaki 211, Japan

[2] Semiconductor Devices Laboratory, Fujitsu Laboratories Ltd., 10-1 Morinosato-Wakamiya, Atsugi 243-01, Japan

ABSTRACT

Near the unperturbed absorptions at about 30, 1100 and 1200 cm^{-1} due to the isolated oxygen, the perturbed absorptions by two kinds of germanium-oxygen complexes were found. While the area-intensities per one oxygen atom (the oscillator strengths) of the perturbed 30 and 1100 cm^{-1} absorptions were almost unchanged from the corresponding unperturbed ones, those of the perturbed 1200 cm^{-1} absorptions were reduced to 17% or less of the unperturbed one. Modification of our previous calculation for the isolated oxygen consistently explained the perturbed oscillator strength and the peak-energies of the Ge-O complexes. The reduction of the 1200 cm^{-1} oscillator strengths was ascribed to the weakening of the anharmonic coupling between the 30 and 1100 cm^{-1} impurity-modes.

1. Introduction

The bond-interstitial oxygen (**Oi**) in silicon causes the impurity absorption bands at 30, 1100 and 1200 cm^{-1} regions.[1-4] These bands attract much interests because they reflect strong anharmonicity of the impurity excitation. The energies of the peaks observed in these bands are listed in Table I.

According to our previous paper,[5] the 30 cm^{-1} band is caused by the 2-dimensional low-energy anharmonic excitation (**2D LEAE**) due to the (r, ϕ)-freedom of motion of the Oi (Fig.1) in the adiabatic potential $\alpha_0 r^2 + \beta r^4$. The 1100 cm^{-1} band is caused by the A_{2u} local mode Q_{A2u} containing the z-displacement of the Oi (Fig.1). The anharmonic coupling $gr^2Q_{A2u}{}^2$ between these two excitations causes the 1200 cm^{-1} band as the 2D LEAE sideband of the 1100 cm^{-1} transition. Figure 2 shows the energy levels of the 2D LEAE belonging to the zero- and one-A_{2u}-phonon states. The observed peaks listed in Table I are assigned to the transitions as shown in Fig.2. Without the coupling (g=0), the shape of the adiabatic potential $\alpha_1 r^2 + \beta r^4 + \hbar\omega$ for the one-A_{2u}-phonon states (Fig.2) became the same as that of the zero-A_{2u}-phonon state potential $\alpha_0 r^2 + \beta r^4$, i.e., $\alpha_1 = \alpha_0$. This leads to the same energy-level-schemes for both of the zero- and one-A_{2u}-phonon states. In this case, the 1100 cm^{-1} band peaks at 1136.4, 1128.2 and 1121.9 cm^{-1} (see Fig.2) are piled up into a single peak. Thus, the separations between these peaks reflect the strength of the coupling. Consequently, in our model, the intensity of the 1200 cm^{-1} band is closely related to the separations of the 1100 cm^{-1} band peaks.

In Sec. 2, we show that the oscillator strengths (the area-intensities per one oxygen atom) of the 1200 cm^{-1} absorptions of the Ge-O complexes are much smaller than that of the isolated Oi, while regarding the 1100 and 30 cm^{-1} absorptions the oscillator strengths of the Ge-O complexes are almost unchanged from those of the isolated Oi. In Sec. 3, using our previous model, we calculate the oscillator strengths. In Sec. 4, we consistently explain both of the measured oscillator strengths and the peak-energies. The weakening of the anharmonic coupling in the Ge-O complexes turns out to be essential for the reduction of the 1200 cm^{-1} oscillator strength.

2. Experiment and results

Four-inch crystal ingots with germanium concentration [Ge] of 13-13600 ppm were grown by the Czochralski method. We measured [Ge] by the inductively coupled plasma (ICP) analysis. The interstitial oxygen concentrations [Oi] of the Ge-doped crystals were 31.1-34.2 ppm. We refer to the crystal without Ge-doping as the *undoped* crystal. All samples were p-type with specific resistivity 8-11 Ω cm. The carbon concentration was under detection limit, 0.1 ppm. The infrared and far-infrared absorption measurements were carried out with the Fourier transform spectro-photometers, JEOL JIR-100 and BRUKER IFS-120HR, respectively. The resolutions were 0.1-0.5 cm^{-1} and 0.1-0.25 cm^{-1} for the low-temperature infrared and far-infrared measurements, respectively.

Table I. Energies of the absorption peaks observed for [Ge] = 0 ppm (undoped) and 13000 ppm. The parameter values adjusted so as to well explain the observed peak energies are also given. The peaks in the undoped case have already been reported by other papers (Ref. 1-4). The values shown are the result of our present measurements. The blanks indicate that the corresponding absorption was not distinguishable as a peak. When adjusting the parameters for Oi-III, the value of β was fixed to that determined for Oi-II.

Types of Oi [Ge] (ppm)	Observed Peak Energies (cm^{-1})							a (meV/Å^2)	β (meV/Å^4)	$\hbar\omega$ (meV)	$\hbar g/\omega$ (meV/Å^2)
Oi-I (undoped)	29.2	37.7	43.3	49	1136.4 1205.7	1128.2 1216.4	1121.9	−15.948	241.95	143.18	−27.364
Oi-I (13000)	30.1	38.3	–	–	1135.9 1206.4	1128.1 1217.2	1121.2	−13.464	239.45	142.93	−25.899
Oi-II (13000)	33	–	–	–	1130.1 1213	1126.2 –	1123	−36.122	372.43	141.12	−15.730
Oi-III (13000)	36	–	–	–	1118.5 –	1115.5 –	–	−27.317	372.43	139.42	−12.386

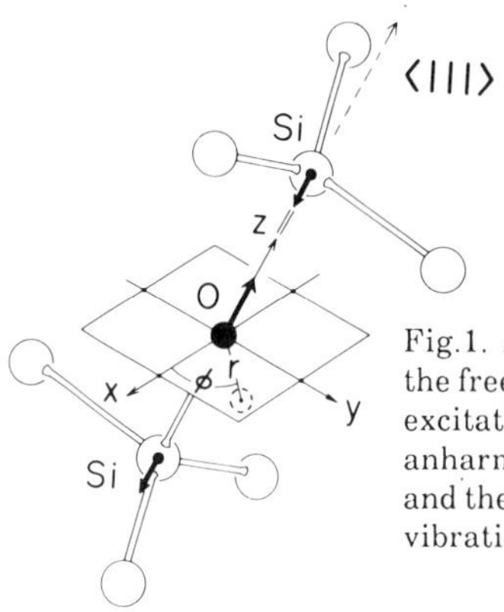

Fig.1. Atomic configuration and the freedoms of impurity excitations, the in-plane anharmonic motion (r, ϕ) and the A_{2u}-symmetric local vibration (bold arrows).

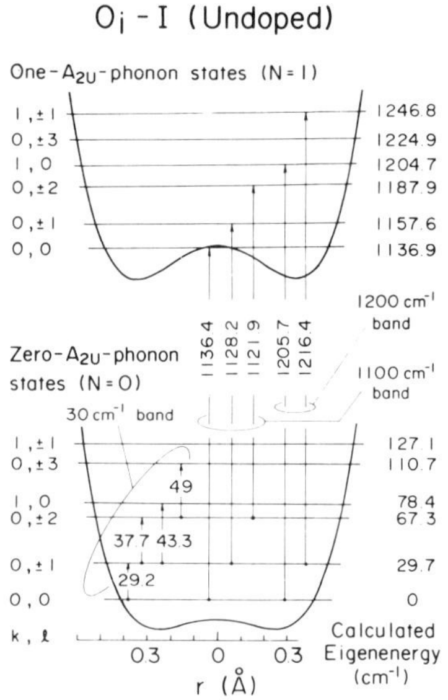

Fig.2. Energy levels of the 2D LEAE in the A_{2u}-local-phonon-renormalized adiabatic potentials calculated for the undoped case. The value in the arrow is the observed energy (Table I) of the peak assigned to each transition.

A. The 1100 cm^{-1} band

As shown in Fig.3, the position and the shape of the room-temperature oxygen peak at 1106 cm^{-1} is almost unchanged by the Ge-doping. The values of [Oi] determined from each peak are given in Fig.3. We used the conversion constant in ASTM F121-79.

The 1106 cm^{-1} peak of the undoped crystal changes to the sharp peak at ~ 1136 cm^{-1} as the temperature lowers.[1,2] Figure 4 shows the bottom of the 1136 cm^{-1} peak measured for various [Ge] at 4.5K. The ordinate of the figure, the *normalized absorbance* (**NA**), is the absorbance normalized by the sample thickness t and the $[Oi]$, i.e., $(-log_{10}[I(\omega)/I_0(\omega)])\,/\,t\,[Oi]$. As shown in Fig.4, new peaks grow up at 1130.1 cm^{-1} and 1118.5 cm^{-1} as [Ge] increases. However, it turned out that the total integral intensities for the range 1100-1150 cm^{-1} were almost the same for all NA spectra in Fig.4, 1.58±0.03 cm^{-2} ppm^{-1}. This suggests that the new peaks originate from two types of the Oi suffering such slight perturbations as to hardly affect the oscillator strength. As shown in Fig.5, the area-intensities of the new peaks increase in proportion to [Ge]. Therefore, the new peaks are attributed to two kinds of complexes involving one Ge atom and the Oi. However, the shifts of the two peaks from the fundamental Oi-peak, ~ 6 cm^{-1} and ~18 cm^{-1} respectively, seem to be too small to be attributed to formation of the direct Ge-O-Si bond. The complexes would be the Si-O-Si units located close to the substitutional Ge atom. Hereafter, we refer to these types of Oi's causing the 1136, 1130.1 and 1118.5 cm^{-1} absorptions as *Oi-I*, *Oi-II* and *Oi-III*, respectively. From the invariability of the oscillator strength suggested above, Fig.5 can be interpreted as [Ge]-dependence of the partition rate of Oi atoms to Oi-I, -II and -III.

Figure 6 (a) and (b) are the NA spectra measured at 55K for [Ge]=13600 and 0 ppm. For [Ge]=13600 ppm, the partition rate of Oi atoms to Oi-I is 66.5% (Fig.5). Figure 6 (c) is the spectrum obtained by subtracting 66.5% of the spectrum (b) from the spectrum (a). This spectrum, in which we can find five absorptions at 1130.1, 1126.2, 1123, 1118.5 and 1115.5 cm^{-1}, approximately expresses the contribution from Oi-II and -III. We performed similar procedure for various temperatures, and found that the temperature dependence of the first three peaks is very similar to that of the 1136, 1128 and 1122 cm^{-1} peaks due to Oi-I. This suggests that these three peaks are due to Oi-II, and are assigned to the similar 1100 cm^{-1} band transitions described in Fig.2. The 1115.5 cm^{-1} peak is considered to be the hot band of the 1118.5 cm^{-1} peak due to Oi-III. According to our model outlined in Sec.1, these assignments mean that also Oi-II and -III have the 2D LEAE and its coupling to the A_{2u} local mode. However, the separations of the peaks belonging to Oi-II and -III (~ 4 cm^{-1} and ~ 3 cm^{-1}, respectively) are much smaller than that of Oi-I (~8 cm^{-1}), indicating much weaker couplings in Oi-II and -III. Therefore, from our model, the oscillator strengths of the 1200 cm^{-1} absorptions of Oi-II and -III are expected to be much smaller than that of Oi-I.

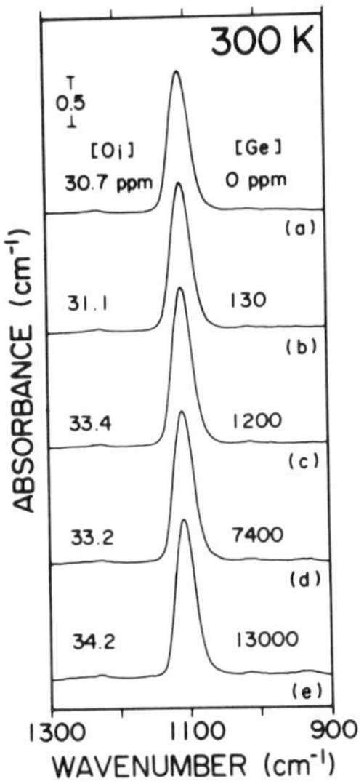

Fig.3. Room-temperature 1100 cm^{-1} band measured for various germanium concentrations [Ge].

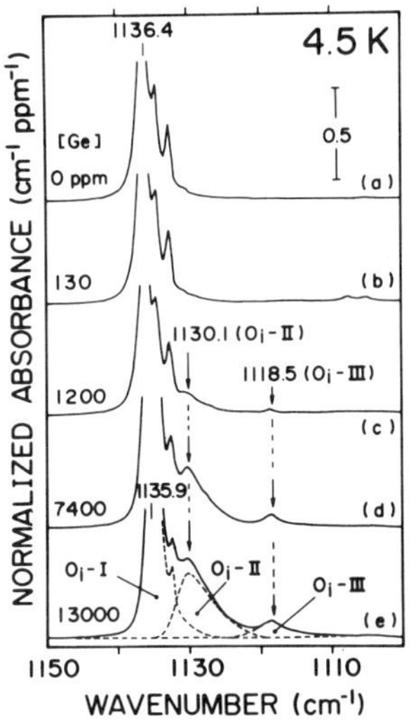

Fig.4. Germanium concentration dependence of the low-temperature 1100 cm^{-1} band.

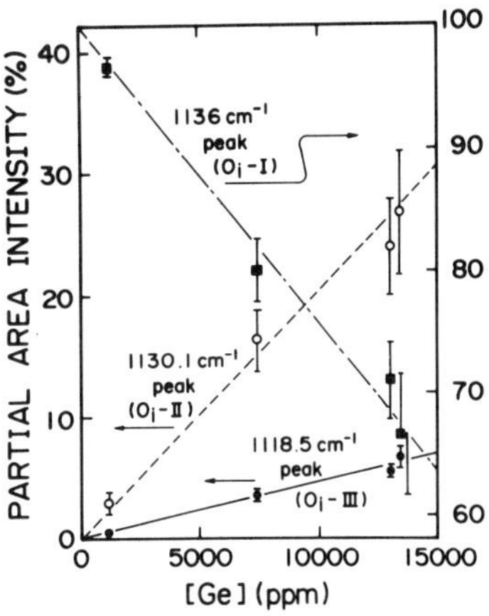

Fig.5. Germanium concentration dependence of the ratio of the area-intensity of each peak to the total integral intensity of the 1100 cm^{-1} band. The ordinate can be regarded as the partition rate of Oi atoms to Oi-I, -II and -III.

B. The 30 cm^{-1} band

Figure 7 shows the NA spectra of the 30 cm^{-1} band measured at 4.5 K. The plateau-like absorption in 32-37 cm^{-1} region becomes distinct as [Ge] exceeds 1200 ppm. This behavior of the plateau coincides with that of the 1130.1 and 1118.5 cm^{-1} peaks due to Oi-II and -III. We have seen above that Oi-II and -III, having the 2D LEAE, may contribute absorptions to the far-infrared region. Thus, we attribute the plateau to Oi-II and -III. Compared to the 1100 cm^{-1} band (Figs.4 (d) and (e)), the peak structure of this band is unclear (Figs.7(b) and (c)). This would be due to the large broadening of each peak caused by the inhomogeneous lattice strain induced by the Ge-doping.

As in the case of the 1100 cm^{-1} band, the total integral intensities of the NA spectra for the range 28-38 cm^{-1} were nearly constant (0.026$\pm$0.004 cm^{-2} ppm^{-1}), independent of [Ge]. This implies the situation analogous to the 1100 cm^{-1} band: the oscillator strengths of Oi-I, -II and -III are nearly the same, and nearly equal to that of Oi-I in the undoped case. From the peak-decomposition consistent with the measured partition rates of Oi atoms (Fig.5), the peaks due to Oi-II and -III are considered to be at ~33 and ~36 cm^{-1}, respectively.

C. The 1200 cm^{-1} band

The NA spectra measured at 4.5 K are shown in Fig.8. The 1200 cm^{-1} band in the heavily doped case consists of only two structures: the main peak at ~ 1206 cm^{-1} due to Oi-I, and the weak absorption at 1213 cm^{-1} which is considered to originate from Oi-II. For [Ge] = 13000 ppm, the total area intensity of the band is 70% of that in the undoped case, 0.0598 cm^{-2} ppm^{-1}: 66% of the main peak, and 4% of the 1213 cm^{-1} absorption. In this case, the partition rates of Oi atoms are 71, 23 and 6%, for Oi-I, -II and -III respectively (see Fig.5). Consequently, for Oi-I and -II respectively, the oscillator strengths are estimated to 93% and 17% of that of Oi-I in the undoped case. The reduction of the oscillator strength in Oi-II is remarkable, and consistent with the weakening of the coupling suggested from the 1100 cm^{-1} absorptions. The absorption by the Oi-III atoms of 6% partition (2.1 ppm) might be observed as a peak, if the oscillator strength of Oi-III were larger than 50%. But this was not the case. The separation (3 cm^{-1}) of the 1100 cm^{-1} band peaks due to Oi-III was smaller than that of Oi-II (~ 4 cm^{-1}). This suggests the oscillator strength of Oi-III to be smaller than that of Oi-II, 17%. The quantitative treatment will be given in Sec.4.

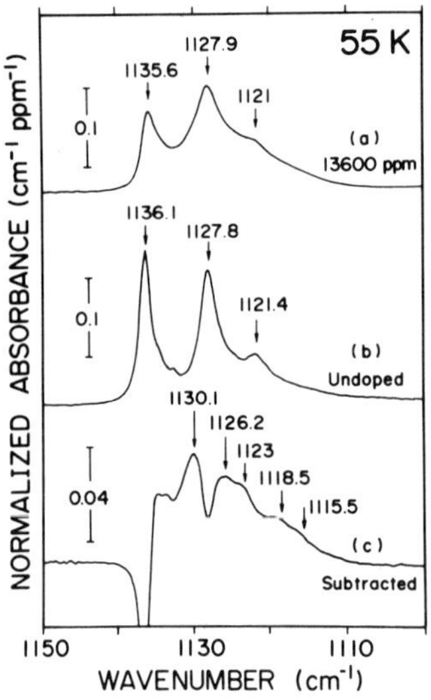

Fig.6. The 1100 cm^{-1} bands measured at 55 K for [Ge] = 13600 ppm (a), and 0 ppm (b). The spectrum (c) was obtained by subtracting 66.5% of the spectrum (b) from the spectrum (a).

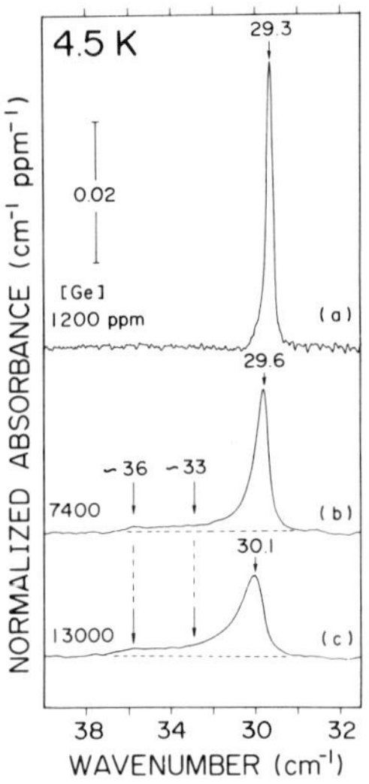

Fig.7. Germanium concentration dependence of the 30 cm^{-1} band measured at 4.5 K.

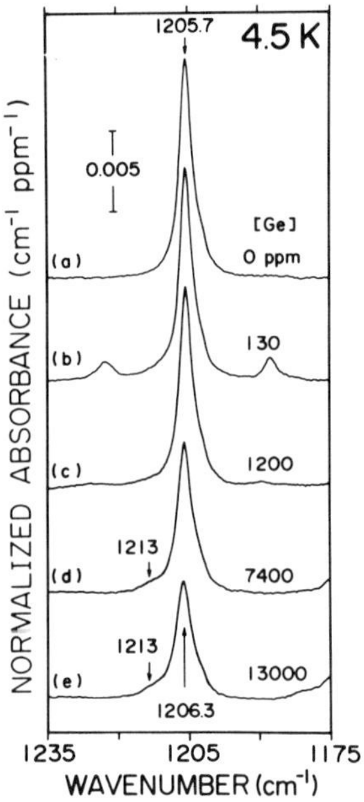

Fig.8. Germanium concentration dependence of the 1200 cm^{-1} band measured at 4.5 K. The small peaks in (b) are irrelevant, which are sometimes observed also for the undoped crystals.

3. Calculation of the oscillator strengths

We first review our treatment[5] for the 2D LEAE coupled to the A_{2u} local mode. The eigenfunction of the coupled excitation is approximated as $\xi_{k|\ell|,N}(r) \cdot (e^{i\ell\phi}/\sqrt{2\pi}) \cdot |N\rangle$, where ℓ and N are respectively the angular momentum quantum number and the A_{2u} phonon number. The label k distinguishes the eigenstates with the same (ℓ, N). The $|N\rangle$ denotes the N-phonon eigenfunction of the A_{2u} local vibration. We express the corresponding eigenstate as $|k, \ell, N\rangle$. The eigenenergy $E_{k\ell,N}$ and the radial part $\xi_{k|\ell|,N}(r)$ is obtained from the following equation in each subspace of (ℓ, N):

$$H_{\ell,N}\, \xi_{k|\ell|,N}(r) = [E_{k\ell,N} - \hbar\omega(N+1/2)]\, \xi_{k|\ell|,N}(r), \tag{1}$$

$$H_{\ell,N} \equiv -\frac{\hbar^2}{2m}\left[\frac{1}{r}\frac{d}{dr}\left(r\frac{d}{dr}\right) - \frac{\ell^2}{r^2}\right] + \alpha_N r^2 + \beta r^4, \tag{2}$$

$$a_N \equiv a + (\hbar g / \omega)\,(N + 1/2). \tag{3}$$

We can regard $a_N r^2 + \beta r^4 + (N+1/2)\,\hbar\omega$ as the A_{2u}-local-phonon-renormalized adiabatic potential for the N-A_{2u}-phonon states. The bare host-lattice potential is expressed by $ar^2 + \beta r^4$. The ω denotes the frequency of the uncoupled A_{2u} local mode. Equation (1) is solved by numerically diagonalizing the Hamiltonian matrix of $H_{\ell, N}$.

For generality, we express $\xi_{k|\ell|, N}(r)$ as $\xi_{k|\ell|}(r; a_N, \beta, m)$ regarding (a_N, β, m) in $H_{\ell, N}$ as the parameters. Using the dimensionless distance $R \equiv r / (\hbar^2 / 2m\beta)^{1/6}$, we rewrite the Hamiltonian (2), and then, divide it by the energy unit $E_{uni} \equiv [\hbar^4 \beta / (2m)^2]^{1/3}$. From this, we see that under the condition $|< \xi_{k|\ell|}(r; a_N, \beta, m) | \xi_{k|\ell|}(r; a_N, \beta, m) >|^2 = 1$, the R-represented eigenfunction

$$\xi_{k|\ell|}(R; A_N) \equiv (\hbar^2/2m\beta)^{1/6}\, \xi_{k|\ell|}((\hbar^2/2m\beta)^{1/6} R; a_N, \beta, m) \tag{4}$$

is specified by only one parameter $A_N \equiv a_N\,[2m / (\hbar\beta)^2]^{1/3}$, as symbolized on the left-hand side. Here, the symbol $< | >$ denotes the radial integral, $< f(\rho) | \rho^n | g(\rho) > \equiv \int_0^\infty d\rho\, \rho^{n+1} f(\rho)\, g(\rho)$. The intensities (the oscillator strengths) of the A_{2u} mode transitions $|0, 0, 0> \to |k, 0, 1>$ (k=0, 1 for the 1136 and 1206cm^{-1} transitions, respectively) at sufficiently low temperatures are expressed as[5]

$$I(|0,0,0> \to |k,0,1>) = (M^*)^2\,(\hbar/2\omega)\,(E_{k0,1} - E_{00,0})\; O_{k0,00}(A_1, A_0), \tag{5}$$

and that of the 2D LEAE transition $|0, 0, 0> \to |0, \pm1, 0>$ (the 30 cm^{-1} transition) as

$$I(|0,0,0> \to |0,\pm1,1>) = (1/2)\,(e^*)^2\,(E_{0\pm1,0} - E_{00,0})\,(\hbar^2/2m\beta)^{1/3}\, T_{01,00}(A_0), \tag{6}$$

where M* and e* are the constants,[5] and

$$O_{k0,00}(A_1, A_0) \equiv |< \xi_{k0}(R; A_1) | \xi_{00}(R; A_0) >|^2, \tag{7}$$

$$T_{01,00}(A_0) \equiv |< \xi_{01}(R; A_0) | R | \xi_{00}(R; A_0) >|^2. \tag{8}$$

The result of the numerical calculations for $O_{00,00}(A_1, A_0)$, $O_{10,00}(A_1, A_0)$ and $T_{01,00}(A_0)$ are diagramed in Figs.9 (a) - (c). From these figures, one can obtain the values of the transition probabilities in various cases of the parameter values (a, β, ω, g, m). The parameter values adjusted so as to well explain the observed peaks in the undoped case are given in Table I. The corresponding position of (A_0, A_1) in Figs.9 (a) and (b) is indicated by a dot with the value of $O_{k0,00}(A_1, A_0)$ (k=0, 1). The position of A_0 in Fig.9 (c) is indicated by an arrow. The relative eigenenergies $E_{k\ell,N} - E_{00,0}$ calculated for the adjusted parameter values are given in Fig.2. Substituting these values into Eq. (5), we obtain $I(|0,0,0> \to |1,0,1>) / I(|0,0,0> \to |0,0,1>) = 0.0432$, as the intensity ratio of the 1205.7 cm^{-1} peak to the 1136.4 cm^{-1} peak. This value agrees well with the measured ratio 0.0378 of the area-intensity between the two peaks.

4. Effect of weakening of the coupling

In Table I, we show the parameter values determined for Oi-I, -II and -III in the case of [Ge] =13000 ppm. It is noticeable that the coupling strengths |g| for Oi-II and -III are much smaller than that for Oi-I. The energy level schemes calculated for these parameter values were similar to that shown in Fig.2. For Oi-II and -III, however, because of the weakened coupling, the difference in potential-curve shape (and thus, the difference in level separations) between the zero- and one-A_{2u}-phonon states became smaller than in Fig.2. The observed peaks corresponding to those in the undoped case (Table I) are assigned to the similar transitions in Fig.2. The adjusting error for each peak-energy was less than 1 cm^{-1}. We show, in Figs.9 (a) and (b), the corresponding position of (A_0, A_1) and the values of $O_{k0,00}(A_1, A_0)$ (k=0, 1). The position of A_0 is indicated in Fig.9 (c) for $T_{01,00}(A_0)$. Using these values and the eigenenergies $E_{k\ell,N}$ calculated for each case of the parameter values, we obtain the values for the relevant factors in Eqs. (5) and (6). The results are listed in Table II, which well explain the experimental results: with regard to the 1100 and 30 cm^{-1} absorptions, the oscillator strengths of Oi-I, -II and -III were almost the same.

However, regarding the 1200 cm^{-1} absorptions, those of Oi-II and -III were reduced to 17% or less of that of Oi-I in the undoped case.

From Figs.9 (a) - (c), it is also qualitatively seen that the weakening of the coupling is an essential factor for these experimental results. In Figs.9 (a) and (b), the contour $A_1 = A_0$ corresponds to the zero-coupling cases; and the contours farther from it roughly correspond to the stronger coupling cases. Thus, we see that even in case of the full reduction of the coupling $|g| \rightarrow 0$, the change of the intensity expected for the 1136 cm^{-1} peak (Oi-I in the undoped case) is negligibly small (~ +4%), while the 1206 cm^{-1} peak vanishes. It is seen from Fig.9 (c), that the full reduction $|g| \rightarrow 0$ reduces the transition probability $(\hbar^2 / 2m\beta)^{1/3}\, T_{01,\,00}\,(A_0)$ by about 14%. However, such the reduction is just compensated by increase of the transition energy $E_{0\pm1,\,0}$-$E_{00,\,0}$, leaving $I\,(|\,0,0,0> \rightarrow |\,0,\pm1,0>)$ almost unchanged.

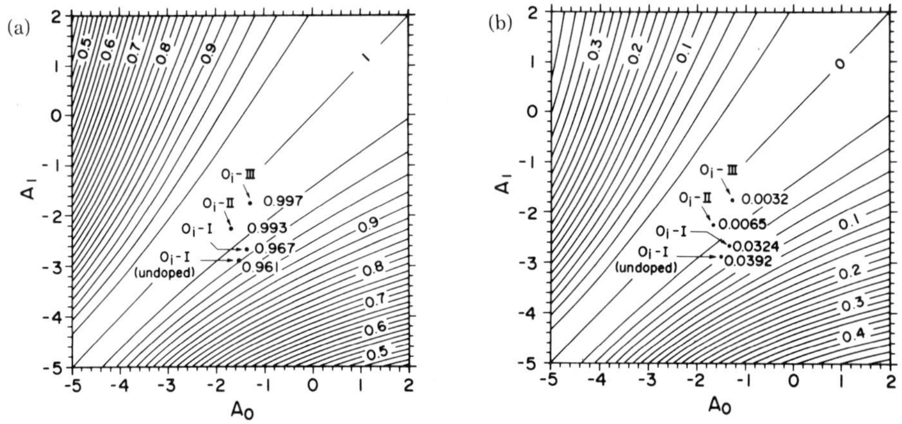

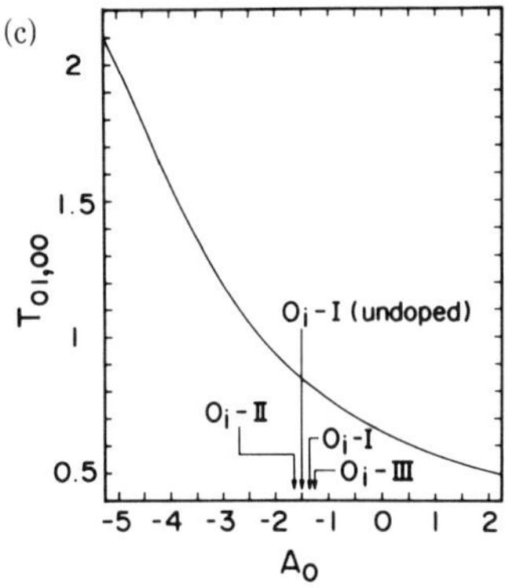

Fig.9. Contour representations for $O_{00,\,00}(A_1, A_0)$ (a), and $O_{10,\,00}(A_1, A_0)$ (b). Their values and the position of (A_0, A_1) are shown for Oi-I in the undoped case, and Oi-I, -II, -III in the case of [Ge] = 13000 ppm. The curve of $T_{01,\,00}(A_0)$ is shown in (c), where the position of A_0 in each case is indicated.

Table II. Result of evaluation for the relevant factors in the oscillator strengths, Eqs. (5) and (6).

Absorptions (cm^{-1})	Relevant Factors in the Oscillator Strengths	Oi-I (undoped)	Oi-I ([Ge]=13000 ppm)	Oi-II	Oi-III
~ 1100	$[(E_{00,1} - E_{00,0})/\hbar\omega]\, O_{00,00}$	0.946	0.954	0.986	0.991
~ 1200	$[(E_{10,1} - E_{000})/\hbar\omega]\, O_{10,00}$	0.0409	0.0340	0.0070	0.0034
~ 30	$(E_{0\pm1,0} - E_{00,0})\,(\hbar/2m\beta)^{1/3}\, T_{01,00}$ (meV/Å^2)	0.253	0.254	0.252	0.256

REFERENCES

1. D.R. Bosomworth, W. Hayes, A.R.L Spray and G.D. Watkins, Proc. R. Soc. London Ser. A 317, 133 (1970).
2. H.J. Hrostovski and R.H. Kaiser, Phys. Rev. 107, 966 (1957).
3. H.J. Hrostovski and B.J. Alder, J. Chem. Phys. 33, 980 (1960).
4. B. Pajot, J. Phys. Chem. Solids 28, 73 (1967).
5. H. Yamada-Kaneta, C. Kaneta and T. Ogawa, Phys. Rev. B 42, 9650 (1991).

Materials Science Forum Vol. 83-87 (1992) pp. 425-432

IR ABSORPTION IN MONOISOTOPIC GERMANIUM

L.I.Khirunenko, V.I.Shakhovtsov, V.K.Shinkarenko

Institute of Physics, Academy of Sciences of the Ukrainian SSR, Kiev-28, Prospekt Nauki, 46, USSR

The structure of oxygen absorption in Ge (vibration ν_3 of quasi-molecule Ge_2O) was studied. Using specially prepared monoisotopic ^{74}Ge single crystal, it was ascertained that four absorption lines, resulting from interaction of vibration ν_3 with low-energy vibration ν_2 are associated with vibration ν_3 of quasi-molecule $^{74}Ge_2O$ (as well as of any other isotopic combination of Ge atoms). Minimum energies of transition for ν_2 (3.2; 9.6; and 16 cm^{-1}) were estimated in the rigid two-dimensional rotator approximation.

Frequencies and intensities of components of all possible isotopic combinations were calculated and the experimentally measured absorption spectrum of natural Ge was identified.

Introduction

Oxygen in Si and Ge is an unavoidable technological impurity and one of elements participating most actively in radiation and thermal defect formation processes. In spite of extensive studies of oxygen in such crystals, however, many points concerning its properties and behavior in defect formation remain debatable so far. It has been particularly little studied in the lattice of Ge, which still remains among basic materials employed in the semiconductor engineering for development of, e.g., high-resolution - spectrometers, photodetectors of a broad IR range, high-power SHF transistors, etc. Studying the properties of O in Ge is therefore one of priority tasks from both scientific and applied standpoints.

When a grown crystal is cooled, oxygen occupies an interstitial position and, breaking the bond between neighboring atoms, bonds covalently with them into a quasi-molecule Si_2O (Ge_2O) having a C_{2v} symmetry / 1,2 /. Such a triatomic molecule exhibits three normal vibrations: symmetric valence vibration ν_1; symmetric bending (with a great share of a rotary component) vibration ν_2; and antisymmetric valence vibration ν_3. For Si the three vibrations have been identified and are related by most authors / 1-5 / to bands of 517 (at 300 K), 29.3 (at 4.2 K) and 1106 cm^{-1} (at 300 K). The most studied of them is vibration ν_3. When the temperature is lowered, the 1106 cm^{-1} band shifts towards higher energies and splits into three components, of 1136.4, 1128.3, and 1122 cm^{-1}, due to interaction of ν_3 with low-frequency vibration ν_2 (Fig. 1, a). The temperature variation of the ratio of intensities of the components correlates with the temperature population of sublevels of vibration ν_2. In addition, the 1136.4 cm^{-1} band with half-width H = 0.6 cm^{-1} has a three-component fine structure, caused by the isotopic composition of Si (^{28}Si, ^{29}Si, ^{30}Si). For Ge, vibration ν_3 (855 cm^{-1}) alone has been identified, but even it has been little studied. It should be noted that the absorption band corresponding to ν_3 is generally used to determine the oxygen concentration in Si and Ge.

The present study dealt with the structure of absorption of vibration ν_3 in germanium.

Experiment and Discussion

The subject of the study was germanium, natural and enriched in the isotope ^{74}Ge (98 %), with an oxygen concentration of $(2\text{-}5)\cdot 10^{16}$ cm^{-3}. The study was conducted on a IFS-113v Fourier spectrometer with a maximum resolution of 0.03 cm^{-1} over a temperature range from 2.5 to 300 K.

Figure 2 (a) shows the absorption spectrum of oxygen in natural Ge at 4 K. As seen, at this temperature the broad 855 cm^{-1} absorption band gets, as for silicon, shifted towards higher energies, but, in contrast to silicon, splits into 25 components.

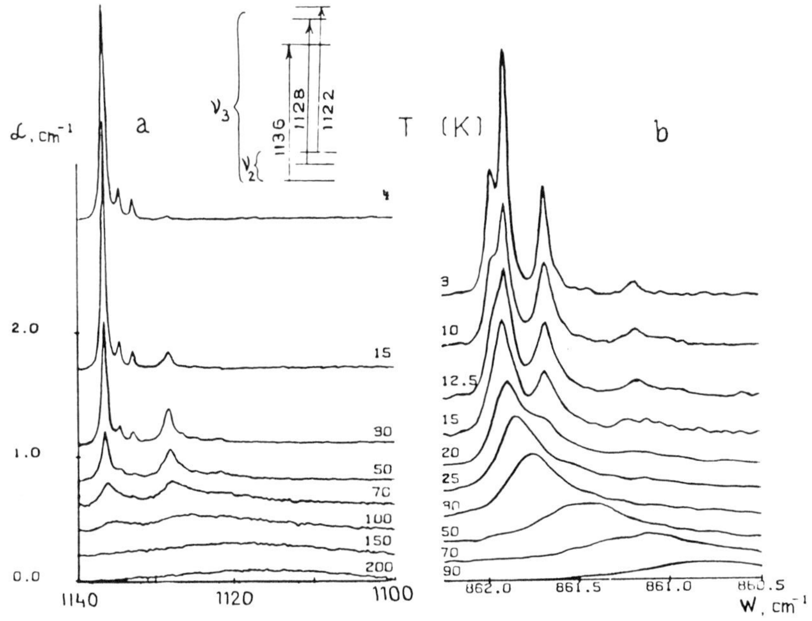

Fig.1. Temperature dependences of absorption spectrum of oxygen (ν_3) in Si (a) and in monoisotopic ^{74}Ge (b). Inset: schematic of transitions / 2 /

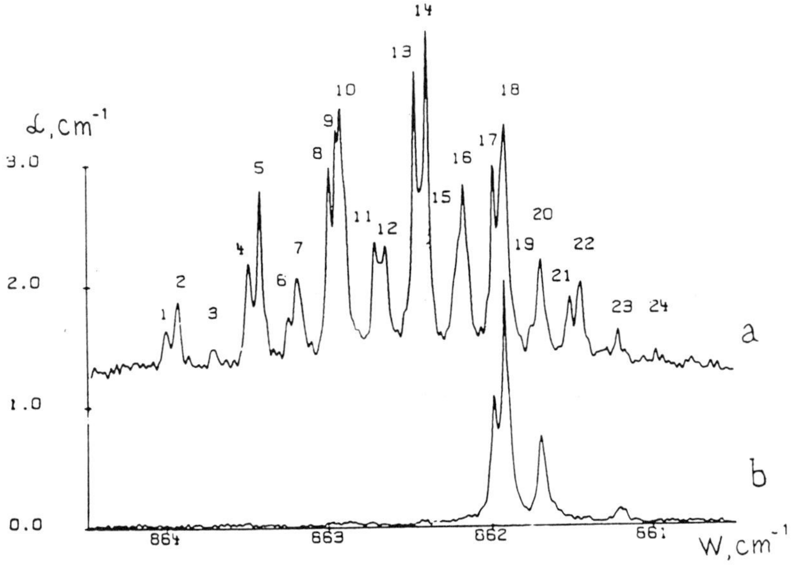

Fig.2. Absorption spectrum of oxygen in natural Ge (a) and in monoisotopic ^{74}Ge (b) at 4 K

Natural Ge is known to consist of five isotopes, ^{70}Ge (20.55 %), ^{72}Ge (27.37 %), ^{73}Ge (7.61 %), ^{74}Ge (36.74 %), and ^{76}Ge (7.67 %), each of which can participate in formation of Ge_2O molecule. The isotopic shift of vibration ν_3 of an isolated bent triatomic molecule X_2Y is described by the following formula / 6 / :

$$\nu_3^2 = \left[1 + \frac{2M_x \sin^2\alpha}{M_y}\right] \frac{k}{4\pi M_x} , \qquad (1)$$

where M_x, M_y are masses of atoms of kinds X and Y, k is the force constant, and 2α-the angle between bonds (X-O and Y-O). For Ge_2O, $\alpha \approx 70°$ / 7 /. It should be pointed out that isotopic combinations having identical total mass of Ge atoms (e.g., ^{74}Ge-O-^{72}Ge and ^{73}Ge-O-^{73}Ge) will exhibit one and the same vibration frequency / 7 /. Allowing for this fact, it follows that the isotopic splitting should give rise to a structure of 11 components. The presence of "excessive" lines in the spectrum and similarity of vibrations of Si_2O and Ge_2O make it reasonable to assume the additional structure in the spectrum to be due to interaction of vibrations of ν_2 and ν_3 The possibility of existence of a complex structure of vibration ν_3 was also noted in / 7 /, but its authors failed to attain a full interpretation of the spectrum because of an overlap of lines belonging to different isotopes and ensuing difficulties in studying the temperature dependence of the absorption spectrum.

A sole possibility for a full interpretation of the structure of the spectrum is investigation of a monoisotopic Ge. To this end we studied the temperature dependence of the absorption spectrum of monoisotopic ^{74}Ge, containing 98 % main isotope. The measured absorption spectrum for such a crystal is shown in Fig. 2 (b), and its temperature dependence, in Fig. 1 (b). As seen, at T =

= 3-4 K the spectrum consists of four bands, A, B, C, and D, spaced at $d_{AB} \approx 0.07$, $d_{BC} = 0.23$, and $d_{CD} = 0.48\ cm^{-1}$, with half-width $H \simeq 0.06\ cm^{-1}$. As the temperature is increased above 5 K, there occur a progressive broadening of the bands and variation of their relative intensities. Next, at $T \geqslant 25$ K the bands merge into a single broad band, which starts shifting gradually towards lower energies up to 855 cm^{-1} at T = 300 K. The similarity of structures of quasi-molecules Si_2O and Ge_3O allows a conclusion that the detected four-component structure of vibration ν_3 for monoisotopic quasimolecule $^{74}Ge_2O$ is similar to that for Si_2O (1136.4, 1128, and 1122 cm^{-1}) and is associated with interaction of vibrations ν_3 and ν_2 and, accordingly, with the temperature population of levels of vibration ν_2 (Fig. 1). It should be noted that the population of levels of ν_2 is affected, apart from the temperature, also by the background thermal radiation penetrating into the cryostat.

Comparing the absorptions in the region of vibration ν_3 for Si and monoisotopic ^{74}Ge reveals the following features: (1) the variation of the ratio of intensities of ν_3 components for Ge_2O in a temperature range of 3-50 K corresponds to that in a temperature range of 40-150 K for Si_2O; (2) Ge_2O exhibits very small spacings of the components (e.g., $d_{AB} \approx 0.07\ cm^{-1}$ as against $\approx 6\ cm^{-1}$ for Si_2O); and (3) even at 3 K the intensity of the transition from the ground state (band A) is less than the intensity of transition from the first excited state (band B). The above features lead to a conclusion that the spacing of energy levels of vibration ν_2 in Ge_2O is much smaller than in Si_2O.

Knowing the structure of the absorption spectrum for vibration ν_3 of monoisotopic ^{74}Ge, ascertained in the present study, as well as the percentage isotopic composition of a natural germanium, one can, using formula (1), identify all bands observed for the natural Ge as well as describe their intensities. It should be taken into account, however, that, since quasi-molecule Ge_2O is disposed in the crystal lattice and is linked to the six nearest Ge atoms, a correction allowing for an effective increase of the mass M_x should be applied, i.e.,

$$M_x = M_{Ge} + m' , \qquad (2)$$

where m' = 11.65 amu / 7 /.

The calculated frequencies and intensities of components of all isotopic combinations are presented in the Table, where also experimental results are given for comparison. The subscripts in component designations indicate the sum of masses of the Ge isotopes that constitute quasi-molecule Ge_2O, minus 140. For example: A_6 is the A band for quasi-molecules $^{70}Ge\text{-}O\text{-}^{76}Ge$, $^{72}Ge\text{-}O\text{-}^{74}Ge$, and $^{73}Ge_2O$.

We also estimated the minimum possible values for transitions of low-frequency vibration ν_2. An earlier study of vibration ν_2 in Si_2O ascertained this bending vibration to contain a great share of a rotary motion. Therefore, using the approximation of a rigid two-dimensional rotator / 8 /, where energies of the levels are calculated from the formula

$$E = \frac{n^2 h^2}{2I} , \qquad (3)$$

where n is the main quantum number and I-the moment of inertia, equal in this case to

$$I = \frac{2(M_{Ge} + m')M_O}{2(M_{Ge} + m') + M_O} r^2 , \qquad (4)$$

where r = 0.6Å is the distance from the oxygen atom to the axis of rotation (Ge-Ge), we obtain a set of levels with energies of

Band No.	Component	Frequency, cm^{-1} experiment	Frequency, cm^{-1} calculation	Relative intensity (calculat.)	Band No.	Component	Frequency, cm^{-1} experiment	Frequency, cm^{-1} calculation	Relative intensity (calculat.)
1	A_0	864.00	864.00	11.18	16	C_6	862.16	862.18	47.3
2	B_0	863.93	863.931	17.66		B_7		862.156	23.8
3	C_0	863.70	863.710	8.35	17	A_8	861.98*	861.98*	46.8
4	A_2	863.48	863.476	29.9		D_5		861.943	1.19
5	B_2	863.41	863.409	47.21	18	C_7	861.91	861.935	11.26
6	D_0	863.24	863.225	1.18		B_8		861.91*	73.86
	A_3		863.22	8.46	19	A_9	861.74	861.744	3.23
7	C_2	863.19	863.188	22.33	20	D_6	861.68	861.695	6.67
	B_3		863.153	13.36		C_8	861	861.69*	34.95
8	A_4	862.99	862.968	59.85		B_9		861.675	5.1
9	C_3	862.94	862.932	6.32	21	A_{10}	861.50	861.507	15.07
10	B_4	862.91	862.899	94.5		C_9		861.454	2.41
11	A_5	862.71	862.718	11.31	22	D_7	861.43	861.45	1.59
	D_2		862.703	3.18		B_{10}		861.438	23.8
	C_4		862.678	44.7	23	C_{10}	861.21	861.217	11.26
12	B_5	862.65	862.649	17.86		D_8		861.205*	4.93
13	A_6	862.47	862.47	63.33	24	A_{12}	860.97	861.042	1.58
	D_3		862.447	0.89		B_{12}		860.973	2.5
	C_5		862.428	8.45		D_9		860.969	0.34
14	B_6	862.39	862.401	100		C_{12}		860.752	1.18
15	A_7	862.23	862.225	15.07		D_{10}		860.732	1.59
	D_4		862.193	6.3		D_{12}		860.267	0.17

* Experimental values for $^{74}Ge_2O$.

0; 3.2; 12.8; and 28.8 cm^{-1}. According to the selection rule for such a rotator, transitions at frequencies of 3.2; 9.6; and 16 cm^{-1} should occur. The allowance for the vibrational component will somewhat increase the values. A low value of the energy of the low-frequency vibration accounts for difficulties of its experimental detection.

Conclusions

Thus, the present study ascertained that four absorption lines are associated with vibration ν_3 of quasi-molecule $^{74}Ge_2O$ (and of any other isotopic combination). Frequencies and intensities of components of all possible isotopic combinations were calculated and the experimentally measured absorption spectrum of natural germanium was identified. The minimum possible frequencies for bending vibration ν_2 were estimated (3.2; 9.6; and 16 cm^{-1}).

References

1. A.M.Stoneham. Theory of Defects in Solids. Clarendon Press, Oxford, 1975.
2. D.R.Bosomworth, W.Hayes, A.R.Spray, G.D.Watkins. Proc. Roy. Soc. (London), 1970, vol. 317, No. 1538, p. 133.
3. W.Kaiser, P.H.Keck, C.F.Lange. Phys. Rev. , 1956, vol. 101, No. 4, p. 1264.
4. W.Hayes, D.R.Bosomworth. Phys. Rev. Lett., 1969, vol. 23, No. 15, p. 851.
5. M.Stavola. Appl. Phys. Lett., 1984, vol. 4, No. 5, p. 514.
6. G.Herzberg. Infrared and Raman Spectra. N.Y., 1949.
7. B.Pajot, P.Claus. Proc. 18 Int. Conf. Phys. Semicond. Stockholm, 1986, vol. 2, p. 911.
8. B.Hrostowski, B.J.Alder. J. Chem. Phys., 1960, vol. 23, No. 4, p. 980.

Materials Science Forum Vol. 83-87 (1992) pp. 433-446

AB-INITIO MOLECULAR DYNAMICS OF SEMICONDUCTOR DEFECTS

R. CAR[1], P. BLÖCHL[2] and E. SMARGIASSI[3]

[1]IRRMA, PHB Ecublens, CH-1015 Lausanne, and Department of Condensed Matter Physics, University of Geneva, CH-1211 Geneva 4, Switzerland

[2]IBM Research Division, Zürich Research Laboratory, CH-8803 Rüschlikon, Switzerland

[3]Physical Chemistry Laboratory, University of Oxford, South Parks Road, Oxford OX1 3QZ, United Kingdom

ABSTRACT

Ab-initio molecular dynamics makes it possible to perform parameter-free calculations of defect motion in real materials. This leads to novel understanding of diffusion processes in silicon, by allowing to identify, without a-priori assumptions, the equilibrium configurations and the important migration paths associated to a given defect mechanism at high temperature. Diffusion coefficients can be calculated quantitatively and compared with experiment. As a consequence, longstanding issues to self-diffusion in silicon, such as the relative importance of different microscopic mechanisms and the so called entropy puzzle have finally become accessible to quantitative theoretical investigations. The emerging picture indicates that both vacancies and self-interstitials have entropies within the experimental range. However, the self-interstitial mechanism dominates self-diffusion because of its lower formation energy.

1. Introduction

Self-diffusion in silicon has been the subject of intense debate over the last decades and even today it is one of the hot topics in the field studying semiconductor defects. In fact, a number of basic questions are still waiting for a satisfactory microscopic explanation. Moreover, understanding self-diffusion is a necessary first step towards understanding the more complex diffusion processes of dopant atoms, which are of central importance to semiconductor technology.

Experimental self-diffusion rates exhibit an Arrhenius behaviour

$$Q(T) = D_0 \exp(-Q/k_B T) \tag{1}$$

with an activation energy Q in the range of 4-5 eV and a pre-exponential D_0 that is much larger than typical numbers for metals[1,2,3,4]. For many years, the debate focussed on whether self-diffusion is mediated primarily or exclusively by vacancies (V) or self-interstitials (I). More recently, the concerted exchange mechanism has been suggested as a new contender[5] (see Fig.1 for a schematic representation of different mechanisms).

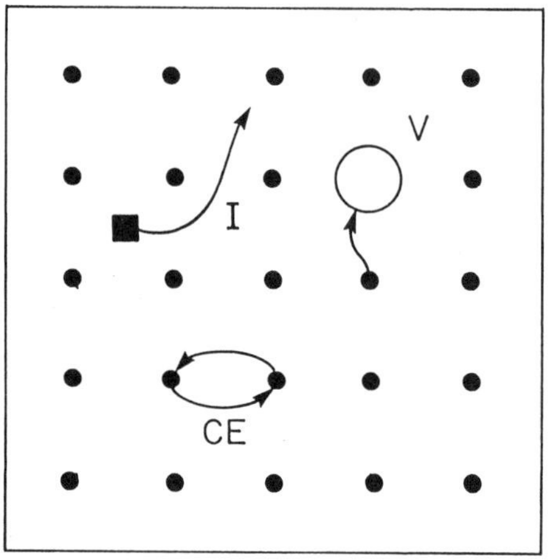

Fig. 1. Schematic illustration of the three mechanisms suggested for self-diffusion in Silicon.

Experimentally it has not been possible to unambiguously determine the relative importance of the different mechanisms, although indirect evidence supports the notion that both vacancies and/or self-interstitials should be present under specific experimental conditions[6]. Theory has focussed so far on the prediction of the activation energies Q for the various mechanisms[7,8,5]. Zero temperature calculations indicate that all of the above mechanisms have activation energies within the experimental range. More recently also the entropy of the concerted exchange mechanism has been calculated using rate theory, resulting in a value of 3.3 k_B, i.e. about 2-3 times lower than observed experimentally[9]. For an unambiguous theoretical determination of the dominant diffusion mechanism, however, one has to calculate the complete self-diffusion constants for all three competing mechanisms.

The high entropy of diffusion, of the order of 10 k_B, which is necessary to account for the observed pre-exponentials D_0 has led to the so-called entropy puzzle, namely a debate on whether or not the diffusion entropy in silicon can be accounted for in terms of simple point-defect mechanisms. Starting from the pioneering work of Seeger and Chik[4], it has been suggested that the defects responsible for self-diffusion at high temperature are extended over several lattice sites, forming disordered local zones having large formation entropies. A detailed microscopic model, however, has not been provided.

These problems can be resolved through computer experiments that have previously been performed on model materials such as 'Lennard-Jonesium'[10]. Diffusion processes in real materials, such as silicon, on the other hand, involve a complex re-arrangement of chemical bonds that go beyond the scope of studies with simplified model interactions. A quantitative basis is provided by the highly accurate electronic structure theory that is used in the present first-principles molecular dynamics calculations (MD)[11]. This approach has paved the way for parameter-free calculations of diffusion rates in real materials as well as for direct dynamical simulations of diffusion processes.

In this paper, we review the results of extensive ab-initio MD studies performed over the last few years on vacancy and interstitial mediated self-diffusion processes in silicon. This work has allowed to determine, for the first time, the complete self-diffusion constants for the two defect mediated mechanisms[12]. The diffusion constant for the exchange mechanism can be directly determined from the energy and entropy of diffusion calculated by Pandey and Kaxiras[9]. One finds that vacancies and interstitials have entropies within the experimental range and give rise to similar induced diffusivities. However, the formation energy favours the interstitial mediated mechanism and as a consequence this dominates self-diffusion. The diffusion constant of the vacancy mechanism is approximately two orders of magnitude smaller and that of the exchange mechanism is a full five orders of magnitude smaller.

A great deal of additional information on atomistic diffusion processes in silicon is obtained from these studies, like e.g. an unbiased determination of the equilibrium defect configurations and of their high temperature migration paths. This is particularly important in the case of the self-interstitial, for which experimental information on the microscopic structure is not available. The present calculations confirm that the (110) split-interstitial structure originally proposed by Bar-Yam and Joannopoulos[8] is the ground-state neutral self-interstitial configuration and is characterized by a relatively small formation energy of 3.3 eV. This is a consequence of the significant amount of rebonding associated with an ad-atom in the silicon crystal. Rebonding effects are important throughout the whole self-interstitial migration path characterized by configurations where the ad-atom is spread over several lattice sites. Similar effects are also found in the case of a vacancy which, at temperatures close to the melting point, may temporarily spread over several lattice sites giving rise to what may be called a dynamically disordered local zone. Rebonding effects facilitate atomic displacements. At the same time they give rise to perturbations larger than expected for a simple point defect, explaining the large diffusion entropies that are observed experimentally.

This paper is organized as follows. In Sect. 2 we discuss the method and the technical details of the calculations, giving also a critical evaluation of their accuracy. In Sect. 3 we present the results for the concentration of native defects in silicon. In Sect. 4 we report defect mediated diffusivities and discuss dynamical effects associated to defect motion at high temperature. Sect. 5 contains a comparison of the calculated diffusion rates with experiment as well as our conclusions.

2. Method and details of the calculations

The technique of ab-initio MD, as introduced by Car and Parrinello[11], allows to perform MD simulations in which the forces acting on the atoms arise directly from the instantaneous electronic ground-state. The electronic structure is obtained from state-of-the-art calculations based on density functional theory within the local density approximation for exchange and correlation[13,14]. This feature is crucial for an accurate description of the subtle chemical effects associated with bond breaking and forming processes in covalent materials.

Direct simulations have been successfully applied to diffusion of positively charged hydrogen atoms in silicon[15]. Here one obtains the diffusion coefficient directly from the mean squared displacement of the atoms in the simulation via the Einstein's relation

$$D = \lim_{t \to \infty} \frac{1}{6t} \langle \Delta R^2(t) \rangle \tag{2}$$

However, since the jump frequency depends exponentially on activation barrier and temperature and since the computer resources limit the absolute length of a simulation, only systems with low barriers and at high temperature can be studied. For systems with large barriers and low temperatures, this approach is complemented by rate theory as described by one of us[16].

As self-diffusion is characterized by large barriers of the order of several eV it seems not apt to direct simulation. This is actually the case for the concerted exchange mechanism that has been studied on the basis of rate theory[9]. For the defect mediated mechanisms, however, the largest contribution is due to creating a native defect whereas the diffusion of a given defect proceeds rapidly and can be studied by simulation. The diffusion constants of the defect-mediated mechanisms, D_I and D_V, can be decomposed into a product of the concentration of native defects (C_I and C_V) that are available for self-diffusion and the diffusivities (d_I and d_V) of silicon atoms in the presence of one such native defect, which are a measure of the speed of diffusion. The total diffusion constant is then given by

$$D = C_I\, d_I + C_V\, d_V + D_X \tag{3}$$

where D_X denotes the diffusion constant for the exchange mechanism. The concentration of native defects requires the evaluation of their free energies of formation

$$F_f = E_f - TS_f \tag{4}$$

where E_f and S_f denote formation energy and entropy respectively. As we will discuss in the next section F_f can be calculated by means of special techniques, which either require a very efficient total energy minimization scheme, when the harmonic approximation is adopted, or require full MD simulations, when complete inclusion of anharmonic effects is needed. In both cases the Car-Parrinello method, which provides an efficient means for relaxing simultaneously electronic and atomic coordinates, has proven to be very useful. As diffusion proceeds rather rapidly it can be directly simulated with ab-initio MD techniques in the high temperature range typical of self-diffusion experiments, i.e. for $T \geq 1100$ K. This approach has been followed to compute both d_I and d_V, as will be discussed in Sect. 4. In principle, the diffusion rate D_X, corresponding to the exchange mechanism, could have been calculated following the same method mentioned above to compute the concentration of native defects, since the basic quantity that determines D_X is the activation free energy characterizing this process at the saddle point. This calculation, however, has not been performed and we will refer to the results of Pandey and Kaxiras for what concerns D_X[9].

All the calculations reported below have been performed at fixed volume, using large supercells containing 64 ± 1 atoms. All defects were assumed to be in their neutral charge state. Norm-conserving pseudopotentials[17] were adopted in combination with plane wave expansions of the electronic orbitals. All of the interstitial and some of the vacancy calculations used smooth pseudopotentials with cutoff radii rc = 1.5, 1.3, 1.5 a.u. for s, p, and d angular momenta respectively. Some of the vacancy calculations were instead performed with the somewhat harder pseudopotentials of Bachelet, Hamann and Schlüter[18] (BHS). Most calculations with the smooth pseudopotentials and some of the calculations with the BHS pseudopotentials used a plane wave cutoff of 8 Ry, while some other calculations with the BHS pseudopotential used a cutoff of 12 Ry. Fewer calculations and some tests used a cutoff of up to 14 Ry. The k-integration has been done with the Γ point (k=0) only. For the formation energies, however, 27 k-points in the reciprocal unit cell have been used. Since we want to compare the relative contribution of different defect mechanisms, the absolute accuracy of the calculations plays a crucial role. The largest contribution to the native defects formation free-energy comes from their formation energy: the convergence of this quantity has been studied in most detail. We estimate the absolute convergence error in our calculated formation energies to be of the order of 0.2 eV, i.e. about 5%. Relative errors are likely to be smaller. However it was not possible to perform the most expensive MD simulations with the larger cutoffs and the most extended sets of k-points. Therefore quantities, like entropies and diffusivities, may have larger errors percentually but smaller in absolute terms.

3. Concentration of native defects

In the dilute (low concentration) limit, the concentration of a native defect is given by:

$$C(T) = C_S \exp(-F_f/k_BT) \tag{5}$$

where C_S = 5 x 10^{22} cm^{-3} is the concentration of lattice sites in the silicon crystal and F_f is the formation free-energy of the defect (Eq. 4).

In order to compute F_f we first adopt the harmonic approximation. In this case we only need to perform static total energy calculations to compute the energy of formation E_f and the formation entropy S_f. Notice that, within the harmonic approximation, the only temperature dependence of the free-energy of formation F_f comes from the explicit T dependence in Eq. 4

The calculation of the defect formation energies requires an accurate determination of their ground-state structures. This was done by minimizing the total energy with respect to all atomic coordinates in the supercell containing one native defect. In order to assure that the minimum energy was found, the system was quenched several times during the high temperature MD runs used to investigate defect induced diffusivities. The structure of the vacancy is relatively well understood[19,20], it consists mainly of a pairing D_{2d} distortion, in agreement with low temperature experiments[20], plus an inward breathing displacement of the neighbouring atoms, in agreement with other recent calculations[21,22]. The self-interstitial has a more complicated structure. This is shown in Fig. 2: it consists of a pair of two silicon atoms, oriented in the (110) direction, that occupies a single lattice site. This structure provides a remarkable illustration of the strong rebonding effects that accompany displaced atoms in the silicon crystal. Interestingly, no states in the gap are associated with this structure, explaining the longstanding ability of the self-interstitial to elude direct experimental observation. The formation energies found were 3.3 eV and 4.1 eV for the self-interstitial and the vacancy, respectively. We notice that the large energy difference of about .8 eV between the vacancy and the interstitial strongly favours the latter mechanism under normal conditions, and it is unlikely to be reversed by finite T effects.

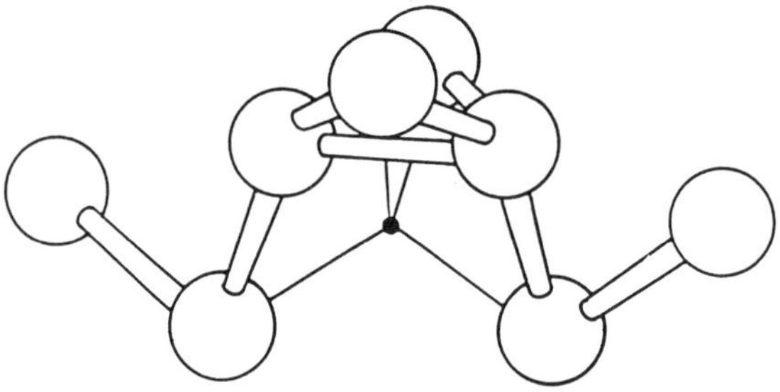

Fig. 2. The self-interstitial ground-state structure.

The entropy of formation is the sum of the configurational entropy and the vibrational entropy. The configurational entropy is simply related to the number of equivalent ways in which the native defect can be incorporated at one particular lattice site: there are 3 possible arrangements for the vacancy and 6 for the split self-interstitial, resulting in configurational entropies of 1.1 k_B and 1.8 k_B respectively. The vibrational entropy is a measure of how tightly the native defects are constrained in their equilibrium configurations. It is rather difficult to calculate it, even within the harmonic approximation, since this needs a lengthy evaluation of the dynamical matrix of the system both in presence of and without a point defect. In order to simplify this calculaton we adopted the local harmonic approximation, recently introduced by LeSar et al.[23], and whose accuracy was amply demonstrated in the case of a simple model for Cu, using truncated pairwise Morse potentials to describe the interatomic interactions. In this approach the vibrational entropy is a sum of local (atomic) contributions. The accuracy of this approximation for silicon can be judged by comparing its results with those of fully anharmonic MD calculations that we will describe below. We find that the contribution from local entropies is already converged beyond the second shell of neighbours of a defect. The resulting entropies of formation, including both configurational and vibrational contributions, are 6 k_B for the self-interstitial and 5 k_B for the vacancy, respectively. The resulting concentrations amount to 10^{12}-10^{13} self-interstitials and 10^8-10^9 vacancies per cm^3 at 1273 K. Since the electronic structure of the vacancy includes empty levels close to the occupied ones in the gap, the electronic contribution to S_f may be not negligible. A rough estimate gives an electronic contribution of the order of 1 to 2 k_B which has to be added to the vacancy formation entropy. This raises the vacancy concentration but still keeps it approximately two orders of magnitude smaller than that of the self-interstitial in the relevant temperature range.

Two questions are in order to assess the accuracy of the above calculations: these are (i) how good is the local harmonic approximation for silicon and (ii) how important are anharmonic effects at high temperature. Both questions can be answered at the same time by performing 'exact' calculations of the formation free-energy F_f. This was done in the case of the silicon vacancy, indicating that (i) the local harmonic approximation is very accurate at low T, (ii) anharmonic contributions are important and tend to increase the vacancy entropy of formation S_f by about 2 k_B at 1000 K.

The exact calculation of an absolute free-energy F for a real material is not possible, since it involves the calculation of a complex multidimensional integral that cannot be cast under the form of an ensemble average, and therefore cannot be calculated using standard statistical mechanical techniques, such as Monte Carlo (MC) or MD[24].

$$F = - k_B T \ln Z \tag{6}$$

where Z is the partition function

$$Z = \mathrm{Tr} \exp(-H/k_B T) \tag{7}$$

Here Tr is a shorthand for phase space integral and H is the (classical) Hamiltonian of the system.

Notice that, if the Hamiltonian depends on a parameter λ, the derivative of the free-energy with respect to λ is given in the form of a canonical ensemble average, as indicated below by the angular brackets.

$$\frac{\partial F}{\partial \lambda} = Z^{-1}\,\mathrm{Tr}\,[-\frac{\partial H}{\partial \lambda} \exp(-H/k_B T)] \;=\; \langle -\partial H/\partial \lambda \rangle \tag{8}$$

It is then possible to use MC or MD techniques to compute a free-energy difference between two reference states characterized by two different values of the parameter λ by evaluating the integral:

$$F(\lambda_2) - F(\lambda_1) = \int_{\lambda_1}^{\lambda_2} < - \frac{\partial H}{\partial \lambda} > d\lambda \qquad (9)$$

If the integrand is a smooth function of λ, as it is usually the case, the integral (9) can be accurately calculated numerically using a Gauss-Legendre quadrature formula with a small set of sampling points (often just 3 or 5 are enough).

This procedure amounts to compute a free-energy difference between two equilibrium states by thermodynamic integration along a reversible path joining the two states of interest; the average over all the possible thermal fluctuations along the path is performed with MC or MD techniques. Essentially the same procedure is used to extract a free-energy difference from experiment, when one integrates the specific heat between two reference states at different temperatures. The advantage of computer experiments over real experiments in this context is that numerical experiments leave a much larger freedom in the choice of the integration parameter λ. For instance, λ can be the coupling constant of one atom to the rest of the crystal, so that when $\lambda = 1$ we have a perfect bulk crystal and when $\lambda = 0$ we have a crystal with one vacancy plus a decoupled atom. In this case the above technique can be used to compute 'exactly' the free-energy of formation of a vacancy, i.e. with full inclusion of anharmonic effects, the only remaining error being the statistical error arising from the finite number of configurations used to sample the thermal fluctuations in (9) with MC or MD. This technique is well known and has been used successfully to compute 'exact' formation free-energies of defects in model systems in which the interatomic interactions are given in terms of classical empirical potentials[10,25].

Within the Car-Parrinello method the above technique can be used in connection with first-principles MD simulations, in which the interatomic potential is derived directly from the electronic ground-state. Such an approach has been followed recently to compute the free-energy of formation of a vacancy in silicon, in the context of plane-wave pseudopotential calculations. The parameter λ was chosen to decouple from the crystal a classical pseudo-ion and its four valence electrons in order to create a neutral vacancy. The calculation required a set of MD runs of several ps each for each one of the intermediate λ values in (9) (a 3-point Gauss-Legendre integration formula was used). Since this is computationally very demanding, only calculations at T=500 K and at T=1000 K were performed. The results are reported in Fig. 3. We remark that: (i) at T=500 K the 'exact' F_f coincides within the statistical error with F_f obtained with the local harmonic approximation, thereby confirming the validity of such a scheme in the case of silicon; (ii) at high temperature anharmonic effects are important, being already of the order of about 2 k_B at T=1000 K. The effect of anharmonicity amounts to a local vibrational softening of the lattice in the defective crystal that reduces the vacancy formation free-energy at high temperature by increasing its formation entropy.

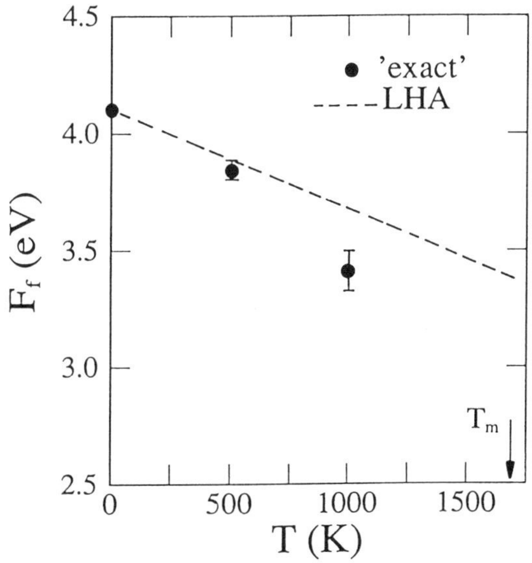

Fig. 3. Vacancy formation free-energy vs. temperature; the dashed line refers to local harmonic calculation, the dots refer to 'exact' calculations via the thermodynamic integration method and the error bars are statistical error estimates.

By adding the various contributions to the vacancy formation entropy we get a high temperature value of about 8-9 k_B. By adding the migration contribution, that will be discussed in the next section, we end up with a total entropy of diffusion of the order of 10 k_B for the vacancy. In the case of the self-interstitial we have not included anharmonic effects and the diffusivity has been calculated for only one temperature, so that the diffusivity cannot be decomposed into an entropy and an energy contribution. However we expect similar effects also for the self-interstitial, whose formation entropy is already of 6 k_B. This indicates that both native defects in silicon have diffusion entropies within the experimental range.

4. Defect induced diffusivities

At the temperatures relevant for self-diffusion experiments, i.e. for T $\geq$ 1100 K, vacancy and interstitial induced diffusivities can be obtained directly from MD simulations, by monitoring over a period of time the motion of the atoms in a supercell containing an artificially created vacancy or self-interstitial. During the simulation the temperature was held constant by means of a Nose' thermostat[26]. Runs were performed at T=1500 K for the interstitial, and at T=1200 K, 1400 K and 1600 K for the vacancy. After sufficient equilibration of each sample at this temperature, we followed the time evolution of the system for up to 20 ps, which corresponds to about 350 phonon oscillations. The diffusivity d was then obtained from the mean squared displacement of all the atoms in the cell, by evaluating the Einstein's relation

$$d = \lim_{t \to \infty} \frac{1}{6t} < \Sigma_I \,|\, R_I(t) - R_I(0) \,|^2 > \tag{10}$$

where $R_I(t)$ is the position of the I-th atom at time t. It is noteworthy that the diffusivities of the vacancy and of the self-interstitial lie in a very similar range. Their magnitudes are of order 10^{-4} cm^2/s at the temperatures of our investigations. The statistical error bars of the diffusivities affect the diffusion constant by less than a factor of two and thus are negligible compared to the uncertainty in the free-energy of formation which is of the order of 0.2 eV.

It is interesting to analyze the atomic trajectories during the diffusion events. In the case of the self-interstitial at T = 1500 K, it is difficult to identify a well defined reaction coordinate, since the diffusion process occurs via a sequence of strongly rebonded configurations in which the ad-atom is always spread over more than one lattice site.

Fig. 4. Atomic trajectories projected on a (110) plane at T=1400 K during a time interval of about 6 ps. The atoms of the upper chain simply vibrate around their equilibrium positions, while those of the lower chain make also occasional first-neighbour jumps due to a vacancy migrating along the chain (from left to right).

A more detailed analysis was attempted in the case the vacancy. Here simulations were performed at different temperatures. At T=1200 K and at T=1400 K diffusion proceeds, as expected, via a sequence of first-neighbour jumps in which the vacancy moves from one site to the next by crossing a barrier of 0.3 eV at the saddle point, corresponding to the so-called split vacancy configuration. This is in good agreement with low temperature experiments that indicate a migration barrier of about 0.4 eV for the neutral vacancy[20]. However, our results show that already at T=1200 K the jumps are not uncorrelated, as they are expected to be at low temperature, but are affected instead by strong dynamical or memory effects which can be inferred by looking at the temporal as well as at the spatial distribution of the jumping events. One of the most characteristic manifestations of such effects is the observed systematic tendency of a vacancy to perform several jumps along the same atomic chain before starting to move along a different one. Consider for instance the atomic trajectories in Fig. 4: only after performing the seven jumps shown in the picture the vacancy began to move along a different tetrahedral chain. This effect was systematically observed both at T=1200 K and at T=1400 K. At the same time we noticed a significant reduction of the rate of return jumps compared to the value of 1/4 expected for a simple random walk.

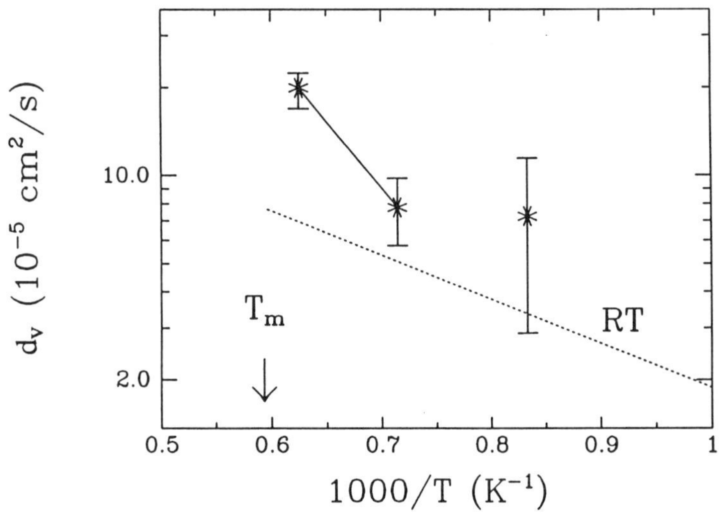

Fig. 5. Arrhenius plot of the vacancy diffusivity d_V. Asterisks: MD results. The error bars are statistical error estimates. Dotted line: simple rate theory estimate using E_m=0.3 eV for the barrier and a frequency of 20 THz for the jump attempt frequency. The continuous line has a slope of 0.9 eV.

Dynamical effects are even more pronounced at T=1600 K, where, in addition to nearest-neighbour jumps, we also observed second- and third-neighbour jumps. These processes are characterized by larger activation barriers but are favoured entropically. Some of these events are associated to the concerted motion of several atoms which create temporarily a disordered local zone. Events of this kind can be considered as dynamical premelting fluctuations and can provide a realistic model of the extended vacancy concept originally introduced by Seeger and Chik to explain the high entropy of diffusion in silicon. Premelting effects associated to bulk point defects have been experimentally detected in the case of lead and are likely to be a common phenomenon in nature[27,28]. In the case of silicon they contribute to increase the vacancy diffusivity at high temperature, as can be seen in Fig. 5. However, their overall effect on the self-diffusion coefficient is modest, since this is dominated by the concentration of native defects rather than by their diffusivity. At this stage we are not able to locate accurately the melting temperature of our model silicon crystal, as it is well known that melting of small sized computer models with periodic boundary conditions is affected by large hysteresis effects. We have observed, however, that vacancy nucleated melting occurs at about 1750 K in our system: this is likely to be an upper bound for the melting temperature of our model silicon crystal (experimentally T_m = 1685 K).

5. Self-diffusion coefficients in silicon

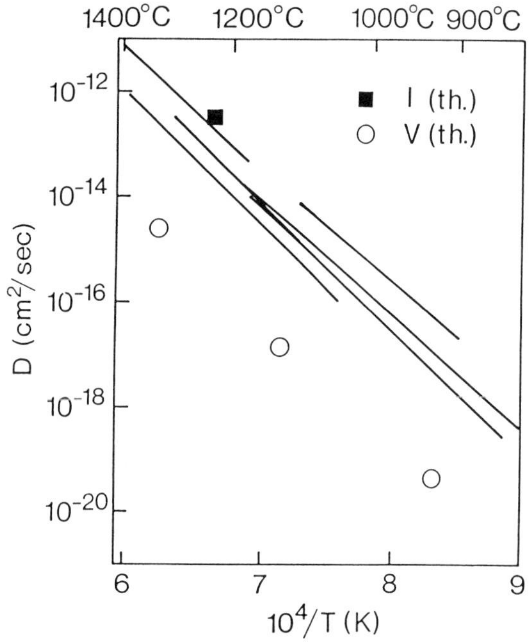

Fig. 6. Calculated self-diffusion coefficients for the interstitial and the vacancy mechanisms. The experimental self-diffusion coefficients are represented by lines.

By combining the results for the concentration of native defects reported in Sect. 3 with those on the diffusivity reported in Sect. 4, the vacancy and the interstitial contributions to the self-diffusion coefficient in silicon are easily evaluated according to Eq. 3. In order to make a meaningful comparison we use here, for both the vacancy and the self-interstitial, the concentration data obtained with the local harmonic approximation. The resulting coefficients are compared with experimental self-diffusion data in Fig. 6. We notice that (i) the self-interstitial mechanism dominates over the vacancy mechanism, (ii) the vacancy mechanism gives a rate that is low compared to experiment, (iii) the diffusion coefficient due to the interstitial mechanism agrees well with experimental data. The diffusion coefficient corresponding to the concerted exchange mechanism can be obtained from the energy and the entropy calculated by Pandey and Kaxiras[9] for this mechanism. The resulting coeffficient is about five orders of magnitude smaller than the coefficient of the interstitial mechanism and is also much too low compared with experimental data. This is because the activation energy corresponding to this mechanism is with 4.7 eV very large according to Pandey's calculation[5], which is comparable in accuracy to our formation energy calculations. On the other hand the entropy calculated by Pandey and Kaxiras for the concerted exchange mechanism is likely to be a lower bound because of their incomplete treatment of relaxation. However it would be necessary to add a very substantial contribution of more than 10 k_B to the presently calculated entropy of the exchange mechanism to bring its rate in the same range of that of the self-interstitial at $T = 1500$ K.

The good agreement found between theory and experiment for the interstitial mechanism should be taken with care, since anharmonic effects have not been included in the calculation of the self-interstitial free-energy of formation. Moreover experiments are done at constant pressure, while the calculations reported here for all the three mechanisms have been performed at constant volume. It is not clear at present how much this will affect the results. Several other important questions remain open. We list here just a few: (i) what is the Arrhenius slope of the interstitial mechanism, (ii) how will doping affect the results for the vacancy. All these questions demand for additional and more extensive investigations. In conclusion, quantitative first-principles calculations of the diffusion constants of real materials just begin to be feasible. A lot more is expected to come in the future.

Acknowledgement. Some of the work reported in this paper was done in collaboration with W. Andreoni, D.B. Laks, S.T. Pantelides and M. Parrinello. One of us (R.C.) would like to acknowledge a longterm stimulating collaboration with S.T. Pantelides on this matter.

REFERENCES

1. S.T. Pantelides, MRS Symp. Proc 63, 511 (1990).
2. U.M. Goesele, in Festkoerperprobleme 16, 89 (1986).
3. W. Frank, U. Goesele, H. Mehrer and A. Seeger, 1984, in *Diffusion in Crystalline Solids*, eds G.E. Murch and A. Nowick (Academic, New York) p. 63.
4. A. Seeger and P. Chik, Phys. Stat. Sol. 29, 455 (1968).
5. K.C. Pandey, Phys. Rev. Lett. 57, 2287 (1986).
6. P.M. Fahey, P.B. Griffin and J.D. Plummer, Rev. Mod. Phys. 61, 289 (1989).
7. R. Car, P.J. Kelly, A. Oshiyama and S.T. Pantelides, Phys. Rev. Lett. 52, 1814 (1984), Phys. Rev. B30, 2260 (1984).
8. Y. Bar-Yam and J. D. Joannopoulos, Phys. Rev. Lett 52, 1129 (1984). Phys. Rev. B30, 2216 (1984).
9. K. C. Pandey and E. Kaxiras, Phys. Rev. Lett. 66, 915 (1991).
10. C. H. Bennet in *Diffusion in Solids*: Recent Develpments, eds A.S. Nowick and J.J. Burton (Academic, New York, 1984) p. 429.
11. R. Car and M. Parrinello, Phys. Rev. Lett. 55, 2471 (1985).

12. P.E. Blöchl, D.B. Laks, S.T. Pantelides, E. Smargiassi, R. Car, W. Andreoni and M. Parrinello in *Proc. of the 20th Int. Conf. on The Physics of Semiconductor*, edited by E. M. Anastassakis and J.D. Joannopoulos (World Scientific,1990) Vol.1, p. 533.
13. P. Hohenberg and W. Kohn, Phys. Rev. 136, B864 (1964).
14. W. Kohn and L. Sham, Phys. Rev. 140, A1133 (1965).
15. F. Buda, G.L. Chiarotti, R. Car and M. Parrinello, Phys. Rev. Lett. 63, 294 (1989).
16. P.E. Blöchl, C.G. Van de Walle, S. T. Pantelides, Phys. Rev. Lett. 64, 1401 (1990).
17. D.R. Hamann, M. Schlueter and C. Chiang, Phys. Rev. Lett. 43, 1494 (1979).
18. G.B. Bachelet, D.R. Hamann and M. Schlüter, Phys. Rev. B26, 4199 (1982).
19. G.A. Baraff, E.O.Kane and M.Schlüter, Phys.Rev. B21, 5662, (1980).
20. G.D. Watkins and J.R. Troxell, Phys. Rev. Lett. 44, 593 (1980); J.R. Troxell and G. D. Watkins, Phys. Rev. B22, 921 (1980).
21. P.J. Kelly, R. Car, and S.T. Pantelides, in *Defects in Semiconductors,* ed. H J. von Bardebelen, Mat. Sci. Forum 10-12 (Trans. Tech. Publications, Aedermannsdorf, CH, 1986).
22. A. Antonelli and J. Bernholc, Phys. Rev. B40, 10643 (1989).
23. R. LeSar, R. Najafabadi and D.J. Srolovitz, Phys. Rev. Lett. 63, 624 (1989).
24. For a review, see D. Frenkel, in *Molecular Dynamics Simulations of Statistical Mechanical Systems,* eds G. Ciccotti and W.G. Hoover (North Holland, Amsterdam, 1986) p. 151.
25. D. Frenkel and A.J.C. Ladd, J. Chem. Phys. 81, 3188 (1984).
26. S. Nose, Mol.Phys. 52, 255 (1984); J. Chem. Phys. 81, 511 (1984).
27. E.A. Stern and Ke Zhang, Phys. Rev. Lett. 60, 1872 (1988).
28. H. Schechter et al., Phys. Rev. Lett. 63, 1400 (1989).

Materials Science Forum Vol. 83-87 (1992) pp. 447-456

IMPURITY AND DEFECT INDUCED METASTABILITIES IN TETRAHEDRAL SEMICONDUCTORS

D. J. Chadi
NEC Research Institute, Princeton, NJ 08540-6620, USA

ABSTRACT

The electronic and structural properties of single and double acceptor and donor centers in tetrahedrally coordinated semiconductors are investigated. The conditions under which a large-lattice relaxation (LLR) leads to a shallow-deep transition of levels are examined for GaAs, ZnSe, and Si. Lattice relaxations characterized by relative atomic displacements of greater than 1 Å and energy changes of about 0.35 eV are found to be common in zincblende semiconductors. For Si, complex formation is generally needed for LLR to occur.

1. Introduction

Impurity and defect induced structural metastabilities in semiconductors have been a subject of considerable interest for many years. Metastability, when present, can strongly affect the electronic, optical, and transport properties of an impurity center. Several examples of such metastabilities are now known in III-V, II-VI and elemental semiconductors and more will probably be discovered in the future as these systems are studied further. In this paper a short review of the theoretical results we have obtained over the last three years in this area is presented. The similarities and differences in the metastabilities exhibited by donor and acceptor centers and the trends in going from covalently bonded to ionic semiconductors are examined. The results obtained from first-principles pseudopotential calculations can be explained in some cases on the basis of a simple orbital energy scheme. In the following we discuss the properties of several acceptor and donor centers in tetrahedrally coordinated semiconductors.

2. Ga-antisite double acceptor in GaAs

This defect occurs when a Ga atom occupies an As lattice site. In *n*-type GaAs, where the Fermi level is near the conduction-band-minimum, the defect is a double acceptor. It captures two electrons from the Fermi level (E_F) and makes four two-electron covalent bonds with its four nearest-neighbor Ga atoms. The total-energy of the doubly charged center varies with E_F as:

$$\Delta E_{tot} = -2\Delta E_F \quad , \tag{1}$$

where E_F is measured from the valence-band-maximum (VBM).[1] As the Fermi level comes down from the CBM, the energy of the center increases. When E_F is close to the VBM, the energy lowering resulting from the transfer of electrons to the center becomes quite small. In this case a new configuration with a neutral charge state for the antisite becomes possible. The justification for considering this state is that the deficiency of two electrons at the center can get localized to one of the four Ga-Ga bonds. With no electrons in this bond, the two Ga atoms are able to undergo a large lattice relaxation. The results of our first-principles pseudopotential calculations show that it is energetically favorable to "break" this zero-electron bond and to move both the antisite and its nearest-neighbor into planar sp^2-bonded configurations.[2] A similar transformation is found to be energetically costly in bulk, defect free, GaAs because the dangling-bonds resulting from bond rupture contain two electrons and their rehybridization from sp^3 to p_z together with the lattice strain energy prevent this distortion. In the case of the Ga-antisite the dangling bonds are completely empty and their rehybridization does not raise the energy. The lattice distortion is driven by the well known tendency of a threefold bonded Ga atom to seek a planar configuration. The bond breaking is predicted to occur when E_F is less than about 0.2 eV above the VBM. Nearly the same results are found to hold for a B antisite in GaAs. These results should be contrasted to those for double and single acceptors in Si, InP, and ZnSe discussed in the following.

3. Be double acceptor in Si

We have examined whether neutral Be impurities in Si undergo the same type of bond breaking metastability as Ga-antisite defects in GaAs and whether this could explain why only 10% of Be atoms are electrically active in Si. The results of our calculations indicate that the broken-bond configuration is unstable in Si.[3] The low electrical activity of Be can be understood in terms of isoelectronic pair formation. The energies of isolated Be atoms, six structurally different Be-Be pairs, and of larger Be complexes were examined by Tarnow et al.[3] and the lowest energy state of a Be pair in p-type Si was identified with that of a relaxed (and "rebonded") substitutional-interstitial pair. The behavior of Be double acceptors in Si is similar to that found for Be single acceptors in InP.

4. Be single acceptor in InP

Gallium-arsenide can be doped heavily p-type and acceptor concentrations in excess of $10^{21}/cm^{-3}$ have been obtained with C impurities. The situation is quite different for InP where the maximum achievable acceptor concentration is apparently only $5x10^{18}/cm^{-3}$. Recent calculations by Zhang[4] on Be dopants in InP suggests that the low hole concentration is most probably caused by Be-Be complex formation in a manner similar to that in Si. A complex consisting of a substitutional-interstitial pair of Be atoms acts as a donor which compensates a normal unpaired Be atom on a Ga substitutional site. This result is consistent with experimental particle-induced-X-ray-emission data on Zn acceptor impurities in InP reported at this Conference by Walukiev icz et al.[5] They found strong evidence for interstitial Zn atoms by performing channeling experiments along the three primary cubic axes.

5. P, As, and N single acceptors in ZnSe

The results of first-principles pseudopotential calculations in conjunction with extensive experimental data, particularly for P dopants, were used to derive a microscopic theory of the properties of column V acceptors in ZnSe.[6] The main results of the present study are that As and P acceptors in ZnSe possess two distinct atomic configurations in a *neutral* charge state. A metastable state with a small lattice relaxation which can be identified with the effective-mass state as well as a more stable A° deep state characterized by a LLR are found. The atomic structure for the A° center is found to be very similar to that derived for the neutral Ga-antisite defect in GaAs in its LLR state, i.e., the acceptor and a nearest-neighbor Zn atom break their bond and move into planar configurations. The electronic structure is different, however. The p_z orbital of the Column V impurity is not empty but contains one electron. The A° state explains key experimental properties of P impurities in ZnSe. In particular, the strong C_{3v} symmetry, the electron-spin-resonance-active ground state, a P atomic displacement of 0.5 Å along the [111] axis, and the localization of the hole on a p_z-orbital of P are all in excellent agreement with experimental data obtained from an analysis of hyperfine splitting data.[6] In addition, other data on the optical and thermal quenching of the 1.9 eV characteristic photoluminescence line can be easily understood. The theoretical results suggest that in the absence of impurities such as H or Li which tend to form complexes and passivate acceptors in Si and GaAs,[7-9] N should give rise to only a shallow effective-mass state in either the LLR or fourfold geometry.

6. Double-donor As-antisite defect in GaAs

This defect is known to be at the core of the much studied EL2 center in GaAs. Theoretical total-energy studies indicate that the neutral isolated As antisite defect has two atomic configurations.[10-12] The stable state corresponds to the normal fourfold coordinated geometry with the As antisite surrounded by four other As atoms. In the metastable configuration the antisite breaks one of its four bonds and moves about 0.2 Å below the plane defined by the remaining three As atoms to which it remains bonded. The total displacement of the antisite is larger than 1 Å. The most remarkable feature of the metastability is the small $\cong 0.25$ eV energy difference between the stable and metastable states. There are several reasons for this. First, in the neutral stable state, the antisite has two tightly bound electrons with energies at midgap.[13] The wavefunctions of these electrons are primarily antibonding in character. The lengthening of an As-As bond during bond rupture *decreases* the energy this state. In the final metastable state, these two electrons lie below the bulk VBM. The resultant decrease of the electronic energy significantly reduces the energy of the metastable state. A second important factor in the stabilization of the metastable state is charge transfer from the threefold coordinated antisite to the remaining threefold coordinated As atom resulting from bond breaking. The transfer is relatively large and is about 0.75 e. As a result, the electron in the high energy dangling-bond p_z orbital of the planar antisite goes into the lower energy sp^3 orbital of the other threefold coordinated As atom. The calculated electronic charge density distributions for the highest occupied and lowest unoccupied states of the As antisite clearly demonstrate the charge transfer.[10] A third factor influencing the energy of the metastable state is that in the nearly planar configuration for the antisite the bond angles are very close to 120°, indicative of the stronger sp^2 type bonding. Altogether, these three factors reduce the energy of the metastable state and nearly overcome the large lattice strain energy required to form this state. The energy barrier to go from the metastable state to the stable state is $\cong 0.34$ eV.[10] For the reverse transition the barrier is close to 0.6 eV. Photoexcitation from the ground A_1 state to the excited T_2 state considerably lowers the barrier for the stable to metastable transition.[11]

There is still strong debate on whether the antisite, by itself, constitutes the EL2 defect. The theoretically derived properties of the isolated As antisite are similar to those observed experimentally for the EL2 center. Positron annihilation experiments[14] support the result that in the metastable state the EL2 defect is vacancylike and has a region of negative charge state. Optical absorption and luminescence experiments[15,16] have also been interpreted[17] as supporting the simple model. But experimental,[18,19] and theoretical[20] evidence that the EL2 center is a complex consisting of an As antisite and an interstitial As (or possibly an intermediate interstitial Ga) atom persists. In the original model for the complex, the interstitial atom was believed to be close to the As antisite, but theoretical studies by Baraff and Schluter[21] showed that it was difficult to explain the binding of such a complex. The problem is that in its ESR active state, the antisite is in a positive charged state; the As interstitial, if present, also has to be in a positive charge state in order to be ESR inactive. The resultant Coulombic repulsion between the two As atoms destabilizes the complex. For this reason it is far more preferable to place the As interstitial near one of the four nearest-neighbors of the antisite. Such a possibility, which turns out to be consistent with the experimental data,[18] removes one of the major obstacles in accepting the antisite-interstitial complex as a candidate for the EL2 defect. The only remaining questions are the nature of the metastability for the complex and the consistency of its calculated electronic structure with experimental data. Our previous study of the interaction of an As-interstitial with a *Ga-antisite* showed a very strong rebonding effect induced by the interstitial.[2] A similar structural change is found for an As-interstitial bound to an antisite pair.[23] It is very likely that similar rebonding effects will play an important role in the antisite-interstitial complex.

7. Donors in Si

We have examined the properties of single and double donors in Si via first-principles pseudopotential calculations.[22,23] In general, we find no metastabilities or LLR effects except in the presence of hydrogen, or when impurities from the first row elements are involved. It is not too difficult to understand the reason why Si is for the most part immune to such metastabilities. Consider first the situation where the "double donor" consists of two nearest-neighbor substitutional As atoms. At first sight this complex looks electronically similar to the As antisite in GaAs. Both defects give rise to two extra electrons that are localized in As-As derived states. However, unlike the As antisite defect in GaAs which has a low energy metastable configuration with a LLR, this state is unstable in Si.[22] A simple reason for why this is the case is as follows. Suppose that the two As atoms in Si did undergo, in fact, a LLR and that as a result both As atoms became threefold coordinated with one atom (labeled As_a) going to a planar like geometry and the other (As_b) staying in its undistorted bulk position. The As_b atom will use three of its five valence electrons to bond to its three nearest-neighbor Si atoms, leaving the remaining two unbonded electrons to fill the lone pair state. This atom is therefore, in a "closed shell" electronic configuration and has no low energy unoccupied states to accept any other electrons. The planar As_a atom also uses three of its valence electrons to bond to its nearest-neighbor Si atoms. However, it ends up with two electrons in a high energy p_z state. Since charge transfer to the other As atom is blocked, the broken bond state ends up to be high in energy and is unstable. We have also done extensive studies on a single As impurity in Si.[23] The results are the same, i.e., the donor does not have any metastabilities accompanied by a LLR when in a neutral, negative, or even a positive charged state. The explanation for this behavior follows the same line of argument given above for the failure of the diatomic As-As complex in Si to exhibit a strong structural metastability.

The only exceptions to this overall tendency against LLR for an isolated donor in Si occurs for N and O.[24-27] The covalent bonding radius of N is sufficiently small that it moves along a [111] axis to strengthen its bond with the other three Si atoms at the cost of breaking a Si-N bond. Similarly, substitutional O moves off-center and forms a bond with two Si atoms with the remaining two Si atoms forming a weak bond.[25,26] Oxygen, in groups of more than two atoms, forms complexes which behave as shallow double donors. Theoretically, a threefold coordination of at least two oxygen atoms seems essential in the formation of these "thermal donors".[28-30] In the threefold geometry, valence electrons are forced into antibonding O-O or Si-O states. Because of the strong O-O and Si-O interactions these states lie considerably above the Si conduction band minimum. The threefold state is stabilized by the ionization of the center i.e., the antibonding electrons drop into the conduction band minimum and become delocalized. Shallow donors results from the Coulombic interaction between the positively charged "core" and these ionized electrons. The precise atomic structure of the center is not yet known with certainty.

In the presence of both H and donor impurities, complexes with a LLR readily occur. Complex formation leads to a passivation of donors[31] (and acceptors[7-9]). The atomic structure and vibrational properties[32] of donor-hydrogen complexes in Si are well understood. For a substitutional P donor, the optimal structure of the P-H complex consists of a broken P-Si bond together with a new Si-H bond.[33] The Si atom involved in the H bonding is the atom undergoing the LLR. The hydrogen atom approaches and bonds to it along the (111) "antibonding" direction. If the (Si-H) unit is regarded as being equivalent in some way to a column V atom, the *stable* structure of the overall complex is seen to be very similar to that of the *metastable* state of the EL2 related As antisite defect in GaAs with (Si-H) playing the role of the antisite. Our calculations indicate that the P-H complex in Si also has a metastable configuration in which the P atom is bonded to four Si atoms and the H atom is loosely bonded along the (111) antibonding direction.[34] This metastable state is the analog of the stable state of the As antisite defect in GaAs.

In the absence of other impurities, H can still induce complexes with a LLR. In particular, we have shown that a nonmolecular H_2 complex in which a Si-Si bond is broken and the resulting dangling bonds are bonded to the two H atoms is one of the lowest energy forms of H in Si.[35] The complex should be especially important in amorphous Si where there are many weak bonds that can be easily broken. If the (Si-H) units are treated as being equivalent to a column V atom, the structure of the H_2 complex is seen to be very similar to that of the metastable state of the As antisite in GaAs.

8. *DX* centers in GaAs and AlGaAs alloys

Substitutional column IV donors on Ga (or Al) sites and column VI donors on As sites in GaAs and AlGaAs alloys possess both shallow and deep states. It is now well known that the relative stability of the two types of states depends on pressure,[36] AlAs mole fraction in an AlGaAs alloy, and on the doping level.[37] The deep state labeled *DX* has been extensively studied. There is strong experimental evidence now that *DX* is a negatively charged center[38-43] and that it involves a LLR.[44-48] For a substitutional Si_{Ga} center, we have proposed that the atomic structure of the *DX* state involves the breaking of one of the four Si-As bonds and the displacement of the Si atom into an "interstitial" threefold coordinated atomic configuration.[49,50] The resulting geometry is very similar to that for the metastable state of the As antisite[10] in GaAs. We have calculated the energy of the Si-induced *DX* center and have shown, in agreement with experiment, that it is metastable in GaAs and that it becomes the stable state under pressure or with alloying as the AlAs mole fraction increases.[50] For a S_{As} center we find that the *DX* formation involves the motion of a nearest-neighbor Ga or Al atom into an interstitial position.[49] The formation energy of the *DX* center is found to be surprisingly small in GaAs and to be approximately 0.16 eV per electron. This is in large part caused by the drop in the electronic energy of the filled antibonding state as the donor-host bond is broken. The highest occupied electronic level of *DX* is optically deep and is located about 0.5-0.6 eV above the bulk VBM. The proposed model explains the large difference between the thermal and optical ionization energies of *DX* centers and the phenomenon of persistent photoconductivity. When an electron from a DX^- center is optically excited into the conduction band, the LLR is no longer stable. The displaced atom moves back to the substitutional position, becomes fourfold coordinated, and gives rise to a shallow donor state. In order to go back to the original LLR position the atom has to overcome an energy barrier (to break a donor-host bond) and this is not possible at sufficiently low temperatures.

A localized noneffective-mass state associated with the donor can occur even in the absence of a LLR.[51] It arises from an antibonding combination of the donor-host orbitals. In GaAs this level is above the conduction band minimum. In AlGaAs alloys the A_1 level associated with this state can for certain alloy concentrations and pressures be lower in energy than both the *DX* center and the effective-mass state.[52] Unlike the *DX* center, the optical and thermal ionization energies of the A_1 center should be nearly equal.

9. Conclusions

A short review of our recent theoretical work on the electronic and structural properties of donor and acceptor centers in zincblende and elemental semiconductors was given above. Isolated impurities or antisite defects are found to be capable of inducing low energy metastable states with large lattice relaxations, particularly in zincblende semiconductors. The association of an impurity with H is often necessary to produce the same type of relaxations in Si. The relaxations generally lead to a shallow-deep transition of impurity levels. In GaAs, the double acceptor Ga and B antisite defects are predicted to exhibit bistability, i.e., their structure depends on the Fermi level position and the charge state of the defects. The double acceptor Be impurity in Si, however, does not exhibit a similar structure. The results of the total-energy calculations indicate that the low electrical activity of Be in Si is instead caused by a pairing of a substitutional and interstitial Be impurities which then act as isoelectronic centers. The same type of Be-Be or Zn-Zn pairing probably occurs in InP. In this case the paired impurities act as donors which compensate substitutional acceptors. In ZnSe, LLR and a shallow-deep transition of acceptor levels already occurs for neutral P and As single acceptors. Nitrogen, because of the high ionization energy of its valence *p* electrons, is expected to have only shallow effective-mass states in either the normal substitutional site or in the large lattice relaxed threefold coordinated state. For the latter state, the p_z orbital of N is ½full and, unlike the case for P and As, it lies below the VBM of ZnSe and is capable of capturing an additional electron. As for donors, isolated arsenic antisite defects in GaAs are found to exhibit metastability and a LLR. Possible effects of an additional As interstitial atom in the vicinity of the antisite on the metastability of the antisite component of the complex remains to be investigated. Such a complex is suggested by recent experimental data on EL2 centers in GaAs and from a comparison of the theoretical and experimental properties of the isolated defect. A LLR is also found for single electron donors in GaAs and AlGaAs alloys (*DX* centers). In Si, LLR around isolated donors (except N) occurs only with the assistance of H which caps dangling bonds resulting from bond breaking. *DX* centers should be absent, therefore, in crystalline Si. These results demonstrate that both lattice relaxation and complexing of impurities or defects need to be considered in general to understand the properties of deep impurity centers.

10. References

1. G.A. Baraff and M. Schluter, Phys. Rev. B **33**, 7346 (1986).
2. S.B. Zhang and D.J. Chadi, Phys. Rev. Lett. **64**, 1789 (1990).
3. E. Tarnow, S.B. Zhang, K.J. Chang, and D.J. Chadi, Phys. Rev. B **42**, 11252 (1990).

4. S.B. Zhang (unpublished).
5. W. Walukiewwicz, K.M. Yu, J. Jaklevic, and E.E. Haller, presented at this Conference.
6. D.J. Chadi (to be published). We had previously found the large-lattice-relaxed geometry only for the positively charged state of the acceptor: D.J. Chadi and K.J. Chang, Appl. Phys. Lett. **55**, 575 (1989).
7. J.I. Pankove, P.J. Zanzucchi, C.W. Magee, and G. Lucovsky, Appl. Phys. Lett. **46**, 421 (1985).
8. B. Clerjaud, D. Cote, F. Gendron, W.S. Hahn, M. Krause, and C. Porte, presented at this Conference.
9. K.J. Chang and D.J. Chadi, Phys. Rev. Lett. **60**, 1422 (1988).
10. D.J. Chadi and K.J. Chang, Phys. Rev. Lett. **60**, 2187 (1989).
11. J. Dabrowski and M. Scheffler, Phys. Rev. Lett. **60**, 2183 (1988); Phys. Rev. B **40**, 10391 (1989).
12. E. Kaxiras and K.C. Pandey, Phys. Rev. B **40**, 10044 (1989).
13. G.A. Baraff and M. Schluter, Phys. Rev. B **35**, 5929 (1987).
14. R. Krause et al., Phys. Rev, Lett. **65**, 3329 (1990).
15. M. Kaminska, M. Skowronski, and W. Kuszko, Phys. Rev. Lett. **55**, 2204 (1985).
16. M.K. Nissen, A. Villemaire, and M.L.W. Thewalt, Phys. Rev. Lett. **67**, 112 (1991).
17. G.A. Baraff, Phys. Rev. B **40**, 1030 (1989).
18. K. Krambrock, H. Söthe, and J.-M. Spaeth, paper presented at this Conference; J.-M. Spaeth, K. Krambrock, and D. Hoffman, in *Proceedings of the Twentieth International Conf. on the Physics of Semiconductors*, edited by E.M. Anastassakis and J.D. Joannopoulos (World Scientific, Singapore, 1990), p. 441.
19. H.J. von Bardeleben, D. Stiévenard, D. Deresmes, A. Huber, and J.C. Bourgoin, Phys. Rev. B **34**, 7192 (1986); C. Delerue, M. Lannoo, D. Stiévenard, H.J. von Bardeleben, and J. C. Bourgoin, Phys. Rev. Lett. **59**, 2875 (1987).
20. M. Lannoo, C. Delerue, and G. Allan, paper presented at this Conference.
21. G.A. Baraff and M. Schluter, Phys. Rev. B **35**, 6154 (1987).
22. S.B. Zhang and D.J. Chadi (unpublished).
23. D.J. Chadi, (unpublished).
24. K.L. Brower, Phys. Rev. B **26**, 6040 (1982).
25. G.D. Watkins and J.W. Corbett, Phys. Rev. **121**, 1001 (1961).
26. G.G. DeLeo, W.B. Fowler, and G.D. Watkins, Phys. Rev. B **29**, 3193(1984)
27. H.P. Hjalmarson and D.R. Jennison, Phys. Rev. B **31**, 1208 (1985); R.P. Messmer and P.A. Schultz, Sol. State Communications **52**, 563 (1984).
28. L.C. Snyder and J.W. Corbett, in *Proceedings of the Thirteenth International Conference on Defects in Semiconductors, Coronado, California, 1984,* edited by L.C. Kimmerling and J.M. Parsey, Jr. (The metallurgical Society of the AIME, Warrendale, 1985), p. 693.
29. D.J. Chadi, Phys. Rev. B **41**, 10595 (1990).
30. P. Deak, M. Heinrich, L.C. Snyder, and J.W. Corbett, paper presented at this Conference.

31. N.M. Johnson, C. Herring, and D.J. Chadi, Phys. Rev. Lett. **56**, 769 (1986).
32. K. Bergman, M. Stavola, S.J. Pearton, and J. Lopata, Phys. Rev. B **37**, 2770 (1988); K. Bergman, M. Stavola, S.J. Pearton, and T. Hayes, Phys. Rev. B **38**, 9643 (1988).
33. S.B. Zhang and D.J. Chadi, Phys. Rev. B **41**, 3882 (1990).
34. K.J. Chang and D.J. Chadi, Phys. Rev. Lett. **60**, 1422 (1988); Phys. Rev. B **40**, 11644 (1989).
35. K.J. Chang and D.J. Chadi, Phys. Rev. Lett. **62**, 937 (1989); Phys.Rev. B **42**, 7651 (1990).
36. M. Mizuta, M. Tachikawa, H. Kukimoto, and S. Minomura, Jpn. J. Appl. Phys. **24**, L143 (1985).
37. T.N. Theis, P.M. Mooney, and S.L. Wright, Phys. Rev. Lett. 60, 361 (1988). For an extensive review of experimental data see: P.M. Mooney, J. Appl. Phys. **67**, R1 (1990).
38. V. Mosser, S. Contreras, J.L. Robert, R. Piotrzkowski, W. Zawadski, and J.F. Rochette, Phys. Rev. Lett. **66**, 1737 (1991).
39. L. Dobaczewski and P. Kaczor, Phys. Rev. Lett. **66**, 68, (1991).
40. J. A. Wolk, M.B. Kruger, J.N. Heyman, W. Walukiewicz, R. Jeanloz, and E.E. Haller, Phys. Rev. Lett. **66**, 774 (1991).
41. T. Fujisawa, J. Yoshino, and H. Kukimoto, J. Cryst. Growth **98**, 243 (1989).
42. T.N. Theis, P.M. Mooney, and B.D. Parker, J. Elec. Materials **20**, 35 (1991).
43. W. Jantsch, Z. Wilamowski, and G. Ostermayer, presented at this Conference.
44. T.N. Morgan, in *Defects in Semiconductors 15*, edited by G. Ferenczi, Vols. 38-41 of *Materials Science Forum* (Trans tech, Aedermannsdorf, Switz., 1989), p. 1079.
45. P.M. Mooney, T.N. Theis, and S.L. Wright, Appl. Phys. Lett. **53**, 2456 (1988).
46. G.A. Northrop and P.M. Mooney, J. Elec. Materials **20**, 13 (1991).
47. P. Gibart, D.L. Williamson, J. Mose, and P. Basmaji, Phys. Rev. Lett. **65**, 1144 (1990).
48. P.M. Mooney, T.N. Theis, and E. Calleja, J. Elec. Materials **20**, 23 (1991).
49. D.J. Chadi and K.J. Chang, Phys. Rev. Lett. **61**, 873 (1988); Phys. Rev. B **39**, 10063 (1989).
50. S.B. Zhang and D.J. Chadi, Phys. Rev. B **42**, 7174 (1990).
51. S.Y. Ren, J.D. Dow, and J. Shen, Phys. Rev. B **38**, 10677 (1988).
52. J.E. Dmochowski, P.D. Wang, and R.A. Stradling, in *Proceedings of the Twentieth International Conf. on the Physics of Semiconductors*, edited by E.M. Anastassakis and J.D. Joannopoulos (World Scientific, Singapore, 1990), p. 658; and paper presented at this Conference.

Materials Science Forum Vol. 83-87 (1992) pp. 457-462

THEORY OF NITROGEN AND PLATELETS IN DIAMOND

P.R. Briddon*, M.I. Heggie‡ and R. Jones

Dept of Physics, University of Exeter, Stocker Road, Exeter, Devon, EX4 4QL. U.K.
EMAIL : pb1@uk.ac.exeter.excc

*Present address : Dept. of Physics, University of Newcastle upon Tyne, Newcastle, NE1 7RU, UK.

‡Present address : Dept. of Computer Science, University of Exeter, Stocker Road, Exeter, Devon, EX4 4QL. U.K.

ABSTRACT

The structural, electronic and vibrational properties of nitrogen in diamond are calculated for the single substitutional atom, the complex of two neighbouring nitrogen atoms and two models of the platelet defects. Our results explain much of the available experimental data. For the isolated substitutional nitrogen atom the calculations reproduce the infrared absorption, giving the resonance at 1130 cm^{-1} and providing an explanation for the so-called anomalous local vibrational mode at 1344 cm^{-1}. The microscopic structure and vibrational properties of the A-centre are then considered and finally the the Lang and Humble CN models of the platelet defects are examined. It is shown that the Lang model cannot reproduce the observed absorption at 1372 cm^{-1} (the B´peak) whereas the CN model would produce several absorptions in this region.

METHOD

The calculations presented here are fully self-consistent and used local density functional theory together with the *ab initio* non-local pseudopotentials of Bachelet, Hamann and Schlüter [1]. Hydrogen-terminated clusters of up to 86 atoms were used with bases of up to 550 s- and p-type Gaussian functions centred on both the atoms and the mid-points of the C–C and C–N bonds. These clusters are sufficiently large to ensure that an atom embedded in the centre of a cluster will feel the same potential as it would in a solid. The main features of our approach that enable such large clusters to be treated are intermediate fits to the charge density (also using bases of Gaussian orbitals) in order to facilitate the evaluation of the Hartree and exchange-correlation energies. The methodology has been presented in detail elsewhere [2,3] and so no further details will be given here.

Using this approach, structural properties can be obtained by calculating the forces on each atom and allowing the atoms to move in the directions of these forces until the energy changes are negligible. For the defects considered here, up to 20 atoms were relaxed. To illustrate this procedure a 71 atom cluster $C_{35}H_{36}$ was constructed, with the atoms arranged in the diamond structure. The innermost core of atoms was then relaxed giving bond lengths and bond angles deviating from the experimental values for diamond by only ~1 %. An *ab initio* interatomic potential for diamond was then determined by calculating the second derivatives of the total energy of the cluster with respect to small displacements of the atoms. The parameters of a Musgrave-Pople interatomic potential [4] were then obtained from these by a least squares fitting procedure. The phonon spectrum of diamond was then calculated from this using the usual procedures of lattice dynamics. However, it should be emphasised that this has been determined from first principles without recourse to empirical fitting parameters. The Raman frequency of diamond was correctly obtained at 1332 cm^{-1}. Full details have been given in [2]. As the method correctly reproduces the structure and vibrational properties of a bulk solid, it may also be expected to give properties of the defective material correctly, as no further approximations are made.

In the vicinity of a defect, the interatomic potential is of course modified, making this valence force potential inadequate. Clearly any elements of the dynamical matrix associated with motion of atoms that are first or second nearest neighbours of a defect atom may be significantly modified by its presence. These modified elements were calculated directly using the *ab initio* program and the remainder found by using the valence force potential determined for pure crystalline diamond as described above, thus giving a complete description of the interatomic potential in the defective material. If desired, a feel for the strengths of the bonds present in the defect complex can then be obtained by fitting the modified double derivatives of energy to a "local" valence force potential with parameters which are allowed to take different values from the bulk values and which vary from atom to atom.

The vibrational frequencies associated with the defect could then be calculated by generating and diagonalising the dynamical matrix describing the motion of several hundred atoms (either with fixed or periodic boundary conditions) containing a single defect. In a previous publication [4] we used exactly this approach. Here we have preferred to use the valence force potential for diamond to generate the unperturbed Green function for the lattice, and then to calculate the perturbed Green function by solving Dyson's equation, with the change in the dynamical matrix forming the perturbation. The local vibrational density of states at frequency ω, $\rho_{i\alpha}(\omega)$, projected onto atom i moving in direction α is then given in the usual way by $\rho_{i\alpha}(\omega) = -[2\omega/\pi]\,\mathrm{Im}\,G_{ii}^{\alpha\alpha}$.

THE ISOLATED SUBSTITUTIONAL NITROGEN ATOM

Information about the structure of the isolated substitutional nitrogen atom is available from ESR experiments on type 1b diamonds [6]. These showed the nitrogen to move off centre forming a defect of C_{3v} symmetry. The exact displacement could only be tentatively estimated, an extension of one of the C–N bonds by 5–10 % being suggested.

Type Ib diamonds also show strong infrared (IR) absorption at 1130 cm^{-1} and 1344 cm^{-1}. These were initially thought to be unrelated as the 1130 cm^{-1} mode shifted to lower frequencies for diamonds containing ^{15}N whereas the 1344 cm^{-1} mode was unaffected. This conclusion was later shown to be incorrect [7] when it was shown that the ratio of the peak heights was reproducible for many samples.

Early optical absorption experiments [8] showed absorption to start at 1.7 eV and to rise rapidly with sharp structure at ~4 eV. Optical and thermal excitation experiments [9] were interpreted as indicating that this was due to transitions from the valence band and conduction bands to a defect level lying 1.7 eV below the conduction band (4 eV above the valence band). However more recent work [10,11] has indicated that an absorption at 4.059 eV is due to a transition between two localised states of A_1 symmetry.

We investigated this defect using an 86 atom cluster (figure 1), the central carbon atom being replaced by nitrogen. It was found that the resulting electronic structure consisted of a (doubly occupied) hyper-deep level 2.5 eV below the valence band and a singly-occupied "A_1" state near mid-gap (in close agreement with [12]). It is therefore clear that the partially occupied level is non-degenerate so that the off-centre distortion cannot be ascribed to the Jahn-Teller mechanism as had been previously thought [13]. The central atoms were then relaxed and a trigonal distortion was obtained. The resulting structure is given in figure 1 (right). We find an extension of 28% - longer than proposed by some workers (5–10% [7,14] and 26% [13] although longer values have also been suggested (30% [15] and 36% [16]). A recent *ab initio* calculation [17] has produced the very similar result of 25%.

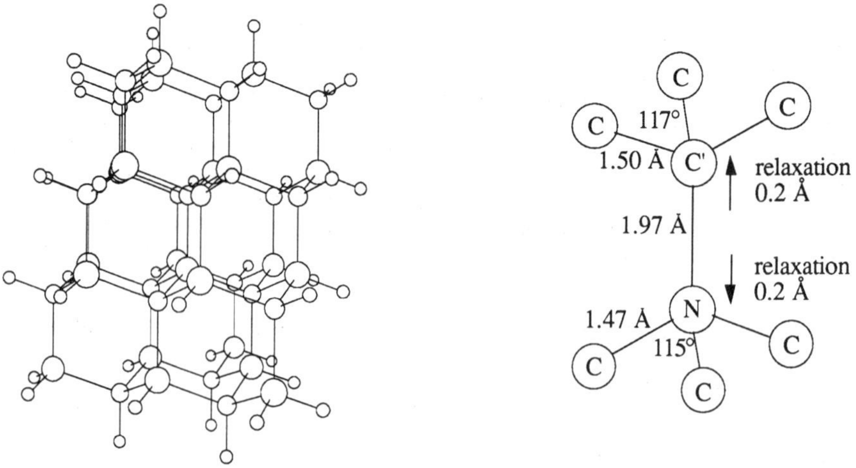

Figure 1 : The 86 atom cluster used to investigate the substitutional N defect (left) and the resulting microscopic structure (right).

Two one-electron energy levels appeared in the band gap. The lower energy level (doubly occupied) was located 1.2 eV above the valence band top and had a wavefunction which was a bonding combination of two *p*-orbitals centred on the nitrogen atom and its unique carbon neighbour (C′ in figure 1). In fact a larger contribution came from the nitrogen atom's *p*-orbital so this may be considered to be derived mainly from the lone pair orbital. The other (partially occupied) energy level fell 0.5 eV below the conduction band and was an antibonding combination of the two *p*-orbitals. The presence of two gap states is consistent with the transitions between two localised states deduced from optical absorption experiments [10,11].

An *ab initio* valence force potential describing the interactions between the atoms in the vicinity of the nitrogen atom was obtained as described above. Eight parameters were found necessary to reproduce the *ab initio* force constants. Further details of the precise choice of fitting functions and parameters will be given in a forthcoming paper. One important result evident from the fit was an increase in the C′– C bond-stretching force constant. This is to be expected as the bonding has changed in character from sp^3 to sp^2 which would be expected to give an increase in strength. An increase in this force constant has recently been assumed in an empirical study of the vibrational properties of this defect [18].

The vibrational modes of this defect were then found using the Green function method outlined above. In figure 2, we plot the local density of states projected onto the nitrogen atom and its unique carbon neighbour. In figure 2(a) this quantity is projected onto a direction perpendicular to that of the elongated bond, and in figure 2(b) it is projected parallel to this bond.

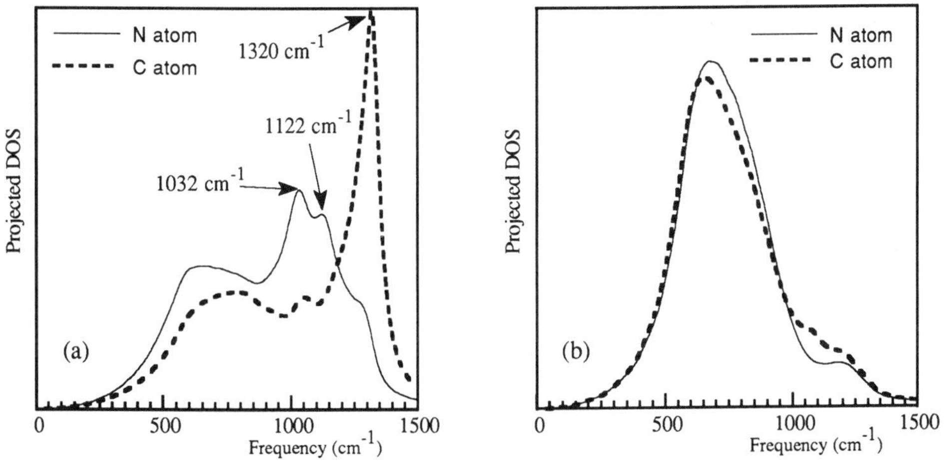

Figure 2 : Vibrational density of states projected onto central N and C atoms.

Four features of interest are evident :

(a) The unique neighbouring carbon atom showed a very sharp resonance at the top of the one-phonon absorption band (see figure 2a). Our calculation predicts this to be 1320 cm^{-1}, in fact, just below the Raman frequency at 1332 cm^{-1}. This mode is doubly degenerate and of symmetry E. It can be assigned to the 1344 cm^{-1} mode observed by Collins, Stanley and Woods [7] - the anomalous mode ! It is clear that the high frequency is associated with the strengthening of the carbon atom's bonds with its neighbours as it relaxes into their plane. The density of states projected onto the nitrogen atom shows only a slight shoulder at this frequency, thus explaining why the mode is insensitive to isotopic replacement of the nitrogen with a ^{15}N atom (we have carried this out and find the frequency of this mode to be unchanged). A similar conclusion has recently been reached by other workers [18]

(b) An peak in the density of states projected on the N atom is seen at 1122 cm^{-1} (figure 2a). This mode is also doubly degenerate and of E–symmetry and can be identified with the observed resonance at 1130 cm^{-1}. This mode dominates the experimental spectrum, presumably because it has a large dipole moment and so couples strongly with infrared radiation.

(c) A further peak is seen at 1032 cm^{-1} in figure 2(a). A similar feature occurs in the unperturbed Green function at a close frequency, but in fact infrared absorption is observed at 1080 cm^{-1}.

(d) A very broad feature is seen in figure 2(b) between 500 and 900 cm^{-1}. This is clearly associated with motion of the nitrogen and unique carbon atoms along the elongated bond. However, this has not yet been reported experimentally, possibly because of its breadth.

THE A-CENTRE : DAVIES' MODEL

Davies [19] has proposed that the "A-centre" consists of two adjacent substitutional N atoms. Using a similar 86 atom cluster, we have investigated this model and found the resulting structure to have the required C_{3v} symmetry with a C–N bond length of 1·46 Å and a very long N–N bond length of 2.14 Å. A doubly occupied electronic level appeared near mid gap (the antibonding combination arising from the overlapping of the lone pair orbitals of the two substitutional nitrogen atoms). In fact a peak in the optical absorption is seen at 3.8 eV. The vibrational frequencies of this defect were calculated to be 547 cm^{-1} and 960 cm^{-1} (associated respectively with in phase and out of phase motion of the two nitrogen atoms), 1150 cm^{-1} and 1324 cm^{-1} (associated with the N atoms moving perpendicular to the long N–N bond. These are in reasonable agreement with the experimental values of 480, 1093, 1203 and 1282 cm^{-1}. This lends support to this model of this defect.

THE PLATELET DEFECTS

The nature of the {100} planar defects known as platelets has been a mystery since they were first observed. Little is known for certain apart from the lattice displacement and several related infrared absorptions, the most prominent being at ~1372 cm^{-1} (the B′peak) and 326 cm^{-1} [20]. There has been much controversy over the composition of the platelets with different workers contradicting one another, but recently, an investigation into the isotopic shift in the B′ peak occurring when synthetic diamond composed of ^{13}C (with both ^{14}N and ^{15}N incorporated) was studied has given a strong indication that nitrogen is involved [21]. Here two models are studied, both involving nitrogen.

The Lang Model

An early model of platelet structure was put forward by Lang [22] who suggested that a (100) plane of carbon atoms should be replaced by a double plane of nitrogen atoms. This results in each "platelet unit cell" containing 6 carbon atoms and 4 nitrogen atoms. However, this composition does not agree with a quantitative analysis of the nitrogen content of platelets which was later carried out by Sumida and Lang [23].

To model this structure, a cluster containing 56 atoms, $C_{18}N_6H_{32}$, was used. The atoms were allowed to relax and the resulting microscopic structure is shown in figure 3 (left). The N-N and C-N bond lengths give rise to a displacement of 1.34 Å, slightly less than some previous experimental measurements 1.424 Å [24]; 1.388 Å [25], but greater than others [26].

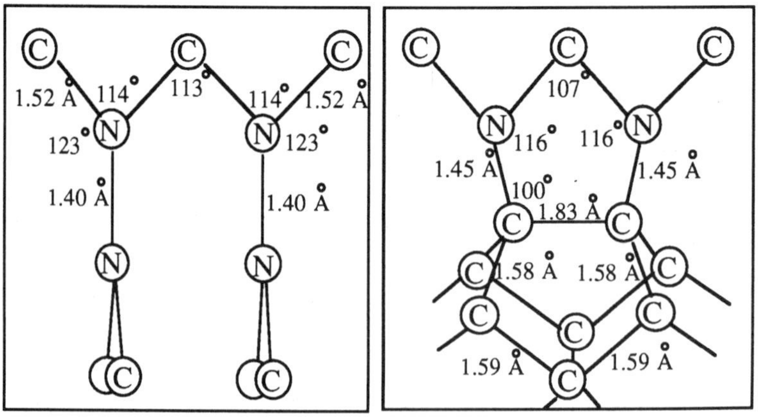

Figure 2 : Structure of Lang (left) and Humble CN (right) models of platelet structure.

The vibrational properties which would correspond to the Lang model were then obtained as previously described [5]. In this case, they were found using a supercell containing the double (100) plane of N atoms and 23 (100) planes of carbon atoms. This is large enough to ensure that the platelets in adjacent unit cells do not interact. No localised modes emerged above the Raman frequency. Projected densities of states on the nitrogen atoms were found as before and these displayed resonances at 276 cm^{-1}, 350 cm^{-1}, 938 cm^{-1} and 1106 cm^{-1}. An analysis of the normal co-ordinates of the low frequency resonances revealed that they were associated with motion of the nitrogen atoms out of the plane of their neighbours. It is possible to associate these frequencies with the observed absorption at 326 cm^{-1} [20]. It is evident however that the Lang model cannot explain the B′ peak that is a signature of the presence of platelets.

Humble's CN Model

An alternative model for the platelet structure is the Humble CN model [25]. This model is obtained from the Lang model by replacing half of the nitrogen atoms with carbon and joining up the extra carbon bonds in pairs. This model contains two atoms of nitrogen per platelet unit cell although the platelet might not have a homogeneous composition.

This structure was modelled using a cluster of 71 atoms. 15 inner atoms were relaxed and the resulting microscopic structure is shown in figure 3 (right). It is seen that the nitrogen-carbon bonds are slightly shorter than found for the Lang model and that one very long carbon-carbon bond is required to complete the five-membered rings of atoms.

The vibrational frequencies were then obtained and revealed the following features :

(a) A low-frequency resonance at 228 cm^{-1} associated with motion of the nitrogen out of the plane of its three neighbours. This mode has considerable in-plane dispersion and the **q=0** mode (which is what would be measured by a photon) occurred at 288 cm^{-1}. This is very similar to what was found for the Lang model - there too the "floppy" mode of the nitrogen atom had a frequency similar to the experimentally observed absorption at 326 cm^{-1}. It seems likely that this frequency is characteristic of a threefold coordinated nitrogen atom moving out of the plane of its neighbours.

(b) This time two bands of modes were obtained with frequencies above the one-phonon band. The maxima in the densities of states occurred at 1350 cm^{-1} and 1460 cm^{-1}. The two high frequency modes both involve stretching of the C–N bonds lying almost parallel to [100]. These bonds are considerably strengthened (in fact, we found its stretching force constant to be 3 eV/$Å^2$ greater that that in the host lattice). This can be understood on chemical grounds : the elongated C–C bond, necessary in order to complete a 5-atom ring, implies that both of the atoms forming it have relaxed towards the plane of their surrounding three atoms, leading to these being strengthened as a consequence of the greater degree of sp^2 bonding.

It is therefore seen that, unlike the Lang model, this 'CN model' can explain the main observed features of the infrared absorption. Clearly the *B′* peak can be explained but there remains one band of modes at a higher frequency. Several infrared absorptions are observed with higher frequencies than the *B′* peak, [27] although these are usually much weaker and broader, the exception being a mode at 1426 cm^{-1} which can have an intensity of up to 10 % of the *B′* peak.

In addition, two filled electronic energy levels fell in the lower half of the band gap, 1.9 and 2.3 eV above the valence band top. The wavefunctions of these were found to be bonding and antibonding combinations of the lone pair orbitals on the two nitrogen atoms in the 5-atom rings. These may be related to the observed luminescence at 1.25 eV [28] (the Lang model didn't produce any gap states).

CONCLUSIONS

To conclude, these results have shown that *ab initio* calculations can give valuable information relating experimental measurements to the microscopic structure of the corresponding defects.

For the isolated substitutional nitrogen atom, all the observed resonances have been assigned to specific atomic displacement patterns. These lead to an explanation for the absence of a shift in the localised vibrational mode at 1344 cm^{-1} when ^{14}N is replaced by ^{15}N. A broad resonance is predicted at around 600 cm^{-1} which has not yet been reported.

Calculations performed on the different models for the platelets have shown that the Lang model, perhaps the best known model, cannot give rise to the high-frequency absorption band that is observed although a low frequency resonance was obtained close to that observed at 326 cm^{-1}. In contrast the 'CN' model can reproduce all the observed absorptions.

The one common feature between both models is the existence of a low-frequency mode, in both cases due to the motion of a nitrogen atom out of the plane of its three neighbours. This is a feature that is unlikely to be reproduced by models such as fully-coordinated aggregated carbon interstitials - any four-fold coordinated atom would not be able to undergo such a low-energy vibration. This suggests either that nitrogen should be present in the platelets or that the carbon atoms take up an unusual bonding configuration (double or triple bonds, for example). This latter suggestion seems unlikely as double or triple bonds would be very susceptible to hydrogenation. Evidence for nitrogen being involved is the shift in the B' peak observed if ^{14}N is replaced by ^{15}N. Our results predict that the 326 cm^{-1} mode also should shift.

ACKNOWLEDGEMENTS
We wish to thank Andrew Lang, Marshall Stoneham and Dimitris Kechrakos for many useful discussions.

REFERENCES
1. G.B.Bachelet, D.R.Hamann and M.Schluter *Phys. Rev.* B **26**, 4199 (1982).
2. R.Jones *J. Phys. C : Solid State Phys.* **21**, 5735 (1988).
3. R.Jones *Mol. Simulation* **4**, 113 (1989).
4. M.J.P.Musgrave, J.A.Pople (1962) *Proc. R. Soc.* A **268**, 474.
5. P.R.Briddon and R. Jones (1990) in *Proceedings of Int. Conf. on New Diamond Science and Technology*, ed. R. Messier.
6. W.V.Smith, P.P.Sorokin, L.L.Gelles, G.J.Lasher *Phys. Rev.* **115**, 1546 (1959).
7. A.T.Collins, M.Stanley, G.S.Woods *Phil. Mag.* A **46**, 77 (1982).
8. H.B.Dyer, F.A.Raal, L.du Preez, J.H.N.Loubser *Phil. Mag.* **11**, 763 (1965).
9. R.G.Farrer *Solid State Commun.* **7**, 685 (1969).
10. J.Koppitz, O.F.Schirmer and M.Seal, *J. Phys. C : Solid State Phys.* **19**, 1123 (1986).
11. M.H.Vaz de Carvalho & A.J.Trindale das Neves *J. Phys. C : Solid State Phys.* **20**, 2713 (1987).
12. G.B.Bachelet, G.A.Baraff and M.Schlüter *Phys. Rev.* B **24**, 4736 (1981).
13. R.P.Messmer and G.P.Watkins *Phys. Rev.* B **7**, 2568 (1973).
14. A.Mainwood *J. Phys. C : Solid State Phys.* **12**, 2543 (1979).
15. A.M.Stoneham, in *The Properties of Diamonds*, edited by J.E.Field (Academic Press, London, 1979), p185.
16. C.A.J.Ammerlaan Inst. Phys. Conf. Ser. No. 59, p81 (1981).
17. J.Bernholc, S.A.Kajihara, A.Antonelli in proceedings of *20th International Conference on the Physics of Semiconductors* **1**, ed. E. M. Anastassakis and J. D. Joannopoulos, published by World Scientific, Singapore , p 332..
18. J.J.Sinai and S.Y.Wu *Phys. Rev.* B **43**, 12419 (1991).
19. G.Davies *J. Phys. C : Solid State Phys.* **9**, L537 (1976).
20. G.S.Woods *Phil. Mag. Lett.* **59**, 339 (1989).
21. T.Evans, G.S.Woods, D.Clark, H.Kanda presented at the 1990 Diamond Conference, held at Reading, 1990 (unpublished).
22. A.R.Lang *Proc. Phys. Soc.* **84**, 871 (1964).
23. N.Sumida and A.R.Lang *Proc. R. Soc. Lond.* A **419**, 235 (1988).
24. P.Humble *Proc. R. Soc. Lond.* A **381**, 65 (1982).
25. P.Humble, D.F.Lynch, A.Olsen *Phil. Mag.* A **52**, 623 (1985).
26. L.A.Bursill, J.L.Hutchison , N.Sumida , A.R.Lang (1981) Nature, **292**, 518.
27. G.S.Woods *Proc. R. Soc. Lond.* A **407**, 219 (1986).
28. D.R.Wright, P.J.Dean , E.C.Lightowlers , C.D.Mobsby, J. Luminescence **3**, 163 (1971).

Materials Science Forum Vol. 83-87 (1992) pp. 463-468

AB-INITIO CLUSTER CALCULATION OF HYPERFINE INTERACTIONS AND TOTAL-ENERGY SURFACES FOR N IN DIAMOND, SILICON AND GERMANIUM

A. FAZZIO[1], C.R. MARTINS DA CUNHA[1] and S. CANUTO[2]
[1]Instituto de Física da Universidade de São Paulo, CP 20516, 01498 São Paulo, SP, Brazil
[2]Departamento de Física da Universidade Federal de Pernambuco, 50 000, Recife, PE, Brazil

ABSTRACT

Using an ab-initio LCAO unrestricted Hartree-Fock cluster model we study the electronic structure of the substitutional N impurity in C, Si and Ge. We find that the stable position for the impurity is off-center in a $<\bar{1}\ \bar{1}\ \bar{1}>$ direction in all cases. The results for the hyperfine interactions over the impurity and first neighbors are in fair agreement with experiment.

1 Introduction

The purpose of this work is to investigate the behavior of the N impurity in group IV semiconductors C, Si and Ge. We may thus follow the trends in electronic structure for the system when the environment of the impurity changes from 2sp → 3sp → 4sp. In spite of the large amount of theoretical studies of N in diamond [1,2] and Si [3,4], this is the first calculation of the electronic structure of N in Ge through an "ab-initio" method. From experiment, also, the situation in diamond and Si is well known: electron paramagnetic resonance (EPR) data indicate that in both cases the center is distorted from T_d symmetry, the impurity displaced along a $<\bar{1}\ \bar{1}\ \bar{1}>$ direction. We show in this work that an "ab-initio" cluster calculation can describe well the known experimental results; we further predict the behavior of N in Ge, for which we obtain the total-energy surface for movement of the central atoms, together with the magnetic hyperfine interactions.

2 Model

We use 17-atoms clusters consisting of a central atom, a first shell of four nearest-neighbors and a second shell of twelve hydrogen atoms to saturate the dangling bonds. The "perfect" crystals are then simulated by clusters in T_d symmetry: C_5H_{12}, Si_5H_{12} and Ge_5H_{12} A N atom replaces the central atom for study of the impurity systems, that is, clusters NX_4H_{12} , with X=C, Si or Ge.

The electronic structure is obtained using an all-electron "ab-initio" LCAO procedure. The molecular orbital expansion is over the primitive gaussian contracted to some sets, and all one-electron, two-electron and overlap integrals are fully calculated, with no pseudopotential. We have used the basis set proposed by Dunning and Hay [7]: 4s contracted 2s for H; 9s5p contracted 3s,2p for C; and 12s,5p contracted 6s,4p for Si. For Ge we used the "Husinaga-MIDI" basis set [8]. The calculation was performed using program GAMESS [9] in both Roothaan Open-shell Hartree-Fock (ROHF) and unrestricted Hartree-Fock (UHF) schemes.

3 Results and Discussion

The N substitutional impurity, for all 3 cases, introduces an a_1 antibonding state in the gap region and, to higher energies, a t_2 state. The ground state configuration in T_d symmetry is then obtained as $|\ldots a_1^1 t_2^0 >$ leading to a non-degenerate ground state $^2A_1$; a Jahn-Teller distortion is not bound to occur. Although we find that the systems distort to a C_{3v} symmetry, we will not enter here in the discussion about the reasons for this distortions, if from a chemical rebonding of a pseudo Jahn-Teller effect.

Using the procedure described in the last section we calculated the total-energy for a large number of positions of the central impurity-coupled to displacement of one of the nearest neighbors-in the impurity clusters.

We show in Fig.1 the results of such a calculation for a grid of 68 coupled displacements. We see that N is unstable in T_d symmetry in the three cases, moving to C_{3v} symmetry. We must stress that, in diamond and silicon, if we displace only the atom, keeping the first neighbors fixed, no off-center minimum is found: for this to happen, it is essential that the C (Si) atom is allowed to move in <111> direction away from the impurity. This finding agree with the results of Schultz and Messmer [4] that if the N atom is inside a rigid Si cage, the stable geometry would be tetrahedral.

The atomic displacements we obtain from our calculation of the minima in the energy surfaces are: for the N-atom, in a $<\bar{1}\,\bar{1}\,\bar{1}>$ direction away from the T_d substitutional site, 0.23Å in diamond, 0.47Å in Si and 0.49Å in Ge; for the ligand atom on the C_{3v} axis in the <111> direction away from the impurity, 0.36Å in diamond, 0.47Å in Si and 0.48Å in Ge. For these coupled displacements, the gain in (cluster) total energy is 1.01 eV in diamond, 0.72 eV in Si and 0.63 eV in Ge.

We now turn to the analysis of bond-charge and spin distribution. The charge distribution is obtained from the bond-order matrix P_{AB}, for atoms A and B. Using the fact that the molecular orbital ψ_a is built in LCAO form

$\psi_a(\vec{r})=\Sigma_\mu C_{a\mu}\chi_\mu(\vec{r})$ from the atomic orbital χ_μ, the bond-order matrix is built from the orbital population on density matrix $P_{\mu\nu}=\Sigma_a^{occ}C^*_{\mu a}C_{\nu a}$ as $P_{AB}=\Sigma_\mu^A\Sigma_\nu^B P_{\mu\nu}$ where μ is centered at A and ν at B. The spin density over center A is defined as $\langle S\rangle_A=\langle S^\alpha\rangle_A-\langle S^\beta\rangle_A$, where $\langle S^\alpha\rangle_A=\Sigma_{\mu\nu}^A P^\alpha_{\mu\nu}$. We show in Table I the results for the bond order between the ligand and the N impurity P_{LN} and the spin density $\langle S\rangle$ over each distinct center: in T_d symmetry all ligands are equivalent, in C_{3v} we distinguish the ligand lying in the C_{3v} axis (A) from the three other ligands lying in a plane perpendicular to the axis (P).

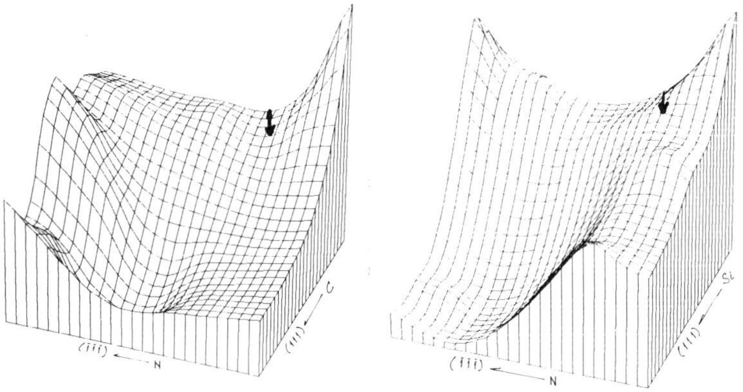

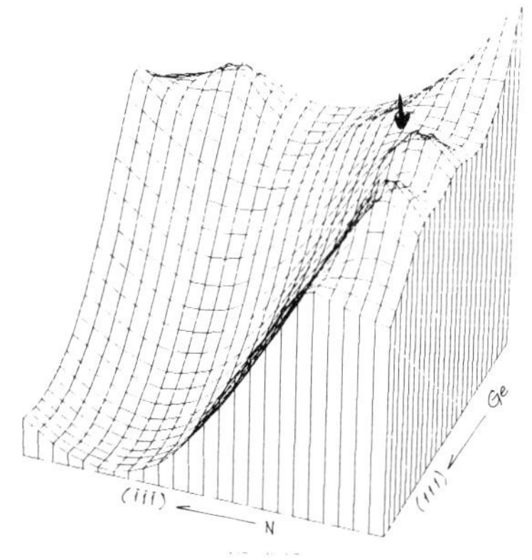

Fig.1-Total energy surface for coupled displacement of N in (a) diamond (b) silicon and (c) germanium. The arrow marks the T_d initial configuration

Table I: Results for bond order P and spin density <S> obtained in this work for the systems C:N, Si:N and Ge:N, in tetrahedral T_d symmetry and in the minimum-energy configuration, of C_{3v} symmetry.

	Substitutional(T_d)			Displaced(C_{3v})				
	$\langle S\rangle_L$	P_{LN}	$\langle S\rangle_N$	$\langle S\rangle_P$	P_{PN}	$\langle S\rangle_N$	P_{AN}	$\langle S\rangle_A$
C	0.3	0.1	0.1	~0.0	0.26	0.08	-0.52	0.12
Si	0.31	0.15	0.4	~0.0	0.29	~0.0	0.01	1.22
Ge	0.49	0.24	0.24	~0.0	0.36	0.01	0.06	1.84

To analyse this results we must consider that the lattice parameter of diamond is very small compared to Si or Ge, and also that the Ge valence orbitals (4s and 4p) are very delocalized compared to the lighter elements. In the T_d geometry the bonds are strenghtened down the column from C to Si to Ge, or, looking from another angle, we have higher overlap population in Ge-N bonds. In C_{3v} symmetry at the minimum-energy configuration we obtain similar trends as in T_d, for the bonds with the three closer ligand neighbors (P_{PN}), with higher overlap population compared to T_d, that is, now we obtain really an sp^2 hybridization. The striking result concerns the bond with the ligand on the axis (P_{AN}): the antibonding character of this bond is very clear for the diamond case, even if for Si and Ge if comes off as non-bonding, with overlap population almost null. The results for spin density indicates a very strong magnetic polarization at the ligand Ge site in the C_{3v} axis, with zero spin density at all other sites.

As said in the introduction, there are EPR [5,6] results for N in Si and Ge which provide detailed pictures of the centers. The usual analysis of experimental data approximates the wave-function ψ_e for the electron in the highest molecular orbital a_1 by a simple LCAO:

$$\psi_e = \Sigma_i \eta_i \ (\alpha_i \chi_{i,s} + \beta_i \chi_{i,p}) \tag{1}$$

where the sum is over valence s and p orbitals at atoms close to the impurity site. We can compare our theoretical results for the isotropic (A_j) and anisotropic (B_j) terms of the hyperfine interation

$$A_j = \frac{8\pi}{3} \ g_e \ \mu_B \mu_N \left(\frac{\mu_j}{I_j}\right) \ |\psi_e(0)|^2 \tag{2}$$

$$B_j = \frac{4}{3} \ \mu_B \mu_N \left(\frac{\mu_j}{I_j}\right) \ \langle r^{-3}_{j,e} \rangle \tag{3}$$

where μ_j is the nuclear magnetic moment of atom j, I_j is nuclear spin. We can write from equation (1)

$$|\psi_e(0)|^2 = |\chi_{j,s}(0)|^2 \ \eta_j^2 \ \alpha_j^2 \tag{4}$$

$$\langle r_e^{-3} \rangle = \langle r^{-3}_{j,p} \rangle \ \eta_j^2 \ \beta_j^2 \tag{5}$$

where now $|\chi(0)|^2$ and $\langle r^{-3}_{j,p} \rangle$ are taken from Hartree-Fock atomic calculations and the coefficients come from our results. We present then in Table II a comparison between our values for A_j and B_j, and results from EPR [5,6]. We may see that we obtain an overall agreement, with the exception of the values for the anisotropic term in Si (which, however, present the observed trend), which is very small and not easy to calculate with a small cluster calculation.

Table II: Isotropic and anisotropic hyperfine terms for N impurity in diamond and Si; all entries in units of $10^{-4}cm^{-4}$, obtained using values for isotopes ^{14}N, ^{13}C and ^{29}Si.

	DIAMOND				SILICON			
	C<111>		N		Si<111>		N	
	A	B	A	B	A	B	A	B
Expt.	68.95	22.17	30.37	3.63	95.7	18.4	13.10	1.00
our	63.7	12.5	16.1	3.9	114.1	5.3	3.7	0.3

There are no available experimental results for Ge:N, hence our results are not included in Table II: A= 12,1 x $10^{-4}cm^{-1}$.

Finally, we find, for the minimum-energy C_{3v} configuration, for all three systems, a level very close to the valence band related to the "lone-pair" N orbital, with strong p-character. This orbital is very localized, with strongest localization for the C:N case. This is another indication that the impurity in this C_{3v} configuration shows sp^2 hybridization.

References

[1] R.P.Messmer and G.D.Watkins, Phys.Rev.B7 , 2568 (1973)

[2] S.A.Kajihara, A.Antonelli, J.Bernholc and R.Car, Phys.Rev.Lett.66, 2010 (1991)

[3] P.A.Schultz and R.P.Messmer, Phys.Rev. B34 , 2532 (1986)

[4] H.P.Hjalmarson, D.R.Jennison and J.S.Binkley, Mat.Res.Soc.Symp.Proc.59, 553 (1986)

[5] a)J.H.N.Loubser and L.Du Preez, Brit.J.Appl.Phys.16, 457 (1965) b)R.J.Cook and D.H.Whiffen, Proc.Roy.Soc.(London) A295, 99(1986)

[6] K.L.Brower, Phys.Rev. B26, 6040(1982)

[7] T.H.Dunning Jr and P.J.Hay, in Modern Theoretical Chemistry Methods of Electronic Structure Theory edited by H.F.Schaefer III, (Plenum, NY - 1977) vol.3 , pp 1

[8] S.Huzinaga, J.Chem.Phys. 42, 1293 (1965)

[9] M.Dupuis, D.Splanger and J.J.Wendowski: NRCC Software Catalog vol.1 Program GG01 (GAMESS) 1980

Materials Science Forum Vol. 83-87 (1992) pp. 469-474

PRESSURE DEPENDENCE OF FORMATION AND MIGRATION ENTHALPIES FOR ATOMIC DIFFUSION IN SI: CONJUGATE GRADIENT MINIMIZATION OF TOTAL ENERGY

Osamu Sugino and Atsushi Oshiyama
Microelectronics Research Laboratories, NEC Corporation, 34 Miyukigaoka, Tsukuba 305, Japan

ABSTRACT

We report our parameter free total energy calculation on atomic diffusion in Si. In order to calculate not only formation but also migration enthalpies, we have implemented a highly efficient conjugate gradient total energy minimization method. This is efficient enough (1) to calculate electronic structures of defects using a very large cell containing up to 216 atoms, and (2) to determine atomic geometry allowing *all* the neighboring atoms fully relaxed using a large 64 site cell. As a result we can determine most favorable migration path among various possible paths, and calculate migration energy. Using this method we have calculated pressure dependence of the activation energies for group V impurities (As and Sb) in Si. We find that each mechanism exhibits its own different pressure dependence; with increasing pressure the activation energy decreases for the vacancy mechanism while it increases for the interstitial(cy) mechanism.

1. Introduction

Impurity diffusion phenomena have long been a main subject in semiconductor physics. Because of the technological importance, diffusion of dopant atoms in Si has been most extensively investigated. Recent studies show that several different mechanisms could contribute to the diffusion. The candidates are the vacancy, interstitial(cy), and concerted exchange mechanisms. So far many theoretical and experimental efforts have been made to determine predominant diffusion mechanism. Yet convincing proof has not been obtained. Fahey *et al.*[1] observed enhanced or retarded impurity diffusion upon defect injection. They were able to show, for several impurities, to what extent each mechanism contributes to the diffusivity with a help of several model assumptions. They found that their analysis leads to a contradiction, however, when applied to arsenic diffusion, i.e. their vacancy injection study and their interstitial Si injection study are inconsistent. Nichols *et al.*[2] formulated equations that relate the change in diffusivity with the concentration of the injected defects on the basis of rate equations that govern the reaction of defects. They showed that the calculated change in the diffusion constant and the activation energy are in agreement with an interstitial injection experiment. From this they concluded that the interstitial(cy) mechanism is dominant for diffusion of B, P, and As. Further comparison with experiment, such as vacancy injection experiment, might confirm their formalism.

Applying pressure is another candidate to resolve this problem because each mechanism could exhibit their own different pressure dependence. The advantage of this study is that the observed pressure coefficient of the activation enthalpy and activation energy can be directly compared with first principles calculation. The purpose of our investigation is to show how each mechanism depends on applied pressure by calculating not only formation but also migration enthalpies using our conjugate gradient total energy minimization method (CG method) (We don't consider the concerted exchange mechanism in this paper following the argument in ref. 2). The reason why we are able to calculate the migration enthalpies is our implementation

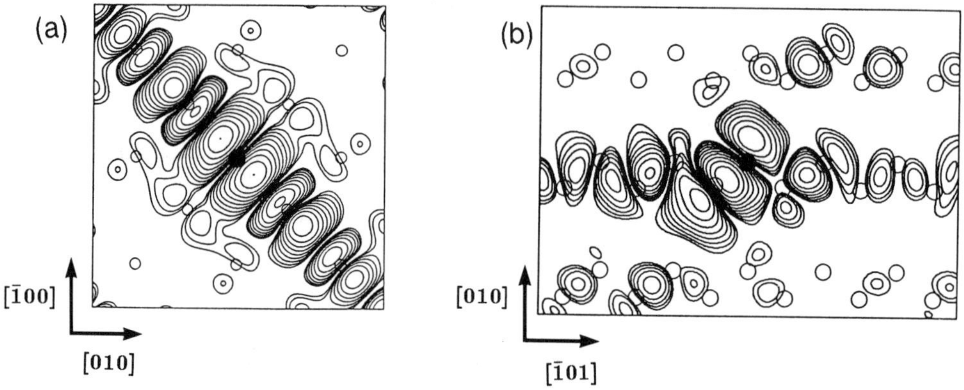

Fig. 1: Charge density contour plot of the gap state of V^-. Position of the Si atom is shown by open circles and the vacancy site is shown by solid circle.

of highly efficient CG method; owing to its efficiency we can determine migration path among various possible paths and thus can calculate migration energy. Such calculations have not been possible with the conventional method. We have investigated the As and Sb impurity diffusion and found that each mechanism exhibits its own different pressure dependence; with increasing pressure the activation energy decreases for the vacancy mechanism while it increases for the interstitial(cy) mechanism.

2. Method of calculation

The calculations are performed within the local density approximation (LDA), using norm-conserving pseudopotentials[3] and a plane-wave basis set. The pseudopotential for Si is approximated by the Kleinman-Bylander form[4], which includes s and p nonlocal terms. In order to prevent the so called "ghost"[5], however, the nonlocal terms for other elements, As and Sb, are strictly calculated using the Štich's method[6]. The electronic structure is obtained by the conjugated gradient (CG) total energy minimization method. The CG method is developed on the basis of the algorithm of Bylander *et al.*[7]. The equilibrium atomic configuration is obtained by alternately minimizing both the electronic and ionic degrees of freedom using the CG method.

3. Advantage of the conjugate gradient method

The conjugate gradient (CG) method is efficient enough (1) to calculate electronic structures of defects using a very large cell containing up to 216 atoms and (2) to determine atomic geometry allowing *all* the neighboring atoms fully relaxed using large 64 site cell. The advantage of this is illustrated in the following two calculations on vacancy in Si.

By using large 216 site cell, we can virtually isolate the point defects, and can obtain wavefunction which is free from the interference from the neighboring cell. In figure 1, we show the gap state wavefunction of the negatively charged vacancy (V^-) in Si. The electronic structure is calculated using the 216 site cell in which neighboring atoms have been relaxed by the independent 64 site-cell calculation: The fully optimized V^- has C_{2v} symmetry.

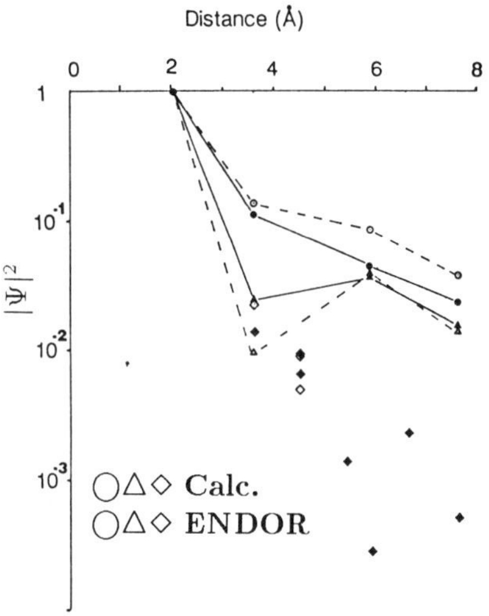

Fig. 2: Charge density at the nuclear site, $|\Psi|^2$, plotted against the distance from the vacancy. The connected data correspond to the zigzag chain of atoms.

Table 1:Migration energy of the vacancy in Si

	V^{2+}	V^{2-}
calc.[a]	0.40 eV	0.10 eV
exp.[b]	0.32 eV	0.18 eV

[a]Calculation was done using 32 site cell.
[b]Watkins[10].

The wavefunction extends in an oscillating manner along coplanar chains of atoms in {110} planes. This behavior is similar to the extension of the charge disturbance in Si crystal caused by a perturbation at the origin[8]. The obtained wavefunction is in good agreement with the experimental data of Sprenger *et al.*[9] as shown in Fig. 2; here the amplitude of the wavefunction at nuclear site is calculated and compared with the experiment. The assignment of the ENDOR data, which is different from the original one, has been done analyzing the anisotropy of the hyperfine tensor.

We can clarify characteristics of the ENDOR data by further analyzing the wavefunction. At most of the atoms located along the chain, the principal axis of the hyperfine tensor is parallel to the direction of the bond. This is because the wavefunction mainly consists of the sp^3 bonding orbital. While at the second nearest neighbor site, the sp^3 bonding orbital is significantly distorted by the large back bond orbital of the 1st nearest neighbor site (see Fig. 1b), which slants the direction of the principal axis. The distortion of the orbital means the enhancement of the p- and d- component of the atomic orbital and reduction of the s- component at this site. As a result the amplitude at the nuclear site is reduced. This is reflected in the dip at $d \simeq 3.5$Å in Fig. 2.

We have shown that the result of the LDA calculation is in good agreement with experiment. This seems to show that, while a simple one-electron model may fail as discussed by Watkins[10], the one-electron picture holds if the electron correlation is effectively included within LDA and the lattice relaxation is rigorously taken into account. So, although Anderson *et al.*[11] stressed the crucial role of the many-electron effect, we believe that the LDA accounts for essential part of the many-electron effect.

The second example is the calculation of migration energy of the vacancy in Si. So far the migration energy has not been calculated because (1) large cell calculation is necessary in order to take into account the long range lattice relaxation and (2) migration path has to be determined among various possible paths. But such calculation is possible with the CG method because we can efficiently calculate potential surfaces allowing *all* the neighboring atoms fully relaxed using large cell (up to 64 site cell). In Fig. 3 we show the obtained migration path and the corresponding potential surface. In this path one of the Si atoms goes nearly straight to the vacancy site. The calculated migration energy is consistent with experimental data within 0.1 eV for various charge states (Table. 1).

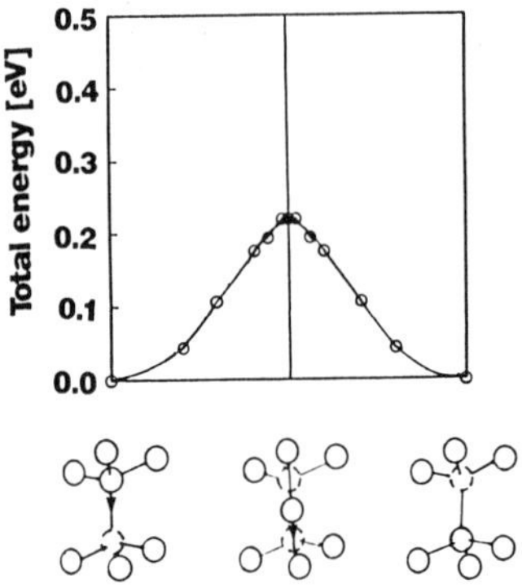

Fig. 3: Potential energy surface for the vacancy migration. Calculation has been done for V^{2+} using 64 site-cell.

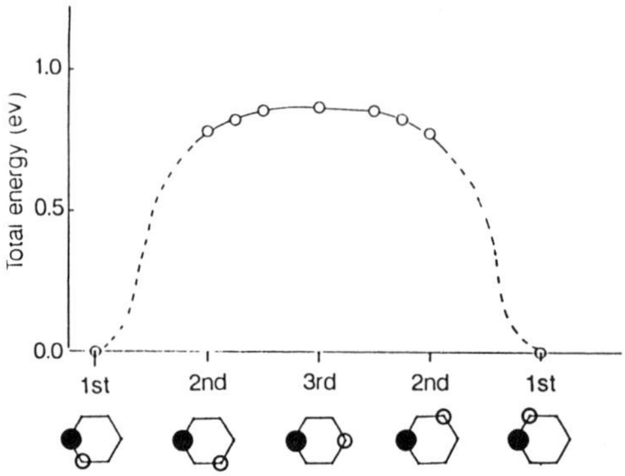

Fig. 4: Potential energy surface of the vacancy migration around the As atom. Solid and open circle correspond to the As atom and the vacancy respectively. Calculation is done using 64 site-cell.

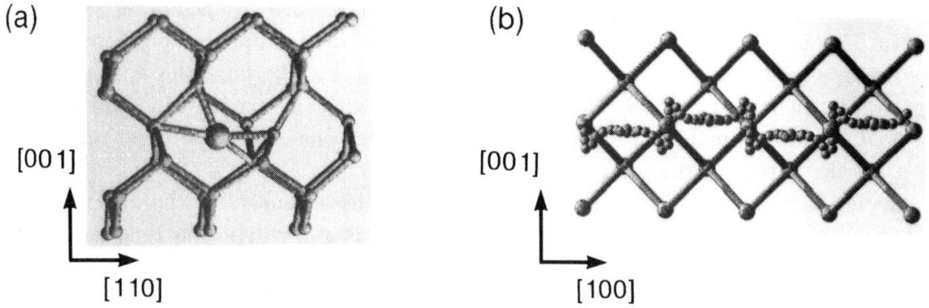

Fig. 5: (a) The most stable As interstitial(cy) structure; larger circle corresponds to the As atom. (b) Snap shot of the migration path of As interstitial(cy); relaxation of the Si atoms are not indicated in the figure.

We now investigate the pressure dependence of the migration energy not only because this is interesting in itself but also this is important for an understanding of the impurity diffusion. The migration energy is found to show a linear dependence on the applied pressure and it disappears at $P \simeq 60$kbar. This means that the vacancy can migrate without surmounting a barrier. The pressure dependence can be intuitively understood as follows. Since the coordination number of the migrating atom is six near the saddle point (see Fig. 3), the gain in bonding energy becomes large here when the neighboring atoms come closer by the applied pressure. This reduces the total energy difference between the vacancy at the saddle point and the vacancy at the lattice site. Thus the migration energy is reduced.

4. Impurity diffusion in Si and its pressure dependence

In this section we investigate the impurity diffusion of group V impurities, As and Sb, in Si. First, migration path and migration energy of arsenic are calculated separately for each diffusion mechanism, i.e. the vacancy and interstitial(cy) mechanism.

In the vacancy mechanism the arsenic first traps a vacancy and then the arsenic–vacancy pair migrates. The pair migration process consists of two steps. First, the vacancy migrates around the arsenic atom from a nearest neighbor site to another nearest neighbor site. Then the vacancy is exchanged with the arsenic atom. In Fig. 4 we show the calculated potential energy surface for the vacancy migration. The potential surface is smooth and at the saddle point vacancy occupies the third nearest neighbor site. This is qualitatively different from the one speculated so far in which it exhibits local minimum every time when the vacancy occupies lattice site and local maximum somewhere in between. The disappearance of the humps in the calculated potential surface can be explained when we notice that neighboring Si atoms are suffered from the local strain by the large arsenic atom and that compressive pressure decreases the migration energy of vacancy (see §3). Due to the local strain, Si atoms relax considerably toward the vacancy. In this situation the six coordinated structure, which is realized when the vacancy is located between the lattice sites, is energetically favored. This reduces the energy barrier height experienced when the vacancy migrates from site to site.

In the interstitial(cy) mechanism the arsenic atom is kicked out and migrates. Figure 5a is the most stable interstitial(cy) structure we obtained. In this structure the arsenic atom is located between the bond center site and the hexagonal site; one Si-Si bond nearest to the arsenic atom is cut and the two Si atoms that lost a bond are pull back into the back bond direction forming a local planar structure. Figure 5b is one of the migration paths with minimum migration energy, in which arsenic atom spirally migrates successively cutting nearby Si-Si bond. The corresponding migration energy is 0.3 eV.

Adding the formation energy to the migration energy we obtained the activation energy of the arsenic diffusion; it is 3.2 eV for the vacancy mechanism and 3.46 eV for the interstitial(cy) mechanism. The energy difference is not large enough to obtain conclusive remarks about the diffusion mechanism.

Next we investigate the pressure dependence of the activation energy for both mechanism. We find that each mechanism shows qualitatively different pressure dependence; the activation energy for the vacancy mechanism decrease by 0.56 eV at P=60 kbar ($\Delta E/\Delta P = -0.93 \times 10^{-2}$ eV/kbar) while that for the interstitial(cy) mechanism increase by 0.72 eV at that pressure ($\Delta E/\Delta P = +1.2 \times 10^{-2}$ eV/kbar). Only for the vacancy mechanism the amount of the decrease is consistent with the experimental data by Nygren *et al.*[12] ($\Delta E/\Delta P = -3.0 \pm 1.3 \times 10^{-2}$ eV/kbar). The calculated activation volume for the vacancy mechanism, -5 Å^3, is also consistent with the experimental data, -9 ± 4 Å^3. This is suggestive of the predominance of the vacancy mechanism.

We have done a parallel calculation for the Sb diffusion. The activation energy at zero pressure is 3.13 eV for the vacancy mechanism and it is 4.1 eV for the interstitial(cy) mechanism. The activation volume is -6Å^3. The pressure dependence of the activation energy for the vacancy mechanism is similar to that for arsenic, $\Delta E/\Delta P = -1.2 \times 10^{-2}$ eV/kbar.

Finally we mention that the sign of the activation volume is not a criterion to determine the diffusion mechanism. Nygren *et al.* discussed that the negative activation volume is an evidence for the interstitial(cy) character of the arsenic diffusion. But this is not true. Since the Si atoms around the vacancy near the arsenic atom is suffered from the local compressive strain, large inward relaxation (0.4 eV) is realized; this reduces the volume around the defect considerably (25Å^3). The activation volume, which is equal to the formation volume of the unrelaxed vacancy (20Å^3) minus the above mentioned relaxation volume, is thus negative.

5. Conclusion

We have implemented a highly efficient conjugate gradient method. This is so efficient that we are able to (1) calculate electronic structures using a large cell containing up to 216 atoms and to (2) calculate migration path and migration energy using 64 site cell, which are impossible with the conventional method. Using the CG method we calculated pressure dependence of the activation energies for group V impurities (As and Sb) in Si. We find that each mechanism exhibits its own different pressure dependence; with increasing pressure the activation energy decreases for the vacancy mechanism while it increases for the interstitial(cy) mechanism.

REFERENCES

1. P. Fahey, G. Barbuscia, M. Moslehi, and R. W. Dutton, Appl. Phys. Lett. **46** 784 (1985); P. Fahey, P. B. Griffin, and J. D. Plummer, Rev. Mod. Phys. **61** 289 (1989).
2. C. S. Nichols, C. G. Van de Walle, and S. T. Pantelides, Phys. Rev. Lett. **62** 1049 (1989); Phys. Rev. **B40** 5484 (1989).
3. G. B. Bachelet, D. R. Hamann, and M. Schlüter, Phys. Rev. **B26**, 4199 (1982).
4. L. Kleinman and D. M. Bylander, Phys. Rev. Lett. **48**, 1425 (1982).
5. X. Gonze, P. Käckell, and M. Scheffler, Phys. Rev. B **41** 12264 (1990).
6. I. Štich, R. Car, M. Parrinello, and S. Baroni, Phys. Rev. B **39**, 4997 (1989).
7. D. M. Bylander, L. Kleinman and S. Lee, Phys. Rev. B **42** 1394 (1990).
8. E. O. Kane, Phys. Rev. B **31** 5199 (1985).
9. M. Sprenger, S. H. Muller, and C. A. J. Ammerlaan, Physica 116B, 224 (1983); M. Sprenger, S. H. Muller, E. G. Sieverts, and C. A. J. Ammerlaan, Phys. Rev. B *35* 1566 (1987).
10. G. D. Watkins, in *Lattice Defects in Semiconductors, 1974*, Institute of Physics Conference Series No. 23, edited by F. A. Huntley (Institute of Physics, London), p. 1.
11. F. G. Anderson and F. S. Ham; the following paper.
12. E. Nygren, M. J. Aziz, D. Turnbull, J. M. Poate, D. C. Jacobson, and R. Hull, Appl. Phys. Lett. **47** 105 (1985).

Materials Science Forum Vol. 83-87 (1992) pp. 475-480

MANY-ELECTRON EFFECTS IN THE NEGATIVE SILICON VACANCY

FREDERICK G. ANDERSON,[1] FRANK S. HAM,[1] and GUNTER GROSSMANN[2]
[1]Physics Department and Sherman Fairchild Laboratory,
Lehigh University, Bethlehem, PA 18015, USA
[2]Department of Solid State Physics
University of Lund, Box 118, S-221 00 Lund, Sweden

ABSTRACT

A reinvestigation of Watkins's LCAO-MO model for the negative silicon vacancy, taking into account linear Jahn-Teller coupling of the one-electron t_2 orbitals, reveals that this model predicts the form of the Jahn-Teller distortion experimentally observed only if many-electron effects (Coulomb interaction) are taken explicitly into account, and then only if Jahn-Teller coupling to τ_2 modes exceeds that to ε modes.

1. Introduction

Watkins's one-electron LCAO-MO (linear combination of atomic orbitals-molecular orbital) model for the vacancy in silicon has been very successful in describing his electron paramagnetic resonance (EPR) spectra.[1] Theorists have for years debated the importance of many-electron effects (Coulomb interaction) for this center.[1-6] These considerations have not led to consensus, especially in the case of the negatively charged vacancy (V^-). The first theoretical prediction for V^-, given by Coulson and Kearsley,[2] was that of a high-spin (S=3/2) ground state, in which case many-electron effects must dominate. Watkins's model, by contrast, takes the Jahn-Teller (JT) coupling to be dominant. Supporting this view, Lannoo[6] has concluded from an explicit consideration of configuration interaction that the ground state for V^- may be correctly obtained in a one-electron approximation. (Lannoo has nevertheless shown that configuration interaction plays a small but significant role within the ground state by modifying the wave function.) In this work, we demonstrate that the EPR results cannot be explained in terms of a one-electron model with linear JT coupling, i.e. that many-electron effects are essential to account for the form of JT distortion found to occur. We show for the case of V^- that the JT coupling to τ_2 modes of distortion must be stronger than the coupling to ε modes. Finally, we determine bounds on the strength of the Coulomb interaction relative to the JT energies.

2. One-Electron Model

Watkins's LCAO-MO model for V^- is reviewed in Fig. 1. Dangling bonds on the four nearest Si neighbors form molecular orbitals belonging to irreducible representations (in T_d symmetry) a_1 and t_2. (We denote by $t_{2\xi}$, $t_{2\eta}$, $t_{2\zeta}$ the t_2 orbitals transforming as yz, zx, xy, respectively.) With the a_1 state doubly occupied and three electrons in the t_2 orbitals (Fig. 1a), the system's energy is lowered and its degeneracy partially removed by a strong tetragonal JT distortion Q_θ (resulting in a defect of D_{2d} symmetry, Fig. 1b) that places two electrons in the lower singlet $t_{2\zeta}$ and one in the higher doublet ($t_{2\xi}, t_{2\eta}$). This remaining orbital degeneracy is removed by a second, weaker (trigonal) JT distortion Q_ζ that splits this doublet, leading to C_{2v} symmetry (Fig. 1c). The Watkins model proposed this simple sequence of JT distortions in a one-electron description in order to account for the spontaneous distortion to C_{2v} symmetry found by EPR to occur for V^-.

The one-electron linear JT coupling of the t_2 states to both ε distortion modes (Q_θ, Q_ε) and τ_2 modes (Q_ξ, Q_η, Q_ζ) is described (in the basis $t_{2\xi}$, $t_{2\eta}$, $t_{2\zeta}$) by

$$H_{JT} = \begin{bmatrix} V_E(Q_\theta - \sqrt{3}Q_\varepsilon)/2 & -V_T Q_\zeta & -V_T Q_\eta \\ -V_T Q_\zeta & V_E(Q_\theta + \sqrt{3}Q_\varepsilon)/2 & -V_T Q_\xi \\ -V_T Q_\eta & -V_T Q_\xi & -V_E Q_\theta \end{bmatrix} , \quad (1)$$

where $V_E(V_T)$ is the one-electron JT coupling coefficient for $\varepsilon(\tau_2)$ modes. Including the elastic energy of these modes,

$$H_{el} = \frac{1}{2} K_E (Q_\theta^2 + Q_\varepsilon^2) + \frac{1}{2} K_T (Q_\xi^2 + Q_\eta^2 + Q_\zeta^2) \quad , \quad (2)$$

where K_E and K_T are force constants, we have for the minimum-energy configuration of the form (Q_θ, Q_ζ) of C_{2v} symmetry, as indicated in Fig. 1c, for three electrons in the t_2 orbitals, a JT stabilization energy (relative to the tetrahedral configuration)

$$E_{JT}(Q_\theta, Q_\zeta) = \frac{9}{4} E_{JT}(\varepsilon) + \frac{3}{4} E_{JT}(\tau_2) \ . \quad (3)$$

Here, $E_{JT}(\varepsilon) = V_E^2/2K_E$ and $E_{JT}(\tau_2) = 2V_T^2/3K_T$ are the one-electron JT energies for a t_2 orbital and ε and τ_2 modes, respectively.

The mode Q_ζ, which splits the doublet ($t_{2\xi}, t_{2\eta}$) into singlets $(t_{2\xi} \pm t_{2\eta})/\sqrt{2}$, is not the only mode that can lift the remaining orbital degeneracy of the tetragonally distorted configuration of Fig. 1b. A nonzero value of Q_ε splits the doublet into the states $t_{2\xi}$ and $t_{2\eta}$ according to Eq. (1) and would yield a configuration of D_2 symmetry in which the principal axes of the defect coincide with the crystal's cubic axes. The minimum energy configuration of this type (Q_θ, Q_ε) has a JT

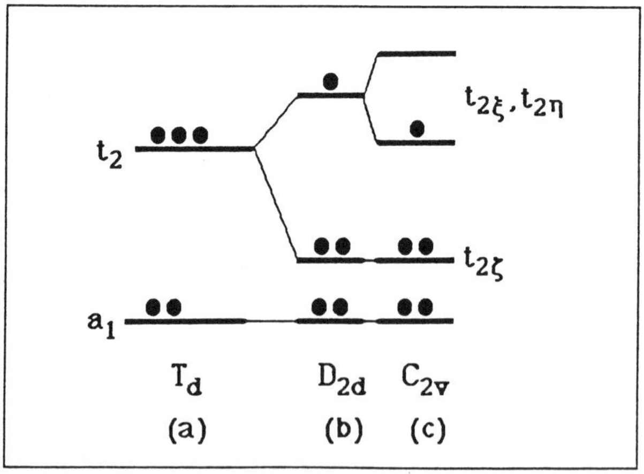

FIG. 1. Splitting of the t_2 manifold in going from (a) T_d to (b) D_{2d} symmetry, and finally to (c) C_{2v} symmetry.

energy

$$E_{JT}(Q_\theta,Q_\varepsilon) = 3E_{JT}(\varepsilon) \quad . \tag{4}$$

Comparing with Eq. (3), we note that this (Q_θ,Q_ε) configuration has a larger JT energy than does the (Q_θ,Q_ζ) configuration if the ε-mode coupling exceeds the τ_2-mode coupling, as measured by $E_{JT}(\varepsilon)$ and $E_{JT}(\tau_2)$, respectively. Under the conditions assumed in the Watkins model, for which the ε-mode coupling is the stronger, the (Q_θ,Q_ε) distortion should therefore be more stable than (Q_θ,Q_ζ). Occurrence of the (Q_θ,Q_ζ) distortion evidently must instead require having

$$E_{JT}(\tau_2) > E_{JT}(\varepsilon) \quad . \tag{5}$$

But in this case there is another form of distortion in which the JT energy exceeds that of Eq. (3), so that configurations of the form (Q_θ,Q_ζ) would not then have the lowest possible energy and thus could not be stable. To see this we assume that only Q_ζ is nonzero, place two electrons in one of the orbitals $(t_{2\xi} \pm t_{2\eta})/\sqrt{2}$ and one in $t_{2\zeta}$, and obtain for the resulting JT energy

$$E_{JT}(Q_\zeta) = 3E_{JT}(\tau_2) \quad . \tag{6}$$

[In fact, there is a continuous path in the space of the τ_2 modes Q_ξ, Q_η, Q_ζ along which the JT energy is given by Eq. (6).] Accordingly, we must conclude that in the one-electron model a distortion of the form (Q_θ,Q_ζ) should never represent a stable configuration, whichever type of JT coupling is the stronger.

3. Many-Electron Energy Levels

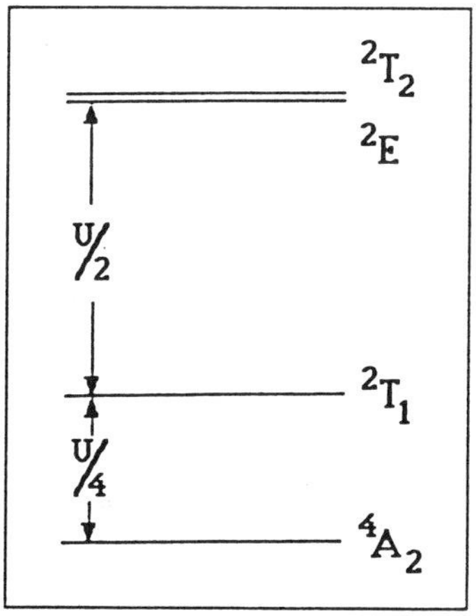

FIG. 2. The ordering of the many-electron states for the $a_1^2t_2^3$ configuration of V^-.

Many-electron effects for V^- in the ground-state configuration $a_1^2t_2^3$ can be handled in just the same way as in standard treatments of the t_2^3 configuration of a transition-metal ion in a cubic host. (See Ref. 7 for energy levels and tables of wave functions.) The many-electron terms in tetrahedral symmetry are $^4A_2$, ^2E, $^2T_1$, and $^2T_2$, with term energies that satisfy the general inequalities $E(^2T_2) > E(^2T_1) > E(^4A_2)$ and $E(^2T_2) > E(^2E) > E(^4A_2)$. These relationships fix $^4A_2$ as the lowest term in general and $^2T_2$ as the highest but do not determine the sequence of $^2T_1$ and ^2E. However, as Lannoo and Bourgoin have shown, with t_2 orbitals formed from dangling bonds that only weakly overlap, the $^2T_2$ and ^2E terms should approximately coincide,[8] as shown in Fig. 2. These two then lie above the $^2T_1$ term by an energy U/2 (in the notation of Ref. 8), where U approximates the direct Coulomb energy difference between charge distributions given by two electrons in (a) a single dangling bond and (b) adjacent dangling bonds.[8]

4. Effect of Jahn-Teller Coupling Among Many-Electron States

The coupling due to the one-electron JT interaction given in Eq. (1) may be shown to be zero within each of the four terms of the $a_1^2t_2^3$ configuration. The JT coupling among the many-electron terms of V^- is thus purely of a pseudo-JT type.[9] (Indeed, this very problem served to motivate the pioneering analysis by Stoneham and Lannoo[9] of such interlevel pseudo-JT coupling.) The ε-mode coupling is only between $^2T_1$ and $^2T_2$ and by symmetry cannot couple these triplets to ^2E; τ_2-mode coupling is between $^2T_2$ and both $^2T_1$ and ^2E (but not between $^2T_1$ and ^2E).

For ε-mode coupling alone, the work of Stoneham and Lannoo[9] shows that we obtain a minimum-energy configuration of the form $(Q_\theta, Q_\varepsilon)$ with energy (relative to the undistorted $^2T_1$ term energy)

$$E(Q_\theta, Q_\varepsilon) = -(1-y_1)^2 E_{JT}(Q_\theta, Q_\varepsilon) , \qquad (7)$$

if we have $y_1 = U/8E_{JT}(Q_\theta, Q_\varepsilon) < 1$, where $E_{JT}(Q_\theta, Q_\varepsilon)$ is given by Eq. (4). On the other hand, with the coupling between $^2T_1$ and $^2T_2$ involving both ε and τ_2 modes, we may have a minimum of the (Q_θ, Q_ζ) type with energy

$$E(Q_\theta, Q_\zeta) = -(1-y_2)^2 E_{JT}(Q_\theta, Q_\zeta) , \qquad (8)$$

if $y_2 = U/8E_{JT}(Q_\theta, Q_\zeta) < 1$, $E_{JT}(Q_\theta, Q_\zeta)$ being given by Eq. (3). The condition that the energy of Eq. (8) be below that of Eq. (7), so that the (Q_θ, Q_ζ) configuration represents the true minimum rather than a saddlepoint, is simply $E_{JT}(Q_\theta, Q_\zeta) > E_{JT}(Q_\theta, Q_\varepsilon)$, requiring that the inequality in Eq. (5) be satisfied. When the inequality of Eq. (5) is reversed, the minimum of the $(Q_\theta, Q_\varepsilon)$ type is instead lower.

A distortion of the pure Q_ζ type with JT energy given by Eq. (6) results from coupling between the ^2E and $^2T_2$ terms, rather than from coupling between $^2T_1$ and $^2T_2$. The energy minimum of this type is

$$E(Q_\zeta) = \frac{U}{2} - E_{JT}(Q_\zeta) , \qquad (9)$$

where $E_{JT}(Q_\zeta)$ is given by Eq. (6). This energy is no longer necessarily below that of Eq. (8) when the τ_2-mode coupling dominates, if U is large enough. The condition that the minimum of (Q_θ, Q_ζ) type be below that of the pure Q_ζ type is then

$$U > 9[E_{JT}(\tau_2) - E_{JT}(\varepsilon)]. \qquad (10)$$

This inequality therefore provides a lower bound for U if the minimum is to be of the (Q_θ, Q_ζ) type. We require, of course, in this case that the inequality of Eq. (5) be satisfied (making the right-hand side of Eq. (10) positive), so that a $(Q_\theta, Q_\varepsilon)$ configuration does not lie at a lower energy than the (Q_θ, Q_ζ) minimum.

Since the $^4A_2$ term is below $^2T_1$ by U/4 in the absence of a JT distortion (Fig. 2), the occurrence of an absolute minimum of (Q_θ, Q_ζ) type, arising from the states with S=1/2, requires that the JT coupling shall be strong enough to place the energy given by Eq. (8) below -U/4. This requirement imposes the further condition

$$U < \frac{1}{2}[9E_{JT}(\varepsilon) + 3E_{JT}(\tau_2)] , \qquad (11)$$

which provides an upper bound for U relative to the strength of the JT coupling.

5. Discussion

We have shown that the linearly coupled one-electron model for V^- should never lead to the (Q_θ, Q_ζ) type of JT distortion, which is observed in EPR, as the minimum-energy configuration. Inclusion of many-electron effects is essential to allow this configuration to have an energy lower than that of competing forms of JT distortion. A lower bound (Eq. 10) for the Coulomb energy U is therefore established by the EPR results in order that the energy of the ^2E term shall be sufficiently far above that of $^2T_1$. An upper bound (Eq. 11) also exists, from the requirement that the energy gain from the JT distortion shall exceed the energy difference U/4 between $^2T_1$ and $^4A_2$, the latter always being the lowest term of the $a_1^2t_2^3$ configuration in T_d symmetry.

The occurrence of the (Q_θ, Q_ζ) distortion also leads to the conclusion that for V^- the JT coupling to τ_2 modes exceeds that to ε modes, as measured by the inequality (5). This result is surprising because the opposite conclusion is unequivocally shown by EPR for the singly positive vacancy V^+, which has a single t_2 electron and reveals a tetragonal distortion. Were the coupling to τ_2 modes stronger for V^+, EPR would have revealed C_{3v} symmetry rather than D_{2d}. We are forced to conclude that the relative strengths of these couplings reverse between V^+ and V^- (more specifically between V^0 and V^- since V^0 can be observed indirectly to have a tetragonal JT distortion).

A value for the difference $[E_{JT}(\tau_2) - E_{JT}(\varepsilon)]$ can be determined from experiment. V^- reorients at T~20 K between the distortions $(Q_\theta, \pm Q_\zeta)$, while the direction of the tetragonal axis remains fixed. The two configurations $(Q_\theta, \pm Q_\varepsilon)$, which have the same orientation for the tetragonal axis as that for $(Q_\theta, \pm Q_\zeta)$, should provide saddlepoints over which this reorientation passes. Thus, the difference between Eqs. (7) and (8) gives the activation energy for this process,

$$\Delta E_{act} = E(Q_\theta, Q_\varepsilon) - E(Q_\theta, Q_\zeta) \cong \frac{3}{4}[E_{JT}(\tau_2) - E_{JT}(\varepsilon)].$$

Watkins measured this activation energy to be 0.008 eV, which indicates nearly equal JT coupling to τ_2 and ε modes. Using Eq. (10), we estimate the lower bound on U to be U > 0.1 eV.

Our work can be extended to several transition-metal defects in silicon. The Ni^-, Pd^-, and Pt^- impurities each have three electrons in t_2 orbitals that are quite vacancy-like in character, according to Watkins's Vacancy Model for these defects.[10,11] In addition, they possess the same C_{2v} symmetry as V^-, also the result of a mixed (Q_θ, Q_ζ) type of distortion. Our results suggest that many-electron effects play an essential role in these transition-metal centers also.

6. Acknowledgment

The portion of this work done at Lehigh University was supported by U.S. Office of Naval Research Grant No. N00014-90-J-1264.

REFERENCES

[1]For a review of Watkins's experimental results and of the theoretical results for the vacancy see G. D. Watkins, in Deep Centers in Semiconductors, edited by S.T. Pantelides (Gordon and Breach, 1986), p. 147.

[2]C.A. Coulson and M.J. Kearsley, Proc. R. Soc., London A241, 433 (1957).

[3]F.P. Larkins, J. Phys. Chem. Solids 32, 965 (1971).

[4]G.D. Watkins and R.P. Messmer, Phys. Rev. Lett. 32, 1244 (1974).

[5]M. Lannoo, G. A. Baraff, and M. Schlüter, Phys. Rev. B 24, 943 (1981); ibid B 24, 955 (1981).

[6]M. Lannoo, Phys. Rev. B 28, 2403 (1983).

[7]S. Sugano, Y. Tanabe, and H. Kamimura, Multiplets of Transition-Metal Ions in Crystals (Academic Press, New York 1970), pp. 66-71, 81-82.

[8]M. Lannoo and J. Bourgoin, Point Defects in Semiconductors I: Theoretical Aspects (Springer-Verlag, Berlin, 1981), pp. 141-154.

[9]A.M. Stoneham and M. Lannoo, J. Phys. Chem. Solids 30, 1769 (1969).

[10]G.D. Watkins, Physica 117B & 118B, 9 (1983).

[11]F.G. Anderson, F.S. Ham and G.D. Watkins, submitted to Phys. Rev. B.

Materials Science Forum Vol. 83-87 (1992) pp. 481-486

DEFECTS AND BANDTAILS IN AMORPHOUS SILICON

by

David A. Drabold[1], P. A. Fedders[2], Stefan Klemm[3] and Otto F. Sankey[4]

[1] Department of Materials Science and Engineering, University of Illinois, Urbana, IL 61801. [2] Physics Department, Washington University, St. Louis, MO 63130. [3] Minnesota Supercomputer Center, Inc., 1200 Washington Ave. South, Minneapolis, MN 55415. [4] Physics Department, Arizona State University, Tempe, AZ 85287.

ABSTRACT

Ab Initio local-orbital molecular dynamics is used to study the zero and finite temperature properties of the best available models of a-Si including the 216 atom cell of Wooten, Weaire and Winer. We show that the existence of coordination defects is neither a necessary or sufficient condition for the existence of localized gap states. We discuss the importance of using accurate interatomic forces in simulations. We connect our calculations to photoemission experiments, and describe a step toward understanding the Staebler-Wronski effect.

INTRODUCTION

Recently developed density functional/molecular dynamics (MD) schemes have made it possible to compute interatomic forces in covalent materials directly from the electronic structure[1,2]. We have shown[3] that this level of accuracy is *essential* to properly describing the structure and dynamics of a-Si. We have even found that simplified models of the electronic structure (such as certain empirical tight-binding formulations) are inadequate for many purposes[4]. In short, we believe that in systems with varied bonding environments, energetics and dynamics must be extracted from accurate electronic structure calculations.

In this paper, we report the application of *ab initio* MD to static and dynamic properties of a-Si. We describe defect conformations, dynamic effects on bandtailing, and simulations involving defects with different charge states.

DEFECTS IN a-SILICON

Wooten Weaire and Winer[5] (WWW) have proposed a 216 atom model structure for a-Si with many desirable properties including a satisfactory radial distribution function and no coordination defects. Not surprisingly, this cell has become almost a "standard" used in theoretical investigations of a-Si. We have studied the WWW structure, and find that the cell is near an energetic minimum according to our MD simulations. However, we find that the relaxed cell has several localized states in the gap. Whether these should be categorized as true "gap states" or "bandtail states" is not altogether clear. No universally accepted criterion exists to differentiate between the two. A quantitative measure of the localization is given in Fig. 1, where we plot inverse participation ratio vs. eigenvalue for states near the gap. At least nine states show appreciable localization. On the other hand, there are only two coordination defects, where coordination is determined by a radius of $R_C = 2.74Å$. This leads us to the conclusion that a coordination criterion is a very incomplete way to characterize defects in a-Si. The number of coordination defects is not very sensitive to R_C. The coordinates of the relaxed WWW cell are available upon request[6].

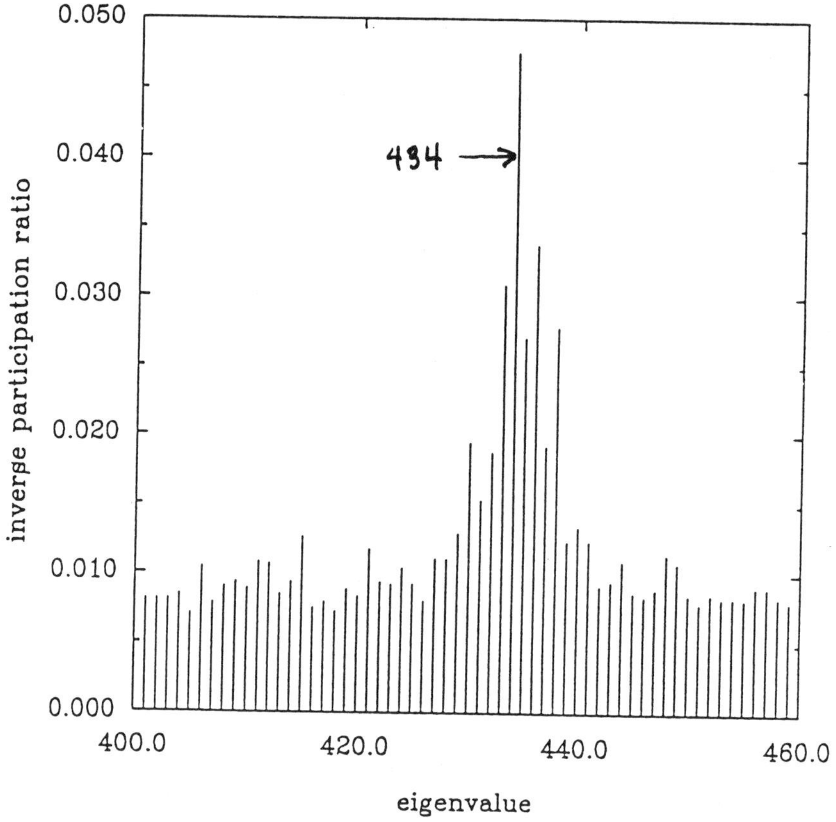

FIGURE 1. Density-functional Inverse Participation Ratios plotted against eigenvalue index for WWW-216 cell. Eigenvalue 432 is the highest occupied state. States near the gap are generally more localized than states deep in the bands. The conduction band edge states are significantly more localized than the valence band edge states.

We have recently proposed a recipe which yields 63 atom model structures of a-Si in agreement with experiment[7]. The vibrational power spectrum is in striking agreement with experiment, and the radial distribution function is in reasonable agreement. The coordination defect concentration is about 3%. A variation of the usual "rapid quenching" method from a highly disordered state was used to obtain this structure. While any reasonable measure of coordination suggests that this cell has two defects. we find that there are *three* localized states in the gap. Simulations at moderate kinetic temperature ($\approx 300K$) show that the number of *instantaneous* coordination defects fluctuate in an apparently random fashion, from two to six. Comparable effects are also observed in the WWW structure. In Fig. 2 we illustrate the three defect conformations giving rise to localized states.

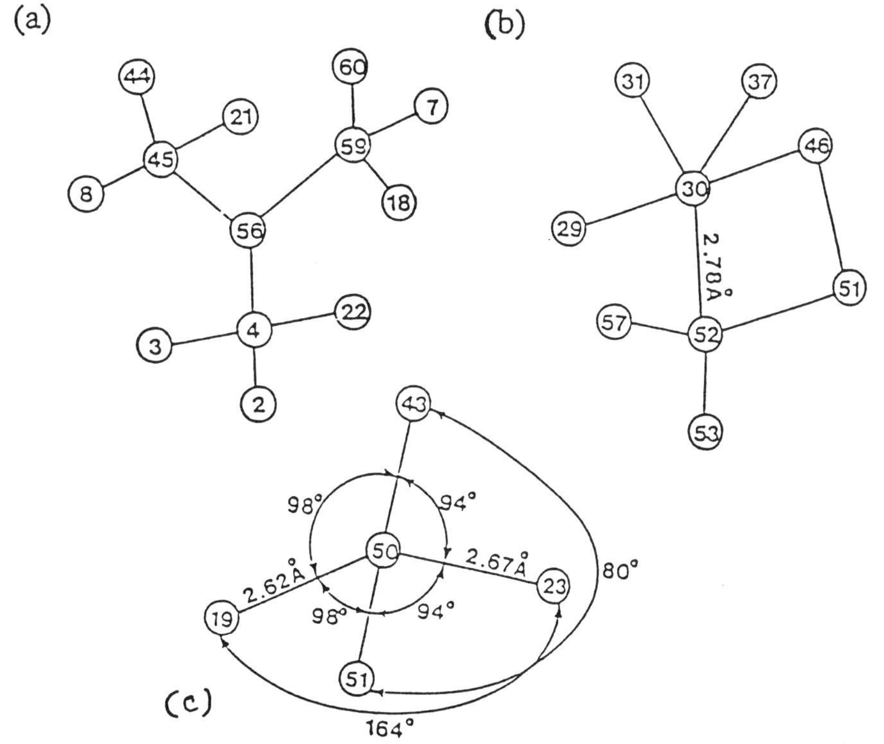

FIGURE 2. Defect conformations for the 63 atom cell. Defect (a) is a simple dangling bond. (b) is either three- or five- fold depending on R_C. (c) is a badly strained, weakly bonded four fold atom that yields a localized level in the gap.

An important *caveat* about this and all other calculations on small cells must be raised. Device-quality a-Si has a low defect concentration (about .1%). The best extant structures (in our opinion, the two we discuss here) have substantially more. Many simulations are performed on model structures with more than 15% defects! It is unlikely that the latter have much in common with "real" a-Si. Even the few percent defect cells we discuss have difficulties: in both the relaxed WWW cell and the 63 atom cell, defect wavefunctions "mix", and it is possible that this mixing affects the geometrical structure of the defects. This is primarily a finite-size effect, and provides us with an incentive to study larger systems. We are also including H in our simulations at present, and expect the situation to be at least partly ameliorated by H defect-passivation. Some related conclusions about finite-size effects have been given by Hollender[8].

THERMAL EFFECTS AND BANDTAILING

Aljishi and coworkers[9] have recently observed interesting and unexpected temperature dependence on bandtailing in a-Si and a-Si:H. In brief, they found that the width of the valence band tail was essentially temperature (T) independent, while the conduction band tail broadened linearly with T. By monitoring the fluctuation of eigenvalues near the gap at a given temperature, we find excellent agreement between simulation and experiment for the conduction tails, and we find that the dynamical component of the valence band tailing is insufficient to explain experiment, implying a fundamental distinction between the valence and conduction tails. The former is both structurally and dynamically broadened; the latter is broadened predominantly by dynamical effects. A much fuller account of this is given elsewhere[10].

STAEBLER-WRONSKI

We have just completed a crude simulation of the dynamical/structural effects of light-induced defects. This work is preliminary and includes no H, and should be treated accordingly. It is nevertheless suggestive.

In the simulations discussed earlier, we assumed that all valence states were occupied. Here, we started with the relaxed 216 atom WWW cell and removed an electron from the highest occupied state. This is a model for the ejection of an electron by light. As this state was a localized dangling bond state, the change in charge state led to substantial local rearrangements. As time passed, the rearrangement became increasingly global and ultimately had some effect on most of the atoms in our 63 or 216 atom cells. Invariably, the rearrangements led to *more* defects. After waiting 100-500 fs, we put the electron back (as with recombination), and let the system equilibrate. The new equilibrium always is found to have more defects than the starting cell. This is consistent with the well known irreversibility of the photodegradation associated with the Staebler-Wronski effect.

It is vital that the electron be ejected from a localized state. If removed from an extended valence state, the forces would change minimally and no major rearrangement would occur. In this simulation there was no H present, though future versions will. Clearly, the inclusion of H is important to a thorough understanding of SW. It is known, for example that the amount of photodegradation is related to the H concentration and also changes if another bond passivator (such as Cl) is employed. We will report these findings elsewhere.

ACKNOWLEDGEMENTS

This work was supported in part by the National Science Foundation under Grant DMR-88-01260, and the Office of Naval Research under Grant N00014-90-J-1304.

REFERENCES

1. R. Car and M. Parrinello, Phys. Rev. Lett. **55** 2471 (1985).
2. O. F. Sankey and D. J. Niklewski, Phys. Rev. B **40**, 3979 (1989).
3. D. A. Drabold, P. A .Fedders, A. E. Carlsson, O. F. Sankey and J. D. Dow, Phys. Rev. B **42** 5345 (1990).
4. P. A. Fedders, D. A. Drabold and S. Klemm, submitted to Phys. Rev. B.
5. F. Wooten and D. Winer, *Solid State Physics: Advances in Research and Applications* **40**, H. Ehrenreich and D. Turnbull, eds., Academic, New York, 1987, p.2.
6. email requests may be directed to klemm@msc.edu.
7. D. A. Drabold, P. A. Fedders and O. F. Sankey, Phys. Rev. B **42** 5135 (1990).
8. J. Hollender et al., J. Phys. C **3** 1947 (1991).
9. S. Aljishi, D. Cohen, S. Jin and L. Ley, Phys. Rev. Lett., **64** 2811 (1990).
10. D. A. Drabold, P. A. Fedders. S. Klemm and O. F. Sankey, submitted to Phys. Rev. Lett.

Materials Science Forum Vol. 83-87 (1992) pp. 487-492

THEORY OF SECOND-ORDER VIBRONIC REDUCTION FACTORS FOR DEEP LEVEL IMPURITIES IN SEMICONDUCTORS

Colin A. Bates[1], Janette L. Dunn[1], Victor Z. Polinger[2], Lee D. Hallam[1], Peter J. Kirk[1] and Sajida Jamila[1]
[1] Physics Department, The University, Nottingham NG7 2RD, UK.
[2] Institute of Chemistry, Academy of Sciences of the Moldavian SSR, Grosul Str. 3, 277028 Kishinev, USSR.

ABSTRACT

Second-order vibronic reduction factors are derived for strongly-coupled orbital triplet systems coupled to t_2-modes of vibration using the recently derived symmetry-related method. This involves the calculation of oscillator overlaps which are projected out of the cubic vibronic states. Details are presented for the examples of second-order spin-orbit coupling and when the perturbations are both of E-type symmetry and when they are both of T_2-type symmetry. The latter calculations are relevant for the modelling of the $^3T_1$ excited state of V^{3+} ions in III-V materials and of the effects of uniaxial stress on the propertries of deep level impurities in semiconductors.

1. Introduction

For many impurity ion systems, the electrons are strongly coupled to the vibrations of their surroundings. The concept of reduction factors is then introduced within such a vibronic system to enable us to describe an electronic perturbation V by a purely electronic Hamiltonian after reduction factors (RF) are inserted within it. First-order RFs are used when V occurs once and second-order RF's are used when V occurs twice. In recent papers[1,2], descriptions have been given of a general method for the derivation of such second-order vibronic or Jahn-Teller (JT) RF's. The analysis was based entirely on symmetry grounds. It was shown that the reduction factors could be deduced from the evaluation of the sums of various oscillator overlaps and explicit expressions were derived. These second-order terms become particularly important in many cases of strong coupling especially in the case of orbital triplets (described by the isomorphic orbital angular momentum operator $\ell = 1$) as they can then dominate the first-order terms. The general principles are discussed in the books by Perlin and Wagner[3] and Bersuker and Polinger[4] which also give references to other work.

Applications of the method using spin-orbit coupling $\lambda \boldsymbol{\ell}.\mathbf{S}$ as an example have been given previously.[1,2]. In this paper we outline the basis of the method, taking spin-orbit coupling again as the example, and extend it to the case when Γ_k and Γ_l are, firstly, both of T_2-smmetry and, secondly, when they are both of E-symmetry. Γ_k and Γ_l denote the symmetries of the two parts of V involved in the calculation. The discussion here will be limited to $T \otimes t_2$ JT systems but equivalent results have also been derived for orthorhombic $T \otimes (e + t_2)$ JT systems[5].

2. Mathematical background for orbital triplets

We consider an isolated orbital T_1 triplet which is strongly coupled to the vibrations of its surroundings. The electronic orbital states may be written in the form $\Psi_{\Gamma\gamma}(r) \equiv |\Gamma\gamma\rangle$; the eigenstates of the system are vibronic states which are written in the form $\Psi_{\Gamma\gamma}^{(N)} \equiv |N\Gamma\gamma\rangle$. The labels $\Gamma\gamma$ give the irreducible representation (ir) and the component of the state while N labels the repeated ir of the vibronic state such that their energies $E_\Gamma^{(N)}$ increase with increasing N. The vibronic eigenstates are written as a sum of products of electronic and vibrational states by using a Clebsch-Gordan convolution form:

$$|N\Gamma\gamma\rangle = \sum_{\sigma\Lambda\lambda} |\Sigma\sigma)|N(\Gamma)\Lambda\lambda\}\ \langle\Sigma\sigma\Lambda\lambda|\Gamma\gamma\rangle \tag{2.1}$$

where $\langle\Sigma\sigma\Lambda\lambda\,|\,\Gamma\gamma\rangle$ are the Clebsch-Gordan coefficients and $|\,N(\Gamma)\Lambda\lambda\} \equiv \chi_{\Lambda\lambda}{}^{N(\Gamma)}(Q)$ are functions of the nuclear coordinates Q and thus represent the phonon states.

A perturbation V can cause a splitting of the ground vibronic state $|\,0\Gamma\gamma\rangle$ in first order. In second order, the additional splitting can be described by the effective Hamiltonian:

$$\mathcal{H}^{(2)} = \sum_{N}\sum_{\Sigma\sigma} V\frac{|N\Sigma\sigma\rangle\langle N\Sigma\sigma|}{E_{\Gamma}^{(0)} - E_{\Sigma}^{(N)}}V \tag{2.2}$$

$\mathcal{H}^{(2)}$ depends upon the nuclear as well as the space coordinates. However, as the contributions from the summation in (2.2) form a scalar, the symmetry of $\mathcal{H}^{(2)}$ depends only on the symmetry within V × V. This means that the effective Hamiltonian to describe the vibronic system can be obtained from the purely electronic Hamiltonian by multiplying (V x V) by $K_M{}^{(2)}$ (Γ_k x Γ_l) where[1]:

$$K_M^{(2)}(\Gamma_k\times\Gamma_l) = \frac{(-1)^{j(\Gamma)}[\Gamma]}{\begin{bmatrix}\Gamma_k & \Gamma_l & M\\ \Gamma & \Gamma & \Gamma\end{bmatrix}}\times\sum_{\Lambda}(-1)^{j(\Lambda)}[\Lambda]\begin{bmatrix}\Gamma_k & \Gamma_l & M\\ \Gamma & \Gamma & \Lambda\end{bmatrix}\times\sum_{N}\left[\frac{S_N(\Gamma_k\Lambda\Gamma)\times S_N(\Gamma_l\Lambda\Gamma)}{E_{\Gamma}^{(0)} - E_{\Lambda}^{(N)}}\right] \tag{2.3}$$

with $$S_N(\Sigma\Lambda\Omega) = \sum_{\Phi}(-1)^{j(\Phi)+j(\Sigma)}\{0(\Omega)\Phi\,||N(\Lambda)\Phi\}\begin{bmatrix}\Gamma & \Gamma & \Sigma\\ \Omega & \Lambda & \Phi\end{bmatrix} \tag{2.4}$$

In the above, the oscillator overlap integrals are $\{0(\Omega)\Phi\,|\,|\,N(\Lambda)\Phi\}$ while the remaining terms give numerical factors. Also the 6Γ symbols are written with large square brackets, the fictitious angular momentum operator as j(Γ) and the dimension of a representation with small square brackets[6].

3. Applications

First we take $V = \lambda\ell.S$ so that $\Gamma_k = \Gamma_l = T_1$. The second-order spin-orbit coupling can be written as an effective Hamiltonian in the form[1,2], where $K_M{}^{(2)}(T_1 \times T_1)$ is the second-order reduction factor:

$$\mathcal{H}_{eff}^{so-so} = \lambda^2\sum_{M\mu} K_M^{(2)}(T_1\times T_1)\ L_{M\mu}^{(2)}\ S_{M\mu}^{(2)} \tag{3.1}$$

In terms of the more usual effective Hamiltonian,

$$\mathcal{H}_{eff}^{so-so} = \lambda^2(b(\underline{\ell}.\underline{S}) + c(E_\theta E_\theta^S + E_\epsilon E_\epsilon^S) + d(L_{yz}S_{yz} + ...) + e\ell(\ell + 1)S(S + 1)),$$

the coefficients b, c, d and e are given by[12]

$$b = K^{(2)}_{T_1} \quad c = \tfrac{2}{3}(K^{(2)}_{E} - b) \quad d = \tfrac{1}{2}(K^{(2)}_{T_2} - b) \quad e = \tfrac{1}{3}(K^{(2)}_{A_1} - b)$$

where $L_{yz} = \ell_y\ell_z + \ell_z\ell_y$ and where $L_{M\mu}{}^{(2)}$ and $S_{M\mu}{}^{(2)}$ are the second-order orbit and spin operators respectively which transform as $M\mu$ with $M = A_1, E, T_1, T_2$. (For example, $L_{E\theta}{}^{(2)} = (1/\sqrt{6})[3\ell_z{}^2 - 2]$.)

Detailed calculations show that:[1]

$$\begin{aligned}
K^{(2)}_{A_1}(T_1\times T_1) &= +3R_{A_1} + 6R_E + 9R_{T_1} + 9R_{T_2} \\
K^{(2)}_{E}(T_1\times T_1) &= -6R_{A_1} -12R_E + 9R_{T_1} + 9R_{T_2} \\
K^{(2)}_{T_1}(T_1\times T_1) &= +6R_{A_1} - 6R_E + 9R_{T_1} - 9R_{T_2} \\
K^{(2)}_{T_2}(T_1\times T_1) &= -6R_{A_1} + 6R_E + 9R_{T_1} - 9R_{T_2}
\end{aligned} \tag{3.2}$$

where the parameters R_Λ involve sums of terms involving overlap integrals (such as $\{0(T_1)T_1\,||\,n(A_1)T_1\}$, for example) divided by an energy. For example[1],

$$R_{A_1} = \sum_n \left[\frac{\{0(T_1)T_1\,||\,n(A_1)T_1\}^2}{9(E^{(0)}_{T_1} - E^{(n)}_{A_1})}\right] \tag{3.3}$$

If we take the orbital part of V to have T_2-symmetry, the second-order effective Hamiltonian equivalent to (3.1) can be written as:

$$\mathcal{H}^{T_2-T_2}_{eff} = v^2 \sum_{M\mu} K^{(2)}_m(T_2\times T_2) L^{(2)}_{M\mu} Q^{(2)}_{M\mu} \tag{3.4}$$

where $Q_{M\mu}{}^{(2)}$ is a non-orbital operator of $M\mu$-symmetry. Detailed calculations show that:

$$\begin{aligned}
K^{(2)}_{A_1}(T_2\times T_2) &= +3R_{A_2} + 6R_E + 9R_{T_1} + 9R_{T_2} \\
K^{(2)}_{E}(T_2\times T_2) &= -6R_{A_2} -12R_E + 9R_{T_1} +9R_{T_2} \\
K^{(2)}_{T_1}(T_2\times T_2) &= -6R_{A_2} + 6R_E + 9R_{T_1} - 9R_{T_2} \\
K^{(2)}_{T_2}(T_2\times T_2) &= +6R_{A_2} - 6R_E + 9R_{T_1} - 9R_{T_2}
\end{aligned} \tag{3.5}$$

Finally, if we take the orbital part of V to have E-symmetry, the same effective Hamiltonian as that given in (3.4) can be used but with the superscripts T_2 replaced by E. The corresponding calculations for the second-order reduction factors are much simpler and give the results:

$$\begin{aligned}
K^{(2)}_{A_1}(E\times E) &= 9R_{T_1} + 9R_{T_2} \\
K^{(2)}_{E}(E\times E) &= 9R_{T_1} - 9R_{T_2}
\end{aligned} \tag{3.6}$$

In the latter expressions (3.5) and (3.6), the R_Λ are different from each other and from those obtained with spin-orbit coupling. The above expressions for the various second-order reduction factors clearly reflect the symmetry of V and of the type of JT problem involved.

4. Calculations

In order to use this method, it is necessary to have expressions for the ground and excited vibronic states in symmetry-adapted form. Such sets of states have been obtained[7] from a unitary transformation method followed by an energy minimisation procedure[8-12].

In this method, the z-component of the ground T_1 vibronic triplet is written in the form:

$$|T_{1z}t\rangle = N_{Tt}(-|a';0\rangle + |b';0\rangle + |c';0\rangle - |d';0\rangle) \tag{4.1}$$

while the inversion level of A_2 symmetry at a relative energy of δ is written in the form:

$$|A_2t\rangle = N_{At}(|a';0\rangle + |b';0\rangle + |c';0\rangle + |d';0\rangle) \tag{4.2}$$

In the above, $|a';0\rangle = U_a|a;0\rangle$ etc., where a labels the trigonal well and simultaneously gives the associated orbital state such that:

$$a = \sqrt{1/3}\,(x + y - z),\; b = \sqrt{1/3}\,(x - y + z),\; c = \sqrt{1/3}\,(-x + y + z),\; d = \sqrt{1/3}\,(-x - y - z) \tag{4.3}$$

Also '0' denotes that there are no excitations in any of the t_2 oscillators with respect to the transformed picture and U_a is the value of the unitary transformation U evaluated at the 'a' minimum etc. and is used to transform the states back to the original basis[1,2]. The other members of the ground triplet $|T_{1x}t\rangle$, $|T_{1y}t\rangle$ are obtained by cyclic permutations of the labels x,y,z and hence of a,b,c. The N's are normalising factors.

The excited vibronic states are much more complicated. A convenient form is[13]:

$$\Psi_i(l,m,n) = N_i(l,m,n)\ \phi_i(l,m,n) \tag{4.4}$$

where l,m,n give the number of t_2 oscillator excitations of symmetry 'yz', 'zx', 'xy' respectively and N_i are normalising factors. The label 'i' goes from 1 to 19; expressions for the ϕ_i are given in table 2 of Dunn[13].

In order to extract the required oscillator parts of the vibronic states, it is necessary to use the expansion formula given by (2.1) using Clebsch-Gordan coefficients[6,14,15] appropriate to the T_d group. Thus $|T_{1z}t\rangle$, $|A_2t\rangle$ and $|N\Gamma\gamma\rangle$ must be expanded in terms of the orbital states $|x\rangle$, $|y\rangle$ and $|z\rangle$ and the oscillator states obtained from a combination of the U_a, U_b, U_c and U_d acting on $|0,0,0\rangle$ for the ground states and on $|4^l5^m6^n\rangle$ for the excited states. On comparing these expressions with those obtained from the expansion formula and equating the coefficients of the equivalent orbital states, the resulting simultaneous equations are easily solved and give the results shown in the figures.

6. Applications

This paper has shown that the second-order reduction factors can be obtained as a sum of various oscillator overlaps which need to be expressed in symmetry-adapted form. The results are particularly important for the spectroscopic study of deep level impurities in semi-conductors.

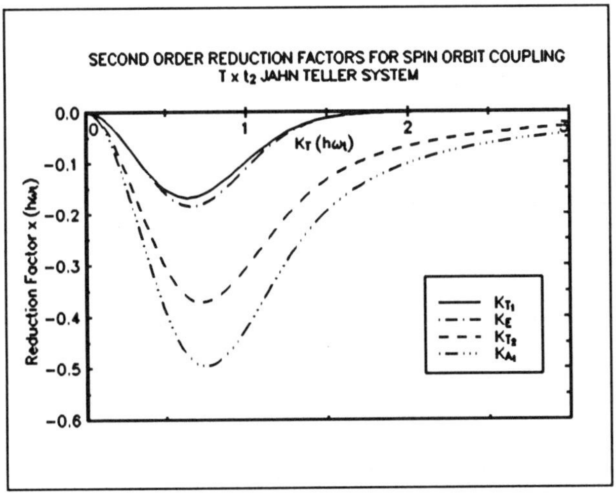
SECOND ORDER REDUCTION FACTORS FOR SPIN ORBIT COUPLING
T x t2 JAHN TELLER SYSTEM
Reduction Factor x (hωι)
0.0
−0.1
−0.2
−0.3
−0.4
−0.5
−0.6
0
1
2
3
Kτ (hωι)
Kτ1
Kε
Kτ2
KA1

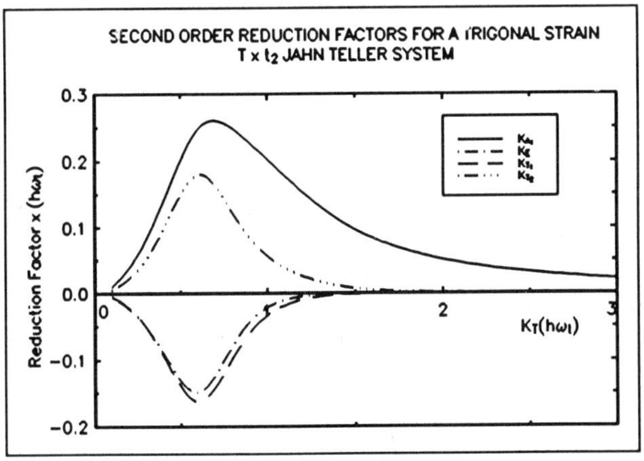
SECOND ORDER REDUCTION FACTORS FOR A TRIGONAL STRAIN
T x t2 JAHN TELLER SYSTEM
Reduction Factor x (hωι)
0.3
0.2
0.1
0.0
−0.1
−0.2
0
2
3
Kτ(hωι)

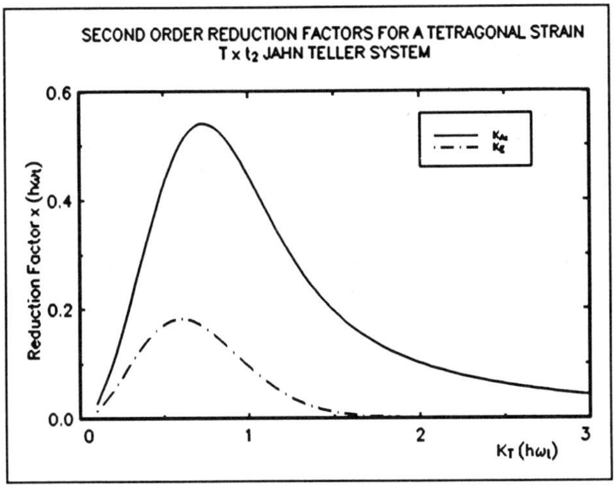
SECOND ORDER REDUCTION FACTORS FOR A TETRAGONAL STRAIN
T x t2 JAHN TELLER SYSTEM
Reduction Factor x (hωι)
0.6
0.4
0.2
0.0
0
1
2
3
Kτ (hωι)

For example, the excited $^3T_1(F)$ state of the V^{3+} ion in III-V materials as observed in zero phonon lines in optical absorption experiments is split mainly by second-order spin-orbit coupling terms wheras the first-order terms are very strongly quenched[16].

Also, it is clear that the T_1 states of almost all such impurities are very strongly coupled to the their surroundings so that uniaxial stress can have a marked effect on the various spectra. The above formula are therefore useful in calculating corrections to first-order stress effects. For example, for a uniaxial stress along a [001] direction of a tetrahedron, (3.4) applies with $v = V_E$ the ion-lattice coupling constant for the E-modes (Q_θ,Q_ϵ) and $Q_{M\mu}{}^{(2)} = Q_\theta{}^2$ for $\mu=\theta$ and $Q_{M\mu}{}^{(2)} = 0$ for $\mu = \epsilon$. Similarly, for a stress along [111], $v = V_T$ and the $Q_{M\mu}{}^{(2)}$ factors are all equal to each other for the t_2-modes (Q_4, Q_5 and Q_6). Further details will be published later.

Acknowledgements

The authors are particularly grateful to the Royal Society of London and The USSR Academy of Sciences for a Protocol agreement which has financed exchange visits between Nottingham and Kishinev. LDH wishes to thank the SERC for a Research Studentship and SJ for a Commonwealth Scholarship during the course of this work.

References

1. V.Z. Polinger, C.A. Bates, and J.L. Dunn, J. Phys.: Condens. Matter **3** 513-27 (1991).
2. C.A. Bates, J.L. Dunn, L. Hallam, P.J. Kirk and V.Z. Polinger, J. Phys.: Condens. Matter **3**, 3441-3453 (1991).
3. Yu.E. Perlin and M. Wagner, *The Dynamical Jahn-Teller Effect in Localised Systems* (North Holland: Amsterdam, 1984).
4. I.B. Bersuker and V.Z. Polinger, *Vibronic Interaction in Molecules and Crystals* (Springer: Berlin, 1989).
5. L. Hallam, C.A. Bates and J.L. Dunn, J. Phys.: Condens. Matter, to be published (1990).
6. J.S. Griffiths, *The Irreducible Tensor Method for Molecular Symmetry Groups* (Prentice Hall: London, 1962).
7. C.A. Bates, J.L. Dunn and E. Sigmund, J. Phys. C: Solid State Phys. **20** 1965-83 (Corrigendum 4013) (1987).
8. J.L. Dunn, J. Phys. C: Solid State Phys. **21** 383-99 (1988).
9. J.L. Dunn and C.A. Bates, J. Phys.: Condens. Matter **1** 375-94 (1989).
10. C.A. Bates and J.L. Dunn, J. Phys.: Condens. Matter **1** 2605-16 (1989).
11. J.L. Dunn and C.A. Bates, J. Phys.: Condens. Matter **1** 2617-29 (1989).
12. J.L. Dunn, C.A. Bates and P.J. Kirk, J. Phys.: Condens. Matter, **2** 0000-00 (1990).
13. J.L. Dunn, J. Phys.: Condens. Matter **1** 7861-81 (1989).
14. S. Sugano, Y. Tanabe and H. Kamimura, *Multiplets of Transition Metal Ions in Crystals* (New York, Academic, 1970).
15. G.F. Koster, J.O. Dimmock, R.G. Wheeler and H. Statz, *Properties of the Thirty-Two Point Groups* (M.I.T. Press: Cambridge, Massachusetts, 1963).
16. C.A. Bates, J.L. Dunn and W. Ulrici, J. Phys: Condens. Matter **2**, 607-612 (1990).

Materials Science Forum Vol. 83-87 (1992) pp. 493-498

DONOR BISTABILITY INDUCED BY ELECTRON-PHONON COUPLING

S. BEDNAREK and J. ADAMOWSKI

Institute of Physics and Nuclear Techniques, Academy of Mining and Metallurgy, PL-30059 Kraków, Poland

ABSTRACT

A microscopic theoretical model for bistable donors in polar crystal is proposed. This approach goes beyond the effective mass approximation and applies the one-band approximation. The dispersion relation for the conduction band is proposed in an analytical form which takes into account the finite width and non-parabolicity of the band and is consistent with the optical absorption data. The polar coupling of the donor center and the electron with *LO* phonons leads to the large and small lattice deformation for the weakly and strongly localized donor states, respectively. The present model allows us to explain the experimental results for donors in cadmium fluoride crystals. It has been shown that the indium impurity is bistable and possesses coexisting donor states with different localization. On the contrary, such impurities like Y and Yb are stable donors and possess only the weakly localized states. For the cadmium fluoride crystal doped with indium, the calculated absorption band shape in the 3-eV region agree well with experiment.

1. Introduction

Several defects in semiconductors exhibit bistable properties [1]. One of them is the *DX* center [2-4], which is created by the group IV or group VI donors in GaAs under large hydrostatic pressure or in AlGaAs with large enough aluminum content. A number of theoretical papers (see, *e.g.*, [5-7]) is devoted to an explanation of the nature of this center. These studies are far away from a final solution owing to a complicated band structure of the host material.

There exists another bistable defect with properties similar to those of the *DX* center. Namely, this is an indium donor in a semiconducting cadmium fluoride crystal [8-12]. The experiments of Langer *et al.* [8, 9] and Dmochowski *et al.* [10-12] show that this defect possesses the bistable properties which are similar to those of the *DX* center. The nature of bistability of both the defects may be different, but we cannot exclude that there are some common causes of the bistable properties of the *DX* center in AlGaAs and In donor in CdF_2. The bistable indium donor in CdF_2 is a very interesting system due to its strong coupling with the lattice [13, 14]. This was the main reason for undertaking the present research. Another reason was the similarity of its bistable behavior to that of the *DX* center. This system is simpler to a theoretical description than the *DX* center due to a direct band gap in CdF_2 and an ionic character of the crystal in which the lattice deformation is mainly due to the Fröhlich coupling with the *LO* phonons [13, 14].

The present paper provides a theoretical description of bistable donors in polar crystals. This approach has been applied to an explanation of the properties of donors in CdF_2. In section 2 we present the variational method proposed to the calculation of energy levels of the donor-lattice system. In section 3 we discuss the problem of optical absorption of CdF_2 crystals. Section 4 includes conclusions and a summary of the results.

2. Calculation of donor energy levels

The present theoretical method is a generalization of the approach proposed to a description of the bound polaron [15]. On the contrary to Ref. [15], we go beyond the effective mass approximation and apply instead the one-band approximation. The first attempt of using this method has been presented in Ref. [14].

The system we describe consists of an electron, a donor center, and a field of *LO* phonons. The Hamiltonian of this system has a form

$$H = H_o + V_C + H_{ph} + H_{int} , \tag{1}$$

where H_o is the one-electron Hamiltonian for the ideal crystal, $V_C = -e^2/\varepsilon_\infty r$ is the Coulomb potential energy of the electron in the field of the donor center,

$$H_{ph} = \hbar\omega \sum_{\mathbf{k}} a^+_{\mathbf{k}} a_{\mathbf{k}} \tag{2}$$

describes the field of *LO* phonons with the energy $\hbar\omega$, where $a^+_{\mathbf{k}}$ $(a_{\mathbf{k}})$ are the *LO* phonon creation (annihilation) operators, and

$$H_{int} = \sum_{\mathbf{k}} \left\{ \left[v^e_{\mathbf{k}} \, exp(i\mathbf{k}\cdot\mathbf{r}) + v^d_{\mathbf{k}} \right] a_{\mathbf{k}} + H.c. \right\} \tag{3}$$

describes the Fröhlich coupling with the *LO* phonons of the electron $(v^e_{\mathbf{k}})$ and the donor center $(v^d_{\mathbf{k}})$. The interaction amplitudes are given by

$$v^e_{\mathbf{k}} = - (i/|\mathbf{k}|)(2\pi e^2/\hbar\omega\varepsilon\Omega)^{1/2} \tag{4}$$

and $v^d_{\mathbf{k}} = - v^e_{\mathbf{k}}$, where $\varepsilon = 1/(1/\varepsilon_\infty - 1/\varepsilon_o)$, ε_∞ (ε_o) is the high-frequency (static) dielectric constant, and Ω is the quantization volume.

We solve the eigenvalue problem for Hamiltonian (1) by variational means. In the one-band approximation, the total wave function of the system is expanded as follows

$$\Psi = \sum_{\mathbf{q}} b_{\mathbf{q}} \, \psi_{c\mathbf{q}}(\mathbf{r}) \, \chi_o , \tag{5}$$

where the sum runs over the first Brillouin zone, $\psi_{c\mathbf{q}}(\mathbf{r})$ is the Bloch function of the conduction band, which fulfils the equation

$$H_o \, \psi_{c\mathbf{q}}(\mathbf{r}) = E_{c\mathbf{q}} \, \psi_{c\mathbf{q}}(\mathbf{r}) , \tag{6}$$

and χ_o is the phonon-vacuum state. We propose the expansion coefficients in the form consistent with the one-band approximation

$$b_{\mathbf{q}} = \frac{1}{N} \sum_n exp(-i\mathbf{q}\cdot\mathbf{R}_n)\ \Phi(\mathbf{R}_n)\ U(\mathbf{R}_n)\ , \tag{7}$$

where $\mathbf{R}_n$ are the lattice vectors ($n = 1,\ldots,N$). The function Φ is proposed in the form

$$\Phi(\mathbf{R}_n) = \sum_{\mathbf{k}} \phi_{\mathbf{k}}\ exp(i\mathbf{k}\cdot\mathbf{R}_n)\ , \tag{8}$$

where the summation is performed over the first Brillouin zone, and U is the operator of unitary transformation

$$U(\mathbf{R}_n) = exp\ \left[\sum_{\mathbf{k}} F_{\mathbf{k}}(\mathbf{R}_n)\ a_{\mathbf{k}} - H.c. \right] . \tag{9}$$

The displacement amplitudes are taken on in the form

$$F_{\mathbf{k}}(\mathbf{R}_n) = f_{\mathbf{k}}\ exp(i\mathbf{k}\cdot\mathbf{R}_n) + g_{\mathbf{k}}\ , \tag{10}$$

where the functions $f_{\mathbf{k}}$ and $g_{\mathbf{k}}$ are determined during a minimization of an expectation value of Hamiltonian (1) calculated with the use of trial wave function (5). The functions $\phi_{\mathbf{k}}$ in equation (8) are proposed as follows

$$\phi_{\mathbf{k}} = \sum_j c_j\ (k^2 + \alpha_j^2)^{-2}\ , \tag{11}$$

where c_j and α_j are the variational parameters. The calculations of energy levels appear to be already convergent for the two-element basis (11).

The variational wave function for the donor-electron-lattice system takes on the simpler form if we introduce the Wannier functions $w_c(\mathbf{r}-\mathbf{R}_n)$ for the conduction band. Then, wave function (5) can be written down as follows

$$\Psi = \frac{1}{\sqrt{N}} \sum_n \Phi(\mathbf{R}_n)\ w_c(\mathbf{r}-\mathbf{R}_n)\ U(\mathbf{R}_n)\ \chi_o\ . \tag{12}$$

Here, we see that for each n the wave function consists of the two parts: the electronic part (the first two terms) and the lattice part (the last two terms).

In order to perform the calculations we have to make some numerical approximations. First, we replace the summation over the first Brillouin zone by the integration over a sphere of the same volume. Next, we neglect the matrix elements of the Coulomb potential between the Bloch waves with reciprocal lattice vectors $\mathbf{G} \neq 0$. These matrix elements lead to a short-range potential, which, nevertheless, has been taken into account in the present calculations by including the potential which is constant in the **k**-space. This potential has the short range in the coordinate space and simulates the action of the short-range part of the Coulomb potential as well as the short-range potential of the impurity core. The strength of this potential is labeled by γ.

The important physical input in our approach is a parametrization of the conduction band. We propose the following analytical form for the dispersion relation of the conduction band

$$E_{c\mathbf{k}} = a\,(k/k_o)^2 + b\left\{ 1 - \left[1 + (k/k_o)^2 \right]^{-2} \right\}, \tag{13}$$

where a and b are determined from the known values of the conduction band width (Δ = 4 eV) [16] and the electron band mass (m_e = 0.21 m_{eo}); k_0 is the fitting parameter. This form of the conduction band agree with the results of the band structure calculations [16, 17]. Moreover, it allows us to reproduce correctly the shape of the optical absorption between the impurity and conduction band states.

We have performed the variational calculations of donor energy levels in the **k**-space using the trial wave function defined by equations (5-11) and the above assumptions. The experimentally measured [8] energy level E_o = -200 meV (with respect to the bottom of the conduction band) of the strongly localized state of indium donor has been reproduced with k_o = 0.235 a_B^{-1} and γ = -0.065 $a.u.$, where a_B is the Bohr radius. In order to calculate the energy of the weakly localized shallow donor state we perform the variational calculation for an excited state with the help of the method given by Löwdin [18]. The energy of this state is E_1 = -90 meV for the indium donor [11]. The optical phonon in CdF_2 has the energy $\hbar\omega$ = 50 meV. This means that between the donor states with the energies E_o and E_1 there are the one- and two-phonon states with energies $E_o + \hbar\omega$ and $E_o + 2\hbar\omega$, respectively. Therefore, the weakly localized donor state should be orthogonal to the zero-phonon ground state with the energy E_0 as well as to these one- and two-phonon states. After perfoming the orthogonalization [18] we find the energy of the shallow donor state in full agreement with the experimental value [11]. The similar calculations have been done for the stable donor, namely yttrium, for which we obtain no strongly localized state and the ground-state energy E_o = -85 meV, which agree with experiment [11], if the short-range potential is repulsive and has the strength γ = 0.01 $a.u.$

From our calculations we also get an estimate of a localization of the donor state. Namely, an expectation value $\lambda = \langle 1/r \rangle$ of the inverse electron-donor distance provides a measure of this localization. For the indium donor, the calculated value of the localization parameter for the ground state is λ_o = 0.226 a_B^{-1}, while for the weakly localized state λ_1 = 0.066 a_B^{-1}.

3. Optical absorption

The optical absorption of CdF_2 doped with In is very characteristic [10]. In this crystal, the two absorption bands are observed [10] with maxima at 0.2 eV and 3 eV, *i.e.*, in the infrared (IR) and visible-ultraviolet range. The 3-eV absorption is due to the transitions between the strongly localized donor state and the conduction band states with many phonons included. However, the IR absorption results from the transitions between the weakly localized shallow donor states and the conduction band.

We have calculated the absorption coefficient for the optical absorption of the CdF_2:In crystal in the 3-eV range. In this process the initial state is the zero-phonon strongly localized state of the donor, the final states are the many-phonon states of the electron in the conduction band. The following formula for the absorption coefficient has been obtained

$$\alpha = (C/h\nu)\, exp(-X) \sum_{\ell} (X^{\ell}/\ell!) \sum_{\mathbf{k}} k^2\, |\phi_{\mathbf{k}}|^2\, \delta(E_{c\mathbf{k}} + \ell\hbar\omega - E_o - h\nu)\ , \qquad (14)$$

where C is a constant, $h\nu$ is the photon energy, and X is given by

$$X = (\hbar\omega)^{-2} \sum_{\mathbf{k}} |v^e_{\mathbf{k}}|^2\, |\rho_{\mathbf{k}}|^2 \qquad (15)$$

with

$$\rho_{\mathbf{k}} = \sum_{\mathbf{q}} \phi^*_{\mathbf{q}}\, \phi_{\mathbf{q+k}}\ . \qquad (16)$$

We insert into equations (14) and (16) the functions $\phi_{\mathbf{k}}$ obtained for the strongly localized donor state. In the calculations of absorption we have only to fit the value of α at the maximum and no additional fitting parameters are used. The calculated absorption coefficient (figure 1) shows good agreement with experimental data [10]. This result provides an *a posteriori* verification of parametrization (13) of the conduction band.

Figure 1. Calculated (solid curve) and measured (dotted curve) absorption coefficient.

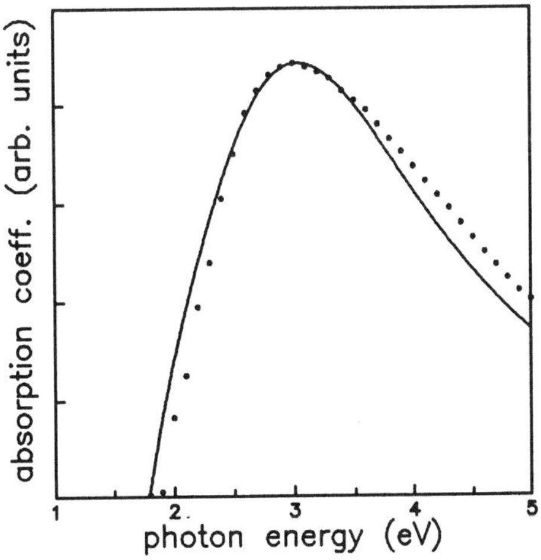

4. Conclusions

The present approach yields the properly orthogonalized strongly and weakly localized donor states. The estimated localization for the ground state of the indium donor amounts 1/3 of the length of the wave vector corresponding to the Brillouin zone boundary. On the other hand, it is only 3 times larger than the localization of the weakly localized state. This means that the ground state of the In donor is not very strongly localized as it was previously suggested [13]. Its "orbital radius" is larger than the distance between the donor center and the nearest ions in the lattice.

The present work shows the possible mechanism of the bistability of donors in polar crystals. This effect results from a combined influence of electronic and lattice properties of the impurity. We were able to describe the influence of electronic states by going beyond the effective mass approximation and applying the one-band approximation. The lattice states contribute to the bistability since the donor states with different localization lead to the different lattice deformations around the defect. In the ionic CdF_2 crystal, the larger deformation is connected with the weakly localized donor state. The different lattice deformations corresponding to the both types of donor states cause their bistability since the direct optical transitions between these states are forbidden. Only the transitions with many phonons involved are allowed, *e.g.*, thermal transitions. The weak repulsive short-range potential destroys the bistability which takes place for the Y donor in CdF_2. This potential shifts upwards the energy of the strongly localized state and changes the localization of the shallow state. The difference in the localization of the shallow donor states is responsible for the shift of the absorption band maxima in the IR range between the stable (Y) and bistable (In) donors [19].

Our results suggest a possibility of a similar mechanism of bistability for the *DX* center in AlGaAs, in which material the conduction band shape is favourable for a creation of donor states with different localization. However, in this case another coupling with phonons should be responsible for the lattice deformation.

REFERENCES

1. G.D. Watkins, Materials Science Forum 38-41, 39 (1989).
2. D.V. Lang and R.A. Logan, Phys. Rev. Lett. 39, 635 (1977).
3. P.M. Mooney, J. Appl. Phys. 67, R1 (1990).
4. L. Dobaczewski and J.M. Langer, Materials Science Forum 65-66, 433 (1990).
5. J.C. Bourgoin and A. Mauger, Appl. Phys. Lett. 53, 749 (1988).
6. D.J. Chadi and K.J. Chang, Phys. Rev. Lett. 61, 873 (1988).
7. M. Lannoo, Diffusion Defect Data, Solid State B 10, 209 (1990).
8. J.M. Langer, U. Ogonowska and A. Iller in *Physics of Semiconductors 1978*, ed. B.L.H. Wilson (IOP Conference Ser. No. 43, Bristol and London, 1979) p. 277.
9. J.M. Langer, J. Phys. Soc. Japan 49, Suppl. A, 207 (1980).
10. J.E. Dmochowski, J.M. Langer, Z. Kaliński and W. Jantsch, Phys. Rev. Lett. 56, 1735 (1986).
11. J.E. Dmochowski, thesis (Institute of Physics, Polish Academy of Sciences, Warsaw, 1986).
12. J.E. Dmochowski, J.M. Langer and W. Jantsch in *Shallow Impurities in Semiconductors 1988*, ed. B. Monemar (IOP Conference Ser. No. 95, Bristol and Philadelphia, 1989) p. 325.
13. S. Bednarek, Acta Phys. Polon. A 77, 115 (1990).
14. S. Bednarek and J. Adamowski, Materials Science Forum 65-66, 427 (1990).
15. J. Adamowski, Phys. Rev. B 32, 2588 (1985).
16. J.P. Albert, C. Jouanin and C. Gout, Phys. Rev. B 16, 4619 (1977).
17. B.A. Orłowski and P. Plenkiewicz, Phys. Stat. Sol. (b) 126, 285 (1984).
18. P.-O. Löwdin, Phys. Rev. 139, A357 (1965).
19. S. Bednarek and J. Adamowski, Acta Phys. Polon. A 79, 393 (1991).

Materials Science Forum Vol. 83-87 (1992) pp. 499-504

THE TEMPERATURE DEPENDENCE OF TRAP CROSS SECTIONS

J. P. LAVINE, E. K. BANGHART, E. T. NELSON, W. F. DesJARDIN AND B. C. BURKEY
Image Acquisition Products Division, Eastman Kodak Company, Rochester, NY 14650-2008, USA

ABSTRACT

The cross section for electron capture by phosphorus donors in silicon is presented over the temperature range of 65 to 90 K. An increase in the temperature leads to a decrease in the cross section in accord with the cubic power law expected from a cascade phonon emission process. The consideration of capture into a high-lying or a low-lying excited state provides a means of reconciling the reported data at lower temperatures.

1. Introduction

Many capture cross sections vary with the temperature T. Power law dependences are common [1,2], especially for shallow levels [3], and exponential dependences have been noted for specific deep levels [4,5]. Phosphorus is a shallow donor in silicon with a ground state at 0.044 eV below the conduction band edge. At temperatures below 100 K phosphorus acts as a trap for electrons. Since phosphorus is commonly used in silicon devices, it is important to understand the low temperature behavior.

Experimental data were obtained by Norton et al. [6] at and below 20 K. The measured cross section decreases with increasing temperature according to T^n with n between 1 and 2. Further measurements were made by Saks and Nordbryhn [7] and by Rosencher et al. [8,9] around 20 K. These results differ from those of Norton et al. This discrepancy prevents an accurate extrapolation of the low temperature results to liquid nitrogen temperatures. The present work provides cross section data over the range of 65 to 90 K for phosphorus in silicon and attempts to resolve the differences between the earlier data.

Charge packet transfer devices known as charge-coupled devices (CCDs) are used to study shallow trap behavior around 77 K. The cross section is found to decrease with increasing temperature according to

$$\sigma (T) \sim 1/T^4 \qquad (1)$$

when the electron's velocity is taken to be the thermal velocity. This result agrees with the prediction of Lax [1] for electron capture accompanied by the emission of phonons as the electron drops down through excited states to the ground state. This cascade model was reexamined by Abakumov and co-workers [3,10] who found

$$\sigma (T) \sim 1/T^3 \qquad (2)$$

The CCD data is consistent with this power law if the thermal velocity is replaced by the drift velocity in the equations for electron capture and emission. The reasonableness of this substitution is considered below.

The second section describes a CCD and outlines the measurements. The third section discusses the dependence of the cross section on temperature and is followed by a summary.

2. Experimental procedures and results

Charge-coupled devices [11,12] are analog devices that pass charge packets from under one gate or electrode to under the next gate by varying the gate voltages. The receiving gate is made more positive than the sending gate and the electrons transfer by electric field drift and diffusion. Figure 1 is a schematic drawing of several gates and the potential wells under them. The charge transfer inefficiency ε is the amount of charge remaining under the sending gate divided by the amount of charge under the sending gate at the beginning of the transfer.

The CCDs are built with a phosphorus-implanted region in a boron-doped substrate. ε is measured for a transfer time of 167 ns, a fixed gate voltage swing of 11 V, and a fixed temperature. Additional details are found in [13]. About two-thirds of the electrons in a charge packet are captured by ionized phosphorus atoms at 77 K. These electrons are then emitted during the transfer time with a spatially-dependent emission time constant

$$\tau = \{g/(\sigma v_{th} N_C)\} \exp\{(E_C - E_t - \delta E)/kT\} \quad , \qquad (3)$$

where g is the degeneracy factor, σ the cross section, v_{th} the thermal velocity, N_C the effective conduction band density of states, $E_C - E_t$ the donor ground state level, and kT the temperature in energy units. δE is the

Poole-Frenkel barrier lowering and depends on the local values of the vertical electric field and the electric field in the direction of charge transfer [13]. The latter electric field is at least 200 V/cm for the present CCDs while the vertical fields vary from 0 to about 10^5 V/cm. The trapped charge that remains after the transfer time is in regions of low vertical electric field.

The measured ε versus T is presented in Fig. 2 for 2 CCDs. The charge transfer improves with an increase in temperature. The temperature dependence of the cross section is assumed to be proportional to $1/T^n$. The calculated ε is shown in Fig. 2 for n = 2, 3, and 4. All three are fit to $\varepsilon(T = 77\ K)$ with $\sigma(T = 77\ K) = 1.8 \times 10^{-16}\ cm^2$. The capture cross section with n = 4 provides the best fit to the data. This result is consistent with the original work of Lax [1], but it is contrary to the later work of Abakumov and co-workers [3,10]. A resolution of this dilemma is proposed in the next section.

3. Discussion of the results

Equation (3) shows that the cross section enters into the transport calculations [13] through the product σv_{th}. The thermal velocity is

$$v_{th} = \{8kT/(\pi m_c)\}^{1/2} \quad (4)$$

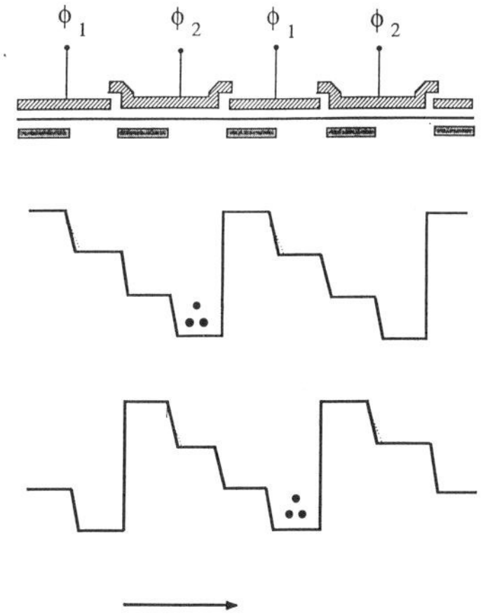

Direction of charge transfer

Fig. 1 Schematic drawing of part of a CCD with potential profiles during charge transfer.

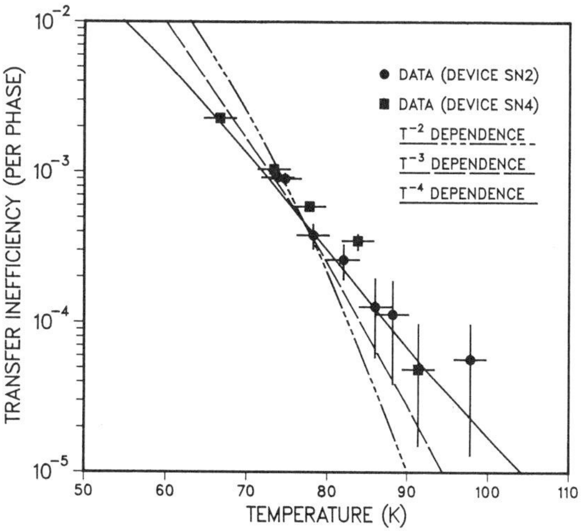

Fig. 2 Measured charge transfer inefficiency vs temperature for two CCDs. The solid curves are calculated using capture cross sections with different temperature dependences.

where m_C is the electron thermal velocity effective mass and is 0.28 times the electron mass [14]. The overall temperature dependence that fits the CCD results is

$$\sigma(T)\, v_{th} \sim T^{-4} * T^{1/2} = T^{-3.5} \quad (5)$$

The same net temperature dependence also arises from

$$\sigma(T)\, v_D \sim T^{-3} * T^{-1/2} = T^{-3.5} \quad (6)$$

where the cross section comes from [3] and v_D is the electron drift velocity. Figure 3 shows that the drift velocity [15] has this temperature dependence around 77 K for an electric field of 200 V/cm. Thus, the present CCD data yield capture cross sections that are explained by the cascade phonon emission theory of Abakumov et al. [3]. It remains to justify the replacement of the thermal velocity with the drift velocity.

Microscopic models for the electron mobility in silicon use acoustic and optical phonon scattering [16], acoustic intravalley and intervalley phonon scattering [17], or combinations thereof [15]. Around and below 77 K the intravalley acoustic phonon scattering dominates [17]. This leads to a rough estimate of 0.6 μm for the mean free path when the standard formulas [18] are used with the thermal velocity. These lengths conflict with the drift velocity results of Canali et al. [15], which translate into scattering lengths an order of magnitude smaller. Consistency is restored if the drift velocity is used in place of the thermal velocity. Both are plotted in Fig. 3. The energy relaxation time is estimated to be about 8 ps at 77 K [16], which is several times the scattering time derived from the drift velocity. Thus, an electron just emitted from a trap probably has a velocity close to the drift velocity.

The CCD data now yield a cross section at 77 K of 1.07×10^{-15} cm^2 when a drift velocity of 0.89×10^6 cm/s is used. This drift velocity [19] is based on an electric field of 200 V/cm and a doping density of $10^{15} cm^{-3}$.

Figure 4 shows the experimental cross sections for electron capture by phosphorus in silicon. Several points should be noted. First, the T^{-3} dependence extrapolates to two data points around 20 K. Secondly, the CCD results and those of [7-9] are about a factor of 30 below the predictions of Abakumov et al. [3] for the phonon cascade mechanism. However, these predictions are in agreement with the results of Norton et al. [6].

All of the data may be understood with the aid of an effective two-level

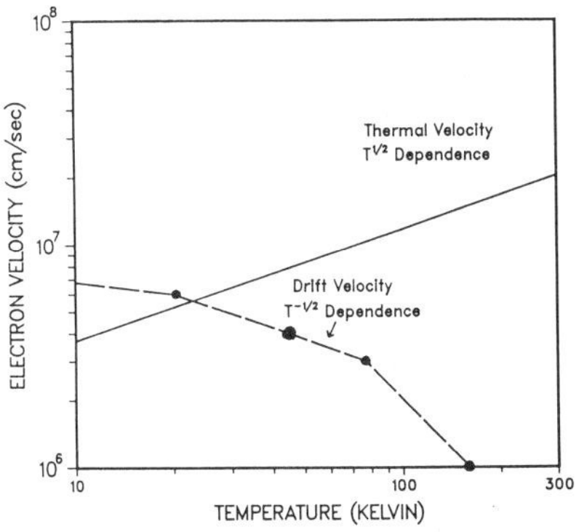

Fig. 3 Drift velocity at 200 V/cm [15] and thermal velocity vs temperature.

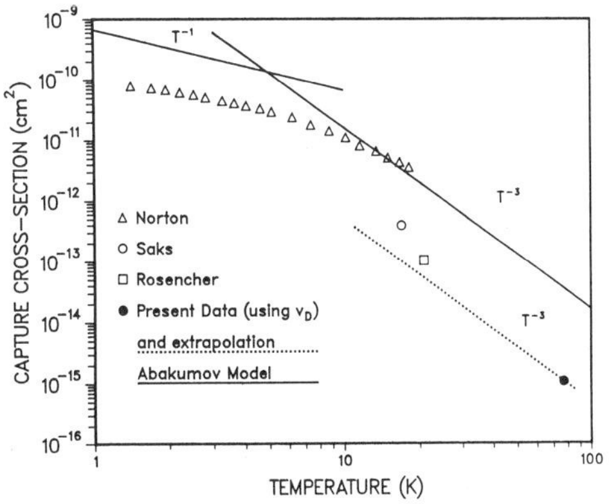

Fig. 4 Experimental electron capture cross sections for phosphorus in silicon. The solid curves use Eqs. 3 and 4 of Ref. 3.

formalism due to Gibb et al. [20] and extended by Landsberg and Dhariwal [21]. The total capture cross section is set equal to the sum of two capture processes

$$\sigma_{TOT} = \sigma_1 + \sigma_2 \nu/(\nu + \tau^{-1}) \quad (7)$$

Here σ_1 represents capture by an excited state and σ_2 capture by a higher-lying excited state. ν is the transition rate from state 2 to state 1 and τ is Eq. (3) for emission from state 2 to the conduction band. If the experiment of Norton et al. [6] probes state 2 as suggested by Rosencher et al. [9], then $\sigma_2 >> \sigma_1$ at low temperatures where $\nu >> \tau^{-1}$ is plausible. At higher temperatures τ^{-1} grows exponentially and Eq. (7) becomes

$$\sigma_{TOT} = \sigma_1 + \sigma_2 \nu \tau \approx \sigma_1 \quad (8)$$

The Poole-Frenkel effect, which plays a role in [7-9] and the present CCD experiments, hastens the growth of τ^{-1}. If σ_1 and σ_2 are based on a similar mechanism, such as cascade phonon emission, then they may be related by a scale factor. A simple estimate for this involves the ratio of the square of the two excited state radii, for example, the 4p or 3p to the 2p, which is approximately 10 [22]. This factor moves the data of [7-9] toward the data of [6] and the theoretical curve. This two-level approach ties together the experimental results for electron capture by phosphorus donors in silicon and shows how the measured cross sections agree with the theoretical predictions of Abakumov et al. [3].

4. Summary

The cross section for electron capture by phosphorus in silicon is extracted from CCD charge transfer inefficiency measurements at 65 to 90 K. The temperature dependence is consistent with the cascade phonon emission mechanism as evaluated by Abakumov et al. [3,10]. It is shown how the present data agree with earlier results at lower temperatures.

REFERENCES

1. M. Lax, J. Phys. Chem. Solids 8, 66 (1959).
2. A. Rose, Concepts in Photoconductivity and Allied Problems (Interscience, New York, 1963) ch. 7.
3. V. N. Abakumov, V. I. Perel' and I. N. Yassievich, Sov. Phys. Semicond. 12, 1 (1978).
4. C. H. Henry and D. V. Lang, Phys. Rev. B 15, 989 (1977).
5. V. N. Abakumov and Z. N. Sokolova, Sov. Phys. Semicond. 12, 962 (1978).
6. P. Norton, T. Braggins and H. Levinstein, Phys. Rev. Lett. 30, 488 (1973).
7. N. S. Saks and A. Nordbryhn, J. Appl. Phys. 50, 6962 (1979).
8. E. Rosencher, V. Mosser and G. Vincent, Solid State Commun. 45, 629 (1983).
9. E. Rosencher, V. Mosser and G. Vincent, Phys. Rev. B 29, 1135 (1984).
10. V. N. Abakumov and I. N. Yassievich, Sov. Phys. JETP 44, 345 (1976).
11. G. S. Hobson, Charge-Transfer Devices (John Wiley, New York, 1978).
12. D. K. Schroder, Advanced MOS Devices (Addison-Wesley, Reading, MA, 1987) ch. 3 and 4.
13. E. K. Banghart, J. P. Lavine, E. A. Trabka, E. T. Nelson and B. C. Burkey, IEEE Trans. Electron Dev. 38, 1162 (1991).
14. M. A. Green, J. Appl. Phys. 67, 2944 (1990).
15. C. Canali, C. Jacoboni, F. Nava, G. Ottaviani and A. Alberigi-Quaranta, Phys. Rev. B 12, 2265 (1975).
16. M. V. Fischetti, IEEE Trans. Electron Dev. 38, 634 (1991).
17. D. L. Rode, phys. stat. solidi b 53, 245 (1972).
18. K. Seeger, Semiconductor Physics (Springer-Verlag, New York, 1973) p. 178.
19. H. De Los Santos and J. L. Gray, IEEE Trans. Electron Dev. 35, 1972 (1988).
20. R. M. Gibb et al. Philos. Mag. 36, 1021 (1977).
21. P. T. Landsberg and S. R. Dhariwal, Phys. Rev. B 39, 91 (1989).
22. L. D. Landau and E. M. Lifshitz, Quantum Mechanics - Non-relativistic Theory (Addison-Wesley, Reading, MA, 1965) p. 119.

Materials Science Forum Vol. 83-87 (1992) pp. 505-510

LATTICE RELAXATION EFFECTS ON DEEP LEVELS: MOLECULAR DYNAMICS CALCULATIONS

CHARLES W. MYLES AND WEI-GANG LI
Department of Physics and Engineering Physics
Texas Tech University, Lubbock, Texas 79409-1051, USA

ABSTRACT

The effects of lattice relaxation on the deep levels due to substitutional impurities are investigated using a recently developed formalism which combines a tight-binding Green's function method with a molecular dynamics technique. Results for several impurities in Si, GaAs and GaP are presented and compared with data and other theories.

1. Introduction

There have been many theories of deep levels in semiconductors,[1,2] and they are of varying degrees of sophistication and accuracy. In this paper, we investigate the effects of lattice relaxation on the deep levels produced by substitutional impurities using a recently developed formalism.[3] Molecular dynamics is used to calculate the relaxation around an impurity[3a] and its effects on the associated deep levels are computed using a tight-binding Green's function technique.[3b] Both first- and second-neighbor relaxation are included in the formalism. As an application, the deep levels produced by several impurities in Si, GaAs and GaP are investigated.

A defect in a semiconductor will interact with the host, displacing the surrounding atoms.[4] Because deep levels are produced by a short-ranged potential,[1,2] this distortion will strongly affect them. First principles theories have been used to study these effects in a few cases.[5] They have also been included phenomenologically within tight-binding theory.[3b,6] While first principles techniques produce the most reliable results, they require considerable computational effort, which is undesirable when the number of defects to be studied is large or when one desires to obtain trends. On the other hand, while tight-binding approaches require comparatively little computational effort and can thus easily be used to study large classes of defects or trends, they are somewhat unsatisfying because of the necessity either to assume an amount of relaxation or to calculate it from a simple model.[3b,6] Our theory is intermediate between these approaches. Since it is tight binding-based, it retains much of the computational simplicity of this approach. However, lattice relaxation is calculated using molecular dynamics, so that the necessity to treat it phenomenologically is circumvented.

2. Formalism

Our approach[3] is a generalization of the Hjalmarson et al.[2] deep level theory to include lattice relaxation. This is a simple, yet widely used[7] theory in which lattice relaxation is neglected. Its extensions have been successful in predicting trends in deep levels in numerous applications.[7] We describe the host using the sp^3s^* tightbinding bandstructure model.[8] However, any suitable choice of tightbinding bandstructures could be used. As the formalism is described in detail elsewhere,[3] only an outline is given here.

We consider the problem of the calculation of the deep levels produced by neutral, sp^3 bonded, substitutional impurities in zinc-blende and diamond structure hosts. The point group for such an impurity is T_d. The deep levels can thus be either of A_1 (s-like) or T_2 (p-like) symmetry. Our technique is applicable for a general distortion of the host atoms around an impurity. Here, however, we consider only T_d symmetry conserving, "breathing mode" distortions, which are thought to dominate for substitutional impurities in the materials of interest.[5,9]

The Koster-Slater[10] theory is convenient for calculating the bound state energies E produced in the bandgap by a defect potential V. In this method, these energies are the solutions to

$$D(E) = det[1 - G^0(E)V] = 0, \tag{1}$$

where $G^0(E) = (E - H_0)^{-1}$ is the host Green's function, and H_0 is the host Hamiltonian. The advantage of this method is that Eq.(1) only needs to be solved in the subspace of V. In the present case, this yields a 68×68 determinant (17 atoms; the impurity, 4 nearest-neighbors, and 12 second-neighbors, with 4 orbitals per atom). For breathing mode distortions, it is shown in Ref. 3c that Eq.(1) factors into the product of four 17×17 determinants, one for the A_1 states and three for the T_2 states.

For an impurity on the anion site in a zinc-blende crystal, the defect potential can be written[3]

$$V = \sum_i (|ia\vec{0} > U_i < ia\vec{0}|) + \sum_i \sum_{\vec{d}} (|ia\vec{0} > \alpha_i < ic\vec{d}| + H.c.) + \sum_i \sum_{\vec{d},\vec{d}'} (|ic\vec{d} > \beta_i < ia\vec{d}'| + H.c.) \quad (2)$$

where the sum on i is over A_1 and T_2 states, a and c refer to anion and cation, $\vec{0}$ is the cell at the origin, the sum on $\vec{d}$ is over the 4 nearest-neighbors, the sum on $\vec{d}'$ is over the 12 second-neighbors, U_i is the diagonal matrix element at the anion site, α_i is the off-diagonal element coupling the anion and cation, β_i is the off-diagonal element coupling first- and second-neighbors, and H.c. denotes the Hermitian conjugate. We have found[3] that the matrix elements coupling the impurity and its second-neighbors are very small in comparison with α_i and β_i, so we neglect them. As is discussed below, α_i and β_i are a measure of the relaxation around the impurity. By treating U_i, α_i and β_i as parameters and computing deep levels as functions of them, their chemical trends (variation with U_i) and trends with varying amounts of relaxation (variation with α_i and β_i) can be explored.

To calculate the deep levels of a particular impurity, U_i, α_i, and β_i must be specified. For U_i, we use the empirical rule[2] of proportionality to atomic energy differences between the impurity and the replaced host atom. To determine α_i and β_i, we use a generalization of the inverse-bond-length-squared scaling rule.[11] This gives $\alpha_i = -C_i[(d_I)^{-2} - (d_H)^{-2}]$ and $\beta_i = -C_i[(d_{II})^{-2} - (d_H)^{-2}]$, where d_I is the impurity bond length, d_{II} is the first- to second-neighbor bond length, d_H is the host bond length, and C_i is a constant determined by the host bandstructures.[12] It is clear from their forms that α_i and β_i represent the effects of first- and second-neighbor relaxation on the defect potential. To calculate them, we use experimental values[11] for d_H and compute d_I and d_{II} by the molecular dynamics scheme we now describe.

For an ideal impurity, the 4 nearest-neighbor and 12 second-neighbor host atoms remain in their perfect crystal positions. In a more realistic treatment, they experience a net force and the lattice distorts to a new configuration where that force is zero. The total force on one of these atoms can be divided into repulsive and attractive parts, $F_x = F_x^r + F_x^a$, where x a bond direction. The origin of F_x^r is the repulsion between electrons in overlapping states. We compute it from a pair potential based on Harrison's overlap interaction,[11] which gives $F_x^r = -A/d^5$, where d is d_I or d_{II} and A is a constant determined by requiring the total force to vanish when the impurity is replaced by a host atom (d_I=d_{II}=d_H). We have found A=199.0, 146.0, and 152.0 eV(Å)^4 for Si, GaAs, and GaP, respectively.

We compute the attractive force, F_x^a, which originates from the occupied one electron levels, from the electronic structure by use of the Hellmann-Feynman theorem.[13] According this theorem, the x component of the force on an atom can be computed from the derivative of the total quantum mechanical energy E_{tot} as

$$F_x^a = -\frac{\partial E_{tot}}{\partial x} = -\frac{\partial}{\partial x} \int_{-\infty}^{E_F} E\rho(E)\,dE, \quad (3)$$

where E_F is the Fermi energy and $\rho(E) = -\frac{1}{\pi} ImTrG(E)$ is the density of states. Here, G is the Green's function, including the impurity, which can be found from G^0 and V using the Dyson equation $G = G^0 + G^0VG = (1 - G^0V)^{-1}G^0$. By combining Eq.(3) with the latter equation and its x derivative, F_x^a can be rewritten as

$$F_x^a = \frac{1}{\pi} ImTr \int_{-\infty}^{E_F} E\frac{\partial G}{\partial x}\,dE = \frac{1}{\pi} ImTr \int_{-\infty}^{E_F} EG\frac{\partial V}{\partial x} G\,dE. \quad (4)$$

Using this formalism and beginning with the first- and second-neighbor host atoms in their perfect crystal positions, one can calculate the motion of each of them for a small time interval,

Δt, using Newton's second law along with standard "molecular dynamics" methods.[14] In this way, new bond lengths d_I and d_{II}, a new V, and a new force can be computed. This process can be repeated for successive intervals Δt until the net force acting on each atom approaches zero. In this way, one can determine d_I, d_{II} and V for the relaxed lattice. Then, deep levels in the bandgap are obtained from Eq.(1).

3. Results for Trends

Our investigations have found that relaxation effects are much more important for A_1-symmetric deep levels than for T_2 levels. Thus, only results for the A_1 levels are discussed here. Further, U_{A_1}, α_{A_1}, and β_{A_1} are denoted below as U, α, and β. In what follows, the zero of energy is taken as the top of the valence band.

Typical results, which illustrate trends in the deep levels with varying amounts of relaxation, are shown in Fig. 1, in which $D(E)$ of Eq.(1) is plotted versus E in the bandgap for various combinations of α and β in the case of the A_1 level produced by Si:Se. In this case, U=-6.5 eV. There are three groups of curves containing three members each. Each group corresponds to a different value of α. From left to right, the groups represent α=0.3 eV, 0.0 eV, and -0.3 eV, respectively, corresponding to d_I/d_H=1.09, 1.00, and 0.93. Within each group, the curves correspond to different values of β. These are β=-0.3 eV (dashed), 0.0 eV (solid), and 0.3 eV (dotted), corresponding to d_{II}/d_H=0.93, 1.00, and 1.09, respectively. The intersections of these curves with $D(E)$=0.0 give the deep levels in the bandgap for these α and β. One can see that, as α becomes more negative (inward nearest-neighbor relaxation), the deep level moves towards the conduction band, while as α becomes more positive (outward relaxation), this level moves towards the valence band. The opposite trend as a function of β can also be seen in Fig. 1, with the level becoming shallower as β becomes more positive (outward second-neighbor relaxation) and deeper as β becomes more negative (inward relaxation). It is also obvious from Fig. 1 that the effects of second-neighbor relaxation are much smaller than those due to first-neighbors. We have found that the latter result and the trends we have illustrated in this case hold qualitatively for all impurities and hosts we have considered.

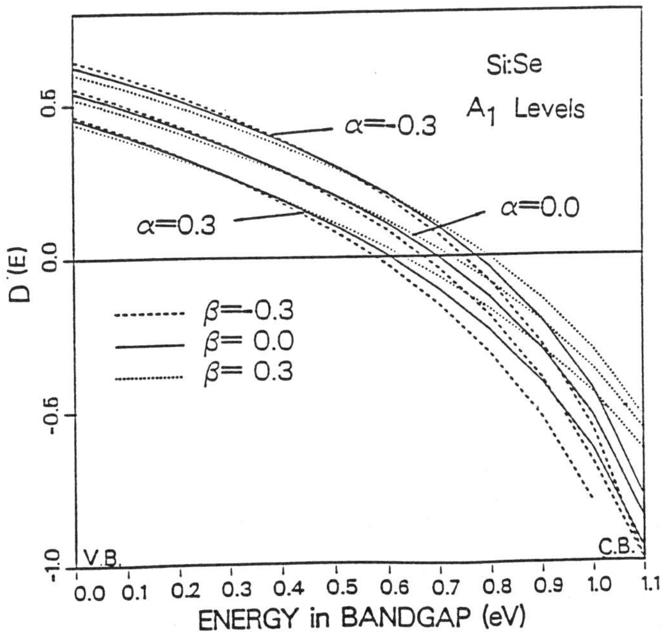

Figure 1: D(E) as a function of the energy E in the bandgap for A_1-symmetric levels due to Si:Se (U=-6.5 eV). The different curves are for the values of α and β indicated.

Figure 2 shows results for A_1 deep levels E produced by impurities substitutional for As in GaAs. The deep level dependence on U is shown for the values of α in the figure label. Since it has only a small effect on E, β has been set to zero. The horizontal scale and the labeled impurities at the top correspond to the value of U calculated from atomic energy differences.[2] Chemical trends (variation in E for fixed α as U is changed) and trends for varying amounts of relaxation (variation in E for fixed U as α is changed) can both be extracted from this figure. For example, it can be seen that the $\alpha \neq 0$ curves have chemical trends which are similar to the α=0 (no relaxation) curve. In addition, it can also be seen that the deep levels for fixed U are shifted towards the conduction band as α becomes increasingly negative (inward relaxation) and towards the valence band as α becomes more positive (outward relaxation), in agreement with the trend discussed above.

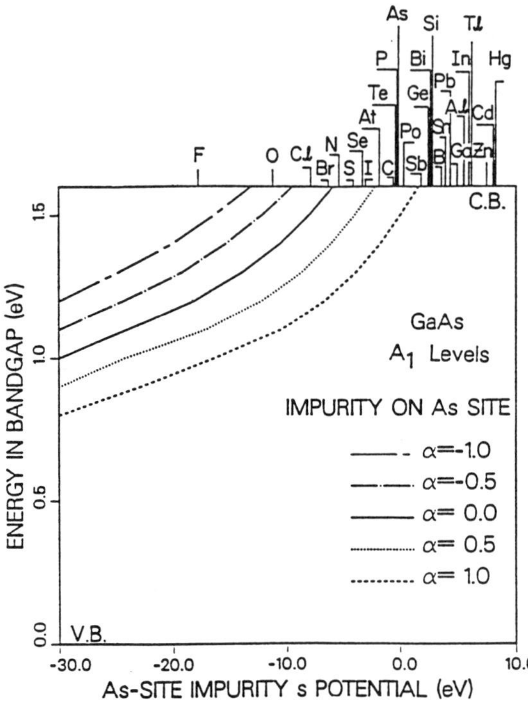

Figure 2: Deep level versus U for A_1 states of an As-site impurity in GaAs for different α (β=0.0). α=1.0, 0.5, 0.0, -0.5, and -1.0 eV corresponds to d_I/d_H=1.53, 1.20, 1.00, 0.90, and 0.81.

4. Molecular Dynamics Results

As an illustration of the molecular dynamics results, in Fig. 3, we show the time dependence of the magnitude of the total force along an impurity atom-host atom bond and of d_I for GaP:N. These are plotted versus $t^* = t/(\Delta t)$, where $\Delta t = 0.35 \times 10^{-14}$s. This choice of Δt was obtained by trial and error. We have found that, while the best choice depends on the particular case, it is always of this order of magnitude. A smaller Δt uses considerably more computer time to achieve similar results and a larger value can cause an "overshoot" of the final equilibrium position and an oscillatory behavior of d_I versus t^*. As can be seen from Fig. 3, after about 16 time steps or about 0.06 ps, the force approaches zero. It can also be seen that, as the atoms approach their new equilibrium positions, the nearest-neighbor Ga atoms move inward by about 0.21 Å, or about 9% of the 2.36 Å GaP bond length. This result is in qualitative agreement with a covalent radius model.[3b]

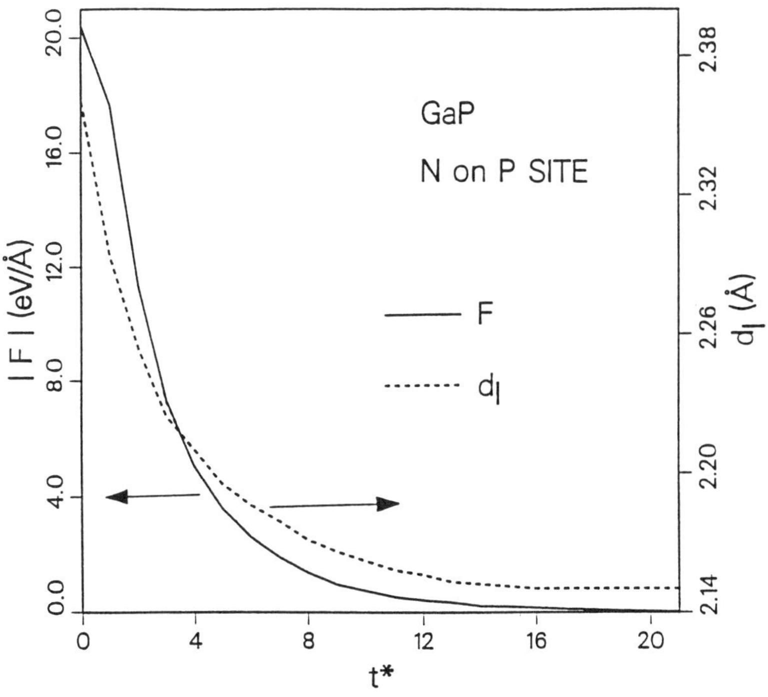

Figure 3: Time dependence of the magnitude of the total force (solid curve) along an impurity atom - host atom bond and of the bond length (dashed curve) for N substitutional for P in GaP.

Results obtained using our formalism to compute A_1 deep levels produced by several impurities in Si, GaAs, and GaP are shown in Table I. Also shown are the experimental deep levels,[1,15-18] and the predictions of the Hjalmarson et al.[2] theory. The last four columns give our results for d_I, $\Delta d = d_I - d_H$, d_{II}, and $\Delta d' = d_{II} - d_H$. As can be seen from the table, we predict an inward relaxation for all cases considered, except for Si:Te, GaAs:O, and GaP:O. Except for the latter two cases, all relaxation results are in qualitative agreement with a covalent radius model.[3b] Also, in all cases considered, our deep level predictions considerably improve upon those of Ref. 2 in comparison with experiment. The remaining discrepancies between theory and experiment might be due to inaccuracies in the host bandstructures and/or charge state effects.

TABLE I: Deep energy levels of A_1 symmetry, bond lengths, and bond length changes for various impurities in Si, GaAs and GaP. All energies are in electron volts, measured from the top of the valence band. The bond lengths are in Å. $\Delta d = d_I - d_H$, $\Delta d' = d_{II} - d_H$.

System	Present Theory	Experiment	Hjalmarson[a]	d_I	Δd	d_{II}	$\Delta d'$
Si:S	0.64	0.85[b]	0.58	2.23	-0.12	2.41	0.06
Si:Se	0.84	0.86[b]	0.65	2.03	-0.32	2.48	0.13
Si:Te	1.04	1.01[b]	1.12	2.60	0.25	2.28	-0.09
GaAs:O	1.16	0.75[c]	1.26	2.73	0.28	2.36	-0.11
GaAs:As_{Ga}	0.74	0.75[c]	0.79	2.39	-0.06	2.47	0.02
GaP:N	2.25	2.34[d]	2.10	2.15	-0.21	2.41	0.05
GaP:O	1.70	1.46[d]	1.85	2.78	0.42	2.28	-0.08
GaP:P_{Ga}	1.09	1.10[e]	1.03	2.28	-0.08	2.40	0.04
GaP:Ge	1.95	2.16[f]	1.85	2.24	-0.12	2.45	0.06

a. Ref. 2; b. Ref. 1; c. Ref. 15; d. Ref. 16; e. Ref. 17; f. Ref. 18

These results depend on a generalization of the inverse-bond-length squared scaling rule[11] for α and β and on a pair potential based on Harrison's overlap interaction,[11]. However, our formalism can be utilized with other reasonable assumptions for these quantities. To test the sensitivity of our results to these assumptions, we have repeated some of these calculations assuming exponential dependences[19] of these quantities on d_I and d_{II}. Results obtained in this manner for d_I, d_{II} and for the deep levels differ by less than 1% from those in Table I.

References

1. M. Jaros, Deep Levels in Semiconductors, (Hilger, Bristol, 1980); S.T. Pantelides, Deep Centers in Semiconductors, (Gordon and Breach, 1986).
2. H.P. Hjalmarson, P. Vogl, D.J. Wolford, and J.D. Dow, Phys. Rev. Lett. 44, 810 (1980).
3. a) W.-G. Li and C.W. Myles, Phys. Rev. B 43, 9947 (1991); b) Phys. Rev. B 43, 2192 (1991); c) W.-G. Li, Ph. D. Dissertation, Texas Tech University, 1991.
4. F.P. Larkins and A.M. Stoneham, J. Phys. C 4, 143 (1971).
5. M. Scheffler, J.P. Vigneron, and G.B. Bachelet, Phys. Rev. B 31, 6541 (1985); U. Lindfelt, Phys. Rev. B 28, 4570 (1982); Phys. Rev. B 30, 1102 (1984).
6. D.N. Talwar, K.S. Suh, and C.S. Tang, Phil. Mag. B 54, 93 (1986); Phil. Mag. B 56, 593 (1987); J.Singh and A. Madhukar, Phys. Rev. B 25, 7700 (1982).
7. For example, O.F. Sankey and J.D. Dow, Phys. Rev. B 22, 7641 (1983); Y.-T. Shen and C.W. Myles, Phys. Rev. B 40, 10425 (1989); E.G. Bylander, C.W. Myles, and Y.-T. Shen, J. Appl. Phys. 67, 7351 (1990); S. Lee, J.D. Dow, and O.F. Sankey, Phys. Rev. B 31, 3910 (1985).
8. P. Vogl, H.P. Hjalmarson, and J.D. Dow, J. Phys. Chem. Solids 44, 365 (1983).
9. T. N. Morgan, Phys. Rev. B 29, 5667 (1984); G.A. Baraff, E.O. Kane, and M. Schlüter, Phys. Phys. Rev. B 25, 548 (1982).
10. G.F. Koster and J.C. Slater, Phys. Rev. 95, 1167 (1957).
11. W. Harrison, Electronic Structure and the Properties of Solids, (Freeman, San Francisco), 1980; Phys. Rev. B 27, 3592 (1983).
12. For the bandstructures of Ref. 8, C_{A_1}= 10.5eV($\AA$)^2 and C_{T_2}=-3eV($\AA$)^2.
13. H. Hellmann, Einfuhrung in Die Quantenchemie, (Deuticke, Vienna, 1937); R.P. Feynman, Phys. Rev. 56, 340 (1939).
14. R. E. Allen, F. W. deWette, and A. Rahman, Phys. Rev. 179, 887 (1969); F. H. Stillinger and A. Rahman, J. Chem. Phys. 60, 1545 (1974).
15. E. Weber, H. Ennen, and U. Kaufmann, J. Appl. Phys. 53, 6142 (1982).
16. A.G.Milnes, Deep Impurities in Semiconductors, (Wiley, New York, 1973).
17. M. Scheffler, J.P. Vigneron, and G.B. Bachelet, Phys. Rev. Lett. 49, 1765 (1982).
18. G.F. Neumark and V.K. Kosai, in Semiconductors and Semimetals, Vol. 19, (Academic, New York, 1983).
19. M.Menon and R.E. Allen, Phys. Rev. B 33, 5611 (1986).

Materials Science Forum Vol. 83-87 (1992) pp. 511-516

RECOMBINATION-INDUCED DEFECT HEATING AND RELATED PHENOMENA

I.N.YASSIEVICH, V.N. ABAKUMOV AND A.A. PAKHOMOV
A.F.Ioffe Physico-Technical Institute,
Politekhnicheskaya 26, 194 021 Leningrad, U.S.S.R.

ABSTRACT

A detailed study of local phonon heating due to multiphonon recombination process is presented. Nonequilibrium distribution of defects by vibration energy is calculated taking into account the dependence of capture and emission rates on the defect energy. It is shown that multiphonon generation and recombination of electron-hole pairs stimulate diffusion of defects in energy space. Thus, "the high energy tail" of defect distribution over the vibration energies essentially increases if there are excess carriers in a semiconductor. The results are used to calculate the rate of recombination-enhanced defect annealing and to interprete the experimental data on inelastic sputtering of semiconductors and insulators by ion bombardment.

1. Introduction

The process of nonradiative recombination of electron-hole pairs is accompanied by energy transfer to the lattice of the order of forbidden gap $\mathcal{E}_g$. If the recombination process is controlled by electron-local vibration interaction this energy is accumulated in the defect vicinity. The forbidden gap width is usually of the order of 1 eV, i.e. of the order of the binding energy of defects in the lattice. On these grounds, the accumulation of such large energy in the defect vicinity is expected to lead to reconstruction of the bonds and thus to induce defect reactions (for example, defect jumps and Frenkel pair recombination).

As early as 1956, Seitz /1/ pointed out that this fact is important for understanding of degradation of semiconductor devices and defect annealing. The first extensive theoretical investigation of this problem was given by Wheeks et al. /2/ who draw attention to essentially nonequilibrium character of defect distribution by vibration energy when the multiphonon recombination of carriers via defect takes place.

In this work we present a detailed study of the relevant nonequilibrium distribution. Contrary to the work /2/ we take into account a strong dependence of carrier capture and emission rates on the defect energy.

We have also shown that simultaneous action of recombination and generation of electron-hole pairs via highly excited defects induces a special kind of defects diffusion in energy space with a step $\mathcal{E}_g$. As a result, in a semiconductor with high concentrations of excess electrons and holes the high-energy tail of defect distribution over the vibration energy is considerably increased.

The results are applied to the consideration of the recombination enhanced defect annealing and inelastic sputtering of solid state under ion irradiation. The modification of the recombination rate due to defect heating is also discussed.

2. Kinetic equations

The distribution of defects by vibration energy E should be characterized by two functions $f_1(E)$, $f_2(E)$ depending on the charge state of the defect. Let state 1 correspond to defect with a bound electron and state 2 to defect without an electron. The functions f_1, f_2 are normalized according to conditions

$$\int_0^\infty f_1(E)\rho_1(E)dE = f \quad ; \quad \int_0^\infty f_2(E)\rho_2(E)dE = 1 - f \qquad (2.1)$$

where f is the population number for the recombination centre, $\rho_{1,2}(E)$ is the density of vibronic states.

The kinetic equations may be written in the form

$$\frac{\partial N_1(E,t)}{\partial t} = nC_n(E-\mathcal{E}_{TC})N_2(E-\mathcal{E}_{TC}) - [pC_p(E) + e_n(E)]N_1(E) + \int_0^\infty d\mathcal{E}\, W_p(E+\mathcal{E}_{TV}+\mathcal{E};\mathcal{E})N_2(E+\mathcal{E}_{TV}+\mathcal{E}) + \frac{\partial}{\partial E}\left\{\frac{E}{\tau_L}\left[N_1(E) + kT\rho_1\frac{\partial}{\partial E}\frac{N_1(E)}{\rho_1}\right]\right\} \qquad (2.2a)$$

$$\frac{\partial N_2(E,t)}{\partial t} = pC_p(E-\mathcal{E}_{TV})N_1(E-\mathcal{E}_{TV}) - [nC_n(E) + e_p(E)]N_2(E) + \int_0^\infty d\mathcal{E}\, W_n(E+\mathcal{E}_{TC}+\mathcal{E};\mathcal{E})N_1(E+\mathcal{E}_{TC}+\mathcal{E}) + \frac{\partial}{\partial E}\left\{\frac{E}{\tau_L}\left[N_2(E) + kT\rho_2\frac{\partial}{\partial E}\frac{N_2(E)}{\rho_2}\right]\right\} \qquad (2.2b)$$

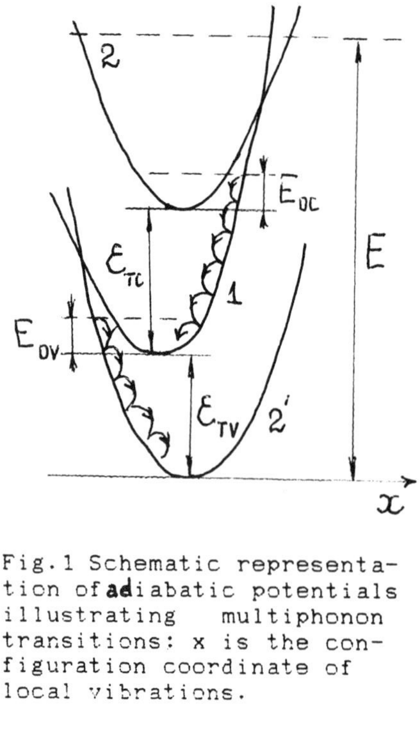

Fig.1 Schematic representation of adiabatic potentials illustrating multiphonon transitions: x is the configuration coordinate of local vibrations.

Here $N_{1,2}(E) = Nf_{1,2}(E)\rho_{1,2}(E)$ is the number of defects in state 1 or 2 with an energy E in a volume unit, N is the defect concentration; $\mathcal{E}_{TC}$, $\mathcal{E}_{TV}$ are electron and hole thermal emission energies (see Fig.1); $C_{n,p}(E)$ are the capture rates of electrons and holes by a defect with vibronic energy E; $e_{n,p}(E)$ are the emission rates of electrons and holes by such a defect. The probabilities $W_{n,p}(E,\mathcal{E})$ of emission of electron or hole with kinetic energy $\mathcal{E}$ by a defect with vibronic energy E are connected with emission rates by

$$e_{n,p}(E) = \int_0^\infty W_{n,p}(E;\mathcal{E})d\mathcal{E} \qquad (2.3)$$

The terms containing derivatives in (2.2) describe energy relaxation of local vibrations via interaction in a Fokker -Planck approximation, is the energy relaxation time τ_L determined by a local phonon decay into lattice phonons. This approximation is justified when the energy losses occur in small steps and is actually valid when the vibration energy E is significantly larger than the energy of a local phonon $\hbar\omega$. While obtaining Eqs. (2.2) we neglected the kinetic energy of the carrier being captured by a defect ($\varepsilon \sim kT_e$) compared to the vibration energy E.

The boundary conditions to the system (2.2) are

$$\left[N_i(E) + kT\rho_i(E)\frac{\partial}{\partial E}\frac{N_i(E)}{\rho_i(E)}\right]_{E=0,\infty} = 0 \qquad (2.4)$$

which correspond to zero flux in the energy space at E = 0 and $E = \infty$.

In one-mode approximation $\rho(E) = 1/\hbar\omega$, where ω is a local vibration frequency.

In a multi-mode case the functions $f_{1,2}$ represent the distributions of defects by total energy of vibrations, i.e. E is the sum energy of all vibration modes of the defect. This description is valid in case of energy redistribution between the modes being faster than the total energy change. Here the density of states depends essentially on energy E.

In a model of s identical oscillators the density of states is given by relation /2/:

$$\rho_s(E) = E^{s-1}/(\hbar\omega)^s(s-1)! \qquad (2.5)$$

Here the emission rate may be presented in the form:

$$W_s(E;\varepsilon) = \int_0^E dE' W_0(E';\varepsilon)\frac{dK_s(E,E')}{dE'} \qquad (2.6)$$

where $W_0(E';\varepsilon)$ is the emission probability when the mode responsible for emission process has accumulated energy E'. To calculate the probability $W_0(E';\varepsilon)$ one can use the results of the theory of multiphonon transitions obtained in the framework of one-mode approximation/3/.The function $K_s(E,E')$ is a probability for the relevant mode to accumulate energy greater than E', when the total vibration energy is E /2/:

$$K_s(E,E') = \rho_s(E-E')/\rho_s(E) \qquad (2.7)$$

3. Defect heating

The system (2.2) can be solved by iterative procedure. If the carrier concentrations n and p are not too great so that inequalities take place

$$nC_n(E),\ pC_p(E) \ll \tau_L^{-1} \qquad (3.1)$$

we get for the distribution function f (E):

$$f_1^{(1)}(E) = \begin{cases} A_1\exp\{-E/kT\}; & E < E_{ov} \\ A_1 e^{-E/kT} + \dfrac{\nu_R\tau_L}{kT}\displaystyle\int_{E_{ov}}^{E} e^{(E'-E)/kT}\frac{dE'}{E'\rho_1(E')}; & E_{ov} \le E < \varepsilon_{Tc} + E_{oc} \\ A_1 e^{-E/kT} + \dfrac{\nu_R\tau_L}{(E_{oc}+\varepsilon_{Tc})\rho_1(E_{oc}+\varepsilon_{Tc})} e^{(E_{oc}+\varepsilon_{Tc}-E)/kT}; & E \ge \varepsilon_{Tc} + E_{oc} \end{cases} \qquad (3.2)$$

where ν_R is the rate of multiphonon recombination via defect:

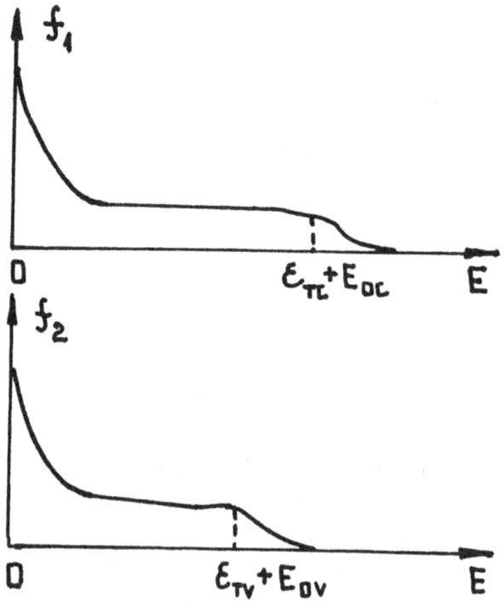

Fig.2 Distribution functions f_1 and f_2 to the first order approximation by the parameter $\nu_R \tau_L$

$$\nu_R = n\langle c_n\rangle p\langle c_p\rangle / (n\langle c_n\rangle + p\langle c_p\rangle) \qquad (3.3)$$

Here $\langle C_{n,p} \rangle$ are the mean values of the capture rates of electrons and holes; E_{0C} is an optimal energy of multiphonon transition corresponding to electron capture and/or emission process (energy of optimal tunneling, see Fig.1). The constant A_1 is determined from the normalization condition (2.1). The function f_2 is given also by the formula (3.2) where ε_{TC} and E_{0C} should be replaced by ε_{TV} and E_{0V}, respectively.

Thus, the distribution functions have nonequilibrium parts in form of an energy plateau extending up to $\varepsilon_{TC} + E_{0C}$ for f_1 and to $\varepsilon_{TV} + E_{0V}$ for f_2 (see Fig.2).

Now we consider the region of high vibration energy (above the crossing points of adiabatic potentials, see Fig.1) where the probabilities of multiphonon transitions are great. If here the inequalities opposite to (3.1) hold, so that the electron and hole capture rates are greater than the rate of energy relaxation of the local-vibration excitation due to the interaction with the lattice, we get a continuous solution to system (2.2) which decreases as $E \to \infty$ in the form:

$$N_1(E) = B\left(\frac{nC_n\, pC_p}{e_n e_p}\right)^{E/\varepsilon_g} ; N_2(E) = \frac{pC_p}{e_p} N_1(E - \varepsilon_{TV}) \qquad (3.4)$$

where B is a coefficient which is independent of the local vibration energy. This solution gives envelope functions to the tails of defect distributions over vibration energies. It is convenient to rewrite these envelope functions in the form:

$$N_{1,2}(E) \sim \exp\left\{- E/kT^*\right\} \qquad (3.5)$$

where we introduce "the effective temperature":

$$T^* = \frac{\varepsilon_g}{k \log(N_c^* N_v^* / np)} \qquad (3.6)$$

which characterizes the scale of decay of nonequilibrium distribution at high energies, the effective densities of states being determined by the relations

$$N_c^* \equiv e_n / C_n \quad , \quad N_v^* \equiv e_p / C_p \qquad (3.7)$$

4. Inelastic sputtering

The inelastic mechanism for fast-ion sputtering of the surface of solids, at which the energy of bombarding particles is transferred to the surface atoms through excitation of the electron subsystem of the solid, was discussed in /4/. It was established that the energy spectrum of sputtered particles can be approximated by a Maxwellian distribution with a temperature on the order of several thousand degrees which, as a rule, greatly exceeds the melting point of the crystal.

Such an energy distribution can be explained if it is assumed that the inelastic sputtering is caused by the heating of local and surface vibrations of the crystal as a result of multiphonon recombination of nonequilibrium carriers through the defects, which are localized at the surface, or through the surface (Tamm) states. The high-energy defect distribution tail, as was shown above, is characterized by an effective temperature T that could easily reach several thousand degrees.

5. Recombination-enhanced annealing

The rate of defect reactions (for example, defect annealing) is proportional to a frequency of defect jumps to neighbour equilibrium positions ν . This frequency is determined by

$$\nu = \nu_0 \int_{E_a}^{\infty} dE \left\{ f_1(E)\rho_{s,1}(E)K_{s,1}(E,E_a) + f_2(E)\rho_{s,2}(E)K_{s,2}(E,E_a) \right\} \qquad (5.1)$$

where E_a is the activation energy of defect reaction.

We define the enhancement coefficient of defect annealing η as a ratio of annealing rates in the presence of excess carriers and without them $\eta = \nu^*/\nu$

The main contribution to the coefficient η comes from the distribution function having a longer high-energy nonequilibrium tail. The difference of f_1 and f_2 functions is due to difference of energies supplied to the system of local vibrations in a process of electron or hole capture.

Let the thermal emission energy for electrons be larger than that for holes ($\varepsilon_{TC} > \varepsilon_{TV}$), then for weak heating (when inequalities (3.1) take place) the distribution function f_1 is given by (3.2).

If $E_a < \varepsilon_{TC}$ the defect reactions become athermal and for the coefficient η we get

$$\eta = \nu_R \tau_L \exp\left\{\frac{E_a}{kT}\right\} \int_{E_a}^{\varepsilon_{TC}+E_{0C}} dE \, \frac{(E-E_a)^{S-1}}{E^S} \qquad (5.2)$$

For $E_a > \varepsilon_{TC} + E_{0C}$ the calculation gives

$$\eta = (s-1)! \, \nu_R \tau_L \left(\frac{kT}{\varepsilon_{TC}+E_{0C}}\right)^S \exp\left\{\frac{\varepsilon_{TC}+E_{0C}}{kT}\right\} \qquad (5.3)$$

Our result differs from that found in /2/ by an additional activation energy $E_{0C}(T)$ that depends, in general, on temperature and also by a strong temperature dependence of preexponential factor.

We present also the coefficient η for heavily doped p-type semiconductor when f_{ℓ} - function has a nonequilibrium plateau extending up to energies $E \approx \mathcal{E}_g$. For $E_a < \mathcal{E}_g$ we have

$$\eta \approx \frac{np\langle c_n\rangle}{N_v^*}\int_{E_a}^{\mathcal{E}_g}\tau^*(E)\frac{(E-E_a)^{S-1}}{E^{S-1}}\,\frac{dE}{E-\mathcal{E}_{TV}} \qquad (5.4)$$

where $\tau^*(E)$ is the total time of energy relaxation of local-vibration excitations:

$$\left[\tau^*(E)\right]^{-1} = \tau_L^{-1} + \tau_p^{-1} \qquad (5.5)$$

Here τ_p is a "hole" time. The new relaxation mechanism appears due to the fact that defects capture cold (thermal) and emit "hot" carriers with the kinetic energy highly exceeding the thermal energy /5/. When the hole concentration is great, so that $\tau^* \simeq \tau_p \sim p^{-1}$, the coefficient η becomes independent of the hole concentration.

In /6/ a saturation of recombination-enhanced annealing of defects in GaAs as a function of injection hole current in n-region of p-n junction under high injection level (p>>n) was observed. We believe that this observation is connected with the fact previously mentioned. A detailed discussion of this experimental result is presented in /7/ where a modification of the recombination rate due to defect heating is also discussed.

References

1. F. Seitz and J. Koehler, in Solid State Physics, edited by F. Seitz and P. Turnbull (Acad. Press, N.Y., 1956) V. 2, p. 307.
2. J.P. Wheeks, J. Tully and L.C. Kimerling, Phys. Rev. B 12, 3286 (1975).
3. V.N. Abakumov, V.I.Perel and I.N. Yassievich, Nonradiative Recombination in Semiconductors (Elsevier, Amsterdam, 1991)
4. I.A.Baranov, Yu.V.Martynenko, S.O.Tsepelevich and Yu.N.Yavlinskii, Sov. Phys. Usp. 31, 1015 (1988).
5. V.N. Abakumov, A.A. Pakhomov, M.K. Sheinkman and I.N. Yassievich, Sov. Phys. Semicond. 23, 1382 (1989).
6. D. Stievenard and J.C. Bourgoin, Phys. Rev. B 33, 8140 (1986).
7. V.N. Abakumov, A.A.Pakhomov and I.N. Yassievich, a review to be published in Sov. Phys. Semicond. (1991).

Materials Science Forum Vol. 83-87 (1992) pp. 517-522

ELECTRON-HOLE MECHANISM OF MIGRATION AND DEFECT INTERACTION

A.B.Gerasimov

Tbilisi State University, Tbilisi, Georgia, USSR

ABSTRACT

A mechanism is proposed in which motion of atoms is interpreted as a result of a change in the quantum state of eletrons which form chemical bonds. This change can be caused by various factors: illumination, injection, heating, etc.

INTRODUCTION

Usually, when describing mechanisms of atomic motion in condensed media such as first order phase transition, migration and diffusion of atoms, defect formation (except for the impact mechanism of defect formation), a concept was used, based on molecular-kinetic theory, according to which the kinetic energy fluctuations near an atom can be large enough for the atom to change its location[1]. Inherent contradictions characteristic of these mechanisms[2] and a large amount of experimental data on stimulated athermic processes of atomic motion[3] made many authors assume that in these processes an essential role is played by the electron subsystem[4]. However, up to now there has been no mechanism which could explain thermal and athermal processes of atomic motion from a unified viewpoint.

This paper presents a mechanism, in which motion of the atoms in condensed media is interpreted as a consequence of change in quantum state of bonding electrons which can be realised in various ways: heating, ionisation, injection, illumination, etc. This approach is based on the consideration of a solid as a large molecule, in which the valence electrons can be in bonding (valence band) and antibonding (conduction band) orbitals[5], and their proportion in the vicinity of a given atom changes the directivity and bonding force of this atom with others.

This concept will be illustrated here by an example of semiconductors with purely covalent bonds.

Let us consider the mechanism of the first order phase transitions in semiconductor crystals[6]. As is known, the bonding orbitals in the p^+ state correspond to the top of the valence band, while the antibonding orbitals in the s^- state correspond to the bottom of the conduction band. Therefore, electron transitions into the conduction band change the spatial directivity and bonding force of those atoms, near which there appears a hole or a free electron (hereinafter they will be referred to as antibonding quasi-particles) in their thermal motion.

There are two ways of electron transition into the conduction band: as a result of the collision of a free electron having sufficient energy with a bound electron (impact mechanism) and as result of collisions of electrons bound with each other (i.e. interatomic collisions).

DEFECT FORMATION

As was shown in Ref. 6, the first order phase transition solid-liquid takes place within the time of one atomic oscillation, the available concentration of antibonding quasi-particles is sufficient for them to run about each atom of the crystal, i.e. $2n_{c2}n_2 = N_A$ where $n_{c2} = N_A\sqrt{m^*/M_A}$. n_2 is the number of atoms, which an antibonding quasi-particle can run about in the time T during its thermal motion; N_A is the concentration of the lattice atoms; m^* is the atomic mass of the substance. n_{c2} does not depend on the way of its realisation (heating, light, etc.). Hence, if we could somehow reach the critical concentration of the excited states at T< T_{melt}, then depending on the quantity of received heat, the state of the substance will be gradually approaching the liquid state, since the whole received heat will be spent on formation of new antibonding quasi-particles. This is confirmed by direct experiments on the influence of light on microhardness[7] and photon recrystallisation of amorphized layers of semiconductors[8].

Since electrons in the conduction band and holes in the valence band (vacant places on bonding orbitals) move throughout the crystal at random, then the probability of occurence of different quantities of antibonding quasi-particles in the vicinity of a certain atom may also be different. There exists a certain probability that near one of the atoms of a crystal (e.g. Ge or Si) four antibonding particles may occur. Then their number will be equal to that of bonding electrons (4 per atom), and this atom will not realise bonding with the neighbouring atoms, and if it is on a free surface, where the number of bonding electrons may be smaller, it may leave the substance, i.e. evaporate (sublime). With an increase in the antibonding particles concentration, the number of atoms able to leave the substance will increase. Due to lack of charge accumulation near a given atom, the most probable situation will be when the state $s^+p^+s^{2-}$ is realized. The leaving atom in this state will result in a decrease of the energy of the crystal, since the atom leaves the substance in an excited state[10].

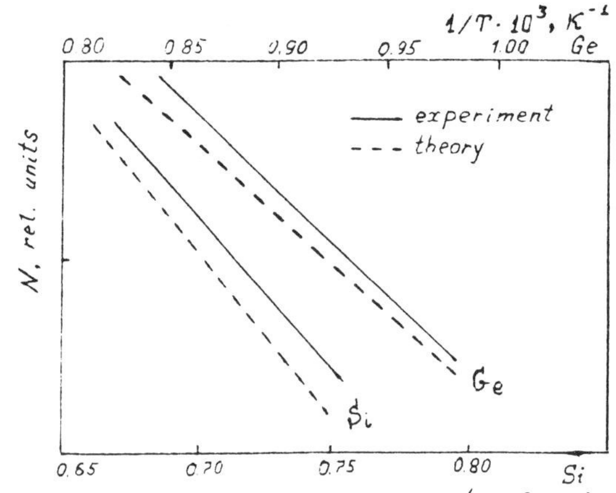

Fig. 1

If near a given atom in the bulk of the crystal the number of antibonding quasi-particles is smaller than that of bonding ones, then spatially this can be presented so that the bonding charge density in one definite direction will be larger than in other directions[9]. That is why displacement of the atom will take place in this direction, i.e. a defect will form. The presence of phonons will contribute to the displacement of the atom.

Defect formation consists of two acts: exit of an atom into an interstitial site and jumping of a neighbouring atom into the vacant place. For the first act it is necessary to have two quasi-particles and for the second act — one, if there are phonons with maximum energy, whose number is equal to their average number at melting temperature[11].

The concentration of the defects formed in the time of one oscillation, T, will be

$$N_{def}(T) = N_A W,$$

where W is the probability of defect formation according to Frenkel, and it is presented as a product of probability of exit of an atom from the site W_{ex} and probability of jumping of another atom into a vacant site W_j; $W_{ex} = \frac{1}{4}n_2^2(2n_i/N_A)^2 p(t)$, where $n_2(n_i/N_A)$ is the probability of existence of one quasi-particle near the given atom in the time T, $p = 1 - \frac{(T-t)}{T}$ is the probability of simultaneous existence of two quasi-particles in the bulk falling at one atom provided that one quasi-particle is waiting for another one for a certain average time t, n_i is the number of antibonding electrons or holes (intrinsic semiconductor). And $W_j = \frac{3}{2} n_2 \frac{n_i}{N_A} W_{ph}$ where $W_{ph} = \bar{n}_{ph}(h_{ph}/h_{ph}!)e^{-n_{ph}}$ where $\bar{n}_{ph} = 3T/t_D$ is the average number of phonons per atom, and $n_{ph} = 3T_{melt}/T_D$ is the average number of phonons per atom at $T = T_{melt}$, T_D being the Debye temperature. Thus,

$$N_{def}(T) = \frac{3}{2} N_A n_2^3 (n_i/N_A)^3 p(t) W_{ph} \quad (1)$$

The stationary defect concentration $N_{def} = (N_{def}(T)/T)T_{lt}$ where T_{lt}, the point defect lifetime is a strong function of the initial defect composition of the condensed medium, its temperature dependence varying only slightly[2,3]. Fig.1 shows temperature dependences of the equilibrium concentrations of point defects in relative units measured during the experiment and calculated by equation (1).

DIFFUSION

In the case of diffusion we shall represent one jump of a labelled atom as the following elementary acts: formation of a point defect; fall of the labelled atom on a site as a result of dissociation of a dumb-bell (a complex of an atom in the interstitial and one of the nearest lattice atoms).

The probability of point defect formation is given by equation (1), and for dissociation of the dumb-bell, the latter must have two antibonding quasi-particles $W_{ab} = \alpha^2 n_2^2 (n_i/N_A)^2 p(t)$, where α is the same parameter characterising an increase of the probability of revealing an antibonding electron or a hole near the defect. In the first approximation α can be expressed as a ratio of the volume of the violated region of the defect to the volume fallng at one lattice atom.

To obtain the self-diffusion coefficient of the label-led atoms, their probability should be multiplied by the mixing coefficient dV_A, where V_A is the thermal velocity of an atom, and d is the shortest distance between the atoms:

$$D = \frac{3}{2} \alpha^2 d V_A n_2^5 (n_i/N_A)^5 p(t)^2 W_{ph} \quad (2)$$

Fig.2 shows temperature dependence of the self-diffusion coefficients of the labelled Ge and Si atoms measured experimentally[12] and calculated by equation (2).

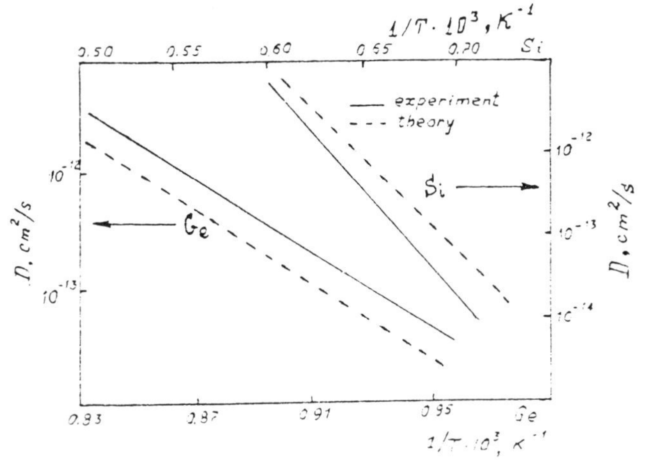

Fig 2

The impurity diffusion coefficient is expressed similarly to equation (2) with the corresponding α for different impurities.

The migration of point defects is also determined by the probability of antibonding quasi-particles existing near the defects. The interstitial atom generates the dumb-bell state which breaks down upon appearance of a quasi-particle in the vicinity of the dumb-bell. As a result one of the atoms can fall on a site and the other one create a new dumb-bell state[14]. Thus the dumb-bell state migration velocity is by many orders of magnitude higher, than that of the labelled atoms, since it is proportional only to $n_2^2(n_i/N_A)^2$, while the labelled atom migration is proportional to $n_2^5(n_i/N_A)^5$.

The vacancy migration is determined by jumps of neighbouring atoms to a vacant site. The electrons captured the vacancies move along the radius equal to several interatomic distances. Unlike the vacant site region, the electrons in this region can only be on loosening orbitals of the lattice atoms, since the bonding orbitals of the latter are occupied and thus loosen the bonds between the atoms surrounding the vacancy, increasing the probability of jumping to a vacant site.

This accounts for the fact that negatively charged vacancies in semiconductors migrate much faster, than non-negatively charged vacancies[15]. In random motion of two negatively charged vacancies, when they approach each other, so that the wave functions captured by electrons overlap to such an extent that they can exchange atoms, which are between these vacancies, is more intensive, than other atoms. As a result, the negatively charged vacancies will tend to each other and form divacancies. Just in the same way polyvacancies and impurity precipitates are formed.

COMPLEX FORMATION

To form a complex, the defects must approach each other. Their motion in the crystal lattice, however, should not be considered as fully chaotic. At certain distances between them there occurs an interaction which correlates their motion. The forces responsible for such interaction can be of various character; but, in our opinion, in the complex formation processes an important role is played by exchange forces, which effectively act at distances specified by an essential overlap of wave functions of the electrons belonging to defects, for example, by an overlap at which hopping conduction can take place.

From the moment of occurence of such overlaps, the molecular orbitals created by the defects can split into antibonding and bonding ones. This splitting will increase with decrease of interdefect distances. So, it will be profitable for the defects to move to each other, thus creating complexes. Just this fact can explain formation of clusters (precipitates) of some impurities in semiconducters. If the energy of bonding between the lattice and atoms of a given impurity is smaller than the energy of bonding between the atoms, then the atoms, occurring at a distance of effective interaction, will accumulate in the form of inclusions of a new phase, i.e. precipitates. And indeed, the precipitates are known to be formed by those impurities which have large diffusion coefficient and small solubilities[16]. Such impurities with weak bonds in the sites can easily go into interstitials and have a high mobility[16]. That is why they can remain in the sites only in such concentrations, when the distances between them are such that the exchange interaction energy is smaller, than the bonding energy of such an atom in the site, and the atoms, which are arranged closer to each other, will go into the precipitate.

During the interaction of an intrinsic interstitial atom (IIA) with an impurity atom in the site, the probability of pressing out the impurity atoms into the site is higher for those atoms, whose bonding orbitals of energy state of the orbitals of the lattice atoms[17]. The same dependence of the energy state of the orbitals exists during dissociation of the complex IIA - impurity atom in the site. In the course of dissociation of the complex vacancy + impurity atom in the site there is a reverse dependence: the more the difference in energy between bonding orbitals of an impurity and lattice atoms, the lower is the probability of leaving the complex by the vacancy[18].

Since for motion and interaction of atoms and defects it is essential that a certain number of antibonding quasi-particles should exist near them, and it is absolutely of no importance how they are formed, one can accomplish almost all technological procedures of making solid state electronic devices using light. Proceeding from these considerations, a low-temperature technology for producing GaAs devices has been developed.

REFERENCES

1) Frenkel, Ya.I. : The Kinetic Teory of Solids and Liquids, 1970, Leningrad (in Russian).
2) Boltaks, B.I. : Diffusion and Point Defects in Semiconductors, 1972, Moscow, 46 (in Russian).
3) Seredin, V.G., Ukrozhenko, V.M. : Zarubezhnaya Elektronmaya Tekhn., 1986, 11, 33.
4) Akhmanov, S.A. et al. : Uspekhi Fiz,Nauk, 1985, 147, 4, 675.
5) Harrison, W.A. : Electronic Structure and the Properties of Solids, 1980, San Francisco.
6) Gverdtsiteli, I.G. et al. : Bull.Acad.Sci. of the Georgian SSR, 1964, 115, 3, 513 (in Russian).
7) Gerasimov, A.B., Jibuti, Z.V., Chiradze, G.D. : Bull. Georgian Acad. Sci., 1991, 142, 1, (in Russian).
8) Gverdtsiteli, I.G. et al. : Poverkhnost, 1985, 11, 132.
9) Lindfelt, U. : Abstracts of the Fourth "Lund" Intern. Conf. on Deep Level Impurities in Semiconductors, Eger, Hungary, 1983; Scheffler, M. et al. : ibid.
10) Gerasimov, A.B., Kimeridze, D.A. : Abstracts of the All-Union Symposium on Emission from semiconductor Surfaces, Including Exoemission, Lvov, October 17-19, 1989, p.12 (in Russian).
11) Gverdtsiteli, I.G. et al. : Bull.Acad.Sci. of the Georgian SSR, 1987, 127, 3, 517 (in Russian).
12) Boltaks, B.I., Budarina, S.I. : Fiz.Tverd.Tela, 1969, 11, 2, 418.
13) Tables of Physical Quantities, Handbook, 1976, Moscow (in Russian).
14) Gerasimov, A.B., Tsertsvadze, A.A.: Fiz.Tekh.Poluprov., 1979, 13, 2, 350.
15) Yemtsev, V.V., Mashovets, T.V. : Impurities and Point Defects in Semiconductors, 1972, Leningrad (in Russian).
16) Milnes, A. : Impurities with Deep Levels in Semiconductors. Moscow, 1977 (in Russian).
17) Alania, N.M. et al. : Proc. of the 3rd Intern. Conf. on Radiation Physics of Semiconductors and Related Materials, 1979, Tbilisi, 570.
18) Gerasimov, A.B., et al. : Fiz. Tekh. Poluprov., 1986, 20, 4, 641.

Materials Science Forum Vol. 83-87 (1992) pp. 523-526

DILUTED MAGNETIC SEMICONDUCTORS WITH Cr^{2+} - UNUSUAL p-d INTERACTION*

J. BLINOWSKI[1] AND P. KACMAN[2]
[1]*Institute of Theoretical Physics, Warsaw University,*
ul. Hoza 69, 00-681 Warsaw, Poland.
[2]*Institute of Physics, Polish Academy of Sciences,*
Al. Lotnikow 32/46, 02-668 Warsaw, Poland

ABSTRACT

In zinc-blende type semimagnetic semiconductors (diluted magnetic semiconductors) with Cr^{2+} content the non-Heisenberg part of the p-d interaction together with the strong, static Jahn-Teller effect is shown to lead to a zero-magnetic field splitting and significant modifications of the spin splittings for the valence band. A maximum effect can be achieved when all Jahn-Teller distortions in the cubic crystal are oriented along one of the [100]-type crystalographic directions.

The interesting properties of diluted magnetic semiconductors (DMS) result from spin dependent exchange interactions of the band electrons with paramagnetic ions. It is commonly believed that these interactions can be described by an effective Heisenberg type Hamiltonian proportional to the product $S \bullet J$, where S is the total spin operator of d-electrons and J is the appropriate quasi-angular momentum for the band electron.

In general, two mechanisms: the direct Coulomb exchange and the hybridization of the band states with the d-orbitals of the ion, are considered. It was rigorously derived for the case of Γ_6 electrons (J=1/2) interacting with any paramagnetic impurity[1] that the interaction induced by the direct Coulomb exchange mechanism can be described by a *ferromagnetic* Heisenberg Hamiltonian. The same was shown for Γ_8 bands (J=3/2) but only in the case of S-state ions (with the total orbital angular momentum of d-electrons L=0)[2]. On the other hand, for Γ_8 bands in DMS doped with S-state paramagnetic impurities (Mn^{2+}, Fe^{3+}), the measurements gave a negative exchange constant β (see ref.2). Since the s-d hybridization mediated interactions were shown to be described by an *antiferromagnetic* Heisenberg Hamiltonian[3,4], the experimental results have been interpreted as an indication that for Γ_8 bands the dominant interactions originate from the p-d hybridization[5,6,7].

The hybridization mechanism can be thought of in terms of virtual transitions, in which an electron hops out of the band into a given impurity d-orbital and vice versa. For S-state

A finite total orbital angular momentum of the d-shell electrons implies, that some of the orbitals are doubly occupied or are empty. As a result, the virtual transitions involving these particular orbitals do not depend on the spin of the band electron. Consequently, a non-trivial modification of the spin-dependent interactions between the mobile and localized electrons can be expected in the case of non-S-state ions.[4,8]

In zinc-blende type DMS with Cr^{2+} (which is a D-state ion) the ground state of the paramagnetic impurity is an orbital triplet T_2 (as long as the weak T_2-E mixing due to the spin-orbit interaction is neglected). For each of the $^5T_2$ states one of the one-electron orbitals t_{2g} is empty. It was demonstrated in Ref.9, that for a Γ_8-electron interacting with four d-electrons of Cr^{2+} ion the usual Heisenberg-like Hamiltonian $H_H = -B\,\mathbf{S}\cdot\mathbf{J}$, which describes the hybridization mediated spin-dependent part of the interaction, has to be completed by additional non-Heisenberg terms:

$$\hat{H}_{NH} = (B+3C)\hat{\mathbf{A}} + (B+2C)\left\{(1/2)\,[\hat{\mathbf{A}},\mathfrak{L}\cdot\mathbf{S} + \mathbf{S}\cdot\mathbf{J}]_+ + \mathbf{S}\cdot\mathbf{J}\right\} \qquad (1)$$

where

$$\hat{\mathbf{A}} = (\mathfrak{L}\cdot\mathbf{J})^2 - (1/2)\mathfrak{L}\cdot\mathbf{J} - (5/2) \qquad (2)$$

and $\mathfrak{L}$ denotes the fictitious orbital angular momentum for the T_2 triplet ($\mathfrak{L}=1$). Both the exchange constants B and C are negative and the formulae for them can be found in ref.10.

According to the Jahn-Teller theorem, the orbital degeneracy makes the ground state of Cr^{2+} unstable against an asymmetric displacement of the neighboring ions. The Cr^{2+} impurities in $A^{II}B^{VI}$ compounds are known to undergo a strong, static Jahn-Teller effect, in which the local distortion reduces the symmetry around each ion to D_{2d}; the ground state becomes an orbital singlet $^5B_2$ separated from the upper states by the Jahn-Teller energy $3E_{JT}$, roughly constant for all $A^{II}B^{VI}$ compounds ($\simeq$400-600 cm^{-1})[11]. Such a static distortion is possible along each of the cubic axes.

For the distortion in z-direction along the [001] crystalographic axis the orbital ground state has the form:

$$|\zeta\rangle = |\mathfrak{M}=0\rangle \qquad (3)$$

where $|\mathfrak{M}=0\rangle$ is the state with the projection of the fictitious angular momentum $\mathfrak{L}$ on the [001] direction equal to zero. The ground states for the remaining two distortions, along [100] and [010] axes, are conventionally denoted by $|\xi\rangle$ and $|\eta\rangle$, respectively.

In DMS crystals the mobile band electrons interact simultaneously with all magnetic ions. Within the mean field approximation this interaction is diagonal in $\vec{k}$ and can be obtained by calculating the thermodynamic average of $H_H + H_{NH}$ for each Cr^{2+} ion and summing over all ions in the unit volume of the crystal. For $kT<3E_{JT}$ the thermodynamic average involves only the

Within the mean field approximation this interaction is diagonal in $\vec{k}$ and can be obtained by calculating the thermodynamic average of $H_H + H_{NH}$ for each Cr^{2+} ion and summing over all ions in the unit volume of the crystal. For $kT<3E_{JT}$ the thermodynamic average involves only the orbital ground singlets $|\xi\rangle$, $|\eta\rangle$ or $|\zeta\rangle$. The final p-d interaction depends strongly on the relative numbers of ions with the Jahn-Teller distortion along each of the three cubic axes.

It is well known that one can change these numbers by, e.g., application of uniaxial stress[11]. In the limit case, with all the Jahn-Teller distortions and the external magnetic field along the [001] axis, the part of the mean field effective mass Hamiltonian representing the p-d interaction is diagonal in the Luttinger m_J,k representation. The eigenvalues of this part, which determine the band splittings at k=0 are:

$$E_{\pm 3/2} = -(B + 3C)xN_0 \pm (3/2)BM$$
$$E_{\pm 1/2} = +(B + 3C)xN_0 \mp (1/2)(B + 4C)M \qquad (4)$$

where M represents the total magnetization of xN_0 Cr^{2+} ions per unit volume (in $g\mu_B$ units). One can see from Eqn. (4) that two important effects of p-d hybridization might be expected in this case. First, there is a zero-field splitting proportional to the concentration of Cr^{2+} ions. Second, the dependence of the valence band Zeeman splittings on the magnetization differs considerably from the form $B_{Mn}MJ_z$ established for DMS with Mn^{2+} content.

In the opposite limit, when none of the three possible axes of the Jahn-Teller distortion is privileged, the effective mass p-d Hamiltonian for the magnetic field along the [001] direction is again diagonal but with quite different eigenvalues:

$$E_{\pm 3/2} = \pm(B - C)M \pm (1/3)(B + 2C)\Delta M$$
$$E_{\pm 1/2} = \pm(1/3)(B - C)M \pm (5/9)(B + 2C)\Delta M \qquad (5)$$

where ΔM is the difference between contributions to the magnetization M from ions with the distortion axes parallel ([001]) and perpendicular ([100] or [010]) to the magnetic field. In this case, there is no zero-field splitting due to p-d hybridization. Still, the expected pattern of the Zeeman splittings and its temperature and magnetic field dependencies are not standard, since in general $\Delta M \neq 0$. It seems that among all zinc-blende type DMS based on II-VI compounds one can put $\Delta M \cong 0$ only for CdTe crystals, where the spin-orbit interaction for Cr^{2+} ions is really very small.

References

*This work was partially supported by the grant G-MEN-128/90

1. S. H. Liu, Phys. Rev. 121, 451 (1961)
2. J. Kossut, *Diluted Magnetic Semiconductors*, (eds. J. K. Furdyna & J. Kossut), Semiconductors and Semimetals, vol.25, p.183, Academic Press, San Diego (1988)
3. J. R. Schrieffer, P. A. Wolff, Phys. Rev. 149, 491 (1966)
4. J. R. Schrieffer, Jour. Appl. Phys. 38, 1143 (1967)
5. T. Dietl, *Physics in High Magnetic Fields*, p.344, (eds. S. Chikazumi & N. Miura), Springer, Berlin (1981).
6. A. K. Bhattacharjee, G. Fishman, B. Coqblin, Physica 117B, 449 (1983)
7. B. E. Larson, K. C. Hass, H. Ehrenreich, A. E. Carlsson, Phys. Rev. B 37, 4137 (1988)
8. R. E. Watson, S. Koide, M. Peter, A. J. Freeman, Phys. Rev. A 139, 167 (1965)
9. J. Blinowski, P. Kacman, H. Przybylinska, Sol. State Commun. (in press)
10. J. Blinowski, P. Kacman, Proc. XX ICPS Thessaloniki, Greece, 1990, (eds. E. M. Anastassakis & J. D. Joannopoulos), World Scientific, (1990), Vol.3, p.1827.
11. J. T. Vallin, G. D. Watkins - Phys. Rev. B, 9, 2051 (1974)

Materials Science Forum Vol. 83-87 (1992) pp. 527-532

STABILITY OF THE POSITIONS OF AN INTERSTITIAL IMPURITY ATOM AND THE ELECTRONIC STATES IN SEMICONDUCTORS

Yuzo SHINOZUKA
Department of Applied Science, Faculty of Engineering, Yamaguchi University
Tokiwadai 2557, Ube, Yamaguchi 755, Japan

ABSTRACT

We apply the Jahn-Teller theorem to an interstitial impurity atom in semiconductors to investigate the stability of the position. On-center and off-center instability, and application to a new impurity diffusion process enhanced by electronic excitation are presented.

1. Introduction

Recently much progress has been done in *ab initio* calculation of electronic states of condensed matter. Even for systems without translational symmetry, the combination of the Local Density Approximation and Molecular Dynamics enable us to obtain the stable atomic positions and electronic states. In order to obtain reliable results, however, it needs enormous computing time. In the present paper, we discuss the position of an interstitial impurity atom in semiconductors. As will be shown below, if we only know the symmetry of the position and the symmetry of the electronic state we can judge the stability of the position.

Let us discuss briefly. An impurity atom may occupy one of the positions with larger empty space in the host lattice, which then possess a high point-symmetry. However, whether they are really stable or not depends on the character of the system and their electronic states. Localized electronic states in highly symmetrical systems are usually degenerate. When electrons are in one of the degenerate states, the Jahn-Teller (J-T) distortion *always* takes place to reduce the symmetry.[1] If the Jahn-Teller distortion mode(s) contains the displacement R_0 of the impurity atom, the position must be unstable for this electronic state. The criterion of this instability depends on the symmetry of positions, as will be shown below.

2. Instability by the Jahn-Teller Effect

Let us discuss stable positions of an interstitial impurity atom in semiconductors. It may occupy a position with high point-symmetry, for example, **A** or **B** in Fig. 1.

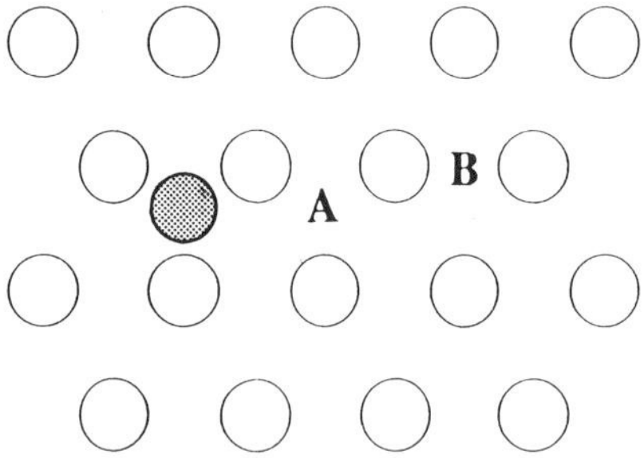

Fig. 1 Position of an interstitial atom.

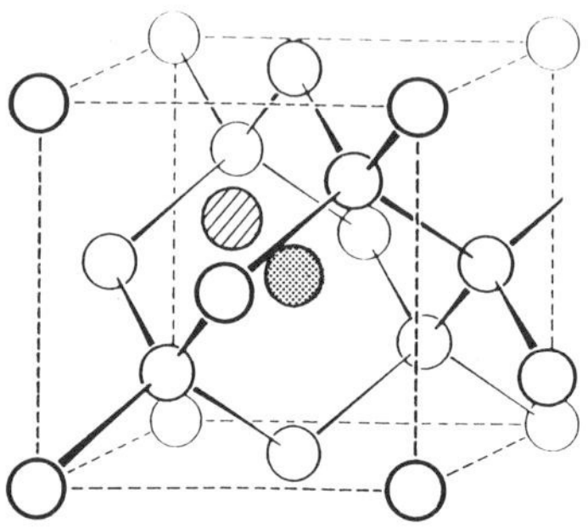

Fig. 2 T_d (●) and D_{3d} (▨) positions in the diamond structure.

When an impurity locates a symmetrical position, surrounding host atoms may be shifted from those in the perfect lattice, while the point symmetry may remain unchanged. We pay attention here only to the symmetry of the system, which will be denoted by **G**. The displacements of the impurity and host atoms from the positions in **G** are denoted as R_j (j = 0, 1,...; j = 0 refers to the impurity).

At first, we discuss a case where an impurity can bind only one electron (hole). The Hamiltonian of the system is given by

$$H = H_e(r) + V(r, R) + U_L(R). \tag{1}$$

Here r denotes the position of an electron and R represents the set $\{R_0, R_1, \ldots\}$. $U_L(R)$ is the elastic energy for the displacements of atoms. The electron-lattice interaction term $V(r, R)$ can be expanded as a power series of R_j's.

$$V(r,R) = V(r,R=0) + \sum_j \frac{\partial V}{\partial R_j} R_j + \ldots \tag{2}$$

Let us start from the case R = 0 (the symmetrical position **G**). An electron is bound to one of the localized electronic states $\Phi(r)$:

$$(H_e(r) + V(r, R = 0))\Phi(r) = E\Phi(r), \tag{3}$$

which can be classified by the irreducible representation Γ of **G** as $\Phi_\Gamma(r)$. Next, we allow the atomic displacement. The generalized force $F_\Gamma{}^j$ for R_j is given by

$$F_\Gamma{}^j = \left\langle \Phi_\Gamma \left| \frac{\partial V}{\partial R_j} \right| \Phi_\Gamma \right\rangle. \tag{4}$$

If $\Phi_\Gamma(r)$ is degenerate and the symmetric representation $[\Gamma \times \Gamma]$ contains the irreducible representation(s) to which the displacement R_0 of the impurity belongs, then $F_\Gamma^j \neq 0$ and we can conclude that the symmetrical position of an impurity in **G** is unstable for the degenerate electronic state Φ_Γ.* Thus, we can obtain useful information on the stable positions of an impurity atom, without calculating the total energy of the system and its derivative with respect to R_0.

* In molecular systems the displacements $\{R_j\}$ are usually transformed into the normal modes $\{Q_\Gamma\}$ in the irreducible representation Γ. In the present case, however, the transformation into $\{Q_\Gamma\}$ is a hard task and sometimes almost impossible. It is much easy to find which Γ('s) the center displacement R_0 belongs to.

(a) *Diamond Structure*
Let us apply the above theorem to the tetragonal interstitial site (T_d) in the diamond structure (Fig. 2). The irreducible representations of T_d symmetry are A_1, A_2, E, T_1, and T_2. The Jahn-Teller distortion modes for degenerated electronic states are given in Table 1. The displacement R_0 of the impurity belongs to T_2. Then we can conclude that in the diamond structure the T_d impurity position is unstable when an electron is bound to a triply degenerate T_1 or T_2 state. How is the hexagonal interstitial site (D_{3d}) in the diamond structure? Because of its inversion symmetry, the irreducible representations are classified into *gerade* and *ungerade* as A_{1g}, A_{2g}, E_g; A_{1u}, A_{2u}, E_u. All the J-T distortion modes are *gerade* (A_{2g}, E_g), while the displacement R_0 of the impurity is *ungerade* (belongs to A_{2u} and E_u). So, the D_{3d} impurity position in the diamond structure does not show the instability caused by the J-T effect.

(b) *Other Lattice Structures*
The last conclusion can be generalized as follows: an impurity atom at any interstitial position which has the inversion symmetry does not show the instability caused by the J-T effect. Another examples of such site are body centered position (O_h) and face centered position (D_{4h}) in the simple cubic lattice.

In the zincblende structure the tetrahedral (T_d) and the hexagonal (C_{3v}) positions are candidates for an interstitial impurity. At the T_d position, the same Jahn-Teller instability takes place as in the diamond structure. The hexagonal (C_{3v}) position, which does not have inversion symmetry, shows the J-T instability to displace the impurity atom when an electron is bound to a doubly degenerate E state (see Table 1).

Wurtzite structure has three candidates for interstitial position, which all have C_{3v} symmetry. The same conclusion can be given as in the zincblende structure.

Table 1. The Jahn-Teller distortion modes and the type of R_0.

T_d		D_{3d}		C_{3v}	
electronic state Φ_Γ	J-T distortion mode Q_Γ	electronic state Φ_Γ	J-T distortion mode Q_Γ	electronic state Φ_Γ	J-T distortion mode Q_Γ
E	E				
T_1	E, T_2	E_g, E_u	A_{2g}, E_g	E	E
T_2	E, T_2				
R_0 belongs to T_2		R_0 belongs to A_{2u} and E_u		R_0 belongs to A_1 and E	
example tetrahedral position in the diamond and zincblende structures		*example* hexagonal position in the in the diamond structure		*example* hexagonal position in the zincblende structure. three positions in the wurtzite structure	

(c) *Many Electrons and Spin*
When an impurity atom binds more than one electron, the above results remain valid if an electronic state Φ_Γ is considered as a many electron state. If an electronic state Φ_Γ contains the spin freedom, the Jahn-Teller instability depends on the number of electrons. The criterion for even number electrons is whether the symmetric representation $[\Gamma \times \Gamma]$ contains the displacement R_0 of the impurity, while that for odd number is whether the antisymmetric representation $\{\Gamma \times \Gamma\}$ contains or not (see ref. 2).

Finally, the above results also can be applied to a *substitutional* impurity. There the present Jahn-Teller instability corresponds to an off-center instability of an impurity atom, as is realized in N in Si.[3,4]

3. New mechanism for electronically enhanced impurity diffusion

An impurity atom can diffuse in a host crystal via interstitial positions. We discuss here an impurity whose stabilities of the interstitial positions are different for different electronic states.
If we excite a bound electron(s) of the impurity to an excited state which shows the present Jahn-Teller instability, the *lattice relaxation* occurs: this means an increase of the kinetic energy of the impurity atom, which may overcome the potential barrier (Fig. 3a). Thus the diffusion process of an impurity can be accelerated. The lattice relaxation of substitutional impurities promotes a defect reaction.

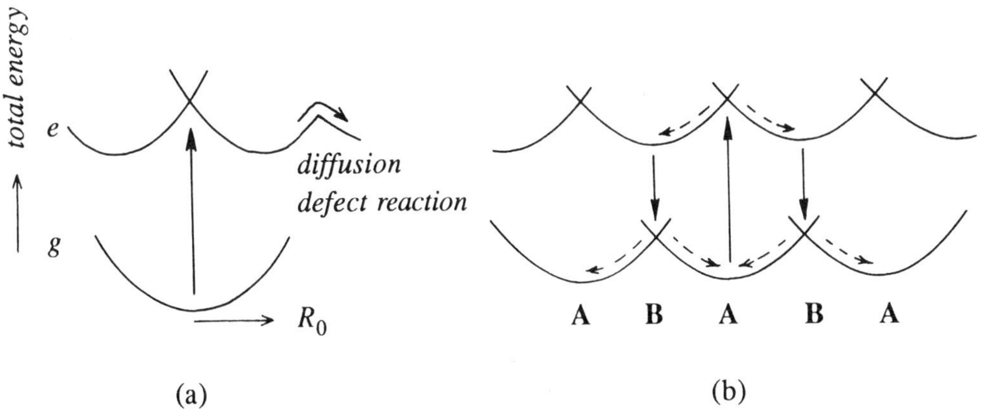

Fig. 3. Electronically enhanced diffusion (defect reaction) of an impurity.

If an electronic state which shows the present Jahn-Teller instability at position **A** continuously changes into a state which does not show at position **B** and *vice versa* (Fig. 3b; self interstitial in Si is shown to be this type[5]), the diffusion process is enhanced by the electronic excitation and the deexcitaion. This is not a kind of enhancement discussed by Sheinkman[6], but is similar to the Bourgoin-Corbett mechanism,[7] however, the number of bound electrons (holes) here remains constant.

4. Summary

We have shown a useful criterion on the stability of positions of an interstitial impurity atom in semiconductors. If we only know the symmetry, the stability of the position can be obtained without doing elaborate calculation by changing the atomic displacement.

In the present paper we have treated the atomic motion classically. When the relaxation energy E_{LR} caused by the J-T effect is smaller than the average phonon energy $\hbar\omega$, transitions between the equivalent minimums of the atomic configuration occur rapidly. If the mean transition time is shorter than characteristic time of the experiment, the average lowering of the symmetry is zero (the *dynamic* J-T effect). On the other hand, when $E_{LR} >> \hbar\omega$ the present J-T instability really takes place to reduce the symmetry. As have been stated in the original paper,[1] the Jahn-Teller theorem does not give us the degree of the atomic distortion and hence that of E_{LR}. However, we hope that the present result might be useful in the physics of defects in semiconductors.

Acknowledgment

The author thanks Professor T. Wada of Nagoya Institute of Technology, Dr. M. Uematsu of NTT LSI Laboratories and Professor F. S. Ham of Lehigh University for valuable discussions. This work is partially supported by a Grant-in-Aid for Scientific Research on Priority Area "New Functional Materials" from the Ministry of Education, Science and Culture of Japan.

References

[1] H. A. Jahn and E. Teller: Proc. Roy. Soc. A**161** (1937) 220, and H. A. Jahn: Proc. Roy. Soc. A**164** (1938) 117.

[2] T. Inui, Y. Tanabe and Y. Onodera: *Group Theory and Its Applications in Physics* (Springer-Verlag, Berlin 1990) Sections 6.5 and 10.4.

[3] K. L. Brower: Phys. Rev. B**26** (1982) 6040.

[4] K. Murakami, H. Kuribayashi and K. Masuda: Phys. Rev. B**38** (1988) 1589.

[5] R. Car, P. J. Kelly, A. Oshiyama and S. T. Pantelides: Phys. Rev. Lett. **52** (1984) 1814.

[6] M. K. Sheinkman: JETP Lett. **38** (1983) 330, and M. K. Sheinkman and L. C. Kimerling: *Defect Control in Semiconductors*, ed. K. Sumino (North-Holland, 1990) p. 97.

[7] J. C. Bourgoin and J. M. Corbett: Phys. Lett. **38**A (1972) 135, and Radiation Effects **36** (1978) 157.

Materials Science Forum Vol. 83-87 (1992) pp. 533-538

LOCAL MODE SPECTROSCOPY AND MODEL STUDY FOR ASSESSING THE ROLE OF LIGHT DEFECTS IN III-V COMPOUND SEMICONDUCTORS

D. N. TALWAR[1], M. O. MANASREH[2], D. W. FISCHER[2], S. J. PEARTON[3], AND G. MATOUS[1]

[1]Department of Physics, Indiana University of Pennsylvania, Indiana, PA, 15705-1087, USA
[2]Wright Laboratory, Wright Patterson AFB, Ohio 45433, USA
[3]AT&T Bell Laboratories, Murray Hill, New Jersey 07974, USA

ABSTRACT

High resolution local-mode spectroscopy is used as a quantitative diagnostic tool to assess the role of light impurities in III-V compounds (GaAs, InP) and related materials such as $Al_xGa_{1-x}As$. In GaAs, the observed localized vibrational modes (LVMs) from Be_{Ga} acceptors, C_{As} and B_{As} acceptors; Si_{Ga} donors; Si_{As} acceptors; Si_{Ga}-Si_{As} pairs, and Si-X(Y) complexes are evaluated in terms of realistic lattice dynamical calculations. New results of the infrared (ir) absorption experiments on proton- and deuteron-implanted InP are reported. To predict the role of hydrogen in III-V compounds, and to provide explanations for the observed oxygen-related LVMs in GaAs, the above theoretical study is found significantly important in our analysis of the experimental data. The effects of local environment on LVMs around light defects in $Al_xGa_{1-x}As$ observed recently in ir experiments are compared and discussed with model calculations.

1. Introduction

The measurements of high frequency localized vibrational mode (LVM) absorptions, due to light substitutional impurities in elemental semiconductors, have long been utilized for impurity identification and concentration determination.[1] In a III-V compound semiconductor (viz., GaAs), the site dependence of the impurity i.e., being substitutional for the cation or anion host lattice atom, is generally inferred by correlation to electrical activation (donor or acceptor behavior) as determined by Hall measurements. Unfortunately, this study becomes frequently complicated if impurities other than the one being investigated are also present. Using additional knowledge available when such LVM are measured with a high resolution Fourier transform infrared (FTIR) spectrometer, the information regarding site selectivity can be obtained directly and unambiguously without the need to resort to supplementary data. Specifically, additional structure in the LVM absorption can be attributed to differences in the isotopic masses of the nearest-neighbor (nn) host lattice isotopes.[2] This aspect of the local-mode spectroscopy is explored here to assess the role of light defects in GaAs with possible extensions to other technologically important materials.

2. Role of light (Si, C, B, Be, and O) defects in GaAs

Background

Silicon is one of the most widely employed shallow dopants in III-V compounds because it has a high incorporation coefficient and low diffusivity. Under appropriate growth conditions one can dope GaAs *n*- or *p*-type with Si atoms. More interestingly, it has been suggested in a recent theoretical work[3] that the donor (Si_{Ga}) impurity in GaAs is metastable. That is, besides forming a shallow level, silicon also forms a deep state as the donor level traps an electron and undergoes a large lattice relaxation. This deep level is identified as a DX center, which is typical of the alloy $Al_xGa_{1-x}As$ and produces a detrimental behavior in the electronic properties of the alloy.

In prior studies, considerable evidence has revealed that $^{28}Si_{Ga}$ donor is responsible for an LVM absorption band near 384 cm^{-1} while $^{28}Si_{As}$ acceptor gives rise to a weaker absorption line near 399 cm^{-1}. Each defect has a tetrahedral T_d symmetry but exhibits a different frequency because of its bonding to different neighbors. At high Si-concentrations, nn "donor-acceptor" pairs $^{28}Si_{Ga}$-$^{28}Si_{As}$ of C_{3v} symmetry are expected. Besides forming nn pairs in GaAs, silicon also enters into a number of complexes if other impurities (e.g., H, Li, Be, Cu, etc.) are present. For nn Si-pairs, three satellite lines were assigned, although, four infrared (ir) active vibrational modes are allowed by symmetry. Two additional defect complexes (Si-X and Si-Y), each involving one silicon atom were also suggested (see: Table - I). Due to the overlapping absorption lines from several next-nearest neighbor (nnn) $^{28}Si_{Ga}$-Cu_{Ga} or $^{28}Si_{Ga}$-Li_{Ga} complexes, the earlier assignments for the LVM of nn pairs were complicated. Subsequent samples, when compensated by high energy (1-2 MeV) electron irradiation, have led to the simplification of the absorption spectra. Consequently, ir lines due to Si-isotopes $^{29}Si_{Ga}$ at 378 cm^{-1} (4.7% abundant), and $^{30}Si_{Ga}$ at 373 cm^{-1} (3%) were identified. Some samples were also grown with dopant enriched with ^{30}Si isotopes. The high resolution FTIR measurements provided positive identification of the lines at 464 cm^{-1} and 393 cm^{-1} as $\{^{28}Si_{Ga}$-$^{28}Si_{As}\}$ nn pairs. On the other hand, no convincing evidence was furnished of the third line at 367 cm^{-1} for a similar interpretation. Absorption strengths of Si-X (at 369 cm^{-1}) and Si-Y (at 367 cm^{-1}) lines when correlated to the pre-compensation values of carrier densities, the mode at 367 cm^{-1} was best fitted by assuming it to be a donor and the line at 369 cm^{-1} was fitted by assuming it to be an acceptor, although this evidence is felt statistically inconclusive. Recent studies in highly doped GaAs:Si samples grown by molecular-beam epitaxy (MBE) and metalorganic chemical vapor deposition (MOCVD) technique, have shown the presence of Si-X defects only and no signature of Si-Y centers is found. Such studies have also confirmed that the Si-X complex is a deep acceptor with the possible structure of Si_{As}-V_{Ga}. Since the absorption lines of Si-Y and Si-X centers overlap in crystals grown by Bridgman technique, therefore, no positive interpretation exists for the structure of Si-Y complex.

In contrast to the *two* LVMs due to isolated Si defects, the ir measurements in ^{12}C-doped GaAs samples have demonstrated the occurrence of only one vibrational mode near 582.4 cm^{-1}. An isotopically shifted band near 561.2 cm^{-1} has also been detected due to ^{13}C. Although, chemical analyses of the samples provided no information about the lattice location of the impurity atoms, however, photoluminescence and other measurements have shown that carbon is a shallow acceptor. Now the observed fine structure of ^{12}C LVM by FTIR and its comparison with $^{28}Si_{As}$ suggests that the mode at 582.4 cm^{-1} arises from an acceptor ($^{12}C_{As}$) and the satellite lines originate from the *two* Ga isotopes in several nn configurations.[2]

The ir absorption due to boron in GaAs was first reported by Newman.[1] Besides the lines due to isoelectronic $^{10}B_{Ga}$ and $^{11}B_{Ga}$ defects observed at 540 and 517 cm^{-1}, two additional modes at 628 and 601 cm^{-1} were also detected in irradiated n-type GaAs. These features were assigned earlier to unidentified but boron related $^{10}B(2)$ ($^{11}B(2)$) defects. Again, two satellite lines for the nn $^{11}B(2)$-Si_{Ga} pairs and three LVMs were observed for the complexes involving B(2) and group VI_{As} donors (see: Table - I). An equally interesting result found was that in each case the mean frequency of the modes of a complex is close to that of an unpaired B(2)-center. Although, these observations suggest that an isolated B(2) center should have a T_d symmetry, however, an ambiguity in the interpretation still remains. If one assumes B(2) center to be an antisite defect (B_{As}), then the nn B_{As}-Si_{Ga} complex would have a C_{3v} symmetry (two modes) and a next nn complex B_{As}-VI_{As} should have a C_s symmetry (three modes) in agreement with experimental observations. If boron atom is considered to be in the tetrahedral interstitial (B_{int}) site, the B_{int}-Si_{Ga} complex would have a C_{3v} symmetry but the complex B_{int}-VI_{As} would now have a C_{2v} symmetry, although three modes are still expected. The available evidences, including recent theoretical studies strongly favour substitutional site ($^{10}B_{As}$, $^{11}B_{As}$) for the B(2) center which has been confirmed by observations of nn Ga isotopic effects similar to $^{12}C_{As}$ and $^{28}Si_{As}$.

Oxygen in GaAs is a technologically important defect, partly because it is present in crystal growth environment (quartz crucible, B_2O_3 encapsulant, etc.) and because it is often intentionally added to the melt to reduce crystal contamination with residual impurities (viz., silicon, carbon, etc.).[4-5] Understanding the role of oxygen has been a subject of intensive investigations for three decades since the discovery that the addition of oxygen to growth ambient can produce semi-insulating (SI) GaAs. The existence of the mid-gap level, referred to as EL2 center, was postulated for the observed SI behavior. At one time it was thought that oxygen might be a component of the mid-gap level, but evidence against this assignment was presented by Huber *et al.*[6] Later, it was proposed that oxygen could constitute a part of another mid-gap level ELO[7] or EL2O.[8] The uncertainty caused by the problems in interpretation of DLTS data in heavily compensated samples proved this assignment inconclusive. Quite recently,[9-11] the atomic structure of oxygen induced defects in GaAs were studied by LVM spectroscopy. By analogy with oxygen related defects in silicon, two different impurity centers have been suggested in GaAs: (i) an isolated interstitial oxygen O_{int} for the LVM located near ~ 845 cm^{-1}, and (ii) V_{As}-O_{int} complexes for the modes observed near ~ 715 and 730 cm^{-1} (see: Table I). The isolated interstitial oxygen (O_{int}) is electrically inactive while the LVMs of the V_{As}-O_{int} complexes are photosensitive. However, there are number of questions remaining to be answered. For instance, it is not clear where the ELO center fit into the proposals made in LVM studies. The level position estimated in Ref. [10] for V_{As}-O_{int} complex seems rather ambiguous and it is not clear which of the variety of traps detected in GaAs is a proper candidate. The metastability of vacancy-oxygen complex (the so-called A-center) in Si is well established. In GaAs optically induced metastability of the two O-related bands is observed, but the processes involved in this interconversion are far from being understood.

Table I Local mode frequency due to light defects in GaAs.

Impurity center	*Linewidths (cm^{-1})*	*Electrical character*	*Local mode (cm^{-1})*
$^{28}Si_{As}$	> 2.1	a^-	398.2
$^{28}Si_{Ga}$	1.4-1.5	d^+	383.7
$^{29}Si_{Ga}$	...	d^+	378.5
$^{30}Si_{Ga}$	...	d^+	373.4
^{28}Si + native defect	> 1.5	donor	367.2
^{28}Si + e^- irr. defect	> 2.0	acceptor	368.5
$^{28}Si_{Ga}$- $^{28}Si_{As}$	> 2.0		464.0, 393.0
$^{12}C_{As}$		acceptor	582.4
$^{10}B_{Ga}$		isoelectronic	540.2
$^{10}B(2)$		acceptor	627.8
$^{10}B(2)$-Si_{Ga}			596.0, 684.8
$^{10}B(2)$-Se_{As}			601.0, 636.5, 649.0
$^{10}B(2)$-Te_{As}			605.0, 633.1, 649.6
O_{int}	0.2		845.9
O_{int}-X_{As}	0.2		794.3
Ga-O-Ga(3)	0.2		604.8
Ga-O-Ga(1)	0.07		730.8
Ga-O-Ga(2)	0.1		733.4

3. New Results of LVM of Hydrogen and Deuterium in InP

A direct exposure of H-plasma in InP is known to cause significant surface degradations due to the formation of phosphine[12-16] which result in forming islands of metallic indium on the surface of the samples. Recently, we have successfully implanted hydrogen and deuterium in InP samples by using 2 MV Van de Graff accelerator.Description of the experimental details will be reported elsewhere. Here the results are presented for the LVMs in proton and deuteron irradiated InP. These measurements were performed at 77 K by Fourier transform infrared spectroscopy using a BOMEM DA3 Michelson interferometer with 0.1 cm^{-1} resolution. The results are displayed in Fig. (1). Four hydrogen-related lines at 2202, 2274, 2286 and 2316 cm^{-1} are observed. Two satellite lines near 2274 and 2286 cm^{-1} are weak compared to the lines observed at 2202 and 2316 cm^{-1}. The deuteron-implanted InP showed also four lines, of about the same width as the proton related ones.

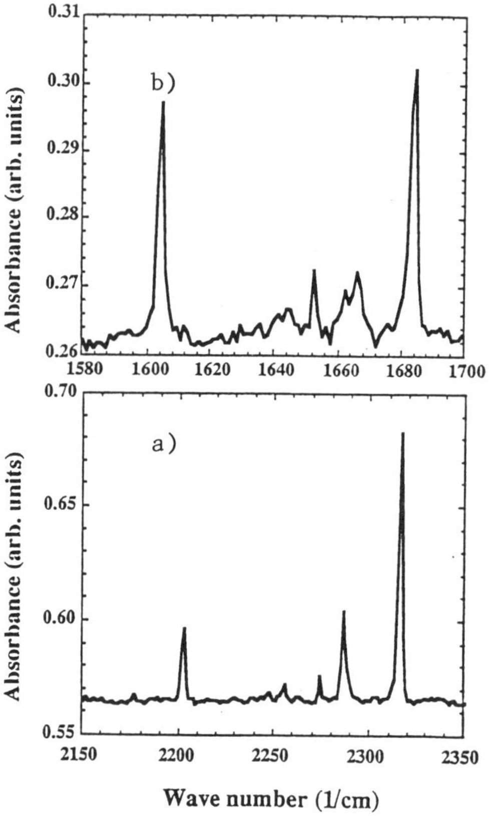

Fig. 1 Localized vibrational modes of (a) H- and (b) D- implanted InP.

4. Model calculations Results and Discussion

To extract the nature of impurity-host bonding in semiconductors from the existing IR data, two equally reliable theoretical efforts have been made in recent years: (a) A microscopic analysis based on a local density functional approach. (b) A macroscopic analysis based on the general treatment of the lattice dynamics in terms of interatomic forces using the Green's function approach. The advantage of the second technique over the first one is that it allows the coupling of the vibrations of the defect to the bulk crystal, and one can really visualize which types of vibrational modes remain localized around the defect. The calculated trends of force variations are found important in understanding the local vibrations of defect complexes of C_{3v}, C_{2v}, and C_s symmetries.

The Green's function matrix elements were computed by incorporating phonons from a lattice dynamical model fitted to the inelastic neutron scattering data of GaAs and InP. The hypotheses adopted in defining the perturbation matrix are those of Ref.[2]. The case of a single substitutional defect involves five atoms with T_d symmetry. The solutions for the LVMs belonging to A_1, E, F_1 and F_2 irreducible representations were carried out with a relative force constant parameter for cases of defect occupying the cation and the anion site, respectively. This serves a good starting point for us to analyze the existing infrared data of LVM's for complex centers and to obtain a set of best-fit parameters which are physically realistic. We discuss the following cases.

In the mass defect approximation (i.e., neglecting changes in force constants) our calculations of the LVM due to a Si_{Ga} donor or Si_{As} acceptor in GaAs provide values smaller than the experimental ones. This requires stiffening in the nearest neighbor (nn) impurity-host bonding to bring the local-mode frequencies into agreement with the experimental data. These results of the force variations and the analysis of LVMs due to several other charged defects in semiconductors suggested a close relationship between the force perturbation and the variation of the covalence of the bond. It should be mentioned here that these trends of force constant changes were found independent of the long range Coulomb forces and we strongly argue that the charge impurities in semiconductors affect only the short-range forces via redistribution of the electron charge density. Using the same values of the force variations for the isolated cases of Si defects, we are able to reproduce the A_1^+ and E^+ modes of the nn Si-pair (C_{3v} symmetry) in excellent agreement with the ir data. With the present choice of the perturbation model, we also predict the E^- mode near 323 cm^{-1} and an A_1^- mode close to the maximum phonon frequency of GaAs. Again, using the Green's function theory we have suggested the possibility of an impurity complex with antisite defect (Si_{As}-As_{Ga}) to be responsible for the Si-Y (367 cm^{-1}) center and a Si_{As}-V_{Ga} pair for the Si-X center. Comparison of the calculated Al-related LVMs[2] with a recent ir data[17] in $Al_xGa_{1-x}As$ samples grown by MBE lend additional support to the reliability of our theoretical technique.

Detailed recent ir measurements and its analyses[2] have led us to believe that the LVMs detected earlier at 601 and 628 cm^{-1} due to B(2) centers actually arise from the defects with antisite structure B_{As}. Further support to our assignment is provided by the calculations of LVMs due to 'B(2)-donor pairs'. In this study we have used the same force constant changes estimated for the isolated cases and the calculated results support the data reasonably well. Again, the substitutional B_{As} is obviously a double acceptor, but the electrical behavior of an essentially unbonded tetrahedral interstitial is much less certain and it would be easier to visualize donor rather than acceptor behavior.

To elucidate the origin of the LVMs in proton and deutron implanted InP we have also performed Greens function calculations. The two lines observed in proton implanted InP near 2202 and 2316 cm^{-1} show one-to-one correspondence with the lines seen in deuteron-implanted materials near 1604 and 1684 cm^{-1}, respectively. The ratios of the observed local mode frequencies for proton- and deuteron related lines are found slightly smaller than the square root of the deuteron to hydrogen mass ratio. This provides additional support that H- and D- are involved in the local mode frequencies. Our preliminary calculations suggest that the highest frequency ~ 2316 cm^{-1} line is a P-H stretching mode of vibration. It seems to us that the impurity-host bond in the defect vibrating at 2202 cm^{-1} may be relatively longer. Therefore, we have fitted this frequency by placing H on an In vacancy which bound to one of the P neighbors.[16] Green's function calculations considering P-H_2 and P-D_2 complexes in InP are in progress which may help us to characterize the additional features observed in our ir data. We are also investigating the microscopic structure of the H-passivating centers (H-Si_{Ga}, H-Si_{As}, H-Be_{Ga}, H-S_{As}, etc.) in *n*- and *p*- type GaAs and other III-V compounds by using sophisticated self-consistent pseudopotential calculations within the local density approximation. Similar studies are also planned to be extended to O-related complexes in GaAs.

Acknowledgements This work was supported in part by a grant from the National Science Foundation to use Pittsburgh's super computer facility and by a grant from the Research Corporation.

References

1. R. C. Newman, *Infrared Studies of Crystal Defects* (Taylor and Francis Ltd, London, 1973).
2. W. M. Theis, K. K. Bajaj, C. W. Litton, and W. G. Spitzer, Appl. Phys. Lett. **41**,70 (1982); W. M. Theis, D. N. Talwar, M. Vandevyver, and W. G. Spitzer, J. Appl. Phys. **58**, 2553, (1985); D. N. Talwar, M. Vandevyver, K. K. Bajaj, and W. M. Theis, Phys. Rev. **B33**, 8525 (1986); D. N. Talwar and M. Vandevyver, Phys. Rev. **B40**, 9779 (1990); C. R. Abernathy, S. J. Pearton, M. O. Manasreh, D. W. Fischer, and D. N. Talwar, Appl. Phys. Lett. **57**, 294 (1990).
3. D. J. Chadi and K. J. Chang, Phys. Rev. **B39**, 10063 (1989).
4. C. Song, B. Pajot, and F. Gendron, J. Appl. Phys.**67**, 7307 (1990).
5. G. M. Martin and S. Makram-Ebeid, in *Deep Centers in Semiconductors*, edited by S. T. Pantelides (Gordon and Breach, New York, 1986), p. 399.
6. A. M. Huber, N. T. Linh, M. Valladon, J. L. Debrun, G. M. Martin, A. Mitonneau, A. Mircea, J. Appl. Phys. **50**,4022 (1979).
7. M. Taniguichi and T. Ikoma, *Semi-Insulating III-V Materials* (Shiva, London, 1982), p. 283; J. Appl. Phys. **54**, 6448 (1983).
8. P. W. Yu, and D. C. Walters, Appl. Phys. Lett. **41**, 863 (1982).
9. G. Song, W. Ge, D. Jiang, and G. Hsu, Appl. Phys. Lett. **50**, 1666 (1987); X. Zhong, D, Jiang, W. Ge, and C. Song, Appl. Phys. Lett. **52**, 628 (1988).
10. H. Ch. Alt, Appl. Phys. Lett. 54, 1445 (1989); H. Ch. Alt and H. Schink, Appl. Phys. Lett. **52**, 1661 (1988).
11. J. Schneider, B. Discler, H. Seelewind, P. M. Mooney, J. Lagowski, M. Matsui, D. R. Beard, and R. C. Newman, Appl. Phys. Lett. **54**, 1442 (1989).
12. S. J. Pearton, J. W. Corbett, and T. S. Shi, Appl. Phys. **A 43**, 153 (1987).
13. J. Chevallier, B. Clerjaud, and B. Pajot, in *Neutralization of Defects and dopants in III-V Semiconductors*, edited by J. I. Pankove and N. M. Johnson (Semiconductors and Semimetals, Academic, New York , 1990).
14. P. R. Briddon and R. Jones, Phys. Rev. Lett. **64**, 2535 (1990).
15. For a recent review, see *Physics of DX centers in GaAs alloys*, guest editor, J. C. Bourgoin, *Diffusion and Defect Data*, Pt. **B10**, 1 (1990)
16. J. Tatarkiewicz, B. Clerjaud, D. Cote, F. Gendron, and A. M. Hennel, Appl. Phys. Lett. **53**, 382 (1982).
17. H. Ono and T. Baba, Phys. Rev. **B42**, 11423 (1990).